Bioengineering Fundamentals

Ann Saterbak

Department of Bioengineering, Rice University
Houston, TX

Ka-Yiu San

Department of Bioengineering, Rice University
Houston, TX

Larry V. McIntire

Department of Biomedical Engineering, Georgia Tech
Atlanta, GA

PEARSON
Prentice Hall

Upper Saddle River, New Jersey 07458

Library of Congress Cataloging-in-Publication Data

Saterbak, Ann.
Bioengineering fundamentals / Ann Saterbak, Ka-Yiu San, Larry V. McIntire.
p. cm.
ISBN 0-13-093838-6
1. Bioengineering. I. San, Ka-Yiu. II. McIntire, Larry V. III. Title.

TA164.S38 2007
660.6—dc22

2005056554

Vice President and Editorial Director, ECS: *Marcia Horton*
Acquisitions Editor: *Holly Stark*
Editorial Assistant: *Jennifer Lonschein*
Executive Managing Editor: *Vince O'Brien*
Managing Editor: *David George*
Production Editor: *Wendy Kopf*
Director of Creative Services: *Paul Belfanti*
Art Director: *Jayne Conte*
Cover Designer: *Bruce Kenselaar*
Art Editor(s): *Greg Dulles and Xiaohong Zhu*
Manufacturing Buyer: *Lisa McDowell*
Senior Marketing Manager: *Tim Galligan*

Pearson Prentice Hall
Pearson Education, Inc.
Upper Saddle River, NJ 07458

Printed in the United States of America

10 9 8 7 6 5 4 3 2 1

ISBN 0-13-093838-6

Pearson Education Ltd., *London*
Pearson Education Australia Pty. Ltd., *Sydney*
Pearson Education Singapore, Pte. Ltd.
Pearson Education North Asia Ltd., *Hong Kong*
Pearson Education Canada Inc., *Toronto*
Pearson Educación de Mexico, S.A. de C.V.
Pearson Education—Japan, *Tokyo*
Pearson Education Malaysia, Pte. Ltd.
Pearson Education, Inc., *Upper Saddle River, New Jersey*

To Miriam
–Ann Saterbak

To my parents
–Ka-Yiu San

To Suzanne Eskin
–Larry V. McIntire

Contents

Preface

The conservation laws of mass, energy, charge, and momentum form the foundation of engineering, including bioengineering. The purpose of *Bioengineering Fundamentals* is to provide a new and unifying approach to teach the conservation laws in an interdisciplinary introductory course for bioengineering students.

The foundation courses in many engineering curricula are based on applications of conservation laws. Conservation of mass and energy is typically the first course in a chemical engineering curriculum. Conservation of momentum including statics and dynamics is often the foundation course in mechanical engineering. Finally, conservation of charge provides the basis for an introduction to electrical circuits. Based on the same fundamental concepts and equations that unify all engineering curricula, *Bioengineering Fundamentals* focuses on these conservation laws and how they apply to biological and medical systems to lay a foundation for beginning bioengineers.

The educational goals of the textbook are designed to help bioengineering students:

1. Develop problem-formulation and problem-solving skills;
2. Develop and understand mass, momentum, charge, and energy conservation equations;
3. Apply the conservation equations to solve problems in the biological and medical sciences and to model biological and physiological systems; and
4. Appreciate the types of technical challenges and opportunities in bioengineering and the rewards of an engineering approach in the life and medical sciences.

The textbook is targeted to first- or second-semester sophomore students for use in a foundation course in a bioengineering, biomedical engineering, or related field. College-level calculus, general chemistry, physics, biology, and some rudimentary computational skills are recommended as prerequisites.

Mastering conservation principles early in the career of an undergraduate bioengineer is essential. These formative years are critical for students as they transition from general courses in science and mathematics (e.g., chemistry, calculus) to upper-level, specialized courses in bioengineering (e.g., biomaterials, bioinstrumentation). Many areas of specialization in bioengineering, such as bioinstrumentation, biomechanics, biochemical engineering, or biomaterials, use forms of accounting or conservation equations as a basis for study or in the derivation of other governing equations. For example, the conservation of momentum is often presented in biomechanics and transport texts. How the accounting and conservation equations are used to derive familiar laws, such as Kirchhoff's current and voltage laws, Newton's laws of motions, Bernoulli's equation, and others, is emphasized in this textbook. This framework of the accounting and conservation principles allows students to build a mental model of how key concepts in engineering and parts of chemistry and physics courses are interrelated. The engineering principles, problem-solving approach, and technical rigor used to model and solve problems make *Bioengineering Fundamentals* an appropriate prerequisite for junior- and senior-level courses.

In addition to conservation principles, the textbook emphasizes engineering problem-solving strategies. For many students, translating written problem statements into a diagram is initially difficult. Appropriately modeling and simplifying complex biological and medical systems into a mathematical description is characteristic of an expert engineer. For a novice, this can be a daunting experience. For this reason, the process of solving a problem is often stressed over the problem itself. The extensive number of example problems that are worked out in detail supports this priority.

Many of these worked example problems are solved using actual numerical parameters. By using actual physical data, students are asked to scrutinize their answers and provide insight into what it signifies in relation to the problem statement and also in the physical and biological world. The example problems span the breadth of modern bioengineering, including physiology, biochemistry, tissue engineering, biotechnology, and instrumentation. This variety of problems is intended to give exposure to bioengineering technology and research and to motivate students.

Chapter 1 provides motivation for a quantitative engineering approach and exposure to different bioengineering technologies and research topics. Physical variables used in accounting equation calculations are introduced in the context of bioengineering technologies and research topics. A methodology or process for solving engineering problems is introduced. This method is similar to those found in other leading engineering textbooks. This methodology for solving problems is used throughout the textbook, and it is recommended that students solve some of their homework problems using this or a similar methodology.

The fundamental framework for the conservation laws is described in Chapter 2. The accounting equation models the movement, generation, consumption, and accumulation of an extensive property in a system of interest. Extensive properties that can be counted include mass, energy, charge, and momentum. The conservation equation is a specialized form of the accounting equation and can be applied to certain extensive properties. Beginning a problem with the appropriate accounting or conservation equation is a familiar process for some types of engineers (e.g., chemical engineers) but not others (e.g., electrical engineers). However, using this framework allows for seemingly disparate introductory concepts in bioengineering (e.g., mass balances, first law of thermodynamics, Bernoulli's equation, and Einthoven's law) to be integrated under one unifying concept. Repeatedly returning to the generic accounting or conservation equation is meant to emphasize the key role these equations play as the foundation of engineering and to reinforce the process of structured problem solving.

The core of the textbook, Chapters 3-6, covers conservation of mass, energy, charge, and momentum, respectively, in biomedical systems. Each chapter opens with a challenge problem or focus that presents a current bioengineering research or design challenge to expose students to the many unanswered questions where further work could make an impact. Within each chapter, basic concepts are reviewed, and the accounting and conservation laws are restated and explicitly formulated for the property of interest. A significant portion is devoted to applying these concepts to solve bioengineering problems and reducing the accounting and conservation equations to other key equations learned in previous courses. Although each of these chapters can stand alone, we highlight how the conservation laws are parallel across the four properties.

Chapter 7 contains three case studies—heart and blood circulation, lungs and a heart-lung bypass machine, and kidneys and dialysis—that are designed to bridge the applications of the mass, energy, charge, and momentum accounting and conservation equations in biomedical systems. We explicitly chose to include these systems since they had physical phenomena at both the cellular and tissue levels. Many of the problems are open-ended and require considerable research on the part of the students. This material could be used as core or supplement material, as a project, or as the basis for large-scale problem-based learning problems.

Structuring the textbook in this fashion allows for considerable flexibility in student learning and faculty teaching. Since Chapters 3-6 can each be self-contained, an instructor can choose to emphasize one or more extensive properties while de-emphasizing the others. In addition, an instructor who wishes to teach from an entirely problem-based learning framework can use the problems presented in Chapter 7 as the basis for the course. Problem-based learning modules based on the Chapter 7 Case Studies are being piloted at Rice University. For students, this textbook lays out problem-solving strategies, covers the fundamental conservation laws, and offers many engaging problems.

The initial concept of *Bioengineering Fundamentals* stemmed from conversations with Rice University faculty and Dr. Jacquelyn Shanks, now at Iowa State University. With the motivation to unify biological phenomena with conservation principles, we wrote this textbook by modeling the same approach as the textbook *Conservation Principles and the Structure of Engineering*, authored by Dr. Charles Glover, Dr. Kevin Lunsford, and Dr. John Fleming at Texas A&M University. Our textbook is also patterned with a similar paradigm as set for in *Transport Phenomena* by Dr. R. Byron Bird, Dr. Warren Stewart, and Dr. Edwin Lightfoot. In their classic textbook, they sought to unify the study of momentum, heat, and mass transfer by using a similar approach to treat each topic. Supplemental material includes an Instructor's Solutions Manual. The National Science Foundation under its Division of Undergraduate Education Course, Curriculum, and Laboratory Instruction program partially funded this work (NSF grant #DUE-0231313). The George R. Brown School of Engineering also provided crucial funding.

We acknowledge with gratitude the many contributions of colleagues and students who have helped us in the preparation of the textbook. Dr. Joseph LeDoux at the Georgia Institute of Technology, Dr. Anita Vasavada at Washington State University, and Dr. Bruce Wheeler at University of Illinois-Champaign provided tremendously helpful feedback on early drafts of the manuscript. In addition, each contributed example or homework problems, together with Dr. Jerry Collins and Dr. Art Overholser at Vanderbilt University. Also contributing to the substantial number of the problems in this book are the numerous undergraduate and graduate students from Rice University: Beth Boulden, Michelle Brock, David Chee, Min-Jye Chen, Stephanie Farrell, Emily Glassinger, Stephen Harder, Elizabeth Hedberg, Heidi Holtorf, Christopher Loo, Amanda Lowery, Sheila Moore, Matthew Murphy, Billy Poon, Thomas Rooney, Adrian Shieh, Aditya Venkataraman, Justin Yang, and the others who we have inadvertently overlooked. Special assistance in the development of Chapter 7 was given by Rice University graduate students Jeremy Blum, Skip Mercier, Mark Sweigart, Lakeshia Taite, and Johnna Temenoff. Dr. Wendy Newstetter at the Georgia Institute of Technology also provided helpful perspective on developing problems for Chapter 7. Technical clarity was provided by tireless editing from Elaine Lee at Rice University. We appreciate the patience of the students at Rice University (taking BIOE 252), the Georgia Institute of Technology, Washington State University, and the University of Illinois-Champaign, who, through their use, improved what is now *Bioengineering Fundamentals*.

On a personal note, Larry McIntire appreciates the support of his spouse, Dr. Suzie Eskin. Ann Saterbak thanks her spouse, Dr. David Ward, for his support during the many, many hours that writing this textbook required.

Introduction to Engineering Calculations

CHAPTER 1

1.1 Instructional Objectives

After completing Chapter 1, you should be able to do the following:

- Perform unit conversions to attain answers in the desired units.
- Distinguish between intensive and extensive properties and list examples of each type.
- Define the physical variables commonly used in accounting and conservation equations. Specifically, you should be familiar with mass, moles, and molecular weight; mass and mole fractions; concentration and molarity; temperature; pressure; density; force and weight; potential, kinetic, and internal energy; heat and work; momentum; charge and current; and flow rates.
- Report answers with an appropriate number of significant figures.
- Adopt a methodology for solving engineering problems; the one described is used to solve many example problems throughout this textbook.
- Begin to develop a sense for the types of engineering problems that bioengineers address.

On December 11, 1998, the National Aeronautics and Space Administration launched the Mars Climate Orbiter, a spacecraft designed to function as an interplanetary weather satellite and a communications relay. It never reached its destination. The loss of the $193 million spacecraft resulted from an embarrassing oversight during the transfer of information between the Mars Climate Orbiter spacecraft team in Colorado and the mission navigation team in California. In calculating an operation critical to maneuvering the spacecraft properly into the Mars orbit, one team used British units while the other used metric units.[1] As a result, instead of the planned 140 kilometers (90 miles), the Mars Climate Orbiter approached the planet at an altitude of about 57 kilometers (35 miles),[2] causing it to either crash in the Martian atmosphere or skip off into space.

Although hundreds of millions of dollars were lost and much hope for scientific advancement was dashed in the failure of this mission, the losses associated with mistakes of this nature in the field of biomedical engineering could be even greater, for human lives are involved. If a bioengineer were to miscalculate the tolerable toxic range of a drug because of unit conversion, a doctor could prescribe an incorrect dosage and cause a patient to die. With so much at stake, the importance of learning basic concepts and giving meticulous attention to applying them cannot be overemphasized.

A thorough understanding of the material presented in Chapter 1 is crucial to your success throughout your bioengineering career. This chapter is an overview of principles and definitions that lay the groundwork for problem solving in bioengineering. In Section 1.5.3, we will demonstrate the relevance of this introductory material in real-life applications.

1.2 Physical Variables, Units, and Dimensions

Being able to measure and quantify physical variables is critical to finding solutions to problems in biological and medical systems. Most of the numbers encountered in engineering calculations represent the magnitude of measurable **physical variables**, which are quantities, properties, or variables that can be measured or calculated by multiplying or dividing other variables. Examples of physical variables include mass, length, temperature, and velocity. Measured physical variables are usually represented with a number or scalar value (e.g., 6) and a unit (e.g., mL/min).

A **unit** is a predetermined quantity of a particular variable that is defined by custom, convention, or law. Numbers used in engineering calculations must be given with the appropriate units. For example, a statement that "The total blood flow in the circulation of an adult human is 5" is meaningless, but "The total blood flow rate in the circulation of an adult human is 5 L/min" quantifies how much blood flows through the adult circulatory system.

A mistake that beginning engineers often make is to write variables without units. Students sometimes claim that they can keep track of the units in their heads and do not need to write them down repeatedly. This attitude leads to many mistakes when calculating solutions, which can lead to significant consequences, as in the Mars Climate Orbiter incident. Experienced engineers rarely omit units.

The basis for measurement of seven physical variables agreed upon internationally is given in Table 1.1. These include length, mass, time, electric current, temperature, amount of substance, and luminous intensity.

In engineering calculations, many other physical variables such as force or energy are commonly used. The units of these variables can be reduced to combinations of the seven base quantities. In this textbook, the term **dimension** can be thought of as a generic unit of a physical variable, which is not scaled to a particular amount for quantitative purposes. The dimensional quantities you will encounter in this textbook are listed in Appendix A. The symbols for the dimensions are given in Table 1.1.

TABLE 1.1

Base Physical Variables

Base quantity	Symbol	Base SI unit	Other unit examples
Length	L	meter (m)	centimeter (cm), foot (ft), inch (in), yard (yd)
Mass	M	kilogram (kg)	gram (g), pound (lb_m), ton (ton)
Time	t	second (s)	minute (min), hour (hr)
Electric current	I	ampere (A)	abampere (abA), biot (Bi)
Temperature	T	Kelvin (K)	Celsius (°C), Fahrenheit (°F)
Amount of substance	N	gram-mole (mol or g-mol)	pound-mole (lb_m-mol)
Luminous intensity	J	candela (cd)	

1.3 Unit Conversion

As discussed in Section 1.2, measured physical variables are usually represented by a number and a unit. The two most commonly used systems are the Système International d'Unités (SI), or metric system, and the British, or English, system. Engineers must be familiar with both systems, since institutions use and publish data in both. Many of the physical variables frequently encountered in bioengineering will be discussed in detail in Section 1.5. The physical variables and corresponding symbols used in this text are listed in Appendix A.

Unit conversion is the process by which the units associated with a physical variable are converted to another set of units by using conversion factors. Common unit conversions are summarized on the inside front cover and in Appendix B. You probably already know some conversion factors, such as that 1 inch is equal to 2.54 cm and 2.2 lb_m is equal to 1 kg. To convert a quantity expressed in terms of one unit to its equivalent in terms of another unit, multiply the given quantity by the conversion factor (new unit/old unit). Just as you would reduce multiples of a number in fractions, cancel out units. For example, you can convert the mass of the standard man in the British system (154 lb_m) to its equivalent in the SI system:

$$154\ \text{lb}_\text{m}\left(\frac{1\ \text{kg}}{2.2\ \text{lb}_\text{m}}\right) = 70\ \text{kg} \qquad [1.3\text{-}1]$$

Because the unit lb_m is present in both the numerator and the denominator, they cancel out. Writing out the units of the conversion factor is critical; if you do not, you may incorrectly scale the physical variable of interest.

Conversion factors are also required to convert within a system of units. For example, within the British system, we convert the mass of a 2200-lb_m car to its equivalent in tons as follows:

$$2200\ \text{lb}_\text{m}\left(\frac{1\ \text{ton}}{2000\ \text{lb}_\text{m}}\right) = 1.1\ \text{ton} \qquad [1.3\text{-}2]$$

Within the SI system, we may convert the length of the average adult's femur, 430 mm, to its equivalent in meters:

$$430\ \text{mm}\left(\frac{1\ \text{m}}{1000\ \text{mm}}\right) = 0.43\ \text{m} \qquad [1.3\text{-}3]$$

A series of prefixes is used to indicate multiples and submultiples of units in the SI system (Table 1.2). The "m" preceding the "m" of meters indicates "milli-" or 10^{-3} of the unit. Often, a series of two or more conversion factors is required to convert a value in a given set of units to the desired one. In situations with several conversions, it is even more critical to write out the units.

EXAMPLE 1.1 **Unit Conversion**

Problem: Convert the force of 50 $\text{lb}_\text{m} \cdot \text{ft/min}^2$ to its equivalent in $\text{mg} \cdot \text{cm/s}^2$.

Solution: From Table 1.2, we see that 1 g consists of 1000 mg and 1 m consists of 100 cm. Appendix B has other conversion factors.

$$50\ \frac{\text{lb}_\text{m}\cdot\text{ft}}{\text{min}^2}\left(\frac{1\ \text{min}}{60\ \text{s}}\right)^2\left(\frac{453.6\ \text{g}}{1\ \text{lb}_\text{m}}\right)\left(\frac{1000\ \text{mg}}{1\ \text{g}}\right)\left(\frac{0.3048\ \text{m}}{1\ \text{ft}}\right)\left(\frac{100\ \text{cm}}{1\ \text{m}}\right) = 1.92 \times 10^5\ \frac{\text{mg}\cdot\text{cm}}{\text{s}^2}$$

■

TABLE 1.2

Prefixes for the SI System

Factor	Prefix	Symbol
10^{12}	tera	T
10^{9}	giga	G
10^{6}	mega	M
10^{3}	kilo	k
10^{2}	hecto	h
10^{1}	deka	da
10^{-1}	deci	d
10^{-2}	centi	c
10^{-3}	milli	m
10^{-6}	micro	μ
10^{-9}	nano	n
10^{-12}	pico	p
10^{-15}	femto	f

TABLE 1.3

Typical Pressure Values*

Pressure	Values
Best laboratory vacuum	1 pPa
Additional pressure on ear drum caused by noise at a rock concert	10 Pa†
Pressure exerted by a penny lying flat on table	85 Pa†
Mars surface air pressure	1 kPa
Earth atmospheric pressure, 10-km altitude (approximate airplane cruising altitude)	26 kPa
Earth atmospheric pressure, sea level	101.3 kPa
Blood pressure	110 kPa
Pressure inside car tire	320 kPa
Household water pressure	350 kPa
Ballet dancer on one foot	2 MPa
Deep trenches in the Pacific Ocean	100 MPa

*Data from Vawter R, "Typical Values of Pressure," www.ac.wwu.edu/~vawter/PhysicsNet/Topics/Pressure/UnitsandValues.html (accessed June 24, 2005).
†Indicates gauge pressure (see Section 1.5.3.3).

As an engineer, it is exceedingly important for you to develop a sense of scale and to be able to tell whether your answer is reasonable (see Section 1.8). Developing a sense of the magnitude of various physical variables is an important goal. Tables 1.3–1.5 give ranges of pressure, length, and current for up to 20 orders of magnitude. Think about the types of bioengineering problems in which you are interested and what their scale is.

1.4 Dimensional Analysis

In high school algebra you learned to manipulate equations to solve for unknown variables. Engineers employ the same fundamental principle to decipher very complex

TABLE 1.4

Typical Length Values

Length	Values
Diameter of carbon nanofibers	100 nm
Human red blood cell diameter*	7.8 μm
Length of smooth muscle fibers in the gastrointestinal tract*	0.4 mm
Diameter of human aorta*	1.8 cm
Height of average man	1.7 m
Length of stretched-out DNA from a human cell†	1.8 m
Deepest ocean point (Marianas Trench in the Pacific Ocean)	11 km
Circumference of Earth	40 Mm
Distance from Earth to the sun	150 Gm

*Data from Guyton AC and Hall JE, *Textbook of Medical Physiology*. Philadelphia: Saunders, 2000.
†Datum from "Deoxyribonucleic Acid (DNA)," *Discovering Science*. Gala Group.

TABLE 1.5

Typical Current Values*

Current	Values
Current between neurons in the brain	1 pA
Current in a memory cell of an integrated circuit	0.01 μA
Current through heart muscle that is deadly	10 μA
Current at threshold of sensation	1 mA
Current across chest that is deadly	0.1 A
Current in typical household devices	10 A
Current from household wall socket	20 A
Typical service to a house	100 A
Typical lightning bolt	10 kA

*Data from Wood S, http://www.ee.scu.edu/classes/1999spring/elen010/LECTS/LECT2/2001lec2.pdf (accessed January 7, 2005).

models and equations. It is a tool to simplify complicated bioengineering problems into smaller, more comprehensible basic tasks in order to find a solution.

Dimensional analysis is an algebraic tool engineers use to manipulate the units in a problem. Numerical values and their corresponding units may be added or subtracted only if the units are the same.

$$5 \text{ m} - 3 \text{ m} = 2 \text{ m} \qquad [1.4\text{-}1]$$

whereas

$$5 \text{ m} - 2 \text{ s} = ?? \qquad [1.4\text{-}2]$$

The units of meters and seconds are not the same, so equation [1.4-2] cannot be executed. On the other hand, multiplication and division always combine numerical values and their corresponding units.

$$(4 \text{ N})(5 \text{ m}) = 20 \text{ N}\cdot\text{m} \qquad [1.4\text{-}3]$$

$$\frac{\left(6\frac{\text{cm}}{\text{s}}\right)}{8\ \text{cm}} = 0.75\frac{1}{\text{s}} \qquad [1.4\text{-}4]$$

Properly constructed equations representing general relationships between physical variables must be dimensionally homogeneous; i.e., the dimensions of terms that are added or subtracted must be the same, and the dimensions of the right-hand side of the equation must be the same as those of the left-hand side. As an example, consider the equation developed by Pennes to relate blood perfusion rate ($\dot{V}/V$ [$\text{L}^{-3}\text{Mt}^{-1}$]) to volumetric heat transfer rate to the tissues (J [$\text{L}^{-1}\text{Mt}^{-3}$]) in the human forearm according to the equation:

$$J = \frac{\dot{V}}{V}C_p(T_a - T_v) \qquad [1.4\text{-}5]$$

where C_p [$\text{L}^2\,\text{t}^{-2}\,\text{T}^{-1}$] is the heat capacity, T_a [T] is the arterial blood temperature, and T_v [T] is the venous blood temperature. We can confirm that the units on each side of equation [1.4-5] reduce to [$\text{L}^{-1}\text{Mt}^{-3}$], and therefore the equation is dimensionally homogeneous:

$$\left[\frac{\text{M}}{\text{Lt}^3}\right] = \left[\frac{\text{M}}{\text{L}^3\text{t}}\right]\left[\frac{\text{L}^2}{\text{t}^2\text{T}}\right][\text{T}] \qquad [1.4\text{-}6]$$

Dimensional analysis is a very powerful method for engineers to calculate quantities. Its basic premise is to cancel out the units appearing in both the numerator and the denominator so that, at the end, the unit for which you are looking on the left-hand side of the equation is matched by the unit on the right-hand side. Given an equation involving a physical quantity, dimensional analysis can be used to determine the dimensions of that quantity. Conversely, it can be used to determine if an equation is dimensionally correct. Thus, dimensional analysis can be used to serve as a check when solving engineering problems to make sure the solution accounts for all necessary variables.

Occasionally, an equation can be specified such that it is **dimensionless.** For these equations, the units in each term cancel to one. An example of this type of equation is an expression for cell growth:

$$\ln\left(\frac{C}{C_0}\right) = \mu t \qquad [1.4\text{-}7]$$

where C [L^{-3}M] is the cell concentration at time t [t], C_0 [L^{-3}M] is the initial cell concentration, and μ [t^{-1}] is the specific growth rate. Note that the units on each side of the equation cancel each other to become dimensionless [−].

Sometimes, groups of variables with dimensions that reduce to one repeatedly show up. These groups become a named dimensionless variable that stands for a specific property or succinctly represents a physical phenomenon. For example, a commonly used dimensionless variable in fluid dynamics is the Reynolds number, Re. For flow in a cylindrical tube or vessel, the Reynolds number is given by the equation:

$$\text{Re} = \frac{Dv\rho}{\mu} \qquad [1.4\text{-}8]$$

where D [L] is the vessel diameter, v [Lt^{-1}] is the fluid velocity, ρ [$L^{-3}M$] is the fluid density, and μ [$L^{-1}Mt^{-1}$] is the fluid viscosity. The Reynolds number is a ratio of the inertial force to the viscous force and describes some flow characteristics of fluid through a tube (see Chapter 6). The Reynolds number for different human blood vessels is shown in Table 1.6.

TABLE 1.6

Typical Reynolds Numbers in Human Circulatory System*

Blood vessel	Reynolds number
Ascending aorta	3600–5800
Descending aorta	1200–1500
Large arteries	110–850
Capillaries	0.0007–0.003
Large veins	210–570
Venae cavae	630–900

*Data from Cooney DO, *Biomedical Engineering Principles: An Introduction to Fluid, Heat, and Mass Transport Processes*. New York: Marcel Dekker, 1976.

1.5 Specific Physical Variables

This section highlights physical variables commonly used to develop and solve systems by means of accounting and conservation equations—concepts that are developed in the remainder of the book. We also briefly introduce extensive and intensive properties and scalar and vector quantities. The physical variables are defined and described in the context of six complex engineering scenarios.

1.5.1 Extensive and Intensive Properties

Physical properties can be classified as either extensive or intensive. **Extensive properties** depend on the size of the system or sample taken. Examples are mass and volume. For example, if you have a system encompassing 1 g of water and you double its size—consequently doubling the amount of water in the system—you have 2 g of water. A partial list of extensive properties is given in Table 1.7. Another characteristic of an extensive property is that it can be counted. Later, you will learn that only extensive properties may be counted in accounting and conservation equations. In this book, extensive properties that are counted include total mass and moles; individual species mass and moles; elemental mass and moles; positive, negative, and net electrical charge; linear and angular momentum; and total, mechanical, and electrical energy.

Intensive properties do not depend on the size of the system or the sample taken. Examples include temperature, pressure, density, mass fractions and mole fractions of individual system components in each phase, and others listed in Table 1.7. If you double the size of a system containing 25°C water, you will not increase the temperature of the water. Later, you will learn that intensive variables are not appropriately used in accounting and conservation equations.

TABLE 1.7

Intensive and Extensive Properties

Intensive properties	Extensive properties
Mass fraction (x)	Mass (m)
Mole fraction (n)	Moles (n)
Molecular weight (M)	Volume (V)
Temperature (T)	Electrical charge (q)
Pressure (P)	Linear momentum ($\vec{p}$)
Specific volume ($\hat{V}$)	Angular momentum ($\vec{L}$)
Density (ρ)	Energy (E_T)
Velocity ($\vec{v}$)	Mechanical energy
Specific energy ($\hat{E}_T$)	Entropy
Saturation (S)	
Humidity (H)	
Boiling point (T_b)	
Melting point (T_m)	

1.5.2 Scalar and Vector Quantities

Physical variables are either scalar or vector quantities. **Scalar** quantities can be defined by a magnitude alone. A **vector** quantity must be defined by both magnitude and direction. The vector must be defined with respect to a reference point to its origin, which can be done by specifying an arbitrary point as an origin and using a coordinate system, such as Cartesian (rectangular), spherical, or cylindrical, to show the direction and magnitude of the vector. To denote a vector quantity in this book, we use an arrow above the variable or symbol that represents the quantity (e.g., $\vec{v}$ for velocity vector).

Two types of vectors are especially important: position and velocity. Position vectors describe the distance and direction of an object's location with respect to an origin; velocity vectors describe the direction with respect to an origin and the distance an object moves per instantaneous time period. To find the magnitude of a vector using the Cartesian system, take the square root of the sum of the squares of each component. For example, a $(45\vec{i} + 45\vec{j})$-km/hr vector in a rectangular coordinate system could describe a car that moves east at 45 km/hr and north at 45 km/hr. However, the same car can be described as moving northeast at a constant 63.6 km/hr. Some examples of scalar and vector quantities are listed in Table 1.8.

TABLE 1.8

Scalar and Vector Quantities

Scalar quantities	Vector quantities
Mass (m)	Force ($\vec{F}$)
Length (L)	Velocity ($\vec{v}$)
Time (t)	Acceleration ($\vec{a}$)
Temperature (T)	Momentum ($\vec{p}$)
Pressure (P)	
Density (ρ)	
Energy (E_T)	
Power (P)	
Charge (q)	

The product of two scalar quantities is still a scalar quantity. The product of a scalar quantity and a vector quantity is a vector that has the same direction as the original vector if the scalar is positive and the opposite direction if the scalar is negative. An example is the multiplication of mass (scalar) and acceleration (vector) to yield a force (vector).

Vectors can be multiplied in two different ways. The scalar product (or dot product) of two vectors is a scalar quantity, as the name indicates. The scalar product is equal to the product of the magnitudes of the two vectors and the cosine of the angle between them:

$$\vec{A} \cdot \vec{B} = |\vec{A}||\vec{B}| \cos \theta \qquad [1.5\text{-}1]$$

Note that if the two vectors are perpendicular, their scalar product is zero. The scalar product is commutative, so $\vec{A} \cdot \vec{B} = \vec{B} \cdot \vec{A}$. The vector product (or cross product) of two vectors is a vector quantity perpendicular to the plane of the two original vectors. Its direction can be found by the so-called right-hand rule. Its magnitude is the product of the magnitudes of the two vectors and the sine of the angle between them:

$$|\vec{A} \times \vec{B}| = |\vec{A}||\vec{B}| \sin \theta \qquad [1.5\text{-}2]$$

This is just a basic outline of some of the properties of vectors. If you need more guidance on vectors, refer to any vector calculus or introductory physics textbook.

1.5.3 Applications

In this section we introduce key concepts about physical variables, presented in open-ended problems. These scenarios will help you master the material presented in Chapter 1, as well as demonstrate and inspire you with real-life challenges that practicing engineers encounter. Five of the scenarios focus on challenging research problems in bioengineering and closely related fields; the sixth captures the splendor of a World Heritage Site.

For some of the problems presented, a method to attack the solution must first be prepared. Each person may attack a problem differently, and we will demonstrate one route to finding an answer, showing how to research background information before formulating a solution. Other problems presented here demonstrate how bioengineers must gather information before they are able to create a solution. In each open-ended scenario, we will show what information physical variables can reveal, using the variables to introduce some concepts that will be developed further in the remainder of the book. Each scenario is significantly more complicated than one engineer can be expected to be solve—particularly one just being introduced to bioengineering. Most of the scenarios are ongoing research problems that still do not have definitive solutions; the calculations presented here are just simple ones that a novice engineer may perform to help set up a problem.

1.5.3.1 Parkinson's Disease Parkinson's disease is a disorder of the central nervous system that affects over 1 million Americans.[3] It is characterized by rigid muscles, involuntary tremor, and difficulty in moving limbs.[4] The disease is caused by the destruction of neurons that secrete dopamine, an inhibitory neurotransmitter that helps regulate the excitation signals for movement. The reduced level of dopamine available in the brain causes the feedback circuits to work improperly, producing the rigidity and tremors associated with Parkinson's disease.

A biotech company has developed a new drug that has the potential to increase dopamine availability in the brain for patients with Parkinson's disease. A potential medication has been determined but has been tested only in animal subjects, who have the drug directly injected through a hole drilled in the skull. This intracranial delivery is hardly a feasible option for human clinical trials, since Parkinson's disease is chronic and the drug will need to be continually administered.

As bioengineering experts at this company, you and your team are asked to formulate a delivery mechanism with proper dosages so that the drug can go to human clinical trials. You must determine the appropriate dose and dosing interval (i.e., how frequently the treatment needs to be administered) as well as the most convenient, safe, and effective manner of drug delivery.

In this problem you need to analyze which task to accomplish first. Because the method of administering the drug will affect how it is formulated, we will decide upon the delivery method first.

In this section we discuss some tools to define this problem, using the following concepts:

- Mass
- Moles
- Mass and mole fraction
- Molecular weight and average molecular weight
- Concentration and molarity

Since direct drug injection through the skull is not a realistic option, other methods must be considered (see box). Of these, only oral administration, which is by far the most convenient and accepted method, is feasible. The other routes require hospital settings (intravenous, intramuscular), have problems with the organ targeted for absorption (rectal, inhalation, topical), or can be interfered with by the symptoms of Parkinson's disease, such as tremors (buccal/sublingual, subcutaneous). A drug taken orally can be absorbed across the membranes of the gastrointestinal tract into the patient's bloodstream and then into the targeted organ.

Drugs can be administered through various routes:

1. *Intravenous:* delivered directly into the bloodstream.
2. *Intramuscular injection:* injected directly into the muscle.
3. *Oral:* taken through the mouth, as with pills.
4. *Buccal/sublingual:* dissolved from small tablets held in the mouth or under the tongue.
5. *Rectal:* administered by a suppository or enema.
6. *Subcutaneous:* injected under the skin, as with insulin.
7. *Inhalation:* contained in an aerosol inhaled by the patient.
8. *Topical:* absorbed through the skin.

To reach the target organ effectively, delivery must overcome limitations involving drugs administered orally, including the first-pass effect, the effect of food on the drug, and the toxic effect of the drug on the gastrointestinal system. However, in developing a drug for patients with Parkinson's disease, the major obstacle is creating a drug that will cross the blood-brain barrier to reach the brain.

The brain has a specialized barrier called the blood-brain barrier, which consists of adjacent endothelial cells tightly fused with one another so that permeability of drugs and other molecules is significantly reduced. Designed to protect the brain from harmful substances, the blood-brain barrier severely restricts the transfer of high-molecular-weight molecules and polar (lipid-insoluble) compounds from the blood to the brain tissue. Lipid-mediated transport is generally proportional to the lipid solubility of the molecule, but is restricted to molecules with a molecular weight lower than approximately 500 g/mol.[5] Currently, 100% of large-molecule drugs and over 98% of small-molecule drugs do not cross the blood-brain barrier.[6] Drug design must recognize and work with this constraint.

To determine the appropriate dose for the drug, you must be comfortable with unit conversion and with the concepts of mass, moles, and molecular weight. Atomic weight and molecular weight should be familiar terms. **Atomic weight** is the mass of an atom relative to 12-carbon (an isotope of carbon with 6 protons and 6 neutrons), which has a mass with a magnitude of exactly 12. The periodic table lists atomic weights for all the elements (Appendix C). The **molecular weight** (M [MN^{-1}]) of a compound is the sum of the atomic weights of the atoms that constitute the molecules of a compound. The molecular weight of a substance can be expressed in a number of units, including daltons, g/mol, kg/kmol, and lb_m/lb-mol. The dalton is a unit used in biology and medicine and is equivalent to g/mol.

One **mole** of a species in the SI system, designated g-mol, is defined to contain the same number of molecules as there are atoms in 12 grams of 12-carbon. This is Avogadro's number or 6.023×10^{23} molecules. The British system uses a similar concept, but the basic mole unit is lb_m-mol. This is defined in an analogous manner: A lb_m-mol is equal to the number of atoms in 12 lb_m of 12-carbon. Because a lb_m is larger than a gram, a lb_m-mol is approximately 450 times larger than a g-mol. In general, you will use g-mol instead of lb_m-mol. In fact, if the units of a quantity are specified as mol, assume g-mol. One way to think of a mole is as the amount of species whose mass (in grams) is equal to its molecular weight. For example, one g-mol of CO_2 contains 44 g of material, since the molecular weight of CO_2 is 44 g/g-mol.

The amount of a material is usually expressed through the physical variables of mass or moles. Both mass (m [M]) and moles (n [N]) are base physical variables (Table 1.1). The **mass** is a measure of the amount of a material, whereas the number of moles present in a sample is calculated. The molecular weight of component A (M_A) is related to the mass of component A (m_A) and the number of moles of component A (n_A) as follows:

$$n_A = \frac{m_A}{M_A} \qquad [1.5\text{-}3]$$

Common biological molecules vary widely in molecular weight. Appendix D lists the molecular weight of common biological molecules (Table D.1).

EXAMPLE 1.2 Mass, Moles, and Molecular Weight

Problem: L-deprenyl prevents the metabolism of dopamine in the brain.[4] With increased dopamine availability in the brain, the symptoms of Parkinson's disease may subside. The chemical formula of deprenyl is $C_{13}H_{17}NHCl$.

Assume a dose of deprenyl of 140 μg/(kg · day) for treating Parkinson's patients. First, calculate the molecular weight of deprenyl and compare it to the threshold of 500 Da, the maximum-size molecule that can cross the blood-brain barrier. Estimate the average person's body mass and calculate how many of each of the following are contained in one day's dose: (a) mol $C_{13}H_{17}NHCl$, (b) lb_m-mol $C_{13}H_{17}NHCl$, (c) mol C, (d) g of C, (e) molecules of $C_{13}H_{17}NHCl$.

Solution: The molecular weight of deprenyl is the sum of the atomic weights of the atoms that constitute the compound. One deprenyl molecule contains 13 atoms of C, 18 atoms of H, and 1 atom each of N and Cl. The atomic weights of the atoms are found in the periodic table (Appendix C).

$$M = 13\left(\frac{12.011\text{ g}}{\text{mol C}}\right) + 18\left(\frac{1.008\text{ g}}{\text{mol H}}\right) + 1\left(\frac{14.007\text{ g}}{\text{mol N}}\right) + 1\left(\frac{35.453\text{ g}}{\text{mol Cl}}\right)$$

$$M = 223.75\,\frac{\text{g}}{\text{mol}} \approx 224\text{ Da}$$

The molecular weight is rounded to 224 Da with three significant figures (see Section 1.6). Thus, deprenyl is a sufficiently small molecule to cross the blood-brain barrier and enter its target region in the brain.

Assume the average person's body weight is 70 kg (154 lb_m). One day's dose of $C_{13}H_{17}NHCl$ is:

$$\text{dose} = \left(\frac{140\ \mu\text{g}}{\text{day}\cdot\text{kg}}\right)(1\text{ day})(70\text{ kg}) = 9800\ \mu\text{g} \approx 10\text{ mg}$$

This dose is consistent with published pharmacological values.[7]

(a) Use molecular weight to convert mass to moles:

$$10\text{ mg } C_{13}H_{17}NHCl\left(\frac{1\text{ mol}}{224\text{ g}}\right)\left(\frac{1\text{ g}}{1000\text{ mg}}\right) = 4.46 \times 10^{-5}\text{ mol } C_{13}H_{17}NHCl$$

(b) Conversion between g and lb_m is needed. Remember that molecular weight can be expressed in units of g/g-mol or lb_m/lb_m-mol:

$$10 \text{ mg } C_{13}H_{17}NHCl\left(\frac{2.2 \text{ lb}_m}{1 \text{ kg}}\right)\left(\frac{1 \text{ kg}}{10^6 \text{ mg}}\right)\left(\frac{1 \text{ lb}_m \text{ mol}}{224 \text{ lb}_m}\right)$$

$$= 9.82 \times 10^{-8} \text{ lb}_m\text{-mol } C_{13}H_{17}NHCl$$

(c) Each molecule of $C_{13}H_{17}NHCl$ contains 13 molecules of C. Therefore, each mole of $C_{13}H_{17}NHCl$ contains 13 moles of C:

$$4.46 \times 10^{-5} \text{ mol } C_{13}H_{17}NHCl\left(\frac{13 \text{ mol C}}{1 \text{ mol } C_{13}H_{17}NHCl}\right) = 5.80 \times 10^{-4} \text{ mol C}$$

(d) Use molecular weight to convert moles to mass. The molecular weight of C is 12 g/mol:

$$5.80 \times 10^{-4} \text{ mol C}\left(\frac{12.011 \text{ g C}}{\text{mol C}}\right)\left(\frac{1000 \text{ mg}}{1 \text{ g}}\right) = 6.97 \text{ mg C}$$

(e) The number of molecules of $C_{13}H_{17}NHCl$ is computed with Avogadro's number:

$$4.46 \times 10^{-5} \text{ mol } C_{13}H_{17}NHCl\left(\frac{6.02 \times 10^{23} \text{ molecules}}{1 \text{ mol}}\right) = 2.68 \times 10^{19} \text{ molecules}$$

■

After calculation of the amount of the active therapeutic agent needed, the drug can be formulated. Many pharmaceutical formulations (e.g., tablet, capsule, lotion, inhaler) contain inert ingredients, such as binders, flavors, colors, and synergists. Some active, therapeutic ingredients may be prescribed in amounts too small for patients to measure or handle, so bioengineers must add these inert ingredients to package the drugs in a form easier to take, such as a pill.

To describe mixtures, the variables of mole and mass fraction are often employed. Suppose that you have a mixture that contains components A, B, C, and D. The **mole fraction** of component A (x_A) in the mixture is defined as:

$$x_A = \frac{\text{moles of A}}{\text{total moles}} = \frac{n_A}{n_A + n_B + n_C + n_D} \quad [1.5\text{-}4]$$

Mole percent is mole fraction multiplied by 100. Within a system of interest, the mole fractions for all the constituents must sum to one:

$$\sum_i x_i = 1 \quad [1.5\text{-}5]$$

For our mixture, this implies that $x_A + x_B + x_C + x_D = 1$.

The **mass fraction** of component A (w_A) in a mixture of compounds A, B, C, and D is defined as:

$$w_A = \frac{\text{mass of A}}{\text{total mass}} = \frac{m_A}{m_A + m_B + m_C + m_D} \quad [1.5\text{-}6]$$

Similarly, **mass percent** is the mass fraction multiplied by 100. The mass fractions for all the constituents in a system must sum to one:

$$\sum_i w_i = 1 \quad [1.5\text{-}7]$$

The term mass fraction is synonymous with **weight fraction.** Both mole fraction and mass fraction are dimensionless [–]. Since the units in the numerator (moles of A, mass of A) and the denominator (total moles, total mass) must be the same, the numerical value of mole or mass fraction does not depend on the selected units. For example, if the mass fraction of O_2 is 0.33, then w_{O_2} equals 0.33 g O_2/g total or 0.33 lb_m O_2/lb_m total, and so on.

EXAMPLE 1.3 Mass and Mole Fraction

Problem: Suppose 10 mg of deprenyl is diluted in 10 mL water. Water has a density of 1.0 g/mL. Calculate (a) the mass fraction of the drug and (b) the mole fraction of the drug.

Solution:

(a) To find the mass fraction of the drug, we first need to find the total mass of the solution, which is the mass of the water plus the mass of deprenyl. Use the density of water to find the mass of a certain volume:

$$m_{\text{water}} = \rho_{\text{water}} V = \left(\frac{1.0 \text{ g}}{\text{mL}}\right)(10 \text{ mL}) = 10 \text{ g}$$

Note that the mass of the drug (10 mg = 0.01 g) is negligible (three orders of magnitude less) when compared to the mass of the water (10 g), so:

$$m_{\text{total}} = m_{\text{water}} + m_{\text{drug}} \approx m_{\text{water}}$$

The mass fraction of the drug is:

$$w_{\text{drug}} = \frac{m_{\text{drug}}}{m_{\text{total}}} \approx \frac{m_{\text{drug}}}{m_{\text{water}}} = \left(\frac{10 \text{ mg}}{10 \text{ g}}\right)\left(\frac{1 \text{ g}}{1000 \text{ mg}}\right) = 1.0 \times 10^{-3}$$

(b) To find the mole fraction of the drug, the number of moles of both the solute (drug) and solvent (water) must be known. The number of moles is calculated using their respective molecular weights. Equation [1.5-3] is used:

$$n_{\text{drug}} = \frac{m_{\text{drug}}}{M_{\text{drug}}} = (10 \text{ mg } C_{13}H_{17}NHCl)\left(\frac{1 \text{ mol}}{224 \text{ g}}\right)\left(\frac{1 \text{ g}}{1000 \text{ mg}}\right) = 4.46 \times 10^{-5} \text{ mol}$$

Similarly, n_{water} is 0.556 mol. Again, the amount of moles of deprenyl present is negligible (four orders of magnitude less) when compared to the amount of moles of water, so:

$$n_{\text{total}} = n_{\text{water}} + n_{\text{drug}} \approx n_{\text{water}}$$

The mole fraction of the drug is:

$$x_{\text{drug}} = \frac{n_{\text{drug}}}{n_{\text{total}}} \approx \frac{n_{\text{drug}}}{n_{\text{water}}} = \frac{4.46 \times 10^{-5} \text{ mol}}{0.556 \text{ mol}} = 8.02 \times 10^{-5}$$

Although the amount of solute dissolved in this scenario is negligible when compared to the amount of solvent present, this is not always the case. The mass and moles of the solute should not be automatically disregarded without careful inspection. ■

A set of mass fractions may be converted to an equivalent set of mole fractions using the following method:

- Set an arbitrary hypothetical mass (e.g., 100 g) of the mixture.
- Calculate the mass of each of the components in the mixture by multiplying the constituent mass fraction by the hypothetical mass.

- Convert the mass of each constituent into the moles of each constituent using its molecular weight.
- Calculate the mole fraction of each constituent as the ratio of the moles of the particular constituent divided by the total moles.

An analogous procedure is followed to convert a set of mole fractions to a set of mass fractions. A hypothetical number of moles (e.g., 100 g-mol) of the mixture must be set. Then calculate the mass of each constituent and the mass fraction in the same manner.

EXAMPLE 1.4 **Conversions Between Mass Fraction and Mole Fraction**

Problem: Suppose a patient with Parkinson's disease is prescribed deprenyl pills. The dosage of the active ingredient, deprenyl, is 5 mg, so to make the tablet large enough to handle, the manufacturer uses inert ingredients to raise its total mass to 200 mg. The inert ingredients are lactose ($w_{\text{lactose}} = 0.475$), cellulose ($w_{\text{cellulose}} = 0.375$), and magnesium stearate ($w_{\text{Mg stearate}} = 0.125$). Find the equivalent mole fractions of these four ingredients in the deprenyl pill. The molecular formulas of lactose and magnesium stearate are $C_{12}H_{22}O_{11}$ and $C_{36}H_{70}MgO_4$, respectively. Cellulose is a polysaccharide composed of glucose monomers with an average molecular weight of 400,000 Da.[8] Recall from Example 1.2 that the molecular weight of deprenyl, $C_{13}H_{17}NHCl$, is 224 Da.

Solution: The mass fraction of deprenyl is given by equation [1.5-6]:

$$w_{\text{deprenyl}} = \frac{\text{mass of deprenyl}}{\text{total mass}} = \frac{5\text{ mg}}{200\text{ mg}} = 0.025$$

The mass fraction of deprenyl could also have been calculated by using $\sum_i w_i = 1$, since all the other mass fractions are known.

To solve this problem, convert the mass fractions to mole fractions using the four-step procedure above. First, set a hypothetical mass of 100 g. Then calculate the mass of each component in the pill by multiplying the constituent mass fraction by the hypothetical mass:

$$m_{\text{deprenyl}} = 0.025(100\text{ g}) = 2.5\text{ g}$$

Similarly,

$$m_{\text{lactose}} = 47.5\text{ g}, \quad m_{\text{cellulose}} = 37.5\text{ g}, \quad \text{and} \quad m_{\text{Mg stearate}} = 12.5\text{ g}$$

Next, use equation [1.5-3] to convert the mass of each ingredient to moles using the corresponding molecular weight:

$$n_{\text{deprenyl}} = 2.5\text{ g}\left(\frac{1\text{ mol}}{224\text{ g}}\right) = 0.0112\text{ mol}$$

Similarly,

$$n_{\text{lactose}} = 0.139\text{ mol}, \quad n_{\text{cellulose}} = 9.38 \times 10^{-5}\text{ mol}, \quad \text{and} \quad n_{\text{Mg stearate}} = 0.0212\text{ mol}$$

The total number of moles is:

$$\begin{aligned} n_{\text{total}} &= n_{\text{deprenyl}} + n_{\text{lactose}} + n_{\text{cellulose}} + n_{\text{Mg stearate}} \\ &= 0.0112\text{ mol} + 0.139\text{ mol} + 9.38 \times 10^{-5}\text{ mol} + 0.0212\text{ mol} = 0.171\text{ mol} \end{aligned}$$

Finally, the mole fraction of each ingredient is calculated as the ratio of the moles of the constituent divided by the total moles, using equation [1.5-4]:

TABLE 1.9

Approximate Composition of Air

Species	Percent composition	Molecular weight (g/mol)
N_2	78.6	28.0
O_2	20.8	32.0
CO_2	0.04	44.0
H_2O	0.5	18.0
Other	0.06	—

$$x_{\text{deprenyl}} = \frac{n_{\text{deprenyl}}}{n_{\text{total}}} = \frac{0.0112 \text{ mol}}{0.171 \text{ mol}} = 0.0655$$

The mole fractions of the other constituents are $x_{\text{lactose}} = 0.813$, $x_{\text{cellulose}} = 0.000549$, and $x_{\text{Mg stearate}} = 0.124$. Note that $x_{\text{cellulose}}$ is much smaller than all the other mole fractions, because cellulose has a much greater molecular weight than the other components.

To check the answers, confirm that the mole fractions sum to 1:

$$\sum_i x_i = 0.0655 + 0.813 + 0.000549 + 0.124 = 1.003$$

This is very close to 1, and the difference can be accounted for by rounding error. ■

The **average molecular weight** of a mixture, M_{avg}, is the ratio of the mass of a sample of the mixture to the number of moles of all species in the sample. If x_i is the mole fraction of the ith component of the mixture and M_i is the molecular weight of this component, M_{avg} is calculated as follows:

$$M_{\text{avg}} = \sum_i x_i M_i \quad [1.5\text{-}8]$$

For a hypothetical mixture of A, B, C, and D, the average molecular weight of the mixture is written as $M_{\text{avg}} = x_A M_A + x_B M_B + x_C M_C + x_D M_D$. One commonly used average molecular weight is that of air, which is 28.8 g/mol. (Convince yourself that this is true using the data in Table 1.9.) We see that the average molecular weight is dominated by N_2 and O_2 because they make up ~99% of air by percent composition.

EXAMPLE 1.5 Average Molecular Weight

Problem: Calculate the average molecular weight of the solution containing 10 mg deprenyl diluted in 10 mL water.

Solution: We need to know the mole fraction of water. Since only two species are present (deprenyl and water), the mole fraction of water is calculated as:

$$x_{\text{water}} = 1 - x_{\text{drug}} = 1 - 8.02 \times 10^{-5}$$

$$M_{\text{avg}} = x_{\text{drug}} M_{\text{drug}} + x_{\text{water}} M_{\text{water}}$$

$$= (8.02 \times 10^{-5})\left(224 \frac{\text{g}}{\text{mol}}\right) + (1 - 8.02 \times 10^{-5})\left(18.0 \frac{\text{g}}{\text{mol}}\right)$$

$$= 18.002 \frac{\text{g}}{\text{mol}} \approx 18.0 \frac{\text{g}}{\text{mol}}$$

It is logical that the average molecular weight of this mixture is much closer to that of the solvent (water) than that of the solute (deprenyl) because the solution is almost 100% water. ■

To establish a suitable dosing regimen, the correct dosages must be calculated to result in the desired concentration of drug in the bloodstream and in the brain. Each drug has a therapeutic range in which it can achieve its remedial effect. The lower end of the therapeutic range is the minimum effective concentration, below which the drug is not concentrated enough to provide any therapeutic value. The upper end of the therapeutic range is the minimum toxic concentration, above which the drug concentration becomes harmful to the patient.

The therapeutic range is often described in terms of mass concentration. The **mass concentration** of a component in a mixture or a solution is the mass of that component (m) per unit volume (V) of the mixture:

$$C = \frac{m}{V} \qquad [1.5\text{-}9]$$

The dimension of mass concentration is $[L^{-3}M]$; common units of concentration include g/cm^3, kg/m^3, and lb_m/ft^3. Depending on the specific drug, therapeutic concentrations vary from the microgram-per-liter levels to the gram-per-liter levels.

The molar concentration of a solution can also be specified. Both mass and molar concentrations are designated as C. The **molar concentration** of a component in a mixture or solution is the number of moles (n) of that substance per unit volume of the mixture:

$$C = \frac{n}{V} \qquad [1.5\text{-}10]$$

The dimension of molar concentration is $[L^{-3}N]$; common units of molar concentration include g-mol/cm^3, g-mol/L, and lb_m-mol/ft^3. The **molarity** (whose units are abbreviated M) of a solution is the value of the molar concentration of the solute expressed in g-mol solute per liter of solution. For example, a 0.1-M fibronectin solution indicates 0.1 g-mol of fibronectin contained per liter of water. The amount (mass or moles) of a substance in a mixture can be determined by multiplying the concentration of the substance by the total volume of the mixture.

EXAMPLE 1.6 Concentration and Molarity

Problem: Using the same solution of 10 mg deprenyl diluted in 10 mL water, calculate (a) the mass concentration of the drug (in g/L) and (b) the molarity of the drug (in g-mol/L).

Solution:

(a) The mass concentration of deprenyl in water is given by equation [1.5-9]:

$$C = \frac{m_{\text{drug}}}{V_{\text{water}}} = \left(\frac{10 \text{ mg}}{10 \text{ mL}}\right)\left(\frac{1 \text{ g}}{1000 \text{ mg}}\right)\left(\frac{1000 \text{ mL}}{1 \text{ L}}\right) = 1\ \frac{\text{g}}{\text{L}}$$

(b) From Example 1.3, we know that 4.46×10^{-5} moles of deprenyl are present. The molarity of deprenyl is given by equation [1.5-10]:

$$C = \frac{n_{\text{drug}}}{V_{\text{water}}} = \left(\frac{4.46 \times 10^{-5} \text{ mol}}{10 \text{ mL}}\right)\left(\frac{1000 \text{ mL}}{1 \text{ L}}\right) = 4.46 \times 10^{-3}\ \frac{\text{mol}}{\text{L}}$$

This is a 4.46×10^{-3} M $C_{13}H_{17}NHCl$ solution. ■

Although deprenyl holds great promise for treating patients with Parkinson's disease, many problems are still associated with the drug, such as hypertension and reduced efficacy with prolonged use. Scientists and engineers must continue to develop new drugs and techniques that target the brain for patients with Parkinson's disease and other neurological disorders.

> One relatively recent technology that holds great promise for treating neurological disorders is buckminster fullerene, a soccer-ball-shaped molecule made of 60 carbon atoms. Buckyball-based drugs boast potential in treating diseases such as Parkinson's, multiple sclerosis, Alzheimer's, and brain cancers because they can slip through the blood-brain barrier and target otherwise inaccessible brain cells.

1.5.3.2 Mars Surface Conditions Outside of the Earth-Moon system, Mars is the most hospitable body in our solar system for humans and is currently the only real candidate for human exploration and colonization. However, many aspects of the surface environment on Mars differ from those on Earth, and these discrepancies must be taken into account when evaluating the feasibility of exploration of Mars. NASA has asked you, as a bioengineer, to compile some data to determine if Mars can support human life.

One part of bioengineering involves researching and establishing the conditions you must improve or compensate. In this problem, you need to analyze the surface conditions on Mars that must be planned for when considering habitation by humans. Only a few conditions will be discussed in the scope of this problem.

In this section, we begin to characterize the surface conditions on Mars using the following concepts:

- Temperature
- Ideal behavior of gases
- Pressure (gas)
- Density
- Saturation and humidity

Compared to Earth, Mars is much farther from the sun (by an average distance of 48.5 million miles), resulting in a significantly colder climate. **Temperature** (T) is a measure of the average kinetic energy of the molecules in a body or a system. The most commonly used temperature scales are Kelvin (K), Celsius (°C), and Fahrenheit (°F). You are probably most familiar with the Fahrenheit scale, since everyday data such as weather reports and cooking temperatures are given in this scale in the United States. Celsius and Kelvin scales are more common in scientific work. The three temperature scales are compared in Figure 1.1.

A temperature scale is arbitrarily defined and its values are assigned by fitting the equation of a line to two known physical quantities, such as the freezing point and boiling point of a substance. In the Celsius scale, the freezing point of water is arbitrarily defined at 0°C and the boiling point of water at 100°C at 1 atmosphere (atm). When the equation of the line is defined for this scale, the lowest theoretical temperature possible is −273.15°C, known as **absolute zero**.

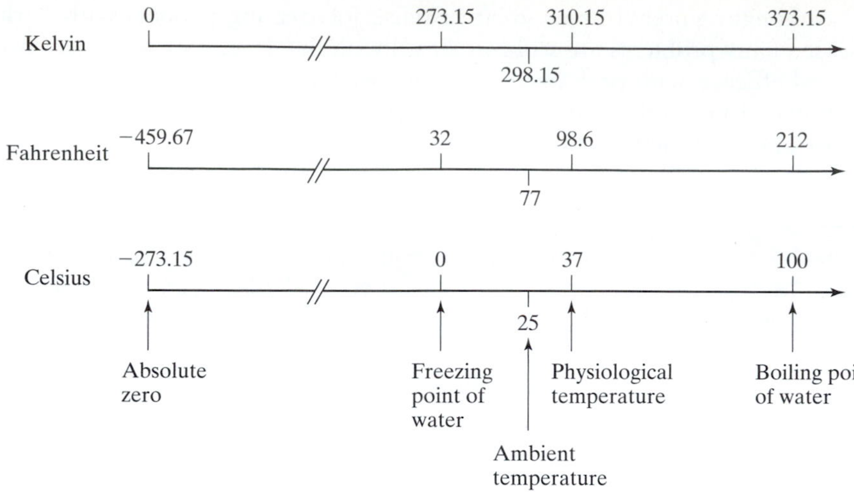

Figure 1.1
Comparison of the Kelvin, Fahrenheit, and Celsius temperature scales. A few temperatures most commonly used in biological problems are labeled. (*Source:* Doran PM, *Bioprocess Engineering Principles,* London: Academic Press, 1999.)

Temperatures at the Martian surface range from −76°C to −10°C.[9] To convert temperature units to different scales, the following equations can be used:

$$T\,(°\text{F}) = 1.8T\,(°\text{C}) + 32 \qquad [1.5\text{-}11]$$

$$T\,(\text{K}) = T\,(°\text{C}) + 273.15 \qquad [1.5\text{-}12]$$

$$T\,(°\text{F}) = 1.8T\,(\text{K}) - 459.67 \qquad [1.5\text{-}13]$$

Using these equations, the highest surface temperature on Mars is equivalent to 14°F or 263 K.

Temperature can affect the behavior of atmospheric gases. The **ideal gas law** describes the relationship between pressure, temperature, number of moles, and temperature of an ideal gas. An **ideal gas** is a hypothetical gas whose individual molecules have negligible volume and negligible intermolecular interaction forces (i.e., all collisions of ideal gas molecules are assumed to be perfectly elastic with each other and with the walls of the container). Many calculations of real gas behavior are approximations based on the assumption of this ideal behavior. The ideal gas law is usually written as follows:

$$PV = nRT \qquad [1.5\text{-}14]$$

where P is the absolute pressure, V is the volume, n is the number of moles, R is the ideal gas constant, and T is the absolute temperature. Equivalent values of R are listed in Table 1.10.

Standard temperature and pressure (STP) is often used when specifying properties of gases, particularly molar volumes, and standard conditions are defined at 273 K (0°C) and 1 atm. **Standard biological temperature and pressure** (BTP) of the human body is defined at 310 K (37°C) and 1 atm.

Real gases deviate from ideal behavior because molecular volume and intermolecular interactions can be significant. To account for these differences, modifications to the ideal gas law have been made, such as the equation developed by Johannes van der Waals in which he introduced two constants specific to each gas to account for the finite volume of gas molecules and the attractive forces between them. For most problems in this text, it is appropriate to assume that a gas has ideal behavior.

TABLE 1.10

Values of Ideal Gas Constant

Equivalent values of ideal gas constant (R)
$1.9872\ \dfrac{\text{cal}}{\text{mol}\cdot\text{K}}$
$8.314\ \dfrac{\text{J}}{\text{mol}\cdot\text{K}}$
$82.057\ \dfrac{\text{cm}^3\cdot\text{atm}}{\text{mol}\cdot\text{K}}$
$10.731\ \dfrac{\text{psi}\cdot\text{ft}^3}{(\text{lb}_\text{m}\text{-mol})\cdot(°\text{R})}$
$0.08206\ \dfrac{\text{L}\cdot\text{atm}}{\text{mol}\cdot\text{K}}$

According to the ideal gas law, one mole of gas at a given temperature and pressure occupies the same volume, regardless of the composition of the gas. At STP, one mole of any gas occupies 22.4 L. For gas, the composition by volume percent is equivalent to its composition by mole percent. In contrast, for a substance in the liquid or solid phase, mole percent is not strictly equivalent to volume percent, and the equality of mole percent and volume percent is rarely encountered.

On Earth, air is composed almost solely of nitrogen (79 vol%) and oxygen (21 vol%), with only trace amounts of other gases (e.g., argon, carbon dioxide, and methane) (see Table 1.9). In contrast, the Martian atmosphere is composed mainly of carbon dioxide (95.3 vol%), nitrogen (2.7 vol%), and argon (1.6 vol%), with small amounts of other gases. Oxygen, which is so important to us on Earth, makes up only 0.13 vol% of the atmosphere on Mars. Based on these percentages, one mole of atmospheric gas on Earth contains 0.21 mol oxygen, while one mole of atmospheric gas on Mars contains just 0.0013 mol oxygen.

Like atmospheric air, many gases are not pure, but contain multiple chemical constituents. The mole fractions of these constituents in the vapor phase can be described as $x_{1v}, x_{2v}, \ldots, x_{iv}$, where the number denotes the constituent and v denotes vapor. For an ideal gas, the total vapor pressure (P) in the vessel containing the gas equals the sum of the partial pressures, P_i, of the constituent gases:

$$P = P_1 + P_2 + \cdots + P_n = \sum_i^n P_i \qquad [1.5\text{-}15]$$

The P_i of each gas in the mixture is defined:

$$P_i = x_{iv}P \qquad [1.5\text{-}16]$$

Thus, for a constituent i, the **partial pressure**—the pressure exerted by a specific component in a mixture of gases—can be determined by multiplying the mole fraction of the constituent by the total vapor pressure of the system.

EXAMPLE 1.7 **Partial Pressures on Earth and Mars**

Problem: Find the partial pressures of nitrogen and oxygen on Earth and Mars. Atmospheric pressure on the surface of Mars is approximately 1% that of Earth.

Solution: Assume that atmospheric air behaves like an ideal gas; this is generally a good assumption. The air pressure on Earth's surface is 1 atm, so the pressure on Mars' surface is approximately 0.01 atm. Equation [1.5-16] is used to calculate partial pressure. On Earth:

$$P_{N_2} = x_{N_2,v}P = 0.79(1 \text{ atm}) = 0.79 \text{ atm}$$

Similarly, $P_{O_2} = 0.21$ atm on Earth. On Mars:

$$P_{N_2} = x_{N_2,v}P = 0.027(0.01 \text{ atm}) = 0.00027 \text{ atm}$$

and $P_{O_2} = 0.000013$ atm.

We see that Mars is almost completely devoid of the oxygen essential for human survival. Oxygen could be transported in limited quantities from Earth to Mars, but this would provide only a finite source, inhibiting extended stays and possible habitation. New technologies that will allow independence from Earth's resources will need to be developed to allow permanent residence on Mars. Future inhabitants of Mars must be able to produce oxygen from resources available in the Martian atmosphere and land. ■

Mars' atmosphere is much thinner than Earth's. When we speak of "thinner" air, as we often do when we talk about traveling in the mountains or reaching higher elevations, we actually refer to the air density. **Density** (ρ) is an intensive property that relates a substance's mass (m) to its volume (V):

$$\rho = \frac{m}{V} \qquad [1.5\text{-}17]$$

The dimension of density is $[L^{-3}M]$; common units include g/cm^3 and lb_m/ft^3. In general, the densities of solids are greater than those of liquids, which are greater than those of gases. An important exception is ice, which is less dense than liquid water. This is why ice floats on bodies of water, enabling plants and aquatic organisms below the surface to survive in cold climates.

Equation [1.5-17] can be rearranged to calculate the volume of 2 lb_m of air using its density at 25°C (0.0012 g/cm^3) as follows:

$$V = \frac{m}{\rho} = \left(\frac{2\ lb_m}{\left(\frac{0.0012\ g}{cm^3}\right)}\right)\left(\frac{1000\ g}{2.2\ lb_m}\right)\left(\frac{1\ L}{1000\ cm^3}\right) = 758\ L \qquad [1.5\text{-}18]$$

The **specific volume** ($\hat{V}$) of a material is the inverse of its density. $\hat{V}$ has a dimension of $[L^3M^{-1}]$ and is an intensive property. The specific volume of air at 25°C is 833.3 cm^3/g.

EXAMPLE 1.8 **Martian Air Density**

Problem: Find the density of air at the Martian equator, given the following air composition: 95.32 vol% carbon dioxide, 2.7 vol% nitrogen, 1.6 vol% argon, 0.13 vol% oxygen, 0.08 vol% carbon monoxide, and 0.03 vol% water. The remaining 0.14 vol% consists of a mixture of gases with an average molecular weight of 30 g/mol. Compare your answer to the density of air on Earth, 1.22 g/L, at Earth's average surface temperature (15°C) and pressure (1 atm).

Solution: Since they are subject to different conditions of pressure and temperature, we need to compare the two gases by using the same quantity, such as 1 mol. Use the ideal gas law [1.5-14] to calculate the volume occupied by 1 mol of gas on Mars. With the given air composition, we calculate the average molecular weight of air to calculate air density.

Recall that the average Martian surface temperature at the equator is −58°C or 215.15 K, and that the absolute surface pressure is 0.01 atm. From the ideal gas law, 1 mol of air on Mars' surface occupies a volume of:

$$V = \frac{nRT}{P} = \left(\frac{(1\ mol)\left(82.057\frac{cm^3 \cdot atm}{mol \cdot K}\right)(215.15\ K)}{(0.01\ atm)}\right)\left(\frac{1\ L}{1000\ cm^3}\right)$$

$$= 1765.5\ L \approx 1770\ L$$

The average molecular weight of Martian air is calculated using equation [1.5-8]:

$$M_{avg} = x_{CO_2}M_{CO_2} + x_{N_2}M_{N_2} + x_{Ar}M_{Ar} + x_{O_2}M_{O_2} + x_{CO}M_{CO} + x_{H_2O}M_{H_2O} + x_{other}M_{other}$$

$$M_{avg} = 43.5\frac{g}{mol}$$

Notice that the molecular weight of atmospheric air on Mars is very close to the molecular weight of CO_2, 44 g/mol, because CO_2 composes the majority of the Martian atmosphere. Therefore, since one mole of air on Mars weighs 43.5 g and occupies 1770 L, the density is:

$$\rho = \frac{m}{V} = \frac{43.5 \text{ g}}{1770 \text{ L}} = 0.0246 \frac{\text{g}}{\text{L}}$$

Thus, the Martian atmosphere is 50 times thinner than Earth's atmosphere. ■

While the density of a gas depends on its temperature and pressure, this dependence is much weaker for substances in the liquid and solid states. The densities of pure solids and liquids are essentially independent of pressure and vary little with temperature. For liquids and solids, specific gravity is used to describe density. The **specific gravity** (SG) of a substance is the dimensionless ratio of the density (ρ) of the substance to the density of a reference substance (ρ_{ref}) at a specified condition:

$$SG = \frac{\rho}{\rho_{ref}} \qquad [1.5\text{-}19]$$

The most common reference material for liquids and solids is water at 4°C, whose $\rho_{ref} = 1.000 \text{ g/cm}^3$.

A gas may contain a vapor capable of condensing to a liquid; the presence of water vapor in air is the most commonly used example. The water vapor in Earth's atmosphere averages 0.8 vol% and can be as great as 4 vol%, but water vapor composes just 0.03 vol% of the Martian atmosphere. In the definitions below, the term **humidity** specifies an air-water system, while **saturation** refers to any other gas-liquid combination.

Suppose a gas at temperature T and pressure P contains a vapor whose partial pressure is P_i and whose saturated vapor pressure is P_i^*. Saturated vapor pressure refers to the maximum pressure a vapor can exert over a pure liquid; P_i^* depends on temperature. For example, when a vapor is saturated with water, it holds all the water possible at that temperature and pressure. Relative saturation (S_R) and relative humidity (H_R) are defined:

$$S_R = H_R = 100\frac{P_i}{P_i^*} \qquad [1.5\text{-}20]$$

A relative humidity of 40% signifies that the partial pressure of water vapor equals 40% of the maximum vapor pressure of water at the system temperature. Relative humidity is usually reported in weather information intended for the general public. Since P_i^* depends on temperature, H_R also depends on temperature. Specifically, hotter air is capable of holding more water vapor than cooler air. Because the atmosphere of Mars is cold and thin, it holds very little water. Although water vapor accounts for 0.03 vol% of the air on Mars, the atmosphere is, at most times and locations, still completely saturated (100% H_R).

Molal saturation (S_M) and molal humidity (H_M) are defined:

$$S_M = H_M = \frac{P_i}{P - P_i} \qquad [1.5\text{-}21]$$

where P is the total pressure of the system. S_M and H_M are the moles of vapor divided by the moles of dry gas. The moles of dry gas are the total moles of the gas minus the moles of the vaporized compound of interest (water, for humidity). The percent saturation (S_P) and percent humidity (H_P) are defined:

$$S_P = H_P = 100\frac{S_M}{S_M^*} = 100\frac{\left(\dfrac{P_i}{P - P_i}\right)}{\left(\dfrac{P_i^*}{P - P_i^*}\right)} \tag{1.5-22}$$

where S_M^* is 100% molal saturation.

If any of these quantities for a gas at a given temperature and pressure are given, you can solve the defining equation to calculate the partial pressure or mole fraction of a constituent in a gas phase.

EXAMPLE 1.9 **Martian Humidity**

Problem: The Martian atmosphere contains approximately 0.03 vol% water vapor, which makes the air 100% saturated at the average surface temperature of −58°C.

(a) What is the partial pressure of water vapor on Mars?

(b) At Earth's average surface temperature of 15°C, the saturated vapor pressure of water is 12.79 mmHg. What is the partial pressure of water vapor on Earth's surface at 15°C if humidity is 90%?

Solution:

(a) We have already established that the atmospheric pressure on Mars is approximately 0.01 atm. Since the volume percent of a gas is equivalent to its mole percent, the mole fraction of water vapor on Mars is 0.0003 (0.03%). Thus, the partial pressure of water vapor on Mars is:

$$P_{H_2O} = x_{H_2O,v}P = (0.0003)(0.01 \text{ atm}) = 3 \times 10^{-6} \text{ atm}$$

(b) Rearranging equation [1.5-22] gives:

$$\left(\frac{P_i}{P - P_i}\right) = \frac{H_P}{100}\left(\frac{P_i^*}{P - P_i^*}\right)$$

$$P_i = \frac{\dfrac{H_P}{100}\left(\dfrac{P_i^*}{P - P_i^*}\right)P}{1 + \dfrac{H_P}{100}\left(\dfrac{P_i^*}{P - P_i^*}\right)}$$

H_P, P_i^*, and atmospheric pressure P are known. Using a conversion factor to make the units consistent for calculation, the saturated vapor pressure at 15°C on Earth is:

$$P_{H_2O}^* = 12.79 \text{ mmHg}\left(\frac{1 \text{ atm}}{760 \text{ mmHg}}\right) = 0.0168 \text{ atm}$$

The partial pressure and mole fraction of water vapor in Earth's atmosphere are:

$$P_{H_2O} = \frac{\dfrac{90}{100}\left(\dfrac{0.0168 \text{ atm}}{1 \text{ atm} - 0.0168 \text{ atm}}\right)(1 \text{ atm})}{1 + \dfrac{90}{100}\left(\dfrac{0.0168 \text{ atm}}{1 \text{ atm} - 0.0168 \text{ atm}}\right)} = 0.0152 \text{ atm}$$

$$x_{H_2O,v} = \frac{p_{H_2O}}{P} = \frac{0.0152 \text{ atm}}{1 \text{ atm}} = 0.0152$$

Note that the partial pressure of water vapor in the Martian atmosphere is over 5000 times lower than that in Earth's atmosphere. ■

Consideration of the surface temperature and air quality as well as many other factors is crucial to designing a closed-system living environment that will make Mars habitable to humans.

1.5.3.3 Getting to Mars From liftoff to arrival, many factors demand consideration in planning long-term travel to Mars. The effects of a zero- or micro-gravity environment during travel time between planets especially pose a great risk to passengers. In space, the head-to-foot arterial blood pressure gradient is severely diminished, causing the regulation and distribution of body fluids to change and possibly resulting in damage to the liver, heart, and other vital organs. Because weightless environments do not allow the bones to be loaded, bone density decreases with extended time periods in space. Bones and muscles become weaker in null gravity, making exercise a necessity.

Other systems of the body also experience detrimental effects when traveling through a microgravity environment. For example, because the human cardiovascular system is adapted to the constant gravitational force of Earth, weaker gravitational forces can cause physiological dysfunction. Unchallenged by gravity, blood vessels lose strength and the ability to dilate and to constrict and send the blood back to the heart, which can cause blood to collect in the lower extremities. The longer the time in microgravity, the weaker the circulatory system becomes.[4,10]

Because of your bioengineering background, NASA has hired you to design a system that will help passengers reduce the risks to their bodies from the effects of microgravity during their travel to Mars. You first need to evaluate the problems the human body encounters during and after extended periods in space.

In this problem, you need to analyze how humans are negatively affected by long-term travel in space. Only a few of the relevant factors will be discussed.

In this section we discuss this problem using the following concepts:

- Force
- Weight
- Potential energy
- Pressure (gas)
- Heat
- Work

Force ($\vec{F}$) is a vector quantity; its influence applied to a free-body object results in acceleration of the object in the direction of the applied force. Four types of forces describe the interactions between particles: electromagnetic, gravitational, strong force, and weak force. This text mostly deals with gravitational forces.

According to **Newton's second law of motion**, force ($\vec{F}$ [LMt^{-2}]) is equal to the product of mass (m [M]) and acceleration ($\vec{a}$ [Lt^{-2}]) as follows:

$$\vec{F} = m\vec{a} \qquad [1.5\text{-}23]$$

As mentioned earlier, units of mass include g, kg, lb_m, and ton. Units of force include the Newton [$N = kg \cdot m/s^2$], dyne [$dyne = g \cdot cm/s^2$], and lb_f [$1\ lb_f = 32.174\ lb_m \cdot ft/s^2$]. Note that lb_m and lb_f are different units that quantitate different physical variables.

Space travelers experience a gravitational acceleration equal to about one-third that of Earth ($g = 9.81\ m/s^2 = 32.174\ ft/s^2$) after they land on Mars ($0.38g$ or $3.72\ m/s^2$). Therefore, the force with which a person is attracted to Mars will be less than that with which the same person is attracted to Earth.

Since the force experienced by a space traveler on Mars is less than that experienced on Earth, she would weigh less on Mars than on Earth. The **weight** of an object is the force of gravity acting on that object. The weight of an object ($\vec{W}$ [LMt^{-2}]) is related to its mass (m) and the free-fall acceleration of the object ($\vec{g}$[Lt^{-2}]) as follows:

$$\vec{W} = m\vec{g} \qquad [1.5\text{-}24]$$

Weight is a force and gravity is an acceleration constant, so both are vector quantities. The magnitudes of weight and the free-fall acceleration constant are often considered by themselves because their direction is assumed to be toward the surface of the Earth and therefore implicit. The acceleration due to gravity varies little with position on the Earth's surface.

The symbol g_c is sometimes used to denote the conversion factor within metric and British force units:

$$g_c = \frac{1\dfrac{kg \cdot m}{s^2}}{1\ N} = \frac{32.174\dfrac{lb_m \cdot ft}{s^2}}{1\ lb_f} \qquad [1.5\text{-}25]$$

Other texts use g_c in formulas and equations; remember it is just a conversion factor, like the others listed in Appendix B.

Consider a 1-kg object on Earth. Using equation [1.5-24], the weight of this object is calculated to be 9.81 N. Consider a lighter 1-lb_m object on Earth. The weight of this object is calculated to be:

$$W = mg = 1\ lb_m\left(32.174\frac{ft}{s^2}\right)\left(\frac{s^2 \cdot lb_f}{32.174\ lb_m \cdot ft}\right) = 1\ lb_f \qquad [1.5\text{-}26]$$

Thus, the weight of a $1-lb_m$ object under Earth's gravity is 1 lb_f. Note that the conversion factor g_c was used. In the British system, because the numerical value of the weight of an object (1 lb_f) is equal to the numerical value of its mass (1 lb_m), there can be confusion that 1 lb_m is equal to 1 lb_f. This is not the case! Mass and weight are not equivalent—they are different physical properties and have different units. Interchanging lb_m and lb_f is like interchanging kg and N. This mistake rarely happens in the metric system because the numerical value of g_c is one; weight and mass have distinct values. However, since g_c has the same numerical value as g in the English system, the numerical value of a mass and its weight are equivalent under Earth's gravitational force. Remember to use g_c when appropriate to convert between lb_m and lb_f.

EXAMPLE 1.10 **Weights on Earth and Mars**

Problem: Calculate the weight in N, dynes, and lb_f of a 70-kg (154-lb_m) astronaut on Earth and Mars.

Solution: On Earth:

$$W = mg = (70\ \text{kg})\left(9.81\frac{\text{m}}{\text{s}^2}\right) = 687\frac{\text{kg}\cdot\text{m}}{\text{s}^2} = 687\ \text{N}$$

$$W = mg = (70\ \text{kg})\left(\frac{1000\ \text{g}}{1\ \text{kg}}\right)\left(9.81\frac{\text{m}}{\text{s}^2}\right)\left(\frac{100\ \text{cm}}{1\ \text{m}}\right)$$

$$W = 6.87 \times 10^7\frac{\text{g}\cdot\text{cm}}{\text{s}^2} = 6.87 \times 10^7\ \text{dynes}$$

$$W = mg = (154\ \text{lb}_\text{m})\left(32.174\frac{\text{ft}}{\text{s}^2}\right)\left(\frac{1\ \text{lb}_\text{f}}{32.174\frac{\text{lb}_\text{m}\cdot\text{ft}}{\text{s}^2}}\right) = 154\ \text{lb}_\text{f}$$

Note that the last conversion required the use of g_c. Although the weight of the astronaut is 154 lb_f and her mass is 154 lb_m, this does not mean that the units of lb_f and lb_m are equivalent.

On Mars:

$$W = mg = (70\ \text{kg})\left(3.72\frac{\text{m}}{\text{s}^2}\right) = 260\frac{\text{kg}\cdot\text{m}}{\text{s}^2} = 260\ \text{N} = 2.60 \times 10^7\ \text{dynes} = 58.4\ \text{lb}_\text{f}$$

■

Any object above the surface of a planet has a potential energy associated with its height that results from the gravitational pull of the planet. **Potential energy** (E_P) results from the position of an object in a potential field, such as a gravitational or electromagnetic field, or from the displacement of a system relative to an equilibrium position (e.g., compression of a spring). The dimension of potential energy is $[L^2Mt^{-2}]$; common units are joule (J), calorie (cal), and British thermal units (BTU). The potential energy of an object of mass m is:

$$E_P = mgz \qquad [1.5\text{-}27]$$

where g is the acceleration of gravity and z is the height of the object above a reference plane at which E_P is arbitrarily defined to be zero.

EXAMPLE 1.11 Potential Energy of a Spaceship

Problem: Suppose that 7 s after launch, a 600-kg spaceship has reached an altitude of 545 ft. What is its potential energy (in joules) with respect to Earth's surface? As the craft approaches Mars, it becomes subject to Mars' gravitational field. At what altitude (in feet) above Mars' surface will it have the same potential energy as it had 7 s after launch?

Solution: In this example, the surface of the planet of interest serves as the reference plane at which E_P is defined to be zero. The potential energy 7 s after launch is:

$$E_P = mgz = (600\ \text{kg})\left(9.81\frac{\text{m}}{\text{s}^2}\right)(545\ \text{ft})\left(\frac{1\ \text{m}}{3.28\ \text{ft}}\right) = 9.78 \times 10^5\ \text{J}$$

To find the altitude at which the ship has a particular potential energy above Mars, rearrange equation [1.5-27] to solve for z:

$$z = \frac{E_P}{mg} = \frac{9.78 \times 10^5\ \text{J}}{(600\ \text{kg})\left(3.72\frac{\text{m}}{\text{s}^2}\right)}\left(\frac{3.28\ \text{ft}}{1\ \text{m}}\right) = 1440\ \text{ft}$$

As expected, the height above Mars' surface at which the spaceship has the same potential energy as in Earth's gravitational field is higher than the height above Earth's surface, because the gravitational force on Mars is lower than on Earth. ■

Absolute pressure is pressure relative to a complete vacuum. Because this reference pressure is independent of location and other meteorological variables, it is a precise and invariant quantity. In general, when pressures of liquids and solids are given, they are absolute pressures. Gas pressure may be given as an absolute pressure or a gauge pressure. **Gauge pressure** (also known as relative pressure) is the difference in pressure between the sample of interest and the surrounding atmosphere. Most pressure-measuring devices detect gauge pressure. Absolute pressure and gauge pressure are related:

$$\text{absolute pressure} = \text{gauge pressure} + \text{atmospheric pressure} \qquad [1.5\text{-}28]$$

Sometimes the units of pressure are identified as absolute or relative. For example, psig is a gauge pressure and psia is an absolute pressure. Blood pressure is a ready example of a measured gauge pressure. The gauge pressure of 70 mmHg in the head corresponds to an absolute pressure of 830 mmHg. On Earth, the blood in the human body experiences gravitational pull, so a pressure gradient is present. In microgravity space, the pressure gradient vanishes, adversely affecting the circulatory system and the body's fluid distribution.

Except on very hot and humid summer days, the air a person inhales, whether on Earth or on Mars, is both cooler and less humidified than the air exhaled. As air is inhaled through the nose and mouth, it is warmed and humidified, taking both heat and water from the body. Thus, the body loses energy in the form of heat during breathing. **Heat** is the flow of energy resulting from a temperature gradient, flowing spontaneously from the body of the higher temperature to the body of the lower temperature. Heat can be used to raise the internal energy of the system or to do work on the system. Heat (Q) and the rate of heat ($\dot{Q}$) have dimensions of $[L^2Mt^{-2}]$ and $[L^2Mt^{-3}]$, respectively.

The flow of energy resulting from any source except for a temperature gradient is termed **work**. Work (W) and rate of work ($\dot{W}$) have dimensions of $[L^2Mt^{-2}]$ and $[L^2Mt^{-3}]$, respectively. Examples of driving forces that generate work include pressure, mechanical force, or an electromagnetic field. When astronauts do work in space or on Mars, the force they exert to perform a certain task is less than the force required to achieve the same task on Earth. Gentle pushes and exertions can move large objects. On Earth, the best simulation of this microgravity environment is immersion in a swimming pool for extended periods. The buoyancy of the water produces fluid shifts and relieves weight-bearing on the bones and muscles as in microgravity.

We have briefly discussed some considerations that engineers and astronauts must bear in mind as they assess the potential for future human travel to the red planet. Many conditions on Mars (e.g., bitterly cold climate, lack of water and oxygen, and reduced gravity) must be overcome to achieve successful human habitation, and engineers and physicians are working to design reliable, long-term solutions to these problems.

1.5.3.4 Gene Transfer Technology Scientists in the 1960s first speculated whether it was possible to treat genetic disorders by introducing functional genes via viral-mediated gene transfer. In 1990, this proposition became a reality when a young girl participated in a clinical trial for the treatment of adenosine deaminase deficiency. Since then, the definition of gene therapy has expanded from replacement of a defective gene to using any nucleic acid (deoxyribonucleic acid [DNA] or ribonucleic acid [RNA]) to treat or prevent a disease. Research developments have indicated possible applications of gene therapy in a wide spectrum of disorders, including cancer, autoimmunodeficiency syndrome (AIDS), cardiopathies, and neurological diseases. However, while it is promising, the success of gene therapy depends on finding a method that efficiently transfers the therapeutic gene to the target cells.

Gene therapy treatment may soon offer relief to patients with cystic fibrosis (CF), a disease characterized by thick mucus buildup in the lungs that hampers breathing, fosters deadly infections, and causes ongoing lung damage. One of the genes in patients with CF incorrectly codes for a membrane channel protein that maintains the balance of water and salts needed to produce the healthy lung secretions that trap and remove harmful bacteria. In clinical trials, the functional gene packaged into a virus is introduced into CF patients either in a saline solution dripped into their nasal passages or in a mist that is inhaled. Unfortunately, the beneficial effects diminishi over time as patients began to generate antibodies against the virus. Despite these setbacks, gene therapy treatment, which attacks CF at its root, may ultimately provide more effective relief than currently available treatments that aim only to control the symptoms of the disease.

A doctor asks you to advise him on the viability and safety of using two relatively inexpensive physical methods of gene transfer—the HELIOS Gene Gun and electroporation.

Bioengineers work in industrial, academic, or patient-care settings, and all of these require interaction not only with other bioengineers, but also with other coworkers, who may range from doctors and surgeons to other types of engineers, to administrative personnel, to patients. You must be able to communicate your expertise clearly on subjects your coworkers are less versed in. In this case, you are working with a doctor who is unsure which physical method of gene transfer will be effective for his patient. You give him a short synopsis of how each method works and the principles it is based on before listing the pros and cons of both methods.

In this section we discuss the physics behind these two methods of gene transfer using the following concepts:

- Momentum
- Kinetic and internal energy
- Charge and current

The HELIOS Gene Gun (Figure 1.2) is a hand-held device that rapidly and directly deploys foreign material, such as DNA or RNA, into nearly any target cell or tissue in vivo in close proximity (within 5–10 cells of the surface or about 1–2 mm deep). At the pull of a trigger, the gun uses a low-pressure helium pulse to shoot DNA- or RNA-coated gold or tungsten particles directly into the target tissue.

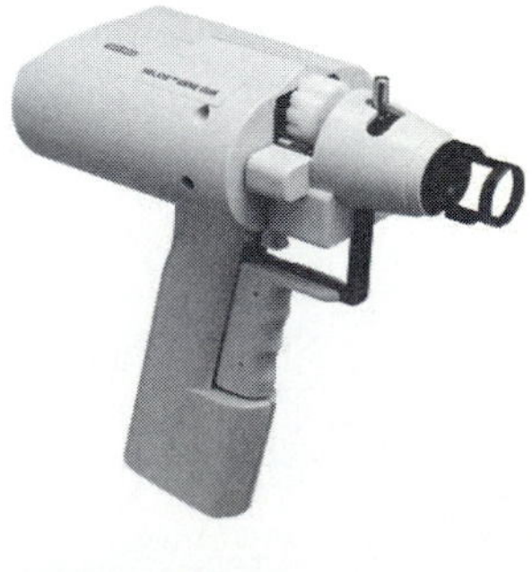

Figure 1.2
The Helios® Gene Gun. Reprinted with permission from Bio-Rad Laboratories.

To deliver the DNA- or RNA-coated particles into the target region, the gene gun uses momentum to forcefully transfer them into the body. **Linear momentum** is an extensive property that quantifies the motion of a particle or a system. The linear momentum, $\vec{p}$ [LMt^{-1}], of a system is the product of the velocity, $\vec{v}$ [Lt^{-1}], and mass, m [M], of the system as follows:

$$\vec{p} = m\vec{v} \qquad [1.5\text{-}29]$$

where velocity and linear momentum are vector quantities.

EXAMPLE 1.12 Calculation of Linear Momentum

Problem: Calculate the momentum of a DNA-coated gold particle exiting a gene gun at $1100\vec{i}$ mph. A typical gold particle is 2 μm in diameter and is coated with 100 plasmids, each with an approximate mass of 6×10^{-18} g. The density of gold is 19.3 g/cm^3.

Solution: To find the momentum of any moving object, we must first know its mass and velocity. Calculate the mass by modeling the particle as a sphere and using the known density of gold:

$$m_{\text{gold}} = \rho_{\text{gold}} V_{\text{gold}} = \rho_{\text{gold}} \frac{4}{3}\pi r^3 = \left(\frac{19.3 \text{ g}}{\text{cm}^3}\right)\frac{4}{3}\pi\left((1 \ \mu\text{m})\left(\frac{1 \text{ cm}}{10^4 \ \mu\text{m}}\right)\right)^3 = 8.09 \times 10^{-11} \text{ g}$$

The mass of the plasmid coating (100 plasmids) is added to the mass of the gold particle to find the total mass of the particle leaving the gene gun:

$$m_{\text{particle}} = 8.09 \times 10^{-11} \text{ g} + 100(6 \times 10^{-18} \text{ g}) = 8.09 \times 10^{-11} \text{ g}$$

Note that the mass of the DNA coating is negligible compared to the mass of the gold particle itself.

The momentum of the gold particle leaving the gene gun can then be calculated:

$$\vec{p} = m\vec{v} = (8.09 \times 10^{-11} \text{ g})\left(\frac{1 \text{ kg}}{1000 \text{ g}}\right)\left(\frac{1100\vec{i} \text{ miles}}{\text{hr}}\right)\left(\frac{1 \text{ hr}}{3600 \text{ s}}\right)\left(\frac{1609.34 \text{ m}}{1 \text{ mile}}\right)$$
$$= 3.98 \times 10^{-11} \vec{i} \ \frac{\text{kg} \cdot \text{m}}{\text{s}}$$

One firing of the gene gun contains about 0.5 mg of particles. Since each particle has a mass of about 8.09×10^{-11} g, the total number of particles ejected per firing is about 6 million, raising the total momentum of one firing to about $2.4 \times 10^{-4} \vec{i}$ kg · m/s. ■

Angular momentum is used to describe the rotational motion and torque on rotating bodies and in static and dynamic analyses of structures. It is an extensive property proportional to the mass of the system and applies to any object undergoing rotational motion about a point. The angular momentum, $\vec{L}$ [L^2Mt^{-1}], of a particle or body is the cross product of the position vector and its momentum and is described by a three-dimensional vector quantity. While the DNA-coated gold particles possess angular momentum, characterizing their linear momentum is more helpful to understanding how the gene gun operates.

Like all moving objects, the gold particles used in the gene gun carry a certain **kinetic energy** (E_K), or energy of motion, due to the translational motion of the system as a whole relative to some frame of reference (usually the Earth's surface). Kinetic energy is a scalar quantity that has the same dimension and units as potential energy. The kinetic energy of an object of mass m, moving with velocity v, is calculated as follows:

$$E_K = \frac{1}{2}mv^2 \qquad [1.5\text{-}30]$$

EXAMPLE 1.13 Calculation of Kinetic Energy

Problem: Calculate the kinetic energy of a DNA-coated gold particle exiting a gene gun. Use the same assumptions as before but note that direction does not matter, as kinetic energy is a scalar term. Report your answers in joules.

Solution: Recall from Example 1.12 that the mass of the DNA-coated gold particle is 8.09×10^{-11} g and that it exits the gene gun with a velocity of 1100 mph. The kinetic energy of the gold particle is:

$$E_K = \frac{1}{2}mv^2 = \frac{1}{2}(8.09 \times 10^{-11}\ \text{g})\left(\frac{1\ \text{kg}}{1000\ \text{g}}\right)\left(\left(\frac{1100\ \text{mi}}{\text{hr}}\right)\left(\frac{1\ \text{hr}}{3600\ \text{s}}\right)\left(\frac{1609.34\ \text{m}}{1\ \text{mi}}\right)\right)^2$$

$$E_K = 9.78 \times 10^{-9}\ \text{J}$$

and the kinetic energy of all the gold particles in one firing is 0.0587 J. ■

Internal energy (U) is the sum of all molecular, atomic, and subatomic energies of matter, including the energy due to the rotational and vibrational motion of the molecules, to the electromagnetic interactions of the molecules, and to the motion and interactions of the atomic and subatomic constituents of the molecules. Internal energy is a scalar quantity with dimension [L^2Mt^{-2}]. Internal energy cannot be measured directly or known in absolute terms; only changes in internal energy can be quantified. For example, the change in internal energy when transferring genes via the gene gun cannot be directly measured but can be calculated if sufficient information is provided. Changes in internal energy can be calculated using the tools presented in Chapter 4.

Electrical energy (E_E) is the energy associated with the flow of electric current, which is discussed in Chapter 5. Electric energy is a scalar quantity that has the same dimension and units as potential energy.

Electric charge (Q) is a physical property of matter that, like mass, is inherent for a particular atom, molecule, or ion. However, unlike mass, charge can be positive or negative. The dimension of charge is [tI], and it is usually expressed in units of coulombs (C). A coulomb is equal to 6.24×10^{18} elementary charges. An elementary charge is equal to the charge on a proton or an electron. A proton is an elementary particle with a positive value of charge, while an electron is one with a negative value. The magnitude of the total charge of one mole of electrons is 96,485 C, which is Faraday's constant.

An **electric current** (I) is the motion of charges and is quantified by the rate of flow of electric charge. Current is a base physical variable; its dimension is [I]. Electric current is usually expressed in units of amperes (1 A = 1 C/s). Some typical levels of current are found in Table 1.5. When charge moves between two points, the energy per unit charge generated is the **potential difference** (v) or **voltage.** The dimension of voltage is [$L^2Mt^{-3}I^{-1}$], and it is usually expressed in units of volts (V).

Electroporation is a gene transfer technology that relies on creating a potential difference across a cell membrane to transfer genes. Cell membranes, which are made up of charged bilayers of phospholipids, have a resting membrane potential; thus, many molecules cannot pass through from outside the cell. Whether in vitro or in vivo, electroporation uses high-voltage pulses to temporarily overcome the target cell membranes' resting potential so that DNA can enter the cell and DNA transfection can occur.

To perform electroporation, the cells are placed between positively charged and negatively charged electrodes to create an electric potential difference. An electric pulse is sent through the electrodes so that the cell's **capacitance**—its ability to store electrical charge when a potential difference is present—can be exceeded to allow molecules in. Capacitance (C) and charge (q) are proportionally related by the potential difference (v):

$$q = Cv \qquad [1.5\text{-}31]$$

The dimension of capacitance is [$L^{-2}M^{-1}t^4T^2$], and it is usually expressed in units of Farad, which is a coulomb/volt.

When a pulse perturbs the cell during a short time frame, the membrane becomes more permeable. Because DNA is negatively charged, the added DNA rushes toward the positive electrode, causing it to enter the cell on the side nearest the negative electrode. Once inside the cell, the DNA stays inside, because it immediately begins to interact with ATP.

EXAMPLE 1.14 **Cell Capacitance**

Problem: Cell membranes are made of phospholipid bilayers. Charge can be stored in the membrane, creating a capacitor. If the magnitude of the potential difference across a cell membrane is 70 mV and the membrane's capacitance is 1 μF/cm^2, calculate the charge stored by a cell membrane that has a diameter of 15 μm.

Solution: We can use the relation between charge and capacitance. We can model the cell as a sphere with the given diameter, so the region the membrane occupies can be modeled as the surface area of the sphere:

$$\text{surface area of a sphere} = 4\pi r^2 = 4\pi(7.5\ \mu\text{m})^2 = 706.85\ \mu\text{m}^2 \approx 707\ \mu\text{m}^2$$

We are given the potential difference of the membrane, as well as the capacitance per square centimeter, so using equation [1.5-31], we get:

$$q = Cv = \left(1\frac{\mu\text{F}}{\text{cm}^2}\right)(707\ \mu\text{m}^2)\left(\frac{1\ \text{cm}}{10^4\ \mu\text{m}}\right)^2(70\ \text{mV}) = 0.0004949\ \mu\text{F}\cdot\text{mV}$$

$$= 0.000495 \times 10^{-6}\frac{\text{C}}{\text{V}}(\text{mV})\left(\frac{\text{V}}{10^3\ \text{mV}}\right) = 4.95 \times 10^{-13}\ \text{C}$$

The charge stored by the cell membrane is equal to 4.95×10^{-13} C. Although the calculated value of stored charge is positive, the sign of the voltage and hence the calculated charge depend on the reference state from which we base our calculation. ■

In addition to the physical methods of insertion, such as the gene gun and electroporation, other methods of gene transfer take advantage of the body's different physical and biochemical mechanisms to transfect genes into target cells. One method is viral-mediated gene transfer, which uses viruses to carry a gene across the cell membrane and then insert it into the host genome. Transfection efficiency is often high because of the virus' natural evolutionary ability to "infect" cells. If the viral vector is replicated during mitotic division, the proteins coded by the gene can be continually expressed. However, viral vectors are expensive and often carry risk of antibody formation against the virus.[11]

Another method employs liposomes to transfer genes. Liposome-mediated gene transfer exploits the tendency of positively and negatively charged particles to interact by complexing cationic (positively charged) liposomes and negatively charged DNA. Because of their structural similarity to cell membranes, liposome-DNA complexes can penetrate cell membrane surfaces. Additionally, liposomes are nonpathogenic, inexpensive, and easy to produce, but their effectiveness in transfection is less than that of viral vectors.

Bioengineers are currently engaged in research to further develop these two methods as well as others to construct a safe and effective gene delivery method for patients.

You advise the doctor on the benefits of the gene gun and electroporation. In addition to being inexpensive and easy to prepare, both can be performed in vivo and ex vivo, allowing direct and targeted gene transfer of both DNA and nucleic acids to a specific area. Neither method uses viral vectors, increasing safety by avoiding risk of pathogenesis and antibody formation. Additionally, both methods can inject nucleic acids into both mitotic and nonmitotic cells. The gene gun produces temporary durations of gene expression with variable effectiveness. In comparison, electroporation has been shown to be highly efficient in some systems. However, the physical method of delivery creates problems for both methods. The gold particle beads the gene gun uses to shoot the DNA or RNA, though they may be innocuous, stay inside the body. During electroporation, cell mortality can occur if the pulsations are too high, too frequent, or too long.

Despite considerable advances in basic research, the therapeutic applications of gene transfer technology remain largely theoretical. Some weaknesses in the engineering of current technology include targeting the gene specifically to the cells of interest, designing the viral vector or other delivery system, regulating the gene, and inhibiting the immune response.

Equally challenging are the ethical issues involved, including safety concerns when testing new therapies in human clinical trials, questions about using gene therapy to enhance traits not related to disease, animal welfare in laboratory testing, and the cost of treatment. As technology advances, engineers, scientists, policy makers, and the public must continue to consider the extent to which they wish to progress human gene transfer research. Only by combining all perspectives can we help human gene transfer advance safely and ethically toward fulfilling its promise.

1.5.3.5 Microsurgical Assistant Microsurgical procedures manipulate extremely small and delicate structures with hand-held surgical tools under stereo-microscopic observation. These procedures require movements on the micron scale and are limited by human sensory-motor skills. During vitreoretinal surgery, a stereo-operating microscope placed in the eye gives visual feedback to the surgeon for guidance on manipulating instruments to achieve very delicate retinal tissue manipulation. Even the slightest errors can cause permanent damage leading to blindness. The Johns Hopkins Microsurgery Advanced Design Lab has demonstrated that the limiting factors of vitreoretinal surgical performance are hand tremor and drift, lack of tactile sensation between the surgical instrument and retinal tissue, and the low-resolution view of the retina through the pupil.[12]

To address these limitations, the Johns Hopkins research group has developed the Microsurgical Assistant, a surgical tool that enhances the information available from the surgical environment and augments the surgeon's ability to manipulate, position, and sense within that environment. Two systems constitute the Microsurgical Assistant: the Information Enhanced Surgery System and the Steady-Hand Augmentation System. The Information Enhanced Surgery System gathers diagnostic information not normally available to the surgeon, such as blood flow rates, temperature, subretinal structure, tissue differentiation, and biomechanical properties. With this extra information gathered in real time, the surgeon can determine the best surgical approach and adjust at a moment's notice. The Steady-Hand Augmentation System (Figure 1.3) uses a robotic platform to increase the surgeon's accuracy and positioning of the instruments on the retinal tissue. As the surgeon guides the surgical tool, the robot simultaneously performs the movements free of tremors, providing long-term positional stability. Because tactile sensation—critical to a surgeon's ability to sense the surgical environment—is lost when replacing the surgeon's hands with a robot, the robot is built with a system that augments tactile sensations. These features of the Steady-Hand Augmentation System allow the surgeon to execute surgical motions with greater precision than with the human hand alone.

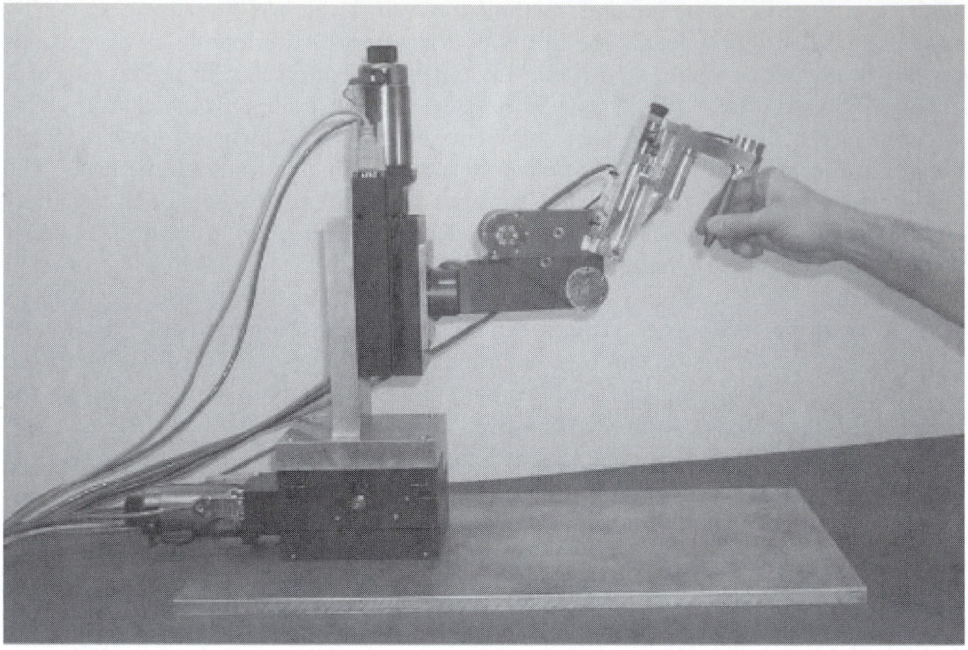

Figure 1.3
The Steady-Hand Augmentation System, the second component of the Microsurgical Assistant. The system allows the surgeon to use the robot to increase his stability and accuracy during surgery.

You are a bioengineer who is working with the Johns Hopkins group to update the software on the Microsurgical Assistant. Because of your knowledge of biological systems and physics and your teammate's knowledge of programming algorithms, you and an electrical engineer must program the robot to measure blood flow rates and the amount of pressure that body tissue can sustain.

Bioengineers must work with all types of people in teams. In addition to sharing your expertise, you must also be able to learn new skills and work collaboratively with others. In this case, if you do not already have a strong background in programming, you must be able to listen so that you can voice clear questions and answers to your teammate to effectively communicate what needs to be added to the program.

In this section we discuss strategies to attack this problem using the following concepts:

- Flow rates
- Pressure (solid)

Your teammate is having difficulty programming an algorithm to calculate the blood flow to the eye. The preliminary experimental runs of the robot output numbers outside the normal range, so you and your teammate reevaluate the assumptions associated with flow rates in the body.

Many systems studied in bioengineering, including the vascular system in the eye, involve the movement of material. The terms **rate** and **flow rate** are used to describe the transport of a physical property over a period of time. **Typically, a dot over a variable signifies a rate.** Material flow rate can be expressed as a mass flow rate ($\dot{m}$ [Mt^{-1}]), molar flow rate ($\dot{n}$ [$t^{-1}N$]), or volumetric flow rate ($\dot{V}$ [L^3t^{-1}]). The molecular weight of the material is used to convert between mass and molar flow rates, while the density of the material allows conversion between mass and volumetric flow rates.

Volumetric flow rate ($\dot{V}$) through a conduit is calculated by multiplying the cross-sectional area of the conduit, A [L^2], by the velocity of the fluid, v [Lt^{-1}]:

$$\dot{V} = Av \quad [1.5\text{-}32]$$

For a cylindrical conduit, the area is that of a circle. Substituting this in equation [1.5-32] gives:

$$\dot{V} = \frac{\pi}{4}D^2 v \quad [1.5\text{-}33]$$

where D [L] is the vessel diameter.

The mass flow rate ($\dot{m}$) is $\dot{V}$ multiplied by the fluid density (ρ [$L^{-3}M$]):

$$\dot{m} = \dot{V}\rho = Av\rho \quad [1.5\text{-}34]$$

The molar flow rate ($\dot{n}$) through a conduit is calculated by dividing $\dot{m}$ by the molecular weight (M [MN^{-1}]) of the moving fluid:

$$\dot{n} = \frac{\dot{m}}{M} \quad [1.5\text{-}35]$$

The concept of material flow rate is illustrated by the flow of blood through a capillary in the eye. A reasonable rate of blood flow through a single capillary is about 8.5×10^{-9} mL/s. Since blood has a density of 1.056 g/mL, that volumetric flow rate corresponds to a mass flow rate of 9.0×10^{-9} g/s. That is, if the vessel is modeled as a cylindrical vessel with a certain cross-sectional area, 8.5×10^{-9} mL or 9.0×10^{-9} g of blood passes through that cross section in the vessel every second.

Flow rates (particularly volumetric flow rates) can often be measured in blood vessels using a number of tools. The laser Doppler velocimeter measures flow rate by quantifying the linear diversion of laser beams off moving platelets in blood vessels. The electromagnetic flowmeter uses a voltmeter to record the electrical potential generated between two electrodes, which is proportional to the rate of blood flow in the vessel of interest. Another apparatus to determine flow rate is the ultrasonic Doppler flowmeter, which transmits sound waves downstream along flowing blood and determines the frequency difference between a transmitted wave and a wave that has been reflected by red blood cells. Both the electromagnetic flowmeter and the ultrasonic Doppler flowmeter boast the advantages of measuring blood flow without opening the vessel and accurately recording both steady flow and rapid, pulsatile changes.

EXAMPLE 1.15 **Central Retinal Artery Flow Rates**

Problem: The Information Enhanced Surgery System gathers information about the retinal tissue prior to and during vitreoretinal surgery. Find the volumetric and mass flow rates in the central retinal artery. Model the artery as a cylindrical vessel that has a diameter of 0.3 mm and a relatively low blood velocity of 25 mm/s.

Solution:

$$\dot{V} = Av = \frac{\pi}{4}D^2 v = \frac{\pi}{4}(0.3\text{ mm})^2\left(\frac{25\text{ mm}}{\text{s}}\right)\left(\frac{1\text{ cm}^3}{10^3\text{ mm}^3}\right)\left(\frac{1\text{ mL}}{1\text{ cm}^3}\right) = 0.00177\ \frac{\text{mL}}{\text{s}}$$

$$\dot{m} = \dot{V}\rho = \left(\frac{0.00177\text{ mL}}{\text{s}}\right)\left(\frac{1.056\text{ g}}{\text{mL}}\right) = 0.00187\ \frac{\text{g}}{\text{s}}$$

■

The Steady-Hand Augmentation System allows a constant and more precise amount of force and pressure to be applied by the surgeon on the retinal tissue, reducing the risk of injury caused by limitations in human motor movements. **Pressure** (P [$L^{-1}Mt^{-2}$]) is the ratio of a force (F [LMt^{-2}]) to the area (A [L^2]) over which the force acts:

$$P = \frac{F}{A} \qquad [1.5\text{-}36]$$

Common units of pressure include N/m^2, atm, dynes/cm^2, and lb$_f$/in^2 or psi (pounds per square inch). The SI pressure unit is Pa (pascal), which is equivalent to N/m^2 or kg/m · s^2. Any substance—liquid, gas, or solid—can exert pressure on another substance or structure. Previously, pressures exerted by gases were discussed. In this section, absolute pressures exerted by solids are highlighted.

The relationship between force, area, and pressure can be explained when considering the steady-hand robot. If the robot or surgeon uses an instrument that is like a needle, the surface area of contact is minuscule. For a specified force, the resulting pressure is great. However, surgical instruments like scalpels that have larger surface areas have their forces distributed over a larger area of contact. Consequently, for the same specified force, the applied pressure is smaller.

EXAMPLE 1.16 **Surgical Instrument**

Problem: Surgeons must carefully decide which instrument can effectively and safely be used for any particular procedure. For instruments that cut tissue open, the force the instrument exerts must exceed a force of F_o. However, to avoid damage to the layers underneath the tissue, the pressure over any given area may not exceed a pressure of P_o. Discuss options for an algorithm to help a surgeon decide which instrument is safe to use.

Solution: Based on equation [1.5-36], with a fixed F_o and P_o, the only free variable is area (A):

$$A > \frac{F_o}{P_o}$$

Instruments with different shapes (e.g., rectangle, oval) are available. A listing of instruments meeting the criteria for area should be developed and presented to the surgeon. Note that this simple calculation will work only if the surgeon applies the pressure uniformly. ■

Instruments like the Microsurgical Assistant are likely to improve surgery success rates and patient safety. Engineers have the opportunity to develop new ways to prevent errors from occurring in surgery by inventing technologies such as computer-aided devices. Additionally, newer generations of current devices can be developed to correct problems. Bioengineers can contribute by designing equipment, writing software, testing equipment, and adapting current equipment for new applications. For example, while the Microsurgical Assistant is currently focused on vitreoretinal surgery, the designers hope to adapt the system for applications in neurosurgery and microvascular, spine, ear, nose, and throat, surgery. By improving current systems and applying more innovative ideas, well-designed robotic processes can compensate for imperfect human motor movements, and the knowledge of a skilled surgeon can compensate for the inevitable fallibilities of robots, allowing safe and successful surgery.

1.5.3.6 Victoria Falls After graduating with your bachelor's degree, you travel to the spectacular Victoria Falls (Figure 1.4). Located on the Zambezi River bordering Zambia and Zimbabwe, Victoria Falls is Africa's most popular tourist attraction.

Figure 1.4
Victoria Falls in Zambia and Zimbabwe.

You continue to be plagued by your scientifically oriented mind and immediately begin analyzing the possibilities of harnessing this stunning natural phenomenon's power.

In this section we discuss the following concepts:

- Rate of momentum
- Rate of kinetic energy
- Rate of potential energy
- Pressure (liquid)

About 1700 meters wide and averaging 100 meters high, Victoria Falls is the largest curtain of falling water on Earth. At the height of the flood season, an estimated 500 million liters of water per minute plummets over the edge into a deep gorge, which means the water must carry great momentum. However, the waterfall is not a single mass (chunk of water) moving at a certain velocity; rather, it is a continuous flow of mass moving at a certain velocity.

The rate of linear momentum ($\dot{\vec{p}}$) due to the movement of mass is equivalent to the rate at which mass travels ($\dot{m}$ [Mt^{-1}]) multiplied by the velocity (or momentum per unit mass) ($\vec{v}$ [Lt^{-1}]):

$$\dot{\vec{p}} = \dot{m}\vec{v} \qquad [1.5\text{-}37]$$

Since the **rate of linear momentum** is a vector quantity, a direction of momentum flow must always be specified in addition to the magnitude. The rate of linear momentum has the dimension of [MLt^{-2}]. Common units are N, lb_f, and dyne.

Analogous to $\dot{\vec{p}}$, the **rate of angular momentum** ($\dot{\vec{L}}$) is used for an object undergoing rotational motion about a point. The rate of angular momentum has the dimension of [L^2Mt^{-2}]. Common units are N · m, lb_f · ft, and dyne · cm.

EXAMPLE 1.17 Rate of Momentum Transfer

Problem: Calculate the rate (in kg · m/s^2, or N) at which momentum carried by flowing water enters the waterfall system across the entire stretch of Victoria Falls. The system, shown in Figure 1.5, is defined by only the free-falling water, from the edge of the cliff to the gorge below. Assume the falling curtain of water is 1700 m long and 10 m deep.

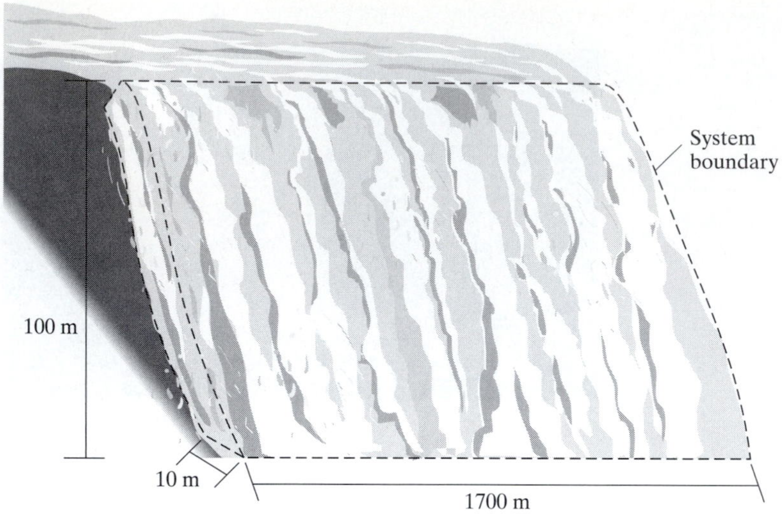

Figure 1.5
System for Victoria Falls.

Solution: To find the rate of momentum entering the waterfall, we need to calculate the mass flow rate and velocity of the water from the volumetric flow rate (500 million L/min). We convert the volumetric flow rate to units of cubic meters per second, then use the density of water (1.0 g/cm^3) to find the mass flow rate:

$$\dot{m} = \dot{V}\rho = \left(\frac{500 \times 10^6 \text{ L}}{\text{min}}\right)\left(\frac{1.0 \text{ g}}{\text{cm}^3}\right)\left(\frac{1000 \text{ cm}^3}{1 \text{ L}}\right)\left(\frac{1 \text{ min}}{60 \text{ s}}\right)\left(\frac{1 \text{ kg}}{1000 \text{ g}}\right) = 8.33 \times 10^6 \frac{\text{kg}}{\text{s}}$$

If we model the blanket of water falling off the cliff as a rectangle with length 1700 m and width 10 m, the cross-sectional area of the waterfall is:

$$A = lw = (1700 \text{ m})(10 \text{ m}) = 17{,}000 \text{ m}^2$$

Thus, the water velocity and rate of momentum are:

$$\vec{v} = \frac{\dot{V}}{A} = \frac{\left(\frac{8.33 \times 10^3 \text{ m}^3}{\text{s}}\right)}{17{,}000 \text{ m}^2} = 0.490 \frac{\text{m}}{\text{s}} \text{ downward}$$

$$\dot{\vec{p}} = \dot{m}\vec{v} = \left(\frac{8.33 \times 10^6 \text{ kg}}{\text{s}}\right)\left(\frac{0.490 \text{ m}}{\text{s}}\right) = 4.08 \times 10^6 \frac{\text{kg} \cdot \text{m}}{\text{s}^2} \text{ downward}$$ ■

Just as momentum flows into the system, kinetic energy must also move at a rate proportional to the mass flow rate and velocity of the water. The relationship between kinetic energy and rate of kinetic energy is similar to that between momentum and rate of momentum. The **rate of kinetic energy** can be formulated for any system with a continuous movement of mass. If a fluid moves with a mass flow rate $\dot{m}$ and uniform velocity v, then the rate at which kinetic energy ($\dot{E}_K$ [L^{2}Mt^{-3}]) is transported is:

$$\dot{E}_K = \frac{1}{2}\dot{m}v^2 \qquad [1.5\text{-}38]$$

Like kinetic energy, the rate of kinetic energy is a scalar quantity. The SI unit for rate of kinetic energy is the watt (W), or J/s. The rate of energy is also known as **power**.

EXAMPLE 1.18 Rate of Kinetic Energy

Problem: Find the rate at which kinetic energy enters the system due to flowing water across the entire stretch of Victoria Falls.

Solution: Recall from Example 1.17 that water flows with a mass flow rate of 8.33×10^6 kg/s and a velocity of 0.490 m/s at the top of Victoria Falls:

$$\dot{E}_K = \frac{1}{2}\dot{m}v^2 = \frac{1}{2}\left(\frac{8.33 \times 10^6 \text{ kg}}{\text{s}}\right)\left(\frac{0.490 \text{ m}}{\text{s}}\right)^2 = 1.0 \times 10^6 \frac{\text{J}}{\text{s}} = 1.0 \times 10^6 \text{ W}$$ ■

As with kinetic energy, potential and electrical energy can enter or leave a system with a flowing material. For a single position in space, knowing the **rate of potential energy** ($\dot{E}_P$) is rarely of practical importance. In contrast, the change in the rate of potential energy when a body or fluid moves from one elevation to another, as in a waterfall, is of interest. The rate of potential energy change is described by:

$$\dot{E}_{P,2} - \dot{E}_{P,1} = \dot{m}g(z_2 - z_1) \qquad [1.5\text{-}39]$$

where g is the gravitational constant, z is position, and the subscripts 1 and 2, respectively, refer to the initial and final positions of the system or material of interest. Electrical potential energy can move into or out of a system at a charge flow rate or current, i. The **rate of electrical energy** ($\dot{E}_E$) is defined as the product of the current and the specific potential energy (voltage) for that current. The rate of potential energy and rate of electrical energy have the same dimension and units as the rate of kinetic energy.

EXAMPLE 1.19 Rate of Potential Energy Change

Problem: Find the rate of change of potential energy as the vast blanket of water falls from the top to the bottom of Victoria Falls.

Solution: Recall that the height of the waterfall is approximately 100 m.

$$\dot{E}_{P,2} - \dot{E}_{P,1} = \dot{m}g(z_2 - z_1) = \left(\frac{8.33 \times 10^6 \text{ kg}}{\text{s}}\right)\left(\frac{9.81 \text{ m}}{\text{s}^2}\right)(0 \text{ m} - 100 \text{ m})$$

$$= -8.17 \times 10^9 \frac{\text{J}}{\text{s}} = -8.17 \times 10^9 \text{ W}$$

The negative sign denotes that, as the water falls, potential energy is lost in the system. The rate of change of potential energy in Victoria Falls is enough to power 80 aircraft carriers or jet airliners, and such power could be harnessed to generate electricity. In comparison, the Grand Coulee Dam in Washington, which is the third largest producer of electricity in the world, generates 7×10^9 W and provides power to the Pacific Northwest.[13] ■

When the water crashes into the gorge at the base of the Falls, it exerts a huge force on the surface of the lake onto which it falls. Like solids, liquids create pressure when forces are exerted over certain areas. Pressures exerted by flowing liquids can be substantial, as when 500 million L/min of water crashes into a gorge at the bottom of a waterfall. Even when the fluid flows through a pipe or other conduit, the fluid exerts pressure on the walls of the vessel as it travels. When a fluid is still, it exerts a pressure called **hydrostatic pressure.** Fluid pressures are measured with a number of devices, including elastic-element methods, liquid-column methods (e.g., manometers), and electric measurements (e.g., strain gauges). The types of pressures and forces exerted by liquids are important in the study of the conservation of momentum.

For the past hundred years, we have been able to harness the power in falling water in streams and rivers to produce electric energy. Hydroelectric power projects provide one of the most efficient means of producing electric energy. In the early 1990s, building a dam above Victoria Falls for this purpose was proposed, but its designation as a World Heritage Site was significant in repealing the proposal.

1.6 Quantitation and Data Presentation

Bioengineers can bring increased quantitation to the fields of biology and medicine. Engineers possess strong problem-solving skills and expertise in modeling, experimental design, and design of devices and equipment. Topics ripe for bioengineering participation include those discussed in Section 1.5 and many problems throughout this book. Increased quantitation in biology and medicine may bring now-unforeseen benefits to scientific understanding and capability and new medical breakthroughs.

In biology and medicine, compact theories of the sort familiar in physics are rare. Rather, explanations of phenomena are often qualitative descriptions. However, to help isolate important interactions or components that dominate the phenomena of interest, engineers often use **quantitative models**. For example, researchers are working on developing models of certain cellular functions (e.g., signal transduction) that involve many different interacting molecules. Quantitative models can help us clarify and understand the complex mechanisms that coordinate the timing of signaling and the cascading reactions that regulate dynamic cell function. For engineers, quantitative models establish a basis for predicting current activity and variation within the system and for understanding the impact of some perturbation on the system.

Developing models to describe the complicated nature of biology and medicine is a critical role for bioengineers. **Mathematical models** are often used to represent biological and physical phenomena. Two general classifications of mathematical models are mechanistic and empirical. **Mechanistic models** are based on theoretical assessment of the phenomenon being measured. When mechanistic models are not available, **empirical models** are developed based on using experimental or computational data to describe complex systems. Both models can accurately account for and predict experimental data.

The quantitative nature of an engineering approach requires appropriate understanding and application of statistical methods (see Schork and Remington, *Statistics with Applications to the Biological and Health Sciences*, 2000). Before an engineer can conclude that a set of data is robust and reliable, the measurements must be carefully examined for possible sources of error. Measurements may contain two types of experimental error: systematic and random. **Systematic error** affects all measurements of the same variable in the same way. For example, an improperly calibrated thermocouple may consistently read values above the actual temperature. If discovered, systematic error may be accounted for in data analysis. On the other hand, **random errors** are due to unknown causes and are present in almost all data. For example, the data generated by using a properly calibrated thermocouple to repeatedly measure the temperature of water in a beaker will contain random error. In this case, the error is manifest as a data set with values that are clustered together, but not exactly the same.

In reporting scientific and engineering data, it is of primary importance to understand the distinction between precision and accuracy. **Precision** refers to the degree of agreement among individual measurements of the same quantity within a set of measurements. In other words, precision is a measure of the repeatability of the measurements. A precise measuring instrument gives very nearly the same value each time it is used to measure the same condition.

Accuracy is a measure of reliability of measurements and indicates the difference between the true value and the measured value of a certain quantity. For example, a balance should read 100 grams if a standard 100-g mass is placed on it. If it does not read 100 grams, then the balance is inaccurate. If a certain data set has high precision but poor accuracy, one may suspect that a systematic bias has been introduced. An example of a systematic error is using an instrument where the zero position is improperly set. In research, you do not often "know" the correct accurate answer, and thus these types of biases are difficult to detect.

Measurements containing random errors but not systematic errors can be analyzed using statistical procedures. Many statistical methods, including sample descriptions, inferences about populations, regression, correlations, and analysis of variance, are available to an educated user. Two customary descriptors of experimental data are arithmetic mean and standard deviation, which can be calculated from a sample of measured data when the data are normally distributed. The **arithmetic mean** ($\bar{x}$) for a variable x measured n times is calculated:

$$\bar{x} = \frac{\sum_{i}^{n} x_i}{n} = \frac{x_1 + x_2 + x_3 + \cdots + x_n}{n} \qquad [1.6\text{-}1]$$

The arithmetic mean is a measure of central tendency of the data set and is often called the average.

The **standard deviation** gives information about the precision of the measurements. For a set of experimental data, the **sample standard deviation** (σ) is calculated:

$$\sigma = \sqrt{\frac{\sum_{i}^{n}\left(x_i - \bar{x}\right)^2}{n - 1}} = \sqrt{\frac{(x_1 - \bar{x})^2 + (x_2 - \bar{x})^2 + \cdots + (x_n - \bar{x})^2}{n - 1}} \qquad [1.6\text{-}2]$$

The standard deviation is a measure of the variation or scatter in the data.

To report results of a repeated measurement, the mean is often given as the best estimate of an experimental variable and the standard deviation as an indication of the variability of the result. The ratio of the standard deviation to the mean gives some indication of the random error in the data. Larger ratio values of the standard deviation to the mean indicate a greater degree of variability in the data, while smaller ratio values indicate a lesser degree of variability. More information on these methods and others can be found in statistics textbooks (e.g., Schork and Remington).

EXAMPLE 1.20 **Plasma Drug Concentration**

Problem: A drug for treating Parkinson's disease is undergoing human clinical trials. A certain Parkinson's patient is administered a 5 mg oral dose of the drug each day for six consecutive days. His plasma drug concentration is measured at regular intervals after each dose. The plasma concentration of the drug one hour after administration is measured on the six days as follows (in mg/L): 0.206, 0.214, 0.211, 0.209, 0.213, 0.205. Determine the mean and standard deviation of the plasma drug concentration.

Solution: The mean of the plasma drug concentration values is calculated from six measurements:

$$\bar{x} = \frac{\sum_{i}^{n} x_i}{n} = \frac{0.206 + 0.214 + 0.211 + 0.209 + 0.213 + 0.205}{6} = 0.210\ \frac{\text{mg}}{\text{L}}$$

The standard deviation is:

$$\sigma = \sqrt{\frac{\sum_{i}^{n}\left(x_i - \bar{x}\right)^2}{n - 1}}$$

$$= \sqrt{\frac{(0.206 - 0.210)^2 + (0.214 - 0.210)^2 + \cdots + (0.205 - 0.210)^2}{6 - 1}}$$

$$= 0.004\ \frac{\text{mg}}{\text{L}}$$

The plasma drug concentration is reported as 0.210 ± 0.004 mg/L. ■

The number of figures used to report a measured or calculated variable is an indirect indication of the precision to which that variable is known. A **significant figure** is any digit, 1–9, used to specify a number; zero may be a significant figure when it is not used merely to locate the position of the decimal point. For example, the following numbers each contain three significant figures: 321, 4.67, 601, 0.0754, and 7.50×10^6. The numbers 340, 8700, 0.0025, and 0.098 have only two significant figures, since the zeros merely serve as placeholders. Rewriting the first two numbers as 3.40×10^2 and 8.700×10^3 would indicate three and four significant figures, respectively.

The diameter of a catheter may be properly reported with three significant figures. On the other hand, an estimate of how much it will cost to develop a new life-support system on Mars is likely known to only one significant figure (if that). Just because your calculator can calculate a number to nine significant figures, it does not mean the number should be reported as such. It is important for engineers to develop a sense of the certainty for measurements and calculations, and to report these properly when presenting data and models.

Rules for rounding numerical values to the correct number of significant figures are widely accepted. A number is rounded to k significant figures using the following rules:

- If the number in the $(k + 1)$ position is less than 5, all figures to the right of the k position should be dropped.
- If the number in the $(k + 1)$ position is 5 or greater, add 1 to the number in the k position and drop all numbers to the right of the k position.

For example, when rounding to two significant figures, 4578 becomes 4600, and 1.43 becomes 1.4.

Often, values with uncertainty are used in a series of calculations with other experimental values. As a general rule, computed values should be reported only to the number of significant figures of the most uncertain value (i.e., the value with the fewest significant digits). The following is a guide:

- After multiplication or division, the number of significant figures in the result should equal the smallest number of significant figures of any quantity in the calculation.
- After addition and subtraction, the position of the last significant figure in the result should be the same as the position of the last digit in the value with the fewest digits to the right of the decimal point.

It is good practice to carry along one or two extra significant figures through your calculations and then round off at the end when the final answer is presented. As a rule of thumb, data and models associated with biological and medical systems are reported to two or three significant figures. Throughout this text, final answers are rounded to two or three significant figures in accordance with the aforementioned rule of thumb.

EXAMPLE 1.21 Correct Number of Significant Figures

Problem: Perform the calculations and report the answers to the correct number of significant figures.

(a) $(4.307 \times 10^4 \text{ kg})(6.2 \times 10^{-3} \text{ m/s}^2)$
(b) $26.127 \text{ A} + 3.9 \text{ A} + 0.0324 \text{ A}$

Solution:

(a) $(4.307 \times 10^4 \text{ kg})(6.2 \times 10^{-3} \text{ m/s}^2) = 267.034 \text{ N}$ using a calculator. The first number, 4.307×10^4 kg, has four significant figures. The second number, 6.2×10^{-3} m/s^2, has two significant figures. Therefore, the answer should be reported to two significant figures, or 270 N or 2.7×10^2 N. (Remember, 270 N has only two significant figures, since the 0 is just a placeholder.)

(b) $26.127 \text{ A} + 3.9 \text{ A} + 0.0324 \text{ A} = 30.0594 \text{ A}$ using a calculator. The first number, 26.127 A, has three digits to the right of the decimal. The second number, 3.9 A, has one digit to the right of the decimal. The third number, 0.0324 A, has four digits to the right of the decimal. Therefore, the answer should be reported to one digit to the right of the decimal, or 30.1 A. Note that 30.0594 A was rounded up to 30.1 A. ■

Learning to effectively use tables, graphs, and models to interpret and present experimental and computational data is critical to successfully demonstrating results. **Tables** report specific experimental, computational, or calculated values. However, tables can easily become too long, and an overall trend of the values may not be readily visualized. Typically, the independent variable—the variable that is controlled or fixed—is listed in the first column, and the dependent variable—the variable that is uncontrolled during the experiment and responds to changes in the independent variable—is listed in subsequent columns. Tables are also valuable when more than one dependent variable is being measured or when graphing is difficult.

Graphical representation uses graphs or plots to aid visualization of relationships between variables. By convention, the independent variable is plotted along the abscissa (x-axis) and the dependent variable(s) is plotted along the ordinate (y-axis). Graphing data helps to identify outliers and trends, and can be used for interpolation between points. A complete graph should include a descriptive title and a legend. The axes should be labeled, and the units should be indicated. Error bars may be included to indicate the variability of the data.

EXAMPLE 1.22 Oxygen Consumption of an Astronaut

Problem: An astronaut undergoes rigorous weight training to help her physically prepare for her time in microgravity during her upcoming space flight. Measurements of the astronaut's oxygen consumption are made at different levels of work output. Work output (kg·m/min) and oxygen consumption (L/min) data are recorded in pairs as follows: (100, 0.55), (1400, 3.00), (225, 0.55), (750, 1.82), (275, 0.75), (375, 0.95), (550, 1.25), (950, 2.10), (110, 0.45), (1200, 2.75), (825, 2.05), (1700, 3.75). (Adapted from Guyton and Hall, 2000.)

TABLE 1.11

Oxygen Consumption of Astronaut

Work output (kg · m/min)	Oxygen consumption (L/min)
100	0.55
110	0.45
225	0.55
275	0.75
375	0.95
550	1.25
750	1.82
825	2.05
950	2.10
1200	2.75
1400	3.00
1700	3.75

(a) Identify the independent and dependent variable.
(b) Present the data in a table.
(c) Present the data in a graph.
(d) Develop a simple model predicting oxygen consumption as a function of work output.

Solution:

(a) Having the astronaut perform at different levels controls the work output. Therefore, the work output is the independent variable. The oxygen consumption is the dependent variable, since consumption depends on the work output.
(b) Data for the astronaut at different work outputs are given in Table 1.11. Note that the data were put in order of increasing values of work output.
(c) A graph of the data is plotted in Figure 1.6. The work output is the x-axis; the oxygen consumption is the y-axis.
(d) A model of the oxygen consumption of an astronaut as a function of work output can be described with a linear relationship as follows:

$$y = \left(0.0021 \frac{\text{L}}{\text{kg} \cdot \text{m}}\right)x + 0.21 \frac{\text{L}}{\text{min}}$$

where y is the oxygen consumption (L/min) and x is the work output (kg · m/min). Note that based on the information provided, there is no physiological basis for this model; it is simply a fit of the experimental data. ■

1.7 Solving Systems of Linear Equations in MATLAB

Most of the problems presented in this text, as well as those you will encounter in the field, involve solving for one or more unknown values. While systems limited to one or two unknown variables often can be solved easily by hand, solving more complicated systems can be considerably more cumbersome. However, for systems described by linear equations, there are computational techniques that can be applied to minimize tedious calculations by hand. The computational tools described below can be applied only to solving sets of independent, linear equations.

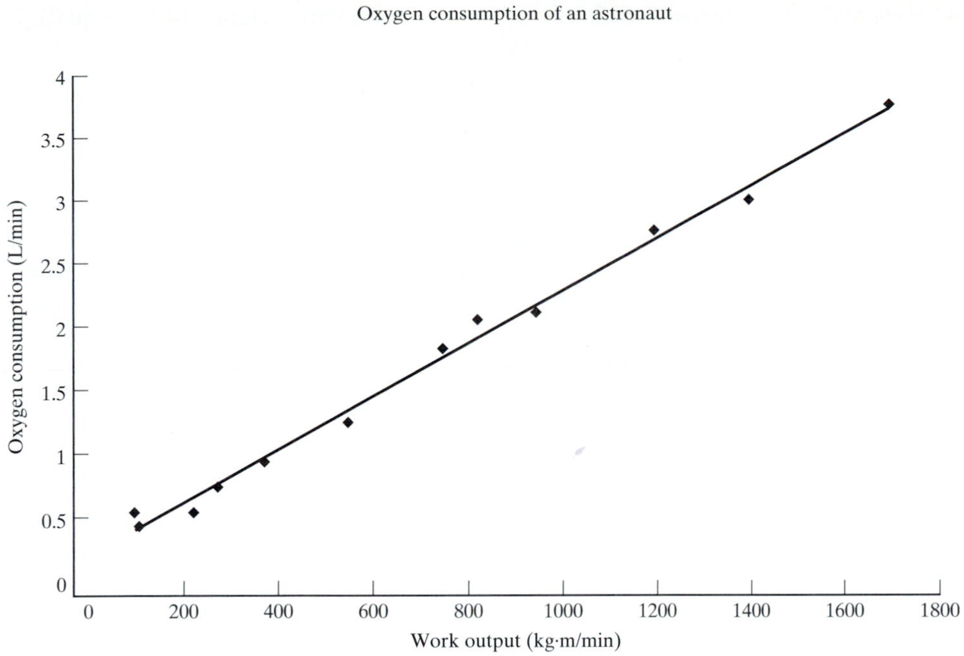

Figure 1.6 Graphical representation of the relationship between work rate done by astronaut and oxygen consumption rate.

A linear equation is an equation of unknown variables of the form:

$$Y = C_1X_1 + C_2X_2 + \cdots + C_nX_n \qquad [1.7\text{-}1]$$

where Y represents a dependent variable, X_i represents an unknown independent variable, and C_i is a constant coefficient, for n variables. If one of the terms is the product of two or more unknown variables (e.g., $C_1X_1X_2$) or if any variable is raised to a power other than one, then the equation is not linear and more complicated solving methods must be used. Likewise, trigonometric and logarithmic equations are nonlinear.

The use of computer software programs such as MATLAB makes solving systems of linear equations relatively easy, since they are designed to handle matrices and vectors. The discussion below assumes some familiarity with MATLAB. A system of linear equations can be represented by a matrix equation. Consider the following example with two linear equations and two unknown variables:

$$\begin{aligned} x_1 + 2x_2 &= 5 \\ 3x_1 + 4x_2 &= 11 \end{aligned} \qquad [1.7\text{-}2]$$

This system of equations is represented by the following matrix equation in the form $A\vec{x} = \vec{y}$ as:

$$\begin{bmatrix} 1 & 2 \\ 3 & 4 \end{bmatrix}\begin{bmatrix} x_1 \\ x_2 \end{bmatrix} = \begin{bmatrix} 5 \\ 11 \end{bmatrix} \qquad [1.7\text{-}3]$$

where A is a 2×2 matrix and $\vec{x}$ and $\vec{y}$ are vectors.

Such a matrix equation is analogous to the following scalar equation:

$$ax = y \qquad [1.7\text{-}4]$$

where a and y are known quantities and x is the unknown variable. In this equation, it is easy to solve for x by simple division:

$$x = \frac{y}{a} = a^{-1}y \qquad [1.7\text{-}5]$$

Although it is desirable to perform an analogous operation on the matrix equation, we cannot simply divide a vector by a matrix. However, the scalar a^{-1} does have a matrix equivalent, defined as the inverse of A or A^{-1}. Thus, the matrix equation can be solved by finding A^{-1} and using it to calculate $\vec{x}$ with basic matrix multiplication. Calculating A^{-1} by hand can be tedious; MATLAB performs this operation quickly.

In MATLAB, all variables, both single values and vectors, are designated by letters. Consequently, an x in MATLAB may represent a single value or it may be the entire vector $\vec{x}$. Therefore, sample MATLAB commands given here do not show any vector arrows. The backward slash (\) is an operator defined by MATLAB for the purpose of solving equations using the inverse matrix. Typing the commandfor example "$x = A\backslash y$" is equivalent to asking the computer to calculate $\vec{x} = A^{-1}\vec{y}$. For example, the matrix A and the vector y are defined:

```
>> A = [1 2; 3 4];
>> y = [5; 11];                [1.7-6]
```

The operation to perform matrix inversion followed by vector multiplication is:

```
>> x = A\y                     [1.7-7]
```

Spaces or commas inside the brackets are used to separate terms within a row; semicolons within the brackets separate rows from one another. When a semicolon is present at the end of a line of code, the program executes the command internally. If there is no semicolon, then MATLAB displays the computed answer on the terminal. By omitting the semicolon in the last line, the program shows the solution for $\vec{x}$, which is:

```
x =
      1
      2                        [1.7-8]
```

The answer can be checked by back-substitution of these values ($x_1 = 1$ and $x_2 = 2$) into the initial system of equations.

EXAMPLE 1.23 Using MATLAB to Solve Three Linear Equations

Problem: Solve the following system of equations:

$$\begin{aligned} x_1 + x_2 + x_3 &= 2 \\ 2x_1 + x_3 &= 4 \\ x_1 + 2x_2 + x_3 &= 1 \end{aligned}$$

Solution: This problem can be solved in the same manner as the two-variable example worked out above. Since the variable x_2 does not appear in the second equation, the coefficient for x_2 in that equation is zero. The corresponding matrix A and vector y are set up:

$$A = \begin{bmatrix} 1 & 1 & 1 \\ 2 & 0 & 1 \\ 1 & 2 & 1 \end{bmatrix}$$

$$y = \begin{bmatrix} 2 \\ 4 \\ 1 \end{bmatrix}$$

Solving with MATLAB gives:

$$
\begin{aligned}
&\gg A = [1\ 1\ 1;\ 2\ 0\ 1;\ 1\ 2\ 1];\\
&\gg y = [2;\ 4;\ 1];\\
&\gg x = A\backslash y\\
&\quad x =\\
&\qquad\quad 1\\
&\qquad -1\\
&\qquad\quad 2
\end{aligned}
$$

Thus, $x_1 = 1, x_2 = -1$, and $x_3 = 2$. ■

1.8 Methodology for Solving Engineering Problems

Developing a pattern or methodology for solving engineering problems is important for consistency and thoroughness. The application of accounting and/ conservation equations (discussed in Chapters 2–7) should be carried out in an organized manner; this makes the solution easy to follow, check, and be used by others. As a new engineer, you may find going through these many steps tedious and excessive for seemingly simple problems. However, when the level of difficulty increases, having a method or process to fall back on will be invaluable. Experienced engineers use most of the steps below when solving real-world problems.

The methodology laid out here or one similar to it should be used to solve problems throughout your bioengineering career. The method outlined below is a general guide of the steps that will be followed to solve problems in Chapters 3–7 of this book, and as in real-world problems, only steps applicable to the problem should be performed. Other methodologies for solving problems are also valid; **the critical issue is that you develop a thorough method and implement it regularly.** As you mature as an engineer, it is appropriate that you develop your own problem-solving method.

1. **Assemble.** Information regarding the problem, including a picture, should be assembled and rewritten.
   (a) The objective of the problem or the answer that you are seeking *to find* should be clearly stated. This is often written as: Find: the flow rate . . .
   (b) Draw a *diagram* showing all relevant information. Often a simple box diagram showing all components entering and leaving the system allows information to be summarized in a convenient way. The system, surroundings, and system boundary should be drawn and labeled. When possible, all known quantitative information should be shown on the diagram.
   (c) Set up a calculation *table*. The known values on your diagram, the components entering and leaving the system, form the foundation for the table. Units should be consistent across the table. The unknown components on the table (blanks) are often the desired answers. As you solve for different components, you can fill in the table. (Developing a table is optional, although useful, especially in multicomponent mass balance problems)
2. **Analyze.** A framework for understanding what is known and what is not known is developed at this stage.
   (a) State any *assumptions* applied to the problem. Biological systems are extremely complex, since many processes and reactions, as well as transport of materials, are often going on simultaneously. Knowing when and where

to make assumptions to simplify the system to a few salient features is the mark of an outstanding engineer. An example of an assumption is that the human forearm can be modeled as a cylinder.

(b) Collect and state any *extra data*. In this step, you may need to research information about a component in your system that was not given in the problem definition. An example of extra data that you may need to look up is the viscosity of plasma.

(c) List the *variables* and *notations* (symbols) adapted for the problem, and select a set of *units* for the problem. Typically, a choice is made to use either metric or British units throughout the problem.

[*Note*: The notations for variables may vary from discipline to discipline, but standard variables that are inherently understood within a subject do not always need to be listed. For example, kinetic energy is defined as E_K in physics, but as T in mechanics. If you are calculating data that will be shared with coworkers of the same discipline, it is not necessary to define standard variables. Rather, variables specific to an aspect of the problem need to be defined (e.g., F_s = force of the astronaut on a chair).]

(d) State a *basis* of calculation. A basis is a specified input or output to a system (usually given as a flow rate or amount). In some problem statements, the basis is given. In other problems, values of components are given relative to one another and not as absolute amounts or rates. Select a basis if one is not given. Mass (Chapter 3) and energy (Chapter 4) problems often need a basis.

(e) If the problem involves chemical *reaction(s)*, list the compounds involved and stoichiometrically balance the equation(s).

3. **Calculate.** Equations are developed and solved in a logical manner.

(a) Write down all appropriate accounting and/or conservation *equations*. Writing down the governing equations and then simplifying them by analyzing the system to eliminate unnecessary terms can be a helpful tool in solving engineering problems. For example, if asking the question, "Is this system at steady-state?" results in a positive response, the governing equation may be simplified by making the Accumulation term equal zero. This concept will be discussed further in Chapter 2. Write down any other essential equations needed to solve the problem.

(b) By applying the appropriate equations, *calculate* the unknown quantities. This is the heart of solving the problem and may require extensive effort. In some cases, the calculation of unknown quantities can be done sequentially. In other cases, it may be best to solve a series of equations using MATLAB or other computer software.

[*Note*: For a multi-unit or complex mass and energy conservation problems, a strategy to solve the problem may be required. A *degree-of-freedom analysis* is a systematic method to demonstrate whether a problem can be solved with the stated information and can help determine the sequence in which the equations should be solved. This process is discussed in chemical engineering textbooks (e.g., Reklaitis, *Introduction to Material and Energy Balances*, 1983).]

4. **Finalize.** Correct answers to the problem statement are stated clearly.

(a) *State the answers clearly* with appropriate significant figures and units. Confirm that you answered the specific questions asked by the problem statement.

(b) Check that your results are *reasonable* and make sense. Three methods to validate a quantitative problem include:
   i. *Back-substitution*: Substitute your solution back into the initial equations and make sure that it works.
   ii. *Order-of-magnitude estimation*: Develop a crude and simple-to-solve approximation of the answer and make sure the more exact solution is reasonably close to it.
   iii. *Test of reasonableness*: Applying a test of reasonableness means verifying that the solution makes sense (e.g., the power needed to operate a pacemaker should be less than that required to operate the facilities at your university).

Beginning engineers may find the last two validation methods difficult, but you will improve with time and practice.

Summary

In this chapter, we defined physical variables, units, and dimensions and showed how to use dimensional analysis and unit conversion. We elaborated on the physical variables in the context of complex engineering applications.

We also discussed why quantitation is important in bioengineering and how to effectively present the quantities and data obtained through experiments and calculations. We demonstrated how MATLAB can be used to solve for unknown variables in a system of linear equations. Finally, we outlined a methodology for solving engineering problems, which is used in solving many problems in the remainder of this book.

References

1. Mars Climate Orbiter Mishap Investigation Board. "Phase 1 report." November 10, 1999:16. ftp://ftp.hq.nasa.gov/pub/pao/reports/1999/MCO_report.pdf (accessed June 24, 2005).
2. Jet Propulsion Laboratory Media Relations Office. California Institute of Technology. "Mars Climate Orbiter mission status." September 24, 1999. http://mars.jpl.nasa.gov/msp98/news/mco990924.html (accessed June 24, 2005).
3. National Parkinson Foundation. "National Parkinson Foundation." http://www.parkinson.org/ (accessed June 24, 2005).
4. Guyton AC and Hall JE. *Textbook of Medical Physiology.* Philadelphia: Saunders, 2000.
5. Miller G. "Drug targeting. Breaking down barriers." *Science* 2002, 297:1116–8.
6. National Institute of Neurological Disorders and Stroke. "The mucopolysaccharidoses: Therapeutic strategies for the central nervous system." September 22, 2004. http://www.ninds.nih.gov/news_and_events/proceedings/mps_2003.htm (accessed June 24, 2005).
7. Beers MH and Berkow R, eds. *The Merck Manual of Diagnosis and Therapy*. Whitehouse Station, NJ: Merck Research Laboratories, 1999.
8. Alberts B, Johnson A, Lewis J, et al. *Molecular Biology of the Cell.* New York: Garland Science, 2002.
9. NASA. "Mars Pathfinder science results: Atmospheric and meteorological properties." http://mpfwww.jpl.nasa.gov/MPF/science/atmospheric.html (accessed June 24, 2005).
10. Schultz J. NASA. "Vascular health in space." http://weboflife.ksc.nasa.gov/currentResearch/currentResearchFlight/vascular.htm (accessed June 24, 2005).
11. Robbins PD and Ghivizzani SC. "Viral vectors for gene therapy." *Pharmacol Ther* 1998, 80:35–47.
12. Jensen P. "Engineered system family #1: A microsurgical assistant for the augmentation of surgical perception and performance." Center for Computer Integrated Surgical Systems and Technology, The Johns Hopkins University. http://cisstweb.cs.jhu.edu/research/MicrosurgicalAssistant/ (accessed December 29, 2004).
13. Federal Energy Regulation Commission. http://www.ferc.gov.

Problems

1.1 Unit conversion.
(a) Convert 10 $lb_m \cdot ft/s^2$ to units of lb_f and dynes.
(b) Convert 20 kPa to units of atm and lb_f/in^2.
(c) Convert 70°F (room temperature) to units of °C and K.
(d) Convert 100 $in^2 \cdot lb_m/s^2$ to units of joules and cal.
(e) If the mass of your roommate is 150 lb_m, what is her weight (lb_f)? If the mass of your father is 70 kg, what is his weight (N)?

1.2 Unit conversion.
(a) Convert 10,000 dynes to units of $lb_m \cdot ft/s^2$ and lb_f.
(b) Convert 0.2 atm to units of kPa and lb_f/in^2.
(c) Convert 37°C (physiological temperature) to units of °F and K.
(d) Convert 50 $in^2 \cdot lb_m/s^2$ to units of joules and cal.

1.3 An 11-lb_m ball is accelerated at 3.4 ft/s^2. Determine the force on the ball in units of lb_f.

TABLE 1.12

Approximate Displaced Volumes of Body Parts in Water

Body part	Volume (in^3)
Head	400
Torso	2000
Arm	350
Leg	700

1.4 Calculate the force and pressure acting across the pelvis and at the feet of a 150-lb_m person. Model the body using cylinders. Neglect external pressures (such as air pressure). Assume that the individual distributes his/her weight equally between two legs. Assume that the cross-sectional area of the foot is approximately that of the leg. Assume that the cross-sectional area of the pelvis is approximately that of the trunk of the body. Several additional assumptions are needed; state them clearly. In your model, is the pressure higher at the pelvis or the feet? Is this consistent with what you expected?

1.5 What is the force (N and lb_f) on a 20.0-kg mass under normal gravity? What is the force (N and lb_f) on a 20.0-lb_m mass under normal gravity?

1.6 According to Archimedes' principle, the mass of a floating object equals the mass of the fluid displaced by the object. A 150-lb_m swimmer is floating in a nearby pool; 95% of her body's volume is in the water while 5% of her body's volume is above water. Determine the density of the swimmer's body. The density of water is 0.036 lb_m/in^3. Does your answer make sense? Why or why not?

1.7 To be a successful water polo player, one needs to have his head, arms, and part of the upper torso out of the water. Since floating alone won't work, players tread water. How much force is the swimmer exerting to keep his head, part of his torso, and his arms above the surface of the water? Use a free-body diagram and a force balance to solve this problem. The buoyant force acts in the direction opposite to gravity and is known to be equal to the weight of the fluid displaced by the object. Use Table 1.12 to estimate the volume of the player that is in the water.

1.8 Liposomes are promising molecules in gene transfer technology because of their similarity to cell membranes and their charge interactions with negatively charged DNA. Estimate the charge on the outside surface of a liposome of diameter 1 μm. Suppose a typical phospholipid head has a diameter of 1 nm and carries a charge of one proton (1.6021×10^{-19} C). Assume that the surface of the spherical liposome is comprised of phospholipid heads packed together as tightly as possible in a configuration known as hexagonal packing, which minimizes the inevitable vacant space that occurs when circles pack against one another (Figure 1.7).

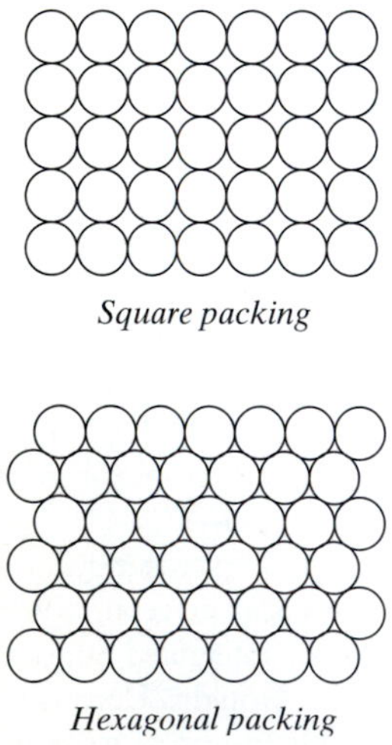

Figure 1.7
Square and hexagonal packing of phospholipid heads.

1.9 Surfactant, a complex mixture of phospholipids, proteins, and ions, plays an important role in decreasing the surface tension of water on the alveolar surface. If surfactant is not present or is present but in less than normal quantities, then the attraction of water molecules for each other (and hence the surface tension) increases. An increased surface tension leads to an increased pressure in the alveoli that can lead to their collapse. The surface tension of normal fluids that line the alveoli with normal amounts of surfactant is 5–30 dyne/cm. The surface tension of normal fluids that line the alveoli without surfactant is 50 dyne/cm. Surface tension is related to the pressure, P, as follows:

$$P = \frac{2\sigma}{r}$$

where σ is the surface tension and r is the radius of the alveolus. Report all answers in units of mmHg.

(a) If the average-sized alveolus has a radius of about 100 μm, what is the surface-tension pressure for an adult when surfactant is present?

(b) What is the pressure for an adult with average-sized alveoli but without surfactant?

(c) Premature babies usually have alveoli with radii one-quarter the size of those of normal adults. In addition, since surfactant does not usually begin to be secreted into the alveoli until the sixth month of gestation, premature babies usually do not have surfactant. Estimate surface-tension pressure for a premature baby.

1.10 You are handed an apple that has a mass of 102 g. To get a better of idea of what different pressures feel like, you rig up several systems where the force of the apple is distributed over objects of different size.

(a) What is the weight of this apple (in units of N and lb_f)?

(b) You stick a square-ended toothpick into the apple and balance it on your finger. What is the pressure of the apple (in units of Pa, psi, and atm) that you feel on your finger?

(c) You cut the apple up into slices and place them on your hand. What is the pressure of the apple on your hand (in units of Pa, psi, and atm)?

(d) You smash the apple into applesauce and spread it over a table. What is the pressure of the applesauce on the table (in units of Pa, psi, and atm)?

1.11 For each of the four tanks of liquid in Figure 1.8, calculate the pressure at the base of each tank. For Fig 1.8(a) and Fig 1.8(b), specify the relationship between P_A and P_B (e.g., $P_A > P_B$, $P_A < P_B$, or $P_A = P_B$). The specific gravity of H_2SO_4 is 1.834.

1.12 Air in the atmosphere is composed primarily of nitrogen, oxygen, and argon. The mole percents of the compounds are as follows: 78% N_2, 21% O_2, 1% Ar. Calculate the mass percents of nitrogen, oxygen, and argon in atmospheric air.

1.13 An alloy used in an artificial hip contains 17 g of Ni, 23 g of Cr, and 40 g of O. Calculate the mole fractions and mass fractions of each element in the alloy. Also, calculate the average molecular weight of the alloy.

1.14 Ti-6Al-4V is a metal alloy used to make biomaterials. Its composition is 90% Ti, 6% Al, and 4% V (mass percents). What are the mass fractions of Ti, Al, and V? What are the mole fractions of Ti, Al, and V? Calculate the average molecular weight of the alloy.

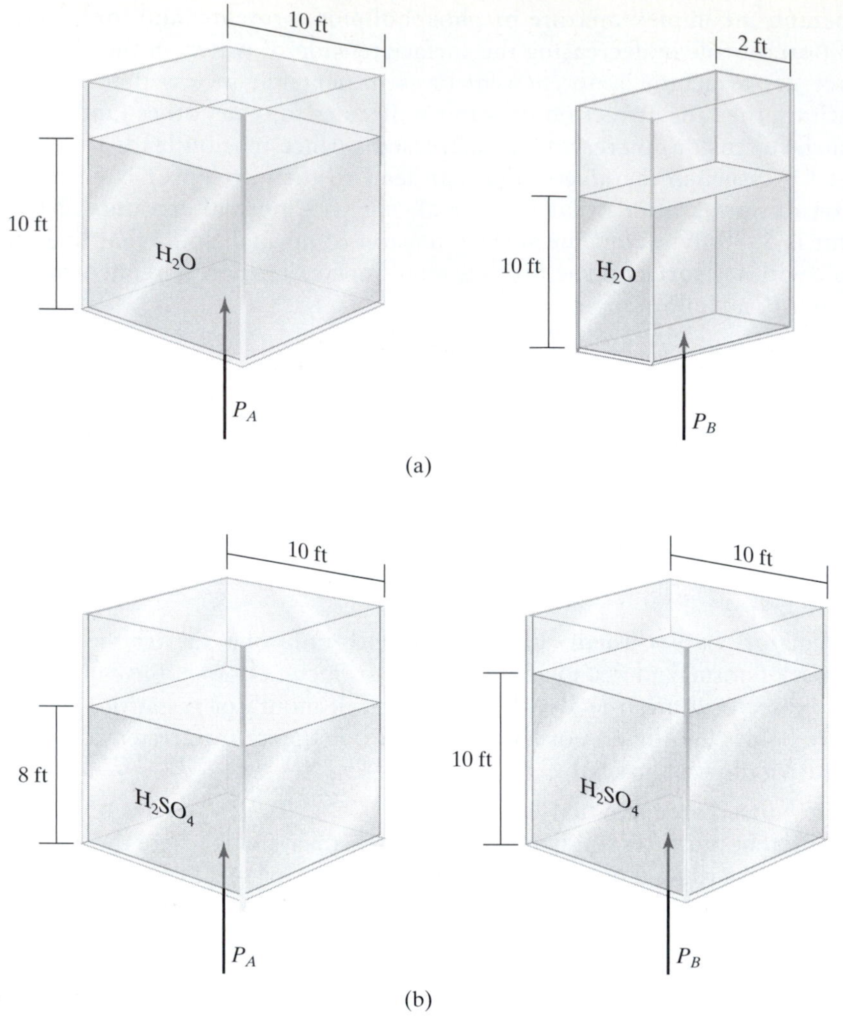

Figure 1.8
Tanks containing (a) water and (b) sulfuric acid.

1.15 A new alloy used for construction of artificial hips is $Co_{20}Cr_{10}Mo$. Calculate the mole fractions and mass fractions of each element in the alloy. Also, calculate the average molecular weight of the alloy.

1.16 The trachea has a diameter of 18 mm; air flows through it at a linear velocity of 80 cm/s. Each small bronchus has a diameter of 1.3 mm; air flows through the small bronchi at a linear velocity of 15 cm/s. Calculate the volumetric flow rate, mass flow rate, and molar flow rate of air through each of these regions of the respiratory system. Also, calculate the Reynolds number for each compartment, given the formula:

$$\text{Re} = \frac{Dv\rho}{\mu}$$

where D is diameter, v is linear velocity, ρ is density, and μ is viscosity. The viscosity of air is 1.84×10^{-4} g/(cm · s).

1.17 In the human circulatory system, large vessels split into two (bifurcate) or more smaller vessels in progression from the aorta to the arterioles and finally the capillaries. In returning blood to the heart, the capillaries join to form venules, and then finally the venae cavae. The diameter of each type of vessel and the blood velocity are given in the Appendix (Table D.9).

(a) Calculate the volumetric flow rate and mass flow rate of blood through each of these regions of the body. (Show your calculations for at least one of the structures. You may use Excel, MATLAB, or another program for the other calculations.)

(b) Can you calculate or estimate a molar flow rate for blood in these different regions? Why or why not?

(c) Calculate the Reynolds number for each of the regions. The density of blood is 1.056 g/mL and the viscosity of blood is 0.040 g/(cm · s).

1.18 Approximate the linear gas velocity in the trachea during normal exhalation. Do this by timing exhalations, measuring volumes of exhaled gas, and looking up or estimating the inner diameter of the trachea. A balloon or paper or plastic bag may be helpful to measure the volume of exhaled gas. Do more than one measurement and compute an average and standard deviation. Repeat the estimate for forced exhalation. Describe the process that you used to make the calculations and estimates. List three potential sources of error in your measurements. Compare your experimental gas velocity in the trachea during normal exhalation with the linear velocity given in Problem 1.16.

1.19 A table showing the composition of the human body is in the Appendix (Table D.4). Calculate the mass fraction, mole fraction, and concentration (mol/L) of each constituent. The molecular weights of fat, protein, and carbohydrate are estimated as 450 g/mol, 60,000 g/mol, and 350 g/mol, respectively. State any assumptions you need to complete the calculations. (You may use Excel, MATLAB, or another program of your choice.)

1.20 You are to prepare a 2.0-mL sample of a diluted drug for injection. The total amount of drug to be injected in this 2.0-mL injection is 0.0210 mg drug/(kg body mass). The patient's body mass is 70.0 kg. The label tells you that the volume of solution in the drug bottle is 30.0 mL. The total mass of drug in the bottle is 294 mg, and the rest is saline (salt solution). In addition to this bottle of concentrated drug, you have an unlimited supply of pure, sterile saline.

(a) What is the concentration (in mg/mL) of drug in the bottle?

(b) What volume of concentrated drug (in mL) and what volume of saline (in mL) will you combine to make 2.0 mL of the drug solution of the necessary concentration?

(c) The molecular weight of the drug is 15,000 g/mol. What is the molarity of the injection?

1.21 A 40-year-old man comes into the hospital complaining of fever, cough, chills, and malaise. Subsequently, he is diagnosed with pneumonia. You decide to treat him with Antibiotic X. Initially you dose him with 5882 mg. In this patient, antibiotic X has a volume of distribution (V_d) of 10 L. The volume of distribution is the volume of blood and plasma in which the drug distributes. The clearance rate (C_L) of the drug is 0.1 L/min. The clearance rate is the volumetric rate of elimination of the drug in the volume of distribution. Antibiotic X has a bioavailability of 85% (i.e., 15% of the drug is not available to be used by the body).

(a) To prepare an injection, you dilute the initial drug dose of 5882 mg in 5 mL of water. What is the drug concentration in mol/L? The molecular weight of Antibiotic X is 372 g/mol.

(b) What is the effective concentration in mg/L of this drug in the body after dosing?
(c) At what rate is the antibiotic eliminated from the body in mg/min?
(d) Typically, after the initial dose, a multiple-day course of Antibiotic X is given to treat pneumonia. How do you think that the clearance rate of the drug would be affected if the patient was an alcoholic? [*Hint*: What organ(s) degrades drugs and other toxins?]

1.22 A laboratory gas bottle used in laboratory studies of hypoxia (low O_2 levels) contains gas of the following composition, expressed as vol%: O_2, 18.0%; N_2, 80.0%; CO_2, 2.0%.
(a) Calculate the partial pressure of each of the gas components. Assume atmospheric pressure (760 mmHg).
(b) Assume that the gas is bottled in a 2.0-L rigid vessel. The temperature in the lab is 23°C. The gauge on the gas bottle reads 1500 psig. How many moles of gas are in the bottle? How many moles of each component are in the gas?
(c) In the pulmonary function lab, a patient exhales a quantity of gas. The technician reports that the volume of gas, measured at 752 mmHg and 22°C, is 1.5 L. What volume would this gas occupy at STP? What volume would it occupy at BTP? Do these answers make sense? Why?

1.23 Laboratory studies and clinical applications require air enriched with high oxygen concentrations. For example, a baby with a compromised heart may need air with a higher oxygen content than regular air to perfuse her entire body adequately. Gas is available at the following composition, expressed as vol%: O_2, 25.0%; N_2, 73.0%; CO_2, 2.0%.
(a) For laboratory studies, the gas is bottled at a pressure of 400 kPa. Calculate the partial pressure of each of the gas components.
(b) Assume that the gas is bottled in a 2.0-L rigid vessel. The temperature in the lab is 22°C. The gauge on the gas bottle reads 1200 psig. How many moles of gas are in the bottle? How many moles of each component are in the gas?
(c) For clinical applications, assume that the gas bottle is at atmospheric temperature and pressure. The dry air must be heated to biological temperature and "wetted" with water to increase its humidity. How will the volume change when the air is heated? If the volume changes, by what percent does it change?

1.24 The concentration of oxygen dissolved in both arterial and venous blood can be determined using Henry's law:

$$P_i = H_i C_i$$

where P_i is the partial pressure of constituent i, H_i is the Henry's law constant for constituent i, and C_i is the dissolved concentration of constituent i. The partial pressure of oxygen is 95 mmHg in the artery and 40 mmHg in the vein. The Henry's law constant of oxygen is 0.74 mmHg/μM. Determine the concentration at arterial and venous condition.

1.25 Sucrose concentration in a fermentation broth is measured using HPLC (high-performance liquid chromatography) (Doran, *Bioprocess Engineering Principles*, 1999). Chromatogram peak areas are measured for five standard sucrose solutions to calibrate the instrument. Measurements are performed in triplicate, with results given in Table 1.13.

TABLE 1.13

Chromatogram Peak Areas of Sucrose Concentrations

Sucrose concentration (g/L)	Peak area
6.0	55.55, 57.01, 57.95
12.0	110.66, 114.76, 113.05
18.0	168.90, 169.44, 173.55
24.0	233.66, 233.89, 230.67
30.0	300.45, 304.56, 301.11

(a) Determine the mean peak area and standard deviation for each sucrose concentration.
(b) Plot the mean peak areas as a function of sucrose concentration. (You may use Excel, MATLAB, or another program.)
(c) Determine an equation for peak area as a function of sucrose concentration.
(d) A sample containing sucrose gives a peak area of 209.86. What is the sucrose concentration?

1.26 Implantable glucose sensors are being developed to aid diabetic patients in monitoring their blood glucose levels. One technology involves polymer spheres that contain dextran (a carbohydrate) and concanavalin A (con A). In the absence of glucose, dextran and con A are weakly bound. However, when glucose is present, it displaces the dextran and binds tightly with con A. Fluorescent molecules can be attached to the dextran and/or con A, which respond to changes in the binding of molecules through con A. Fluorescent signal intensity can be correlated to glucose concentration, providing the rudimentary aspects of a sensor. Since light can travel though several millimeters of skin tissue, polymer spheres containing dextran and con A can be placed subcutaneously in some parts of the body. Work is ongoing to determine the feasibility of this method.

In an effort to assess the function of the sensor, the following experimental data were collected. The glucose concentration was estimated using the sensor. Then, a standard chemical assay was performed to determine definitively the glucose concentration. The data are in Table 1.14.

TABLE 1.14

Measurements of Glucose Concentration Using Sensors and Chemical Assays

Glucose concentration (mg/dL) Chemical assay	Glucose concentration (mg/dL) Sensor
4	5
10	12
24	28
65	64
95	100
147	150
256	240
407	352
601	425
786	465
982	500

(a) Plot the data. Over what range of the curve does the sensor respond in a linear fashion to the concentration of glucose measured by the chemical assay? Over what range of the curve does the sensor lose its sensitivity to changes in glucose concentration measured by the chemical assay? Justify.

(b) What does the term "hypoglycemic" mean? What is the hypoglycemic range for diabetics?

(c) Assume that the polymer spheres have a diameter of 5 nm. Estimate the number of beads in a patch that is 1 mm in height and has a cross-sectional area of 4 mm^2.

(d) Assume that the patch of polymer beads has a binding capacity of 1 μmol of glucose. Estimate the number of glucose binding sites per bead.

(e) What would be the advantages and disadvantages of using a sensor like this in a diabetic patient?

Foundations of Conservation Principles

CHAPTER 2

2.1 Instructional Objectives

After completing Chapter 2, you should be able to do the following:

- Identify an extensive property that can be counted in a system of interest.
- Appropriately define a system, system boundary, and surroundings for a system of interest.
- Specify a time period of interest for a given system, and identify systems that involve continuous or indefinite time periods.
- Know the theory and scope of the conservation laws.
- Explain the differences between an accounting equation and a conservation equation.
- Explain the differences between algebraic, differential, and integral accounting and conservation equations, and apply the appropriate equation to a system.
- Describe the differences between open, closed, and isolated systems; reacting and nonreacting systems; systems with and without energy interconversions; and steady-state and dynamic systems. Correctly evaluate systems using these definitions.

2.2 Introduction to the Conservation Laws

Conservation—the preservation of a physical quantity during movement, transformation, or reaction—is a fundamental concept in engineering and science. Along with the second law of thermodynamics (the disorder in the universe spontaneously increases), a handful of conservation principles provide the governing laws for virtually all physical behavior. Engineers mathematically describe these laws in conjunction with initial or boundary conditions. The foundations of many fields of engineering, including bioengineering, are based on understanding and applying conservation laws.

For centuries, scientists and engineers have recognized that certain physical quantities could be described differently than other physical quantities. For example, Sir Isaac Newton's second law of motion, developed in the late 1600s to relate the net force on an object to its mass and acceleration, is a special case of the law of conservation of linear momentum. Formulated in 1845, Kirchhoff's current law, which states that the total charge flowing into a node must equal the total charge flowing out of the node, is an equation developed on the concept of conservation of charge. Although these specific applications of the conservation laws were developed centuries ago, scientists and engineers more recently have generalized the conservation of multiple extensive properties into a few governing conservation laws. These laws can be applied to total mass, mass and moles of elements, linear and angular momentum, net electrical charge, and total energy. These laws have become axioms that serve as the fundamental basis for problem solving across engineering disciplines.

The conservation laws can be mathematically described by and formulated into conservation equations, which are described in Section 2.4. Conservation laws have been used in a wide variety of applications across all fields of engineering:

- Refining crude oil into gasoline
- Finding bending moments and loads in building structures
- Designing and constructing circuits and computers
- Estimating ground-water contamination
- Designing and producing microchips
- Modeling the carbon cycle in the environment
- Designing and building aircraft
- Developing life support systems

In this text we explore the application of the conservation laws to many systems, such as the following:

- Human kidney
- Blood circulation
- Cell metabolism
- Cell membrane ion pumps
- Human exertion
- Air flow in lungs
- Platelet adhesion
- Hypothermia
- Hemodialysis
- Batteries
- Basal metabolic rate
- Stenotic blood vessels
- Electrical circuits
- Acid/base buffering
- Pumping heart
- Membrane potentials
- Biomaterials
- Tissue engineering
- Medical device design

Not all extensive physical properties are conserved. Those that are not must be described using an accounting equation, a more generalized version of the conservation equation. This textbook is structured to cover how each governing conservation law is applied, and provides numerous examples of the application of accounting and conservation equations to the diverse fields in bioengineering.

2.3 Counting Extensive Properties in a System

Counting objects or quantities of extensive properties is the basis of **accounting**. In engineering systems, the counting of objects can be reduced to a few **accounting statements**. Specifically, many extensive properties—but no intensive properties—can be counted using accounting statements. Recall that extensive properties depend on the size of the system (see Section 1.5.1). Below is a list of extensive properties that can be counted:

- Total mass
- Mass of individual species
- Mass of individual elements
- Total moles
- Moles of individual species
- Moles of individual elements
- Total energy
- Thermal energy
- Mechanical energy
- Electrical energy
- Net electrical charge
- Positive electrical charge
- Negative electrical charge
- Linear momentum
- Angular momentum

All the extensive properties listed above can be counted in accounting equations, but only a subset of these extensive properties is always conserved. Below is a complete list of extensive properties that are conserved in all situations (except nuclear reactions):

- Total mass
- Mass of individual elements
- Moles of individual elements
- Total energy
- Net charge
- Linear momentum
- Angular momentum

To set up an accounting or conservation statement, **the property to be counted must be defined**. You are probably familiar with the concept and application of using an organized accounting scheme to keep track of items or objects. For example, the manager at a university bookstore needs to count all the books, school supplies, and clothing items that enter and leave the bookstore. The manager wants to keep track of each type of item and the type of client that is purchasing the various items, as shown in Table 2.1. Developing a simple spreadsheet like the one depicted

TABLE 2.1

Inventory at University Bookstore

Date	Item	Inventory at bookstore	Delivered to bookstore	Sold to students	Sold to faculty and staff	Sold to nonuniversity individuals
8/20	Books	13,000	800	4,900	100	0
8/20	Supplies	1,000	150	300	25	10
8/20	T-shirts	400	0	15	25	100
8/20	Sweatshirts	400	0	15	25	100
8/21	Books	8,800	200	4,000	100	5
8/21	Supplies	815	0	300	0	0

in Table 2.1 allows the manager to keep track of the inventory items that can be counted. Accounting statements capture the process this manager uses in mathematical equations by counting items that enter and leave the bookstore.

When we consider this example, some features of accounting equations become apparent. First, **the same property must be counted in all terms in a single equation.** If you were writing an accounting equation for the total number of books in the university bookstore, school supplies such as rulers and notebooks would not be included. However, if you were interested in developing a "total clothing" accounting equation, you would include both the number of T-shirts and the number of sweatshirts in your count. When accounting a particular property, the units of all the items must be the same. For example, when accounting mass, all quantities must be a dimension of mass; when accounting energy, all quantities must be a dimension of energy; when accounting a particular chemical species, all quantities must be that specific chemical species; and so on for other extensive properties.

The extensive property to be counted must be included in the system of interest. A **system** consists of matter identified for investigation. The system is set apart from the **surroundings,** defined as the remainder of the universe (Figure 2.1). The **system boundary** separates the system from the surroundings.

Defining the system of interest before you begin to solve a problem is important, since it may change your conditions and assumptions. **Systems are defined by the problem solver based on the needs of the problem.** The system can be quite large or quite small. For example, when investigating a biochemical reaction in the human body, the system may be defined as the entire body, a specific organ in the body, one cell in that organ, an organelle within the cell (e.g., nucleus), or in a number of other ways (e.g., an arbitrary volume of cytoplasm in the cell) (Figure 2.2).

The system is determined by the system boundary. Boundaries are of two types. The first type of boundary is real and tangible, meaning the boundary naturally exists and has a definitive border. Such boundaries may enclose the entire object of

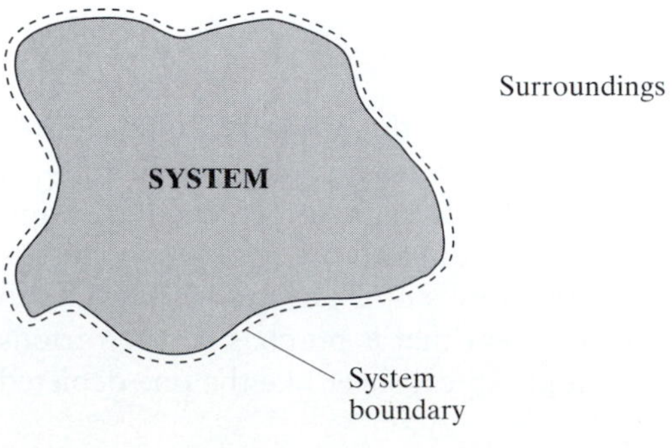

Figure 2.1
System and surroundings.

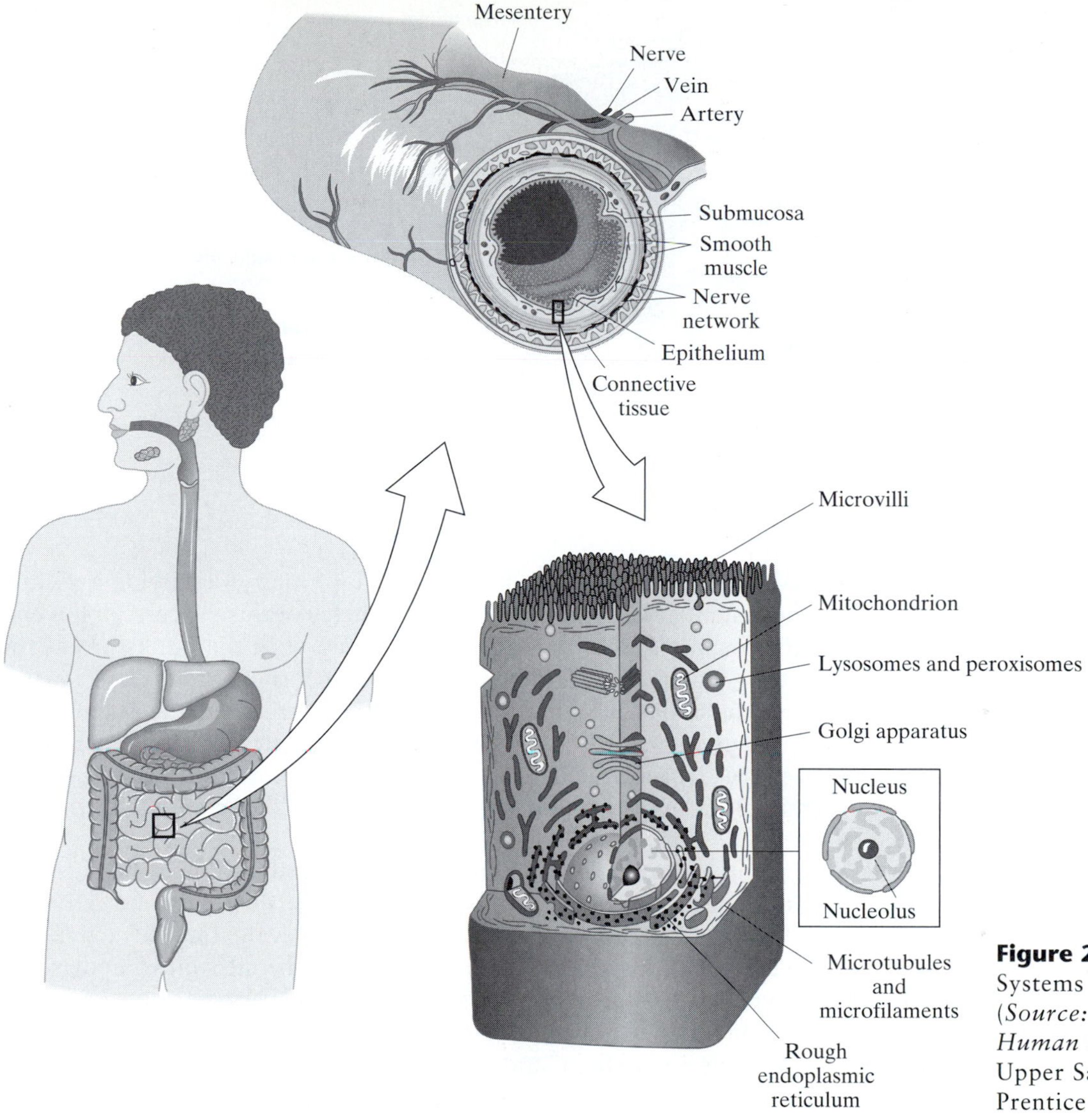

Figure 2.2
Systems of various scales. (*Source:* Silverthorn DU, *Human Physiology*, 2d ed. Upper Saddle River, NJ: Prentice Hall, 2001.)

interest. Some examples are the walls of a glass beaker, where the system is the liquid contained in the beaker; the casing of a total artificial heart, where the artificial heart is the system; or the plasma membrane of a cell, where the cell is the system. The second type of boundary is arbitrarily defined by the problem solver. It can be a sectional area of an object that accurately models the larger object or one that captures all relevant elements including the movement of the property of interest across the boundary. An example of an arbitrary boundary is a 1 cm × 1 cm patch of the villi wall; modeling the villi wall as such a patch is equivalent to looking at how all the villi in the small intestine act. A second example is isolating two malfunctioning units of a seven-unit bioreactor process. Other examples are a hypothetical line in space or a plane in a vessel. For the list of systems mentioned in the previous paragraph, the system boundaries are defined by the skin, the organ wall, the cell membrane, the nuclear membrane, or some hypothetical boundary line in the cytoplasm.

As a problem solver, it is extremely important that you carefully define the system such that the unit(s) or item(s) of interest are isolated for study. (We will demonstrate in Examples 2.15 and 2.16 how changing the boundary of a system can

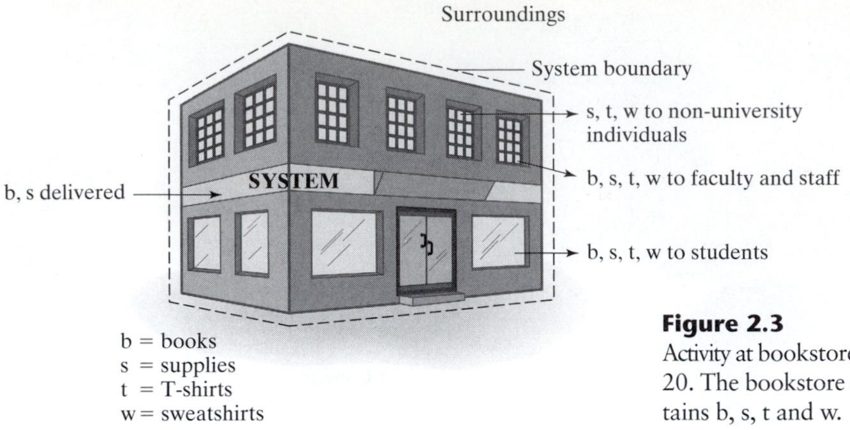

Figure 2.3
Activity at bookstore on August 20. The bookstore system contains b, s, t and w.

change how we view the movement of an extensive property.) Drawing a picture and labeling the system and its surroundings often helps in this process. In the university bookstore example, the problem can be set up with the bookstore as the system of interest (Figure 2.3). A dotted line separates the system from its surroundings. When the books, clothing, and supplies are purchased and taken out of the bookstore, the items leave the system and become part of the surroundings; when a delivery truck brings items to the bookstore, the items enter the system from the surroundings. Arrows indicate the movement of the items across the system boundary.

Finally, to set up an accounting statement, **a time period must be defined**. All elements of the accounting statement must be evaluated over the same time period. In the university bookstore example, if we are interested in evaluating the activity on August 20, the time period is one day. Sometimes it may be difficult to understand the distinction between a system that has a defined, specific period of time and a system that operates indefinitely. In some systems, a clear "beginning" and "end" are apparent. The time period is finite and is calculated as the difference between the start time and the end time, such as the one-day period in the bookstore example. In other systems, no beginning or end has been defined, and the system operates on an ongoing or continuous basis. It is appropriate in these cases to describe the time period as ongoing or continuous. An example is the beating of a heart; while one heartbeat has a distinct beginning and end, the heart continues beating over an indeterminate span of time. Ongoing and continuous systems are described mathematically using rates and differential equations.

In review, to be able to write an accounting statement, three items are necessary:

- The *property* to be counted must be specified.
- The *system* and its *surroundings* must be defined by specifying a *boundary*.
- A *time period* must be specified.

EXAMPLE 2.1 Pacemaker

Problem: A pacemaker is a small electrical unit that uses electrical impulses to initiate heart contraction when the heartbeat is irregular. Implanted in the chest cavity just under the collarbone, the pacemaker runs on a battery and is connected to the heart by leads, usually

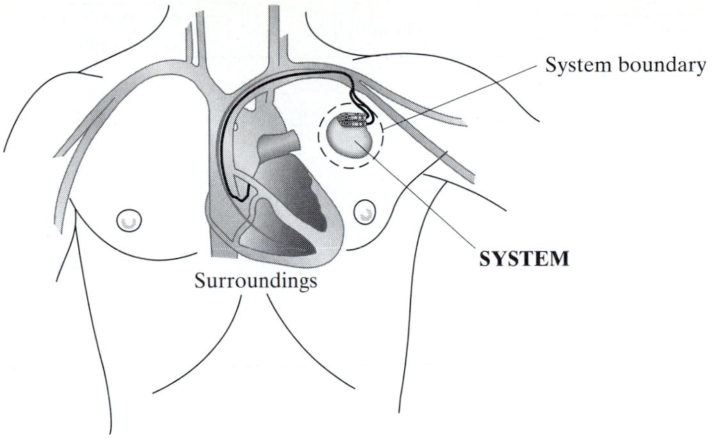

Figure 2.4
Pacemaker connected to heart.

wires. The wires carry electric charges from the battery to an electrode touching the inner wall of the heart, where an electrical potential difference stimulates a heartbeat.

Consider the electrical charges from the pacemaker used to stimulate one heartbeat. Name the property to be counted. Draw a picture and label the system, system boundary, and its surroundings. State the time period of interest.

Solution: The property to be counted is electrical charge. The system is defined as the pacemaker (excluding the wire running from the pacemaker to the heart), and the boundary is defined by the perimeter of the pacemaker (Figure 2.4). The surroundings include everything outside of the pacemaker, including the wires, the heart, and the remainder of the body. The time period is one heartbeat, since we want to know the charge required to initiate one beat. ■

In 1950, electrical engineer John Hopps accidentally discovered that a cooled heart could be started again by stimulating it with electrical impulses. With this discovery, Hopps inadvertently made the world's first cardiac pacemaker. The device's large size and limited battery life were impractical for use in patients, and not until 1960 did a team of surgeons successfully insert the first implantable pacemaker into a human.

The engineer who improved Hopps' design was Wilson Greatbatch. While working to create a circuit to record fast heart sounds, he accidentally used the wrong resistor—but discovered that it generated electrical impulses in the unique "lub-dub" rhythm of the heart. For the next several years, Greatbatch worked to develop a long-lasting, corrosion-free lithium-ion battery and to reduce the device to the size of a matchbook. Greatbatch's innovation has helped millions of patients maintain a regular heartbeat, allowing them to regain a normal quality of life and to live a lifespan comparable to those of healthy individuals. In 1985, the National Society of Professional Engineers named the pacemaker as one of the ten greatest engineering contributions in the past 50 years.

EXAMPLE 2.2 **Drug Delivery**

Problem: A patient with chronic obstructive pulmonary disease is treated with theophylline by continuous intravenous (IV) infusion at a rate of 0.5 mg/min. Since IV drug administration directly infuses the drug into the bloodstream, the blood carries the theophylline and exchanges it with the target organ, the lungs.

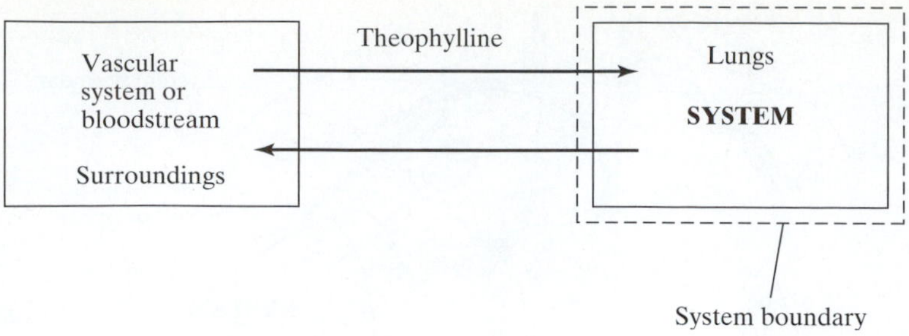

Figure 2.5
Theophylline drug delivery for the treatment of chronic obstructive pulmonary disease.

A physician studying the distribution of theophylline in the body evaluates the dose proportion that reaches the target organ. Name the property to be counted. Draw a picture and label the system, system boundary, and its surroundings. State the time period of interest.

Solution: The property to be counted is the mass of theophylline in the patient's lung tissue. The system is defined as the patient's lungs (Figure 2.5), since this is the target organ. The vascular system is intertwined with the alveoli of the lungs. Although it would be difficult to actually dissect all the blood vessels and separate them from the alveoli, you can abstractly think about the vascular system as a compartment. The vascular system is the surroundings of interest. Simplifying a complex structure, such as the bloodstream, by modeling it as a box is a common engineering practice.

Since the drug is constantly administered by IV infusion, no discrete time period of interest can be defined. Rather, we are looking at the ongoing distribution of drug over an extended, undefined period time. ■

2.4 Accounting and Conservation Equations

Accounting equations are used to track extensive properties and their movement across system boundaries (Figure 2.6). Specifically, an **accounting equation** is a mathematical description of the movement, generation, consumption, and accumulation of an extensive property in a system of interest. Accounting equations can be written for extensive properties and the rates of those extensive properties that can be counted. In this text, we discuss how accounting equations are used to track the extensive properties of:

- Total mass
- Mass of individual species
- Mass of individual elements
- Total moles
- Moles of individual species

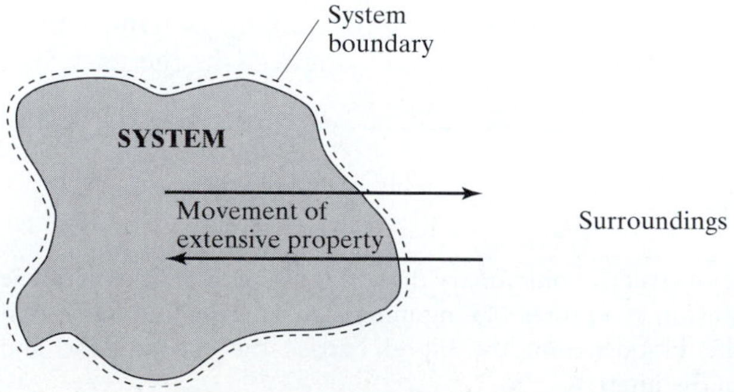

Figure 2.6
Movement of material across system boundary within a universe.

- Moles of individual elements
- Total energy
- Thermal energy
- Mechanical energy
- Electrical energy
- Net electrical charge
- Positive electrical charge
- Negative electrical charge
- Linear momentum
- Angular momentum

A specialized version of the accounting equation tracks only extensive properties and the rates of extensive properties that are conserved. By definition, a **conserved property** is one that is neither created nor destroyed. The **conservation law** states: **When an extensive property is conserved, that property is neither created nor destroyed despite changes in or to the system or the surroundings.** Not all extensive properties are conserved, but several key ones are:

- Total mass
- Mass of individual elements
- Moles of individual elements
- Total energy
- Net charge
- Linear momentum
- Angular momentum

The **conservation equation** is a mathematical description of the movement and accumulation of an extensive property in a system of interest, eliminating any terms that account for creation or destruction of an extensive property. When an extensive property is conserved, the amount or rate of that extensive property is unchanged in the universe at all times (except for nuclear reactions), where the **universe** is defined as the system and its surroundings. Thus, there is no net change of the amount of that conserved extensive property in the universe. In other words, the total amount of a conserved extensive property in the universe is a constant. This is slightly different from how conservation is commonly defined in physics, where it means that the quantity of an extensive property is a constant value. In physics, when a property—such as momentum—in a system is conserved, the quantity in the system does not change. For this textbook, the term *conserved* means that the quantity in the universe (both the system and the surroundings) is constant, and the extensive property may accumulate in a system when material is transferred between the surroundings and the system.

While it cannot be created or destroyed in a system, a conserved extensive property can be exchanged with the system's surroundings. Changes in the amount of a conserved property in a system can happen only by a one-for-one exchange with its surroundings. Consider the conservation of total mass: The amount of total mass added to a system is equivalent to the amount of total mass that the system's surroundings gave up. In this case, the net amount of total mass in the system increased, the net amount of total mass in the surroundings decreased, and the amount of total mass in the universe is unchanged. As an example, consider a system defined to be a person. If he eats a candy bar, the mass of the system increases by the mass of the candy bar. However, the mass of the surroundings decreases by the same amount, since the candy bar is no longer included in the surroundings. The total mass of the universe (i.e., person, candy bar, and everything else) is unaffected by the transfer of the candy bar into the system.

In contrast to total mass, species mass is not conserved and should be described using an accounting equation. When a reaction occurs in a system to produce a particular chemical species, the net amount of that species mass increases—both in the system and in the universe. Since there is a net change in species mass in the universe, the species mass is not conserved. It is important to understand the criteria that separate conserved and nonconserved properties. To reiterate, all extensive properties that can be counted can be described with an accounting equation. Only those extensive properties that are neither created nor destroyed in the universe may be described using a conservation equation.

Mathematically, we illustrate both these concepts—accounting and conservation—in governing equations:

Accounting Equation

$$\text{Input} - \text{Output} + \text{Generation} - \text{Consumption} = \text{Accumulation} \quad [2.4\text{-}1]$$

Conservation Law Equation

$$\text{Input} - \text{Output} = \text{Accumulation} \quad [2.4\text{-}2]$$

The final and initial amounts in the system mathematically describe the Accumulation term in both the accounting and conservation equations:

$$\text{Final Condition} - \text{Initial Condition} = \text{Accumulation} \quad [2.4\text{-}3]$$

Figure 2.7 schematically represents the generic accounting equation situation for any extensive property. For a system in which the extensive property is conserved, no generation or consumption occurs.

The **Input** and **Output** terms capture the exchange or transfer of an extensive property into and out of a system. The Input and Output terms describe all exchanges in which the amount of extensive property transferred to or from the system is equal to that lost or gained by the surroundings, respectively. The terms can also describe situations in which the quantity of extensive property exchanged is balanced across the system boundary. Exchange or transfer of extensive property can occur by several different modes.

- An extensive property can be transported by *bulk movement*. In this case, an extensive property is physically transferred or moved across the system boundary. For example, if we define a system as a baseball catcher's mitt and the air in

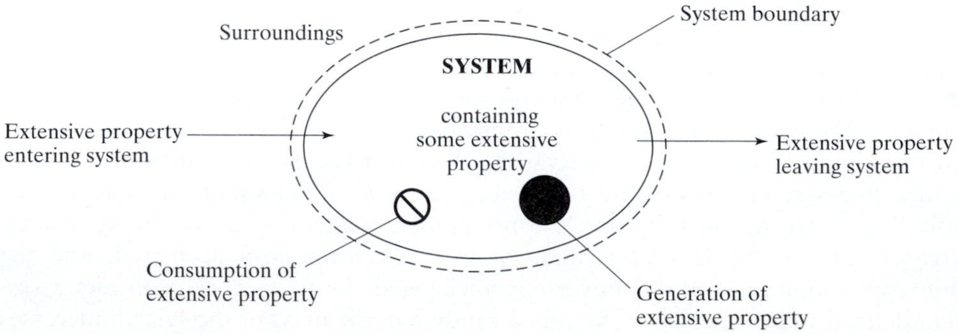

Figure 2.7
Schematic diagram of the principle of accounting extensive properties.

proximity surrounding the glove, moving a baseball across the boundary into the mitt system adds mass, energy, and momentum to the system through the bulk movement of the baseball.

- An extensive property can be transferred by *direct contact*. In this case, an extensive property is transferred to or from an object that physically touches the system without the object or any material crossing the system boundary. An example is the transfer of heat by a heating jacket touching the system boundary around a bioreactor. Heat is not transferred through movement of a hot material, such as a molten liquid; the heat is energy moving down a temperature gradient. In this case, the energy gained by the bioreactor through heat is equal to the energy lost by the surroundings (the heating jacket).
- An extensive property can be transferred through *nondirect contacts*. In this case, the system is acted on across a distance. The most frequently applied type of this transfer is a potential field. For example, when considering the momentum conservation law, gravitational forces are included in the Input and Output terms. The gravitational forces that act on a system are balanced by other forces in the surroundings.

The last two types of exchanges are a bit more abstract and difficult to visualize. In summary, the Input includes all extensive property that is added to the system across the system boundary. Likewise, the Output term describes the amount of extensive property that is lost by the system across the system boundary.

The **Generation** term describes the quantity of an extensive property that is created by the system, while the **Consumption** term describes the quantity that is used or destroyed by the system. The Generation and Consumption terms capture the production and elimination of an extensive property within the system. When a Generation or Consumption term is present, there is also a net production or destruction of that specific extensive property in the universe (i.e., the system and its surroundings). In some textbooks, momentum and energy that are not transferred through bulk material are considered Generation or Consumption terms. This textbook does not adopt this convention.

A chemical reaction is the easiest example to illustrate this concept of the Generation or Consumption of an extensive property in a system. When a chemical reaction occurs in a system, new chemical species—the products—are formed as the old chemical species—the reactants—are consumed. Consider the summary equation for photosynthesis:

$$6\ CO_2 + 12\ H_2O + \text{light} \longrightarrow C_6H_{12}O_6 + 6\ O_2 + 6\ H_2O \qquad [2.4\text{-}4]$$

In this case, the new chemical species—1 mole of glucose, 6 moles each of oxygen and water, and energy—are counted as generated, since their total amount has increased in the system and in the universe. On the other hand, the old chemical species—6 moles of carbon dioxide and 12 moles of water—have been consumed simultaneously, because their total amount has decreased in the system and in the universe. However, the *total mass* (products and reactants) in the universe has remained the same since total mass is conserved.

Thus, when one quantity of extensive property increased (was created) in the universe, something else must have decreased (was destroyed). Mass and charge accounting equations may contain Generation and Consumption terms that stem from chemical reactions. Another example of extensive properties that are generated or consumed is the conversion between different forms of energy, such as mechanical energy and thermal energy, when considering the mechanical energy accounting equation.

The **Accumulation** term describes the quantity that has collected in the system within the system boundary during the defined time period. In other words, the Accumulation term quantifies the gain or loss of an extensive property within a system. The Accumulation term can be determined by finding the difference in quantity present in the system between the initial condition and the final condition, which are used to define the time interval during which the system is examined. The **initial condition** of a system is the state at the start of the interval, and the **final condition** of a system is the state at the end. For example, the Accumulation term can measure the change in the mass of a blood clot from when the skin is cut to when the wound is sealed.

Accounting and conservation equations can be represented in three forms:

- Algebraic
- Differential
- Integral

The derivation of the differential and integral equations is covered in other textbooks (e.g., Bird, Stewart, and Lightfoot, *Transport Phenomena*, 2002). For this textbook, it is important that you understand all three equations and how to apply them appropriately. You should also be able to recognize that each equation is dimensionally homogenous.

2.4.1 Algebraic Accounting Statements

Algebraic accounting equations are generally applied to extensive properties within a defined system during a defined time period. **Algebraic equations can be applied when discrete quantities or "chunks" of extensive property are involved.** They cannot be applied when rates or time-dependent terms are involved. Based on equation [2.4-1], the generic algebraic accounting equation is written:

$$\psi_{in} - \psi_{out} + \psi_{gen} - \psi_{cons} = \psi_{acc} \qquad [2.4\text{-}5]$$

where:

ψ = Any extensive property
ψ_{in} = Input, the quantity of extensive property that enters the system during the time period
ψ_{out} = Output, the quantity of extensive property that leaves the system during the time period
ψ_{gen} = Generation, the quantity of extensive property that is generated in the system during the time period
ψ_{cons} = Consumption, the quantity of extensive property that is consumed in the system during the time period
ψ_{acc} = Accumulation, the difference between the quantities of extensive property contained within the system at the end of the time period as compared to those at the beginning

The initial and final conditions can also be used to define Accumulation:

$$\psi_f - \psi_0 = \psi_{acc} \qquad [2.4\text{-}6]$$

where:

ψ_f = final condition, the quantity of extensive property contained in the system at the end of the specified time period, and

ψ_0 = initial condition, the quantity of extensive property contained in the system at the beginning of the specified time period.

The dimension of the terms in equations [2.4-5] and [2.4-6] is that of the extensive property.

2.4.2 Differential Accounting Statements

Consider a system where the extensive property of interest moves into and out of the system continuously over time by inlet and outlet streams. A **stream** is the pathway by which an extensive property enters or leaves a system. We can measure the amount of an extensive property that moves continuously into or out of the system by using rates. An example is the rate of air exhalation at 6 g/min. In addition to transfer across a system boundary, an extensive property can be generated, consumed, or accumulated at a specified rate in a system. An example of consumption in a biological reaction is the rate of oxygen metabolism in tissues at 0.64 mg/s. Note that each rate given is an extensive property, in this case mass, divided by time. **The differential form of the accounting statement is most appropriate when the extensive properties are specified as rates.** The differential form of the accounting equation is written:

$$\dot{\psi}_{in} - \dot{\psi}_{out} + \dot{\psi}_{gen} - \dot{\psi}_{cons} = \dot{\psi}_{acc} = \frac{d\psi}{dt} \qquad [2.4\text{-}7]$$

where:

$\dot{\psi}_{in}$ = rate at which an extensive property enters the system,

$\dot{\psi}_{out}$ = rate at which an extensive property leaves the system,

$\dot{\psi}_{gen}$ = rate at which an extensive property is generated in the system,

$\dot{\psi}_{cons}$ = rate at which an extensive property is consumed in the system, and

$\dot{\psi}_{acc}$ or $\frac{d\psi}{dt}$ = rate at which an extensive property accumulates within the system.

Note that $\dot{\psi}_{in}$, $\dot{\psi}_{out}$, $\dot{\psi}_{gen}$, $\dot{\psi}_{cons}$, and $\dot{\psi}_{acc}$ are all rates. The dot over the symbol ψ signifies a **rate**, a change in the extensive property with respect to time, which is the mathematical definition of a derivative. For example, if ψ is defined as the extensive property mass, then $\dot{\psi}$ is the change of mass in the system over time or the mass flow rate. If ψ is defined as charge, then $\dot{\psi}$ is the time derivative of charge, or current. The accumulation term, $d\psi/dt$, is often expressed as the instantaneous rate of change of the extensive property in the system. The dimension of the terms in equation [2.4-7] is that of the extensive property divided by time.

Equation [2.4-7] is written for a differential time period (dt); hence this form of the accounting statement is called a differential balance. You can think of the differential accounting statement as describing what is happening in the system at an instant in time. Differential equations are often used when the system is operating on an ongoing or continuous basis. To solve a differential equation in general, a boundary or initial condition must be specified. Depending on the problem, the dependent variable, ψ, may be specified at some value of the independent variable (in this case, time). Frequently, the initial condition of the system at $t = 0$ is specified.

2.4.3 Integral Accounting Statements

Finally, the accounting statement can be written in an integral form. **Integral balances are most useful when trying to evaluate conditions between two discrete time points.** Integral accounting equations can be written to incorporate rates of change of an extensive property. When developing an integral balance, you can write the differential balance equation and integrate it between the initial and final times. The integral accounting statement is:

$$\int_{t_0}^{t_f} \dot{\psi}_{\text{in}}\,dt - \int_{t_0}^{t_f} \dot{\psi}_{\text{out}}\,dt + \int_{t_0}^{t_f} \dot{\psi}_{\text{gen}}\,dt - \int_{t_0}^{t_f} \dot{\psi}_{\text{cons}}\,dt = \int_{t_0}^{t_f} \dot{\psi}_{\text{acc}}\,dt \qquad [2.4\text{-}8]$$

where:

$\int_{t_0}^{t_f} \dot{\psi}_{\text{in}}\,dt$ = total extensive property that enters the system between the initial time t_0 and final time t_f,

$\int_{t_0}^{t_f} \dot{\psi}_{\text{out}}\,dt$ = total extensive property that leaves the system between t_0 and t_f,

$\int_{t_0}^{t_f} \dot{\psi}_{\text{gen}}\,dt$ = total extensive property that is generated in the system between t_0 and t_f,

$\int_{t_0}^{t_f} \dot{\psi}_{\text{cons}}\,dt$ = total extensive property that is consumed in the system between t_0 and t_f, and

$\int_{t_0}^{t_f} \dot{\psi}_{\text{acc}}\,dt$ = total extensive property that accumulates in the system between t_0 and t_f.

The Accumulation term can also be written as $\int_{t_0}^{t_f} \frac{d\psi}{dt}\,dt$ or $\int_{\psi_0}^{\psi_f} d\psi$, where ψ_f and ψ_0 are values of the extensive property at the final and initial conditions, respectively.

Like the terms in differential accounting equations, $\dot{\psi}_{\text{in}}$, $\dot{\psi}_{\text{out}}$, $\dot{\psi}_{\text{gen}}$, $\dot{\psi}_{\text{cons}}$, and $\dot{\psi}_{\text{acc}}$ are rates. The $\dot{\psi}$ terms can be specified functions of time. The dimension of the terms in equation [2.4-8] is that of the extensive property. Information on the conditions of the system at either t_0 or t_f or both is usually needed to solve problems using the integral equation. For simple systems whose terms do not change with time, the integral equation [2.4-8] can be reduced to the algebraic equation [2.4-5].

The differential and integral forms are particularly useful, since rates are often given to specify the operation of bioengineering systems. The characteristics of the algebraic, differential, and integral accounting equations are summarized in Table 2.2.

TABLE 2.2

Characteristics of Accounting Equations

	Algebraic	Differential	Integral
Can it incorporate discrete quantity of extensive property?	Yes	No	Sometimes
Time interval	Finite	Instantaneous	Finite
Can it incorporate rates?	No	Yes	Yes
Dimension of equation	Extensive property	$\frac{\text{Extensive property}}{\text{time}}$	Extensive property

EXAMPLE 2.3 **Population of the United States**

Problem: The United States is one of the fastest-growing countries in the world. Between July 1, 2000, and July 1, 2001, 1,279,800 people immigrated to the United States, 214,800 U.S. residents emigrated to other countries, 4,052,800 babies were born, and 2,435,300 people died (Figure 2.8).[1] Considering the population change in this year, would an algebraic, differential, or integral balance be most useful? Write a balance on the U.S. population during this time period.

Solution: The property to be counted is people. The system is the United States, and the system boundary is defined by the national border and includes other entry or exit points using various transportation modes, such as airplanes. The time period is one year. Because a person can be considered a "chunk" of extensive property, an algebraic accounting equation is appropriate to use. The term ψ_{in} is 1,279,800 people, since this many people crossed the system boundary (immigrated) into the United States; and ψ_{out} is 214,800 people, since this many people left the system across its boundary (emigrated). Because births increase the total number of people in the system and the universe, ψ_{gen} is 4,052,800 people; and because

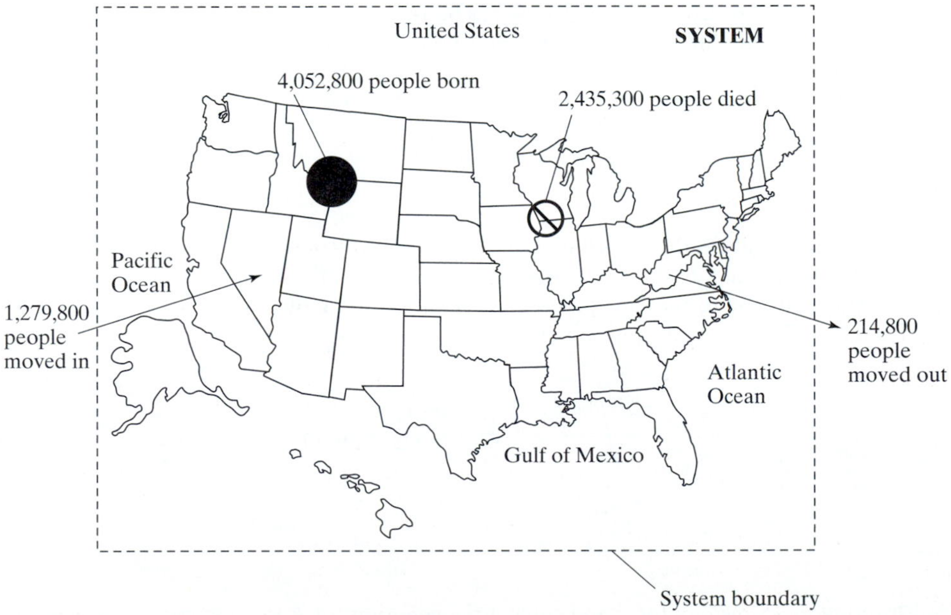

Figure 2.8
Population changes in the United States between July 1, 2000, and July 1, 2001.

deaths decrease the total number of people in the system and in the universe, ψ_{cons} is 2,435,300 people. Therefore,

$$\psi_{\text{acc}} = 1{,}279{,}800 \text{ people} - 214{,}800 \text{ people} + 4{,}052{,}800 \text{ people} - 2{,}435{,}300 \text{ people}$$

$$\psi_{\text{acc}} = 2{,}682{,}500 \text{ people}$$

A net gain (i.e., positive accumulation) of 2,682,500 people occurred in the United States between July 1, 2000, and July 1, 2001. ■

EXAMPLE 2.4 Diabetes Medication

Problem: Jean is a type II diabetic, which means her cells are insensitive to insulin, a hormone that helps cells take in glucose. Her doctor has prescribed one 30-mg Pioglitazone pill a day to promote the uptake of glucose into muscle cells. Her body metabolizes the drug at a rate dependent on the time since she ingested the pill:

$$\text{rate of drug metabolism} = ae^{-kt}$$

where a is a rate constant equal to 45 mg/day and k is an unknown rate constant. At the end of the 24-hour dosage period, Jean's physician finds that 2 mg of unmetabolized drug remains in her bloodstream. Assume that the rest of the drug is completely metabolized and no Pioglitazone is created or leaves her body through other means (e.g., urine). Is it best to use an algebraic, differential, or integral accounting equation to calculate k? Why? Consider Jean's body as the system. Calculate the metabolic rate constant k for Pioglitazone.

Solution: The drug consumption is given as a rate. Since the drug Pioglitazone is converted into another chemical, the net amount of Pioglitazone in the universe has decreased. Thus, the drug is not conserved in this problem. The rate of metabolism should be considered a Consumption term. Since the consumption rate is time dependent and a finite period of time (1 day) is given, the integral accounting equation is most appropriate.

If we define that the 1-day time period starts immediately after Jean ingests the pill, the Input term is zero. Because no drug leaves her body unmetabolized, the Output term is zero. No drug is generated in Jean's body, so the Generation term is eliminated. Thus, equation [2.4-8] reduces to:

$$-\int_{t_0}^{t_f} \dot{\psi}_{\text{cons}}\, dt = \int_{\psi_0}^{\psi_f} d\psi = \psi_f - \psi_0$$

where $\dot{\psi}_{\text{cons}}$ is the rate of drug consumption (ae^{-kt}), ψ_0 is the amount of drug present in Jean's body at the beginning of the time period (30 mg), and ψ_f is the amount of drug remaining in her system at the end of the time period (2 mg). Substituting the known values and integrating from $t_0 = 0$ to $t_f = 1$ day, we obtain:

$$-\int_{0}^{1\text{ day}} ae^{-kt}\, dt = 2\text{ mg} - 30\text{ mg}$$

$$\left.\frac{a}{k}e^{-kt}\right|_{0}^{1\text{ day}} = \frac{a}{k}e^{-k(1\text{ day})} - \frac{a}{k} = -28\text{ mg}$$

Substituting in 45 mg/day for a and solving the equation for k in a computer software program such as MATLAB or Excel yields $k = 1.04\text{ day}^{-1}$. Therefore, the rate constant associated with Jean's metabolism of Pioglitazone is $k = 1.04\text{ day}^{-1}$. ■

Before 1921, a patient diagnosed with diabetes mellitus, detected by excessive amounts of sugar in the urine, could hope to live for only a few months before starving to death. Although the disease had been known since the time of the ancient Egyptians and Greeks, no treatment was found until Canadians F. Banting and C. Best isolated insulin from the pancreas of a dog. They injected their extract into diabetic dogs, which then demonstrated normal glucose uptake. Another group member, J. B. Collip, found a way to purify the extracted insulin. In 1922, the group tested their sample on a 14-year-old boy near death, and he improved after the injection.

To keep up with the demand for insulin, extractions from pigs and cows sent to the slaughterhouse were purified for patient use. In the 1980s, recombinant DNA technology allowed scientists to engineer human insulin.

Insulin is not a cure for diabetes. A few medical trials to transplant diabetic patients with islets of Langerhans, which normally produce insulin, have had limited success. Until a cure—possibly one involving tissue engineering—is found, Banting and Best's revolutionary medical discovery of insulin remains the best treatment for diabetic patients.

Diabetes and its treatment provide an example of physiological constraints on an accounting equation. Insulin is consumed during normal metabolism. However, for a diabetic person, there is no Generation term. Consequently, the insulin Consumption term must be balanced by regular injections of insulin (an Inlet term) or the implantation of islets of Langerhans (a Generation term).

2.4.4 Algebraic Conservation Equation

To write an algebraic conservation equation, we rewrite equation [2.4-2] as follows:

$$\psi_{in} - \psi_{out} = \psi_{acc} \qquad [2.4\text{-}9]$$

where ψ is any extensive property, ψ_{in} is the Input of that property, ψ_{out} is the Output, and ψ_{acc} is the Accumulation during the defined time period. (Refer to the algebraic accounting equation in Section 2.4.1 for more complete definitions of these variables.) Recall that a conserved extensive property is neither created nor destroyed, so the algebraic accounting equation is reduced by setting ψ_{gen} and ψ_{cons} equal to zero.

The initial and final conditions can be used to define Accumulation:

$$\psi_f - \psi_0 = \psi_{acc} \qquad [2.4\text{-}10]$$

where ψ_f is the quantity of property of the system present during the final condition, and ψ_0 is the quantity present during the initial condition.

2.4.5 Differential Conservation Equation

Similarly, the conservation law is written in differential form:

$$\dot{\psi}_{in} - \dot{\psi}_{out} = \dot{\psi}_{acc} = \frac{d\psi}{dt} \qquad [2.4\text{-}11]$$

where $\dot{\psi}_{in}$ is the rate of Input, $\dot{\psi}_{out}$ is the rate of Output, and $\dot{\psi}_{acc}$ or $d\psi/dt$ is the rate of Accumulation. For more complete definitions of these variables, refer to the differential accounting equation in Section 2.4.2.

2.4.6 Integral Conservation Equation

Finally, the conservation equation is written in integral form as follows:

$$\int_{t_0}^{t_f} \dot{\psi}_{\text{in}}\, dt - \int_{t_0}^{t_f} \dot{\psi}_{\text{out}}\, dt = \int_{t_0}^{t_f} \dot{\psi}_{\text{acc}}\, dt \qquad [2.4\text{-}12]$$

where:

$\int_{t_0}^{t_f} \dot{\psi}_{\text{in}}\, dt =$ Input between t_0 and t_f, and

$\int_{t_0}^{t_f} \dot{\psi}_{\text{out}}\, dt =$ Output between t_0 and t_f.

The Accumulation between t_0 and t_f can be written as:

$$\int_{t_0}^{t_f} \left(d\psi/dt\right) dt, \quad \int_{\psi_0}^{\psi_f} d\psi, \quad \text{or} \quad \int_{t_0}^{t_f} \dot{\psi}_{\text{acc}}\, dt.$$

For more complete definitions of these variables, refer to the integral accounting equation in Section 2.4.3.

Information on the conditions of the system at either t_0 or t_f or both is usually needed to solve problems using the integral equation. The integral equation [2.4-12] can be reduced to the algebraic equation [2.4-9] for simple systems whose terms do not change with time.

It is critical to understand the differences between the accounting and conservation equations and to recognize when to apply each. Much of the rest of the book is devoted to setting up conservation and accounting equations for particular systems and to solving the appropriate equations for unknown parameters. **If you are uncertain of which equation to use, write down the accounting equation first.** For particular systems or extensive properties or both, the accounting equation may then reduce to the conservation equation.

EXAMPLE 2.5 **Water in a Bathtub**

Problem: Your showerhead sprays water into your bathtub at a rate of 5 kg/min, and water accumulates in the tub at 1.5 kg/min. At what rate does water drain from the tub? If 15 kg of water accumulates in the tub, how long was your shower? After you turn the shower off, how long does it take for the remaining water to drain?

Solution: As shown in Figure 2.9, the system is the tub holding the water, and the extensive property of interest is the quantity of water. No water enters or leaves the system except as shown (e.g., no water spills over the side of the tub). It is assumed the drain works at maximum capacity for the entire duration of the shower. Because the quantity of water is given in flow rates, a differential equation should be used. The total mass of water is conserved, since water is neither generated nor consumed in a chemical reaction, so the differential conservation equation [2.4.11] can be used.

The rate of the inlet stream of water, $\dot{\psi}_{\text{in}}$, is 5 kg/min. Water accumulates in the bathtub at a rate, $\dot{\psi}_{\text{acc}}$, of 1.5 kg/min.

$$\dot{\psi}_{\text{in}} - \dot{\psi}_{\text{out}} = \dot{\psi}_{\text{acc}}$$

$$\frac{5\text{ kg}}{\text{min}} - \dot{\psi}_{\text{out}} = \frac{1.5\text{ kg}}{\text{min}}$$

$$\dot{\psi}_{\text{out}} = 3.5\,\frac{\text{kg}}{\text{min}}$$

Thus, the drain rate is 3.5 kg/min.

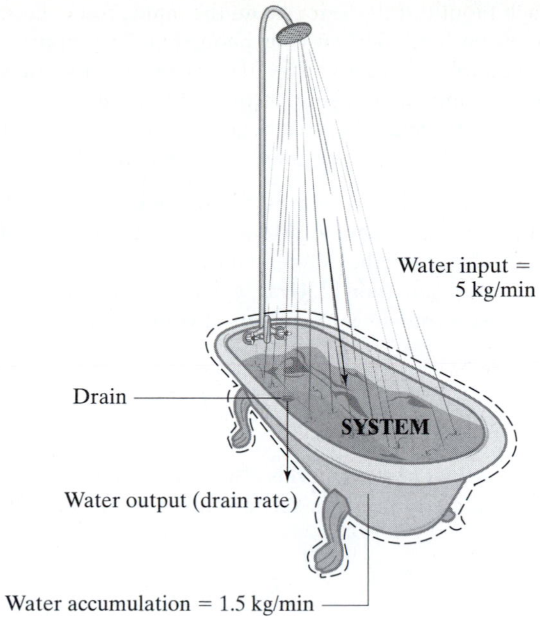

Figure 2.9
Accumulation of water in a bathtub.

To calculate the length of the shower, we use the integral mass conservation equation [2.4-12]. The difference in the amount of water present in the bathtub between the end and beginning of the shower ($\psi_f - \psi_0$) is 15 kg:

$$\int_{t_0}^{t_f} \dot{\psi}_{in}\, dt - \int_{t_0}^{t_f} \dot{\psi}_{out}\, dt = \int_{\psi_0}^{\psi_f} d\psi = \psi_f - \psi_0$$

$$\int_{t_0}^{t_f} \left(\frac{5 \text{ kg}}{\text{min}}\right) dt - \int_{t_0}^{t_f} \left(\frac{3.5 \text{ kg}}{\text{min}}\right) dt = 15 \text{ kg}$$

When $t_0 = 0$ and the equation is integrated with respect to time:

$$\left(1.5 \frac{\text{kg}}{\text{min}}\right) t_f = 15 \text{ kg}$$

$$t_f = 10 \text{ min}$$

Therefore, the shower lasted 10 minutes.

After the shower is turned off, $\dot{\psi}_{in} = 0$. To calculate the length of time necessary to drain the remaining water, we again use the integral conservation equation, which reduces to:

$$-\int_{t_0}^{t_f} \dot{\psi}_{out}\, dt = \int_{\psi_0}^{\psi_f} d\psi = \psi_f - \psi_0$$

Again, when $t_0 = 0$ and the equation is integrated with respect to time:

$$-\left(\frac{3.5 \text{ kg}}{\text{min}}\right) t_f = 15 \text{ kg}$$

$$t_f = 4.3 \text{ min}$$

Therefore, the accumulated water continues to drain for 4.3 minutes after the showerhead is turned off. ■

EXAMPLE 2.6 **Bank Account**

Problem: On the first day of every month, you receive a bank statement for your checking account that lists the previous month's activity and your current balance. Transactions for each month of the year remain the same: You collect \$5 of interest and spend \$75 on books, \$150 on food, \$40 on your phone bill, \$50 on utilities, and \$400 on rent. Fortunately, you have a job that pays you \$450 every two weeks, and you are very diligent about depositing all of your money into your account. To calculate your savings rate, would an algebraic, differential, or integral balance be most appropriate? Write a balance on the money in your bank account. Assume there are 4 weeks in a month.

Solution: The property to be counted is money. The system is your bank account, and the time period is ongoing. Money goes into and out of your account each month at a certain rate. Therefore, the differential conservation equation is most appropriate.

Using equation [2.4-11], you can figure out your savings rate by writing a balance on the transfer of money into and out of your account. The money you deposit from your paycheck is an Input term:

$$\dot{\psi}_{\text{in}} = \left(\frac{\$450}{2\text{ weeks}}\right)\left(\frac{4\text{ weeks}}{\text{month}}\right) = \frac{\$900}{\text{month}}$$

The interest of \$5/month is also an Input term.

All the money you spend is taken out of the system (your account), so these are Output terms:

$$\dot{\psi}_{\text{out}} = \frac{\$75}{\text{month}}(\text{books}) + \frac{\$150}{\text{month}}(\text{food}) + \frac{\$40}{\text{month}}(\text{phone}) + \frac{\$50}{\text{month}}(\text{utilities}) + \frac{\$400}{\text{month}}(\text{rent}) = \frac{\$715}{\text{month}}$$

The net amount of money in the universe (system and surroundings) is constant. No money is generated or consumed; it is just moved between your account and different institutions. You might think that the interest term is a Generation term; however, this is really just the transfer of money from the surroundings (the bank) to the system (your account). The printing of money by the government would be an example of a Generation term, since it would increase the net amount of money in the universe.

Substituting these terms into the differential conservation equation gives:

$$\dot{\psi}_{\text{acc}} = \frac{\$900}{\text{month}} + \frac{\$5}{\text{month}} - \frac{\$715}{\text{month}} = \frac{\$190}{\text{month}}$$

Therefore, you are accumulating (saving) \$190 per month in your bank account. ■

EXAMPLE 2.7 Discharge of a Capacitor

Problem: Capacitors are devices used in biomedical instrumentation to store charge (Figure 2.10). The positive plate of a capacitor has a charge of 10 mC initially ($t = 0$). The plate discharges at a rate i proportional to the net charge on the positive plate (q^{sys}):

$$i = kq^{\text{sys}}$$

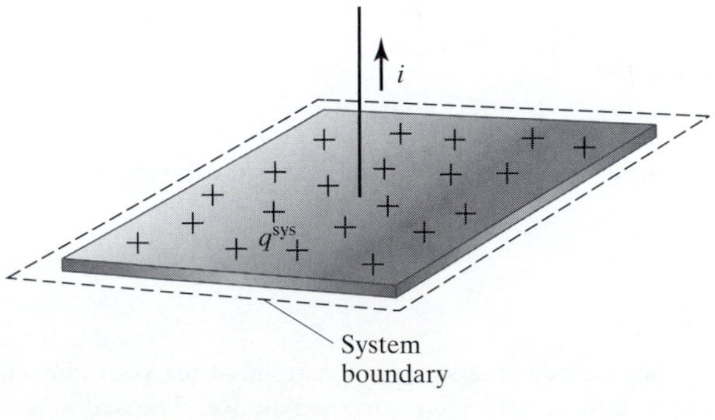

Figure 2.10
Discharging capacitor.

where k is the proportionality constant 0.5 s^{-1}. To calculate the net charge on the plate at a specified time, should you use an accounting or conservation equation? Is it better to use an algebraic, differential, or integral equation? Consider the capacitor as the system.

Solution: No charge enters the system; however, charge leaves the system. Net charge is neither generated nor consumed in the system, since it is a conserved property. Therefore, a conservation equation can be used. Since a rate of movement of charge is specified, the algebraic conservation equation cannot be used. Either the differential or integral conservation equation could be used. Since the problem indicates interest in a specified time, an integral equation is most appropriate. ■

It is very important to understand that the accounting and conservation equations are parallel across the discussed extensive properties. The same mathematical and computational tools can be applied to the accounting and conservation equations. **The central theme of this book is that two key general equations—accounting and conservation—can be applied to the four main properties of mass, energy, charge, and momentum to solve problems across all fields of bioengineering.** The remainder of the book is organized around the four main properties: mass in Chapter 3, energy in Chapter 4, charge in Chapter 5, and momentum in Chapter 6. In Chapter 7, the four properties are integrated in the study of three physiological systems. Mastering the general accounting and conservation equations and learning how to apply them to one extensive property will enable you to transfer your knowledge to other extensive properties.

2.5 System Descriptions

A system or process under investigation may be described using terms that characterize a system. Labeling a system accurately will help you identify the correct governing equation, make appropriate assumptions, and incorporate the correct terms of the accounting and conservation equations.

2.5.1 Describing the Input and Output Terms

Recall that Input and Output terms describe the transfer of extensive properties across the system boundary. These terms also encapsulate all types of transfers of conserved extensive properties. When an extensive property crosses the boundary, it is exchanged between the system and the surroundings.

An **open system** is one that exchanges an extensive property with its surroundings through **bulk material transfer**, which is the transfer of extensive properties across a system boundary through quantities of mass. The movement of mass into or out of an open system can transfer mass, energy, charge, or momentum. The extensive property enters or leaves the system by crossing the system boundary. In an open system, either the Input or Output terms or both are nonzero. No universally applied reductions to the accounting or conservation equations can be made. Open systems are very common in bioengineering applications. Open systems are dealt with in Sections 3.4–3.9, 4.5–4.10, 5.5–5.10, and 6.8–6.11.

A **closed system** is one that exchanges an extensive property with its surroundings through means other than bulk material transfer. In a closed system, extensive properties do not cross the system boundary through the transfer of mass. However, both energy and momentum can be exchanged in a closed system through direct contact interactions, such as heat, or noncontact interactions, such as gravity. In a closed system describing energy or momentum, no universally applied reductions to the accounting or conservation equations can be made. Closed systems are somewhat common in bioengineering applications. Closed systems are highlighted in Sections 4.4, 5.6, 5.9, and 6.5–6.6.

An **isolated system** is one that does not exchange any extensive properties by any means with its surroundings. No extensive property enters or leaves the system. In an isolated system, both the Input and Output terms are zero. Truly isolated systems are uncommon in biological and medical applications. Isolated systems are explored in Sections 4.4 and 6.7.

The terms open and closed are used to describe the accounting of mass and charge. The terms open, closed, and isolated are used to describe the accounting of energy and momentum. Typically, the term isolated is used to describe systems in which momentum or energy—but not mass or charge—is counted.

EXAMPLE 2.8 Penicillin Production in a Bioreactor

Problem: Bioreactors are widely used in bioengineering to produce large quantities of vaccines, monoclonal antibodies, antibiotics, pharmaceutical products, and other products. Bioreactors may be operated in a batch, semibatch, or continuous mode.

In a **batch** process, the feed materials are added to the bioreactor before the process or reaction begins. No reactants or additional substances are added during operation. Similarly, no products or materials are removed until the process is complete. In a **continuous** bioreactor process, the feed is continuously supplied in input streams, and products and wastes are continuously removed via output streams. In a **semibatch** process, either continuous inlet or outlet streams, but not both, are present.

Figure 2.11 shows a bioreactor designed to produce penicillin. Consider mass and energy as the extensive properties of interest and the bioreactor tank as the system. Materials

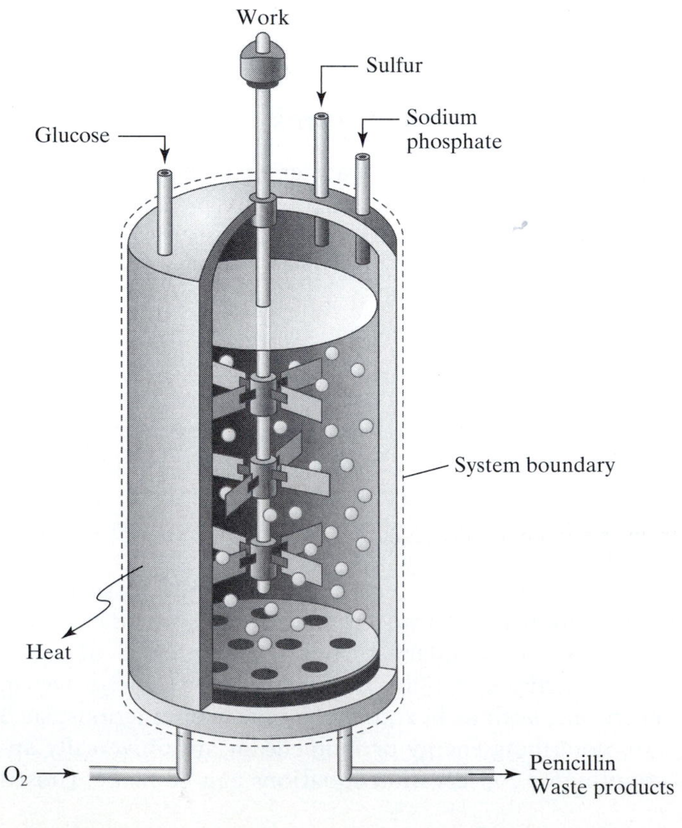

Figure 2.11
Penicillin bioreactor.

including glucose, sodium phosphate, sulfur, and oxygen are necessary for penicillin production. Work in the form of stirring is added, and heat is removed from the bioreactor during operation. Identify and describe each bioreactor process as an open, closed, or isolated system.

Solution: The batch process is considered closed when mass is the extensive property because no mass crosses the system boundary (vessel wall) during operation. Penicillin and waste products are removed from the bioreactor after operation is complete. No energy enters or leaves the system through bulk material transfer, since no mass crosses the system boundary. Since the biochemical reaction produces energy, heat is removed from the bioreactor to maintain a constant operating temperature. Also, work is added when stirring. Thus, with respect to the accounting of energy, the batch process is described as closed.

In the continuous and semibatch processes, mass and energy (possessed by the mass) continuously cross the system boundary during operation. Bioreactors that operate in these modes are open systems, since bulk material transfer occurs. These two modes of operation are the most common in bioprocessing.

A bioreactor configuration that is isolated with respect to energy is uncommon, because controlling the temperature in the bioreactor is critical to successful operation. ■

While it is well known that Sir Alexander Fleming was the first scientist to recognize the importance of a specific strain of mold in 1928, it is not so well known that the team of Florey, Chain, and Moyer engineered the revolutionary drug penicillin and forever altered how antibiotics were used and produced to treat bacterial infections.

In 1939, Florey and Chain were the first to show that mice infected with multiple strains of bacteria could be successfully cured with penicillin, implicating its potential in treating disease. Their team soon developed a powder form of penicillin—the first antibiotic. In 1941, Moyer developed a method to increase yields of the mold, which was incredibly difficult to purify. Moyer's method allowed massive quantities of the drug to be produced and later led to large-scale production.

Penicillin's success in treating millions of wounded American soldiers during World War II paved the way for antibiotics research, as well as the industrialization of U.S. pharmaceutical companies. As methods of purifying the drug improved, the price per dose dropped rapidly, from $20 in 1943 to 55 cents in 1946. It still remains one of the most economical drugs to manufacture and one of the best for combating many different types of infections.

EXAMPLE 2.9 Momentum on Earth and in Space

Problem: An astronaut is standing completely still on the Earth's surface while she waits to board the shuttle on her mission to Mars. Except for Earth's gravitational field, no other external forces act on her body (Figure 2.12a). Assume that during the mission to Mars, the effect of gravitational fields from celestial bodies is negligible, and that her body does not feel the effects of any other contact or noncontact forces (Figure 2.12b).

Consider her body as the system and momentum as the extensive property of interest. Is the system open, closed, or isolated on Earth? in space? Neglect unmentioned sources of momentum transfer (e.g., electrical and magnetic fields) to her body.

Solution: On Earth's surface, no momentum carried through bulk material transfer crosses the system boundary (the surface of her body). However, Earth's gravitational field exerts a noncontact force on the astronaut's body. In this way, the system exchanges momentum with the surroundings. Thus, the system is closed but not isolated.

Figure 2.12a
Astronauts experiencing a force caused by Earth's gravitational field. The force her body exerts on the ground ($\vec{W}$) is equal in magnitude to and opposite in direction from the normal force the ground exerts on her body ($\vec{n}$). Photo courtesy of NASA.

Figure 2.12b
Astronaut free-floating in space. The gravitational forces of celestial objects are negligible, and she does not experience any contact forces. Photo courtesy of NASA.

For a similar analysis in space, no bulk material transfer into or out of the system occurs. We assume that between Earth and Mars, the astronaut is sufficiently far from each planet that any gravitational force acting on her body is negligible. Furthermore, in the absence of gravity, the astronaut floats around freely within the space vessel, so no other contact forces act on her body. When neglecting other sources of momentum transfer, her body may be considered an isolated system with respect to momentum. ■

2.5.2 Describing the Generation and Consumption Terms

Recall that Generation and Consumption terms describe the creation and destruction of an extensive property within a system. These terms are present in accounting equations describing nonconserved properties, such as moles of a chemical species or mechanical energy. When an extensive property is generated or consumed in a system, the surroundings do not lose or gain, respectively, the equivalent amount of that property. Instead, the extensive property is created or destroyed in both the system and the universe. This is the main criterion for characterizing a term as an Input or Output term or as a Generation or Consumption term. When a Generation or Consumption term is present, an accounting equation must be used.

Two main processes constitute generation or consumption of an extensive property. The first is a chemical reaction. When a chemical species reacts to produce a new product, a quantity of its mass is destroyed in the system and in the universe. Chemical reactions are considered in the accounting of mass, charge, and energy. Chemical reactions are defined broadly in this book to include both the rearrangement of molecules between compounds and species dissociation or electrochemical reactions, as well as the transfer of electrons and other atomic particles in nuclear reactions.

A **reacting system** is one in which at least one biochemical or chemical reaction is taking place. When a system is reacting, accounting equations are required to deal with nonconserved extensive properties, such as individual species moles. Conservation equations are appropriate only when applied to conserved properties, such as total mass. Reacting systems are discussed in Sections 3.8–3.9, 4.8–4.9, and 5.9–5.10.

In a **nonreacting system,** no biochemical, chemical, or other reactions are taking place. The Generation and Consumption terms in the accounting equation can be set to zero if a system is both nonreacting and without energy interconversions. Nonreacting systems are discussed in Sections 3.4–3.7, 3.9, 4.4–4.7, 4.10, 5.5–5.8, and 6.5–6.11.

The second process constituting the generation or consumption of an extensive property is the interconversion of different types of energy (e.g., mechanical, thermal, electrical). In a system with **energy interconversion,** one form of energy is converted to another form, so the first form of energy has depleted in the system and in the universe. For example, when mechanical energy is converted to heat, such as when frictional loss occurs, the total amount of mechanical energy in both the system and the universe has decreased. Therefore, these types of energy interconversions are considered in the Generation and Consumption terms. When tracking for a specific type of energy (but not total energy), the accounting equation must be used. Systems with energy interconversion are discussed in Sections 5.6, 5.8, 5.10, and 6.11.

Table 2.3 summarizes the types and classification of Input/Output and Generation/Consumption terms covered in this textbook. Notice that all extensive properties without any Generation or Consumption terms are conserved. That is, the extensive properties of total mass, elemental mass, elemental moles, total energy, net charge, linear momentum, and angular momentum can be neither created nor destroyed in the system and in the universe. **It is appropriate to use the conservation equation for conserved properties. For all other extensive properties, the accounting equation, which contains the Generation and Consumption terms, must be used.**

TABLE 2.3

Summary of Classifications of Terms in Accounting Equation

Accumulation	Input − Output		+ Generation − Consumption	
Extensive property	Bulk material transfer	Direct and non-direct contacts	Chemical reactions	Energy interconversions
Total mass	X			
Species mass	X		X	
Elemental mass	X			
Total moles	X		X	
Species moles	X		X	
Elemental moles	X			
Total energy	X	X		
Thermal energy	X	X		X
Mechanical energy	X	X		X
Electrical energy	X	X		X
Net charge	X			
Positive charge	X		X	
Negative charge	X		X	
Linear momentum	X	X		
Angular momentum	X	X		

EXAMPLE 2.10 **Penicillin Production in a Bioreactor II**

Problem: As discussed in Example 2.8, bioreactors may be used for the production of a wide variety of biological and pharmaceutical products. A multistep process isolates the product after it leaves the bioreactor by using physical separation methods to remove waste. Identify the processes in both the bioreactor and separation systems as reacting or nonreacting.

Solution: In the bioreactor, different chemical constituents of the feed, such as glucose and oxygen, undergo biochemical reactions and are converted to penicillin and waste products. Thus, the bioreactor is a reacting system.

Methods of product isolation are numerous, including liquid-liquid extraction, vacuum distillation, and precipitation. Product isolation usually involves the physical separation of substances, not any chemical reactions between components; thus, the separation system is nonreacting. ■

EXAMPLE 2.11 **Solutions in a Beaker**

Problem: A chemistry student has three beakers (Figure 2.13). In the first, she adds a block of inert polymer to water. In the second, she mixes two salts, NaCl and KCl, in water. In the third, she mixes NaCl and $AgNO_3$ in water. Consider the charge in each beaker as the extensive property of interest. Identify each beaker as a reacting or nonreacting system.

Solution: In the first beaker, the polymer does not undergo any type of chemical or dissociation reaction. Therefore, this system is nonreacting.

In the second beaker, the two salts dissolve in water and their molecules dissociate to form Na^+, Cl^-, and K^+ ions. These ions mix and intersperse throughout the solution. Because the salts dissociate into charged species, this system is considered reacting.

In the third beaker, a double replacement reaction occurs in which the cations exchange anionic partners:

$$\mathrm{NaCl(aq)} + \mathrm{AgNO_3(aq)} \longrightarrow \mathrm{NaNO_3(aq)} + \mathrm{AgCl(s)}$$

These specific compounds chemically react to form a silver chloride precipitate, so the third beaker is also a reacting system. ■

2.5.3 Describing the Accumulation Term

The Accumulation term describes the net gain or loss of an extensive property contained within a system. When an Accumulation term is present, the amount of extensive property has changed in the system during the time period of interest.

Figure 2.13
Beakers containing chemical species, some of which react.

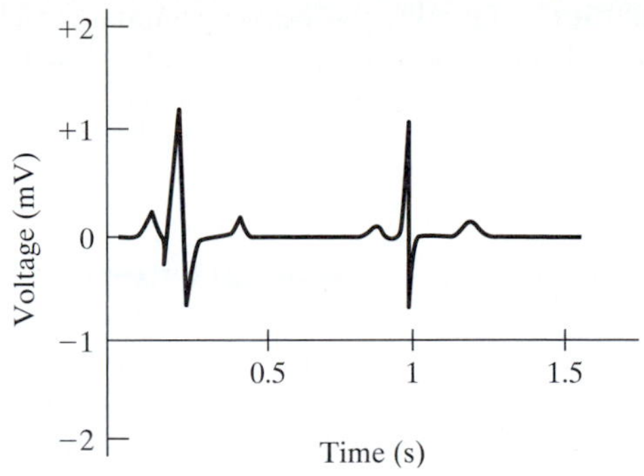

Figure 2.14
One cardiac cycle in an electrocardiogram.

Steady-state is a condition in which the values of all the variables in a system (e.g., temperature, pressure, volume, flow rate) do not change with time, although minor fluctuations about constant mean values may occur. Let us illustrate this by using a photography analogy. If you take multiple, imaginary "snapshots" of a steady-state system over a time period, each snapshot should look the same as the previous one. The initial and final conditions and all the intermediate snapshots of the system are identical or nearly so. The snapshots show that no quantity of the extensive property has collected in the system. If the extensive property is continuously flowing into the system at the same rate it flows out, such as when water flows at a constant rate through a pipe, the amount of extensive property in the system remains the same, and thus the snapshots also look the same. In a steady-state system, the Accumulation term is zero. Steady-state systems are highlighted in Sections 3.4–3.8, 4.5–4.9, 5.5–5.6, 5.10, 6.5–6.8, and 6.11.

Dynamic state, also called **unsteady-state** or **transient**, is a condition in which the values of at least one variable in a system change with time. Using the photography analogy, if you take multiple, imaginary snapshots of an unsteady-state system, the snapshots should look different from one another. Because the initial and final conditions of the system are not equivalent, the Accumulation term is nonzero. Thus, a positive accumulation (gain) or a negative accumulation (loss) of an extensive property means that no universally applied reductions to the accounting or conservation equation can be made. Dynamic systems are highlighted in Sections 3.9, 4.4, 4.10, 5.7–5.9, and 6.9.

Whether a system is steady-state or dynamic depends greatly on the time scale over which you are examining it. Consider the heartbeat of a person during his teenage years. If you looked at snapshots of his heartbeat on an electrocardiograph (Figure 2.14), where each wave corresponds to the electrical depolarization or repolarization of a compartment of the heart, the snapshots would look virtually identical over the course of a year. Thus, the system (the heart) is in a steady-state condition during the course of one year. However, if you took several snapshots over the course of a single heartbeat, each snapshot would look drastically different from the previous one; the heart is in a dynamic state during the course of one heartbeat.

EXAMPLE 2.12 **Solutions in a Beaker II**

Problem: Consider the mixing of two different salts, NaCl and KCl, in a beaker of water (Example 2.11). Consider the time period over which the chemistry student adds and mixes the salts. If charge is the extensive property of interest in the beaker system, is the system steady-state or dynamic?

Solution: Initially, the beaker contains no charged species, only water, NaCl, and KCl. After dissociation (final condition), the beaker contains water and Na^+, Cl^-, and K^+ ions. Using the photography analogy, comparing the snapshots over the time period of interest reveals that the charge contents in the beaker change with time. Thus, the system is dynamic. ■

EXAMPLE 2.13 Freshman Fifteen

Problem: Students matriculating into college are often challenged by the "freshman fifteen," when they gain an average of 15 lb_m by the end of their freshman year. The mass gain is often attributed to the lack of exercise and the late-night pizzas, sodas, and snacks needed to cram for exams.

Josh's mass is 175 lb_m when he matriculates. By the end of his freshman year, he has gained 15 lb_m. During the summer Josh works out every day and makes a conscious effort to eat healthy foods. By the beginning of his sophomore year, he is back to 175 lb_m, and he is able to maintain that mass until graduation.

Consider Josh as the system and his mass as the extensive property of interest. Consider his freshman year. Is Josh a steady-state or dynamic system? What about for the time period of a single day during his freshman year? Is he a steady-state or dynamic system over his collegiate career?

Solution: This is one example of how changing the time period can change your assumptions about how the system is described.

At matriculation (initial condition), Josh's mass is 175 lb_m. At the end of his freshman year (final condition), Josh's mass is 190 lb_m. Because the initial and final conditions of the system are different and mass has collected in the system, Accumulation is nonzero and the system is dynamic over Josh's first year of college.

For a single day in his freshman year, Josh's mass may fluctuate slightly as he eats, drinks, and excretes over a 24-hour period. However, using the photography analogy, snapshots of Josh during this time period would not change drastically and would look virtually identical. His overall mass is not expected to change much; that is, the change in mass in one day is negligible in comparison to his overall mass. Thus, Josh is considered a steady-state system over a 24-hour period.

Over his collegiate career, Josh's mass mostly remains the same due to his attention to diet and his regular exercise routine. At his matriculation (initial condition) and at his graduation (final condition), he is 175 lb_m. Although he experienced mass accumulation during his freshman year and mass loss the following summer, taking multiple snapshots of Josh over the time period of interest would reveal snapshots that mostly looked the same. Thus, Josh's "freshman fifteen" can be considered a fluctuation about his mean mass value. Josh is considered a steady-state system over his collegiate career. ■

2.5.4 Changing Your Assumptions Changes How a System Is Described

When analyzing complex systems, it is essential to examine the three categorical distinctions discussed in the previous sections. Learning to assess systems (e.g., open, closed, or isolated; steady-state or dynamic) is crucial to accurate and thorough problem solving in bioengineering. The assumptions you make about a system may profoundly affect how the governing accounting and conservation equations are applied, and may change the final answer. Learning to make good engineering assumptions is a difficult skill; it is one that you will master as you mature as an engineer. The following examples illustrate the importance of how the system is set up and how the assumptions affect the solution.

EXAMPLE 2.14 Training for a Cycling Event

Problem: Robbie decides to enter a cycling event to help raise money for a charity that aids multiple sclerosis patients and funds research for treatment and cures. In the summer, he trains for 45 minutes each day and stops to drink 200 mL of water every 15 minutes. Consider Robbie's body as the system. Consider the extensive properties of mass and energy. Is the system open or closed? Is he at steady-state? During warm-up, how would you best describe Robbie as a system?

Solution: *Mass* (Figure 2.15a): When Robbie drinks water, mass enters the system. Robbie loses mass (e.g., water and salts) through perspiration. Because mass enters and leaves the system, the system is open. Since Robbie only drinks every 15 minutes during the 45-minute training ride, the Input term is not continuous with time; therefore, the accumulation of water in Robbie's system on a short time scale (e.g., 5 min) is not constant with time. Because Robbie's mass at the initial and final conditions of the ride is different, he is in transient mode. The same analysis holds true during Robbie's warm-up. If Robbie drank one sip of water every few seconds and his physiological rates (e.g., heart, blood pressure, perspiration) were constant for a period of time, you might choose to assume that he is at steady-state.

Energy (Figure 2.15b): Stored energy in Robbie's body is converted to work to generate power for him to cycle. Assuming the water he drinks has no caloric value, energy does not enter his system from food or other sources during cycling, but Robbie does lose a tremendous amount of heat during exercise. The work and heat we have just described, which are forms of energy, leave the system, but not through bulk material transfer. From this perspective, the system describing energy is closed, but not isolated. However, when Robbie perspires, energy is lost in the bulk material transfer of water through the skin's surface. Additionally, Robbie loses energy when he breathes. With these considerations, the system should be classified as open. If Robbie's vital signs (e.g., body temperature) and his rate of perspiration do not change with time, you might think that he is at steady-state. However, the initial and final conditions of Robbie's energy status are definitely different, since his energy reserves have been depleted. Therefore, the best assumption is that Robbie is in transient mode. During warm-up, his physiological characteristics change with time; therefore, he is definitely in transient mode.

The description of a system depends on the extensive property of interest (e.g., mass or energy) as well as the time scale. In Chapters 3 and 4, quantitative examples of mass and energy transfer in the human body are presented. Also, you will learn how cells metabolize food to generate heat, work, and stored energy. ■

Recall that a system is defined by its boundary. Defining the system of interest allows you to make certain assumptions before you begin to solve a problem. Thus, changing a system by moving its boundary can change your assumptions, such as whether a system is steady-state or dynamic.

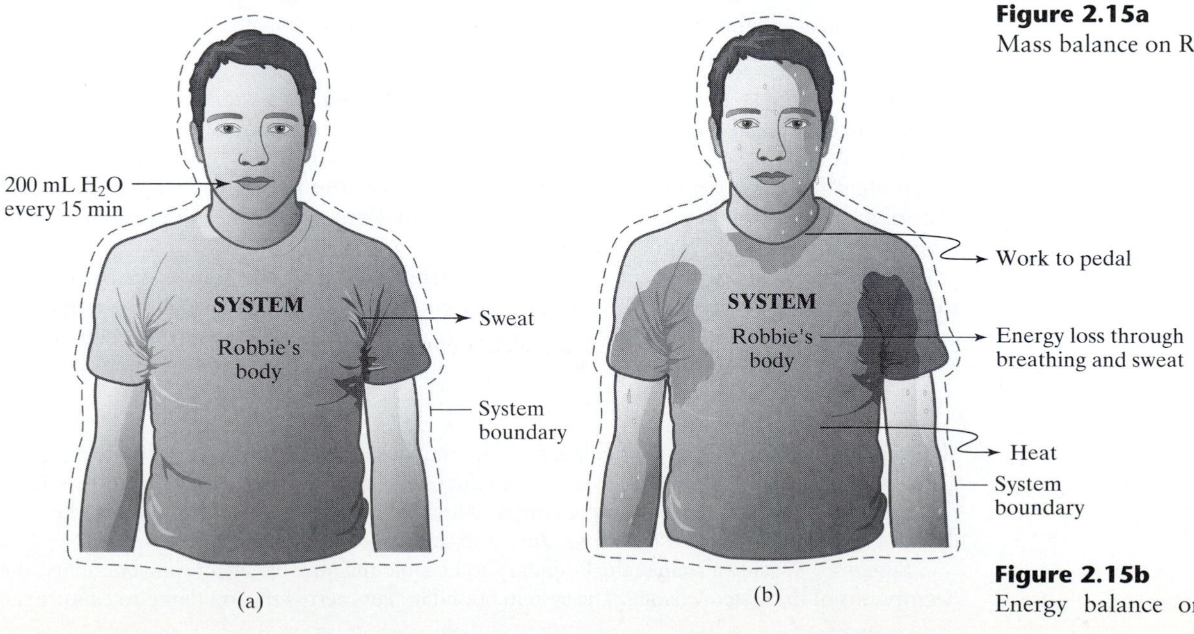

Figure 2.15a
Mass balance on Robbie.

Figure 2.15b
Energy balance on Robbie.

How you define the system characterizes the system itself. This in turn affects which governing equations are appropriate to use, as well as what reductions to these equations are suitable to the system and the problem. In most cases, the system should be defined such that the movement of the extensive property being examined can be tracked across the system boundary. Often, **the system boundary should be drawn to cut across any inputs and outputs to the system.** The importance of placement of the system boundary relative to movement of extensive property is highlighted in the following two examples.

EXAMPLE 2.15 Action Potential in Neurons

Problem: Across the plasma membrane of most neuronal cells, there is a charge gradient. This gradient is established because the volume inside the cell membrane has a negative electrical charge relative to the extracellular space adjacent to the membrane. The gradient keeps the cell at its resting electrical potential, approximately −90 mV in neurons, which is necessary for signal transduction. To maintain this membrane potential, ion pumps and channels facilitate the movement of charge into and out of the cell. For example, the sodium/potassium pump pushes two potassium ions into the cell for every three sodium ions it pumps out, resulting in a net loss of positive charges from the intracellular space. Potassium channels in the membrane allow K^+ ions to diffuse back out of the cell in response to the excess of K^+ ions inside the cell.

Certain external stimuli, such as electrical stimuli (e.g., sensory events, such as touch or temperature sensation), mechanical stimuli (e.g., stretching), and neurotransmitters (e.g., acetylcholine), cause the rapid influx of sodium and potassium ions into the cell. If the membrane potential reaches a threshold voltage of about −65 mV, an action potential is generated and the membrane completely depolarizes in a period of less than 1 ms. When the membrane depolarizes, it becomes highly permeable to sodium ions, which rapidly move into the cell to neutralize the charge inside, collapsing the charge gradient across the membrane. This ability to rapidly alter the membrane potential gives neurons the ability to transmit signals across the body.

Consider the three system boundaries for the Na^+/K^+ pump in the neuron as shown in Figure 2.16. For systems A, B, and C, analyze the movement of positive charge by characterizing each system as open, closed, or isolated; reacting or nonreacting; steady-state or dynamic; and deciding which governing equation to use to describe the activity of the system. Consider a time period when the Na^+/K^+ pump is operational. For system A, consider the system to include one neuronal cell, as well as the extracellular fluid surrounding the membrane (Figure 2.16a). Assume the ions remain in the extracellular space just next to the membrane. For system B, consider the system boundary to cut across the ion pump of the neuronal cell (Figure 2.16b). For system C, consider the system boundary to include only one ion pump (Figure 2.16c).

Solution: *System A:* When the system boundary encloses the intra- and extracellular spaces of the neuronal cell, no ions cross the system boundary by any means. Thus, the system is best described as closed. Since no chemical reactions with the positively charged ions of interest are taking place, the system is nonreacting. Even though charge has moved around within the system, the net amount of positive charge in the system is the same at the initial and final conditions. Thus, the system is at steady-state.

Since the system is nonreacting, you can eliminate the Generation and Consumption terms from the accounting equation. This reduces the accounting equation to the conservation equation. You can also reduce the Input and Output terms to zero, since the system is closed. The Accumulation term is zero, which is consistent with the reductions to the governing equation:

$$\psi_{acc} = 0$$

While this equation is a true statement about the system of interest, very little insight is gleaned from it. This choice of a system boundary that includes the cell and regions outside the cell gives the result that nothing changes, which counters what we know about the very dynamic movement of positive charges during maintenance of a membrane potential.

System B: When we change the boundary to become the tangible membrane boundary, the descriptions of the system change. The system boundary cuts across the ion pump to capture the transfer of positive charge between the intracellular and extracellular space. Since the ions cross the system boundary of the membrane, the system is open. The system is still nonreacting, since no charges are created or destroyed within it through chemical reactions. When the pump is active, snapshots show the loss of Na^+ ions from the system and the gain of K^+ ions in the system. Because its initial and final conditions are different, the system is dynamic. During the time period of interest, there has been a negative accumulation (or loss) of positive charge in the system.

As in system A, you can eliminate the Generation and Consumption terms from the governing accounting equation, reducing it to the conservation equation. However, because the system is open and dynamic, the Input, Output, and Accumulation terms are nonzero:

$$\psi_{in} - \psi_{out} = \psi_{acc}$$

Thus, the accumulation of positive charge in the system is equal to the difference of the Input and the Output. This choice of system boundary gives us some useful information, namely that the difference between the influx and outflow of ions is the amount of charge that accumulates in the cell. An actual numerical value for the accumulation of charge can be calculated when measured experimental values of net influx and outflow of charge are known.

System C: Last, we examine how the system descriptions change when the system boundary includes only an ion pump. Again, the nonreacting system is open, since positive charges enter and leave the ion-pump boundary. Since the charges move through the pump and do not remain there, no charges accumulate within the system, the initial and final conditions are the same, and the system is at steady-state.

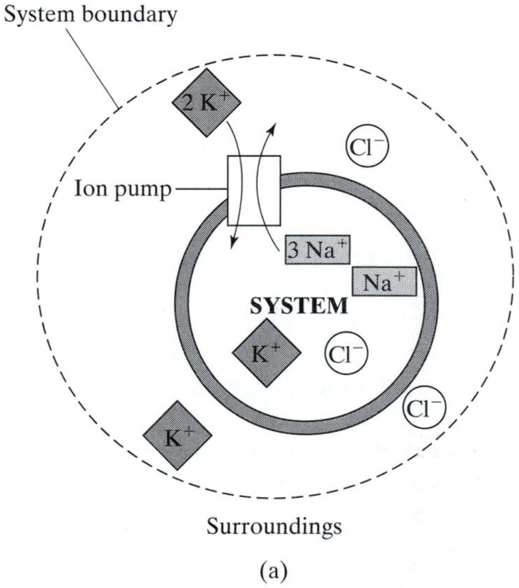

(a)

Figure 2.16a
System that includes cell and immediate extracellular space.

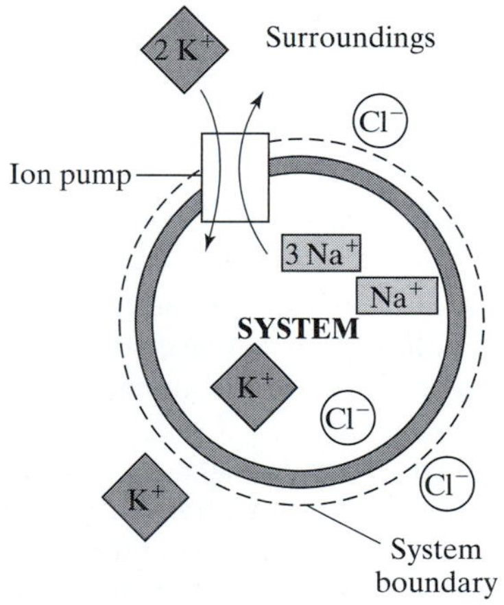

(b)

Figure 2.16b
System containing only the cell.

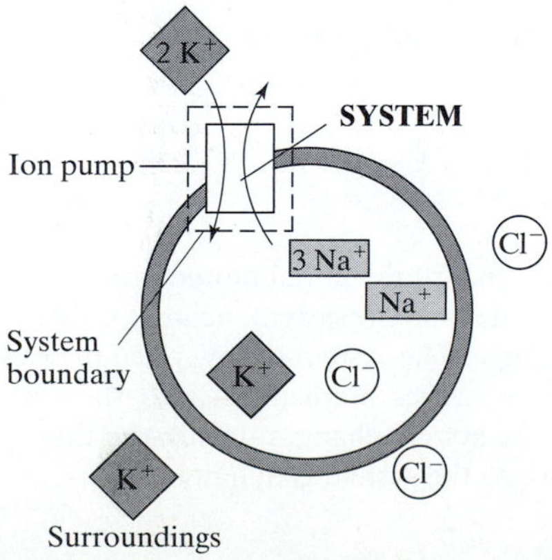

(c)

Figure 2.16c
System containing one ion pump.

As in the other two systems, the conservation equation is appropriate. As in system A, you can also set the Accumulation term to zero, since the system is at steady-state:

$$\psi_{in} - \psi_{out} = \psi_{acc} = 0$$
$$\psi_{in} = \psi_{out}$$

This equation states that all positive charges entering the pump also leave it. This statement certainly makes sense, although it provides little useful information about how the movement of positive ions affects the membrane potential of the cell.

To reiterate, where you define the system boundary makes a significant impact on the reductions to the accounting equation. In systems A and C, true statements describing the systems were derived, but neither was useful for understanding membrane potential. In contrast, the system boundary defining system B enabled the derivation of an equation that captured the behavior of positive ions through a Na^+/K^+ pump. ■

EXAMPLE 2.16 Collision of Plaque in an Atherosclerotic Vessel

Problem: Atherosclerosis is the buildup of plaque—fatty deposits, cholesterol, calcium, and other substances—that eventually blocks blood flow through the arteries. Usually when a material collides with an atherosclerotic lesion, it sticks to the plaque buildup already present and hardens over time. Suppose a fatty deposit is carried by the blood at a known velocity to the lesion site, where it gets trapped by the plaque buildup. Ignore the effects of gravity.

Consider linear momentum as the extensive property being tracked. Consider the entire artery (Figure 2.17a), which includes the fatty deposit and atherosclerotic site. How would you characterize the system? What governing equation is appropriate to use to find the linear momentum after the collision? What reductions to the governing equation can you apply? What happens if you change the system boundary so that the system is only the atherosclerotic lesion site (Figure 2.17b)?

Solution: When the system boundary encloses the entire artery as the system, momentum is not transferred to the system by the movement of mass across the system boundary. Because the extensive property of interest does not enter or leave through bulk material transfer, the system can be characterized as closed. Also, no external forces, such as gravity, act on the system. In the absence of any momentum transfer mechanism, the system is considered isolated.

When a fatty deposit collides with the plaque, it does not chemically react with the substance buildup already present at the atherosclerotic region. Since no reaction takes place, the system can be classified as nonreacting. The momentum of the system does not change between the initial and final conditions, making the system steady-state.

By definition, linear momentum is a conserved property, and no Generation or Consumption terms are present. Because the system is isolated, both the Input and Output terms are eliminated. Thus, the Accumulation term is equal to zero, and the system is at steady-state:

$$\psi_{acc} = \psi_f - \psi_0 = 0$$
$$\psi_0 = \psi_f$$

Thus, the final momentum is equal to the initial momentum of the system.

When we change the boundary so the system includes only the atherosclerotic region, the descriptions of the system change. The system is now open instead of isolated, since the fatty deposit crossing the boundary carries momentum into the system through bulk material transfer. The momentum of the system changes during the time period, since the initial and final conditions are not equal, so the system is dynamic.

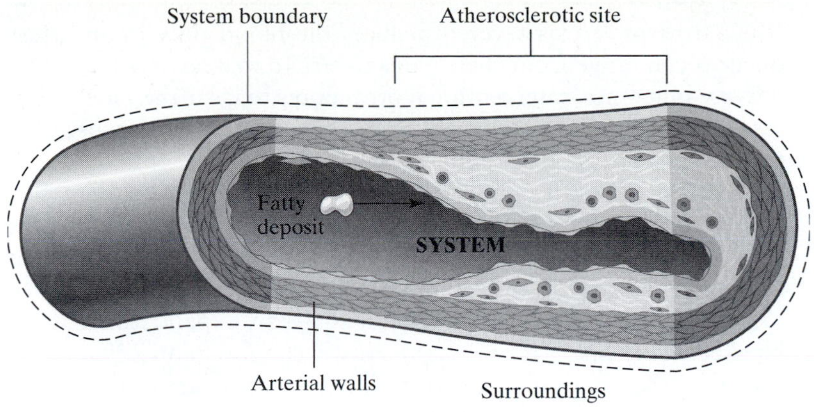

(a)

Figure 2.17a
System that includes fatty deposit and atherosclerotic site.

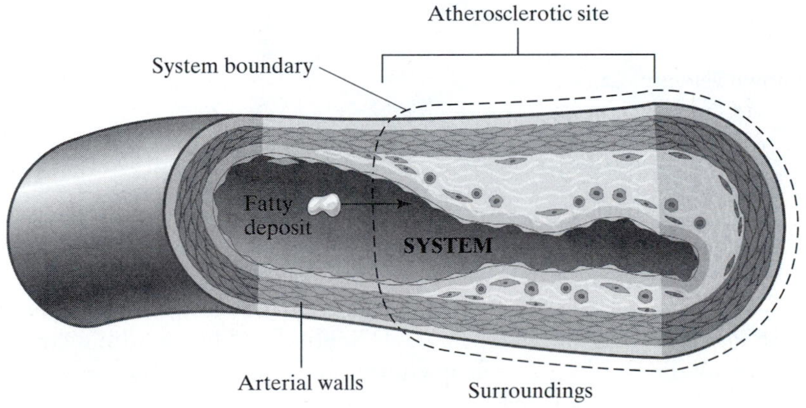

(b)

Figure 2.17b
System that includes only the atherosclerotic site.

As for the other system, we can use the conservation law as the governing equation. Since this system is open and dynamic, the Input, Output, and Accumulation terms may be nonzero. Given that no mass, and hence no momentum, leaves the system boundary, the Output term is zero. The initial momentum of the system is zero, since it has no initial velocity:

$$\psi_{in} - \psi_{out} = \psi_{acc} = \psi_f - \psi_0$$
$$\psi_{in} = \psi_f$$

Thus, the momentum of the system at the final condition is equal to the momentum added to the system.

If you were to assign masses to the fatty deposit and the plaque buildup and a velocity to the fatty deposit, the governing equations would reduce similarly. In this case, the two proposed system boundaries yield similar information, even though the system boundaries were established at different places. ■

When atherosclerosis severely reduces the blood flow in an artery, the consequences can range from chest pains to stroke to a heart attack. To prevent such adverse outcomes, interventional procedures to clear the blocked artery are necessary. These procedures range from oral medicines to transcatheter intervention to surgery.

One of the most common treatment procedures is minimally invasive angioplasty. A deflated balloon-tip on a catheter is inserted into a vein in the patient's groin, guided through the vessels to the blockage, and inflated to flatten out the plaque against the artery wall. Often surgeons also use the balloon-tip to insert a stent, a wire-mesh skeleton that can expand to hold the newly unblocked artery open (Figure 2.18). However, more than 25% of patients experience tissue scarring caused by abrasion from the edges of the metal stent, which can cause plaque to build up and block the artery again.

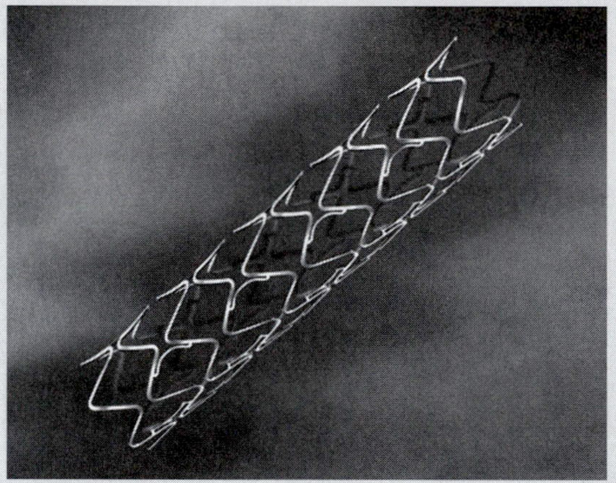

Figure 2.18
Fully expanded stent.

To combat this problem, bioengineers and surgeons have combined efforts to improve the stent. In 2003, the U.S. Food and Drug Administration approved one potential solution: a drug-eluting stent that prevents scar tissue from growing. With fewer adverse outcomes, early results are promising; hopefully this new generation of stents will perform well in the long term.

2.6 Summary of Use of Accounting and Conservation Equations

As discussed in Section 2.3, some extensive properties should be counted using accounting equations, whereas others can be counted using conservation equations. Table 2.4 summarizes which equations can be used to count specific extensive properties. Note that the accounting equation is *always* valid, while the conservation equation is appropriate only for those extensive properties that are conserved. Because the accounting equation is always valid, we use it as the starting point for a majority of the example problems in this book.

Determining when and how to employ conservation and accounting equations in a variety of medical and biological systems is the major focus of Chapters 3–7. Frequently, solving a problem is not what makes bioengineering difficult; rather, it is setting up the system, evaluating the system parameters, and writing the appropriate accounting and conservation equations that is challenging. Regardless of your

TABLE 2.4

Use of Accounting and Conservation Equations

Name of property	Accounting equation universally valid?	Conservation equation universally valid?	Number of scalar equations
Total mass	Yes	Yes	1
Species mass	Yes	No	m, for m species
Elemental mass	Yes	Yes	n, for n elements
Total moles	Yes	No	1
Species moles	Yes	No	m, for m species
Elemental moles	Yes	Yes	n, for n elements
Total energy	Yes	Yes	1
Thermal energy	Yes	No	1
Mechanical energy	Yes	No	1
Electrical energy	Yes	No	1
Net charge	Yes	Yes	1
Positive charge	Yes	No	1
Negative charge	Yes	No	1
Linear momentum	Yes	Yes	3
Angular momentum	Yes	Yes	3

Adapted from Glover C, Lunsford KM, and Fleming JA, *Conservation Principles and the Structure of Engineering*, 4th ed. New York: McGraw-Hill, Inc., 1994.

mathematical and problem-solving skills, an inappropriately drawn system or an invalid equation will lead to incorrect answers. Thus, making you comfortable with defining a system and writing its corresponding accounting or conservation equation is a major goal of this book.

The number of **scalar equations** is the number of equations that can be written for an extensive property in a system. For example, with elemental mass, one equation can be written for each element n in the system. Linear and angular momentum each have three scalar equations, since there are three directional axes (e.g., x, y, z in Cartesian; r, θ, ϕ in spherical). **Independent equations** are those that are mathematically linearly independent from one another. (If one equation in a group can be formed by a linear combination of the other equations, then that equation is not linearly independent.) The number of independent equations for a system is equal to the number of linearly independent equations that can be written for a system.

Most biological systems contain many different components or chemical constituents. Many different types of mass and mole accounting equations can be written to describe a system. One species mass accounting equation can be written for each of m species in a multicomponent situation. One conservation equation for total mass can also be written. Thus, for a system with m species, $m + 1$ mass balance equations can be written. Of these, only m equations are linearly independent. One elemental mole accounting equation can be written for each of n elements in a system. One total mole accounting equation can also be written. Of these $n + 1$ mole balance equations, only n are linearly independent. The situation is similar when considering all forms of species and elemental mass and mole balance equations.

One scalar accounting or conservation equation can be written for the properties total energy, thermal energy, mechanical energy and electrical energy. Although they are linearly independent from one another, these equations are typically not used together to solve a problem.

One scalar conservation equation can be written for net electric charge, and one accounting equation can be written each for positive and negative electric charge. However, only two of these three charge equations are linearly independent. As will be shown in Chapter 5, the positive and negative electric charge equations can be added together to generate the net charge equation.

Linear and angular momentum are three-dimensional vector quantities. Thus, three scalar, linearly independent equations can be written each for linear momentum and angular momentum.

Summary

In this chapter, we defined how engineering processes can be captured with mathematical statements in the forms of the accounting and conservation equations. To describe the system concerning the particular extensive property of interest, we characterized the system parameters and used these assumptions to find reductions to the governing mathematical equations, which can be in the algebraic, differential, or integral form. We also introduced the concepts of conservation; system, system boundary, and surroundings; open, closed, and isolated; reacting and nonreacting; energy interconversion; and steady-state and dynamic. These concepts and terms help define and reduce the Input, Output, Generation, Consumption, and Accumulation terms found in the accounting equation.

Reference

1. Population Reference Bureau. "Population Reference Bureau." 2004. www.prb.org.

Problems

2.1 For problem parts (a) through (q), do the following:

- Draw a picture of the system.
- Name an extensive property that can be counted.
- Label the system, surroundings, and system boundary.
- State the time period of interest. Explain.
- Identify the system as open, closed, or isolated and state why.
- Identify the system as steady-state or dynamic and state why.
- Identify whether the system has a reaction and/or energy interconversion and state why.
- Should an algebraic, differential, or integral equation be used to describe the system? State why.
- Is the selected extensive property conserved in this system? State why.

(*Note*: There can be more than one correct answer; it depends on how you set up the system.)

(a) Blood is flowing through the heart. Consider the inlets to the heart as the pulmonary vein and the vena cava and the outlets from the heart as the pulmonary artery and the aorta. Ignore the coronary artery and cardiac veins. You want to write a model of the blood flowing through the heart that considers changes that occur during one second or less (i.e., a model that looks at different points in the cardiac cycle).

(b) Blood is flowing through the left side of the heart. The inlet is the pulmonary vein; the outlet is the aorta. Momentum is carried by the blood as it flows. Momentum is also added to the fluid through the force exerted by the heart as it pumps. You want to write a model that quantifies the momentum of the left side of the heart on the time scale of minutes to hours.

(c) A new drug is produced in a bioreactor. The process is termed a batch operation, since the reactants are added all at the beginning, the system is sealed and the reaction takes place, and then the products are removed. Within this processing system, the concentration of drug in the reactor increases as the reaction proceeds. Consider an accounting equation that tracks the drug only during the time that the reaction is occurring.

(d) The liver converts toxins to more innocuous constituents. Arteries and veins transport the blood containing the toxins into and out of the liver. The liver works continuously to detoxify materials, and the conditions of the liver do not change with time. Toxins do not accumulate in the liver. You are interested in writing an accounting equation on the toxins that are transformed in the liver.

(e) A cell in the kidney works constantly to keep the ion balance in the blood correct. The cell membrane contains ion pumps and channels that move Na^+ ions across the membrane. Looking specifically at one type of pump, Na^+ ions are transported from the inside of the cell to the outside of the cell. Assume that the cell does not generate or consume Na^+ ions. You are interested in writing a model on the positive charge contributed by Na^+ ions as they move across the cell membrane through the pumps.

(f) A calorimeter can determine a person's metabolic rate by measuring the quantity of heat liberated from the body in a given period of time. The calorimeter consists of a large air chamber with well-insulated walls that prevent energy transfer into or out of the calorimeter. As the person's body produces heat, the air temperature in the chamber is maintained constant by forcing the air through pipes in a cool water bath. A thermometer measures the temperature increase, which is used to calculate the rate of heat gain by the water bath, which equals the rate of heat release from the person's body. You want to measure your metabolic rate as you exercise in the calorimeter.

(g) A heart-lung bypass machine is used to circulate blood through the body during open heart surgery when the heart is stopped. Flowing blood enters and leaves the bypass machine. Special biocompatible material is used to line the walls of the machine so that no reactions occur in the blood. Energy in forms of heat and mechanical work enters the bypass machine. During an operation, the bypass machine is maintained at constant temperature and other operating conditions. You are interested in writing an accounting equation on the total energy of the bypass machine.

(h) The effects of osmotic water shifts on red blood cells (RBC) can be observed experimentally by exposing the cells to hypertonic and hypotonic saline solutions. The inner fluid of RBCs is isotonic with 0.15 M NaCl. When RBCs are placed in a hypertonic solution, water leaves the cells, causing them to shrink. When RBCs are placed in a hypotonic solution, water enters the cells, causing rapid swelling, which may result in the bursting of some cells. An aliquot of 10^5 RBCs is added to 1 L of water with 0.05 M NaCl. Assume that no metabolic activity is occurring in the

RBCs to generate water. You are interested in writing an accounting equation on the water in the RBC as a function of time. Will the size of this system change?

(i) Poly(lactic acid), p(LA), is a biodegradable polymer currently approved by the FDA for use in humans as suture material. One limitation is that p(LA) undergoes biochemical degradation in the body to lactic acid (LA), which is acidic. If lactic acid is not cleared from the sutured area quickly enough, the local pH will be altered, potentially causing damage to the surrounding tissue. You design an in vitro experiment to determine the fluid flow rate needed to remove enough lactic acid from the p(LA) site so as to alter the pH only one pH unit. Assume p(LA) undergoes bulk erosion, which means that the degradation rate is not constant. Assume that the local flow rate of solution is constant. Answer the above questions for the properties of LA and p(LA).

(j) A hydrogel is a unique polymer network that absorbs water. To determine the equilibrium swelling content, the gel is first weighed dry. Then it is submerged in water. Water is absorbed into the gel in a time-dependent manner. After one hour, the gel is removed, the surface is blotted dry, and the sample is reweighed. Absorption of water is not a chemical reaction. Answer the above questions for the properties of water and hydrogel during the process of determining the equilibrium swelling content.

(k) A burn patient comes into a hospital emergency room. She is immediately connected to an IV bag to replace the fluids she has lost during an accident and is currently losing by evaporation from her skin. Initially, you must administer IV fluid at a higher rate than she is currently losing it to replace the fluid lost in the accident. Once the patient is stabilized, you want to administer the fluid at the same rate that she is losing it through evaporation. You are interested in writing an accounting equation to determine the rate of IV fluid replacement. Answer the above questions for the initial time period and for the time after she is stabilized.

(l) When a person gives blood, the blood flows from his body into a collection bag for about half an hour. You are interested in writing a model to determine the blood in the person's body.

(m) Soon after you wake up on a mid-December morning, you grab a cup of hot coffee. You instantly feel the warmth of the mug in your hand. You would like to know how the rate of energy from the cup to your hand changes with time.

(n) Sprinters and long-distance runners rely on muscles to propel them as they run. Muscles require oxygen and glucose for contraction. Metabolism of these reactants generates carbon dioxide and lactate. Consider chemical compounds in the adductor longus muscle in the leg of a sprinter during a 100-m race.

(o) Your roommate, a sociology major, does not understand the principles of heat conduction and grabs the metal handle of a pot of boiling water. The pain begins when outer nerve cells in the skin depolarize. This triggers an action potential along the axons of nerves until the signal reaches the brain. You are interested in modeling the signal conduction as a current from the hand to the brain during the time period before your roommate screams.

(p) Tumor cells are known for their rapid replication times. Using high-resolution imaging equipment, you are able to estimate the number of cells in the tumor. You are asked to determine the rate of growth of the tumor. Will the system boundary change size and/or shape?

(q) Nerve growth factor (NGF) is from the neurotrophin family of proteins that has been shown to have trophic effects on some cholinergic systems of the brain. Since it helps prevent neurons from dying, work has been done to see if NGF is viable as a therapeutic agent in the treatment of Alzheimer's disease or other neurodegenerative disorders. One possible treatment involves fabricating polymers into porous implants loaded with NGF powder and surgically placing them within the brain. Consider a model to track implanted NGF in the brain.

2.2 Water and solids enter and leave the human body through various means. The masses of water and solids for an average man for one day are shown in Figure 2.19.

(a) Write an algebraic accounting statement for the total solids in the system. Is the system open or closed, steady-state or dynamic, reacting or nonreacting? Calculate the total solids entering and leaving the system. Since Input does not equal Output, postulate an explanation.

(b) Write an algebraic accounting statement for water in the system. Is the system open or closed, steady-state or dynamic, reacting or nonreacting? Plug the mass of water entering, leaving, and being generated in the system into the algebraic accounting statement. Can all the water be accounted for?

2.3 A patient in the hospital is being given a saline solution through an IV. Each day, she receives 1200 g of water through the IV. She receives water through no other means. A catheter collects all the urine leaving her bladder. It is determined that the daily water loss in the urine is 1600 g. Assume that water leaves her body through no other means. Metabolic activity is known to be normal.

While the patient in the hospital for a week, the doctor notices that she loses no weight. (From this, assume that the mass of water in the body does

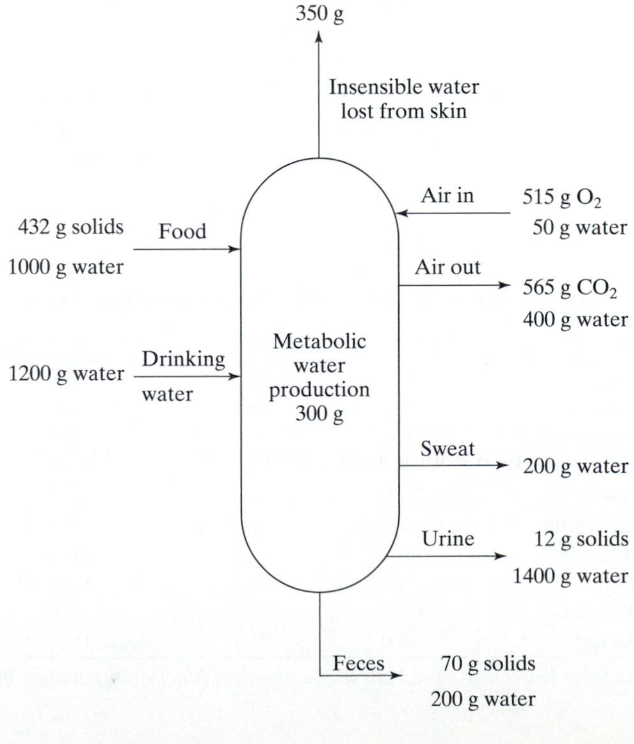

Figure 2.19 Water and solid matter production, consumption, and waste of an average man. (*Source*: Cooney DO, *Biomedical Engineering Principles: An Introduction to Fluid, Heat, and Mass Transport Processes*. New York: Marcel Dekker, 1976.)

not change with time.) Yet, a quick mass balance on water that incorporates only the Inlet and Outlet terms doesn't make sense. Help the doctor figure out what is going on by defining whether the system is open or closed, steady-state or dynamic, reacting or nonreacting. What is going on? Try to find some biological evidence to support your hypothesis. Finally, write the accounting equation with appropriate terms to describe the water balance in the patient.

2.4 For breakfast, Joe had 200 g oatmeal, 75 g milk, and 1 orange (225 g). For lunch, he had 1 apple (100 g), 4 pieces of bread (100 g), 90 g bacon, and 40 g cheese. For dinner, Joe had 350 g pork, 150 g asparagus, 150 g potatoes, and 2 pieces of bread (50 g). The protein, fat, and carbohydrate content as well as the energy content of the different foods are given in Table 2.5. You may use Excel, MATLAB, or another program of your choice for calculations.

(a) Assuming that the solids content is 30% and the water content is 70%, what is the total solids and water intake through food for Joe in one day? How does this value compare with that of the average man (Problem 2.2)?

(b) Calculate the total grams of protein, fat, and carbohydrates derived from each food.

(c) Calculate the energy derived from each food based on the fuel values.

(d) The physiologically available energy in food is as follows: carbohydrates–4 calories/gram, fat–9 calories/gram, and protein–4 calories/gram. Using the total amount of protein, fat, and carbohydrates in Joe's diet, calculate the available energy from protein, fat, and carbohydrates. Average Americans receive 15% of their energy from protein, 40% from fat, and 45% from carbohydrates. Compare Joe to this average.

(e) How similar are the total energy values calculated in parts c and d?

2.5 Mammalian cells are cultured (grown) in a bioreactor. The chemical building blocks of cells are carbon, hydrogen, nitrogen, and oxygen; cells are often modeled as $CH_{\alpha}N_{\beta}O_{\delta}$.

(a) To begin a batch process, 50 L of cells at 100 g/L are initially added to the reactor. After operation, the cell concentration is 25 g/L and the cell mass fills the entire reactor volume of 1000 L. Determine the amount of $(NH_4)_2SO_4$ to be supplied, assuming that the cells are 12 wt% nitrogen and that $(NH_4)_2SO_4$ is the only nitrogen source.

(b) During a continuous operation, the steady-state cell concentration in the reactor is 20 g/L, and the cell mass fills the entire reactor volume of

TABLE 2.5

Protein, Fat, Carbohydrate, and Energy Content of Food*

Food	% Protein	% Fat	% Carbohydrate	Fuel value /100 g (kcal)
Apples	0.3	0.4	14.9	64
Asparagus	2.2	0.2	3.9	26
Bacon, broiled	25.0	55.0	1.0	599
Bread, white	9.0	3.6	49.8	268
Cheese	23.9	32.3	1.7	393
Milk, whole	3.5	3.9	4.9	69
Oatmeal	14.2	7.4	68.2	396
Orange	0.9	0.2	11.2	50
Pork, ham	15.2	31.0	1.0	340
Potatoes	2.0	0.1	19.1	85

*Data from Guyton AC and Hall JE, *Textbook of Medical Physiology*. Philadelphia: Saunders, 2000.

1000 L. Assume that no cells enter the reactor and that the product stream containing the cells leaves the reactor at a rate of 20 L/day. Determine the mass flow rate of $(NH_4)_2SO_4$ to be supplied in the feed stream, assuming that the cells are 12 wt% nitrogen and that $(NH_4)_2SO_4$ is the only nitrogen source.

(c) During batch and continuous operations in the real world, the added nitrogen source is in 20% excess of stoichiometric needs. Redo the calculation for (b), assuming that a 20% excess of nitrogen is supplied in the feed stream.

2.6 The body needs a constant supply of energy in order to survive. The minimum level of energy required just to perform chemical reactions in the body and maintain essential activities of the central nervous system, heart, kidney, and other organs is known as the basal metabolic rate (BMR). However, if an individual is to engage in such activities as eating and walking, additional energy must be available. On average, an individual performing normal daily activities expends 2750 kcal/day. The daily energy expenditure is comprised of maintaining the BMR, digesting and processing food (220 kcal), nonexercise activities such as maintaining body temperature (190 kcal), and purposeful physical activity (690 kcal).

(a) Given that breathing accounts for 5% of the BMR, calculate the energy required for an individual at rest to breathe. Report your answers in units of joules/breath.

(b) Heavy exercise can increase the daily energy expenditure to 7000 kcal. In addition, exercise can increase the energy requirements for breathing about 20-fold. Calculate the energy expended for purposeful physical activity when exercising. Report your answer in units of joules/breath.

2.7 Consider Problem 2.1, part (q). It has been experimentally determined that the implant can release 0.24 μg NGF/day. In the brain, 0.11 μg NGF/day is eliminated due to metabolic processes and 0.023 μg NGF/day is lost due to nonspecific binding.

(a) Write a generic accounting equation to track NGF. Which (if any) terms can be eliminated?

(b) Neurons do not exhibit any reaction to NGF until the concentration reaches 2.0 ng/mL. What is the rate of accumulation of NGF in the brain? How long does it take before the entire brain (volume $\cong$ 1400 cm^3) reaches a therapeutic level? Assume perfect mixing and that the volume taken up by the implant is negligible.

(c) More realistically, you would want to treat only the diseased part of the brain. Alzheimer's disease is often associated with the death of cholinergic neurons in the basal forebrain (volume $\cong$ 400 cm^3). If the implant were placed in that region, how long would it take to reach 5.0 ng/mL?

2.8 Perform a "candle in the jar" experiment. In order to set up this experiment, light a candlestick whose base is in a pool of water. Cover the candle with a glass jar and observe what happens. The water level should be steady at first and should rise significantly when the flame goes out. Your goal is to quantitatively describe the total change in the height of the water in the jar as a function of system parameters. Predict only the final change in water height, not how the height of the water changes as a function of time. Develop an explicit mathematical relationship between the change in water height and system parameters; resist the temptation to describe the phenomenon only qualitatively.

(a) Perform an engineering analysis that quantitatively predicts how much the water will rise as a function of the key parameters of the system. Report the water rise in nondimensional terms such that the height of the water is 1.0 if the jar becomes completely filled and 0.0 if the jar is empty. Include in your analysis:

- Description of the experiment.
- Identification of all of the relevant physical laws and principles.
- Identification of the key system parameters that control how high the water rises.
- List of hypotheses of the possible main causes of the water rise. Include a systematic evaluation of each of these hypotheses. Your findings may help you simplify the model that you develop in your model.
- Mathematical formulation (model) that quantitatively relates the level of water rise to the key parameters of the system. Describe the model graphically with at least one graph that relates the change in the height of the water to one or more of the key system parameters.
- List of all assumptions and simplifications made to complete your model.
- Check(s) of your model. Did your model give reasonable results? Pay particularly close attention to the behavior of the model at extreme parameter values.
- Discussion of the limitations of your model and what steps you could take to improve the model.

(b) Summarize the process by which you "solved" this particular open-ended problem. Include a critique of the approach you took, identifying any limitations or weaknesses in your problem-solving approach, and a critique of your own personal knowledge that you used or needed to solve this problem. What principles and facts did you know before this problem that helped you? What principles did you have to learn or refamiliarize yourself with in order to solve the problem? What principles would you still have to learn if you were asked to further improve the quality of your model?

Conservation of Mass

CHAPTER 3

3.1 Instructional Objectives and Motivation

After completing Chapter 3, you should be able to do the following:

- Explain the different types of flow rates.
- Write the algebraic, differential, and integral mass accounting and conservation equations.
- Apply mass accounting and conservation equations correctly.
- Justify why conservation equations cannot be universally applied when counting species mass, total moles, and species moles.
- Explain the meaning and significance of a basis of calculation and how to select a proper one.
- Set up and solve mass accounting and conservation equations for systems with multiple streams and compounds.
- Understand the strategy of degree-of-freedom analysis for handling multi-unit systems.
- Isolate a small system or unit within a large system.
- Balance a complex chemical reaction.
- Know the definition and application of the reaction rate and the fractional conversion of a reaction, and the meaning and significance of the limiting reactant.
- Set up and solve mass accounting and conservation equations for reacting systems.
- Set up and solve mass accounting and conservation equations for dynamic systems.
- Comfortably use the Methodology for Solving Engineering Problems.

3.1.1 Tissue Engineering

Mass accounting and conservation equations are used widely in the field of bioengineering. When tracking or monitoring the mass of a particular compound or material, mass balance equations are helpful. Mass accounting and conservation equations are prevalent in systems involving chemical and biochemical reactions, such as the human body and bioreactors. In this chapter the conservation of mass is applied to a wide range of example and homework problems.

In this introduction we highlight tissue engineering, with a specific emphasis on bone. Tissue engineering is a diverse and growing field where conservation principles are routinely applied in order to model systems and to solve problems. The complex challenge below serves to motivate our discussion of mass accounting and conservation equations.

In the late 1980s, the National Science Foundation (NSF) formally debuted a new and exciting medical field:

> Tissue engineering is the application of principles and methods of engineering and life sciences toward fundamental understanding of structure-function relationships in normal and pathological mammalian tissues and the development of biological substitutes to restore, maintain, or improve tissue functions.[1]

Although it informally existed before its definition, the field of tissue engineering has expanded dramatically in recent decades. The widespread and successful applications of tissue engineering to blood substitutes, bone and cartilage replacements, and even neuronal and organ replacements have invigorated new prospects and technologies in a rapidly expanding field.

Tissue engineering involves the use of artificial or foreign biomaterials generated in the laboratory to restore or replace human tissues lost or damaged by disease, trauma, age, or congenital abnormalities. Its expansive domain requires the cooperation of many medical and technical disciplines, including cell biology, molecular biology, cellular and tissue biomechanics, biomaterials engineering, computer-aided design, and robotics engineering. Successful application often requires a balanced and skilled team, ranging from bioengineers, to chemical engineers, to molecular biologists, to bioreactor technicians, and a variety of other specialists.

Tissue engineers have endeavored to grow virtually every type of human tissue, including skin, cartilage, tendon, bone, muscle, blood vessels, cardiovascular valves, liver tissue, urinary bladders, nerves, and pancreatic islets. Synthetic skin was the first commercially produced tissue to reach the market and is used to treat patients with burns and diabetic ulcers. Tissue-engineered cartilage has become especially valuable, since damaged adult cartilage does not naturally regenerate or heal. The vast number of bone fractures and the prevalence of osteoporosis has drawn considerable attention to using tissue engineering to improve bone structure.

Each year, over 200,000 people in the United States undergo hip replacement surgery with synthetic prostheses to decrease pain and restore mobility.[2] An estimated 800,000 patients are hospitalized annually with severe bone fractures. Many of these fractures heal improperly or incompletely, thus requiring supplemental procedures like bone grafts. Some commonly practiced procedures include autografts (relocating a patient's healthy tissue to replace tissue loss in another area), allograft (using a human donor's tissue, usually from a cadaver, to replace damaged or degraded tissue), or synthetic materials (metal implants, such as plates and screws). Each year, over 500,000 patients receive bone grafts, with approximately half of these procedures related to spine fusion.[3]

Patients receiving autografts and allografts must cope with possible negative immune responses, and they often still experience bone loss if the grafts do not take hold. Metal or alloy implants, while structurally sound, often fail or degrade after a time period, since the materials cannot mimic the continuous cyclic process of bone formation and resorption, so crucial to sustaining the interaction of the implant with the surrounding native bones. However, tissue engineering offers a new paradigm in these applications by synthesizing a biomaterial to replace the bone mass, thereby alleviating some of the complications associated with grafts and implants.

Resorbable polymers, which the body naturally absorbs as the tissue heals and rebuilds, are promising, since the wide range of polymer building blocks allows engineers to design specific chemical and mechanical characteristics, as well as the degradation rate. The construct is initially designed to replace the load-bearing capability of the lost bone. As the implanted polymer decays, new bone tissue is synthesized at the same rate.

While researchers continue testing new options, bone substitutes, such as VITOSS (Orthovita; Malvern, PA), are being used. VITOSS is an example of a biodegradable, absorbable scaffold that facilitates interconnection with the host bone, bone remodeling, and vascularization. After the natural bone has regenerated, the minerals that compose the scaffold are incorporated into the body. Although it behaves like a biodegradable polymer, VITOSS is actually composed of nanoparticles of calcium and phosphate, the primary constituents of bone.

Researchers are also designing polymer constructs with cells or molecular signals that promote bone formation. For example, a biodegradable polymer is shaped and seeded with living cells to recreate its intended tissue function. The polymer is then bathed in growth factors in vitro to stimulate proliferation, creating a three-dimensional tissue as the cells multiply throughout the scaffold. Upon implantation, the scaffold either dissolves or is absorbed, and blood vessels innervate the implanted tissue, allowing nutrients and other materials to be transferred to and from the tissue, a process called vascularization. Thus, the newly grown tissue eventually assumes the same structural and functional role as the original, native tissue.

Despite the progress in tissue engineering, the potential for this technology is not fully matured. Although several synthetic biomaterials have been approved for human use, such procedures are still mainly in the clinical testing stage. As the insufficiencies of current technologies become clearer and the feasibility of using a synthetic biomaterial to replace damaged tissue increases, the landscape of clinical options for tissue repair or replacement continues to broaden. The success of this innovative technology depends on the ability of engineers and scientists to overcome the technical hurdles in designing synthetic tissue replacements. The following list focuses on a few issues specific to bone tissue:

- *Knowledge base*: Thorough comprehension of the essential elements of osteogenesis, such as cell distribution and growth factors, is necessary.
- *Vascularization of bone substitutes*: To achieve proper cell growth and tissue formation, the biomaterial must accommodate blood flow similar to that of the native tissue.
- *Tissue architecture*: Although tissues grown in vitro secrete the correct biochemicals necessary to maintain structure, engineers have yet to design tissues to form into the proper architecture, which is imperative for proper tissue function.
- *Degradation rate*: The material must degrade at the same rate as bone regenerates.
- *Mechanical properties*: The biomaterial must be porous to facilitate natural tissue regeneration but simultaneously strong enough to support routine applied forces in the native tissue.
- *Toxicity*: The degradation products of the biomaterial should not harm the patient and should ideally elicit no foreign-body immune response.

Multidisciplinary teams around the world tackle these research challenges in industry, government labs, and academia. Along with many unique assays and computational tools, bioengineers use mass balances to help them model different aspects of tissue engineering. To tie this material into the chapter, we examine how mass accounting and conservation equations can be used to evaluate bone grafts in Examples 3.6, 3.19, and 3.21. Remember that tissue engineering is only one of many exciting areas where mass accounting and conservation equations can be applied to bioengineering and its related fields.

This chapter opens with an overview of basic mass concepts and then discusses how system definitions can be applied to solve systems involving mass. We also discuss how to solve systems with multiple components or units. Systems involving chemical reactions change how the governing accounting equation is applied. Finally, we show how to use the governing equations to solve dynamic systems.

The conservation of mass is presented first in this text because it is used to solve more complex problems in conjunction with the conservation of total energy (Chapter 4) and linear momentum (Chapter 6), as well as the accounting of electrical energy (Chapter 5) and mechanical energy (Chapter 6).

3.2 Basic Mass Concepts

The amount of a material is expressed through the base physical variables of mass or mole. **Mass** (m [M]) is a quantity of matter that has weight in a gravitational field. The **mole** (n [N]) is a base unit describing an amount of any substance containing Avogadro's number of molecules of that substance. One mole contains 6.02×10^{23} atoms of that element, and has a mass, in grams, equal to the atomic weight of the element. For example, a single molecule of O_2 has an atomic weight of 32.0 amu; in one mole of O_2, there are 6.02×10^{23} molecules of O_2, having a mass of 32.0 g collectively. Common units of mass are g, kg, and lb_m. Common units of mole are g-mol (usually written as mol) and lb_m-mol.

The molecular weight (M) of component A is related to the mass (m) and the number of moles (n) of that component:

$$n_A = \frac{m_A}{M_A} \qquad [3.2\text{-}1]$$

Molecular weight has the dimension [MN^{-1}]. Common units are g/mol and lb_m/lb_m-mol. Molecular weights of elements are on the periodic table (Appendix C).

A **flow rate** describes the transport of material over a period of time. For example, suppose a conduit has a volumetric flow rate of 25 L/hr. A detector set up at a specific point along the conduit would measure 25 L of material passing the detector point in a one-hour period. Accounting and conservation equations are developed with three types of flow rates: mass flow rate, volumetric flow rate, and molar flow rate.

The **mass flow rate** ($\dot{m}$ [Mt^{-1}]) is the rate of movement of mass and is calculated:

$$\dot{m} = Av\rho \qquad [3.2\text{-}2]$$

where A is the cross-sectional area of the conduit, v is the velocity of the fluid, and ρ is the fluid density. Because $\dot{m}$ is a scalar quantity, it is not necessary to denote a direction for the velocity. For a cylindrical conduit, the cross-sectional area is a circle, giving:

$$A = \frac{\pi}{4}D^2 = \pi r^2 \qquad [3.2\text{-}3]$$

where D is the vessel diameter and r is the vessel radius.

The **volumetric flow rate** ($\dot{V}$ [L^3t^{-1}]) is the rate at which a volume of material flows and is described by:

$$\dot{V} = Av = \frac{\dot{m}}{\rho} \qquad [3.2\text{-}4]$$

EXAMPLE 3.1 Density Calculation

Problem: Using only a scale, a graduated cylinder, and a stopwatch, devise a way to determine the density of a fluid flowing in a pipe.

Solution: The density of the fluid can be determined by rearranging equation [3.2-4]:

$$\rho = \frac{\dot{m}}{\dot{V}}$$

By determining the time required to collect an arbitrary volume of fluid (e.g., 1 L), the volumetric flow rate can be calculated by dividing the volume by the collection time. The mass of the sample can be found using the scale. The mass flow rate is determined by dividing the mass by the collection time. The mass and volumetric flow rates are then used to find the density of the fluid. (*Note*: The density can also be calculated without considering the collection time using $\rho = m/V$.) ■

Finally, the **molar flow rate** ($\dot{n}$ [Nt^{-1}]) through a conduit is calculated by dividing the mass flow rate $\dot{m}$ by the molecular weight M of the moving fluid:

$$\dot{n} = \frac{\dot{m}}{M} \qquad [3.2\text{-}5]$$

It is valuable to understand the relationship between variables in a system and to determine how a change in one variable affects other variables in the system. For example, we can investigate the relationship between any two of the variables in equation [3.2-2] if we hold the other variables constant. Suppose a fluid of constant density ρ moves at a constant $\dot{m}$ through a conduit with varying cross-sectional areas along the path of fluid flow. In this case, the velocity of the fluid must be changing along the path of the fluid flow to maintain a constant $\dot{m}$. If we assume a cylindrical vessel with an initial radius r_0 and a fluid flowing at velocity v_0, equation [3.2-2] can be rewritten as:

$$\dot{m} = \pi r_0^2 v_0 \rho \qquad [3.2\text{-}6]$$

If the vessel radius reduces to half its original value ($r_1 = r_0/2$), equation [3.2-6] becomes:

$$\dot{m} = \pi r_1^2 v_1 \rho = \pi\left(\frac{r_0}{2}\right)^2 v_1 \rho = \frac{\pi}{4} r_0^2 v_1 \rho \qquad [3.2\text{-}7]$$

For constant ρ and $\dot{m}$, the fluid velocity must increase by a factor of four. That is, in order for the mass flow rate to remain constant as the radius of the conduit reduces by half, the velocity of the fluid must be four times faster than the initial velocity (i.e., $v_1 = 4v_0$). Conversely, if the vessel radius doubles ($r_2 = 2r_0$), equation [3.2-6] becomes:

$$\dot{m} = \pi r_2^2 v_2 \rho = \pi(2r_0)^2 v_2 \rho = 4\pi r_0^2 v_2 \rho \qquad [3.2\text{-}8]$$

For a constant ρ and $\dot{m}$, the fluid velocity must decrease by a factor of four. That is, in order for the mass flow rate to remain constant as the radius of the conduit doubles, the velocity of the fluid must be four times slower than the initial velocity (i.e., $v_2 = v_0/4$).

EXAMPLE 3.2 **Constriction of a Blood Vessel**

Problem: Atherosclerosis is a dangerous condition characterized by the accumulation of fatty deposits on arterial walls, forming plaques. As the fatty tissue thickens and hardens, blood flow is hindered, and the arterial wall erodes and diminishes in elasticity. Blood clots can begin to form around the plaques, posing additional danger if the clots rupture and cause debris to travel to the heart, lungs, or brain, frequently resulting in heart attack or stroke. Patients with diabetes, obesity, or high cholesterol are at the greatest risk of developing atherosclerosis.

Suppose an overweight diabetic patient has plaque buildup in his coronary artery, decreasing the vessel diameter by two-thirds and the blood flow velocity by 25%. Compare the mass flow rates in the coronary artery of a healthy patient with one who has atherosclerosis. If the blood has the same mass flow rate in both patients, what velocity must the blood flow be when the diameter is reduced? The average diameter of a healthy coronary artery is 2.5 mm, with a blood flow rate of 6.4 cm/s. The density of blood is 1.056 g/cm^3.

Solution: Since we model the coronary artery as a cylindrical vessel, we can substitute the cross-sectional area of a circle into equation [3.2-2]. For a healthy individual, the mass flow rate is:

$$\dot{m} = Av\rho = \frac{\pi}{4}D^2 v\rho = \frac{\pi}{4}(0.25\text{ cm})^2\left(6.4\frac{\text{cm}}{\text{s}}\right)\left(1.056\frac{\text{g}}{\text{cm}^3}\right) = 0.332\ \frac{\text{g}}{\text{s}}$$

For the diabetic patient with atherosclerosis, the diameter of the artery is reduced by 67% to 0.825 mm, and the blood flow velocity is reduced 25% to 4.8 cm/s. Solving for the mass flow rate of blood through his coronary artery using equation [3.2-2] gives 0.027 g/s, which is only about 8% of blood flow through a healthy coronary artery.

To calculate the velocity required to maintain the healthy mass flow rate with the reduced diameter, equation [3.2-8] is rearranged:

$$v = \frac{4\dot{m}}{\pi D^2\rho} = \frac{4\left(0.332\frac{\text{g}}{\text{s}}\right)}{\pi(0.0825\text{ cm})^2\left(1.056\frac{\text{g}}{\text{cm}^3}\right)} = 58.8\ \frac{\text{cm}}{\text{s}}$$

Thus, to maintain the same mass flow rate as that in a healthy coronary artery, blood velocity through the diseased coronary artery is nine times greater. ■

Material balance equations are performed on the basis of any convenient amount or flow rate, and the consequent results can be scaled. A **basis of calculation** is an amount (mass or moles) or a flow rate (mass or molar) of one stream or stream component in a system upon which the material balance is solved. The succeeding calculations of other variables in the system are solved from this basis. Recall that step 2d in the Methodology for Solving Engineering Problems (Section 1.8) is to state a basis of calculation. An explicit basis is usually required for mass accounting and conservation equations.

If a specific amount or flow rate (either inlet or outlet) is given in a problem statement, it is convenient to use this quantity as a basis. When an amount or flow rate is already specified, it is unnecessary (in fact, wrong) to assign a numerical value to components entering or exiting the system that have unspecified values. Doing so can result in an overspecified problem statement or an incorrect solution or both. However, if no amounts or flow rates are known, you need to assume one by picking an arbitrary amount or flow rate for a specific component or stream in the system of interest. If possible, select an amount or flow rate for part of the system where the composition is known. If mass fractions are known, choose a total mass or mass flow rate (e.g., 100 kg or 100 kg/hr) as a basis. If mole fractions are known, choose a total number of moles or a molar flow rate (e.g., 100 mol or 100 mol/hr).

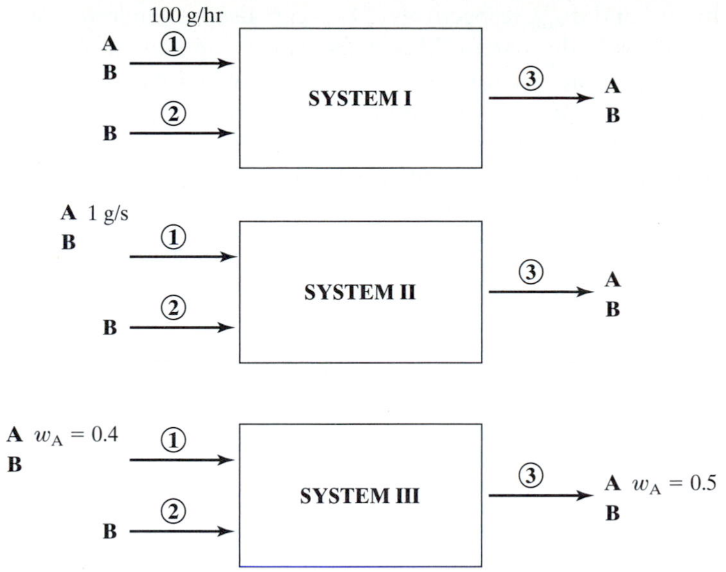

Figure 3.1
Mass flows in three different systems. A and B are compounds in the streams.

EXAMPLE 3.3 **Determination of a Basis**

Problem: Determine a basis for each of the systems in Figure 3.1.

Solution: In System I, a mass flow rate of 100 g/hr is given for stream 1. Therefore, the basis of 100 g/hr is selected for System I to solve for the flow rates of streams 2 and 3, as well as the flow rates of the individual components A and B in all three streams.

In System II, the mass flow rate of compound A in stream 1 is given. We can use this flow rate of 1 g/s of compound A in stream 1 as the basis for all other calculations.

In System III, no amount or flow rate is given. However, the mass fraction of compound A is given in streams 1 and 3. We can choose either stream—but not both—to serve as the basis. For example, we can arbitrarily define our basis such that stream 3 has a mass flow rate of 10 lb_m/hr. Therefore, the flow rate of compound A in stream 3 is 5 lb_m/hr. We could also have chosen a basis of 100 lb_m/hr for stream 3. In this case, all of the calculated flow rates in the system would have been a factor of 10 higher. Note that the ratios of any two streams to each other remain constant. This is an example of how the results can be scaled. ■

3.3 Review of Mass Accounting and Conservation Statements

Mass accounting equations mathematically describe the movement, generation, consumption, and accumulation of mass in a system of interest, and can be used to analyze any mass descriptor (e.g., total mass, total moles, species moles). Consider the system shown in Figure 3.2. Masses entering and leaving the system are represented

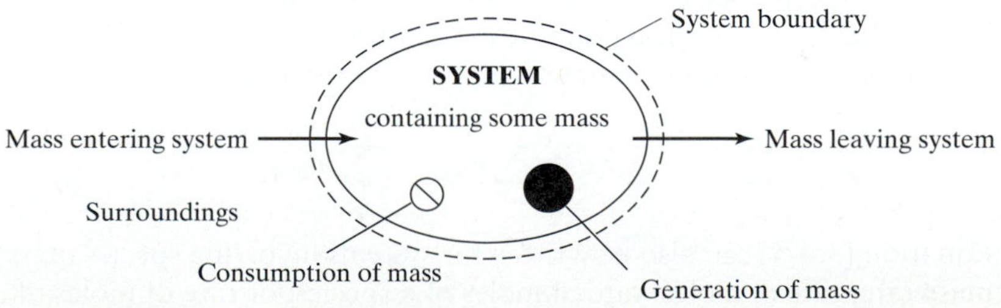

Figure 3.2
Graphical representation of mass accounting equation.

by m_{in} and m_{out}, respectively. Mass can also be generated or consumed by chemical reactions in the system. Mass may also accumulate in the system.

Algebraic equations can be applied when discrete quantities or "chunks" of mass (e.g., 10 kg of penicillin) are specified. Algebraic accounting equations can be applied when specific quantities of mass are given, but not when rates or time-dependent terms are involved. To review, the generic algebraic accounting equation is written:

$$\psi_{\text{in}} - \psi_{\text{out}} + \psi_{\text{gen}} - \psi_{\text{cons}} = \psi_{\text{acc}} \qquad [3.3\text{-}1]$$

ψ_{acc} can be defined by:

$$\psi_{\text{acc}} = \psi_{\text{f}} - \psi_0 \qquad [3.3\text{-}2]$$

See Section 2.4.1 for review of variable definitions.

When applying the generic accounting equation [3.3-1], the extensive property ψ can be replaced to solve for systems involving mass (m), species mass (m_s), element mass (m_p), moles (n), moles of a species (n_s), or moles of an element (n_p). For example, when solving the extensive property m, equation [3.3-1] is written as:

$$\sum_i m_i - \sum_j m_j + \sum m_{\text{gen}} - \sum m_{\text{cons}} = m_{\text{acc}}^{\text{sys}} \qquad [3.3\text{-}3]$$

For the moles, n, of a system, equation [3.3-1] is written as:

$$\sum_i n_i - \sum_j n_j + \sum n_{\text{gen}} - \sum n_{\text{cons}} = n_{\text{acc}}^{\text{sys}} \qquad [3.3\text{-}4]$$

The indices i and j represent the numbered inlet and outlet amounts, respectively. Summation signs indicate that every amount and/or process should be included. Equation [3.3-1] can be written for systems involving individual species mass, element mass, moles of a species, and moles of an element in a similar fashion to equations [3.3-3] and [3.3-4]. Terms in algebraic mass accounting equations have the dimension of [M] or [N].

The differential form of the accounting statement is more appropriate when rates are specified:

$$\dot{\psi}_{\text{in}} - \dot{\psi}_{\text{out}} + \dot{\psi}_{\text{gen}} - \dot{\psi}_{\text{cons}} = \dot{\psi}_{\text{acc}} = \frac{d\psi}{dt} \qquad [3.3\text{-}5]$$

See Section 2.4.2 for review of variable definitions. The Accumulation term is usually expressed as the instantaneous rate of change of the extensive property of the system, while all $\dot{\psi}$ terms in equation [3.3-5] are rates. The $\dot{\psi}_{\text{in}}$ and $\dot{\psi}_{\text{out}}$ terms are material flow rates entering and leaving the system, respectively, across the system boundary.

When applying the differential form of the accounting equation to mass, $\dot{\psi}$, the rate of the extensive property, can be replaced to solve for systems involving mass rate ($\dot{m}$), species mass rate ($\dot{m}_s$), element mass rate ($\dot{m}_p$), rate of moles ($\dot{n}$), rate of moles of a species ($\dot{n}_s$), or rate of moles of an element ($\dot{n}_p$). For example, for the mass rate $\dot{m}$ of a system, equation [3.3-5] is written as:

$$\sum_i \dot{m}_i - \sum_j \dot{m}_j + \sum \dot{m}_{\text{gen}} - \sum \dot{m}_{\text{cons}} = \dot{m}_{\text{acc}}^{\text{sys}} = \frac{dm^{\text{sys}}}{dt} \qquad [3.3\text{-}6]$$

Equation [3.3-5] can also be written for systems involving species mass rate, element mass rate, rate of moles, rate of moles of a species, or rate of moles of an element in a similar fashion to equation [3.3-6]. Terms in differential mass accounting equations have the dimension of [Mt^{-1}] or [Nt^{-1}].

The integral accounting equation is most useful when trying to evaluate conditions between two discrete time points:

$$\int_{t_0}^{t_f} \dot{\psi}_{\text{in}}\, dt - \int_{t_0}^{t_f} \dot{\psi}_{\text{out}}\, dt + \int_{t_0}^{t_f} \dot{\psi}_{\text{gen}}\, dt - \int_{t_0}^{t_f} \dot{\psi}_{\text{cons}}\, dt$$
$$= \int_{t_0}^{t_f} \frac{d\psi}{dt} dt = \int_{\psi_0}^{\psi_f} d\psi = \int_{t_0}^{t_f} \dot{\psi}_{\text{acc}}\, dt \qquad [3.3\text{-}7]$$

See Section 2.4.3 for variable definitions.

When applying the integral accounting equation to mass, $\dot{\psi}$ can be replaced by $\dot{m}$, $\dot{m}_s$, $\dot{m}_p$, $\dot{n}$, $\dot{n}_s$, or $\dot{n}_p$. For example, when solving for $\dot{m}$, equation [3.3-7] is written as:

$$\int_{t_0}^{t_f} \sum_i \dot{m}_i\, dt - \int_{t_0}^{t_f} \sum_j \dot{m}_j\, dt + \int_{t_0}^{t_f} \sum \dot{m}_{\text{gen}}\, dt - \int_{t_0}^{t_f} \sum \dot{m}_{\text{cons}}\, dt = \int_{t_0}^{t_f} \frac{dm^{\text{sys}}}{dt} dt \qquad [3.3\text{-}8]$$

Equation [3.3-7] can also be written for systems involving species mass rate, element mass rate, rate of moles, rate of moles of a species, or rate of moles of an element in a similar fashion to equation [3.3-8]. Terms in integral mass accounting equations have the dimension of [M] or [N].

Recall that the law of the conservation of mass states that total mass can be neither created nor destroyed, so total mass of a system is always conserved. (The one exception to this law is nuclear reactions that interconvert mass and energy, which is governed by the equation $E = mc^2$, where E is energy, m is mass, and c is the speed of light.) Since total mass is conserved, Generation and Consumption terms are both set to zero in that equation. Element mass and element moles are also conserved in all systems. Even in cases with chemical reactions, specific chemical elements are neither created nor destroyed. Thus, **the conservation of mass equation mathematically describes phenomena where mass is neither created nor destroyed and can be universally applied only when counting total mass, element mass, and element moles.**

To illustrate this mathematically, equation [3.3-1] is rewritten as the generic algebraic mass conservation equation as:

$$\psi_{\text{in}} - \psi_{\text{out}} = \psi_{\text{acc}} \qquad [3.3\text{-}9]$$

When applying the conservation equation to mass, ψ can be replaced for systems involving m, m_s, m_p, n, n_s, or n_p when no chemical reactions are occurring. In systems with chemical reactions, only m, m_p, and n_p may be substituted in equation [3.3-9].

Similarly, the conservation equation is written in differential form as:

$$\dot{\psi}_{\text{in}} - \dot{\psi}_{\text{out}} = \dot{\psi}_{\text{acc}} = \frac{d\psi}{dt} \qquad [3.3\text{-}10]$$

and in integral form as:

$$\int_{t_0}^{t_f} \dot{\psi}_{\text{in}}\, dt - \int_{t_0}^{t_f} \dot{\psi}_{\text{out}}\, dt = \int_{t_0}^{t_f} \frac{d\psi}{dt} dt = \int_{\psi_0}^{\psi_f} d\psi = \int_{t_0}^{t_f} \dot{\psi}_{\text{acc}}\, dt \qquad [3.3\text{-}11]$$

In equations [3.3-10] and [3.3-11], $\dot{\psi}$ can be replaced for nonreacting systems involving $\dot{m}$, $\dot{m}_s$, $\dot{m}_p$, $\dot{n}$, $\dot{n}_s$, or $\dot{n}_p$. In systems with chemical reactions, only $\dot{m}$, $\dot{m}_p$, and $\dot{n}_p$ may be substituted into equations [3.3-10] and [3.3-11].

TABLE 3.1

Appropriate Usages of Conservation Equations for Reacting Systems

	Conservation equation valid?
Total mass	Yes
Species mass	No
Element mass	Yes
Total moles	No
Species moles	No
Element moles	Yes

Accounting statements can be used for *all* mass and mole balances, but are necessary for mass and molar species balances and total mole balances when chemical reactions are involved. On the other hand, conservation statements can always be applied for mass and mole element balances and total mass balances regardless of the presence of chemical reactions. Finally, conservation statements can be applied for species mass and moles and total mole balances when no chemical reactions are involved. In other words, **an accounting statement can always be used for all representations of mass and moles. A conservation equation can always be used for nonreacting systems, but only for certain representations of mass and moles for reacting systems.** Table 3.1 summarizes when conservation equations are appropriate to use in a reacting system.

EXAMPLE 3.4 Bacterial Production of Acetic Acid

Problem: Under aerobic conditions (i.e., oxygen is present), *Acetobacter aceti* bacteria convert ethanol to acetic acid (vinegar). A bioreactor with a continuous fermentation process for vinegar production is shown in Figure 3.3. The conversion reaction is:

$$C_2H_5OH\ (\text{ethanol}) + O_2 \longrightarrow CH_3COOH\ (\text{acetic acid}) + H_2O$$

A feed stream containing ethanol enters the reactor, and air continuously bubbles into the reactor. An off-gas stream and a liquid product stream containing acetic acid leave the reactor.

Characterize the bioreactor system using definitions from Chapter 2 (e.g., open or closed, steady-state or dynamic, reacting or nonreacting). For what species or elements can you write conservation equations? For what species must you write accounting statements?

Solution: The diagram of the system clearly shows two inlet and two outlet streams crossing the boundary, so the system is open. The term "continuous" in the problem statement suggests the fermentation process is at steady-state. Since vinegar is produced, as described by the reaction in the problem statement, the system is reacting.

The following species are in this system: C_2H_5OH, O_2, CH_3COOH, H_2O, and N_2. (Remember, air, not oxygen, is pumped into the system.) A conservation equation can be written for the total mass of the system. Also, conservation equations can be written for the mass and moles of each individual element (i.e., C, H, O, and N). Because N_2 is a nonreacting component of the system, a conservation equation can be written for N_2.

Since a biochemical reaction occurs in the system, accounting equations must be written for the compounds C_2H_5OH, O_2, CH_3COOH, and H_2O. Because the system is reacting, accounting equations are appropriate for total moles and species mass and moles for these compounds. ■

Remember that the key difference between the accounting and conservation equations is the presence of the reaction terms (i.e., Generation and Consumption terms).

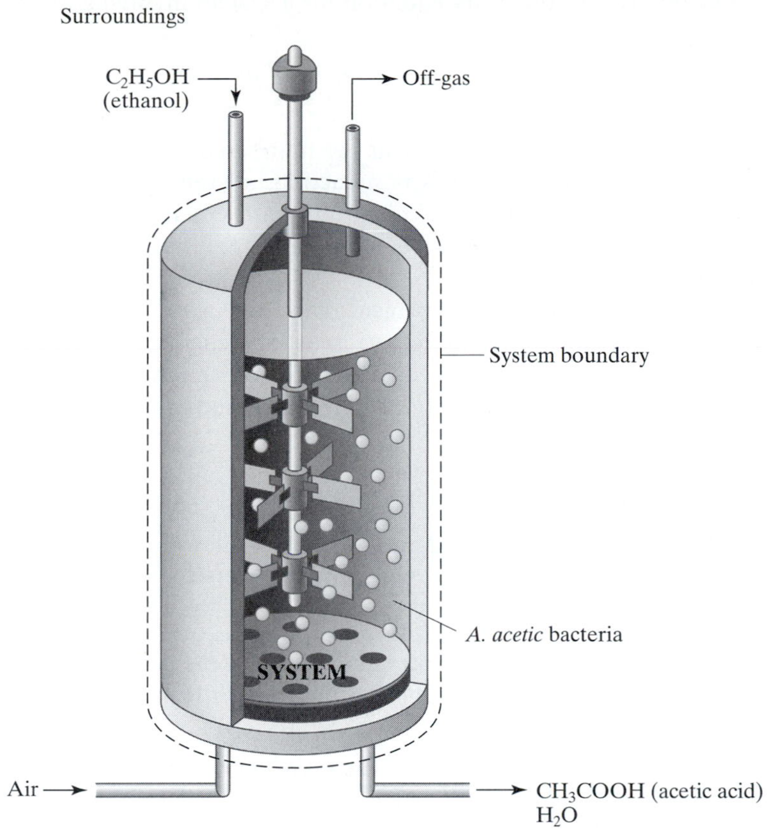

Figure 3.3
A bioreactor with a continuous fermentation process for vinegar production.

If you are confused about which equation should be used (i.e., accounting or conservation), you can always begin with the accounting statement and simplify it according to the assumptions you make about the system. In Sections 3.4–3.7, the systems are nonreacting, so only applications of the conservation equation are presented.

3.4 Open, Nonreacting, Steady-State Systems

Systems with open, steady-state conditions are very common in bioengineering. For example, some organs in the human body can be modeled as open, nonreacting, steady-state systems. These systems involve the movement of material across the system boundary. The variables characterizing the system do not change with time, and no mass accumulates in the system. Additionally, many systems are nonreacting, allowing for additional simplifications of the accounting equation.

In applications such as biomixers, an open, steady-state system is often termed **continuous**, since materials are constantly fed into it and mixed products are constantly removed. Such continuous flow of the inlet and outlet streams creates an unchanging system, and the continuous nature of the flow means the Input and Output terms are usually specified as rates, making the differential equation most appropriate to use.

For open, steady-state, nonreacting systems, the differential accounting equation reduces to the corresponding **continuity equation:**

$$\dot{\psi}_{in} - \dot{\psi}_{out} = 0 \quad [3.4\text{-}1]$$

$$\dot{\psi}_{in} = \dot{\psi}_{out} \quad [3.4\text{-}2]$$

For example, the continuity equation for a system involving the flow of mass is:

$$\sum_i \dot{m}_i - \sum_j \dot{m}_j = 0 \qquad [3.4\text{-}3]$$

where the indices i and j represent the numbered inlet and outlet rates of flow, respectively. Since the system is nonreacting, all representations of mass and moles ($\dot{m}$, $\dot{m}_s$, $\dot{m}_p$, $\dot{n}$, $\dot{n}_s$, $\dot{n}_p$) may be substituted into equation [3.4-1]. The algebraic and integral equations can also be formulated for open, nonreacting, steady-state systems.

There is a large class of problems with one inlet rate and one outlet rate requiring the application of the conservation of total energy (Chapter 4), the conservation of linear momentum (Chapter 6), or the accounting of mechanical energy (Chapter 6). In these systems, the continuity equation is often required as well.

EXAMPLE 3.5 **Blood Flow in the Heart**

Problem: The heart is divided into two sides, each with two chambers. Blood enters the left atrium from the pulmonary veins and drains into the left ventricle, where the blood is pumped out at an average rate of 60 beats per minute. The stroke volume (the volume of blood discharged from the ventricle to the aorta) is 70 mL for each contraction. Assuming that no reactions with blood occur in the heart and the heart chambers do not accumulate blood, determine the inlet and outlet volumetric flow rates for the left side of the heart (Figure 3.4).

Solution: The left side of the heart is an open, nonreacting, steady-state system, since blood flows in and out, does not react in the heart, and does not accumulate. Thus, the total mass of blood is conserved. Because the problem asks for rates, the differential conservation equation is most appropriate. Since blood does not accumulate in the heart, the continuity equation involving mass flow [3.4-3] is most appropriate.

The stroke volume gives the amount of blood that exits the system. Volume is not a conserved extensive property, but mass and mass flow rate are. To convert the volume of blood flowing out of the system, the volumetric flow rate ($\dot{V}$) must be calculated

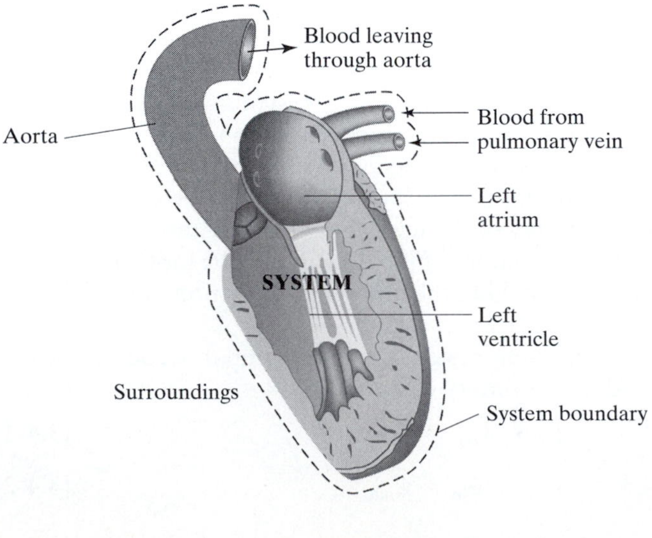

Figure 3.4
Blood flow through the left side of the heart.

and converted to a mass flow rate ($\dot{m}$). The outlet flow rate from the ventricle is calculated as follows:

$$\dot{V}_{out} = \left(70\frac{\text{mL}}{\text{beat}}\right)\left(60\frac{\text{beats}}{\text{min}}\right) = 4200\frac{\text{mL}}{\text{min}}$$

Recall from equation [3.2-4] that $\dot{m}$ and $\dot{V}$ are related by fluid density (ρ). Assuming no change in the density of the blood as it passes through the heart ($\rho_{in} = \rho_{out}$), the inlet and outlet volumetric flow rates of blood must be equal:

$$\dot{m}_{in} - \dot{m}_{out} = \dot{V}_{in}\rho_{in} - \dot{V}_{out}\rho_{out} = 0$$

$$\dot{V}_{in} - \dot{V}_{out} = \dot{V}_{in} - 4200\frac{\text{mL}}{\text{min}} = 0$$

$$\dot{V}_{in} = 4200\frac{\text{mL}}{\text{min}}$$

Therefore, the volumetric flow rate of the blood entering the left side of the heart is 4200 mL/min, the same volumetric flow rate that leaves the left side of the heart. See Case Study 7B for a much more extensive analysis of the heart. ■

In the example above, it might appear that you could perform a balance on volumetric flow rate. In a certain subset of problems involving incompressible fluids (fluids with constant density) in nonreacting systems, the continuity equation [3.4-1] can often be reduced such that the volumetric flow into the system is equal to the flow out. In fact, some chemical engineering textbooks illustrate examples with volumetric flow rate balances.

However, the volumetric flow rate is often misapplied in balance equations. The reduction of equation [3.4-1] to one where volumetric flows into and out of the system are equal cannot be made when dealing with many reacting systems or with systems containing compressible fluids such as gases. **In general, it is always best to use mass or molar flow rates, rather than volumetric flow rates, when solving accounting and conservation equations.**

EXAMPLE 3.6 Flow Through a Bone Graft

Problem: To achieve proper cell and tissue growth, a tissue-engineered product must accommodate blood flow similar to that of the native tissue. Without blood flow to deliver oxygen, glucose, and other essential nutrients while removing waste materials, the size of the implant is limited.

A company develops a porous bone graft that permits glucose to flow into the interior. A 50 g/min buffer solution containing 5 mg/mL of glucose flows into the bone graft in a laboratory experiment.

(a) What are the expected total mass flow rate and mass flow rate of glucose out of the system?
(b) Laboratory results show the total outlet mass flow rate is 50 g/min. The outlet glucose mass flow rate is 225 mg/min. Make conjectures about what has happened in this experiment.

Solution:

(a) The porous bone scaffold system is shown in Figure 3.5. To find the mass flow rates, a number of assumptions must be made for the experimental setup:

- The system is at steady-state.
- The bone graft is modeled as a cylindrical vessel.

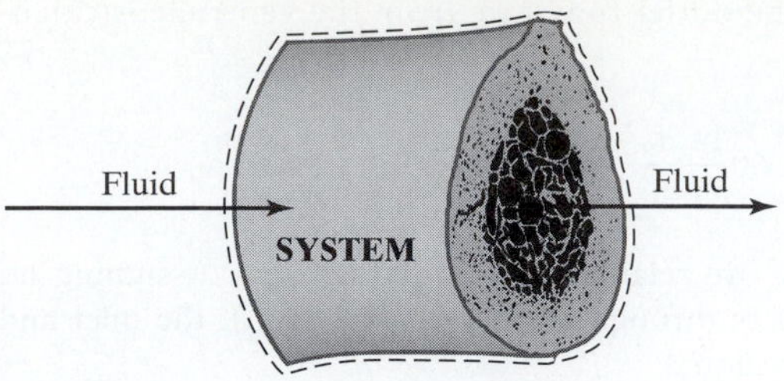

Figure 3.5
Porous bone scaffold system.

- The buffer is an incompressible fluid.
- The buffer density is equal to that of water (1.0 g/mL).
- No reactions occur in the system.

Since the system is open, nonreacting, and at steady-state, the differential mass continuity equation [3.4-3] can be used. Note that the accounting and conservation equations are often applied to flow through unobstructed tubes, pipes, or vessels, but they are also valid for flow through a porous bone graft. The presence of tissue inside the bone through which blood flows does not nullify the standard accounting and conservation equations.

The mass flow rate into the system is given, so we can find the expected total mass flow rate out using equation [3.4-3]:

$$\dot{m}_{\text{in}} - \dot{m}_{\text{out}} = 50\frac{\text{g}}{\text{min}} - \dot{m}_{\text{out}} = 0$$

$$\dot{m}_{\text{out}} = 50\frac{\text{g}}{\text{min}}$$

The amount of glucose (G) entering the system is:

$$\dot{m}_{\text{in,G}} = 5\frac{\text{mg}}{\text{mL}}\left(50\frac{\text{g}}{\text{min}}\right)\left(\frac{\text{mL}}{1.0\text{ g}}\right) = 250\ \frac{\text{mg}}{\text{min}}$$

Using equation [3.3-10] written for the species mass rate, the expected mass flow rate of glucose out of the steady-state, nonreacting system is:

$$\dot{m}_{\text{in,G}} - \dot{m}_{\text{out,G}} = 250\frac{\text{mg}}{\text{min}} - \dot{m}_{\text{out,G}} = 0$$

$$\dot{m}_{\text{out,G}} = 250\frac{\text{mg}}{\text{min}}$$

Thus, for a nonreacting, steady-state system, the expected total outlet mass flow rate is 50 g/min and the expected outgoing mass flow rate of glucose is 250 mg/min.

(b) The lab results confirm the calculated total outlet mass flow rate of 50 g/min. As expected, total mass is conserved. However, since the experimental outlet mass flow rate of glucose of 225 mg/min differs from the expected 250 mg/min, the assumption that the system is nonreacting or at steady-state may not be valid. Since the amount of glucose is not conserved, an accounting equation—not a conservation equation—must be used to model the system.

Let us suppose the system is reacting—for example, suppose the graft is seeded with metabolizing cells that consume the glucose, reducing the amount in the outlet stream. Even if glucose is consumed, some types of waste products are generated. Therefore, the total mass inlet and outlet should remain equal, since total mass is conserved.

Another conjecture is that glucose may be accumulating in the system, making the system dynamic. The glucose might bind nonspecifically to the scaffold or through another mechanism and build up in the system, causing the outlet mass flow rate of glucose and the total mass flow rate to drop. In experimental measurements, the change in glucose concentration may be detectable, but the drop in total mass flow rate would be very small ($<0.5\%$) and might be undetectable. ■

3.5 Open, Nonreacting, Steady-State Systems with Multiple Inlets and Outlets

Systems often have multiple inlets and outlets crossing the system boundary, regardless of whether the system is reacting or nonreacting, at steady-state or dynamic. In this section we focus on analyzing nonreacting, steady-state systems with multiple inlet and outlet streams.

Biomedical and bioprocessing situations often have multiple streams that cross the system boundaries. A physiological example is the bifurcation (splitting into two branches) of vessels in the body. In the lungs, the airflow in the trachea is split to the left and right main stem bronchi, creating two outlet streams from one inlet (see Example 3.8). Bifurcation and branching continue throughout the lungs until reaching the alveoli, where gas exchange occurs.

Another example of a system with multiple inlets and outlets is a bioprocess operation that has multiple units, each of which can contain multiple inlet or outlet streams or both. For example, one compartment of the operation could be a mixing tank, where different streams containing glucose (carbon source), ammonia (nitrogen source), oxygen, water, and other nutrients flow into the tank system and are mixed. A second example is a separation tank, where the product is sorted from the waste, and each leaves through a different outlet. The mixing tank has several inlets and one outlet, while the separation tank has one inlet and two outlet streams. A system can also contain multiple inlets and multiple outlets.

Many applications require the differential form of the conservation equation [3.3-10] to develop mass balances. The Input term $\dot{\psi}_{\text{in}}$ encompasses *all* the rates of mass that enter the system, while the Output term $\dot{\psi}_{\text{out}}$ encompasses *all* exiting mass rates. The differential conservation equation describing mass flow rate ($\dot{m}$) for an open, steady-state system is rewritten as:

$$\sum_i \dot{m}_i - \sum_j \dot{m}_j = 0 \qquad [3.5\text{-}1]$$

where i corresponds to the index number of the inlet streams and j to the index number of the outlet streams. Algebraic and integral conservation equations can also be written for systems with multiple streams. Similar equations can be written for species mass rate, element mass rate, mole rate, species mole rate, and element mole rate under open, nonreacting, steady-state conditions.

EXAMPLE 3.7 Lymph Vessel Collection

Problem: In the lymphatic system, lymphatic capillaries collect and filter excess fluid away from the interstitial spaces before returning it to the bloodstream. Near the armpit, three lymphatic vessels come together to a collecting vessel, as shown in Figure 3.6. Write the appropriate mass conservation equation for the fluid. Assume the system is at steady-state.

Solution: Since there are multiple streams in the open, nonreacting, steady-state system, Equation [3.5-1] is appropriate. The three mass inlet streams and one outlet stream are described for this lymphatic system as follows:

$$\sum_i \dot{m}_i - \sum_j \dot{m}_j = \dot{m}_1 + \dot{m}_2 + \dot{m}_3 - \dot{m}_4 = 0$$

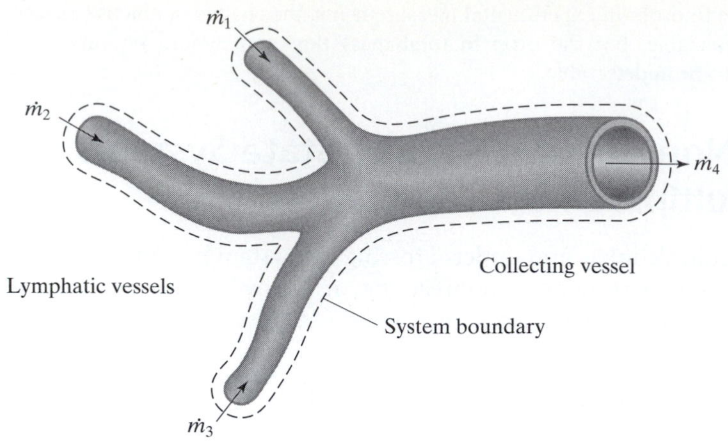

Figure 3.6
Lymphatic vessels combining into a collecting vessel.

where $\dot{m}_1$, $\dot{m}_2$, and $\dot{m}_3$ are the total mass flow rates of the inlet streams, and $\dot{m}_4$ is the total mass flow rate of the outlet stream. ■

EXAMPLE 3.8 Air Flow in the Respiratory Pathway

Problem: Air inhaled through the nostrils travels down the trachea before splitting into the two main bronchi, which then split into several generations of bronchi, becoming bronchioles. The path of airflow terminates at air sacs called alveoli at the ends of the bronchioles. The alveoli transfer oxygen from the inhaled air to the pulmonary capillaries and pick up carbon dioxide for exhalation.

Suppose a person inhales 0.5 L of air in an average breath over a period of 2 seconds. Write a mass conservation equation for airflow through the trachea and main bronchi (Figure 3.7). What is the mass flow rate of air through the trachea? What are the velocities of air through the two main bronchi if the mass flow rate in each bronchus is assumed to be the same? Assume the air does not humidify in the respiratory tract. The trachea has a diameter of approximately 2 cm. The right main bronchus is slightly larger than the left, with diameters of approximately 12 mm and 10 mm, respectively. The density of air at room temperature is 1.2 g/L.

Solution: First we assume that the system during one inhaled breath is steady-state and nonreacting. We model the trachea and bronchi as cylindrical conduits. Since no humidification occurs, we can assume a constant air density.

As drawn, the system has multiple outlets with one given inlet flow rate. We can use the differential form of the conservation equation for a system with multiple streams [3.5-1]:

$$\dot{m}_t - \dot{m}_r - \dot{m}_l = 0$$

where t is the trachea, r is the right main bronchus, and l is the left main bronchus.

To solve for the unknown airflow velocity in the trachea given the volume of incoming air, we can use equation [3.2-4], using the volume to find the volumetric flow rate and dividing by the cross-sectional area (equation [3.2-3]) of the airflow. The average volumetric flow rate of air through the trachea during the time period of inhalation is:

$$\dot{V}_t = \frac{V}{t} = \frac{0.5 \text{ L}}{2 \text{ s}} = 0.25 \ \frac{\text{L}}{\text{s}}$$

The velocity of air in the trachea is then calculated:

$$v_t = \frac{\dot{V}_t}{A_t} = \left(\frac{0.25\frac{\text{L}}{\text{s}}}{\frac{\pi}{4}(2 \text{ cm})^2}\right)\left(\frac{1000 \text{ cm}^3}{\text{L}}\right) = 79.6 \ \frac{\text{cm}}{\text{s}}$$

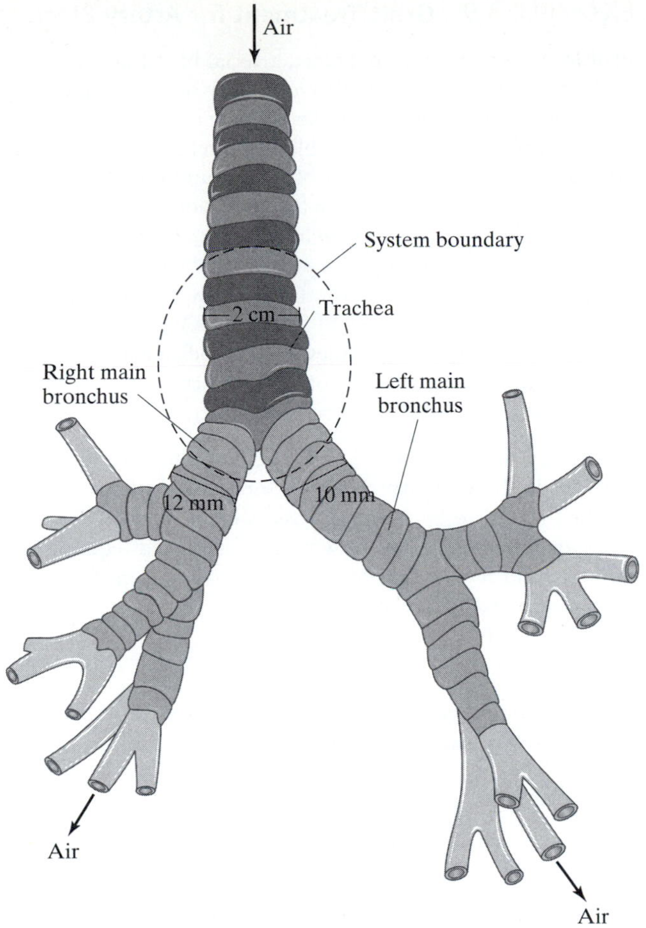

Figure 3.7
Trachea and main bronchi in respiratory pathway.

Thus, the mass flow rate in the trachea is calculated using equation [3.2-2]:

$$\dot{m}_t = A_t v_t \rho = \frac{\pi}{4} D_t^2 v_t \rho = \frac{\pi}{4}(2\text{ cm})^2\left(79.6\frac{\text{cm}}{\text{s}}\right)\left(1.2\frac{\text{g}}{\text{L}}\right)\left(\frac{1\text{ L}}{1000\text{ cm}^3}\right) = 0.30\ \frac{\text{g}}{\text{s}}$$

Since the mass flow rate in each bronchus is the same, $\dot{m}_r = \dot{m}_l$. Using our differential mass conservation equation for this problem:

$$\dot{m}_r = \dot{m}_l = \frac{\dot{m}_t}{2} = 0.15\ \frac{\text{g}}{\text{s}}$$

To find the velocity of air in each of the main bronchi:

$$v_r = \frac{4\dot{m}_r}{\pi D_r^2 \rho} = \left(\frac{4\left(0.15\frac{\text{g}}{\text{s}}\right)}{\pi(1.2\text{ cm})^2\left(1.2\frac{\text{g}}{\text{L}}\right)}\right)\left(\frac{1000\text{ cm}^3}{\text{L}}\right) = 111\ \frac{\text{cm}}{\text{s}}$$

Thus, the airflow velocity in the right main bronchus is 111 cm/s. Solving in a similar manner for the airflow in the left main bronchus yields 159 cm/s. It should make sense that for equivalent mass flow rates through the left and right main bronchi the velocity through the right bronchus is lower than through the left bronchus, since the vessel diameter of the right bronchus is higher. See Case Study 7A for a much more extensive analysis of the lungs. ■

EXAMPLE 3.9 Graft Treatment for Artery Blockage

Problem: In coronary ischemia, arterial blood flow is obstructed at a few discrete locations in the coronary arteries, while blood flow on either side of these blocked points remains normal. Patients with coronary ischemia may undergo a surgical treatment called coronary artery bypass grafting, which uses a blood-vessel graft to connect the segments of the vessel with normal blood flow to redirect blood around the blocked points. While synthetic vessels can be used, surgeons generally use the saphenous vein from the leg for the graft.

Consider again the patient in Example 3.2 whose coronary artery diameter has decreased by 67%. His surgeon implants a graft to bypass the constricted region (Figure 3.8). If the velocity of the blood and the total mass flow rate are the same on either side of the blockage, what must be the diameter of the graft? Assume the cross-sectional area of the coronary artery is the same on both sides of the blockage and the curvature in the graft does not alter blood flow. Assume the sections of the artery with normal blood flow have a diameter of 2.5 mm and a blood flow velocity of 6.4 cm/s.

Solution: In Example 3.2, we calculate the mass flow rate of blood through the healthy portion of the coronary artery before the blockage to be 0.332 g/s and that through the blocked region to be 0.027 g/s (assuming a reduced velocity of 4.8 cm/s). Since no reactions occur in the blood, mass flow rate is conserved (equation [3.3-10]). If we draw our system around point B where the blocked artery and the graft reconvene and assume steady-state, the conservation equation becomes:

$$\dot{m}_{\text{block}} + \dot{m}_{\text{graft}} - \dot{m}_{\text{out}} = 0$$

where *block* denotes blood flow from the constricted artery, *graft* denotes blood flow from the graft, and *out* denotes the outlet flow beyond the bypass. Since the mass flow rate out of the blockage must equal that before the blockage, the desired mass flow rate through the graft is:

$$\dot{m}_{\text{graft}} = \dot{m}_{\text{out}} - \dot{m}_{\text{block}} = 0.332\frac{\text{g}}{\text{s}} - 0.027\frac{\text{g}}{\text{s}} = 0.305\frac{\text{g}}{\text{s}}$$

The diameter of the graft is a variable, although typically it is that of the saphenous vein from the patient. If we assume that the blood velocity in the graft is the same as in the healthy portion of the artery (6.4 cm/s), we can then rearrange equations [3.2-2] and [3.2-3] to find the diameter of the graft vessel to achieve these conditions:

$$D_{\text{graft}} = \sqrt{\frac{4\dot{m}_{\text{graft}}}{\pi v \rho}} = \sqrt{\frac{4\left(0.305\frac{\text{g}}{\text{s}}\right)}{\pi\left(6.4\frac{\text{cm}}{\text{s}}\right)\left(1.056\frac{\text{g}}{\text{cm}^3}\right)}} = 0.24 \text{ cm} = 2.4 \text{ mm}$$

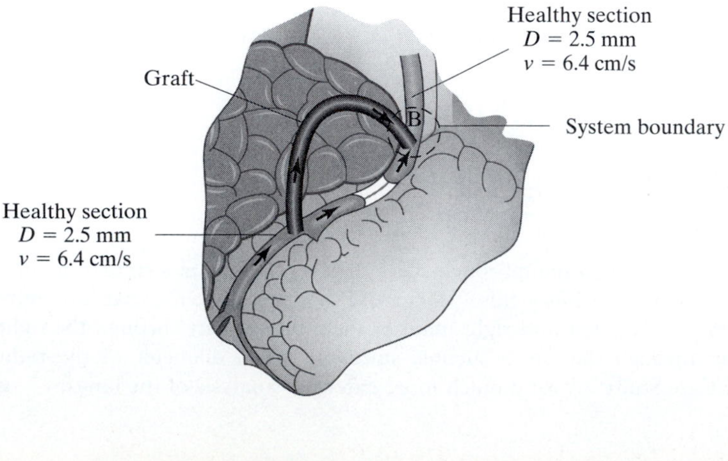

Figure 3.8
Using a coronary artery bypass graft to redirect blood flow around a constricted vessel.

Thus, to have the same mass flow rate and blood velocity on either side of the atherosclerotic region, the graft diameter must be 2.4 mm, which is nearly the size of the healthy coronary artery.

■

3.6 Systems with Multicomponent Mixtures

The equations in Sections 3.4 and 3.5 can be generalized further for systems where the inlets and outlets may contain multiple species or compounds. This process for solving such systems is very common in bioengineering. Most fluids in the body (e.g., blood, air), as well as streams in a biotechnological processing unit (e.g., insulin product), are multicomponent mixtures.

Consider systems with mass entering and leaving in streams. Each chemical species or compound s in each stream is associated with its species flow rate ($\dot{n}_s$ [moles of s/time] or $\dot{m}_s$ [mass of s/time]). The total flow rate of the stream, in either moles or mass, can be calculated by summing the individual species flows over all species s present in the stream:

$$\dot{n} = \sum_s \dot{n}_s \qquad [3.6\text{-}1]$$

$$\dot{m} = \sum_s \dot{m}_s \qquad [3.6\text{-}2]$$

An alternative way of representing a stream is to give its total flow, in rate of either moles or mass, together with the composition of the stream. Two convenient measures of composition of a species are the mass or weight fraction (w_s) and the mole fraction (x_s). All mass and mole fractions of all species s in a stream must sum to 1:

$$\sum_s w_s = 1 \qquad [3.6\text{-}3]$$

$$\sum_s x_s = 1 \qquad [3.6\text{-}4]$$

Mass and mole fractions are related to mass and molar flow rates, respectively:

$$w_s = \frac{\dot{m}_s}{\dot{m}} \qquad [3.6\text{-}5]$$

$$x_s = \frac{\dot{n}_s}{\dot{n}} \qquad [3.6\text{-}6]$$

Mass and molar flow rates for a particular compound s are related to one another through the molecular weight as follows:

$$\dot{n}_s = \frac{\dot{m}_s}{M_s} \qquad [3.6\text{-}7]$$

where M_s is the molecular weight of compound s. Either (a) molar flow rates and mole fractions or (b) mass flow rates and mass fractions can be used to characterize a stream. If the molecular weight (M) is known for each of the species in the stream, interconversions between the mass and molar units can be obtained:

$$\dot{n} = \sum_s \frac{w_s \dot{m}}{M_s} = \dot{m} \sum_s \frac{w_s}{M_s} \qquad [3.6\text{-}8]$$

$$x_s = \frac{w_s\dot{m}}{M_s\dot{n}} \qquad [3.6\text{-}9]$$

The derived equations [3.6-1] through [3.6-9] can also be formulated for chemical elements in a system. The equations would contain a p subscript in place of the s subscript.

The differential conservation equations from Section 3.5 for systems containing multiple inlets and outlets can be extended to systems that contain multiple species in each stream. For open, nonreacting, steady-state systems, the following mass conservation equations are written for:

$$\text{Species mass:} \quad \sum_i \dot{m}_{i,s} - \sum_j \dot{m}_{j,s} = 0 \qquad [3.6\text{-}10]$$

$$\text{Element mass:} \quad \sum_i \dot{m}_{i,p} - \sum_j \dot{m}_{j,p} = 0 \qquad [3.6\text{-}11]$$

$$\text{Species moles:} \quad \sum_i \dot{n}_{i,s} - \sum_j \dot{n}_{j,s} = 0 \qquad [3.6\text{-}12]$$

$$\text{Element moles:} \quad \sum_i \dot{n}_{i,p} - \sum_j \dot{n}_{j,p} = 0 \qquad [3.6\text{-}13]$$

where s is the species or compound of interest, p is the chemical element of interest, and i and j are indexed inlet and outlet streams, respectively. Note that when both a stream and species element are specified, the stream index is always listed first.

In some systems containing multiple components within their boundaries, an individual component can come from a single inlet source, mix together with components from other inlet streams, and exit through a single outlet. For example, consider a simplified model of the stomach, where one of the primary functions is mixing. Neglecting any chemical reactions facilitated by gastric juices (i.e., no digestion or breakdown of food), the stomach can be modeled as a nonreacting system. The stomach has two inlets, the esophagus and the gastric glands, and one outlet, the duodenum (Figure 3.9). Food, which is modeled to be comprised of proteins, fats, and carbohydrates, enters the stomach from the esophagus. Gastric juices, which are modeled to be comprised of enzymes, mucus, and pH regulators, are secreted by the gastric glands. All individual components (e.g., protein) enter by either the esophagus or the gastric glands (but not both), mix together in the stomach, and exit the duodenum as chyme, a multicomponent mixture of all the substances.

When one inlet is the only source of a specific component in a multicomponent mixture and the mixture leaves through only one outlet, we can use equations [3.6-5] and [3.6-10] to find the mass flow rates into and out of the system for a specific component or species s:

$$\sum_i \dot{m}_{i,s} - \sum_j \dot{m}_{j,s} = w_{\text{in},s}\dot{m}_{\text{in}} - w_{\text{out},s}\dot{m}_{\text{out}} = 0 \qquad [3.6\text{-}14]$$

We can rearrange equation [3.6-14] to solve for the mass fraction of species s in the outlet stream in terms of the mass fraction of species s in the inlet stream and the mass flow rates:

$$w_{\text{out},s} = \frac{w_{\text{in},s}\dot{m}_{\text{in}}}{\dot{m}_{\text{out}}} \qquad [3.6\text{-}15]$$

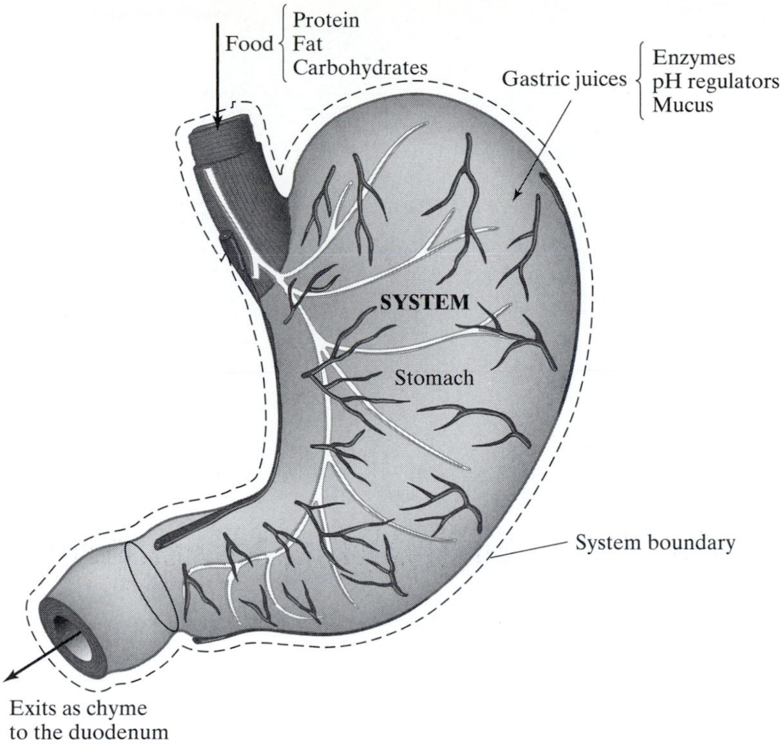

Figure 3.9
Model of stomach with multiple components entering and leaving the system.

Since species s is contained in just one inlet and one outlet stream, the mass flow rate of species s in the outlet stream equals the mass flow rate of species s in the inlet stream containing the species:

$$\dot{m}_{\text{out},s} = \dot{m}_{\text{in},s} \qquad [3.6\text{-}16]$$

For example, in the stomach mixing case, $\dot{m}_{\text{chyme,protein}} = \dot{m}_{\text{food,protein}}$ and $\dot{m}_{\text{chyme,mucus}} = \dot{m}_{\text{gastric,mucus}}$.

For each species or compound in a system, a material balance equation can be written. Therefore, for s species, s species mass (or mole) equations can be written. Also, one overall or total mass (or mole) balance equation can be written. Therefore, $s + 1$ mass (or mole) equations can be written for a system with s species. However, only s of those $s + 1$ equations will be linearly independent. (See Section 2.6 and Table 2.4.)

To illustrate the process of setting up steady-state conservation equations for particular compounds and for the overall system in a multicomponent-stream system, consider a mixing tank (Figure 3.10). Stream 1 contains compounds A and B and has a total mass flow of $\dot{m}_1$. The mass fractions of the compounds in stream 1 are $w_{1,\text{A}}$ and $w_{1,\text{B}}$, respectively. Stream 2 contains compounds B and C and has a total mass flow of $\dot{m}_2$. The mass fractions of the compounds in stream 2 are $w_{2,\text{B}}$ and $w_{2,\text{C}}$, respectively. Stream 3 contains compounds A, B, and C and has a total mass flow of $\dot{m}_3$. The mass fractions of the compounds in stream 3 are $w_{3,\text{A}}$, $w_{3,\text{B}}$, and $w_{3,\text{C}}$, respectively. For each compound in this nonreacting, steady-state system, a species mass conservation equation can be written. Equation [3.6-10] can be generalized for species s to:

$$\sum_i w_{i,s}\dot{m}_i - \sum_j w_{j,s}\dot{m}_j = 0 \qquad [3.6\text{-}17]$$

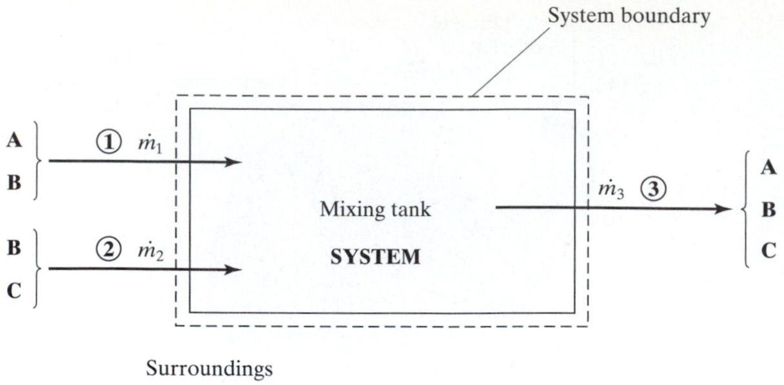

Figure 3.10
A mixing tank with inlet and outlet streams composed of A, B, and C.

Thus, an equation can be written for each of the compounds A, B, and C using equation [3.6-17]:

$$\text{A:} \quad w_{1,A}\dot{m}_1 - w_{3,A}\dot{m}_3 = 0 \qquad [3.6\text{-}18a]$$

$$\text{B:} \quad w_{1,B}\dot{m}_1 + w_{2,B}\dot{m}_2 - w_{3,B}\dot{m}_3 = 0 \qquad [3.6\text{-}18b]$$

$$\text{C:} \quad w_{2,C}\dot{m}_2 - w_{3,C}\dot{m}_3 = 0 \qquad [3.6\text{-}18c]$$

The total mass balance on the system is:

$$\text{total:} \quad \dot{m}_1 + \dot{m}_2 - \dot{m}_3 = 0 \qquad [3.6\text{-}18d]$$

Overall, four equations are written for this system containing three compounds. Only three of the above equations are linearly independent. [Recall that the sum of the mass fractions of all the components in any stream equals one (equation [3.6-3]).] Equations [3.6-18a], [3.6-18b], and [3.6-18c] add up to equation [3.6-18d]. To solve the system above, any three of the four equations could be used to solve for the unknowns. While the example above is developed from equation [3.6-10] with mass fractions and mass flow rates, it is equally valid to develop equation [3.6-12] for a system described by mole fractions and molar flow rates.

EXAMPLE 3.10 **Blood Flow in Two Joining Venules**

Problem: Two cylindrical venous capillary vessels join to form a larger cylindrical venule to return blood to the heart. The first vessel has a diameter of 0.0006 cm, an average blood flow velocity of 0.07 cm/s, and a hematocrit of 0.43. [Hematocrit is the ratio of the volume of packed red blood cells (RBC) to the volume of the whole blood.] The second vessel has a diameter of 0.0007 cm, an average blood flow velocity of 0.08 cm/s, and a hematocrit of 0.46. The diameter of the collecting vessel is 0.0008 cm. Find the hematocrit of the collecting vessel.

Solution:

1. Assemble
   (a) Find: hematocrit of the venule.
   (b) Diagram: The diagram is shown in Figure 3.11, which indicates the direction of blood flow and the system boundary.
2. Analyze
   (a) Assume:
   - Blood flow through the vessels is smooth and nonpulsatile.
   - Blood-vessel walls are stiff and do not expand or contract.
   - Although the hematocrit is slightly different among the vessels, the density of the blood is assumed to be constant throughout the system.

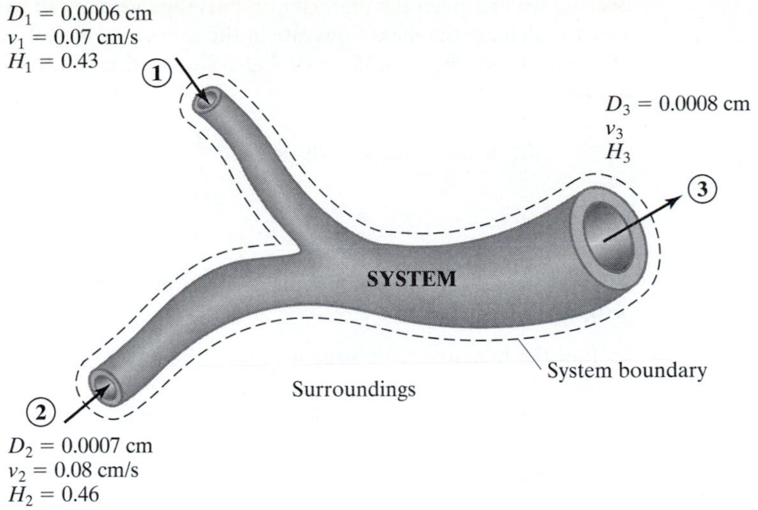

Figure 3.11
Blood flow through joining venous vessels.

- No blood accumulates in the vessels.
- The blood does not react.

(b) Extra data:
- The density of whole blood is 1.056 g/cm^3.

(c) Variables, notations, units:
- The two smaller vessels are labeled streams 1 and 2, and they join to create a larger venule, labeled stream 3.
- H = hematocrit, which is dimensionless.
- Use cm, s, g.

(d) Basis: A basis is not given explicitly, but a flow rate can be calculated using the density, diameter, and average velocity of blood in one of the inlet streams. The basis is set as stream 1 and is calculated as follows:

$$\dot{m}_1 = \frac{\pi}{4} D_1^2 v_1 \rho = \frac{\pi}{4}(0.0006 \text{ cm})^2\left(0.07\frac{\text{cm}}{\text{s}}\right)\left(1.056\frac{\text{g}}{\text{cm}^3}\right) = 2.09 \times 10^{-8}\ \frac{\text{g}}{\text{s}}$$

3. Calculate

(a) Equations: The differential form of the mass accounting statement [3.3-6] is appropriate, since mass flow rates can be calculated. Blood enters and leaves the system, so the system is open. No reactions occur within the system, so the Generation and Consumption terms in the governing equation are zero, and the total mass of blood is conserved in the system. Since blood does not accumulate, the system is also at steady-state. Because the open, nonreacting, steady-state system has multiple streams, we can use the mass conservation continuity equation [3.4-3]:

$$\sum_i \dot{m}_i - \sum_j \dot{m}_j = 0$$

To solve for the hematocrit, since whole blood consists of multiple components (e.g., RBC, plasma), we can use equation [3.6-10]:

$$\sum_i \dot{m}_{i,s} - \sum_j \dot{m}_{j,s} = 0$$

(b) Calculate:
- The mass conservation equation specific to this problem is:

$$\sum_i \dot{m}_i - \sum_j \dot{m}_j = \dot{m}_1 + \dot{m}_2 - \dot{m}_3 = 0$$

Because we are given the diameter and average velocity of the blood in stream 2, we can calculate the mass flow rate in the same way we solved the mass flow rate of stream 1. So, $\dot{m}_2 = 3.25 \times 10^{-8}$ g/s. We use this to find the mass flow rate of stream 3:

$$\dot{m}_1 + \dot{m}_2 - \dot{m}_3 = 2.09 \times 10^{-8}\,\frac{\text{g}}{\text{s}} + 3.25 \times 10^{-8}\,\frac{\text{g}}{\text{s}} - \dot{m}_3 = 0$$

$$\dot{m}_3 = 5.34 \times 10^{-8}\,\frac{\text{g}}{\text{s}}$$

With the mass flow rate and diameter of stream 3, we calculate the velocity, v_3, to be 0.10 cm/s.

- To find the hematocrit of stream 3, we need the RBC volume and whole blood volume of stream 3. While hematocrit is traditionally defined as the volume fraction of RBC, it may also be calculated as the ratio of the volumetric flow rate of RBC to the total volumetric flow rate. Using the relationships in equation [3.2-4], we can find the volumetric flow rate. Given the hematocrit in stream 1, the RBC volumetric flow rate in stream 1 is:

$$H_1 = \frac{\dot{V}_{1,\text{RBC}}}{\dot{V}_1} = \frac{\dot{V}_{1,\text{RBC}}}{A_1 v_1} = \frac{\dot{V}_{1,\text{RBC}}}{\frac{\pi}{4}D_1^2 v_1} = \frac{\dot{V}_{1,\text{RBC}}}{\frac{\pi}{4}(0.0006\text{ cm})^2\, 0.07\frac{\text{cm}}{\text{s}}} = 0.43$$

$$\dot{V}_{1,\text{RBC}} = 8.51 \times 10^{-9}\,\frac{\text{cm}^3}{\text{s}}$$

A similar calculation for stream 2 gives $\dot{V}_{2,\text{RBC}} = 1.42 \times 10^{-8}$ cm^3/s.

- As with the mass flow rates, we can perform a mass conservation balance on the RBC component of the blood to find the RBC mass flow rate of stream 3:

$$\sum_i \dot{m}_{i,s} - \sum_j \dot{m}_{j,s} = \dot{m}_{1,\text{RBC}} + \dot{m}_{2,\text{RBC}} - \dot{m}_{3,\text{RBC}} = 0$$

- Again using the relationships in equation [3.2-4], we can solve for the RBC volumetric flow rate:

$$\dot{m}_{1,\text{RBC}} + \dot{m}_{2,\text{RBC}} - \dot{m}_{3,\text{RBC}} = \rho_{1,\text{RBC}}\dot{V}_{1,\text{RBC}} + \rho_{2,\text{RBC}}\dot{V}_{2,\text{RBC}} - \rho_{3,\text{RBC}}\dot{V}_{3,\text{RBC}} = 0$$

Since we assume the blood density is constant despite the differences in hematocrit in the three streams, the above equation simplifies to:

$$\dot{V}_{1,\text{RBC}} + \dot{V}_{2,\text{RBC}} - \dot{V}_{3,\text{RBC}} = 8.51 \times 10^{-9}\frac{\text{cm}^3}{\text{s}} + 1.42 \times 10^{-8}\frac{\text{cm}^3}{\text{s}} - \dot{V}_{3,\text{RBC}} = 0$$

$$\dot{V}_{3,\text{RBC}} = 2.27 \times 10^{-8}\,\frac{\text{cm}^3}{\text{s}}$$

- Using the RBC volumetric flow rate and average velocity of blood in stream 3, we can calculate the hematocrit:

$$H_3 = \frac{\dot{V}_{3,\text{RBC}}}{\dot{V}_3} = \frac{\dot{V}_{3,\text{RBC}}}{\frac{\pi}{4}D_3^2 v_3} = \frac{2.27 \times 10^{-6}\,\frac{\text{cm}^3}{\text{s}}}{\frac{\pi}{4}(0.0008\text{ cm})^2(0.10\frac{\text{cm}}{\text{s}})} = 0.45$$

Thus, the hematocrit of the venule is 0.45.

4. Finalize
   (a) Answer: In the venule, the average velocity of blood is 0.10 cm/s and the hematocrit is 0.45.

(b) Check: It makes sense that the velocity of the outlet stream is higher than that of the two inlet streams, since the diameter increases only slightly but must still accommodate the combined mass flows of the two venous vessels. It also makes sense that the hematocrit in the outlet venule falls between the hematocrit values in the inlet capillary vessels, since the constituents other than RBC (e.g., plasma) in the first vessel would decrease the RBC concentration of the second vessel when combined in the venule. See Case Study 7B for a much more extensive analysis of blood flow. ■

EXAMPLE 3.11 Blood Detoxification in the Liver

Problem: One of the vital functions of the liver is detoxifying blood contaminants, such as drugs, alcohol, food additives, and metabolic end products. The liver returns detoxified blood to the heart through three hepatic veins, which drain into the inferior vena cava (IVC). The right hepatic vein drains separately into the IVC, while the left and middle hepatic veins usually join before emptying into the IVC.

Consider the system where detoxified blood from the liver drains into the inferior vena cava (Figure 3.12). Although blood from the liver enters the IVC through two separate vessels, we can assume they contain blood of nearly the same composition, so we can model them as one inlet stream.

The liver receives a total of about 1.35 L blood per minute. The only other outlet stream from the liver drains bile to the intestines, but it contains no blood and has a negligible volumetric flow rate when compared to the outflow of blood through the hepatic veins. Thus, we can assume blood leaves the liver only through the hepatic veins at a total rate of 1.35 L/min. The inferior vena cava returns 3.33 L/min of blood, about two-thirds of the cardiac output, to the heart.

Assume a patient must clear the following concentrations of toxins from his body: 0.5 mg/L of ammonia (NH_3), 0.6 mg/L of cyanide (CN), and 0.25 mg/L of lead (Pb). The toxins have the following known relationships between their mass fractions in streams 1 and 3:

$$3w_{1,NH_3} = 5w_{1,CN} \qquad w_{3,NH_3} = 6.34w_{1,NH_3} \qquad w_{3,CN} = 2.46w_{3,Pb}$$

Find the mass fractions of the three toxins in each of the streams. What percents of ammonia, cyanide, and lead are cleared by the patient's liver?

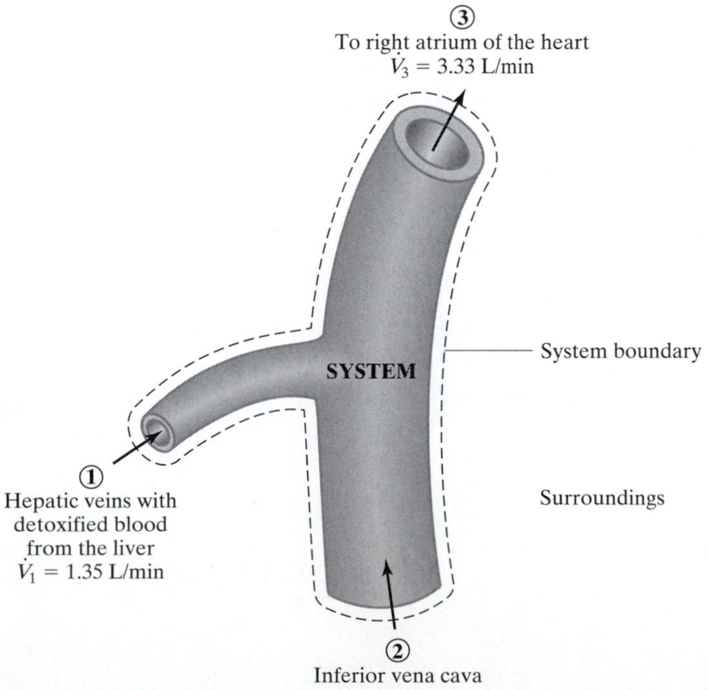

Figure 3.12
Mixing of regular blood and detoxified blood from the liver.

Solution: While blood is detoxified (i.e., reacted) in the liver, the blood leaving the liver through the hepatic veins and entering the system in Figure 3.12 is nonreacting, as is the blood entering from the rest of the body. Thus, Generation and Consumption are zero, and the conservation equation is most appropriate to use. Since we are given rates of blood flow, we use the differential form of the conservation equation. Because the blood flows continuously, the system is at steady-state. The open, nonreacting, steady-state system has multiple streams composed of multiple components. The total mass conservation equation for systems with multiple streams [3.5-1] can be used:

$$\sum_i \dot{m}_i - \sum_j \dot{m}_j = \dot{m}_1 + \dot{m}_2 - \dot{m}_3 = 0$$

We use equation [3.2-4] to relate the known volumetric flow rates in streams 1 and 3 to their mass flow rates. We assume the difference in blood density when toxins are present is negligible, so the blood density is 1.056 g/mL in all three streams.

$$\dot{m}_1 = \dot{V}_1\rho = \left(1.35\frac{\text{L}}{\text{min}}\right)\left(1.056\frac{\text{g}}{\text{mL}}\right)\left(1000\frac{\text{mL}}{\text{L}}\right) = 1430\ \frac{\text{g}}{\text{min}}$$

The mass flow rate in stream 3 is calculated similarly, giving $\dot{m}_3 = 3520$ g/min. Now we solve for the mass flow rate in stream 2:

$$\dot{m}_2 = \dot{m}_3 - \dot{m}_1 = 3520\frac{\text{g}}{\text{min}} - 1430\frac{\text{g}}{\text{min}} = 2090\ \frac{\text{g}}{\text{min}}$$

To find the mass fractions of the three toxins in streams 1 and 3, equation [3.6-10] is rewritten for each of the toxins ammonia, cyanide, and lead:

$$\text{NH}_3: \quad w_{1,\text{NH}_3}\dot{m}_1 + w_{2,\text{NH}_3}\dot{m}_2 - w_{3,\text{NH}_3}\dot{m}_3 = 0$$

$$\text{CN}: \quad w_{1,\text{CN}}\dot{m}_1 + w_{2,\text{CN}}\dot{m}_2 - w_{3,\text{CN}}\dot{m}_3 = 0$$

$$\text{Pb}: \quad w_{1,\text{Pb}}\dot{m}_1 + w_{2,\text{Pb}}\dot{m}_2 - w_{3,\text{Pb}}\dot{m}_3 = 0$$

Since the blood in stream 2 has not been detoxified by the liver, the toxin concentrations are those stated in the problem: $C_{2,\text{NH}_3} = 0.5$ mg/L, $C_{2,\text{CN}} = 0.6$ mg/L, and $C_{2,\text{Pb}} = 0.25$ mg/L. The mass fractions of the toxins in stream 2 are found using the corresponding concentrations and the density of blood. The mass fraction of ammonia in stream 2 is calculated:

$$w_{2,\text{NH}_3} = \frac{C_{2,\text{NH}_3}}{\rho_{\text{blood}}} = \left(\frac{0.5\dfrac{\text{mg NH}_3}{\text{L}}}{1.056\dfrac{\text{g blood}}{\text{mL}}}\right)\left(\frac{1\ \text{L}}{1000\ \text{mL}}\right)\left(\frac{1\ \text{g}}{1000\ \text{mg}}\right) = 4.73 \times 10^{-7}$$

The mass fractions of cyanide and lead in stream 2 are calculated similarly, giving $w_{2,\text{CN}} = 5.68 \times 10^{-7}$ and $w_{2,\text{Pb}} = 2.37 \times 10^{-7}$.

We can substitute the known values for the mass flow rates and the mass fractions in stream 2:

$$\text{NH}_3: \quad w_{1,\text{NH}_3}\left(1430\frac{\text{g}}{\text{min}}\right) + (4.73 \times 10^{-7})\left(2090\frac{\text{g}}{\text{min}}\right) - w_{3,\text{NH}_3}\left(3520\frac{\text{g}}{\text{min}}\right) = 0$$

$$\text{CN:} \quad w_{1,CN}\left(1430\frac{\text{g}}{\text{min}}\right) + (5.68 \times 10^{-7})\left(2090\frac{\text{g}}{\text{min}}\right) - w_{3,CN}\left(3520\frac{\text{g}}{\text{min}}\right) = 0$$

$$\text{Pb:} \quad w_{1,Pb}\left(1430\frac{\text{g}}{\text{min}}\right) + (2.37 \times 10^{-7})\left(2090\frac{\text{g}}{\text{min}}\right) - w_{3,Pb}\left(3520\frac{\text{g}}{\text{min}}\right) = 0$$

Thus, we have three equations for the system containing six unknown variables: w_{1,NH_3}, $w_{1,CN}$, $w_{1,Pb}$, w_{3,NH_3}, $w_{3,CN}$, and $w_{3,Pb}$. Recall that three additional relationships between these unknowns are specified in the problem statement. This gives us a total of six equations for the six unknown variables. This is a system that can be set up as a matrix equation to solve in MATLAB:

$$3w_{1,NH_3} - 5w_{1,CN} = 0$$

$$w_{3,NH_3} - 6.34w_{1,NH_3} = 0$$

$$w_{3,CN} - 2.46w_{3,Pb} = 0$$

$$\text{NH}_3\text{:} \quad w_{1,NH_3}\left(1430\frac{\text{g}}{\text{min}}\right) - w_{3,NH_3}\left(3520\frac{\text{g}}{\text{min}}\right) = -9.89 \times 10^{-4}\frac{\text{g}}{\text{min}}$$

$$\text{CN:} \quad w_{1,CN}\left(1430\frac{\text{g}}{\text{min}}\right) - w_{3,CN}\left(3520\frac{\text{g}}{\text{min}}\right) = -1.19 \times 10^{-3}\frac{\text{g}}{\text{min}}$$

$$\text{Pb:} \quad w_{1,Pb}\left(1430\frac{\text{g}}{\text{min}}\right) - w_{3,Pb}\left(3520\frac{\text{g}}{\text{min}}\right) = -4.95 \times 10^{-4}\frac{\text{g}}{\text{min}}$$

Arranging these linearly independent, scalar equations into the matrix equation form $A\vec{x} = \vec{y}$ gives:

$$\begin{bmatrix} 3 & -5 & 0 & 0 & 0 & 0 \\ -6.34 & 0 & 0 & 1 & 0 & 0 \\ 0 & 0 & 0 & 0 & 1 & -2.46 \\ 1430 & 0 & 0 & -3520 & 0 & 0 \\ 0 & 1430 & 0 & 0 & -3520 & 0 \\ 0 & 0 & 1430 & 0 & 0 & -3520 \end{bmatrix} \begin{bmatrix} w_{1,NH_3} \\ w_{1,CN} \\ w_{1,Pb} \\ w_{3,NH_3} \\ w_{3,CN} \\ w_{3,Pb} \end{bmatrix} = \begin{bmatrix} 0 \\ 0 \\ 0 \\ -9.89 \times 10^{-4} \\ -1.19 \times 10^{-3} \\ -4.95 \times 10^{-4} \end{bmatrix}$$

Putting this matrix equation into MATLAB yields the mass fraction for each component in each stream:

$$x = \begin{bmatrix} w_{1,NH_3} \\ w_{1,CN} \\ w_{1,Pb} \\ w_{3,NH_3} \\ w_{3,CN} \\ w_{3,Pb} \end{bmatrix} = \begin{bmatrix} 4.735e - 008 \\ 2.841e - 008 \\ 3.6747e - 009 \\ 3.002e - 007 \\ 3.4961e - 007 \\ 1.4212e - 007 \end{bmatrix} \approx \begin{bmatrix} 4.74 \times 10^{-8} \\ 2.84 \times 10^{-8} \\ 3.67 \times 10^{-9} \\ 3.00 \times 10^{-7} \\ 3.50 \times 10^{-7} \\ 1.42 \times 10^{-7} \end{bmatrix}$$

One method to calculate the fraction of toxin present after the blood has been detoxified is to divide the mass fraction of the toxin in the detoxified blood (stream 1) by the mass fraction of the same toxin in uncleared blood (stream 2). This result is then subtracted from 1.0 to find the clearance of the toxin by the patient's liver. For ammonia,

$$\text{NH}_3 \text{ clearance} = 1 - \frac{w_{1,NH_3}}{w_{2,NH_3}} = 1 - \frac{4.74 \times 10^{-8}}{4.73 \times 10^{-7}} = 0.90$$

Solving similarly for cyanide and lead clearance gives values of 0.95 and 0.985, respectively. Thus, as the blood passes through the liver it clears 90% of the ammonia, 95% of the cyanide, and 98.5% of the lead. ■

3.7 Systems with Multiple Units

All the systems discussed up to this point were represented as a single unit, acting like a black box that communicates with the surrounding environment through inputs and outputs. In other words, one system and one surrounding were defined for each problem. However, many engineering systems consist of multiple complex units acting in a certain sequence, sometimes making it difficult to analyze how the big picture works.

Multistep processes, such as creating pharmaceutical drugs in biochemical engineering, can be too complicated to solve in one series of material balance equations. Isolating individual units allows the problem to be simplified so that the mathematical and engineering tools discussed in previous sections can be used. For systems with multi-unit processes, isolating units and writing balances on several subsystems of the process allows enough equations to be obtained for determining all unknown variables. In multi-unit systems, understanding how more microscopic (unit-specific) details, such as intermediate stream flows, affect other units allows us to analyze the macroscopic system more thoroughly and accurately.

Defining a unit system within a larger system for microscopic analysis is to some degree arbitrary. What a system boundary encloses depends on what variables and properties need to be evaluated. **The system boundary should be drawn to cut across the inlets and outlets that isolate the unit of interest and contain the unknown variables.** For a system with two units, labeled I and II, a system boundary can be drawn around unit I, unit II, or both units together. For a system with three or more units, many more system boundaries can be drawn. An overall balance can always be written with the system boundary enclosing all units.

Consider the kidney as an example of a multi-unit system. The kidney separates wastes from blood while conserving water. It is often modeled with two or more compartments. The nephron, the functional unit of the kidney, can be divided into many smaller units and systems, since it contains distinct regions that perform different and specific functions (Figure 3.13a). For example, the nephron may be divided into the Bowman's capsule, which separates a cell-free filtrate from the blood, and the tubules, which concentrate the urine from the filtrate (Figure 3.13b).

In the kidney model, three system boundaries (I, II, and III) can be drawn. System Boundary I cuts across streams 1, 2, and 3; System Boundary II cuts across streams 3, 4, and 5; and System Boundary III cuts across streams 1, 2, 4, and 5. Given information on streams 4 and 5, you can find information on stream 3 using System Boundary II. Given information about stream 2 and the ratio of the amount of material in streams 1 and 5, you can use System Boundary III to analyze stream 4. If you want to analyze how the inlet flow of the Bowman's capsule affects the outlet flow from the tubules, you would first need to analyze the system enclosed by System Boundary I to find information on stream 3. Example 3.14 shows a detailed calculation for a two-unit kidney model.

For a more microscopic analysis, the tubules may be further divided into the proximal convoluted tubule, the loop of Henle, the distal convoluted tubule, and the collecting duct. These components may be considered as individual units or grouped together in various ways when drawing a system boundary to analyze different parts of the kidney. The kidneys are analyzed in more detail in Case Study 7C.

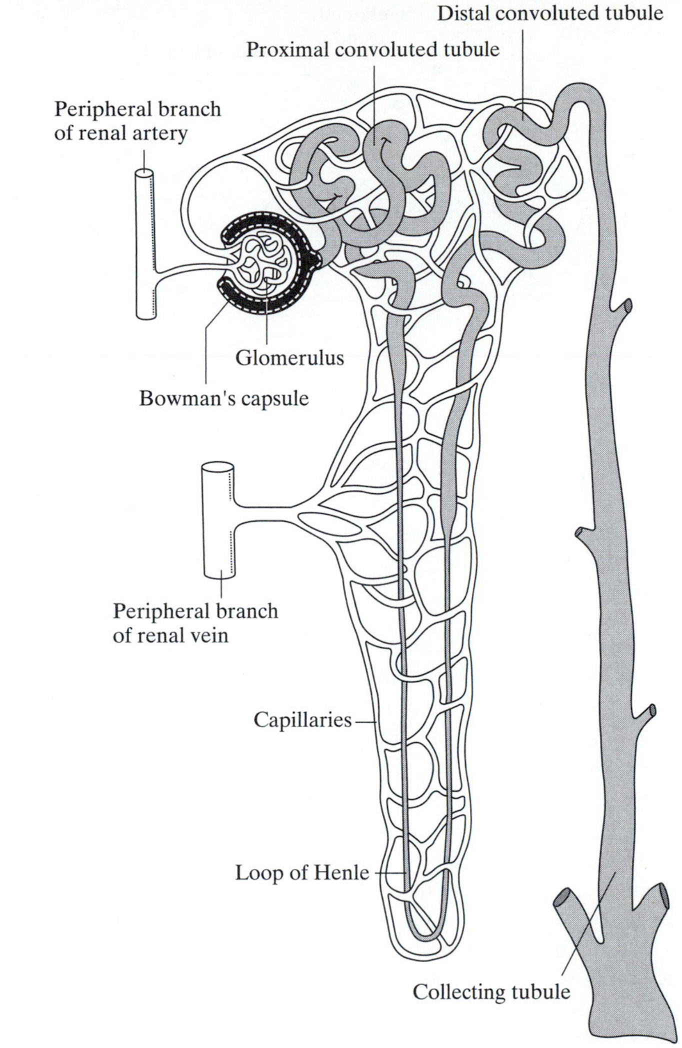

Figure 3.13a
The nephron has many distinct section. (*Source*: Keeton WT and Gould JL, *Biological Science*, 4th ed. New York: W. W. Norton and Company, Inc., 1986.)

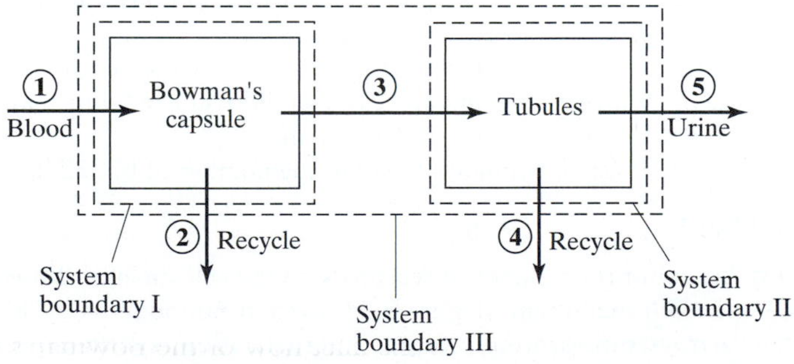

Figure 3.13b
Two-compartment model of the human kidney.

EXAMPLE 3.12 **Cell Receptor Trafficking**

Problem: Cell surface receptors covering a cell membrane bind ligands in the extracellular matrix to allow communication between the outside and inside of the cell. Under normal physiological conditions, dynamic trafficking events, such as receptor synthesis, degradation, internalization, and recycling, occur concurrently with receptor/ligand binding on the cell surface, and can alter the number of receptors present and ligands remaining in the extracellular matrix.

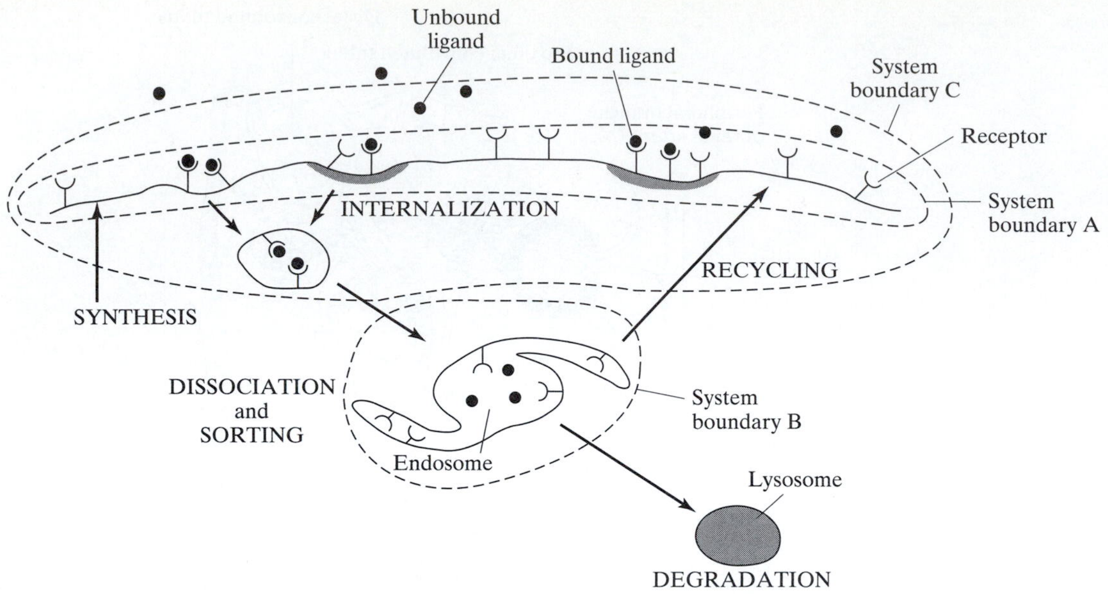

Figure 3.14
Simplified model of receptor trafficking. (*Source*: Adapted from Lauffenburger DA and Linderman JJ, *Receptors: Models for Binding, Trafficking, and Signaling*. New York: Oxford University Press, 1993.)

Once internalized from the cell membrane into the intracellular space, a receptor travels to an endosome, where receptor-ligand dissociation and sorting of the receptor occur. The endosome can recycle the receptor and send it back to the surface, or traffic the receptor to the lysosome for degradation. To increase the cell's ability to respond to a particular ligand, the cell can also synthesize receptors to increase the number of cell surface receptors. Figure 3.14 shows a simplified model of receptor trafficking, where a single endosome directs all the receptors in the cell.

(a) Using Figure 3.14, draw a system boundary designed to count the receptors on the cell surface. Which trafficking processes would you need information about to find the rate of receptor internalization in this system?
(b) Draw a system boundary designed to count the receptors in the endosomal compartment. Which trafficking processes would you need information about to find the rate of receptor internalization in this system?
(c) Draw a system boundary to count the number of bound ligands.

Solution:

(a) To count the number of receptors on the cell surface, draw a system boundary around the cell membrane (Figure 3.14, System Boundary A). This boundary cuts across the arrows representing receptor internalization, synthesis, and recycling. Thus, to find the rate of receptor internalization using a balance on receptor movement into and out of the system, we would need information on the rates of receptor synthesis and recycling.
(b) To count the receptors in the endosomal compartment, draw the system boundary around the endosome (Figure 3.14, System Boundary B). The boundary cuts across the arrows for receptor internalization, recycling, and degradation. Thus, to find the rate of receptor internalization using a balance on receptor movement into and out of this system, we would need information on the rates of receptor recycling and degradation.

(c) Ligands are bound to receptors both on the cell surface and after they have been internalized. However, when the internalized ligand/receptor complexes reach the endosome, they dissociate. To count the number of bound ligands, the boundary must include the ligands near the cell surface so that ligands from the extracellular matrix that bind and unbind to the receptors are counted. The boundary must also include endocytosed intracellular vesicles before they reach the endosome (Figure 3.14, System boundary C). ■

EXAMPLE 3.13 **Wastewater Treatment**

Problem: Wastewater from sewage, municipal water supplies, manufacturing facilities, and other sources must be purified at wastewater treatment plants before the water can be used again. Treated water discharged from treatment facilities may be used for drinking water, irrigation, or recreational uses (e.g., swimming). Before reuse, the wastewater is treated by both physical and biological processes. Figure 3.15a shows a simplified model of one design scheme for wastewater treatment.

In the aeration tank, bacteria degrade unwanted wastes. In the secondary settling tank, the bacteria and other solids settle to the bottom for recycling to the aeration tank and to the waste sludge; the remediated liquid stream is further processed. The section isolated for the model involves bacteria recycling from the secondary settling tank to the aeration tank (Figure 3.15b). When analyzed together, the system contains two process units, a splitter, and a recycle stream. The splitter separates the recycled bacteria (stream *r*) without changing the composition to two streams: one entering the aeration tank (stream *b*) and one moving to the sludge management unit (stream *s*).

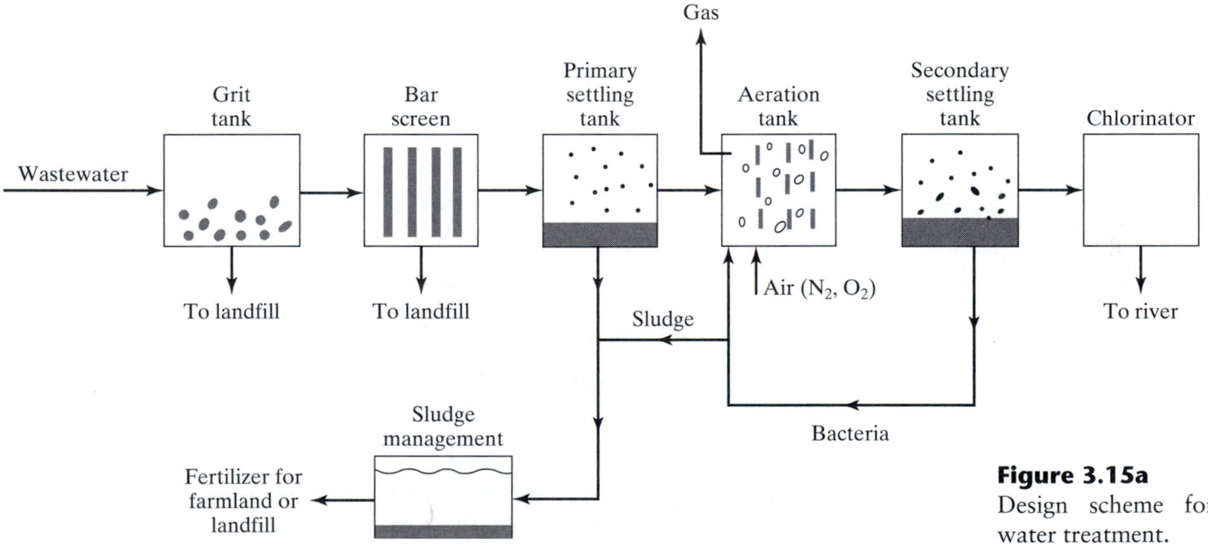

Figure 3.15a
Design scheme for wastewater treatment.

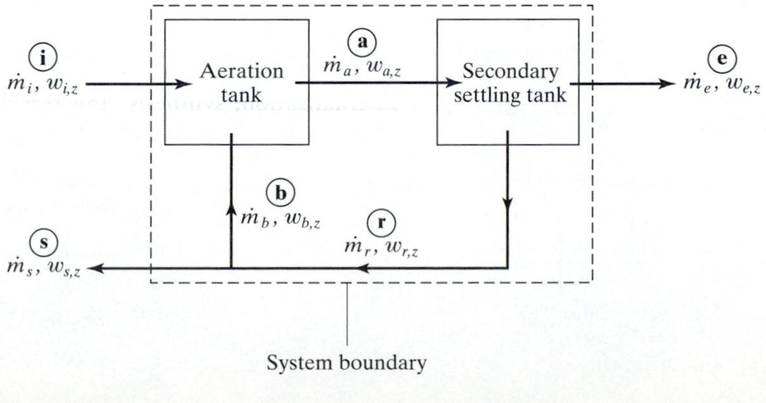

Figure 3.15b
Model isolating the aeration and secondary settling tanks.

A hypothetical nonreacting compound z is contained in each stream. Given that $\dot{m}_s/\dot{m}_i = 0.1$, $\dot{m}_b/\dot{m}_i = 0.05$, and $w_{e,z}/w_{i,z} = 0.95$, find the total mass flow rate in each stream ($\dot{m}_a$, $\dot{m}_e$, $\dot{m}_r$, $\dot{m}_b$, and $\dot{m}_s$) in terms of the total mass flow rate of the inlet stream ($\dot{m}_i$). Also solve for the mass fraction of the inert compound z in each stream in terms of the mass fraction of z in the inlet stream ($w_{i,z}$). Assume the system is operating at steady-state.

Solution: The mass rate $\dot{m}_i$ is the basis. As in all systems, total mass is conserved. The steady-state conservation equation [3.5-1] is applied for the isolated section in Figure 3.15b:

$$\sum_i \dot{m}_i - \sum_j \dot{m}_j = \dot{m}_i - \dot{m}_e - \dot{m}_s = 0$$

From the given information, we know that $\dot{m}_s = 0.1\dot{m}_i$. Substituting this into the overall mass balance for the system gives:

$$\dot{m}_i - \dot{m}_e - \dot{m}_s = \dot{m}_i - \dot{m}_e - 0.1\dot{m}_i = 0$$
$$\dot{m}_e = 0.9\dot{m}_i$$

To find the mass flow rates of the other streams, however, we need to know how the streams acting between the two process units relate. So we isolate each unit as a system. Since we have information about streams i and b, we first isolate the aeration tank as a system to solve for information about stream a (Figure 3.15c). The steady-state mass conservation equation for the aeration tank is:

$$\dot{m}_i + \dot{m}_b - \dot{m}_a = 0$$

From the given information, we know that $\dot{m}_b = 0.05\dot{m}_i$. Substituting this into the overall mass balance for the aeration tank gives:

$$\dot{m}_i + \dot{m}_b - \dot{m}_a = \dot{m}_i + 0.05\dot{m}_i - \dot{m}_a = 0$$
$$\dot{m}_a = 1.05\dot{m}_i$$

We now have information about streams a and e, so we can isolate the secondary settling tank as a system to obtain information about stream r (Figure 3.15d). The overall mass conservation equation for the secondary settling tank is:

$$\dot{m}_a - \dot{m}_e - \dot{m}_r = 0$$

Substituting the known mass flow rates for streams a and e gives:

$$\dot{m}_a - \dot{m}_e - \dot{m}_r = 1.05\dot{m}_i - 0.9\dot{m}_i - \dot{m}_r = 0$$
$$\dot{m}_r = 0.15\dot{m}_i$$

The mass flow rate for each stream is given in Table 3.2.

To solve for the mass fraction of compound z in each stream, recall equation [3.6-5], which relates the mass flow of a specific compound to the mass flow of the stream. Since compound z

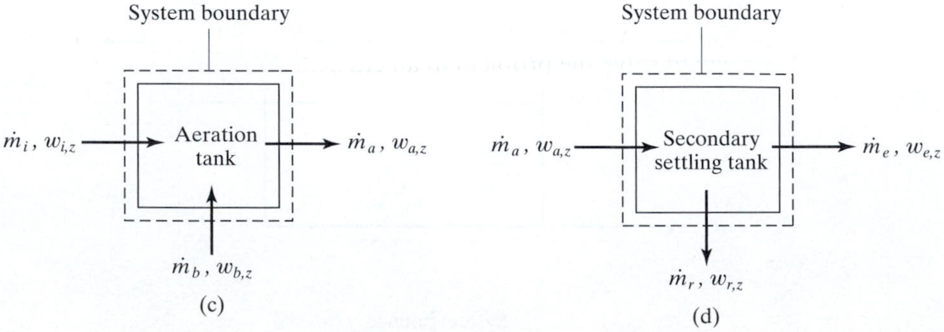

Figure 3.15c
Isolated aeration tank system.

Figure 3.15d
Isolated secondary settling tank.

TABLE 3.2

Mass Flow Rates and Mass Fractions in Wastewater Treatment Facility

Stream	Mass flow rate	Mass fraction of compound z
a	$1.05\dot{m}_i$	$1.02w_{i,z}$
b	$0.05\dot{m}_i$	$1.45w_{i,z}$
e	$0.9\dot{m}_i$	$0.95w_{i,z}$
i	$\dot{m}_i$	$w_{i,z}$
r	$0.15\dot{m}_i$	$1.45w_{i,z}$
s	$0.1\dot{m}_i$	$1.45w_{i,z}$

is nonreacting, it is neither generated nor consumed. Since the system is also at steady-state, conservation equations for compound z can be written for each unit system and the overall system:

Overall: $\dot{m}_i w_{i,z} - \dot{m}_e w_{e,z} - \dot{m}_s w_{s,z} = 0$

Aeration tank: $\dot{m}_i w_{i,z} + \dot{m}_b w_{b,z} - \dot{m}_a w_{a,z} = 0$

Secondary settling tank: $\dot{m}_a w_{a,z} - \dot{m}_e w_{e,z} - \dot{m}_r w_{r,z} = 0$

From the given information, we know that $w_{e,z} = 0.95w_{i,z}$. Substituting this and the values of $\dot{m}_e$ and $\dot{m}_s$ into the compound z balance for the overall system gives:

$$\dot{m}_i w_{i,z} - 0.9\dot{m}_i(0.95w_{i,z}) - 0.1\dot{m}_i w_{s,z} = 0$$
$$w_{s,z} = 1.45w_{i,z}$$

Since the splitter at the intersection of streams r, b, and s divides stream r into streams b and s without changing the composition of stream r, the mass fractions of all three streams must be the same:

$$w_{r,z} = w_{b,z} = w_{s,z} = 1.45w_{i,z}$$

The mass fraction of compound z is known in every stream except for stream a. Either of the two unit system compound conservation equations can be used to find $w_{a,z}$. Using the isolated aeration tank as the unit system and substituting in the known variables, the mass balance of compound z can be used to find the mass fraction:

$$\dot{m}_i w_{i,z} + \dot{m}_b w_{b,z} - \dot{m}_a w_{a,z} = \dot{m}_i w_{i,z} + 0.05\dot{m}_i(1.45w_{i,z}) - 1.05\dot{m}_i w_{a,z} = 0$$
$$w_{a,z} = 1.02w_{i,z}$$

The mass fraction of compound z in each stream is given in Table 3.2. ■

In the above example, we solved for the unknown variables without much difficulty. For other problems, however, a formal strategy, such as the **degree-of-freedom analysis**, may be needed. A degree-of-freedom analysis can help answer two questions: (1) Is the correct amount of information available to solve the problem? (2) What is a good strategy to solve the problem in an efficient manner? The latter question is particularly important when multiple compartments exist in a problem statement. The degree-of-freedom analysis is readily used to solve complex and multicompartment mass and energy balance problems.

The degree-of-freedom analysis is a systematic mechanism for counting all the variables, accounting and/or conservation equations, and relations involved. To solve a set of equations with N unknowns, the set must consist of N independent equations. If fewer than N independent equations are available, no solution is possible, resulting in an **underspecified** solution set. If more than N equations are available, then any N equations can be used to find the solution. This situation, called an **overspecified** solution set, always has the risk of errors or inconsistencies, since the solution obtained will depend on the N equations chosen. Hence, the only reliable solution set contains the same number of variables as equations (i.e., there are N unknowns for N independent equations) prior to solving, a system which is called **correctly specified.** The degree-of-freedom analysis is an index that measures the balance of known equations and information to unknown variables.

A degree-of-freedom analysis can also be used to develop a strategy to solve the problem in an efficient manner. Within a multi-unit system, a degree-of-freedom analysis can be conducted on each unit, on groups of units, and around the whole system. Often, one or two units are correctly specified, but the others are underspecified. An appropriate strategy is to first solve for information in correctly specified units, then repeat the degree-of-freedom analysis to decide which units are now solvable with the newly calculated information. Since it is outside the scope of this text to teach degree-of-freedom analysis, a solution strategy for each multi-unit system is presented along with the problem statement. Further information about degree-of-freedom analysis can be found in chemical engineering textbooks (e.g., Reklaitis, *Introduction to Material and Energy Balances*, 1983; Felder and Rousseau, *Elementary Principles of Chemical Processes*, 2000).

EXAMPLE 3.14 Two-Compartment Model of Kidneys

Problem: The functional unit of the kidney is the nephron (Figure 3.13a). Each nephron consists of a fully invaginated bulb called the Bowman's capsule and a long coiled tubule (consisting of the proximal convoluted tubule, the loop of Henle, and the distal convoluted tubule). Wastes, salts, and water move across the nephron walls to the interstitial space, which drains material to the peripheral branch of the renal vein.

In a healthy person, an average of 1200 mL/min of blood flows to the kidney to be filtered. In the Bowman's capsule, 125 mL/min is filtered based on size and collected as the filtrate, while the remainder drains to the renal vein and exits the kidneys. Only small molecules (≤69,000 g/mol), such as salts, urea, and creatinine, pass through to the filtrate. Few proteins and no cells move into the filtrate. After water resorption in the tubules, 0.69 mL/min of urine exits to the bladder, and the remainder of the filtrate moves into the interstitial space for drainage to the renal vein.

The kidney filtration process can be modeled with two units: the Bowman's capsule and the tubules. The blood that enters the kidney can be modeled to include red blood cells (RBC), proteins, urea, creatinine, uric acid, and water. (Other constituents are present, but not modeled in this problem.) RBCs compose 45% of the volume of blood. Plasma is defined as the part of blood that is not RBCs (i.e., water and other components). The composition of the small molecules is assumed to be the same in the filtrate and the plasma fraction of the blood. The urine contains only urea, creatinine, uric acid, and water. The ratios of the constituents in the urine as compared to the filtrate entering the tubules have been determined as follows:

$$\frac{C_{\text{urine,ur}}}{C_{\text{filt,ur}}} = 70 \qquad \frac{C_{\text{urine,cr}}}{C_{\text{filt,cr}}} = 140 \qquad \frac{C_{\text{urine,ua}}}{C_{\text{filt,ua}}} = 14$$

where ur is urea, cr is creatinine, ua is uric acid, and filt is filtrate. The concentration of each constituent in the urine can be easily measured:

$$C_{\text{urine,ur}} = 18.2\,\frac{\text{mg}}{\text{mL}} \qquad C_{\text{urine,cr}} = 1.96\,\frac{\text{mg}}{\text{mL}} \qquad C_{\text{urine,ua}} = 0.42\,\frac{\text{mg}}{\text{mL}}$$

Compare the concentrations and mass flow rates of urine, creatinine, uric acid, and water in the urine to those leaving the tubules to drain to the renal vein. Also compare these concentrations and mass flow rates with those of the blood entering the Bowman's capsule. Discuss the significance of these comparisons.

Solution:

1. Assemble
   (a) Find: concentrations and mass flow rates of all constituents in all inlet and outlet streams to the kidney. Compare these values to each other and analyze the significance of the differences.
   (b) Diagram: The system is modeled with two units: the Bowman's capsule and the tubules (Figure 3.16). The streams are numbered 1–5. The volumetric flow rates given for streams 1, 2, and 4 are listed on the figure. The constituents in each stream are labeled. Note that cells and proteins are present only in streams 1 and 3.
   (c) Table: Using a table and filling it in as you solve for variables can help you keep track of the different constituent concentrations and mass flow rates. Table 3.3A is an example of how to fill out such a chart.
2. Analyze
   (a) Assume:
   - Only the constituents listed in the problem (urine, creatinine, uric acid, water, protein, and cells) are found in the blood.
   - Filtration in the Bowman's capsule excludes 100% of the proteins and cells from the filtrate (Stream 2).
   - All cells are RBCs.
   - The concentrations of small molecules in streams 1, 2, and 3 are in equilibrium (i.e., the Bowman's capsule is assumed to be a passive separation/filtration device).
   - The density of the plasma before filtration is equal to the density of plasma and filtrate after filtration.
   - Proteins, urine, creatinine, and uric acid are soluble in water.

   (b) Extra data:
   - The concentration of proteins in the plasma phase is 82.18 mg/mL.[4]
   - The densities of blood, plasma, and RBCs are 1.056 g/mL, 1.0239 g/mL, and 1.098 g/mL, respectively (values from Table D.6).

   (c) Variables, notations, units:
   - ur = urea
   - cr = creatinine
   - ua = uric acid
   - H_2O = water
   - pr = protein
   - cell = cells
   - pl = plasma
   - filt = filtrate = stream 2

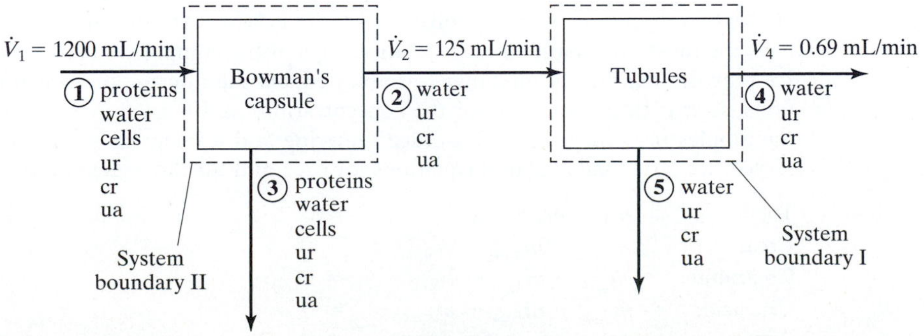

Figure 3.16
Two-unit model of the Bowman's capsule and the tubules in the kidney.

TABLE 3.3A

Setup of Concentrations and Mass Flow Rates of Constituents in Nephron

	Concentration (mg/mL)				
	Stream 1 Blood into Bowman's capsule	Stream 2 Filtrate into tubules	Stream 3 Exit Bowman's capsule to renal vein	Stream 4 Urine leaving tubules	Stream 5 Exit tubules to renal vein
Urea				18.2	
Creatinine				1.96	
Uric acid				0.42	
Proteins		0.0		0.0	0.0
Cells		0.0		0.0	0.0

	Mass flow rate (mg/min)				
	Stream 1	Stream 2	Stream 3	Stream 4	Stream 5
Urea					
Creatinine					
Uric acid					
Water					
Proteins		0.0		0.0	0.0
Cells		0.0		0.0	0.0
Total (mg/min)					
Total (mL/min)	1200	125		0.69	

- urine = stream 4
- Use mg, mL, min.

(d) Basis: The basis is calculated using the inlet blood flow rate of 1200 mL/min:

$$\dot{m}_1 = \dot{V}_1\rho_{\text{blood}} = \left(1200\frac{\text{mL}}{\text{min}}\right)\left(1.056\frac{\text{g}}{\text{cm}^3}\right) = 1267\frac{\text{g}}{\text{min}} = 1.27 \times 10^6\ \frac{\text{mg}}{\text{min}}$$

3. Calculate

(a) Equations: Because rates are given, we can use the differential accounting equation [3.3-6]. Since no chemical reaction or accumulation occurs in the system, we can reduce the equation to the steady-state conservation of mass [3.4-3]:

$$\sum_i \dot{m}_i - \sum_j \dot{m}_j = 0$$

(b) Calculate:

- Since the problem can be modeled using multiple units, using a degree-of-freedom analysis to figure out what to solve for first is helpful. Around the tubules, we have the most information on the relationships of the composition and flow. We see that the number of unknown variables equals the number of known equations, so it is easiest to solve for the concentrations and mass flow rates around the tubules first. Since each constituent entering and leaving the tubules is conserved, we write mass balance equations for each and for the overall unit:

Total: $\dot{m}_2 - \dot{m}_4 - \dot{m}_5 = 0$

Urea: $\dot{m}_{2,\text{ur}} - \dot{m}_{4,\text{ur}} - \dot{m}_{5,\text{ur}} = 0$

Creatinine: $\dot{m}_{2,\text{cr}} - \dot{m}_{4,\text{cr}} - \dot{m}_{5,\text{cr}} = 0$

Uric acid: $\dot{m}_{2,\text{ua}} - \dot{m}_{4,\text{ua}} - \dot{m}_{5,\text{ua}} = 0$

Water: $\dot{m}_{2,\text{H}_2\text{O}} - \dot{m}_{4,\text{H}_2\text{O}} - \dot{m}_{5,\text{H}_2\text{O}} = 0$

Only four of these five equations are linearly independent.

- We can find the mass flow rates of stream 4, since we are given the most information about its constituents. The overall mass flow rate for stream 4 is:

$$\dot{m}_4 = \dot{m}_{4,ur} + \dot{m}_{4,cr} + \dot{m}_{4,ua} + \dot{m}_{4,H_2O}$$

Using the given volumetric flow rate and the density of plasma (since no cells are present), we use equation [3.2-4] to find the mass flow rate of stream 4:

$$\dot{m}_4 = \dot{V}_4\rho_{plasma} = \left(0.69\frac{\text{mL}}{\text{min}}\right)\left(1.0239\frac{\text{g}}{\text{mL}}\right) = 0.706\frac{\text{g}}{\text{min}} = 706\ \frac{\text{mg}}{\text{min}}$$

Using the given concentration and the volumetric flow rate, we find the mass flow rate of urea in stream 4:

$$\text{Urea:} \qquad \dot{m}_{4,ur} = C_{4,ur}\dot{V}_4 = \left(18.2\frac{\text{mg}}{\text{mL}}\right)\left(0.69\frac{\text{mL}}{\text{min}}\right) = 12.6\ \frac{\text{mg}}{\text{min}}$$

Similarly, the mass flow rates for creatinine and uric acid in stream 4 are 1.35 mg/min and 0.29 mg/min, respectively. Substituting in all these values into the overall mass conservation equation for stream 4 gives the mass flow rate of water:

$$\begin{aligned}\text{Water:} \qquad \dot{m}_{4,H_2O} &= \dot{m}_4 - (\dot{m}_{4,ur} + \dot{m}_{4,cr} + \dot{m}_{4,ua}) \\ &= 706\frac{\text{mg}}{\text{min}} - \left(12.6\frac{\text{mg}}{\text{min}} + 1.35\frac{\text{mg}}{\text{min}} + 0.29\frac{\text{mg}}{\text{min}}\right) \\ &= 692\ \frac{\text{mg}}{\text{min}}\end{aligned}$$

- Using the given relationships between the concentrations of the constituents in the filtrate and urine streams, we can find the mass concentration of each constituent in stream 2. For urine, the concentration in stream 2 is:

$$\text{Urea:} \qquad C_{2,ur} = \frac{C_{4,ur}}{70} = \frac{18.2\frac{\text{mg}}{\text{mL}}}{70} = 0.26\ \frac{\text{mg}}{\text{mL}}$$

Solving in a similar manner for the concentrations of creatinine and uric acid in stream 2 gives 0.014 mg/mL and 0.03 mg/mL, respectively. We can then use these concentration values and the volumetric flow rate of filtrate to find the total mass flow rate of stream 2, as well as the mass flow rate of each constituent in the stream; we can do this in the same manner as we did for stream 4. The mass flow rate of stream 2, $\dot{m}_2$, is 1.28×10^5 mg/min; and the mass flow rates for the constituents are: $\dot{m}_{2,ur} = 32.5$ mg/min, $\dot{m}_{2,cr} = 1.75$ mg/min, $\dot{m}_{2,ua} = 3.75$ mg/min, and $\dot{m}_{2,H_2O} = 1.28 \times 10^5$ mg/min. Note that in stream 2 the mass flow rate of water is approximately the same as the total mass flow rate; this is because the small molecules are present in such minute amounts.
- With the mass flow rate values for streams 2 and 4, we can now solve for the constituent mass flow rates in stream 5 using the overall mass conservation equation for each constituent in the tubule system. For urea:

$$\text{Urea:} \qquad \dot{m}_{2,ur} - \dot{m}_{4,ur} - \dot{m}_{5,ur} = 0$$

$$\begin{aligned}\dot{m}_{5,ur} &= \dot{m}_{2,ur} - \dot{m}_{4,ur} \\ &= 32.5\frac{\text{mg}}{\text{min}} - 12.6\frac{\text{mg}}{\text{min}} = 19.9\ \frac{\text{mg}}{\text{min}}\end{aligned}$$

Similarly, the total mass flow rate for stream 5, $\dot{m}_5$, is 1.27×10^5 mg/min, and the remaining constituent flow rates are: $\dot{m}_{5,cr} = 0.40$ mg/min, $\dot{m}_{5,ua} = 3.46$ mg/min, and $\dot{m}_{5,H_2O} = 1.27 \times 10^5$ mg/min. To solve for the concentration of urea in stream 5, we first find the volumetric flow rate using equation [3.2-4]:

$$\dot{V}_5 = \frac{\dot{m}_5}{\rho_{plasma}} = \frac{1.27 \times 10^5 \frac{mg}{min}}{1.0239 \frac{g}{mL}} = 124.3 \ \frac{mL}{min}$$

$$\text{Urea:} \qquad C_{5,ur} = \frac{\dot{m}_{5,ur}}{\dot{V}_5} = \frac{19.9 \frac{mg}{min}}{124.3 \frac{mL}{min}} = 0.16 \ \frac{mg}{mL}$$

The concentrations for creatinine and uric acid are found in a similar fashion, giving $C_{5,cr} = 0.0032$ mg/mL and $C_{5,ua} = 0.0278$ mg/mL, respectively.

- With the information for the inlet and outlet flows of the tubule unit system, we now solve the Bowman's capsule unit. As with the tubule system, all constituents are conserved in the Bowman's capsule, so a total mass conservation equation can be written, as well as one for each constituent:

Total	$\dot{m}_1 - \dot{m}_2 - \dot{m}_3 = 0$
Urea:	$\dot{m}_{1,ur} - \dot{m}_{2,ur} - \dot{m}_{3,ur} = 0$
Creatinine:	$\dot{m}_{1,cr} - \dot{m}_{2,cr} - \dot{m}_{3,cr} = 0$
Uric acid:	$\dot{m}_{1,ua} - \dot{m}_{2,ua} - \dot{m}_{3,ua} = 0$
Water:	$\dot{m}_{1,H_2O} - \dot{m}_{2,H_2O} - \dot{m}_{3,H_2O} = 0$
Protein:	$\dot{m}_{1,pr} - \dot{m}_{3,pr} = 0$
Cells:	$\dot{m}_{1,cell} - \dot{m}_{3,cell} = 0$

Only six of these seven equations are linearly independent.

- We find the total mass flow rate of stream 1, $\dot{m}_1$, using the volumetric inlet flow rate and the density of blood:

$$\dot{m}_1 = \dot{V}_1\rho_{blood} = \left(1200 \frac{mL}{min}\right)\left(1.056 \frac{g}{cm^3}\right) = 1267 \frac{g}{min} = 1.27 \times 10^6 \ \frac{mg}{min}$$

- Using the total mass conservation equation for the Bowman's capsule, we find the mass flow rate of stream 3:

$$\text{Total:} \qquad \dot{m}_1 - \dot{m}_2 - \dot{m}_3 = 0$$

$$\dot{m}_3 = \dot{m}_1 - \dot{m}_2 = 1.27 \times 10^6 \ \frac{mg}{min} - 1.28 \times 10^5 \ \frac{mg}{min}$$

$$= 1.14 \times 10^6 \ \frac{mg}{min}$$

- To find the concentrations of the constituents in stream 3, we first need to find the volumetric flow rate. The density in stream 3 is not that of blood, since some of the small molecules are filtered into stream 2, which is similar in consistency to plasma. To estimate the density in stream 3, we perform a simple thought experiment to determine the new ratio of cells to plasma.

  Assume 1000 mL of blood enters the Bowman's capsule. Of that, 450 mL is cells and 550 mL is plasma. Since about 10% (1.28×10^5 mg/min of 1.267×10^6 mg/min) of the blood volume entering the capsule is filtered, 100 mL of plasma goes to stream 2. No volume of cells moves to stream 2, since the cells cannot cross the filter, which leaves 450 mL cells and 450 mL plasma, or 50 vol% cells and 50 vol% plasma mixture, in stream 3. Thus, the density of stream 3 can be estimated:

$$\rho_3 = 0.5\rho_{cells} + 0.5\rho_{plasma} = 0.5\left(1.098\frac{g}{mL}\right) + 0.5\left(1.0239\frac{g}{mL}\right) = 1.061\ \frac{g}{mL}$$

The density of stream 3 is close to that of normal blood, 1.056 g/mL. In fact, it might have been acceptable to make an engineering approximation and assume the density of stream 3 equals that of stream 1. Thus, the volumetric flow rate of stream 3 is:

$$\dot{V}_3 = \frac{\dot{m}_3}{\rho_3} = \frac{1.14 \times 10^6 \frac{mg}{min}}{1.061\frac{g}{mL}\left(1000\frac{mg}{g}\right)} = 1075\ \frac{mL}{min}$$

- We can now calculate the concentrations of the constituents in streams 1 and 3. For proteins, recall from the problem statement that the concentration of proteins in plasma is 82.18 mg/mL and the volume percent of plasma is 55%. Thus, the concentration of proteins in stream 1 can be used to find the mass flow rate of proteins in stream 1:

$$\dot{m}_{1,pr} = C_{1,pr}\dot{V}_1 = 0.55\left(82.18\frac{mg}{mL}\right)\left(1200\frac{mL}{min}\right) = 5.42 \times 10^4\frac{mg}{min}$$

Using the overall mass balance equation for proteins in the Bowman's capsule, we find the mass flow rate and concentration of proteins in stream 3:

Protein: $\dot{m}_{1,pr} - \dot{m}_{3,pr} = 0$

$$\dot{m}_{3,pr} = \dot{m}_{1,pr} = 5.42 \times 10^4\ \frac{mg}{min}$$

$$C_{3,pr} = \frac{\dot{m}_{3,pr}}{\dot{V}_3} = \frac{5.42 \times 10^4\frac{mg}{min}}{1075\frac{mL}{min}} = 50.46\ \frac{mg}{mL}$$

The mass flow rates and concentrations of cells in streams 1 and 3 are calculated in the same manner, except that the volume percent of cells is 45%. The calculated values are: $\dot{m}_{1,cell} = \dot{m}_{3,cell} = 5.93 \times 10^5$ mg/min, $C_{1,cell} = 494$ mg/mL, and $C_{3,cell} = 552$ mg/mL. It makes sense that the protein and cell concentrations are slightly higher in stream 3 than in stream 1, since the material is effectively concentrated in the Bowman's capsule.

- Since the concentrations of urea, creatinine, and uric acid in the filtered and plasma phases are in equilibrium with one another (i.e., the concentrations are equal), the concentrations of these constituents in streams 1, 2, and 3 are equal. Stream 2 is entirely filtered liquid (i.e., no cells), but streams 1 and 3 contain both plasma and cells. Since the currently calculated concentrations are for constituents dissolved in the plasma phases, those concentrations need to be scaled to represent the total stream. To do this, we multiply the concentrations by the fraction of the total stream that the plasma phase comprises (0.55 and 0.50 for streams 1 and 3, respectively):

Urea: $C_{1,ur} = 0.55C_{1/pl,ur} = 0.55C_{2,ur} = 0.55\left(0.26\frac{mg}{mL}\right) = 0.143\frac{mg}{mL}$

TABLE 3.3B

Concentrations and Mass Flow Rates of Constituents in Nephron

	Concentration (mg/mL)				
	Stream 1 Blood into Bowman's capsule	Stream 2 Filtrate into tubules	Stream 3 Exit Bowman's capsule to renal vein	Stream 4 Urine leaving tubules	Stream 5 Exit tubules to renal vein
Urea	0.143	0.26	0.13	18.2	0.16
Creatinine	0.0077	0.014	0.007	1.96	0.0032
Uric acid	0.0165	0.03	0.015	0.42	0.0278
Proteins	45.2	0.0	50.5	0.0	0.0
Cells	494	0.0	552	0.0	0.0

	Mass flow rate (mg/min)				
	Stream 1	Stream 2	Stream 3	Stream 4	Stream 5
Urea	172	32.5	140	12.6	19.9
Creatinine	9.24	1.75	7.53	1.35	0.40
Uric acid	19.8	3.75	16.1	0.29	3.46
Water	6.20×10^5	1.28×10^5	4.93×10^5	692	1.27×10^5
Proteins	5.42×10^4	0.0	5.42×10^4	0.0	0.0
Cells	5.93×10^5	0.0	5.93×10^5	0.0	0.0
Total (mg/min)	1.27×10^6	1.28×10^5	1.14×10^6	706	1.27×10^5
Total (mL/min)	1200	125	1075	0.69	124.3

The other concentrations of urea, creatinine, and uric acid in streams 1 and 3 are calculated similarly and are shown in Table 3.3B.

- The mass flow rates for the soluble constituents in streams 1 and 3 can be solved in the same manner as for the other streams. For urea in stream 1:

$$\dot{m}_{1,\text{ur}} = C_{1,\text{ur}}\dot{V}_1 = \left(0.143\,\frac{\text{mg}}{\text{mL}}\right)\left(1200\,\frac{\text{mL}}{\text{min}}\right) = 171.6\,\frac{\text{mg}}{\text{min}}$$

The remaining constituent mass flow rates in streams 1 and 3 are shown in Table 3.3B.

- To calculate the mass flow rate of water in stream 1, we need to write an overall mass balance equation for stream 1:

$$\dot{m}_1 = \dot{m}_{1,\text{ur}} + \dot{m}_{1,\text{cr}} + \dot{m}_{1,\text{ua}} + \dot{m}_{1,\text{H}_2\text{O}} + \dot{m}_{1,\text{pr}} + \dot{m}_{1,\text{cell}}$$

$$\begin{aligned}\dot{m}_{1,\text{H}_2\text{O}} &= \dot{m}_1 - (\dot{m}_{1,\text{ur}} + \dot{m}_{1,\text{cr}} + \dot{m}_{1,\text{ua}} + \dot{m}_{1,\text{pr}} + \dot{m}_{1,\text{cell}}) \\ &= 1.267 \times 10^6\,\frac{\text{mg}}{\text{min}} \\ &\quad - \left(171.6\,\frac{\text{mg}}{\text{min}} + 9.24\,\frac{\text{mg}}{\text{min}} + 19.8\,\frac{\text{mg}}{\text{min}}\right. \\ &\quad \left. + 5.42 \times 10^4\,\frac{\text{mg}}{\text{min}} + 5.93 \times 10^5\,\frac{\text{mg}}{\text{min}}\right) \\ &= 6.196 \times 10^5\,\frac{\text{mg}}{\text{min}}\end{aligned}$$

The mass flow rate of water in stream 3 is calculated in the same fashion, giving a flow rate of 4.93×10^5 mg/min.

4. Finalize

(a) Answer: The answers are given in Table 3.3B. In comparing the relative concentrations of urea, creatinine, and uric acid in the urine in all streams of the system, we see that the patterns of change are the same. The kidney is highly efficient at concentrating waste products and conserving water.

Using urea as an example, the entering concentration is 0.143 mg/mL. After passive separation in the Bowman's capsule, the concentration entering the renal vein is nearly unchanged (0.13 mg/mL) and the concentration entering the tubules is two times greater (0.26 mg/mL). With the active transport mechanism in the tubules, the concentration of urea in the urine increases nearly 100-fold (to 18.2 mg/mL), while the concentration entering the renal vein is comparable to the inlet concentration (0.16 mg/mL).

Looking at the total mass flow rate of urea entering the Bowman's capsule (172 mg/min), we see that about 80% (140 mg/min) is shunted off to the renal vein, while only 20% (32.5 mg/min) flows into the tubules, where about 60% (19.9 mg/min) exits in the urine and 40% (12.6 mg/min) leaves through the renal vein. Overall, only 11.5% of the urea mass entering the kidneys exits in the urine.

Looking at the total mass flow rate of water entering the kidney (6.20×10^5 mg/min), 80% (4.93×10^5 mg/min) is shunted off to the renal vein, while 20% (1.28×10^5 mg/min) flows into the tubules. (Note that these percentages are the same as urea.) In the tubules, 0.54% of water (692 mg/min) exits in the urine and 99.5% (1.27×10^5 mg/min) leaves through the renal vein. Overall, 99.9% of the water mass entering the kidneys remains in the body.

(b) Check: The numerical results can be checked in a number of ways. For example, one could construct an overall mass conservation equation on the system as follows:

$$\dot{m}_1 - \dot{m}_3 - \dot{m}_4 - \dot{m}_5 = 0$$

You can check the equation for total mass flow rates, as well as for individual constituent mass flows. You can also confirm that constituent flow rates in each stream add up to the total flow rate. (This will work as a check as long as you didn't use the overall equation for a particular stream to determine the last of the unknown flow rates.) The details on checking the numerical results are not shown. ■

3.8 Systems with Chemical Reactions

Chemical reactions are present in many biological systems. For this reason, we need a formal way to treat reactions in these systems. **The accounting equations introduced earlier contain Generation and Consumption terms, which are used to rigorously solve systems involving chemical and biochemical reactions.** Before we can solve mass accounting equations that handle chemical reactions, concepts such as stoichiometry, fractional conversion, and reaction rates must be mastered.

3.8.1 Balancing Chemical Reactions

Stoichiometry is the theory of how specific chemical species are proportioned in chemical reactions and is based on the conservation of elemental mass. The **stoichiometric equation** of a chemical reaction is a statement of the relative number of molecules or moles of reactants and products that participate in the reaction. An example of a stoichiometric equation is the conversion of glucose into ethanol and carbon dioxide during fermentation:

$$C_6H_{12}O_6 \longrightarrow 2\ C_2H_6O + 2\ CO_2 \qquad [3.8\text{-}1]$$

For a stoichiometric equation to be valid, it must be *balanced* to meet the restrictions of the conservation of elemental mass. A chemical equation is balanced when the number of atoms of each atomic species is the same on both sides of the equation. In the example above, both sides of the equation have 12 H atoms, 6 C atoms, and 6 O atoms, which indicate the equation is balanced.

Recall in the conservation equation describing element mass and moles that the Generation and Consumption terms were never present, since atoms, and therefore elements, can be neither created nor destroyed. (This statement does not hold for nuclear reactions, but they are not considered in this chapter.) This is why the conservation equation is always valid for total mass. On the other hand, the Generation and Consumption terms are nonzero in the accounting equations describing mass and moles of specific chemical species, since chemical species can be transformed during a reaction.

In general, a stoichiometric reaction equation is written:

$$aA + bB + cC + dD + \cdots \rightarrow pP + qQ + rR + \cdots \qquad [3.8\text{-}2]$$

where a, b, c, d, p, q, and r are the stoichiometric coefficients (σ) and A, B, C, D, P, Q, and R are the chemical compounds. **Stoichiometric coefficients** are numbers that precede a chemical species in a reaction that ensure the reaction is balanced. To solve for the stoichiometric coefficients, the following system is employed:

- Index the compounds, corresponding each reactant to A, B, C, and so on, and each product to P, Q, R, and so on.
- Index the elements present in the reaction, numbered 1, 2, 3, and so on.
- For each element, the number of atoms in each compound can be written k_{ij}, where i is the element number index and j is the compound index.
- Construct a balanced equation using the stochiometric coefficents, labeled a, b, c, d, p, q, r, and so on, for each element i:

$$\text{element } i: \quad -ak_{iA} - bk_{iB} - ck_{iC} - dk_{iD} - \cdots + pk_{iP} + qk_{iQ} + rk_{iR} + \cdots = 0 \qquad [3.8\text{-}3]$$

- Solve the set of equations for the unknown stoichiometric coefficients.

Coefficients associated with compounds consumed in the reaction are negative in sign, whereas coefficients associated with compounds generated are positive in sign. Note that to solve for n stoichiometric coefficients, there must be n element equations.

For the fermentation reaction above, the compounds are $C_6H_{12}O_6$, C_2H_6O, and CO_2. The compounds are indexed such that $C_6H_{12}O_6$ is A, C_2H_6O is P, and CO_2 is Q. The elements carbon, hydrogen, and oxygen are indexed 1, 2, and 3, respectively. Looking at carbon, the term k_{1A} is 6, since there are 6 carbon atoms (element 1) in $C_6H_{12}O_6$ (compound A). The numbers of carbon atoms in C_2H_6O and CO_2 are 2 and 1, respectively. The stoichiometric coefficients are: a is 1 for $C_6H_{12}O_6$, b is 2 for C_2H_6O, and c is 2 for CO_2. Thus, the balanced element equation for carbon is:

$$\begin{aligned} \text{C:} \quad & -ak_{1A} + pk_{1P} + qk_{1Q} = 0 \\ & -1(6) + 2(2) + 2(1) = 0 \end{aligned} \qquad [3.8\text{-}4]$$

Similar stoichiometric balances can be written for oxygen and hydrogen.

Often the stoichiometric coefficients of a chemical reaction are unknown, resulting in unknowns in the element balance equations. Some chemical reactions, like the one described above, can be balanced by inspection. In fact, this process may seem to be overkill for simple chemical reactions. However, some classes of chemical reactions involve production of biomass and/or other organic products that require this process tool to balance complex chemical reactions.

Aerobic biochemical reactions that produce organic products, such as biomass, frequently involve the consumption of oxygen and the release of carbon dioxide. The ratio of the amount of carbon dioxide (in moles) released or emitted by a system to the amount of oxygen (in moles) consumed during a given period of time is defined as the **respiratory quotient** (RQ), which is an experimental datum commonly gathered when operating a bioreactor:

$$RQ = \frac{n_{CO_2}}{n_{O_2}} \qquad [3.8\text{-}5]$$

In situations where the number of element balance equations is not sufficient to solve for the number of unknown stoichiometric coefficients (i.e., the system is underspecified), the RQ may be added as an additional equation to facilitate a solution.

> The respiratory quotient is also used to estimate carbohydrate and fat use in the human body. When the body metabolizes carbohydrates for energy, it consumes exactly one oxygen molecule for every carbon dioxide molecule produced, resulting in a respiratory quotient of 1.0. On the other hand, fat metabolism has an average respiratory quotient of 0.70, since an average of 70 carbon dioxide molecules form for every 100 oxygen molecules used. This is because fat molecules contain excess hydrogen atoms that combine with a portion of the oxygen metabolized in foods. Thus, the relative amount of carbon dioxide produced is lower for fat metabolism than for carbohydrate metabolism, resulting in a lower respiratory quotient.

EXAMPLE 3.15 **Cell Growth on Hexadecane**

Problem: The conversion of hexadecane ($C_{16}H_{34}$) into cell mass and CO_2 is described by the following reaction equation:

$$C_{16}H_{34} + aO_2 + bNH_3 \longrightarrow pCH_{1.66}O_{0.27}N_{0.20} + qCO_2 + rH_2O$$

where $CH_{1.66}O_{0.27}N_{0.20}$ represents the produced biomass. In laboratory experiments, the respiratory quotient (RQ) is determined to be 0.43. Find the stoichiometric coefficients. (Adapted from Doran PM, *Bioprocess Engineering Principles*, 1999.)

Solution: Using the formulation [3.8-3], the following element balance equations are written:

Carbon (element 1): $-1(16) + 1p + 1q = 0$

Hydrogen (element 2): $-1(34) - 3b + 1.66p + 2r = 0$

Oxygen (element 3): $-2a + 0.27p + 2q + 1r = 0$

Nitrogen (element 4): $-1b + 0.20p = 0$

Note there are four equations (for C, H, O, and N), but five unknown stoichiometric coefficients (a, b, p, q, and r). To solve for these unknowns, a fifth equation is required, which is given by the respiratory quotient:

$$RQ = \frac{n_{CO_2}}{n_{O_2}} = \frac{q}{a} = 0.43$$

The RQ equation is transformed to the following:

$$RQ: \quad -0.43a + q = 0$$

to give five equations to solve for the five unknowns. This series of equations can be solved using variable elimination, Cramer's rule, or MATLAB. Arranging these scalar equations into the matrix equation form $A\vec{x} = \vec{y}$ gives:

$$\begin{bmatrix} 0 & 0 & 1 & 1 & 0 \\ 0 & -3 & 1.66 & 0 & 2 \\ -2 & 0 & 0.27 & 2 & 1 \\ 0 & -1 & 0.2 & 0 & 0 \\ -0.43 & 0 & 0 & 1 & 0 \end{bmatrix} \begin{bmatrix} a \\ b \\ p \\ q \\ r \end{bmatrix} = \begin{bmatrix} 16 \\ 34 \\ 0 \\ 0 \\ 0 \end{bmatrix}$$

Plugging this matrix equation into MATLAB yields the stoichiometric coefficients of the chemical reaction:

$$x = \begin{bmatrix} a \\ b \\ p \\ q \\ r \end{bmatrix} = \begin{bmatrix} 12.4878 \\ 2.1260 \\ 10.6302 \\ 5.3698 \\ 11.3660 \end{bmatrix}$$

Thus, the balanced equation is written as follows:

$$C_{16}H_{34} + 12.49\ O_2 + 2.13\ NH_3 \longrightarrow$$
$$10.63\ CH_{1.66}O_{0.27}N_{0.20} + 5.37\ CO_2 + 11.37\ H_2O$$

Each element can be checked to make sure the equation is balanced correctly. For example, there are 24.98 moles of oxygen on both sides of the equation. Because the stoichiometric coefficients are noninteger values, it is much easier to do this problem using a computer or calculator than to balance it by inspection. ■

To balance a stoichiometric reaction equation, the stoichiometric coefficient of one compound may be arbitrarily set to one, and all other calculated compound coefficients must be scaled according to the compound with the set coefficient. In Example 3.15, the stoichiometric coefficient of $C_{16}H_{34}$ was set to one, and the other coefficients were calculated subsequently.

Often, sources of carbon (e.g., glucose), nitrogen (e.g., ammonia), and oxygen (e.g., oxygen gas) are present as reactants in biochemical reactions. These biochemical reactions frequently involve compounds with only the four elements of carbon, hydrogen, oxygen, and nitrogen. The respiratory quotient provides an additional equation. Another experimentally measured value that can provide another equation is the **yield**, which is the ratio of the amount of organic product formed (in moles) to the amount of glucose or another carbon-containing reactant consumed (in moles):

$$\text{yield} = \frac{n_{\text{organic product}}}{n_{\text{organic reactant}}} \qquad [3.8\text{-}6]$$

The yield can be used to solve for stoichiometric coefficients in a chemical reaction when glucose or another carbon-containing reactant is the only carbon source for the organic product formed.

EXAMPLE 3.16 **Production of Citric Acid**

Problem: Citric acid ($C_6H_8O_7$) is a natural preservative that prevents discoloration of foods and can be manufactured as a food additive industrially in a bioreactor containing *Asperigillus niger*:

$$C_6H_{12}O_6 + aNH_3 + bO_2 \longrightarrow pCH_{1.79}N_{0.2}O_{0.5} + qH_2O + rCO_2 + sC_6H_8O_7$$

For this reaction, the respiratory quotient is 0.45. The yield of citric acid per mole of glucose consumed is 0.70. The cell mass is given as $CH_{1.79}N_{0.2}O_{0.5}$.

Solution: Using the formulation [3.8-3], the following element balance equations are written:

$$\begin{aligned}
&\text{Carbon:} && -6 + p + r + 6s = 0 \\
&\text{Hydrogen:} && -12 - 3a + 1.79p + 2q + 8s = 0 \\
&\text{Oxygen:} && -6 - 2b + 0.50p + q + 2r + 7s = 0 \\
&\text{Nitrogen:} && -a + 0.2p = 0
\end{aligned}$$

Note there are four equations (for C, H, O, and N) but six unknowns (a, b, p, q, r, and s). The respiratory quotient and yield provide two additional linearly independent equations. The organic product of interest is citric acid.

$$\text{RQ} = \frac{n_{CO_2}}{n_{O_2}} = \frac{r}{b} = 0.45$$

$$\text{yield} = \frac{n_{C_6H_8O_7}}{n_{C_6H_{12}O_6}} = \frac{s}{1} = s = 0.70$$

Having solved for s, we are left with five equations and five unknowns. Again, this series of equations can be solved using your method of choice. Arranging these scalar equations into the matrix equation form $A\vec{x} = \vec{y}$ gives:

$$\begin{bmatrix} 0 & 0 & 1 & 0 & 1 \\ -3 & 0 & 1.79 & 2 & 0 \\ 0 & -2 & 0.5 & 1 & 2 \\ -1 & 0 & 0.2 & 0 & 0 \\ 0 & -0.45 & 0 & 0 & 1 \end{bmatrix} \begin{bmatrix} a \\ b \\ p \\ q \\ r \end{bmatrix} = \begin{bmatrix} 1.8 \\ 6.4 \\ 1.1 \\ 0 \\ 0 \end{bmatrix}$$

Solving this matrix equation in MATLAB yields the stoichiometric coefficients of the chemical reaction:

$$x = \begin{bmatrix} a \\ b \\ p \\ q \\ r \end{bmatrix} = \begin{bmatrix} 0.196 \\ 1.82 \\ 0.979 \\ 0.821 \\ 2.62 \end{bmatrix}$$

Thus, the balanced equation is written:

$$\begin{aligned} C_6H_{12}O_6 + 0.196\ NH_3 + 1.82\ O_2 &\longrightarrow 0.979\ CH_{1.79}N_{0.2}O_{0.5} \\ &+ 0.821\ H_2O + 2.62\ CO_2 + 0.70\ C_6H_8O_7 \end{aligned}$$

■

3.8.2 Using Reaction Rates in the Accounting Equation

When approaching a problem involving a chemical reaction, you must write out the reaction and balance the equation before moving any further into the problem. **Stoichiometric balances and reaction rates must always be worked in units of moles or molecules.** Recall the example of the fermentation of glucose to ethanol:

$$C_6H_{12}O_6 \longrightarrow 2\ C_2H_6O + 2\ CO_2 \qquad [3.8\text{-}7]$$

Assume that 100 kg/day of glucose enters the fermentation vessel and completely converts to ethanol and carbon dioxide. Clearly, 200 kg/day of C_2H_6O and 200 kg/day of CO_2 are not formed (although this might be the answer if you forgot that chemical reactions are stoichiometrically balanced in moles, not mass).

Instead, a total mass balance can be used to solve this problem. Looking at the total mass balance, only 100 kg/day of material enters the system. Assuming the system is at steady-state, total mass is conserved (equation [3.4-3]):

$$\dot{m}_i - \dot{m}_j = 0 \qquad [3.8\text{-}8]$$

$$\dot{m}_i = \dot{m}_j = 100\frac{\text{kg}}{\text{day}} \qquad [3.8\text{-}9]$$

Therefore, the outlet mass flow rate is 100 kg/day.

To determine the outlet mass flow rate of each of the constituents, the rate of moles of glucose entering the system is calculated using equation [3.2-5]:

$$\text{Glucose:} \quad \dot{n}_{i,C_6H_{12}O_6} = \frac{\dot{m}_{i,C_6H_{12}O_6}}{M_{C_6H_{12}O_6}}$$

$$= \left(100\,\frac{\text{kg}}{\text{day}}\right)\left(\frac{\text{mol}}{180\text{ g}}\right)\left(\frac{1000\text{ g}}{1\text{ kg}}\right) = 555\,\frac{\text{mol}}{\text{day}} \qquad [3.8\text{-}10]$$

To determine the rate of moles of particular compounds exiting the system, the stoichiometric coefficients must be known. In this problem, the stoichiometric coefficients of glucose, ethanol, and carbon dioxide are 1, 2, and 2, respectively. Thus, with 555 mol/day of glucose entering the system, 1110 mol/day of ethanol and 1110 mol/day of carbon dioxide are generated. The mass flow rate for ethanol can then be determined using equation [3.2-5]:

$$\text{Ethanol:} \quad \dot{m}_{j,C_2H_6O} = \dot{n}_{j,C_2H_6O}M_{C_2H_6O}$$

$$= \left(1110\,\frac{\text{mol}}{\text{day}}\right)\left(\frac{46\text{ g}}{\text{mol}}\right)\left(\frac{1\text{ kg}}{1000\text{ g}}\right) = 51.1\,\frac{\text{kg}}{\text{day}} \qquad [3.8\text{-}11]$$

Similarly, the mass flow rate for carbon dioxide is calculated to be 48.9 kg/day. Note that the total mass leaving the system (both ethanol and carbon dioxide) sums to 100 kg/day (Table 3.4).

If reactants are fed to a system in stoichiometric proportions and the reaction proceeds to completion, then all of the reactants are consumed. In reality, both of these conditions rarely occur. If reactants are present in stoichiometric proportions,

TABLE 3.4

Glucose Fermentation Under Conditions with Different Fractional Conversions

	$f = 1$		$f = 0.5$	
	Inlet (kg/day)	Outlet (kg/day)	Inlet (kg/day)	Outlet (kg/day)
Glucose	100	0	100	49.95
Ethanol	—	51.1	—	25.53
Carbon dioxide	—	48.9	—	24.42
Total	100	100	100	100

the molar ratios of the reactants are equivalent to the ratios of the stoichiometric coefficients. Often, however, one compound is the limiting reactant and the others are present in excess. A **limiting reactant** is a compound that is present in less than its stoichiometric proportion. **Excess reactants** are compounds present in greater than their stoichiometric proportions. If a limiting reactant is completely consumed in a reaction, excess reactants will still be present. **Note that just because a compound is the limiting reactant, it does not mean that the compound is completely consumed in the reaction.**

One common mistake is to assume that a limiting reactant is completely consumed; in reality, this is rarely the case. In the case where the limiting reactant is completely consumed, the products and all excess reactants are present after the reaction. In the case where the limiting reactant is not completely consumed, the products and all reactants are present after the reaction. Depending on the extent of the reaction, the quantities of reactants are present in different amounts.

The **reaction rate** (R) characterizes the extent to which a chemical reaction proceeds. Reaction rate is expressed in moles or moles/time. (*Note*: Using units of mass, such as g or lb_m, will generally not work when reactions are taking place.) R is a constant for a stoichiometric equation and is not tied to a specific species and/or compound in a reacting system. R may be given, deduced, or calculated using equations [3.8-13] and [3.8-15] below.

To illustrate the concept of reaction rate, consider two beakers with the same reactants. A catalyst is added to one beaker. The contents in both beakers begin reacting at the same time. After an hour, we examine the beakers and find that very few of the reactants remain in the beaker with the catalyst, while most of the reactants remain in the beaker without the catalyst. The reaction rate associated with the reaction in the beaker with the catalyst is higher than that associated with the reaction in the beaker without the catalyst.

Recall the overall differential accounting equation [3.3-5]. In the reacting systems discussed in this section, no accumulation occurs in the system. The appropriately reduced steady-state differential accounting equation is:

$$\dot{\psi}_{\text{in}} - \dot{\psi}_{\text{out}} + \dot{\psi}_{\text{gen}} - \dot{\psi}_{\text{cons}} = 0 \qquad [3.8\text{-}12]$$

Within a system, a particular compound is either consumed or generated. (A compound can be both consumed and generated in a system with multiple simultaneous, chemical reactions; however, this case is outside the scope of this text.) The Generation and Consumption terms are lumped together as the term $\sigma_s R$, where σ_s is the stoichiometric coefficient of compound s and R is the reaction rate. For reactants, $\sigma_s < 0$; for products, $\sigma_s > 0$; and for inerts, $\sigma_s = 0$. Thus, in the glucose fermentation example above, σ for $C_6H_{12}O_6$ is -1, and σ for CO_2 is $+2$.

Because the reaction process must be analyzed on a molar basis, the overall rate of extensive property ($\dot{\psi}$) is replaced with the molar flow rate ($\dot{n}$). For one inlet and one outlet in a steady-state, reacting system, equation [3.8-12] is rewritten for a compound s:

$$\dot{n}_{i,s} - \dot{n}_{j,s} + \sigma_s R = 0 \qquad [3.8\text{-}13]$$

Equation [3.8-13] can be generalized for a system with multiple inlets and outlets and multiple, simultaneous reactions:

$$\sum_i \dot{n}_{i,s} - \sum_j \dot{n}_{j,s} + \sum_n \sigma_{n,s} \mathrm{R}_n = 0 \qquad [3.8\text{-}14]$$

where n is the index for the chemical reaction.

Rearranging equation [3.8-13] gives R:

$$R = \left(\frac{\dot{n}_{i,s} - \dot{n}_{j,s}}{-\sigma_s}\right) \qquad [3.8\text{-}15]$$

for a system with one inlet and one outlet. R can be calculated using different species or compounds; however, **for a particular chemical reaction in a system, R is constant**. When calculating R, use a species for which the inlet and outlet species molar rates are known. If one species is completely consumed, $\dot{n}_{j,s}$ is equal to zero and R is easily calculated.

The **fractional conversion** (f_s) of a reactant is the fraction of reactant s that reacts in the system relative to the total amount of s introduced into the system. The value of f is defined here on the basis of moles or molar rates for one inlet and one outlet. It is assumed the reactant is only consumed (i.e., not generated). The fractional conversion for a system described by molar rates is expressed mathematically:

$$f_s = \frac{\dot{n}_{\text{cons},s}}{\dot{n}_{i,s}} = \frac{\dot{n}_{i,s} - \dot{n}_{j,s}}{\dot{n}_{i,s}} \qquad [3.8\text{-}16]$$

The fractional conversion value should be higher for the limiting reactant than for the excess reactants. R can also be written in terms of fractional conversion:

$$R = \frac{\dot{n}_{i,s} f_s}{-\sigma_s} \qquad [3.8\text{-}17]$$

Finally, the limiting reactant is mathematically defined as the minimum of:

$$\left\{\frac{\dot{n}_{i,s}}{-\sigma_s}\right\} \qquad [3.8\text{-}18]$$

Equations [3.8-13] to [3.8-18] are developed based on molar flow rate. An alternative is to adapt the algebraic accounting equation [3.3-1] and develop similar equations by replacing ψ with moles (n). In this case R, f_s, and other variables are defined in terms of n_s rather than $\dot{n}_s$. Integral accounting equations incorporating reaction terms can also be written.

For example, in the conversion of glucose to ethanol, assume the fractional conversion of glucose is 50%. R is calculated:

$$R = \frac{\dot{n}_{i,C_6H_{12}O_6} f_{C_6H_{12}O_6}}{-\sigma_{C_6H_{12}O_6}} = \frac{\left(555\frac{\text{mol}}{\text{day}}\right)(0.5)}{-(-1)} = 277.5\frac{\text{mol}}{\text{day}} \qquad [3.8\text{-}19]$$

With this new constraint on the conversion, the amount of glucose leaving the system is calculated using the differential accounting equation [3.8-13] written for glucose:

$$\text{Glucose:} \quad \dot{n}_{i,C_6H_{12}O_6} - \dot{n}_{j,C_6H_{12}O_6} + \sigma_{C_6H_{12}O_6} R = 0 \qquad [3.8\text{-}20]$$

$$\dot{n}_{j,C_6H_{12}O_6} = 555\frac{\text{mol}}{\text{day}} + (-1)\left(277.5\frac{\text{mol}}{\text{day}}\right) = 277.5\frac{\text{mol}}{\text{day}} \qquad [3.8\text{-}21]$$

In similar calculations, the molar outflow rate for both ethanol and carbon dioxide is 555 mol/day. The mass leaving the system is calculated using molecular weight conversions, so $\dot{m}_{j,C_6H_{12}O_6} = 49.95$ kg/day, $\dot{m}_{j,C_2H_6O} = 25.53$ kg/day, and $\dot{m}_{j,CO_2} =$ 24.42 kg/day. Note that in contrast to the situation described earlier, both the reactants and products leave the reactor. However, the total outlet mass is still 100 kg/day (Table 3.4). Remember that total mass is conserved regardless of reaction rate and fractional conversion.

EXAMPLE 3.17 Glucose Metabolism in the Cell

Problem: Nutrients obtained from food are needed to fuel the human body. Food is broken down in the digestive system into amino acids, sugars, salts, and other materials, and transported by the circulatory network to individual cells, where metabolism occurs at the cellular level. The cellular metabolism of sugars into carbon dioxide and the conversion of oxygen into water require many enzymes. The details of this process are given in biochemistry textbooks (e.g., Nelson and Cox, *Lehninger Principles of Biochemistry*, 2004) and are sketched in Figure 3.17a. The energy exchanges involving ATP, NADH, and FADH are not included in this analysis.

Assume carbohydrates are present as glucose sugars at the cellular level at a rate of 200 g/day, and 200 g of oxygen per day are available for combustion. Calculate the rate of carbon dioxide and other by-products released. Also determine the rate of carbon, hydrogen, and oxygen in this metabolic reaction before and after combustion with oxygen. Assume the limiting reactant is completely consumed.

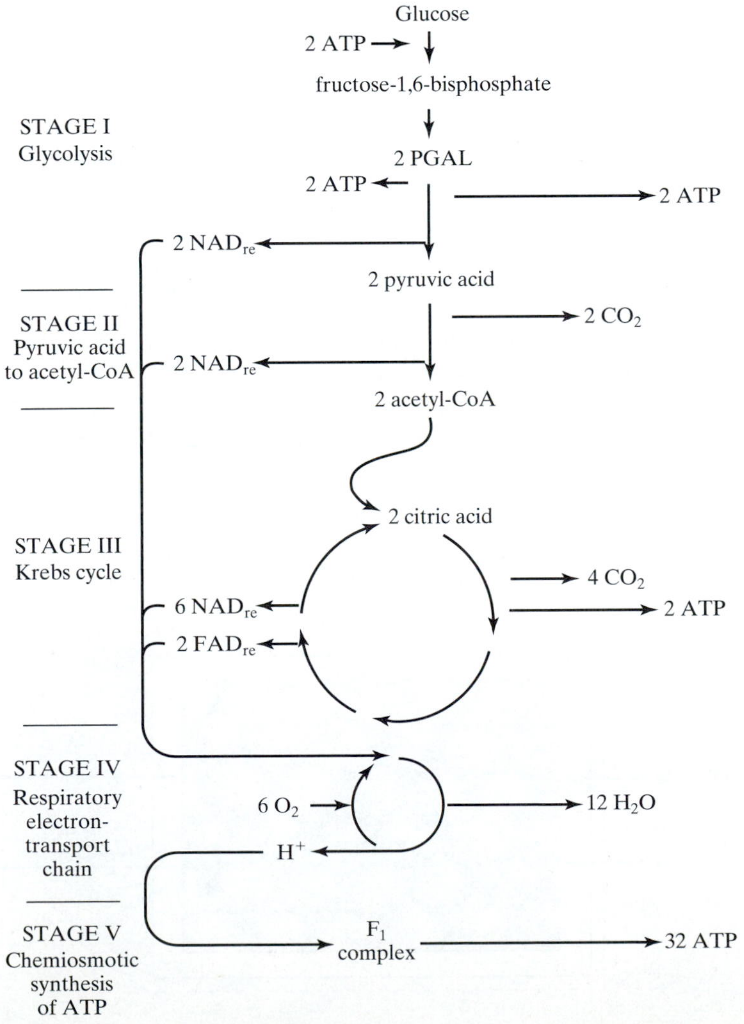

Figure 3.17a
Pathway of glucose metabolism in a cell. (*Source*: Keeton WT, Gould JL. *Biological Science, 4th ed.* New York: WW Norton and Co. Inc., 1986.)

Solution:

1. Assemble
   (a) Find: rate of carbon dioxide and other by-products released; the rate of carbon, hydrogen, and oxygen before and after combustion.
   (b) Diagram: Based on Figure 3.17a, it appears that glucose and oxygen are inputs to respiration and carbon dioxide and water are outputs (Figure 3.17b). The system is defined as the group of cells undergoing metabolism. The surroundings include all space outside the cells.
   (c) Table: A table is used to keep track of the mass rates of the compounds (Table 3.5).
2. Analyze
   (a) Assume:
   - All the intermediate constituents and enzymes needed for the listed reactions are present in the cells at sufficient concentrations.
   - The system is at steady-state.
   - Limiting reactant is completely consumed in the reaction.

   (b) Extra data:
   - The molecular weights of glucose and other constituents are needed.

   (c) Variables, notations, units:
   - Use g, day, mol.

   (d) Basis: 200 g/day of glucose into the system.
   (e) Reaction: The balanced, cellular metabolism combustion equation is:

$$C_6H_{12}O_6 + 6\ O_2 \longrightarrow 6\ CO_2 + 6\ H_2O$$

   Since one of the two reactants is known to be limiting, the other reactant will be an output.
3. Calculate
   (a) Equations: Since rates are given and no discrete time interval is specified, the differential form of the mass accounting equation is most appropriate for this steady-state system. Since total mass is conserved, equation [3.4-3] is appropriate:

$$\sum_i \dot{m}_i - \sum_j \dot{m}_j = 0$$

   Because elements in a reaction are conserved, we can use equation [3.6-11] to write mass balances for each element p:

$$\sum_i \dot{m}_{i,p} - \sum_j \dot{m}_{j,p} = 0$$

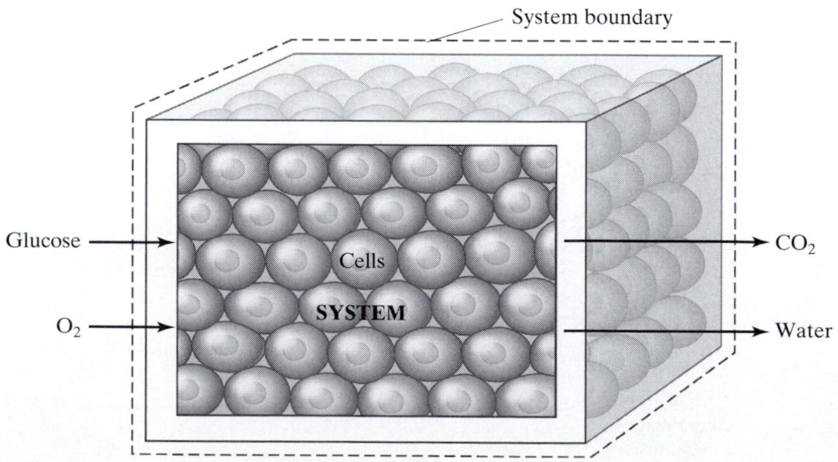

Figure 3.17b
A simplified schematic of glucose metabolism in the system.

To determine the limiting reactant, molar rates, and reaction rate, we need the following formulas:

$$\text{Minimum of} \left\{ \frac{\dot{n}_{i,s}}{-\sigma_s} \right\} = \text{limiting reactant}$$

$$\dot{n}_{i,s} - \dot{n}_{j,s} + \sigma_s R = 0$$

(b) Calculate:

- We first determine which of the two reactants is limiting. To do this for glucose, we first convert the mass rate to a molar rate, and then substitute this value into the formula:

$$\dot{n}_{i,C_6H_{12}O_6} = \frac{\dot{m}_{i,C_6H_{12}O_6}}{M_{C_6H_{12}O_6}} = \frac{200 \frac{\text{g}}{\text{day}}}{180 \frac{\text{g}}{\text{mol}}} = 1.11 \frac{\text{mol}}{\text{day}}$$

$$\text{Glucose:} \quad \left\{ \frac{\dot{n}_{i,C_6H_{12}O_6}}{-\sigma_{C_6H_{12}O_6}} \right\} = \left\{ \frac{1.11 \frac{\text{mol}}{\text{day}}}{-(-1)} \right\} = 1.11 \frac{\text{mol}}{\text{day}}$$

  Similarly for oxygen, the inlet molar rate is 6.25 mol/day and the limiting reactant calculation gives 1.04 mol/day, the smaller of the two calculated values. Thus, oxygen is the limiting reactant.

- Since the limiting reactant is completely consumed in this problem, the outlet rate of oxygen is equal to zero. The reaction rate R is calculated:

$$R = \left(\frac{\dot{n}_{i,O_2} - \dot{n}_{j,O_2}}{-\sigma_{O_2}} \right) = \left(\frac{6.25 \frac{\text{mol}}{\text{day}} - 0}{-(-6)} \right) = 1.04 \frac{\text{mol}}{\text{day}}$$

  If we had used the molar rates of glucose instead to calculate R, the value would be identical.

- Given the stoichiometrically balanced equation, the molar rates of carbon dioxide and water should be six times the reaction rate, so:

$$\dot{n}_{j,CO_2} = \dot{n}_{j,H_2O} = \sigma_s R = 6\left(1.04 \frac{\text{mol}}{\text{day}} \right) = 6.24 \frac{\text{mol}}{\text{day}}$$

  Thus, 6.24 mol/day of carbon dioxide and water are each produced.

- Because glucose is the excess reactant, the outlet stream has unreacted glucose:

$$\dot{n}_{i,C_6H_{12}O_6} - \dot{n}_{j,C_6H_{12}O_6} + \sigma_{C_6H_{12}O_6} R = 0$$

$$\dot{n}_{j,C_6H_{12}O_6} = \dot{n}_{i,C_6H_{12}O_6} + \sigma_{C_6H_{12}O_6} R = 1.11 \frac{\text{mol}}{\text{day}} + (-1)\left(1.04 \frac{\text{mol}}{\text{day}} \right) = 0.07 \frac{\text{mol}}{\text{day}}$$

- The outlet mass rates of carbon dioxide, water, and glucose are calculated by multiplying the molar rates by the appropriate molecular weights. For carbon dioxide:

$$\dot{m}_{j,CO_2} = \dot{n}_{j,CO_2} M_{CO_2} = \left(6.24 \frac{\text{mol}}{\text{day}} \right)\left(44 \frac{\text{g}}{\text{mol}} \right) = 275 \frac{\text{g}}{\text{day}}$$

  Similarly, the outlet mass rates for water and glucose are 112 g/day and 13 g/day, respectively. No oxygen flows out, since it is completely consumed.

TABLE 3.5

Compound Mass and Elemental Mass Rates in Cellular Metabolism

Compound	In (g/day)	Out (g/day)
$C_6H_{12}O_6$	200	13
O_2	200	0
CO_2	–	112
H_2O	–	275

Element	In (g/day)	Out (g/day)
Carbon	80	80
Hydrogen	13.4	13.4
Oxygen	307	307

- To find the rates of carbon, hydrogen, and oxygen before and after combustion, we need to find the mass rate of each individual element. For carbon:

$$\text{Carbon:} \quad \sum_i \dot{m}_{i,p} - \sum_j \dot{m}_{j,p} = \dot{m}_{i,\mathrm{C}} - \dot{m}_{j,\mathrm{C}} = 0$$

$$\dot{m}_{i,\mathrm{C}} = \dot{m}_{j,\mathrm{C}} = \dot{m}_{i,\mathrm{C_6/C_6H_{12}O_6}}$$

$$= \dot{m}_{i,\mathrm{C_6H_{12}O_6}}\left(\frac{M_{\mathrm{C_6}}}{M_{\mathrm{C_6H_{12}O_6}}}\right)$$

$$= 200\,\frac{\mathrm{g}}{\mathrm{day}}\left(\frac{72\,\frac{\mathrm{g}}{\mathrm{mol}}}{180\,\frac{\mathrm{g}}{\mathrm{mol}}}\right) = 80\,\frac{\mathrm{g}}{\mathrm{day}}$$

where the notation $C_6/C_6H_{12}O_6$ is the amount of carbon in glucose. Note that it is possible to calculate the elemental mass outlet instead of the elemental mass inlet, since the two quantities are equivalent. Element mass balance equations can also be written for hydrogen and oxygen. Calculating the elemental mass rates for hydrogen and oxygen gives 13.4 g/day and 307 g/day, respectively.

4. Finalize
   (a) Answer: The answers are given in Table 3.5.
   (b) Check: One way to check the results is to look at the overall inlet and outlet mass rates. Since total mass is neither created nor destroyed, the sum of the inlets should equal the sum of the outlets:

$$\sum_i \dot{m}_i = \sum_j \dot{m}_j$$

$$\dot{m}_{i,\mathrm{C_6H_{12}O_6}} + \dot{m}_{i,\mathrm{O_2}} = \dot{m}_{j,\mathrm{C_6H_{12}O_6}} + \dot{m}_{j,\mathrm{CO_2}} + \dot{m}_{j,\mathrm{H_2O}}$$

$$200\frac{\mathrm{g}}{\mathrm{day}} + 200\frac{\mathrm{g}}{\mathrm{day}} = 275\frac{\mathrm{g}}{\mathrm{day}} + 112\frac{\mathrm{g}}{\mathrm{day}} + 13\frac{\mathrm{g}}{\mathrm{day}} = 400\frac{\mathrm{g}}{\mathrm{day}}$$

■

EXAMPLE 3.18 **Artificial Liver Device**

Problem: Liver failure can cause a variety of life-threatening abnormalities, including the accumulation of ammonia and bilirubin in the plasma, and decreased levels of albumin and clotting

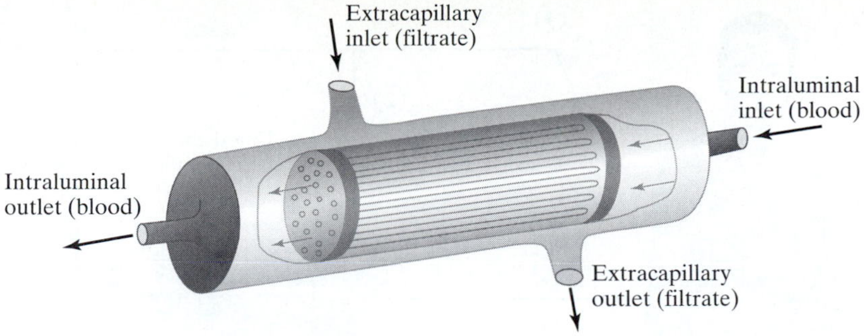

Figure 3.18a
A hollow-fiber membrane for use in an artificial liver device. (*Source*: Nyberg SL, Shatford RA, Peshwa MV, et al., "Evaluation of a hepatocyte entrapment hollow fiber bioreactor: a potential bioartificial liver." *Biotechnol Bioeng* 1993, 41:194–203.)

factors in the plasma. In addition, toxins build up and the hormonal system becomes overactive. The only relatively successful long-term treatment for liver failure is transplantation.

You want to design a hollow-fiber membrane device to be used as an artificial liver to bridge a patient from liver failure to transplantation (Figure 3.18a). Blood enters the device and branches into thousands of smaller fiber membranes. In between the fibers are hepatocyte cells. The membranes retain all large compounds (>100,000 g/mol; i.e., all cells and antibodies) and permit all small compounds (<100,000 g/mol; i.e., many proteins and toxins) to pass into the space containing the hepatocytes. Material trapped in the fibers exits the device without further processing. However, when the filtrate contacts the hepatocytes, the toxins are processed before leaving the device and are remixed with the exit stream containing the unprocessed blood. The remixed blood is then returned to the patient.

Determine the concentration of bilirubin and albumin in the filtrate flowing out of the device and into the patient. Blood flows into the device at 150 mL/min, and filtrate flows out at 20 mL/min. The volume of the device is 500 mL. The entering bilirubin concentration is 10 μg/mL, and its fractional conversion in the device is 83.4%. The entering serum albumin concentration is 2 μg/mL, and its rate of production by the hepatocytes in the device is 5 g/day.

Solution:

1. Assemble
   (a) Find: concentration of bilirubin and albumin in the filtrate flowing out of the device and flowing into the patient.
   (b) Diagram: A simplified model is shown in Figure 3.18b. The device has one inlet (stream 1) and two outlets, the filtrate (stream 2) and all other material (stream 3). The two outlet streams are rejoined (stream 4) before reentering the patient.
2. Analyze
   (a) Assume:
   - The device does not accumulate blood constituents.
   - The changes in other constituent concentrations in the blood do not affect the composition of the constituents of interest.
   - The material in the system contains only cells, plasma (filtrate), bilirubin, and albumin.
   - No small molecules (e.g., bilirubin, albumin) are in stream 3.
   - No cells are in stream 2.
   - The system is at steady-state.

   (b) Extra data:
   - The molecular weights of bilirubin and albumin are 474 g/mol and 66,000 g/mol, respectively.

   (c) Variables, notations, units:
   - bili = bilirubin
   - alb = albumin
   - Use mL, min, μg, mol.

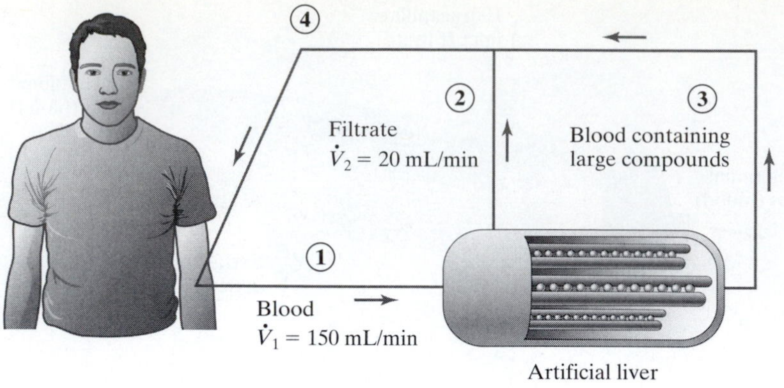

Figure 3.18b
A schematic drawing of blood flow between a patient and an artificial liver device.

(d) Basis: An inlet flow rate for stream 1 of 150 mL/min is given. Since the density of blood is approximately 1.0 g/mL, we use an inlet basis of 150 g/min.

(e) Reactions: Albumin is generated, and bilirubin is consumed in the device. Explicit chemical reactions are not given, but a fractional conversion and a reaction rate are.

3. Calculate

(a) Equations: Since rates are given and no discrete time interval is specified, the differential form of the mass accounting equation [3.3-5] is most appropriate. Compound-specific equations are needed on bilirubin and albumin. Neither of these constituents is conserved, since each is involved in a chemical reaction. However, the system is at steady-state, and since a chemical reaction occurs, molar flow rates are used; thus, we can use equation [3.8-13] for a reacting system:

$$\dot{n}_{i,s} - \dot{n}_{j,s} + \sigma_s R = 0$$

To find the fractional conversion, we use:

$$f_s = \frac{\dot{n}_{i,s} - \dot{n}_{j,s}}{\dot{n}_{i,s}}$$

(b) Calculate:

- Since we assume bilirubin is not present in stream 3, we need to include only streams 1 and 2 in a mass balance equation for the reacting, steady-state system:

$$\dot{n}_{1,\text{bili}} - \dot{n}_{2,\text{bili}} + \sigma_{\text{bili}} R = 0$$

- We can find the inlet molar flow rate of bilirubin since we are given the entering concentration (10 μg/mL). Since molar flow rate is given by the molar concentration multiplied by the volumetric flow rate, the molar flow rate is:

$$\dot{n}_{1,\text{bili}} = \frac{C_{1,\text{bili}}\dot{V}_1}{M_{\text{bili}}} = \left(10\,\frac{\mu\text{g}}{\text{mL}}\right)\left(150\,\frac{\text{mL}}{\text{min}}\right)\left(\frac{\text{mol}}{474\text{ g}}\right)\left(\frac{\text{g}}{10^6\,\mu\text{g}}\right)$$

$$= 3.16 \times 10^{-6}\,\frac{\text{mol}}{\text{min}}$$

- We can find the bilirubin molar flow rate out of the device by using the fractional conversion of bilirubin (83.4%):

$$f = 0.834 = \frac{\dot{n}_{1,\text{bili}} - \dot{n}_{2,\text{bili}}}{\dot{n}_{1,\text{bili}}} = \frac{3.16 \times 10^{-6}\,\frac{\text{mol}}{\text{min}} - \dot{n}_{2,\text{bili}}}{3.16 \times 10^{-6}\,\frac{\text{mol}}{\text{min}}}$$

$$\dot{n}_{2,\text{bili}} = 5.21 \times 10^{-7} \frac{\text{mol}}{\text{min}}$$

- The mass flow rate and concentration of bilirubin in the filtrate (stream 2) can then be calculated:

$$\dot{m}_{2,\text{bili}} = \dot{n}_{2,\text{bili}} M_{\text{bili}} = \left(\frac{5.21 \times 10^{-7}\ \text{mol}}{\text{min}}\right)\left(\frac{474\ \text{g}}{\text{mol}}\right)\left(\frac{10^{6}\ \mu\text{g}}{\text{g}}\right) = 247 \frac{\mu\text{g}}{\text{min}}$$

$$C_{2,\text{bili}} = \frac{\dot{m}_{2,\text{bili}}}{\dot{V}_2} = \frac{247 \frac{\mu\text{g}}{\text{min}}}{20 \frac{\text{mL}}{\text{min}}} = 12.4 \frac{\mu\text{g}}{\text{mL}}$$

- Streams 2 and 3 combine before returning the blood to the patient (stream 4). To find the mass flow rate of bilirubin in stream 4, we combine the bilirubin mass flow rates of the two streams. Since no bilirubin is in stream 3, the bilirubin mass flow rate of stream 4 must equal the bilirubin mass flow rate of stream 2 by the conservation of mass ($\dot{m}_{2,\text{bili}} = \dot{m}_{4,\text{bili}} = 247\ \mu\text{g/min}$). Since the fluid maintains a constant density, we know the volumetric flow rate going into and out of the patient is 150 mL/min. We can then find the concentration of bilirubin returned to the patient:

$$C_{4,\text{bili}} = \frac{\dot{m}_{4,\text{bili}}}{\dot{V}_4} = \frac{247 \frac{\mu\text{g}}{\text{min}}}{150 \frac{\text{mL}}{\text{min}}} = 1.65 \frac{\mu\text{g}}{\text{mL}}$$

- We can find the mass flow rates and concentrations of albumin in streams 2 and 4 in the same way. Albumin is also a small molecule, so no albumin is present in stream 3. Writing out the equation for the reacting, steady-state system:

$$\dot{n}_{1,\text{alb}} - \dot{n}_{2,\text{alb}} + \sigma_{\text{alb}} R = 0$$

  The entering albumin concentration is 2 μg/mL, and the molar inlet flow rate is calculated in the same way as it was for bilirubin, giving 4.54×10^{-9} mol/min.
- Given the reaction rate (5 g/day), we can find the molar flow rate of albumin in stream 2:

$$\sigma_{\text{alb}} R = \left(5 \frac{\text{g}}{\text{day}}\right)\left(\frac{1\ \text{day}}{24\ \text{hr}}\right)\left(\frac{1\ \text{hr}}{60\ \text{min}}\right)\left(\frac{\text{mol}}{66{,}000\ \text{g}}\right) = 5.26 \times 10^{-8} \frac{\text{mol}}{\text{min}}$$

$$\dot{n}_{2,\text{alb}} = \dot{n}_{1,\text{alb}} + \sigma_{\text{alb}} R = 4.54 \times 10^{-9} \frac{\text{mol}}{\text{min}} + 5.26 \times 10^{-8} \frac{\text{mol}}{\text{min}}$$

$$= 5.71 \times 10^{-8} \frac{\text{mol}}{\text{min}}$$

  Note that since no stoichiometrically balanced equation is given for the generation of albumin, the stoichiometric coefficient is not explicitly defined and is therefore assumed to be equal to one.
- The concentrations of albumin in the filtrate and in the stream returning blood to the patient are calculated similarly: $C_{2,\text{alb}} = 189\ \mu\text{g/mL}$ and $C_{4,\text{alb}} = 25.1\ \mu\text{g/mL}$.

4. Finalize
   (a) Answer: The concentrations of bilirubin are 12.4 μg/mL in the filtrate and 1.65 μg/mL in the bloodstream returning to the patient. The concentrations of

albumin are 189 μg/mL in the filtrate and 25.1 μg/mL in the bloodstream returning to the patient.

(b) Check: Looking at bilirubin, the inlet concentration to the artificial liver is 10 μg/mL. While the concentration in the filtrate is higher at 12.4 μg/mL, the concentration in the stream returning to the patient is much lower at 1.65 μg/mL. This overall drop in bilirubin concentration is expected, since bilirubin is consumed in the device. Looking at the albumin, the inlet concentration to the artificial liver is 2 μg/mL; the concentration returning to the patient is 25.1 μg/mL. This 10-fold increase makes sense, since albumin is generated in the device. ■

EXAMPLE 3.19 **Oxygen Consumption in Bone**

Problem: One challenge in designing tissue-engineered bone is the requirement that the new tissue must be vascularized, so that cells in the new bone can obtain the oxygen necessary for respiration. Hemoglobin (Hb) in red blood cells binds oxygen for transport to the cells. Each Hb molecule can hold four oxygen molecules. The Hb concentration in whole blood is 0.158 g/mL. The molecular weight of Hb is 64,500 g/mol. You need to complete a rough estimate for the oxygen consumption in a bone prior to building an implant.

Consider the femur as a steady-state system with arterial blood flow in and venous blood flow out. What is the concentration of oxygen in the blood flowing out of the femur? The rate of blood flow in the femur is estimated as 34 mL/min. Assume the hemoglobin is 100% saturated, and the bone cells receive oxygen only from hemoglobin. The oxygen consumption of the femur is estimated to be 4.0×10^{-2} mg/s.

Solution: The femur system is modeled with one inlet flow and one outlet flow (Figure 3.19). Since the system is at steady-state, oxygen does not accumulate in the femur. We assume that the Hb flowing in the arterial blood is totally saturated with oxygen. We assume no oxygen is produced by bone tissue. We can then simplify the differential accounting equation for molar rate to:

$$\sum_i \dot{n}_{i,s} - \sum_j \dot{n}_{j,s} + \sum \dot{n}_{\text{gen},s} - \sum \dot{n}_{\text{cons},s} = \dot{n}^{\text{sys}}_{\text{acc},s}$$

$$\dot{n}_{i,O_2} - \dot{n}_{j,O_2} - \dot{n}_{\text{cons},O_2} = 0$$

To find the molar flow rate of oxygen into the system, we need to find the number of moles of Hb per unit blood volume:

$$C_{i,\text{Hb}} = \left(\frac{0.158\text{ g Hb}}{\text{mL blood}}\right)\left(\frac{\text{mol Hb}}{64{,}500\text{ g Hb}}\right) = 2.45 \times 10^{-6}\ \frac{\text{mol Hb}}{\text{mL blood}}$$

If four molecules of O_2 are bound to one molecule of hemoglobin, then the number of moles of O_2 per unit volume of blood is four times the number of Hb moles per unit volume of blood. Thus, the inlet molar flow of oxygen in blood is:

$$\dot{n}_{i,O_2} = 4C_{i,\text{Hb}}\dot{V}_{\text{blood}} = \left(\frac{4\text{ mol O}_2}{1\text{ mol Hb}}\right)\left(\frac{2.45 \times 10^{-6}\text{ mol Hb}}{\text{mL blood}}\right)\left(34\frac{\text{mL}}{\text{min}}\right)$$

$$= 3.33 \times 10^{-4}\ \frac{\text{mol}}{\text{min}}$$

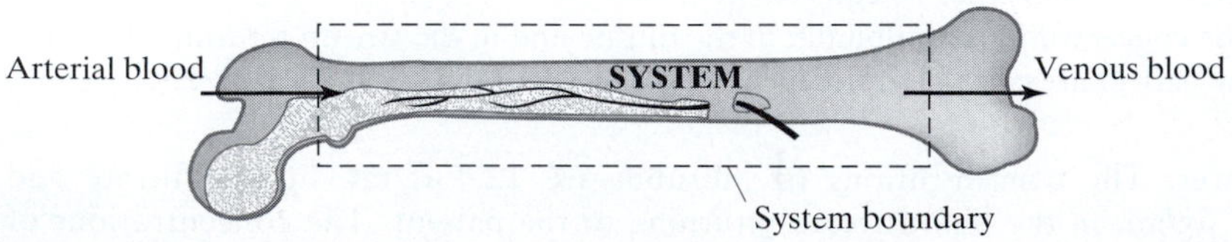

Figure 3.19
Femur system with steady-state blood flow.

To calculate the molar flow rate of oxygen consumed, we use the mass flow rate of oxygen consumed:

$$\dot{n}_{\text{cons,O}_2} = \frac{\dot{m}_{\text{cons,O}_2}}{M_{\text{O}_2}} = \left(\frac{4.0 \times 10^{-2}\,\frac{\text{mg}}{\text{s}}}{32\,\frac{\text{g}}{\text{mol}}}\right)\left(\frac{60\text{ s}}{\text{min}}\right)\left(\frac{1\text{ g}}{1000\text{ mg}}\right) = 7.5 \times 10^{-5}\,\frac{\text{mol}}{\text{min}}$$

We can now calculate the molar flow rate of oxygen exiting the femur:

$$\dot{n}_{j,\text{O}_2} = \dot{n}_{i,\text{O}_2} - \dot{n}_{\text{cons,O}_2} = 3.33 \times 10^{-4}\,\frac{\text{mol}}{\text{min}} - 7.56 \times 10^{-5}\,\frac{\text{mol}}{\text{min}} = 2.58 \times 10^{-4}\,\frac{\text{mol}}{\text{min}}$$

The outlet O_2 concentration is calculated from the outlet molar flow rate of O_2 and the volumetric flow rate of blood through the femur:

$$C_{j,\text{O}_2} = \frac{\dot{n}_{j,\text{O}_2}}{\dot{V}_{\text{blood}}} = \left(\frac{2.58 \times 10^{-4}\,\frac{\text{mol}}{\text{min}}}{34\,\frac{\text{mL}}{\text{min}}}\right)\left(\frac{32\text{ g}}{\text{mol}}\right)\left(\frac{1000\text{ mg}}{\text{g}}\right) = 0.242\,\frac{\text{mg}}{\text{mL}}$$

Thus, the concentration of oxygen leaving the femur is 0.242 mg/mL. Given that the inlet oxygen concentration is 0.314 mg/mL, the femur consumes about 23% of the available oxygen. Normally, bone tissue consumes about 25% of the oxygen bound to hemoglobin, so this number is reasonable. ■

3.9 Dynamic Systems

Recall that in a dynamic system, the variables describing the system (e.g., temperature, pressure) may change with time. In addition, the amount or rate of extensive property at the initial and final conditions of the system are not equal, always making the Accumulation term nonzero. **The Accumulation term tracks the change in the extensive property** (e.g., mass, moles) **contained in the system.**

The differential form of the unsteady-state accounting equation is as follows:

$$\dot{\psi}_{\text{in}} - \dot{\psi}_{\text{out}} + \dot{\psi}_{\text{gen}} - \dot{\psi}_{\text{cons}} = \dot{\psi}_{\text{acc}} = \frac{d\psi}{dt} \qquad [3.9\text{-}1]$$

For unsteady-state, nonreacting systems, the governing differential accounting equation becomes the unsteady-state differential conservation equation:

$$\dot{\psi}_{\text{in}} - \dot{\psi}_{\text{out}} = \dot{\psi}_{\text{acc}} = \frac{d\psi}{dt} \qquad [3.9\text{-}2]$$

Consider the system of a tank that is half full of a liquid. You begin to fill up the tank with more liquid at a constant rate ($\dot{\psi}_{\text{in}}$); assume that the tank also drains the liquid ($\dot{\psi}_{\text{out}}$). Assuming no reactions, equation [3.9-2] is appropriate to use to determine at what rate the liquid accumulates in the system. For the case where $\dot{\psi}_{\text{in}}$ is greater than $\dot{\psi}_{\text{out}}$, then $\dot{\psi}_{\text{acc}}$ is greater than zero, and the tank accumulates liquid. For the case where $\dot{\psi}_{\text{in}}$ is less than $\dot{\psi}_{\text{out}}$, then $\dot{\psi}_{\text{acc}}$ is less than zero, and the tank loses liquid.

Both the differential and integral unsteady-state accounting and conservation equations are commonly used to solve transient systems. Further information to help discriminate between appropriate uses of the differential and integral forms of the equation is provided in Section 3.3 and in Section 2.4. In general, the integral form of the equation is used when a fixed time period (i.e., having a t_0 and a t_f) is specified. The integral form of the unsteady-state accounting equation is:

$$\int_{t_0}^{t_f} \dot{\psi}_{\text{in}}\,dt - \int_{t_0}^{t_f} \dot{\psi}_{\text{out}}\,dt + \int_{t_0}^{t_f} \dot{\psi}_{\text{gen}}\,dt - \int_{t_0}^{t_f} \dot{\psi}_{\text{cons}}\,dt = \int_{t_0}^{t_f} \dot{\psi}_{\text{acc}}\,dt$$

$$= \int_{t_0}^{t_f} \frac{d\psi}{dt}\,dt = \int_{\psi_0}^{\psi_f} d\psi \quad [3.9\text{-}3]$$

For unsteady-state, nonreacting systems, the governing integral accounting equation becomes the unsteady-state integral conservation equation:

$$\int_{t_0}^{t_f} \dot{\psi}_{\text{in}}\,dt - \int_{t_0}^{t_f} \dot{\psi}_{\text{out}}\,dt = \int_{t_0}^{t_f} \dot{\psi}_{\text{acc}}\,dt = \int_{t_0}^{t_f} \frac{d\psi}{dt}\,dt = \int_{\psi_0}^{\psi_f} d\psi \quad [3.9\text{-}4]$$

Note that the accumulation term is written with different notations. The option of $\dot{\psi}_{\text{acc}}$ or $\int_{t_0}^{t_f} \dot{\psi}_{\text{acc}}\,dt$ may be preferable when you are given a rate of accumulation. For the case where $\dot{\psi}_{\text{acc}}$ is not a function of time, the integral becomes $\dot{\psi}_{\text{acc}}(t_f - t_0)$. The option of $\frac{d\psi}{dt}$ or $\int_{t_0}^{t_f} \frac{d\psi}{dt}\,dt$ may be preferable when the extensive property of the system, ψ, is a function of time. Finally, the option of $\int_{\psi_0}^{\psi_f} d\psi$ may be helpful when the accumulated amount of the extensive property is specified. When integrated to $\psi_f - \psi_0$, the amount of extensive property at the beginning and end of the process is being directly evaluated.

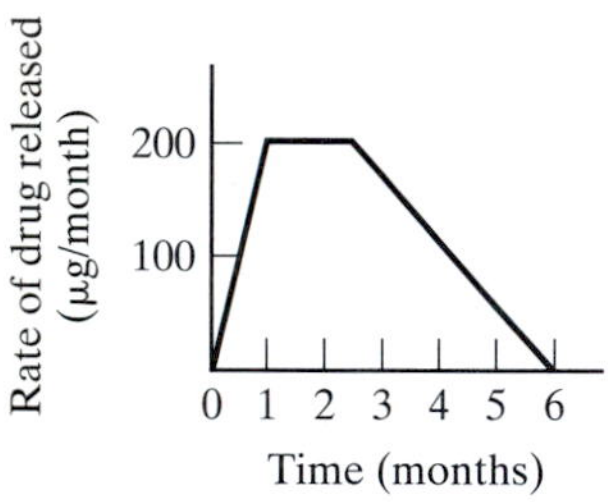

Figure 3.20a
Profile of drug release over a 6-month period.

EXAMPLE 3.20 Drug Delivery

Problem: Innovative drug delivery methods using synthetic polymers are being explored. One such drug delivery system is implanted subcutaneously (under the skin), and the drug is released passively from the device into the tissue for the necessary time period. You are designing a polymer from which drug is released for 6 months (Figure 3.20a). Determine the mass of drug released during the 6-month period for your new design.

Solution:

1. Assemble
   (a) Find: the mass of drug released during the 6-month period.
   (b) Diagram: Figure 3.20b shows the system as the polymer and the encapsulated drug. Since we are interested in the amount of drug released, we define the system such that it encloses the polymer, and the drug travels across the system boundary out into the surroundings of the body.

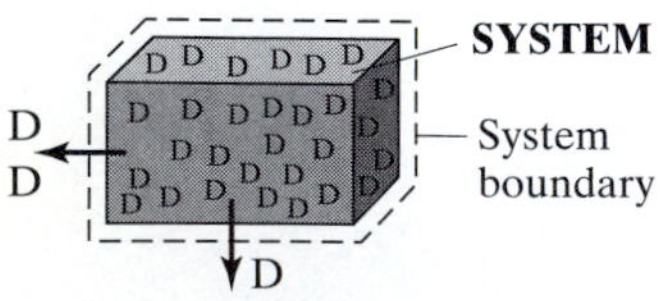

Figure 3.20b
Polymer containing a drug (D) that is released over time.

2. Analyze
   (a) Assume:
   - The drug release profile is given in Figure 3.20a.
   - No drug reenters the polymer from the surrounding tissue.
   - Drug release can be modeled by three different linear relationships in three different time periods (0–1 month, 1–2.5 months, and 2.5–6 months).
   - At the end of the 6-month period, no more drug remains in the polymer.

   (b) No extra data are needed.
   (c) Variables, notations, units:

- m = slope of line
- b = intercept of line
- Use μg, mo.

(d) Basis: Although it changes with time, the outlet flow of drug from the polymer is the basis.

3. Calculate

(a) Equations: We use the integral mass accounting equation, since discrete time points are given:

$$\int_{t_0}^{t_f} \dot{\psi}_{in}\, dt - \int_{t_0}^{t_f} \dot{\psi}_{out}\, dt + \int_{t_0}^{t_f} \dot{\psi}_{gen}\, dt - \int_{t_0}^{t_f} \dot{\psi}_{cons}\, dt = \int_{\psi_0}^{\psi_f} d\psi$$

(b) Calculate:

- The drug does not react with any constituents inside the polymer, so the Generation and Consumption terms can be eliminated from the governing equation. Because drug is only being released from the polymer and is not reabsorbed, the Input term is also zero. The Accumulation term can be evaluated at the initial and final time points:

$$-\int_{t_0}^{t_f} \dot{\psi}_{out}\, dt = \psi_f - \psi_0$$

- Each discrete time period is treated separately. Between the start time and the first month ($t_0 = 0$ and $t_f = 1$ mo), we develop a linear equation describing the amount of drug released:

$$\frac{y_2 - y_1}{x_2 - x_1} = m = \frac{200\dfrac{\mu g}{mo} - 0\dfrac{\mu g}{mo}}{1\text{ mo} - 0\text{ mo}} = 200\,\frac{\mu g}{mo^2}$$

$$y = mt + b = \left(200\frac{\mu g}{mo^2}\right)t + 0 = \left(200\frac{\mu g}{mo^2}\right)t$$

- We can then solve for the amount of drug released between the start time and the first month:

$$\int_{0}^{1\text{ mo}} \dot{\psi}_{out}\, dt = \int_{0}^{1\text{ mo}} \left(200\frac{\mu g}{mo^2}\right)t\, dt = 100\ \mu g$$

- We can repeat this process to solve for the amounts of drug released during the other two time periods. During the time period between 1 month and 2.5 months, 300 μg is released; between 2.5 months and 6 months, 350 μg is released.
- Finally, we apply the mass integral accounting equation for the overall system over the entire 6 months. The amount of drug present in the polymer at 6 months (ψ_f) is zero:

$$-\int_{t_0}^{t_f} \dot{\psi}_{out}\, dt = \psi_f - \psi_0$$

$$-\int_{0}^{1\text{ mo}} \dot{\psi}_{out}\, dt - \int_{1\text{ mo}}^{2.5\text{ mo}} \dot{\psi}_{out}\, dt - \int_{2.5\text{ mo}}^{6\text{ mo}} \dot{\psi}_{out}\, dt = \psi_f - \psi_0$$

$$\psi_0 = \psi_f + \int_{0\text{ mo}}^{1\text{ mo}} \dot{\psi}_{out}\, dt + \int_{1\text{ mo}}^{2.5\text{ mo}} \dot{\psi}_{out}\, d.t + \int_{2.5\text{ mo}}^{6\text{ mo}} \dot{\psi}_{out}\, dt$$

$$\psi_0 = 0 + 100\ \mu g + 300\ \mu g + 350\ \mu g = 750\ \mu g$$

4. Finalize

(a) Answer: During the 6-month period, the drug-filled polymer implant releases 750 μg of drug.

(b) Check: We can check that this answer is valid by calculating the area under the curve in Figure 3.20a, which equals 750 μg. ■

EXAMPLE 3.21 Toxin Accumulation in a Laboratory Bone Implant

Problem: Engineers need to design biodegradable bone tissue replacements such that the degradation products do not harm the patient. Potentially toxic levels of degraded by-products must be assessed in laboratory and animal models before human clinical trials can begin. Most polymers examined for in vivo uses are composed of carbon, hydrogen, oxygen, and sometimes nitrogen; while these elements are prevalent in the body, they can also be toxic when formed into certain chemical structures at certain concentrations. You are testing for the toxic concentrations of a degrading porous polymeric biomaterial.

Test 1: The polymer is 10.0 g in mass and nontoxic in its injected (i.e., nondegraded) form. From previous studies, you know the degradation rate is constant, and it takes 8.0 weeks to completely degrade into monomers in a controlled setting with a buffered saline solution flowing through the material at a constant rate. However, one of the degradation products is known to be toxic to bone tissue. You design an experiment to explore the relationship between the concentration of toxic material in the saline solution leaving the implant and the volumetric flow rate of the saline solution. You run the saline solution through the porous polymer at a flow rate $\dot{V}$ to simulate blood flow in the implant in vivo. You measure the concentration of the toxic product after the fluid stream has passed through the polymer. The toxin concentration in the outlet decreases as flow rate increases, indicating an inverse relationship between $\dot{V}$ and concentration. $\dot{V}$ multiplied by the toxin concentration gives a constant generation rate of toxin. These data are given in Table 3.6A and graphed in Figure 3.21a.

Test 2: Your colleague criticizes your experimental design and suggests testing volumetric flow rates less than 40 mL/min. Additional testing shows that the toxin concentration does not follow the predicted curve seen in Figure 3.21a for flow rates below 40 mL/min. The new experimental design yields the results in Table 3.6B, and the complete data are graphed in Figure 3.21b. For $\dot{V}$ less than or approximately equal to 30 mL/min, the concentration is constant and independent of flow rate. Since the polymer still degrades at the same rate, you wonder if the saline solution is unable to dissolve all of the toxic product.

Use the data from the two tests to answer the following:

(a) Perform a mass balance on the degraded product. What minimum flow rate ensures that the implant is safe (i.e., no toxin accumulation)?

(b) The implant irreversibly damages tissue when 0.10 g of the degraded polymer is concentrated in the implant area. Derive an equation for the time until enough toxins are released such that the tissue becomes irreversibly damaged.

Solution:

(a) The differential mass accounting equation for the degraded toxic product is:

$$\dot{m}_{\text{in,toxin}} - \dot{m}_{\text{out,toxin}} + \dot{m}_{\text{gen,toxin}} - \dot{m}_{\text{cons,toxin}} = \dot{m}^{\text{sys}}_{\text{acc,toxin}}$$

We assume no other source of toxic material enters the system and no metabolic process destroys any of the toxins, so the Input and Consumption terms become zero. We also assume the toxic material is not generated in the system by any method other than polymer degradation. Using the data from Table 3.6A, the toxin is generated at a rate of 62 μg/min. To avoid tissue damage, no toxin should accumulate, so the Accumulation term becomes zero:

$$-\dot{m}_{\text{out,toxin}} + 62\frac{\mu\text{g}}{\text{min}} = 0$$

$$\dot{m}_{\text{out,toxin}} = 62\frac{\mu\text{g}}{\text{min}}$$

TABLE 3.6A

Test 1 Data

$\dot{V}$ (mL/min)	Concentration of toxin (μg/mL)	$\dot{V}$ (mL/min)	Concentration of toxin (μg/mL)
40	1.54	90	0.691
50	1.26	100	0.623
60	1.03	110	0.564
70	0.89	120	0.517
80	0.776	130	0.478

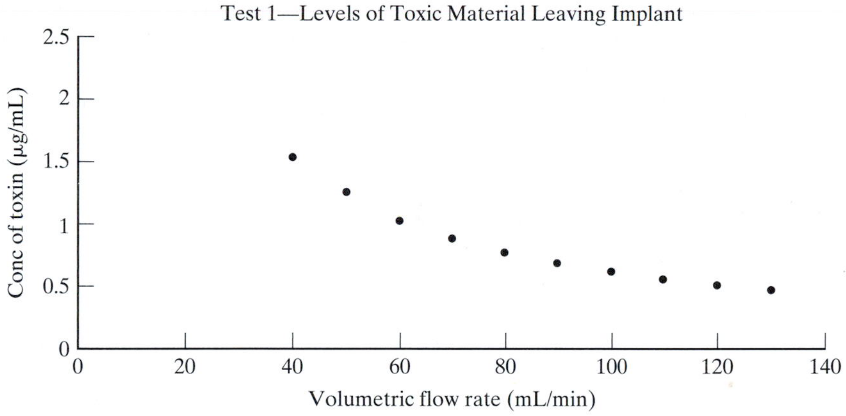

Figure 3.21a
Test 1: Levels of toxic material leaving implant (volumetric flow rates ≥40 mL/min).

The minimum volumetric flow rate of the saline solution can then be calculated from the mass flow rate:

$$\dot{V} = \frac{\dot{m}_{\text{out,toxin}}}{C_{\text{out,toxin}}} = \frac{62\dfrac{\mu\text{g}}{\text{min}}}{2\dfrac{\mu\text{g}}{\text{mL}}} = 31\ \frac{\text{mL}}{\text{min}}$$

where $C_{\text{out,toxin}}$ is 2 μg/mL, the maximum concentration of soluble toxin in the buffered saline. For flow rates less than 31 mL/min, toxic by-products accumulate in the system. From Figure 3.21b, we can verify that 31 mL/min is around the "break point."

(b) To derive an equation finding the time it takes for enough toxins to accumulate to cause irreversible damage, the differential mass accounting equation is written for the system of interest:

$$-\dot{m}_{\text{out,toxin}} + \dot{m}_{\text{gen,toxin}} = \frac{dm^{\text{sys}}_{\text{toxin}}}{dt}$$

where $dm^{\text{sys}}_{\text{toxin}}/dt$ is the time derivative of the mass of toxin in the system (or change in the mass of the toxin per unit time). For flow rates less than 31 mL/min, the outlet mass flow rate of toxin is equal to the volumetric flow rate $\dot{V}$ multiplied by the maximum concentration of soluble toxic material in the saline solution (2 μg/mL). Thus, for a constant toxin generation rate, the differential mass accounting equation above can be rewritten:

$$-\left(2\frac{\mu\text{g}}{\text{mL}}\right)\dot{V} + 62\frac{\mu\text{g}}{\text{min}} = \frac{dm^{\text{sys}}_{\text{toxin}}}{dt}$$

TABLE 3.6B

Test 2 Data

$\dot{V}$ (mL/min)	Concentration of toxin (μg/mL)	$\dot{V}$ (mL/min)	Concentration of toxin(μg/mL)
1	2.01	10	2.01
2	1.99	20	2.00
3	2.00	30	1.98
4	1.98	35	1.77
5	2.00		

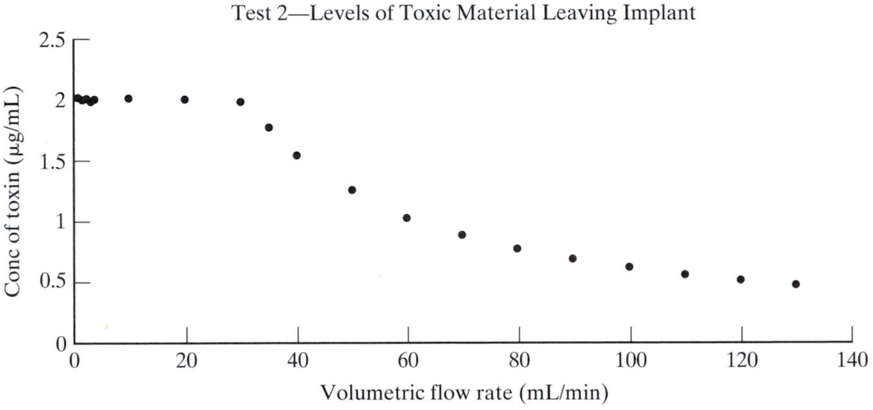

Figure 3.21b
Test 2: Levels of toxic material leaving implant (all test flow rates).

This is applicable only for flow rates less than 31 mL/min. Integrating this over a defined time period of toxin accumulation gives:

$$\int_0^t \left(-\left(2\frac{\mu\text{g}}{\text{mL}}\right)\dot{V} + 62\frac{\mu\text{g}}{\text{min}}\right) dt = \int_{m_0}^{m_f} dm_{\text{toxin}}^{\text{sys}}$$

$$\left(-\left(2\frac{\mu\text{g}}{\text{mL}}\right)\dot{V} + 62\frac{\mu\text{g}}{\text{min}}\right)t = m_{\text{toxin},f}^{\text{sys}} - m_{\text{toxin},0}^{\text{sys}}$$

Since no mass of toxic by-product is accumulated in the system at $t = 0$, $m_{\text{toxin},0}^{\text{sys}}$ also equals zero. Rearranging the equation for the time to irreversible damage gives:

$$t = \frac{m_{\text{toxin,f}}^{\text{sys}}}{-\left(2\frac{\mu\text{g}}{\text{mL}}\right)\dot{V} + 62\frac{\mu\text{g}}{\text{min}}}$$

Irreversible damage occurs when the mass of toxin in the system reaches 0.10 g. Thus, the equation for time to irreversible damage as a function of volumetric flow rate is:

$$t = \frac{0.10\text{ g}}{-\left(2\frac{\mu\text{g}}{\text{mL}}\right)\dot{V} + 62\frac{\mu\text{g}}{\text{min}}}$$

Note that as $\dot{V}$ increases, the time for toxin buildup to reach a damaging level decreases.

■

EXAMPLE 3.22 Culture of Plant Roots

Problem: Plant roots produce valuable chemicals that are often captured for in vitro uses. A batch culture of *Atropa belladonna* roots at 25°C is established in an air-driven reactor (Figure 3.22). During operation, roots cannot be removed, so their growth is monitored using mass balances.

The bioreactor operates for a 10-day period. A feed of 1425 g of nutrient media containing 3 wt% glucose ($C_6H_{12}O_6$) and 1.75 wt% ammonia (NH_3) goes into the reactor; water constitutes the remainder of the media. Air at 25°C and 1 atm pressure is continuously sparged into the reactor at a rate of 22 cm^3/min. Oxygen (O_2), carbon dioxide (CO_2), and nitrogen (N_2) are collected continuously in the off-gas and expunged from the reactor. After 10 days, the reactor is drained of depleted media containing 0.699 g of glucose, as well as water and NH_3. The wet-weight to dry-weight ratio of plant tissue is known to be 14:1.

In the reactor, $C_6H_{12}O_6$ is converted to CO_2, H_2O, and plant mass:

$$C_6H_{12}O_6 + 3.43\ O_2 + 1.53\ NH_3 \longrightarrow 3.37\ CH_{1.27}O_{0.43}N_{0.45} + 2.63\ CO_2 + 6.16\ H_2O$$

The chemical formula for plant mass is given as $CH_{1.27}O_{0.43}N_{0.45}$, which is determined empirically from experimental data.

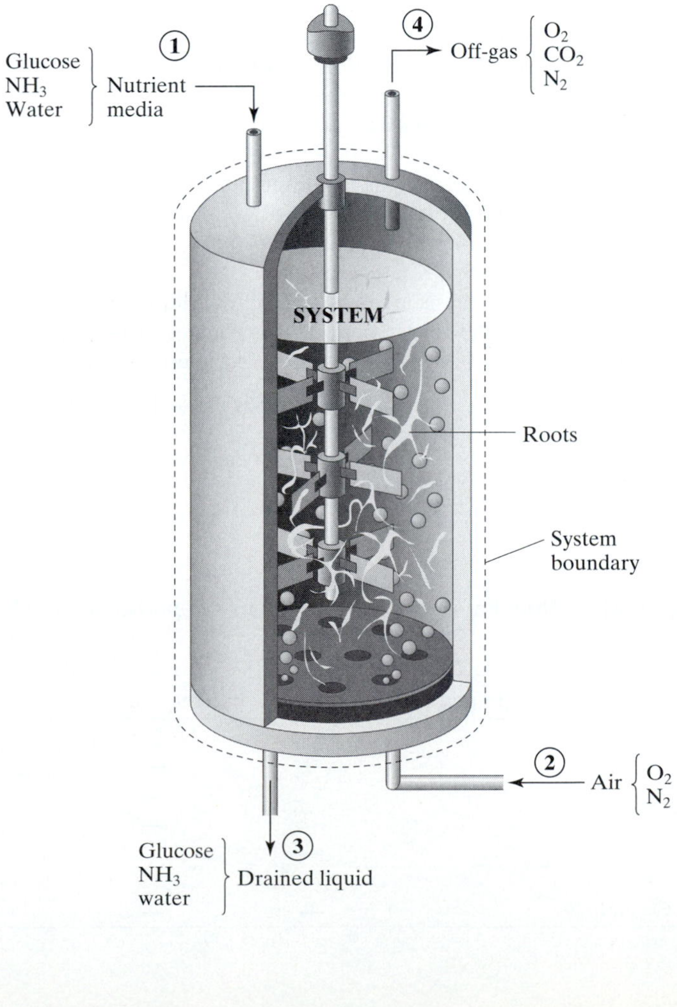

Figure 3.22
Culture reactor with *Atropa belladonna* roots.

For a single batch run, determine the limiting reactant, the rate of reaction, and the outlet masses of $C_6H_{12}O_6$, O_2, N_2, NH_3, CO_2, and H_2O. What mass of *Atropa belladonna* roots accumulates in the system at the end of the 10-day run? (Adapted from Doran, Bioprocessing Principles, 1991).

Solution:

1. Assemble
   (a) Find:
   - Limiting reactant.
   - Rate of reaction (R).
   - Individual masses for glucose, oxygen, nitrogen, ammonia, carbon dioxide, and water in the outlet.
   - Mass (dry weight) of roots at end of the 10-day run.

   (b) Diagram: A diagram is given in Figure 3.22. The inlet gas stream 2 contains O_2 and N_2. The outlet gas stream 4 contains CO_2, N_2, and O_2. The liquid media into (inlet 1) and of (outlet 3) the reactor contains glucose, ammonia, and water.

   (c) Table: The numbers in parentheses in the heading row indicate labeled inlets and outlets (Table 3.7A).

2. Analyze
   (a) Assume:
   - The gases (O_2, N_2, and CO_2) and the nutrients ($C_6H_{12}O_6$ and NH_3) do not accumulate in the reactor.
   - Plant mass accumulates in the reactor. Since water is a substantial component of plant biomass, water also accumulates in the reactor.
   - No leaks in the system.
   - The inlet air and off-gas are dry (i.e., gases have no water vapor; humidity is zero). This assumption ensures that water in the liquid phase is not transferred to the gas phase in the reactor.
   - All CO_2 produced leaves in the off-gas (i.e., is not dissolved in the liquid).

   (b) Extra data:
   - Molecular weights of compounds.
   - Composition of air is 79 vol% N_2 and 21 vol% O_2.

   (c) Variables, notations, units:
   - Use g, mol, K, atm, day, cm^3.

   (d) Basis: The basis is the 1425 g of total liquid nutrient feed entering the system at the onset of the 10-day period. This equates to 1425 g/run added to the system.

   (e) Reaction: The reaction is given in the problem statement:

$$C_6H_{12}O_6 + 3.43\ O_2 + 1.53\ NH_3 \longrightarrow$$

$$3.37\ CH_{1.27}O_{0.43}N_{0.45} + 2.63\ CO_2 + 6.16\ H_2O$$

TABLE 3.7A

Setup of Mass Flow Rates of Components in *Atropa belladonna* Bioreactor

	Inlet (g/run)		Outlet (g/run)		Accumulation (g/run)
	Liquid (1)	Gas (2)	Liquid (3)	Gas (4)	In system
$C_6H_{12}O_6$		—	0.699	—	
CO_2	—	0	—		
O_2	—		—		
N_2	—		—		
NH_3		—		—	
H_2O		—		—	
$CH_{1.27}O_{0.43}N_{0.45}$	—	—	—	—	

Because noninteger proportions of elements constitute the chemical formula for cell mass, the stoichiometric coefficients are also noninteger. Take a minute to convince yourself that this reaction is balanced.

3. Calculate
   (a) Equations: Since rates of material transfer are given, the differential form of the accounting statement is appropriate:

$$\dot{\psi}_{in} - \dot{\psi}_{out} + \dot{\psi}_{gen} - \dot{\psi}_{cons} = \dot{\psi}_{acc} = \frac{d\psi}{dt}$$

For the gases (O_2, N_2, and CO_2) and nutrients ($C_6H_{12}O_6$ and NH_3), the Accumulation term is zero. Thus, for a steady-state, reacting system, we can use equation [3.8-13], which is simplified from the governing accounting equation given above:

$$\dot{n}_{i,s} - \dot{n}_{j,s} + \sigma_s R = 0$$

Plant biomass and water accumulate in the system, so the governing accounting equation becomes:

$$\dot{n}_{i,s} - \dot{n}_{j,s} + \sigma_s R = \dot{n}^{sys}_{acc,s}$$

   (b) Calculate:
   - Using our basis of the nutrient media (1425 g/run) and the information given in the problem, we can calculate the mass and molar rates of glucose, ammonia, and water into the bioreactor. For glucose:

$$C_6H_{12}O_6: \qquad \dot{m}_{1,C_6H_{12}O_6} = 0.03\left(1425\,\frac{\text{g}}{\text{run}}\right) = 42.75\,\frac{\text{g}}{\text{run}}$$

$$\dot{n}_{1,C_6H_{12}O_6} = \frac{\dot{m}_{1,C_6H_{12}O_6}}{M_{C_6H_{12}O_6}} = \frac{42.75\,\frac{\text{g}}{\text{run}}}{180\,\frac{\text{g}}{\text{mol}}} = 0.2375\,\frac{\text{mol}}{\text{run}}$$

Similar calculations are performed for ammonia and water. For ammonia, the mass rate is 24.94 g/run, and the molar rate is 1.47 mol/run. For water, the mass and molar rates are 1357 g/run and 75.4 mol/run, respectively. Remember that these materials are added in batch fashion at the beginning of the 10-day period.
   - Air is continuously fed to the reactor at 22 cm^3/min. After finding the volume of air flowing in during the 10-day run, we can then find the volumetric flow rate of O_2 and N_2, which compose air in quantities 21 vol% and 79 vol%, respectively:

$$\text{Air:} \qquad \dot{V}_2 = \left(22\,\frac{\text{cm}^3}{\text{min}}\right)\left(\frac{60\text{ min}}{\text{hr}}\right)\left(\frac{24\text{ hr}}{\text{day}}\right)\left(\frac{10\text{ day}}{\text{run}}\right) = 316{,}800\,\frac{\text{cm}^3}{\text{run}}$$

$$O_2: \qquad \dot{V}_{2,O_2} = 0.21\left(316{,}800\,\frac{\text{cm}^3}{\text{run}}\right) = 66{,}500\,\frac{\text{cm}^3}{\text{run}}$$

The volumetric flow rate for N_2 is calculated in the same way to be 250,000 cm^3/run.
   - Using the ideal gas law, we can convert the volumetric flow rates into molar flow rates, and then into mass flow rates. For O_2:

$$O_2: \qquad \dot{n}_{2,O_2} = \frac{P\dot{V}_{2,O_2}}{RT} = \frac{(1.0\text{ atm})\left(66{,}500\,\frac{\text{cm}^3}{\text{run}}\right)}{\left(82.06\,\frac{\text{atm}\cdot\text{cm}^3}{\text{mol}\cdot\text{K}}\right)(298\text{ K})} = 2.72\,\frac{\text{mol}}{\text{run}}$$

$$\dot{m}_{2,O_2} = \dot{n}_{2,O_2}M_{O_2} = \left(2.72\,\frac{\text{mol}}{\text{run}}\right)\left(32\,\frac{\text{g}}{\text{mol}}\right) = 87.04\,\frac{\text{g}}{\text{run}}$$

Calculating for N_2 in the same manner gives molar and mass inflow rates of 10.23 mol/run and 286.4 g/run, respectively.

- We can now find the limiting reactant using equation [3.8-18]. For glucose:

$$C_6H_{12}O_6: \quad \left\{\frac{\dot{n}_{i,s}}{-\sigma_s}\right\} = \left\{\frac{0.2375 \dfrac{\text{mol}}{\text{run}}}{-(-1)}\right\} = 0.2375 \frac{\text{mol}}{\text{run}}$$

Similar calculations for O_2 and NH_3 give 0.793 mol/run and 0.96 mol/run, respectively. Since glucose has the minimum value, it is the limiting reactant.

- With the inlet and outlet rates for glucose, we can calculate R. To do this, we must first convert the given mass outlet rate for glucose to a molar rate:

$$\dot{n}_{3,C_6H_{12}O_6} = \frac{\dot{m}_{3,C_6H_{12}O_6}}{M_{C_6H_{12}O_6}} = \left(0.699 \frac{\text{g}}{\text{run}}\right)\left(\frac{\text{mol}}{180 \text{ g}}\right) = 0.00388 \frac{\text{mol}}{\text{run}}$$

We can then use the molar rates in and out of the system to find a fractional conversion (equation [3.8-16]), which can then be used to find R (equation [3.8-15]):

$$f_{C_6H_{12}O_6} = \frac{\dot{n}_{1,C_6H_{12}O_6} - \dot{n}_{3,C_6H_{12}O_6}}{\dot{n}_{1,C_6H_{12}O_6}} = \frac{0.2375 \dfrac{\text{mol}}{\text{run}} - 0.00388 \dfrac{\text{mol}}{\text{run}}}{0.2375 \dfrac{\text{mol}}{\text{run}}} = 0.98$$

Note that even though glucose is the limiting reactant, it is not completely consumed.

$$R = \frac{\dot{n}_{1,C_6H_{12}O_6} f_{C_6H_{12}O_6}}{-\sigma_{C_6H_{12}O_6}} = \frac{\left(0.2375 \dfrac{\text{mol}}{\text{run}}\right)(0.98)}{-(-1)} = 0.2336 \frac{\text{mol}}{\text{run}}$$

- With R and the previously calculated molar inlet rate of NH_3, we can calculate the molar and mass outlet rates using the governing accounting equation for steady-state, reacting systems:

$$NH_3: \quad \dot{n}_{1,NH_3} - \dot{n}_{3,NH_3} + \sigma_{NH_3} R = 0$$

$$\dot{n}_{3,NH_3} = \dot{n}_{1,NH_3} + \sigma_{NH_3} R = 1.47 \frac{\text{mol}}{\text{run}} + (-1.53)\left(0.2336 \frac{\text{mol}}{\text{run}}\right)$$

$$= 1.11 \frac{\text{mol}}{\text{run}}$$

$$\dot{m}_{3,NH_3} = \dot{n}_{3,NH_3} M_{NH_3} = \left(1.11 \frac{\text{mol}}{\text{run}}\right)\left(17 \frac{\text{g}}{\text{mol}}\right) = 18.9 \frac{\text{g}}{\text{run}}$$

We can perform similar calculations to find the mass flow rates of O_2 and CO_2 out of the system. For O_2, the mass flow rate out is 61.44 g/run; for CO_2, 27.0 g/run. The inlet flow rate of N_2 should equal that coming out, since it does not react, so the mass flow rate out is 286.4 g/run.

- Since the bioreactor is sealed during the 10-day run, the plant biomass cannot enter or leave the system; however, its mass does increase as it grows. Therefore, the material produced in the reaction accumulates in the system. To find the amount of accumulation, we use the governing accounting equation for a reacting, dynamic system:

$$\text{Biomass:} \quad \dot{n}_{i,\text{biomass}} - \dot{n}_{j,\text{biomass}} + \sigma_{\text{biomass}} R = \dot{n}^{\text{sys}}_{\text{acc,biomass}}$$

$$0 - 0 + 3.37\left(0.2336 \frac{\text{mol}}{\text{run}}\right) = \dot{n}^{\text{sys}}_{\text{acc,biomass}} = 0.787 \frac{\text{mol}}{\text{run}}$$

The accumulation of dry plant biomass in molar units is converted to mass units using the molecular weight of the plant biomass ($CH_{1.27}O_{0.43}N_{0.45}$), which is 26.45 g/mol:

$$\dot{m}^{sys}_{acc,\ biomass} = \left(0.787\ \frac{mol}{run}\right)\left(26.45\ \frac{g}{mol}\right) = 20.8\ \frac{g}{run}$$

The problem statement indicated the wet-weight to dry-weight ratio of plant tissue is known to be 14:1. Thus, the accumulation of wet biomass and water in the reactor is:

$$\dot{m}^{sys}_{acc,wet\ biomass} = 14\left(20.8\ \frac{g}{run}\right) = 291.2\ \frac{g}{run}$$

$$\dot{m}^{sys}_{acc,H_2O} = \dot{m}^{sys}_{acc,wet\ biomass} - \dot{m}^{sys}_{acc,\ biomass}$$

$$= 291.2\ \frac{g}{run} - 20.8\ \frac{g}{run} = 270.4\ \frac{g}{run}$$

- We can now use the differential mass accounting equation for a reacting, dynamic system to find the outlet water rate:

$$H_2O: \quad \dot{n}_{1,H_2O} - \dot{n}_{3,H_2O} + \sigma_{H_2O}R = \dot{n}^{sys}_{acc,H_2O}$$

$$\dot{n}_{3,H_2O} = \dot{n}_{1,H_2O} + \sigma_{H_2O}R - \dot{n}^{sys}_{acc,H_2O}$$

$$= 75.4\frac{mol}{run} + 6.16\left(0.2336\ \frac{mol}{run}\right) - \left(270.4\ \frac{g}{run}\right)\left(\frac{mol}{18\ g}\right)$$

$$= 61.82\ \frac{mol}{run}$$

$$\dot{m}_{3,H_2O} = \dot{n}_{3,H_2O}M_{H_2O}$$

$$= \left(61.82\ \frac{mol}{run}\right)\left(18\ \frac{g}{mol}\right) = 1113\ \frac{g}{run}$$

4. Finalize

(**a**) Answer: The limiting reactant is glucose. The rate of reaction is 0.234 mol/run. The outlet mass rates for each of the compounds are given in Table 3.7B, where all numerical values are given with three significant figures. The dry-weight mass of *Atropa belladonna* roots in the system at 10 days is 20.8 g. The wet-weight mass is the sum of the water (270.4 g) and the plant dry mass (20.8 g), which is equal to 291 g, the total amount of accumulation in the system at the end of 10 days.

(**b**) Check: We can perform an overall total mass balance to check the solutions using the unsteady-state conservation equation [3.9-2]. If we use the given 1425 g/run of entering liquid and the calculated 373 g/run of gas (Table 3.7B), the total inflow is 1798 g/run. For the outlet, the calculated total outflow is 1504 g/run. Thus, the overall total mass balance is:

$$\sum_i \dot{m}_i - \sum_j \dot{m}_j = \dot{m}^{sys}_{acc}$$

$$1798\ \frac{g}{run} - 1504\ \frac{g}{run} = 294\ \frac{g}{run} \approx 291\ \frac{g}{run}$$

This is close to the calculated accumulation of 291 g/run. The differences can be accounted to rounding errors. ■

TABLE 3.7B

Mass Flow Rates of Components in *Atropa belladonna* Bioreactor

	Inlet (g/run)		Outlet (g/run)		Accumulation (g/run)
	Liquid (1)	Gas (2)	Liquid (3)	Gas (4)	In system
$C_6H_{12}O_6$	42.8	—	0.699	—	0
CO_2	—	0	—	27.0	0
O_2	—	87.0	—	61.4	0
N_2	—	286	—	286	0
NH_3	24.9	—	18.9	—	0
H_2O	1360	—	1110	—	270
$CH_{1.27}O_{0.43}N_{0.45}$	—	—	—	—	20.8

Summary

In this chapter, we described basic mass concepts, which included definitions for mass; moles; mass, molar and volumetric flow rates; and mass and mole fractions. We also described how the accounting and conservation statements can be applied to the extensive properties of total mass, species mass, elemental mass, total moles, species moles, and elemental moles.

We focused our analysis on how we can simplify and reduce the accounting and conservation equations for a variety of systems, such as open, nonreacting, steady-state systems. The equations were also applied in cases with multiple streams entering and exiting the system or with multiple components flowing in a stream or both. We also examined how to isolate simple systems from complex, multi-unit systems with multiple streams to solve for individual constituents and variables. We demonstrated a method to stoichiometrically balance complex biochemical reactions and showed how to apply the accounting and conservation statements to systems with reactions. Finally, we analyzed how the equations can be used to solve for variables in dynamic systems.

Table 3.8 reinforces that mass may accumulate in a system because of bulk material transfer across the system boundary or because of the generation or consumption of mass through chemical reactions. See the tables concluding other chapters for comparison. The conservation of mass is presented first in this text because it is used to solve more complex problems in conjunction with the conservation of total energy (Chapter 4) and of linear and angular momentum (Chapter 6), as well as the accounting of electrical energy (Chapter 5) and of mechanical energy (Chapter 6).

TABLE 3.8

Summary of Movement, Generation, Consumption, and Accumulation in Mass Accounting Equations

Accumulation	Input − Output		+ Generation − Consumption	
Extensive property	Bulk material transfer	Direct and non-direct contacts	Chemical reactions	Energy interconversions
Total mass	X			
Species mass	X		X	
Elemental mass	X			
Total moles	X		X	
Species moles	X		X	
Elemental moles	X			

References

1. Lewis R. "A compelling need." *Scientist* 1995, 9:12.
2. DePuy Orthopaedics I. "JointReplacement.com: Restoringthe Joy of Motion." 2000. http://www.jointreplacement.com/xq/ASP.default/mn.local/pg.header/joint_id.5/newFont.2/joint_nm.Hip/qx/default.htm (accessed July 15, 2005).
3. Greenwald AS, Boden SD, Goldberg VM, et al. "Bone-graft substitutes: Facts, fictions, and applications." *J Bone Joint Surg Am* 2001, 83-A Suppl 2 Pt 2:98–103.
4. Cooney DO. *Biomedical Engineering Principles: An Introduction to Fluid, Heat, and Mass Transport Processes*. New York: Marcel Dekker, 1976.

Problems

3.1 Arterioles bifurcate (i.e., split into) capillaries in the circulatory system. Blood flows at a velocity of 20 cm/s through an arteriole with a diameter of 0.20 cm. This vessel bifurcates into two vessels: one with a diameter of 0.17 cm and a blood flow velocity of 18 cm/sec, and one with a diameter of 0.15 cm. Each of these two vessels splits again. The 0.17-cm diameter vessel splits into two vessels, each with a diameter of 0.15 cm. The 0.15-cm diameter vessel splits into two vessels, each with a diameter of 0.12 cm. Determine the mass flow rate and velocity of blood in each of the four vessels at the end of the arteriole bifurcations. You may need to set up several systems, each with a different system boundary, in order to solve this problem.

3.2 You are interested in modeling the air flow in the lungs. The trachea is represented as Generation 0. The diameter of the trachea, D_0, is 1.8 cm. The volumetric flow rate in the trachea is 200 mL/sec. The density of air is 0.0012 g/cm^3 and the viscosity of air is 0.00018 g/(cm · s).

The trachea splits into two vessels, represented as Generation 1 (see Figure 3.23). These two vessels have the same diameter; each diameter is 75% of the diameter of the trachea. Each of the Generation 1 vessels splits into two vessels, represented as Generation 2. The four vessels in Generation 2 have the same diameter; each diameter is 75% of the diameter of a vessel in Generation 1.

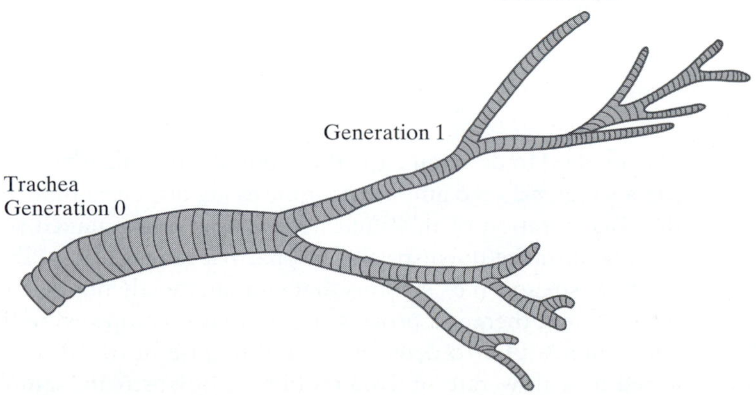

Figure 3.23
Bifurcations of the trachea.

This pattern continues with each vessel bifurcating (splitting into two) at each new generation, and the diameter of the vessel of the new generation is always 75% of the diameter of the preceding generation.

(a) Write an equation for the linear air velocity, v_n, for Generation n, in terms of $\dot{m}_0$ (mass flow rate in Generation 0), n (generation number), D_0 (vessel diameter of Generation 0) and ρ (density of air). Calculate the linear air velocity at Generations 6 and 12.

(b) Write an equation for the Reynolds number, Re_n, for Generation n, in terms of $\dot{m}_0$ (mass flow rate in Generation 0), n (generation number), D_0 (vessel diameter of Generation 0) and μ (viscosity of air). Calculate the Reynolds number at Generations 6 and 12.

3.3 Corn-steep liquor contains 2.5 wt% dextrose and 50 wt% water; the rest of the liquor stream is solids. Beet molasses contains 50 wt% sucrose, 1.0 wt% dextrose, and 18 wt% water; the rest of the molasses stream contains solids. Beet molasses is mixed with corn-steep liquor and water in a mixing tank to produce a dilute sugar mixture. The exit stream contains 2.0 wt% dextrose and 12.6 wt% sucrose, and is ready to be fed into a fermentation unit. See Figure 3.24. (Adapted from Doran PM, *Bioprocess Engineering Principles*, 1999.)

(a) What is the basis in your solution to this problem?

(b) What are the weight percents (wt%) of dextrose, sucrose, solids, and water in the exit stream?

(c) What is the ratio of the mass flow rate of the water stream to the mass flow rate of the corn-steep liquor stream?

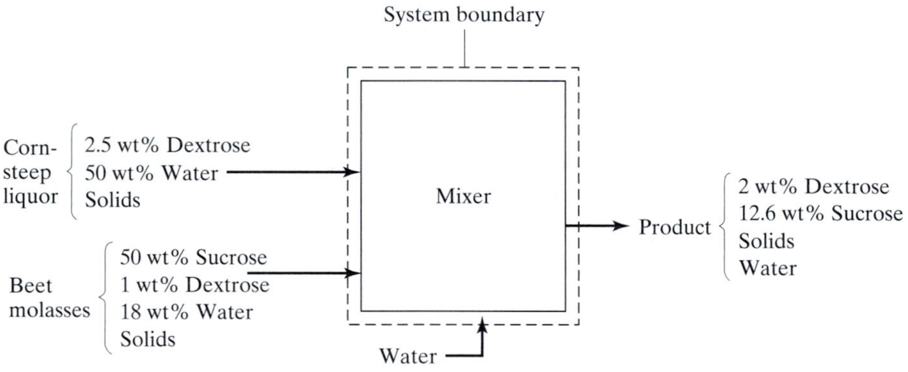

Figure 3.24
Corn-steep liquor and beet molasses mixer.

3.4 One method to determine the flow rate of a turbulently flowing stream is to inject a small, metered amount of some easily dispersed fluid and then to measure the concentration of this fluid in a sample of the mixed stream withdrawn a suitable distance downstream. In pharmaceutical plants, there are often many inert gas streams (i.e., streams that contain mostly nonreacting gases). In a particular plant, there is a process stream that is composed of 95 mol% N_2 (inert) and 5 mol% O_2. To determine the flow rate of this process stream, O_2 is injected at a flow rate of 16.3 mol/hr. A downstream sample analyzes the O_2 content at 10 mol%. You can assume that no reactions occur in the pipe and that the flows are operating at steady-state.

(a) How many balance equations can you write? How many of those are linearly independent?
(b) Calculate the flow rate of the process stream containing 95 mol% N_2 and 5 mol% O_2.

3.5 You need to prepare blood for a transfusion. You have the following three processed blood packs available:

PACK A—ENRICHED IN RED BLOOD CELLS (RBCs)
Contents: 2.5 wt% white blood cells (WBCs); 50.0 wt% isotonic fluid; the rest is RBCs.
PACK B—ENRICHED IN SERUM PROTEINS
Contents: 50.0 wt% serum proteins; 1.0 wt% WBCs; 18.0 wt% isotonic fluid; the rest is RBCs.
PACK C
Contents: 100.0 wt% isotonic fluid.

All three packs must be mixed in the correct proportions to generate blood for a transfusion pack. The transfusion pack needs to have the following composition: 2.0 wt% WBCs and 12.6 wt% serum proteins.
(a) Write out mass conservation equations for RBCs, WBCs, isotonic fluid, serum proteins, and total mass.
(b) Calculate the weight percents (wt%) of RBCs and isotonic fluid in the transfusion pack.
(c) What is the ratio of the mass of pure isotonic fluid (Pack C) to the mass of Pack A? What is the ratio of the mass of Pack B to the mass of Pack A?

3.6 In an industrial process to make alcohol, bacteria, sugar, and water are fed into a bioreactor. The bacteria make alcohol out of the sugar, and the stream leaving the bioreactor contains bacteria, alcohol, and water as well as leftover sugar. We desire to remove all the cells from the process stream so we may purify our alcohol product. The process stream enters a separator where the cellular components are separated from the rest of the stream. The entering process stream contains 30 wt% alcohol, 5 wt% sugar, 10 wt% cells, and the rest water. Two product streams, a cell-rich stream and a cell-free stream, leave the separator. The cell-rich stream is 90 wt% cells, 2.5 wt% sugar, 0.5 wt% alcohol, and 7 wt% water.
(a) Write out species mass conservation equations for alcohol, bacteria, sugar, and water. Write a total mass conservation equation.
(b) How many of the mass balance equations are linearly independent?
(c) Determine the composition of the cell-free stream.

3.7 A synthetic hemoglobin-based blood substitute would be invaluable in situations where donated blood supplies run low. In early substitutes, the hemoglobin molecule was genetically modified to improve its affinity for oxygen.

Modified hemoglobin is dried with sodium chloride (1.0 wt%) and potassium phosphate (1.0 wt%). The dried hemoglobin solution is combined with a solid salt mixture containing sodium bicarbonate (50.0 wt%), sodium chloride (20.0 wt%), and potassium phosphate. Each "blood bag" contains 2.0×10^2 g total, which includes the dried modified hemoglobin mixture and the dried salt mixture. When the blood substitute is needed, water is added to the dried mixture. To reconstitute, 8.0×10^2 g of water is added to each bag. In the reconstituted solution (containing water, hemoglobin, and salts), the modified hemoglobin must be at least 19 wt%.

(a) Write a conservation equation for each of the four chemical constituents (modified hemoglobin, sodium chloride, potassium phosphate, sodium bicarbonate) that are mixed as a dried powder.
(b) Determine the wt% of each of the four constituents in the dried mixture.
(c) Determine the wt% of each of the five constituents after reconstitution with water.
(d) There are occasionally supply problems with the dried salt mixture. In the event that you run out of the salt mixture, how much extra water do you need to add to maintain modified hemoglobin at 19 wt% in the reconstituted solution?

3.8 The drug streptomycin is produced on a large scale in the United States. After purification, streptomycin is 50 wt% in water. For applications, including IV drips, the streptomycin must be diluted and preservative must be added. The diluent stream contains 2 wt% NaCl in water. The preservative stream contains 10 wt% preservative and 5 wt% NaCl in water. These three streams are mixed together in a mixing tank. The outlet stream is ready for packaging in IV bags.
(a) Determine the ratio of the stream containing the drug to the outlet stream, given that the drug is 10 wt% in the outlet stream.
(b) Determine the ratio of the stream containing the preservative to the outlet stream, given that the preservative is 3 wt% in the outlet stream.
(c) Determine the ratio of the diluent stream to the outlet stream, given that the drug is 10 wt% and the preservative is 3 wt% in the outlet stream.
(d) What is the basis in this problem?
(e) Calculate the mass flow rates of the three inlet streams.
(f) What is the wt% of the NaCl in the outlet stream?

3.9 One role of the kidneys is to remove toxins that build up as a result of metabolism. When people experience kidney failure, a machine called a dialyzer must remove the toxins. In the dialyzer, the blood passes through thin-walled tubes (or membranes) in one direction while the dialysate flows along the outside of the tubes in the opposite direction. Small pores in the tubes allow small molecules to pass back and forth between the two streams, but prevent larger molecules (such as proteins and cells) from passing through. Given the composition of the blood entering the machine and the dialysate entering and leaving, as shown in Table 3.9, calculate the concentration of each small molecule

TABLE 3.9

Concentration of Species in Dialysis Unit

Species	Molecular weight (g/mol)	Conc. of blood entering dialyzer (mM)	Conc. of dialysate entering dialyzer (mM)	Conc. of dialysate leaving dialyzer (mM)
Na^+	23.0	142	133	133
K^+	39.1	7	1	2
HCO_3^-	61.2	14	35.7	29.2
HPO_4^{2-}	96.0	9	0	3
Glucose	180.2	100	125	125
Urea	60.1	200	0	87

in the detoxified blood in units of mM. Assume there are no reactions going on inside the machine. Blood flows at 200 mL/min while dialysate flows at 400 mL/min.

3.10 You are collecting plasma from a patient with a pheresis machine. The machine takes whole blood out of the body. It separates and collects 80% (by weight) of the plasma. The remainder of the blood is returned to the body. The composition of blood is modeled to contain erythrocytes, leukocytes, and plasma. The mass fractions in whole blood are: erythrocytes $w_E = 0.40$, leukocytes $w_L = 0.05$, and plasma $w_P = 0.55$. Assume a person donates 2 pints (1895 g) of plasma. What are the mass fractions of erythrocytes, leukocytes, and plasma in the material that reenters the body?

3.11 A pharmaceutical company is nearly ready to market an allergy drug for individuals sensitive to hay fever. The product is primarily a mixture of IgG and IgM antibodies in water, with trace amounts of IgE and reagent to keep the antibodies from reacting with other in solution (i.e., the stabilizer). Your goal is to design a process to purify IgM antibodies at a high concentration. The stream from your designed separation unit will be mixed with a stream with a high concentration of IgG antibodies and a stream containing the stabilizer to produce the product stream.

The process to purify high concentrations of IgG antibodies has been refined. This material is delivered to the mixer with the following composition: $w_{IgG} = 0.40$, $w_{IgM} = 0.025$, $w_{IgE} = 0.0030$, and the rest water. The stream containing the stabilizer has the following composition: $w_{stab} = 0.10$, and the rest water. The target composition for the product stream is as follows: $w_{IgG} = 0.15$, $w_{IgM} = 0.20$, $w_{IgE} = 0.0040$, $w_{stab} = 0.01$, and the rest water. The only constraint that you have on your design for the stream containing a high concentration of IgM antibodies is that $w_{IgE} = 0.0050$.

(a) Write a mass conservation equation for each of the five constituents (IgG, IgM, IgE, stabilizer, and water) in the system.
(b) Determine the flow rate of each of the four streams.
(c) Determine the mass fractions of each constituent (IgG, IgM, IgE, and water) in the stream leaving the separation unit that you designed for the stream containing a high concentration of IgM antibodies.

3.12 To generate the biodegradable polymer, poly-L-lactic acid (PLA), a continuous production method is used. The monomer, lactic acid (LA), is reacted with a catalyst in the presence of water to generate PLA. One unit of this multi-unit process is a mixer (Figure 3.25). Streams 1, 2, and 3 are inlets; stream 4 is an outlet. Stream 1 contains water and catalyst. The mass fraction of catalyst in stream 1 is 0.40. Stream 2 contains water and LA. Stream 3 is a recycle stream from further down the operation and contains water, LA, catalyst and PLA. In stream 3, the mass fraction of PLA is 0.050, the mass fraction of catalyst is 0.020, and the mass fraction of LA is 0.150. The contents of the streams are well mixed and leave in stream 4. In stream 4, the mass fraction of catalyst is 0.10 and the mass fraction of PLA is 0.010. The mass flow rate of LA in stream 2 is ten times the mass flow rate of LA in stream 3.

(a) Write a mass conservation equation for each of the four constituents (water, LA, catalyst and PLA) in the system.
(b) Determine the flow rate of each of the four streams.
(c) Determine the mass fractions of each of the constituents (water, LA, catalyst and PLA, as appropriate) in streams 2 and 4.

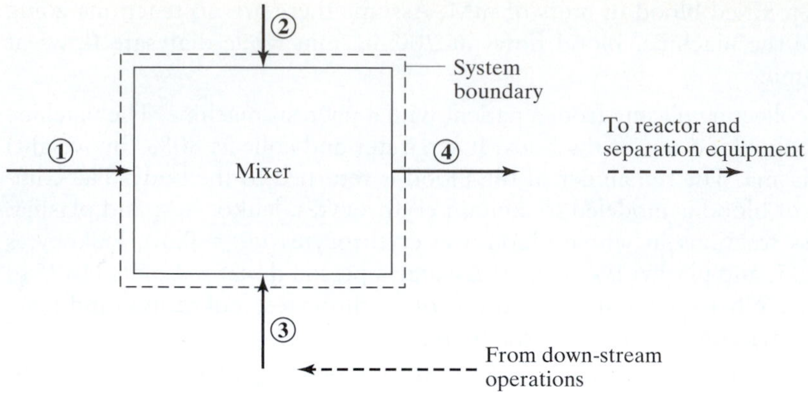

Figure 3.25
Isolated mixer from a multi-unit process to make PLA.

3.13 Hollow-fiber membrane devices are used in a number of applications in bioengineering and biochemical engineering. A typical unit consists of thousands of small hollow fiber tubes packed in a tubular device (Figure 3.26a). Components within the fibers can be isolated from components outside the fibers based on solubility and/or size restrictions. Some materials can easily diffuse across the membrane between the fibers to the annular space. In this problem, the hollow-fiber membrane device is modeled as an inner tube, representing the membrane fibers, and an outer tube, representing the outer (annular) space (Figure 3.26b).

A hollow-fiber membrane device is operated to concentrate a bacterial suspension. The flow rate of cell suspension into the fibers is 350 kg/min. The inlet cell suspension is comprised of 1.0 wt% bacteria; the rest of the suspension can be considered water. An aqueous buffer solution enters the annular space at a flow rate of 80.0 kg/min. Because the cell suspension in the membrane tubes is under pressure, water is forced from the tubes, across the membrane, and into the buffer. Bacteria in the cell suspension are too large to pass through the membrane, and thus they remain in the membrane tubes throughout the device. The outlet cell suspension is comprised of 6.0 wt% bacteria. Assume that the cells do not grow. Also assume that the membrane does not allow any molecules other than water to pass across it. (Adapted from Doran PM, *Bioprocessing Engineering Principles*, 1999.)

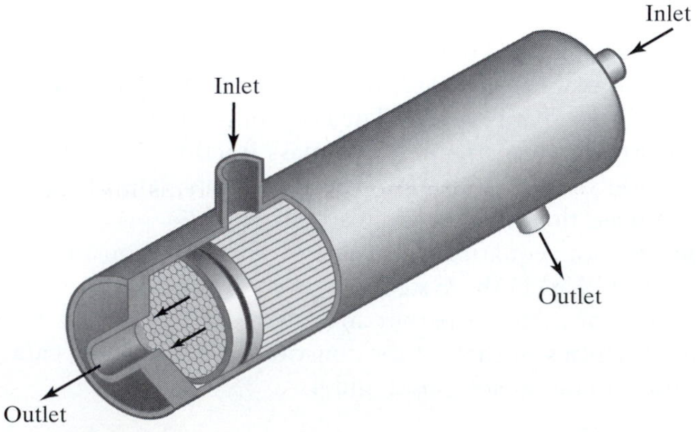

Figure 3.26a
Hollow-fiber membrane device.

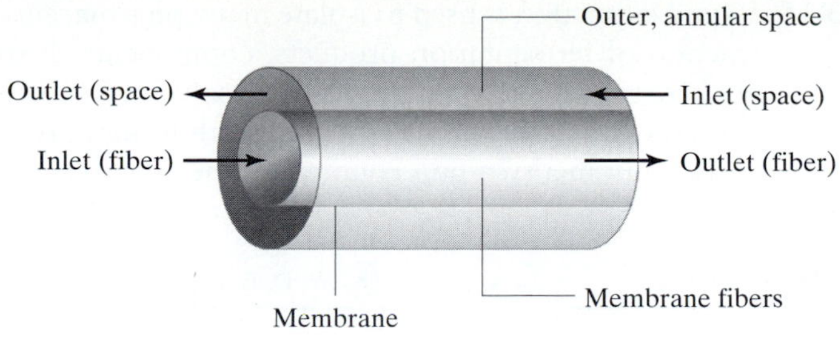

Figure 3.26b
Simplified model of a hollow-fiber membrane device.

(a) Determine the mass flow rates of the outlet cell suspension stream and the outlet buffer stream.
(b) Determine the mass flow rate of the water across the membrane.
(c) Determine the mass flow rate of the cells in the outlet cell suspension stream.

3.14 A membrane system is used to filter waste products from the bloodstream (Figure 3.27). The blood can be thought of being comprised of "waste" and "all other blood constituents." The membrane can extract 30.0 mg/min of pure waste (stream W) without removing any blood. The unfiltered entering bloodstream (stream U) contains 0.17 wt% waste, and the mass flow rate of the entering bloodstream is 25 g/min. After exiting the membrane, the blood is split into two streams: one (stream R) is recycled to join with the unfiltered bloodstream before entering the membrane and one (stream F) leaves the system as filtered blood. The recycle mass flow rate (stream R) is known to be twice that of the filtered mass flow rate (stream F). Calculate the mass flow rate and the wt% of waste in streams A, B, F, and R.

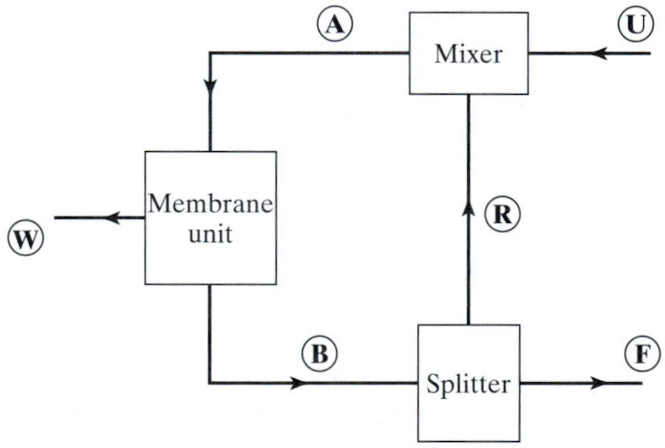

Figure 3.27
Multi-unit membrane system to filter waste products from the bloodstream.

In order to solve for the requested information, several systems must be sequentially specified. First, draw a system boundary around the whole operation that cuts across streams W, F, and U. Solve for the flow rates and compositions of those streams. Then, set your system boundary around the mixer; solve for the required information. Finally, set your system boundary around the splitter; solve for the required information. (Adapted from Glover C, Lunsford KM, Fleming JA, *Conservation Principles and the Structure of Engineering*, 1994.)

3.15 Liquid extraction is used to isolate many pharmaceutical products. In liquid extraction of fermentation products, components dissolved in liquid are recovered by transfer into an appropriate solvent. In the isolation of penicillin, the drug is extracted from its aqueous broth using butyl acetate. This separation is carried out in a two-unit countercurrent design, as shown in Figure 3.28.

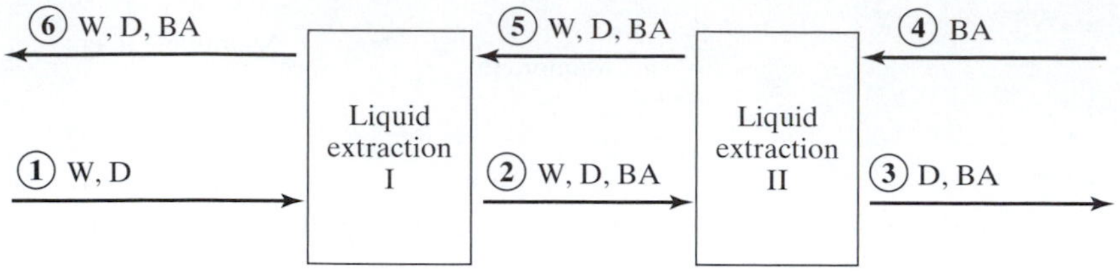

Figure 3.28
Two-unit countercurrent design for liquid extraction of penicillin. BA is butyl acetate; W is water; D is drug.

1.00×10^3 lb_m/hr of a dilute aqueous penicillin stream (stream 1) is extracted using butyl acetate in twounits. The inlet penicillin stream (stream 1) contains 0.50 wt% penicillin; the rest of the stream is water. The inlet butyl acetate mass flow rate (stream 4) is 30.0% of the inlet aqueous penicillin mass flow rate (stream 1). One outlet stream (stream 3) contains 3.0 wt% penicillin; the rest of the stream is butyl acetate. The other outlet flow (stream 6) contains water, penicillin, and butyl acetate. The mass fraction of penicillin in stream 6 is 1/4000 of the mass fraction of water in that stream. The separation of penicillin in the first stage is 98%; in other words, 98% of the penicillin mass entering Unit I is retained in stream 2. The compositions in stream 2 of penicillin and water are 1.7 wt% and 2.0 wt%, respectively.

To solve this problem, an overall system encompassing both liquid extraction units must be drawn first. Solve for the required information in streams 1, 3, 4, and 6. Then, draw a system around Unit I and solve for the required information. You are encouraged to carry four or five significant figures for each number through this problem until you present the final answer.

(a) Determine the total mass flow rate in each stream.
(b) Determine the weight percents of butyl acetate, penicillin, and water, as appropriate, in each of the six streams.

3.16 Poly(propylene fumerate) is a promising polymer for implants for orthopedic applications. It is synthesized in a methylene chloride solvent using a zinc chloride ($ZnCl_2$) catalyst. Since $ZnCl_2$ is potentially toxic to human cells, it needs to be rinsed from the polymer solution. After processing, both the polymer precipitate and the catalyst are dissolved in methylene chloride. The polymer stream is washed in two sequential units with the solvent methylene chloride, as shown in Figure 3.29. The system should operate such that there is a reduction of an order of magnitude (factor of 10) in the wt% of catalyst in the recovered polymer stream after processing by the two wash units.

The unprocessed polymer solution contains 40.0 wt% polymer, 10.0 wt% catalyst, and 50.0 wt% solvent. 80% of the catalyst fed to each unit leaves in the waste solution (which contains only solvent and catalyst). At each unit, the catalyst concentration in the waste solution is the same as the catalyst concentration in the polymer mixture leaving that unit. The units are operated such that the stream between the two wash units has 65.0 wt% polymer and the product stream leaving Wash Unit II has 80.0 wt% polymer. Your first system should be drawn around Wash Unit I.

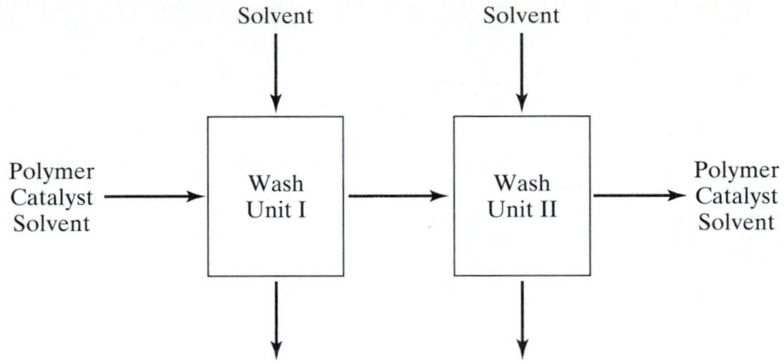

Figure 3.29
Two wash units used in the purification of poly(propylene fumerate).

(a) Determine the mass flow rates of the two inlet solvent wash streams.
(b) Determine the catalyst weight percent in the final outlet polymer product stream.
(c) Does the design as given achieve the order-of-magnitude reduction of catalyst weight percent in the polymer stream?

3.17 A bioinstrumentation manufacturer mixes four alloy feeds to continuously produce desired alloys to cast into scalpels and other surgical equipment. (Adapted from Reklatis GV, *Introduction to Material and Energy Balances*, 1983.)

(a) Inlet Alloy Feeds 1, 2, 3, and 4 are combined in one mixing unit (Figure 3.30a). The Target Alloy I outlet mass flow rate is 1.00×10^4 lb_m/hr. F, G, H, and K are hypothetical compounds. The weight fractions of components F, G, H, and K in the Alloy Feeds and Target Alloy I are given in Table 3.10. Calculate the mass flow rates at which the four Alloy Feeds should be supplied to mix to produce the Target Alloy I stream.

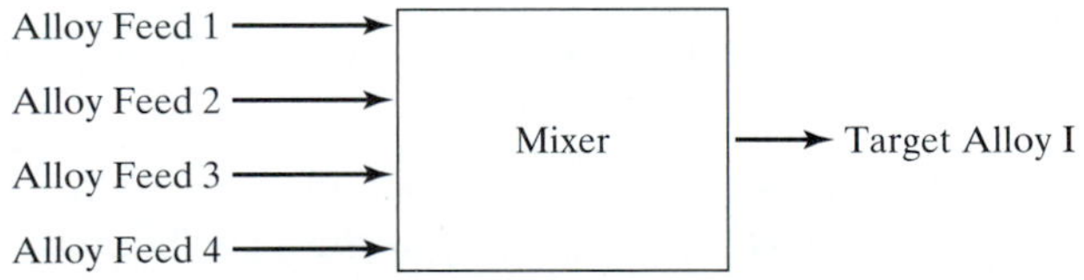

Figure 3.30a
Alloy feeds in a mixing unit.

TABLE 3.10

Alloy Feeds and Target Compositions

	Component weight fractions			
	F	G	H	K
Alloy Feed 1	0.60	0.20	0.20	0
Alloy Feed 2	0.20	0.60	0	0.20
Alloy Feed 3	0.20	0	0.60	0.20
Alloy Feed 4	0	0.20	0.20	0.60
Target Alloy I	0.25	0.25	0.25	0.25

(b) For a different application, Alloy Feed 1 and Alloy Feed 2 are combined in a mixing tank labeled Mixer 1 (Figure 3.30b). The weight fractions of components F, G, H, and K in Alloy Feeds 1 and 2 are given in Table 3.10. The mass fraction of F in the outlet stream from Mixer 1 is 0.50. The outlet stream from Mixer 1 is then combined with Alloy Feed 5 in a second mixing tank labeled Mixer 2 to produce Target Alloy II. Alloy Feed 5 contains only

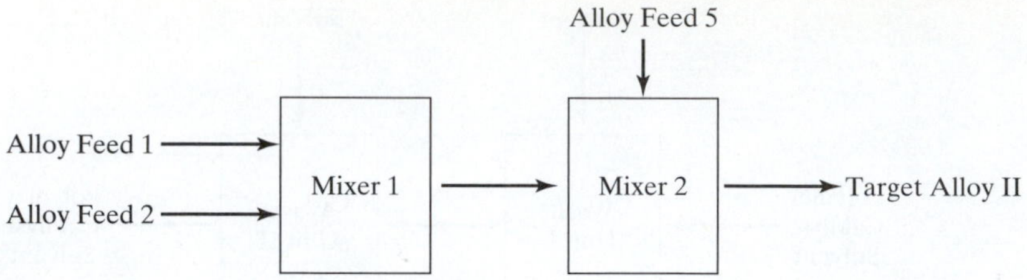

Figure 3.30b
Two mixer units in an alloy processing system.

F, H, and K; the mass fraction of H is one-half the mass fraction of F. The mass fraction of F in Target Alloy II is 0.40. In Target Alloy II, the mass fraction of G is equal to the mass fraction of H. The Target Alloy II outlet mass flow rate is 1.00×10^4 lb_m/hr.

- Set up and solve mass conservation equations around Mixer 1. Determine the mass flow rates of all streams and the mass fractions of all components in all streams entering and leaving Mixer 1. Report the final answers so that the outlet mass flow of Target Alloy II is 1.00×10^4 lb_m/hr.
- Set up and solve mass conservation equations around Mixer 2. Determine the mass flow rates of all streams and the mass fractions of all components in all streams entering and leaving Mixer 2. Report the final answers so that the outlet mass flow of Target Alloy II is 1.00×10^4 lb_m/hr.

3.18 Balance the following equations by solving for the appropriate unknowns. Using MATLAB may facilitate the solution to several parts of this problem.

(a) $ZrCl_4 + aH_2O \rightarrow pZrO_2 + qHCl$

(b) $C_6H_{12}O_6 + aNH_3 + bO_2 \rightarrow pC_5H_9NO_4 + qCO_2 + rH_2O$, RQ = 0.45

(c) $CH_2O + aO_2 + bNH_3 \rightarrow pCH_{1.8}N_{0.2}O_{0.75} + qH_2O + rCO_2$, RQ = 0.30

(d) $C_2H_5OH + aNa_2Cr_2O_7 + bH_2SO_4 \rightarrow pCH_3COOH + qCr_2(SO_4)_3 + rNa_2SO_4 + sH_2O$

(e) Aerobic (i.e., with O_2) growth of *S. cerevisiae* (yeast) on ethanol. $CH_{1.704}N_{0.149}O_{0.408}$ is the yeast/biomass.

$C_2H_5OH + aO_2 + bNH_3 \rightarrow pCH_{1.704}N_{0.149}O_{0.408} + qCO_2 + rH_2O$, RQ = 0.66

(f) Anaerobic (i.e., without O_2) growth of *S. cerevisiae* (yeast) on glucose. $CH_{1.74}N_{0.2}O_{0.45}$ is the yeast/biomass.

$C_6H_{12}O_6 + aNH_3 \rightarrow 0.59\ CH_{1.74}N_{0.2}O_{0.45} + pC_3H_8O_3 + qCO_2 + 1.3\ C_2H_5OH + rH_2O$

3.19 The vitamin company you work for produces alanine. Alanine is a nonessential amino acid synthesized by the body. It is important as a source of energy for muscle tissue, the brain, and the central nervous system. Alanine also helps in the metabolism of sugars and organic acids.

Alanine is produced in a reactor in a continuous process. There are two separate inlet streams that contain glutamine (100 mol/min) and pyruvic acid (50 mol/min), respectively. The ratio of the molar flow rate of pyruvic acid in the outlet stream to that in the inlet stream is 0.6.

$$C_5H_{10}N_2O_3\ \text{(glutamine)} + C_3H_4O_3\ \text{(pyruvic acid)} \longrightarrow C_5H_7NO_4\ (\alpha\text{-ketoglutamic acid}) + C_3H_7NO_2\ \text{(alanine)}$$

(a) Balance the equation.
(b) What is the reaction rate, R, of glutamine? pyruvic acid? Find the limiting reagent. What are the fractional conversions of glutamine and pyruvic acid?
(c) Find the outlet molar flow rates of alanine, α-ketoglutamic acid, and any excess reactants.

3.20 *Acetobacter aceti* bacteria convert ethanol to acetic acid (vinegar) under aerobic (i.e., with oxygen) conditions. A continuous fermentation process for acetic acid production is shown in Figure 3.31. The conversion reaction is as follows:

$$C_2H_5OH \text{ (ethanol)} + O_2 \longrightarrow CH_3COOH \text{ (acetic acid)} + H_2O$$

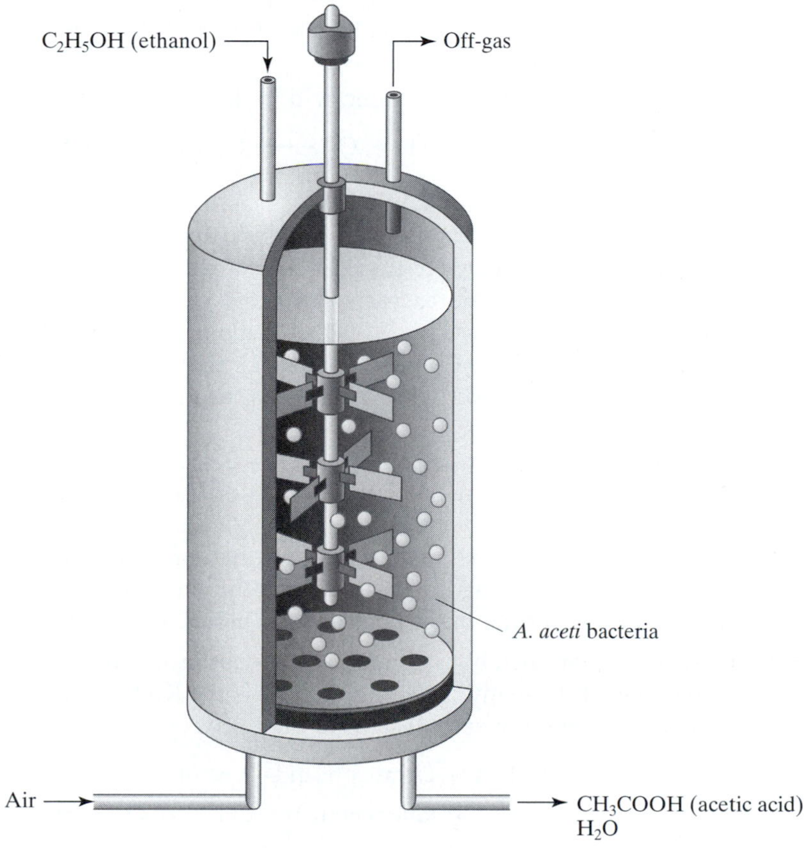

Figure 3.31
Acetobacter aceti bacteria in a bioreactor for acetic acid production.

The feed stream containing ethanol enters the reactor at a rate of 1.0 kg/hr. Also, air bubbles into the reactor at a rate of 40.0 L/min. An exit off-gas stream as well as the liquid product stream containing acetic acid and water leave the reactor.
(a) Check that the above reaction is properly balanced.
(b) What is the reaction rate, R, for this reaction? What is the limiting reactant? What are the fractional conversions of C_2H_5OH and O_2?
(c) Determine the outlet flow rates of the elements C, H, O in the acetic acid product stream. Also, determine the outlet mass flow rates of all compounds in the liquid product stream and the volumetric flow rates of all compounds in the off-gas stream.

3.21 In an attempt to solve the world's energy shortage, engineers have discovered a new type of microbial cell that converts carbon dioxide to propane in the presence of water. A simple continuously stirred tank reactor (CSTR) is designed for the reaction. After months of optimization, it is discovered that the rates of cell division and cell death are equal when the reactor is held at 25°C. Furthermore, complete conversion of CO_2 to propane is achieved in 10% excess water (on a molar basis). CO_2 is bubbled through the reactor at 1680 L/hr. No cells are lost in the liquid product stream. The density of CO_2 is 0.00197 g/cm^3.

(a) Write the balanced chemical equation.
(b) Sketch the reactor setup including all reactant and product streams.
(c) How much feed water is required (mol/hr)?
(d) What is the daily output of propane (kg/day)?
(e) Is the concentration of propane higher in the reactor vessel or the product stream? Explain.

3.22 Glucose is converted into glutamic acid in the following reaction:

$$C_6H_{12}O_6 + NH_3 + O_2 \longrightarrow C_5H_9NO_4 + CO_2 + H_2O$$

The production of glutamic acid via this reaction occurs in many of your cells. Also, mammalian cells can be placed in bioreactors and the biochemical conditions optimized to convert glucose to glutamic acid.

Assume a simple bioreactor system that contains mammalian cells. The inlet flow rate of $C_6H_{12}O_6$ is 1.00×10^2 mol/day. NH_3 is delivered at a rate of 1.20×10^2 mol/day. O_2 is provided such that 1.10×10^2 mol/day are dissolved in liquid (i.e., accessible to the cells for uptake). Assume that the reaction goes to completion.

(a) Balance the reaction, given that RQ = 0.45. Determine the limiting reactant, the reaction rate, R, and the fractional conversions of O_2, NH_3, and $C_6H_{12}O_6$.
(b) Calculate the molar and mass flow rates of all constituents leaving the bioreactor including the products and the excess reactants.
(c) Confirm that total mass, but not total moles, is conserved.

3.23 During cellular metabolism, glucose is combusted to carbon dioxide and water. One of the many steps in glycolysis is the Krebs cycle. A simplified biochemical summary of several steps in the Krebs cycle is as follows:

$$1\ C_6H_5O_7\,(\text{citrate}) + aH_2O + bPO_4 \longrightarrow$$
$$pC_4H_2O_5\,(\text{oxaloacetate}) + qH + rCO_2 + sPO_3$$

Note that this equation captures only the exchange of chemical species, not the exchange of their respective charges. It is known through biochemical experiments that for every molecule of citrate consumed, one molecule of oxaloacetate is generated. A tissue mass is comprised of many cells, each which conducts the process of glycolysis including the Krebs cycle. Assume a molar flow rate of 0.10 mol/day of $C_6H_5O_7$ into this tissue.

(a) Balance the above equation. Determine the stoichiometric coefficients a, p, q, r,and s.
(b) What is the minimum flow rate of water needed for the fractional conversion of $C_6H_5O_7$ to be 1.0?
(c) Assume that the fractional conversion of water is 0.80 and the fractional conversion of $C_6H_5O_7$ is 1.0. What is the limiting reactant? Calculate the reaction rate (R) and the inlet molar flow rate of H_2O. Determine the

molar flow rates of the products and the excess reactant leaving the tissue excluding PO_3 and PO_4.

3.24 Genetically engineered strains of *Escherichia coli* have become essential tools in the production of recombinant human peptides and proteins. One of the first substances synthesized using engineered *E. coli* was recombinant human insulin, or humulin, for the treatment of people suffering from type I diabetes mellitus. A simple reaction scheme for the production of humulin is described below. Bacteria consume glucose under aerobic conditions, and produce humulin and biomass.

$$C_6H_{12}O_6\,(\text{glucose}) + O_2 + NH_3\,(\text{ammonia}) \longrightarrow$$
$$C_{2.3}H_{2.8}O_{1.8}N\,(\text{humulin}) + CH_{1.9}O_{0.3}N_{0.3}\,(\text{biomass}) + CO_2 + H_2O$$

A typical humulin production scheme consists of *E. coli* cultured in a large bioreactor. A continuous stream of media is fed into the reactor. A continuous stream of products and unused reactants is removed from the bioreactor for further processing, including purification of the humulin for therapeutic use. Media containing glucose and ammonia is fed into the reactor at a rate of 100 L/hr; the concentrations of glucose and ammonia in that stream are 150 mM and 50 mM, respectively. Pure oxygen gas is sparged (i.e., bubbled) into the reactor at a rate of 100 mL/min. The exit flow rate of liquid containing biomass, product, and excess reactant is 100 L/hr. Assume that there is no accumulation in the system. Assume that the reaction goes to completion.

(a) Write element balances for C, H, O, and N. Write two additional balance equations given the following information:
- RQ = 0.5.
- The ratio of humulin to biomass production is 1:5.

Solve for the stoichiometric coefficients to balance the equation.

(b) Determine the inlet molar flow rates of glucose, oxygen, and ammonia in mol/hr. In this bioreactor, temperature is 310 K and pressure is 1 atm.

(c) Determine the limiting reactant, the reaction rate (R), and the fractional conversion of glucose.

(d) Calculate the molar flow rates of all constituents leaving the bioreactor.

(e) The desired output of the bioreactor is 1 kg/day of humulin. Can this output be achieved by increasing the oxygen flow rate? Justify.

3.25 Acetic acid (vinegar) can be produced via the reaction:

$$3\,C_2H_5OH + 2\,Na_2Cr_2O_7 + 8\,H_2SO_4 \longrightarrow$$
$$3\,CH_3COOH + 2\,Cr_2(SO_4)_3 + 2\,Na_2SO_4 + 11\,H_2O$$

A diagram of the process is shown in Figure 3.32. Fresh C_2H_5OH is fed in one inlet stream; fresh $Na_2Cr_2O_7$ and H_2SO_4 are fed in a second inlet stream. A recycle stream meets these two inlet streams to mix before entering the reactor. After leaving the reactor, the stream enters a separator from which three streams exit: one that contains only CH_3COOH (acetic acid), one that contains excess H_2SO_4 and C_2H_5OH, which are recycled, and one that contains all the other waste products and excess reactants (including C_2H_5OH, $Na_2Cr_2O_7$, H_2SO_4, and other compounds, but not CH_3COOH).

The overall system fractional conversion of C_2H_5OH is 90.0%. (*Note*: This conversion relates streams 1 and 7.) The recycle mass flow rate is equal to the fresh inlet mass flow rate of C_2H_5OH. The flow rates of fresh H_2SO_4 and $Na_2Cr_2O_7$ are in excess of the stoichiometric amounts required by the fresh flow rate of C_2H_5OH by 20.0% and 10.0%, respectively. The recycle stream contains 94.0 wt% H_2SO_4, and the rest is C_2H_5OH.

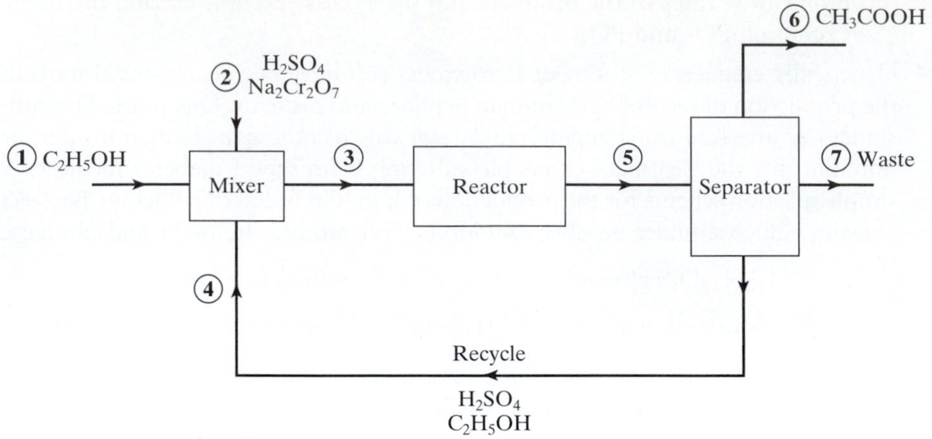

Figure 3.32
Process for acetic acid production.

First, draw a system boundary around and solve for unknowns on the overall system; then, isolate the mixer as a system. Once mass balances around the overall system and mixer are solved, the separator and reactor can be solved. (Adapted from Reklatis GV, *Introduction to Material and Energy Balances*, 1983.)

(a) Label all streams with the compounds they contain.
(b) Determine the reaction rate, R, for the overall system. (*Hint*: Set up a mass accounting equation for the overall system.)
(c) Determine the molar flow rates of each compound in each stream.
(d) Determine the mole fraction of each compound in the exiting waste stream.
(e) Determine the reaction rate, R, and the fractional conversion of C_2H_5OH for the reactor. (*Hint*: Use a mass balance just around the reactor.) Is this fractional conversion higher or lower than for the overall system? Does this offer a reason to utilize recycle streams in chemical and biochemical processing?

3.26 A stream containing compound A_2B mixes and reacts with a stream containing CD in a reactor (Figure 3.33). All products and excess reactants leave in one exit stream. The reactor is operated at steady-state. The primary reaction of A_2B with CD is as follows:

$$A_2B + CD \longrightarrow A_3CD + B_2 \qquad \text{Reaction [1]}$$

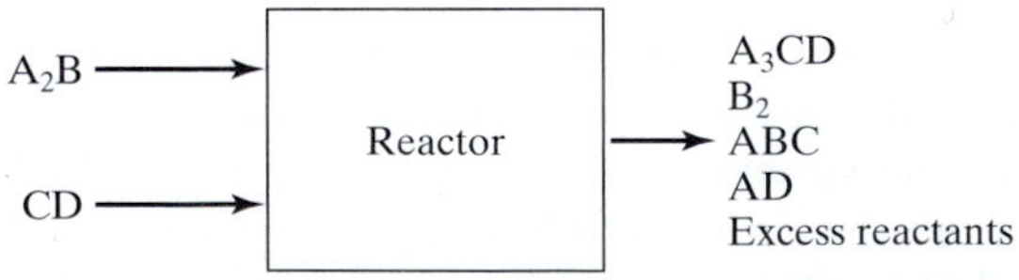

Figure 3.33
A reactor with two stream inlets containing A_2B and CD.

The compound A_3CD is what you are trying to produce. Unfortunately, there is a competing side reaction as follows:

$$A_2B + CD \longrightarrow ABC + AD \qquad \text{Reaction [2]}$$

The inlet mass flow rate of CD is 90.0% of the inlet mass flow rate of A_2B. The mass fraction of B_2 in the outlet stream is 0.2105, and the mass fraction of

AD in the outlet stream is 0.0614. The molecular weights of the compounds are 10 g/mol for A, 20 g/mol for B, 30 g/mol for C, and 15 g/mol for D.

(a) Develop a generic mass accounting equation that can be used to describe an open, steady-state system that contains two or more simultaneous reactions.

(b) Find the reaction rates of the two reactions.

(c) Determine the outlet mass flow rates of each of the compounds (products and excess reactants) in the outlet stream.

3.27 A stream containing compound A_2B mixes and reacts with a stream containing CD in a reactor. All products and excess reactants leave the reactor in one outlet stream. The reactor is operated continuously and at steady-state. The primary reaction of A_2B with CD is as follows:

$$A_2B + CD \longrightarrow A_3CD + B_2 \qquad \text{Reaction [1]}$$

The inlet mass flow rate of CD is 90.0% of the inlet mass flow rate of A_2B. The mass fraction of B_2 in the outlet stream is 0.2105. Assume that Reaction [1] is the only reaction for parts (a) and (b). The molecular weights of the compounds are 10 g/mol for A, 20 g/mol for B, 30 g/mol for C, and 15 g/mol for D.

(a) Find the reaction rate, R.

(b) Determine the outlet molar flow rates of each of the compounds (products and excess reactants) in the outlet stream.

The compound A_3CD is what you are trying to produce. Unfortunately, there is a competing reaction as follows:

$$A_2B + 2CD \rightleftharpoons 2AC + BD_2 \qquad \text{Reaction [2]}$$

This reaction is an equilibrium reaction. The equilibrium constant, K, is defined as follows:

$$K = \frac{x_{AC}^2 x_{BD_2}}{x_{A_2B} x_{CD}^2}$$

where x_s is the mole fraction of species s at equilibrium. You study this equilibrium reaction in a batch reactor. To begin the study, you add 100.0 mol of A_2B and 80.0 mol of CD to the reactor. The equilibrium constant, K, is known to be 0.50.

(c) Determine the moles of A_2B, CD, AC, and BD_2 in the reactor at equilibrium. Assume that Reaction [2] is the only reaction for this part. Recall that the mole fraction of a compound can be written as the number of moles of that compound divided by the total moles in the system.

3.28 The following chemical reaction takes place in a bioreactor:

$$3A + 2B_2 \longrightarrow 2AB + AB_2 \qquad \text{Reaction [1]}$$

The molecular weight of A is 10.0 g/mol and that of B is 15 g/mol.

(a) Based on work from your friend, you assume that the fractional conversion of A is 0.50. Calculate the outlet molar flow rate of A.

(b) Calculate the reaction rate, R_1, for Reaction [1].

(c) You also know that A and B_2 are fed to the reactor in stoichiometric amounts; therefore you assume that the fractional conversion of B_2 is also 0.50. Given this information, determine the inlet and outlet molar flow rates of B_2.

(d) Calculate the outlet flow rates of the two products, AB and AB_2.

(e) Using a detector, you are able to measure the following compounds and their respective mass fractions: $w_{AB} = 0.211$, $w_{AB_2} = 0.155$, and $w_{A_2B} = 0.094$. From this, you suspect that a second reaction, consuming the products from Reaction [1], is occurring as follows:

$$AB + AB_2 \longrightarrow A_2B + B_2 \qquad \text{Reaction [2]}$$

Determine the mass fractions of A and B_2 (w_A and w_{B_2}) in the outlet stream, given both reactions. (*Note:* It is not appropriate to continue to assume that the fractional conversions of A and B_2 are 0.50.)

(f) Determine the outlet molar flow rates for A, B_2, AB, AB_2, and A_2B, given the information from the detector. Calculate the reaction rate, R_2, for Reaction [2].

(g) Calculate the fractional conversion for B_2 that includes both reactions. Is this value higher or lower than the fractional conversion for Reaction [1] alone, which is 0.50? Explain.

3.29 Cell biomass, represented as $C_\alpha H_\beta N_\gamma O_\delta$, is grown in a bioreactor. α, β, γ, and δ are the numbers specifying the molecular formula. The molecular weight of $C_\alpha H_\beta N_\gamma O_\delta$ is 91.34 g/mol. The volume of the bioreactor is 100 L.

There are two inlet streams to the bioreactor (Figure 3.34). The first inlet stream contains glucose and ammonia; the second contains air. There are two outlet streams from the bioreactor. One outlet stream contains $C_\alpha H_\beta N_\gamma O_\delta$, excess $C_6H_{12}O_6$, and H_2O; the other contains the gases O_2, N_2, and CO_2. The outlet gas flow rate is 1.13×10^5 cm^3/min. Assume that the gas streams are dry (i.e., contain no H_2O). Assume that the density of air, O_2, N_2, and CO_2 is 0.0012 g/cm^3.

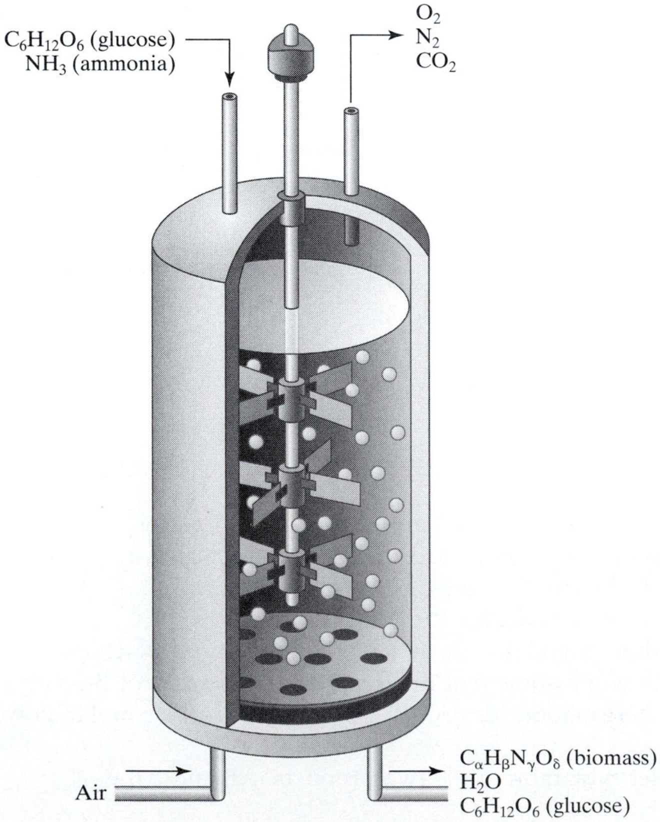

Figure 3.34
Bioreactor for production of cell biomass.

The following biochemical reaction takes place in the bioreactor:

$$C_6H_{12}O_6 + aO_2 + bNH_3 \longrightarrow pC_\alpha H_\beta N_\gamma O_\delta + qH_2O + rCO_2$$

Ammonia is the limiting reactant and is completely consumed in the reaction. The bioreactor is in steady-state operation. The mass and molar flow rates of some of the compounds are summarized in Table 3.11; others must be calculated or deduced.

(a) Determine the outlet molar flow rate of biomass ($C_\alpha H_\beta N_\gamma O_\delta$).
(b) Solve for the stoichiometric coefficients (a, b, p, q, r) that correctly balance the biochemical reaction.
(c) Solve for the values of α, β, γ, and δ.

TABLE 3.11

Setup of Material Flows in Cell Biomass Production

	Inlet rate (mol/min)	Inlet rate (g/min)	Outlet rate (mol/min)	Outlet rate (g/min)	Mol. wt. (g/mol)
O_2	0.7875	25.2	0.221	7.072	32
N_2	3.386	94.81			28
CO_2	–	–	0.768	33.79	44
Glucose ($C_6H_{12}O_6$)	0.80	144	0.416	74.88	180
Ammonia (NH_3)	0.30	5.1			17
Biomass ($C_\alpha H_\beta N_\gamma O_\delta$)	–	–			91.34
H_2O	–	–	1.478	26.60	18

3.30 In your new position at NASA, you are asked to design a life support system. You must devote considerable attention to the supply of air, water, and food as well as the disposal of respiratory and bodily wastes. To begin, you consider the consumption of food by the astronauts (Figure 3.35). Food is modeled as C_2H_2, since the carbon to hydrogen ratio in the average diet is about unity. Food is metabolized (i.e., oxidized) by the astronauts to form CO_2 and H_2O using the O_2 in the cabin atmosphere (which contains 25 vol% O_2 and 75 vol% N_2) in the following reaction:

$$C_2H_2 + O_2 \longrightarrow CO_2 + H_2O$$

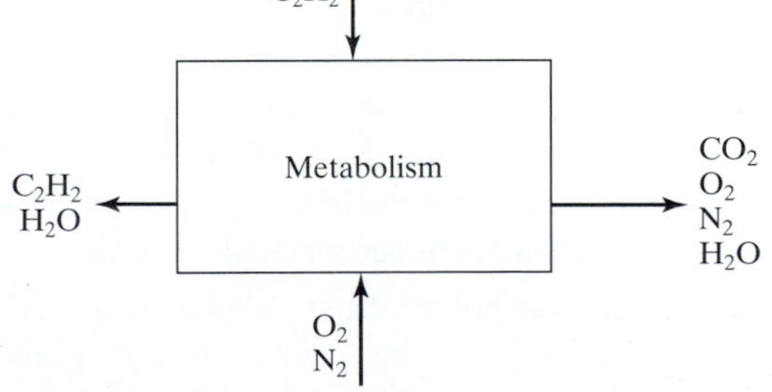

Figure 3.35
Metabolism activity of astronauts in a NASA cabin.

CO_2, O_2, N_2, and a partial amount of the H_2O leave in one exit stream. The mole fraction of water leaving in this stream is 0.050. The rest of the H_2O and unreacted C_2H_2 (couldn't eat all that freeze-dried food ...) leave in a second exit stream.

The fractional conversion of O_2 is 0.80. The inlet molar flow rate of O_2 is 100.0 mol/day. The outlet molar flow rate of C_2H_2 is 0.10 times the outlet molar flow rate of O_2. Although the food is consumed in discrete quantities, assume that the process can be modeled as steady-state. (Adapted from Reklaitis GV, *Introduction to Material and Energy Balances*, 1983.)

(a) Determine the reaction rate, R. Determine the fractional conversion of C_2H_2.

(b) Determine the outlet molar flow rates of CO_2, O_2, N_2, and H_2O in the first exit stream.

(c) Determine the outlet molar flow rates of C_2H_2 and H_2O in the second exit stream.

3.31 Recycling of resources is essential for long-term space missions. For example, water is distilled from many sources, including the astronauts' urine. Some of this water from the distillation system is used to produce oxygen and hydrogen gases by electrolysis. Oxygen is returned to the cabin; in current designs, the hydrogen gas is vented from the spacecraft. Another gas that is currently vented from the spacecraft is carbon dioxide. Research is ongoing to recycle both the hydrogen and carbon dioxide gases.

One design under development is a Methane Formation System (MFS) (Figure 3.36). With an appropriate catalyst, CO_2 and H_2 react to form water and methane gas (CH_4). The water can then be sent to the electrolysis unit for O_2 recovery. The H_2 gas from the electrolysis unit, rather than being vented, is sent to the MFS as the exclusive source of H_2.

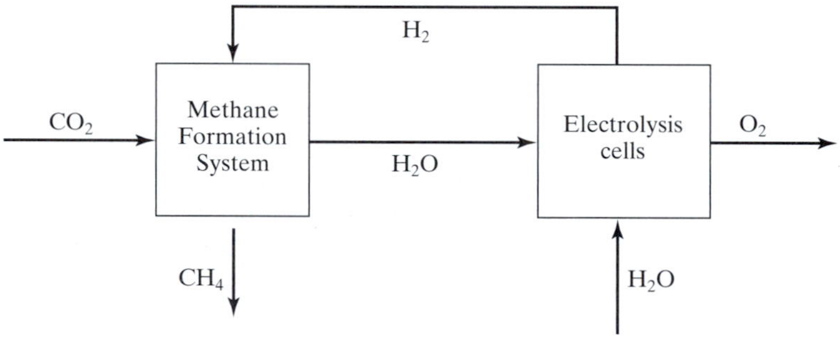

Figure 3.36
Methane formation and electrolysis systems for recycling resources in space.

The goal is to produce 10.0 mol O_2/hr in the electrolysis unit from water sent directly from the MFS. Assume 20.0 mol/hr of CO_2 is supplied to the MFS and that all water in the electrolysis unit is converted to O_2 and H_2. For this problem, assume that any excess reagent can be vented from the MFS.

(a) Write out balanced chemical equations for the electrolysis reaction and the methane formation reaction.

(b) Determine the limiting reactant, reaction rate, and molar flow for all constituents for the MFS unit.

(c) Determine the limiting reactant, reaction rate, and molar flow for all constituents in the electrolysis unit.

3.32 A similar hollow-fiber membrane device as described in Problem 3.13 is operated to achieve the fermentation of glucose to ethanol using yeast cells. Yeast cells are immobilized on the outside walls of the hollow fibers (i.e., the yeast are in the outer, annular space). Once immobilized, the yeast cells cannot reproduce, but they can convert glucose ($C_6H_{12}O_6$) to ethanol (C_2H_6O) according to the equation:

$$C_6H_{12}O_6 \longrightarrow C_2H_6O + CO_2$$

The aqueous feed stream for the yeast cells contains 10.0 wt% glucose; the rest of the stream can be considered water. The aqueous feed for yeast cells enters the annular space in the reactor at 40.0 kg/min. An organic solvent enters the fiber membranes at a mass flow rate of 40.0 kg/min.

The membrane is constructed of a polymer that repels organic solvents. Thus, the solvent cannot penetrate the membrane, and the yeast cells are relatively unaffected by its toxicity. Glucose and water are insoluble in the solvent, and these compounds remain in the annular space (i.e., they do not cross the membrane into the solvent). On the other hand, ethanol is soluble in the solvent; much of the ethanol passes through the membrane into the solvent and leaves dissolved in the solvent stream in the membrane fibers. The by-product CO_2 gas exits from the annular region through an escape valve. The aqueous stream leaving the annular space contains 0.20 wt% glucose and 0.50 wt% ethanol. (Adapted from Doran PM, *Bioprocessing Engineering Principles*, 1999.)

(a) What is the fractional conversion of glucose?
(b) What is the reaction rate, R, of the system?
(c) Determine the mass flow rate of ethanol across the membrane.
(d) Determine the outlet mass flow rates of glucose in the aqueous stream and of ethanol in the aqueous and solvent streams.
(e) Determine the outlet mass and volumetric flow rates of CO_2.

3.33 You are starting up a bioreactor containing mammalian cells to do the following chemical conversion:

$$A_2B + BC \longrightarrow AB + C$$

In addition to the cells, the bioreactor contains many charcoal pellets. Based on previous research, you know that the mammalian cells have better long-term stability when attached to the charcoal pellets as compared to just being in suspension. (Cells that exhibit this pattern are known as anchorage-dependent cells.)

Water containing A_2B and BC flows into the bioreactor at a flow rate of 0.10 L/min. The concentration of A_2B in the input stream is 70.0 g/L. The concentration of BC in the input stream is 140 g/L. A_2B, BC, AB and C are all fully dissolved in water and do not contribute substantially to the density of the solution. The molecular weights of A, B, and C are 2.0 g/mol, 3.0 g/mol, and 4.0 g/mol, respectively.

The reactor runs continuously. The output stream empties into a very large container (Figure 3.37). Since there is no on-line detection for the outlet stream, samples are taken from the container, and the concentrations of the various compounds are determined. Assume that the large container is well mixed.

(a) You run the bioreactor for 4.0 days. During that period, the entire outlet is captured in the large container; none is emptied. You sample the container after 4.0 days and determine that the concentration of A_2B is 3.5 g/L. Based on this information, determine the outlet flow rate of A_2B.
(b) Determine the reaction rate, R, of the system. What are the fractional conversions of A_2B and BC?
(c) You manage to borrow an instrument to do on-line detection right at the end of your 4.0-day experiment. You sample the outlet stream (not the container) and determine that the outlet concentration of AB is 90.0 g/L. Is this measurement consistent with your results from parts (a) and (b)? Why or why not?

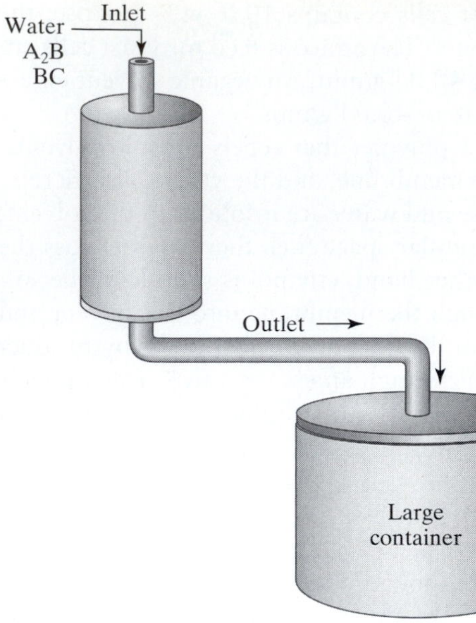

Figure 3.37
Bioreactor containing mammalian cells and charcoal pellets. The outlet empties into a large container.

You decide to rerun the whole experiment. You throw out all the charcoal pellets and mammalian cells. You reload the bioreactor with fresh charcoal and cells. Before you begin this new experiment, you also empty out your large container that was capturing the outlet flow. During this run, you decide to sample from the large container every 12 hours. The concentrations of A_2B are recorded in Table 3.12. Note that the large container is not emptied during the 4.0-day run. Instead, all of the outlet liquid is collected and mixed in the large container.

TABLE 3.12

Samples of A_2B Taken from Large Container

Time (hr)	Conc. of A_2B (g/L)
12	0.0
24 (1.0 day)	0.0
36	0.0
48 (2.0 days)	0.0
60	1.40
72 (3.0 days)	2.33
84	3.00
96 (4.0 days)	3.50

(d) Based on the data in Table 3.12, write an equation (or equations) describing the outlet mass flow rate of A_2B.
(e) What kind of physical phenomena might account for the form of the equation derived in part (d)?
(f) Is the equation from part (d) consistent with your measurement of the outlet concentration of AB of 90.0 g/L at 4.0 days (part (c))? Why or why not?
(g) If you had an infinitely large container and the reacting system proceeded forever, at what concentration would A_2B plateau in the large container? Calculate the time at which the concentration will be at 99% of the plateau value.

3.34 Snorzin is a hypothetical protein produced by the body at a rate dependent on the time of day. Production of the protein (in units of mass/time) follows the equation:

$$\text{Production} = k\{1 + \sin[A(t + 5\text{ hr})]\}$$

where $k = 10$ g/hr, $A = \pi/(12\text{ hr})$, and t = time of day in military hours (00:01 to 24:00).

(a) At what time is Snorzin production at a maximum? When is it at a minimum? Calculate the production rate at these two extremes.

(b) How much Snorzin accumulates in the body between 7 a.m. and 11 p.m.?

3.35 Transfecting *Escherichia coli* with a desired gene has become a common practice in molecular biology. Because *E. coli*'s population growth is rapid, a desired gene or protein can be synthesized more quickly than by many other methods.

(a) Assume that the doubling time of *E. coli* is 20 min. Write an equation to model the generation of *E. coli*, assuming no nutrient or cell density restrictions.

(b) *E. coli* is grown in a 10-L bioreactor. Nutrients enter at a rate of 1.0 L/min. An outlet stream containing waste and *E. coli* leaves the bioreactor at a rate of 1.0 L/min. Assuming the volume of material in the bioreactor stays constant, write an equation describing the concentration of *E. coli* in the outlet stream as a function of time. (*Hint:* The outlet concentration of *E. coli* is equal to that in the reactor.)

(c) Assume that the reactor is seeded with 1×10^2 cells/mL. How long can the bioreactor be run before the cell concentration exceeds 1×10^8 cells/mL? For this part, assume that no cells leave the bioreactor in the outlet stream.

3.36 The cell membrane is covered with protein complexes called Na^+-K^+ pumps. Each pump moves 3 Na^+ ions from the intracellular space to the extracellular milieu for every 2 K^+ ions it moves into the cell. During normal function, the pumps work constantly. Under normal conditions, there is a gradient of Na^+ and K^+ ions between the cell and the extracellular space; the concentrations of these ions in these spaces can be found in Table 3.13. Since the ions are pumped against their respective gradients, energy in the form of ATP is required.

TABLE 3.13

Normal Extracellular and Intracellular Ion Concentrations

	Extracellular concentration (mM)	Intracellular concentration (mM)
Na^+	145	15
K^+	5.0	140

An experiment is conducted using ouabain, which blocks Na^+-K^+ pumps in cells. During this time the gradient collapses; the intracellular Na^+ concentration becomes 80 mM and the intracellular K^+ concentration 72.5 mM. After the experiment, ouabain is removed from the cells by rinsing with PBS (phosphate buffered saline). The pumps begin to operate again to reestablish the gradients. The recovery phase of the experiment lasts 4.0 hr, during which time the cells work to regain the previous ion balance. Model the ion pumps on a cell membrane during the recovery phase.

Assume that the volume of the cell is 65.4 μm^3. Assume there are 1.0×10^5 ion pumps per cell. Assume that the pumping rate is constant (i.e., it is not dependent on the gradient of the ions). Assume that there is no diffusion of Na^+ or K^+ ions across the cell membrane and that no other ion pumps or channels are active.

(a) Calculate the Na^+ pumping rate for one cell (molecules/pump · s) needed to reestablish the intracellular Na^+ concentration in 4.0 hr without consideration of the K^+ pumping rate.

(b) Calculate the K^+ pumping rate for one cell (molecules/pump · s) needed to reestablish the intracellular K^+ concentration in 4.0 hr without consideration of the Na^+ pumping rate.

(c) Will the cell be able to reestablish the ion balance for both Na^+ and K^+ to the equilibrium intracellular concentrations listed in Table 3.13 with just the described Na^+-K^+ pump? Why or why not?

(d) In a different experiment, you calculate a Na^+ pumping rate of 1.6 molecules/pump · s What intracellular K^+ concentration (mM) can be established in 3.0 hr? Assume the collapsed intracellular conditions above as the starting point for the recovery phase.

3.37 A newly manufactured biomaterial needs to be dried before it can be sterilized for transfer to a patient (Figure 3.38). Immediately after processing, the biomaterial contains 30.0 wt% water. To begin sterilization, the biomaterial needs to have a maximum of 20.0 wt% water. The biomaterial is placed on a solid silica gel, which absorbs the water at the following rate:

$$\text{water absorption} = be^{-at}$$

where a is 1 1/min, b is 0.13 lb_m/min, and t is time. 3.2 lb_m of silica gel has the capacity to absorb 1.0 lb_m of water. Assume a basis of 1 lb_m of biomaterial. (From Glover C, Lunsford KM, and Fleming JA, *Conservation Principles and the Structure of Engineering*, 1994.)

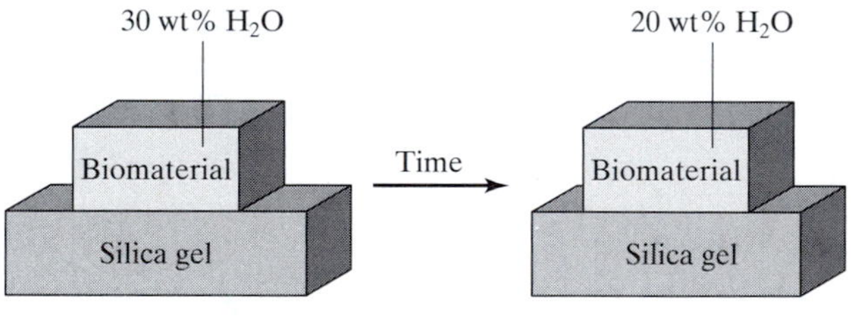

Figure 3.38
Absorption of water in biomaterials by silica gel over time.

(a) Determine the mass (lb_m) of silica gel required per mass (lb_m) of wet biomaterial (containing 30.0 wt% water) needed to absorb the water lost from the biomaterial in the transition from 30.0 wt% water to 20.0 wt% water.

(b) Find the time required for the silica gel to absorb the water from the biomaterial in the transition from 30.0 wt% water to 20.0 wt% water if the silica gel absorbs the water at the above rate.

3.38 A polymer slab dissolves when in contact with water. You are assigned to develop a model to predict the length of time that it will take the polymer to dissolve when placed in a beaker of water. Assume that you begin with 1.00×10^2 g of polymer. Since there is no chemical stoichiometry in this problem, you do not have to use molar rates in order to solve the problem; you can use mass rates.

(a) You start simple with your modeling predictions. You assume that the polymer does not degrade (i.e., it undergoes no chemical transformations to another polymer or smaller monomer units). Rather, you assume that the dissolution rate of the polymer is proportional to a constant, k, and the surface area of the polymer, A. This can be modeled as follows:

$$\text{dissolution rate} = kA$$

where:

$$k = 2\frac{\text{g}}{\text{hr}\cdot\text{cm}^2}$$

$$A = 10\ \text{cm}^2$$

Using this predictive model, how long will it take for the 1.00×10^2 g polymer slab to dissolve completely in water?

(b) Realizing that your first assumption is too simple, you try to model the dissolution rate as a function of the square root of time as follows:

$$\text{dissolution rate} = kA_0t^{1/2}$$

where:

$$k = 2.0\frac{\text{g}}{\text{hr}\cdot\text{cm}^2}$$

$$A_0 = 10.0\frac{\text{cm}^2}{\text{hr}^{1/2}}$$

Using this predictive model, how long will it take for the 1.00×10^2 g polymer slab to dissolve completely in water?

(c) After talking with a colleague, you realize that in addition to dissolving, the polymer also degrades to a monomer. Expecting that taking account of this might significantly improve your model, you investigate further. You conduct a series of experiments to determine the polymer's degradation rate. During the first hour the rate increases linearly to a value of 10.0 g/hr, then plateaus and is a constant at 10.0 g/hr, as shown in Figure 3.39.

Using the model for the dissolution rate from part (a), you combine the dissolution and degradation terms into one model. For this experiment, you begin the trial with a 1.00×10^2 g polymer slab, yet stop the experiment when the mass of the polymer is 20.0 g. Using this model, how long will it take for the polymer to reduce in mass from 1.00×10^2 g to 20.0 g in water?

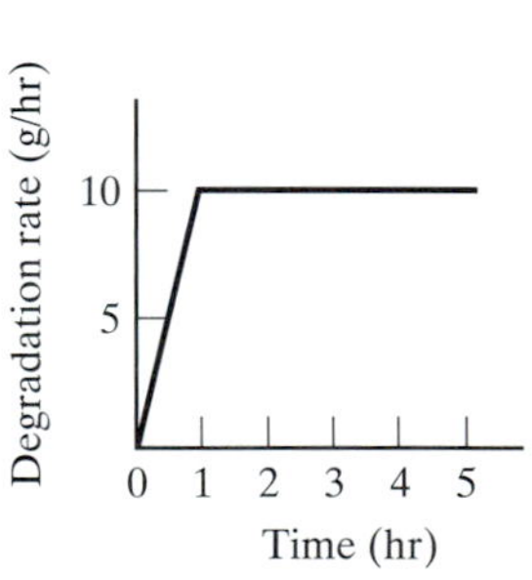

Figure 3.39
Degradation of a polymer into monomers in water.

3.39 Biodegradable synthetic materials are now being explored for use as carriers for drug delivery. Poly(lactic-co-glycolic) acid (PLGA) is one such material currently being explored for this purpose, as it is already approved by the FDA for use in the human body. Microspheres loaded with drug can be fabricated. By varying the properties of the polymer comprising the microsphere, the release profile of the drug can be altered systematically.

You run an experiment to determine the effects of microsphere diameter on the release of the model drug, FITC-BSA (fluorescently labeled bovine serum albumin). The release curve is shown in Figure 3.40. After fitting a curve to your data, you find that the release of FITC-BSA can be modeled as follows:

$$\text{release} = \left(\frac{1\ \text{mg}}{20\ \text{day}}\right)\exp\left(\frac{-t}{20\ \text{day}}\right)$$

Initially, the mass of the model drug FITC-BSA in the microsphere is 1 mg. Determine the amount of drug released after 30 days.

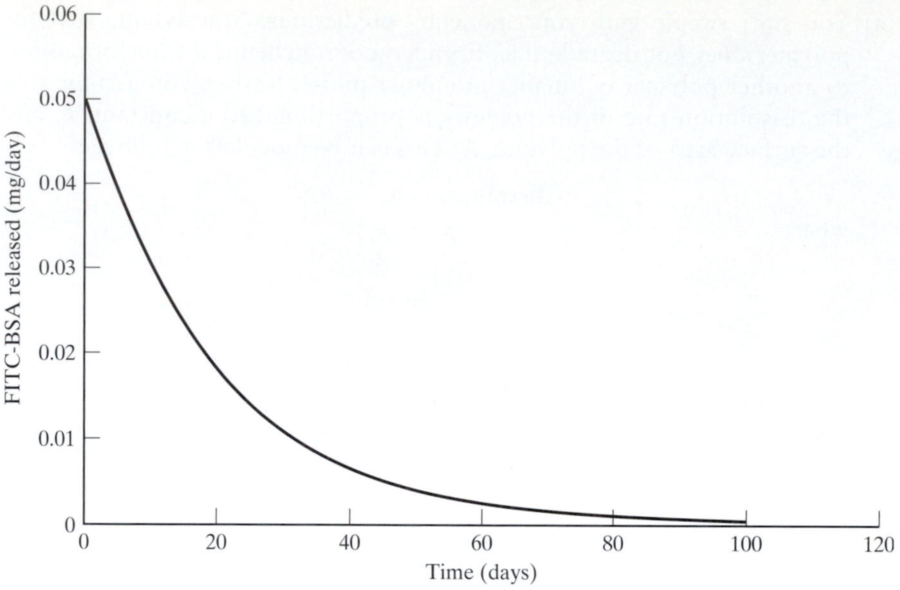

Figure 3.40
Drug release of FITC-BSA over time.

3.40 The cell membrane is covered with a diverse array of receptors. Most receptors are transmembrane proteins. Receptors serve to facilitate communication between the extracellular matrix and the intracellular space. Soluble ligands in the extracellular matrix bind with high specificity to particular receptors. When this binding occurs, an intracellular signal can be promulgated and/or the receptors can be internalized and processed in the cell. Receptors are found on the cell surface, in the endosome, and in transit in the intracellular space. Receptors move around within the cell and on the cell membrane and are constantly in flux.

Once internalized, a receptor moves toward an endosome. Endosomes are compartments in the cell where receptors, proteins, ligands and other small molecules are sorted and targeted for their future destination in the cell. In the endosome, some fraction of the receptors (f_R) is targeted to be degraded, while the remaining fraction is recycled to the membrane surface. Assume that the rate of movement of receptors in the cell is not dependent on the density of receptors on the cell or in the endosome.

Through protein synthesis, new receptors are generated in the cell. These receptors are transported from the intracellular space to the cell membrane. A simplified model of receptor trafficking (i.e., movement) is shown in Figure 3.41.

Nomenclature:

R_S = total number of receptors on the cell surface [#]
R_E = total number of receptors in the endosome [#]
V_s = rate of receptor synthesis [#/min]
k_{rec} = receptor recycle rate constant [1/min]
k_{deg} = receptor degradation rate constant [1/min]

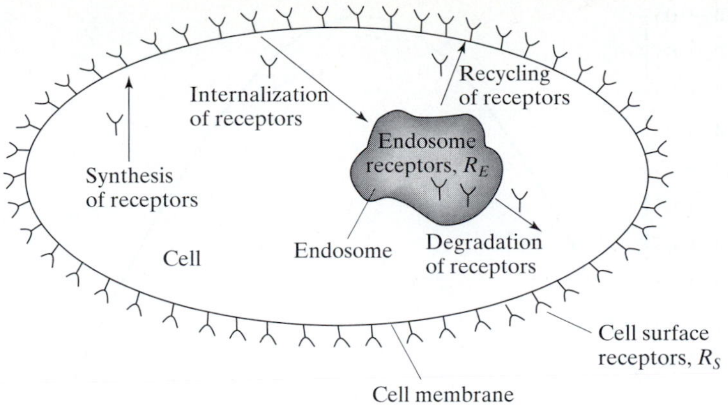

Figure 3.41
A simplified model of receptor trafficking.

f_R = fraction of receptors that are targeted to be degraded [−]

k_e = receptor internalization (endocytosis) rate constant [1/min]

t = time [min]

(a) Draw a system with its system boundary designed to count the receptors on the surface (R_S). Determine whether the system is open or closed, reacting or nonreacting, steady-state or dynamic. Write the appropriate differential equation that describes the rate of change of receptors on the surface (R_S). Internalization of receptors, synthesis of receptors, and recycling of receptors should be included in this balance. (*Hint*: The rate of internalization of cell surface receptors can be written $k_e R_S$. The units of rate are [#/time]).

(b) Draw a system with its system boundary designed to count the receptors in the endosomal compartment (R_E). Determine whether the system is open or closed, reacting or nonreacting, steady-state or dynamic. Write the appropriate differential equation that describes the rate of change of receptors in the endosomal compartment (R_E).

(c) Assume that no accumulation of receptors occurs on the membrane or in the endosome. Solve for the steady-state value of R_S in terms of k_{deg}, k_e, V_s, k_{rec} and f_R.

(d) Using a graphical analysis, show how R_S varies as the magnitudes of the variables k_{deg}, k_e, and f_R vary over the given ranges, assuming the fixed values in Table 3.14 for the other variables. You should have three graphs: R_S vs. k_{deg} (k_e, f_R, V_s, k_{rec} fixed), R_S vs. k_e (k_{deg}, f_R, V_s, k_{rec} fixed), and R_S vs. f_R (k_{deg}, k_e, V_s, k_{rec} fixed). Do these graphs make sense to you? Why or why not?

TABLE 3.14

Values to Model Receptor Trafficking

Variable [unit]	Fixed value	Range
V_s [#/min]	130	
k_{deg} [1/min]	0.010	0.0020–0.050
k_e [1/min]	0.030	0.030–3.0
k_{rec} [1/min]	0.058	
f_R [−]	0.010	0.010–1.0

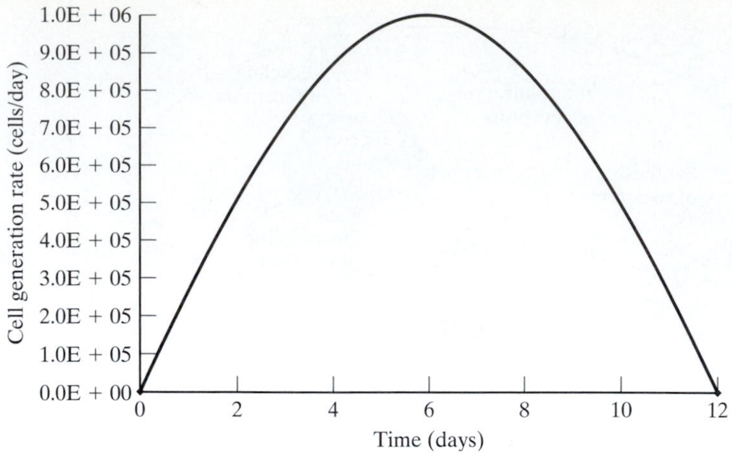

Figure 3.42
Cell reproduction rate.

3.41 A hollow-fiber membrane device like the one described in Problem 3.13 is operated to achieve the fermentation of glucose to ethanol using yeast cells. Yeast cells are immobilized on the outside walls of the hollow fibers (i.e., the yeast cells are in the outer, annular space). To seed the unit, 1.0×10^5 cells are added to the reactor. The cells attach to the fibers. Initially after seeding, the cell generation rate increases. However, as the cells begin to cover the fibers, the generation rate slows. The change in cell reproduction rate occurs as shown in Figure 3.42. The generation rate, $\dot{\psi}_{gen}$, is modeled as follows:

$$\dot{\psi}_{gen} = 1.0 \times 10^6 \frac{\text{cells}}{\text{day}} \sin\left[\frac{t\pi}{12 \text{ days}}\right]$$

where t is the time in units of days. Cells die at a constant rate of 1.0×10^4 cells/day. Determine the number of cells in the reactor at 12 days.

3.42 In gene therapy research, it is common to use reporter genes to quantify the ability of a cell population to produce a foreign protein. These reporter genes typically encode for either a protein that fluoresces or luminesces or an enzyme that will convert a substrate into a colored, fluorescent, or luminescent product. One such gene encodes for the enzyme β-galactosidase that turns the substrate ONPG (*o*-nitrophenyl-β-D-galactopyranoside) into the yellow product *o*-nitrophenol. Measuring the light absorbance of the cell lysate at 420 nm can quantitate the amount of this product.

The consumption of the substrate, ONPG, can be modeled using the Michaelis-Menton equation:

$$\frac{-d[S]}{dt} = \frac{[E]_0 k_2 [S]}{K_m + [S]}$$

(a) Develop an equation for reaction time in terms of $[E]_0$, k_2, $[S]_0$, $[S]$, and K_m. The units and definitions of each variable are shown in Table 3.15.
(b) Given that $[E]_0 = 3.0$ μg/mL and $[S]_0 = 2$ mM, find the time it takes for the substrate concentration to drop to half its initial value, given that $K_m = 0.161$ mM and $k_2 = 0.006$ μmol/(μg enzyme·min).
(c) If $[E]_0$ is decreased by one order of magnitude (i.e., $[E]_0 = 0.3$ μg/mL), find $[S]$ after 30 min. (*Hint*: You cannot explicitly solve for $[S]$ in terms of other variables.)

TABLE 3.15

Variables Used in Michaelis-Menton Equation

Variable	Units	Definition
$[S]$	mM	Concentration of substrate (ONPG)
$[S]_0$	mM	Initial concentration of substrate
$[E]_0$	μg/mL	Initial enzyme concentration (β-galactosidase)
k_2	μmol/(μg enzyme · min)	Reaction rate constant
K_m	mM	Equilibrium constant
t	min	Time

3.43 You are working to design a skin patch that delivers a drug to the body. Examples of patches on the market today include the Nicoderm patch (for helping individuals to quit smoking) and patches that contain hormones, including estrogen and testosterone. The skin patch is thin and flat and has a surface area of 1 in^2 (Figure 3.43a). It is placed on the arm with one side up against the skin and the other side exposed to the air. Your intention is to design a patch that delivers a pain medication that might be used after surgery or traumatic injury. To minimize the risk of addiction, the amount of drug delivered decreases as a function of time. You conduct a number of tests, as described below, to help design and characterize the patch. In all tests below, ignore drug and water losses from the edges (sides) of the patch.

Test A. As noted above, the amount of drug delivered decreases as a function of time. Your research has shown that the rate the drug leaves the patch, y, is described by the following formula:

$$y = -1\frac{\mu\text{g}}{\text{day}^2}t + 40\frac{\mu\text{g}}{\text{day}}$$

where t is the time. Assume that the drug is delivered from the patch to the skin and that no drug is lost to the air. Assuming that the patch is loaded with 800 μg of drug, how long does it take before the drug is exhausted from the patch?

Test B. To try to minimize changes in its structure and size, the patch is designed so that water from the skin enters it to replace the drug it loses. For this test, assume that for every 1 μg of drug lost from the patch, 1 μg of water is

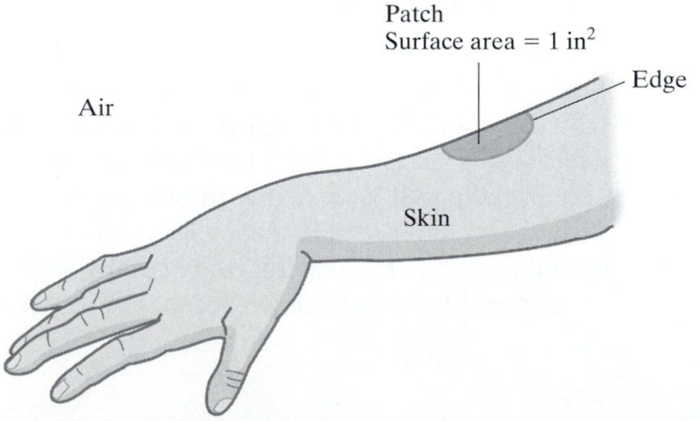

Figure 3.43a
Skin patch with a surface area of 1 in^2.

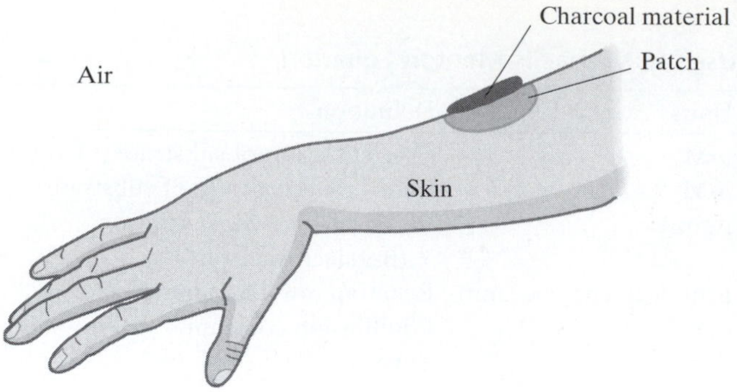

Figure 3.43b
Skin patch with charcoal material to capture water.

absorbed into the patch from the body (i.e., an equal mass exchange of drug and water). Assuming that the patch is initially loaded with 800 μg of drug and 600 μg water, what is the mass of drug in the patch after 20 days? What is the mass of water in the patch after 20 days?

Test C. Your colleague has begun tests on the mobility of water through the patch. She discovers that water is readily absorbed from the skin into the patch and is also readily evaporated from the patch to the air. She sets up a test similar to that which would be found in a clinical application, in which one side of the patch is attached to skin and the other side is exposed to air. On top of the patch, she attaches a novel charcoal material that captures the water that evaporates from the patch (Figure 3.43b). (*Note*: This charcoal material does not absorb water from the surrounding air or accelerate the loss of water from the patch; it only captures the water that has evaporated from the patch.) She samples the charcoal material at 5 days and measures 500 μg of water. She assumes a constant rate of water loss from the patch. Develop an equation describing the rate of water loss from the patch.

Test D. Being the sharp biomedical engineer that you are, you question her assumption of a constant rate of water loss from the patch. You ask her to repeat her experiment and sample the charcoal material after 10 days. Your colleague repeats the test described above, samples the charcoal material at 10 days, and measures 950 μg of water. (*Note*: She does not sample or remove any water at 5 days during this test.) With this second data point, you know that the rate of water loss from the patch is not constant. Develop a linear equation describing the rate of water loss from the patch, using the two data points collected in Test C and Test D.

Test E. Your colleague conducts tests on the patch and determines that water is absorbed from the skin into the patch at a constant rate of 100 μg/day, which is more than enough to replace the drug lost from the patch. The capacity of the patch is 3000 μg of total mass (water and drug combined). As in Test B, the patch is loaded with 800 μg drug and 600 μg water. To model the water loss from the patch, use the equation you determined in Test D. At what time does the patch reach its capacity of 3000 μg? Will the drug be completely delivered before the patch reaches its capacity?

Conservation of Energy

CHAPTER 4

4.1 Instructional Objectives and Motivation

After completing Chapter 4, you should be able to perform the following:

- List and explain the types or forms of energy.
- Explain how heat and work relate to energy.
- Write the algebraic, differential, and integral forms of the conservation of energy.
- Appropriately apply the first law of thermodynamics.
- Describe the concepts of enthalpy and heat capacity.
- Calculate changes in enthalpy due to mixing and from temperature, pressure, and phase changes.
- Apply the conservation of total energy equation to open, nonreacting systems.
- Calculate the heat of reaction using heats of formation and heats of combustion data.
- Apply the conservation of total energy equation to open, reacting systems.
- Apply the conservation of total energy equation to dynamic systems.

4.1.1 Bioenergy

Energy accounting and conservation equations are used widely by engineers to design systems to harness and conserve energy. To track or monitor the energy of a particular system or process, you often will need to apply energy balance equations. To fully understand the human body, biomedical equipment, bioprocessing applications such as biofuels, and many other bioengineering systems, you need to be facile in manipulating the conservation of energy equation. Energy accounting and conservation equations are prevalent in systems involving chemical reactions, as well as pressure and temperature changes. In this chapter, the conservation of energy is applied to a wide range of example and homework problems.

In this introductory section we highlight global energy needs, with a specific emphasis on renewable energy resources and biofuels. Harnessing and conserving energy is a critical issue for the human population, and many diverse solutions are being proposed. In all cases, the conservation of energy is fundamental to developing and designing strategies.

Figure 4.1
Amber waves of bioenergy. (*Source:* http://news.bbc.co.uk/2/hi/science/nature/2523241.stm.)

Bioengineers have a unique role to play in developing bioenergy, since they bridge the engineering and biological worlds. The complex challenge below serves to motivate our discussion of the energy conservation equation.

Without a continuous supply of energy, life as we know it would cease. Even as you read this paragraph, you are breathing, your optical nerves are firing, your blood is pulsing. Each of these processes and many others in the human body require energy, which is obtained from the food you eat. An intricate relationship between solar energy, plant photosynthesis, and aerobic metabolism enables these complex physical processes to occur.

The sun is our major source of energy, emitting about 4.2×10^{22} watts. A very small fraction of that energy, roughly 10^{17} watts, reaches Earth's surface. The energy from solar radiation is available to many biological systems on Earth, supporting the photosynthetic processes of terrestrial plants and oceanic algae and microorganisms. Each year, photosynthesizing organisms fix approximately 10^{11} tons of atmospheric carbon by a photosynthetic reaction that combines atmospheric carbon dioxide, water, and sunlight to form organic compounds and oxygen.[1] The organic compounds, primarily glucose, are then incorporated into the structure of the organism or its progeny. Overall, about 1.1×10^{14} watts of solar energy are converted annually by photosynthesis into organic mass.

Most creatures do not directly capture the sunlight as energy; instead, they obtain their energy by ingesting photosynthetic organisms or organisms that eat photosynthetic organisms. For example, humans eat plants or other animals that eat plants or both to recover the stored energy from photosynthesis. In humans, the metabolism of carbohydrates, fats, and proteins from food generates energy that is stored in a chemical compound called adenosine triphosphate (ATP), a molecule that drives most cellular processes such as nerve conduction, muscle contraction, and active transport.

If we assume an individual basal metabolic rate of 70 kcal/hour and a world population of 6.3 billion, the total energy requirement of all people on Earth is about 5.1×10^{11} watts. This figure is less than 0.5% of the energy provided by photosynthesizing plants. Thus, stored energy from plants can meet the metabolic needs of humans. However, the human population consumes about 1.4×10^{13} watts in daily activities like cooking, transportation, and lighting and heating homes and buildings. Photosynthetic organisms do not supply most of humans' nonmetabolic energy requirements. Instead, humans have developed methods of harnessing energy from other sources.

Formed millions of years ago from the decomposed remains of dead plants and animals, fossil fuels are a major nonrenewable energy source for industrialized countries. We transport these materials from the subsurface for our energy needs. The production and refining of fossil fuels provides about 85% of Earth's energy supply or about 1.2×10^{13} watts.

While fossil fuels are the most common energy source in industrialized nations, renewable energy sources, such as wind, solar, hydropower, ocean, and geothermal, are becoming more popular. Wind turbines can harness wind energy to generate electricity or pump water. Solar energy devices use energy that reaches Earth from the sun to supply buildings with heat, light, hot water, electricity, and even cooling. Hydropower installations capture energy created by flowing water and convert it into electricity, currently generating nearly 10% of the electricity used in the United States. Ocean energy can be drawn from differences in height between high and low tides and differences in temperature between surface water and deep ocean water. Energy in renewable forms is abundant, but designing and refining new methods of converting it to meet our needs is a major engineering challenge.

An exciting new field in renewable energy sources is bioenergy, which exploits renewable biomass (i.e., any plant-derived organic matter), which can be replenished by cultivating fast-growing trees and grasses. Many biomass materials, such as plants, agricultural by-products, and organic components of municipal and industrial wastes, are now being used to produce biofuels and power. Biofuels, such as ethanol and biodiesel, serve transportation needs. Ethanol is an alcohol made by fermenting any biomass high in carbohydrates, such as corn. Biodiesel is an ester made from vegetable oils, animal fats, algae, or recycled cooking grease. Biomass can be burned to produce steam for electricity production, or it can be chemically converted into a fuel oil, which can then be burned to generate electricity. Gasification systems use heat to convert biomass into a gas composed of hydrogen, carbon monoxide, and methane for use in generating electricity. Biomass decay in landfills produces methane gas, which can also be burned to produce steam for electricity generation.

Although technologies for using biological energy resources are still being developed, the eventual benefits should be significant. In addition to the technical hurdles involved in designing efficient systems, the economical, social, and environmental consequences of bioenergy must be taken into account. The following issues associated with bioenergy systems face bioengineers today:

- *Source evaluation*: Engineers must use analytical methods to evaluate and compare the relative availability of various renewable energy sources, the economical and political ease with which they can be exploited for practical uses, their efficiencies, and their impacts on the environment. Material and energy balances help quantify resource depletion, emissions, and energy consumption in all steps of the process.
- *Design*: Processes and equipment must be designed, built, and operated.
- *Sustainable development*: Biomass and bioenergy technologies can move the United States and world economy to a more sustainable basis by reducing our dependence on nonrenewable fossil fuels. Government policies and business practices must reflect long-term commitment to sustainable development.
- *Land use*: Land must support and preserve agricultural, forestry, biomass production, biota, and human populations. Biomass production raises concerns regarding soil erosion control, nutrient retention, and carbon sequestration. Changing the land use to support increased biomass production may destroy some species' native habitat and cause changes in biodiversity.
- *Water conservation*: Bioenergy technologies may impact watershed stability, ground-water quality, surface-water runoff and quality, and local water supplies.

- *Safety*: All aspects of alternative energy production must be engineered to ensure the highest level of safety. All steps of each process must undergo stringent design and testing. Codes and standards for equipment and procedures must be outlined and strictly followed.

Multidisciplinary teams around the world tackle the problem of the best way to generate and harness bioenergy. Bioengineers use energy balances to help them model and evaluate the feasibility of various proposals for renewable energy. In Examples 4.10, 4.11, and 4.15 we examine how energy balances are used to evaluate photosynthesis and hydroelectric power. Example and homework problems in this chapter demonstrate many other exciting applications of the conservation of energy equation.

This chapter opens with an overview of basic energy concepts and then discusses how system definitions can be applied to solve systems involving energy. We discuss how to calculate changes in enthalpy with changes in temperature, pressure, and phase and from reactions. We then use the governing equations to solve open, reacting, and dynamic systems.

4.2 Basic Energy Concepts

Accounting for energy is important in a number of bioengineering applications, such as modeling energy gains and losses from the body, analyzing biochemical reactions, and designing and operating bioreactors. Conceptually, the conservation of energy for a system is very similar to the conservations of mass and momentum. We analyze the transfer of energy of various forms across the system boundary and the accumulation of energy within the system.

We begin by reviewing some definitions. An *open system* allows the exchange of an extensive property through bulk material transfer with its surroundings. In an open system, energy is exchanged through movement of material; an example is the net loss of energy from the body as air is exhaled from the lungs. A *closed system* allows for the transfer of an extensive property through means other than bulk material transfer. Heat and work are forms of energy that transit across the system boundary in the absence of any material. Removing heat by placing a cold pack on a person's forehead is an example of energy transfer in a closed system. Finally, an *isolated system* is enclosed by a boundary that does not allow the transfer of any extensive property by any means. Some types of calorimeters mimic isolated systems. The concepts of open, closed, and isolated systems are defined in greater depth in Chapter 2.

4.2.1 Energy Possessed by Mass

All mass possesses energy. The dimension of energy is $[L^2Mt^{-2}]$. Common units of energy are joule, cal, Btu (British thermal unit), $ft \cdot lb_f$, and $kW \cdot hr$. The dimension of the rate of energy is $[L^2Mt^{-3}]$. Common units of rate of energy are watt, cal/s, and Btu/s. Remember that energy is a scalar (not vector) quantity. The total energy of a system is the sum of three different forms of energy: potential, kinetic, and internal.

A body possesses **potential energy** as a result of its position in a potential field. The gravitational field and the electromagnetic field are the two most common potential fields in bioengineering applications. Both of these fields are conservative. One attribute of a **conservative** field is that the energy required for an object to move across it is independent of the path the object takes; in other words, potential energy and the rate of potential energy are state functions (see Section 4.5.1).

Potential energy can also be thought of as the stored energy of a body relative to a reference state.

The **gravitational potential energy** (E_P) of an object of mass m must be defined relative to a reference plane. The absolute potential energy is rarely needed or calculated; more commonly the change in potential energy is used, and it is incorporated into the conservation of energy equation. To consider the change in potential energy of a mass between two different positions or heights, the following equation is used:

$$E_{P,2} - E_{P,1} = mg(h_2 - h_1) \qquad [4.2\text{-}1]$$

where g is the gravitational acceleration constant, h is the height relative to a reference plane, and 1 and 2 denote the two different positions in space.

Gravitational potential energy can also move into and out of a system at a mass flow rate described by $\dot{m}$. The change in the **rate of gravitational potential energy** ($\dot{E}_P$) can be calculated when material crosses the system boundary as:

$$\dot{E}_{P,2} - \dot{E}_{P,1} = \dot{m}g(h_2 - h_1) \qquad [4.2\text{-}2]$$

The change in **electromagnetic potential energy** (E_E) is given as:

$$E_{E,2} - E_{E,1} = q(v_2 - v_1) \qquad [4.2\text{-}3]$$

where q is the net charge, v is the electric potential energy per unit charge, and 1 and 2 denote two different positions in space. The difference in potential energy per unit charge is commonly called **voltage** and has the dimension of energy per charge [$L^2Mt^{-3}I^{-1}$]. Thus, the change in electromagnetic potential energy is calculated by multiplying the net charge of the species by the voltage difference through which the charged species moves. By definition, if a positive change in potential energy is accomplished when moving a test charge from position 1 to position 2, then the electric potential is higher at position 2 than at position 1 and the voltage ($v_2 - v_1$) is positive.

The change in the **rate of electromagnetic potential energy** ($\dot{E}_E$) is defined as:

$$\dot{E}_{E,2} - \dot{E}_{E,1} = i(v_2 - v_1) \qquad [4.2\text{-}4]$$

where i is the charge flow rate or current. The potential energy change and the rate of potential energy change of an object moving between two arbitrary positions in an electric field are independent of the path of the object (see Section 4.5.1).

A body or mass possesses **kinetic energy** as a result of its translational or rotational motion. Translational motion is the movement of the center of mass of a rigid body as a whole or the movement of a fluid relative to a reference frame (usually the Earth's surface). Rotational motion is the rotational movement of a body relative to an axis or the center of mass of an object. Rotational motion is primarily applicable when dealing with rigid bodies and will not be discussed further in this chapter. For more information, other textbooks have more in-depth discussions (e.g., Glover C, Lunsford KM, and Fleming JA, *Conservation Principles and the Structure of Engineering*, 1996).

The kinetic energy (E_K) of a system is calculated as:

$$E_K = \frac{1}{2}mv^2 \qquad [4.2\text{-}5]$$

where m is the mass of a body and v is the velocity. Since kinetic energy is a scalar quantity, a direction for the velocity does not need to be denoted. Kinetic energy can also move into and out of a system with a flow rate described by $\dot{m}$. The **rate of kinetic energy,** $\dot{E}_K$, is calculated as:

$$\dot{E}_K = \frac{1}{2}\dot{m}v^2 \qquad [4.2\text{-}6]$$

EXAMPLE 4.1 Change in Rate of Kinetic Energy of Blood

Problem: Blood travels from the heart to the body's tissues and organs through blood vessels that continuously branch off each other and become smaller in diameter. In the capillaries, the smallest blood vessels, the exchange of nutrients and other substances between the blood and interstitial fluid takes place. Oxygenated blood from the heart starts in the aorta, which has a diameter of about 2 cm and through which blood travels at a velocity of 33 cm/s. In contrast, an average capillary has a diameter of 8 μm, and blood travels through it at a velocity of about 0.3 mm/s. What is the difference in the rate of kinetic energy of blood between the aorta and a capillary? Calculate the rates of kinetic energy for blood in these vessels in units of W and Btu/s. The density of blood is 1.056 g/cm^3.

Solution: The mass flow rate in the aorta is calculated:

$$\dot{m} = \rho v A = \rho v \frac{\pi}{4} D^2 = \left(1.056 \frac{\text{g}}{\text{cm}^3}\right)\left(33 \frac{\text{cm}}{\text{s}}\right)\frac{\pi}{4}(2 \text{ cm})^2 = 109 \frac{\text{g}}{\text{s}}$$

A similar calculation for the capillary gives a mass flow rate of 1.59×10^{-8} g/s. The mass flow rate of blood ranges over 10 orders of magnitude between the aorta and a capillary.

The rate of kinetic energy in the aorta is:

$$\dot{E}_K = \frac{1}{2}\dot{m}v^2 = \frac{1}{2}\left(109 \frac{\text{g}}{\text{s}}\right)\left(33 \frac{\text{cm}}{\text{s}}\right)^2\left(\frac{1 \text{ kg}}{1000 \text{ g}}\right)\left(\frac{1 \text{ m}}{100 \text{ cm}}\right)^2 = 5.94 \times 10^{-3} \text{ W}$$

$$\dot{E}_K = \left(5.94 \times 10^{-3} \frac{\text{J}}{\text{s}}\right)\left(9.486 \times 10^{-4} \frac{\text{Btu}}{\text{J}}\right) = 5.63 \times 10^{-6} \frac{\text{Btu}}{\text{s}}$$

A similar calculation for the capillary gives the rate of kinetic energy as 7.16×10^{-19} W, which equals 6.79×10^{-22} Btu/s. The rate of kinetic energy in the blood vessels ranges 16 orders of magnitude between the aorta and a single capillary. ■

Mass possesses **internal energy** (U) by virtue of atomic and molecular interactions. The electromagnetic interaction of molecules, the motion of molecules relative to the center of mass of the system, the rotational and vibrational motion of molecules, and other sources contribute to the internal energy of matter. All energy possessed by a mass that is neither kinetic nor potential energy is internal energy.

Internal energy cannot be measured directly or known in absolute terms. Like potential energy, it is calculated relative to a reference point or state. The internal energy of a system is a function of temperature, pressure, chemical composition, phase (vapor, liquid, solid, or crystal), and other characteristics. **While internal energy cannot be known in absolute terms, the change in internal energy can often be quantified.**

The **rate of internal energy** ($\dot{U}$) is the rate at which internal energy enters or leaves a system with a fluid or other material that crosses the system boundary. Again, it cannot be known in absolute terms, but changes in the rate of internal energy can often be quantified.

A system's **total energy** (E_T) is defined as the sum of potential, kinetic, and internal energies:

$$E_T = E_P + E_K + U \qquad [4.2\text{-}7]$$

These three forms of energy may be possessed by the system or may move into or out of the system with bulk material transfer. An analogous equation can be written for the **rate of total energy** ($\dot{E}_T$) in a system:

$$\dot{E}_T = \dot{E}_P + \dot{E}_K + \dot{U} \qquad [4.2\text{-}8]$$

Contributions to the total energy from electromagnetic potential energy are not considered in this chapter. Accounting and conservation equations that explicitly include electromagnetic potential energy are in Chapter 5.

We can convert extensive variables to specific variables, which are designated by a circumflex accent or "hat," by dividing by another extensive variable such as mass or moles. In this chapter, the term **specific** refers strictly to the quantity of a variable per unit mass or moles. Different types of specific energy (kinetic, potential, internal), specific enthalpy, and specific volume, as well as the rates of these listed variables, are used. Specific variables are intensive variables, since they are independent of the size of the system.

Specific energy is energy per unit mass or per unit mole; specific energy rate is the rate of energy per unit mass or per unit mole. As an illustration, suppose the rate of kinetic energy, $\dot{E}_K$, of a stream is 400 kcal/hr and the mass flow rate, $\dot{m}$, is 100 kg/hr. Using the formula:

$$\dot{E}_K = \hat{E}_K \dot{m} \qquad [4.2\text{-}9]$$

the specific kinetic energy, $\hat{E}_K$, of the stream is 4 kcal/kg.

The total energy, E_T, and the rate of total energy, $\dot{E}_T$, are written in terms of specific variables (per unit mass):

$$E_T = m\hat{E}_T = m(\hat{E}_P + \hat{E}_K + \hat{U}) \qquad [4.2\text{-}10]$$

$$\dot{E}_T = \dot{m}\hat{E}_T = \dot{m}(\hat{E}_P + \hat{E}_K + \hat{U}) \qquad [4.2\text{-}11]$$

where m is mass; $\hat{E}_T$, $\hat{E}_P$, $\hat{E}_K$, and $\hat{U}$ are specific energies; and $\dot{m}$ is the mass flow rate. $\hat{E}_T$, $\hat{E}_P$, $\hat{E}_K$, and $\hat{U}$ have the dimension of $[L^2t^{-2}]$.

4.2.2 Energy in Transition

Heat and work are energy in transit across the boundary between the system and its surroundings. Heat and work only exist when a system, and hence a system boundary, has been established. **Heat and work can only be understood as the movement of energy through direct and nondirect contacts—not bulk material transfer.** Heat transfer is a result of a temperature driving force; work is energy in transit as the result of any other driving force (e.g., pressure). Neither heat nor work can be stored in or possessed by a body or system. When looking for contributions from work and heat, look for energy transfer at the system boundary.

Heat (Q) is energy that flows due to a difference in temperature. The direction of heat flow is always from a region of higher temperature to one of lower temperature. Heat can be transferred across all or part of the system boundary. Heat transfer is apparent in simple processes like the flow of heat from a heating pad to an aching area of your body. The dimensions of heat and the **rate of heat** ($\dot{Q}$) are $[L^2Mt^{-2}]$ and $[L^2Mt^{-3}]$, respectively.

Sometimes the heat or rate of heat is specified. Other times, heat transfer must be estimated. One of the simpler methods of estimation relates the rate of heat, $\dot{Q}$, to the effective surface area over which the transfer occurs, A, and the temperature gradient:

$$\dot{Q} = hA(T_{\text{surr}} - T_{\text{sys}}) \qquad [4.2\text{-}12]$$

where h is the heat transfer coefficient (per unit area), T_{surr} is the temperature of the surroundings, and T_{sys} is the temperature of the system. Depending on the definition and complexity of the system, the term U, the overall heat transfer coefficient (per unit area), is used instead of the term h. Typically, when the system consists of multiple layers of material through which the heat passes, the term U is used. Since the variable U stands for internal energy in this chapter, we use h for the heat transfer coefficient in all problems. The dimension of the heat transfer coefficient is $[\text{Mt}^{-3}\text{T}^{-1}]$. Common units are $\text{W}/(\text{m}^2 \cdot \text{K})$ and $\text{Btu}/(\text{ft}^2 \cdot \text{hr} \cdot {}^\circ\text{F})$. The value of h is dependent on the geometry of the system, the types of materials across which the heat transfer occurs, whether those materials are moving or stationary, and other factors. Heat transfer coefficients are estimated or measured; this calculation is outside the scope of this textbook. Other textbooks discuss this in further detail (e.g., Bird RB, Stewart WE, and Lightfoot EN, *Transport Phenomena*, 2002; Johnson AT, *Biological Process Engineering*, 1999).

Heat is defined as a positive value when it is transferred to the system from the surroundings. In other words, heat is a positive value when it is added to the system. Heat is a negative value when energy leaves the system due to a difference in temperature. (*Note*: Different textbooks define the direction of heat differently. The sign convention for the direction of heat is arbitrary; double-check all definitions before comparing equations.) If heat does not transfer into or out of the system, then the system or process is considered **adiabatic**. For example, if an adiabatic wall encloses a system, the system cannot extract heat from or release heat to the surroundings.

Work (W) is energy that flows across a system boundary as a result of a driving force other than temperature. The dimensions of work and **rate of work** ($\dot{W}$) are $[\text{L}^2\text{Mt}^{-2}]$ and $[\text{L}^2\text{Mt}^{-3}]$, respectively. Some driving forces that generate work include pressure, mechanical force, or an electromagnetic field. In all cases, the force acts on the system or part of the system to move it across a distance. Common examples of work include the motion of a piston against a resisting force, the rotation of a shaft (e.g., mixer), and the passage of electrical current across the system boundary. Work is also done when matter flows into and out of a system. Work is typically classified into two types: shaft (or nonflow) work and flow work.

The **rate of shaft work** or **nonflow work** ($\dot{W}_{\text{nonflow}}$) includes the rate of work done by a moving part (e.g., rotor or mixer) on the system. Devices such as motors, pumps, and compressors apply nonflow work on a system. Nonflow work also includes the work associated with the expansion of the system volume against an external force or pressure, the work associated with electrical current, and surface tension forces. Shaft work is all work that is not flow work. **We adopt the convention that $\dot{W}_{\text{nonflow}}$ is positive when work is done on the system by the surroundings.** (*Note*: Different textbooks define the direction of work differently. Double-check all definitions before comparing equations.)

One type of nonflow work often encountered in the study of thermodynamics is associated with the contraction or expansion of a system volume against an external force. When the surroundings exert a force in the x-direction, F_x, on the system, the differential work on the system, dW, is written:

$$dW = F_x\,dx \qquad [4.2\text{-}13]$$

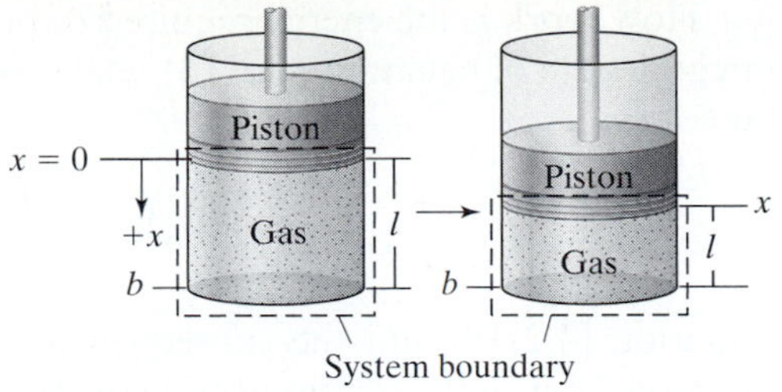

Figure 4.2
Expansion of a gas against a frictionless piston.

where dx is the differential distance or the displacement. In this textbook, the force F_x and the displacement dx are defined to be in the same direction. (*Note*: This may differ in other textbooks.)

A classic example of nonflow work is the contraction of gas within a container against a frictionless piston (Figure 4.2). Work is done on the system by the surroundings ($dW > 0$), causing the volume of the system to decrease ($dV < 0$). From equation [4.2-13], we can derive an equation that describes this relationship. Recall that the definition of **pressure** is force divided by area; thus, the force F_x is equal to pressure times area. Substituting this into equation [4.2-13] gives:

$$dW = PA\,dx \qquad [4.2\text{-}14]$$

The change in volume or differential volume, dV, is the cross-sectional area, A, times the differential distance, dx. To find an equation describing dV, imagine a system with a fixed end b and a piston at position x, which is initially positioned at $x = 0$. The coordinate system is defined such that x is positive as the piston moves toward the fixed end b. The distance between b and the position of the piston x is the length l. As the piston contracts, the distance x increases; therefore:

$$l = b - x \qquad [4.2\text{-}15]$$

The volume of the system, V, is the cross-sectional area A multiplied by the length l, so:

$$V = Al = Ab - Ax \qquad [4.2\text{-}16]$$

Thus, when the piston moves by a distance dx, the change in the volume of the system, dV, is:

$$dV = d(Ab - Ax) = -A\,dx \qquad [4.2\text{-}17]$$

where $d(Ab)$ is equal to zero because the term is constant. Thus, for the work of contraction, W, equation [4.2-13] is transformed for a reversible, closed process to:

$$W = -\int_{V_1}^{V_2} P\,dV \qquad [4.2\text{-}18]$$

where P is pressure of the system, dV is the differential volume change of the system, and V_1 and V_2 are the initial and final volumes, respectively. When the volume of the system expands ($V_2 > V_1$), work is negative; thus, work is being done by the system on the surroundings. When the volume of the system contracts ($V_2 < V_1$), work is positive; thus, work is being done on the system by the surroundings. The work done by the expansion of a gas is described by equation [4.2-18] and is examined in Example 4.3.

Flow work is the energy required to push matter into or out of the system. The integral form of equation [4.2-13] relates work, W, to force and the differential distance, dx:

$$W = \int_{x_1}^{x_2} F_x \, dx \qquad [4.2\text{-}19]$$

Equation [4.2-19] and the subsequent derivation can be generalized for three dimensions. Given the relationship between work and the rate of work, $\dot{W}$:

$$W = \int_{t_0}^{t_f} \dot{W} \, dt \qquad [4.2\text{-}20]$$

it can be shown that:

$$\dot{W} = F_x \frac{dx}{dt} = F_x v \qquad [4.2\text{-}21]$$

where t_0 is the initial time, t_f is the final time, and v is the velocity in the x-direction. The rate of work is also known as **power** (P), which has dimension $[L^2Mt^{-3}]$. Common units of power are horsepower (hp), $ft \cdot lb_f/s$, and watt (W), which is equivalent to J/s.

Consider a system with fluid flow entering the system with velocity v. Because force is pressure multiplied by area, the rate of work, $\dot{W}$, flowing into the system is:

$$\dot{W} = F_x v = PAv \qquad [4.2\text{-}22]$$

where P is the pressure where the fluid crosses the system boundary and A is the cross-sectional area of conduit carrying the fluid (Figure 4.3). Note that since $\dot{W}$ is a scalar, the other variables in this equation are written as scalars as well. Since the rate of work is considered only at the system boundary, only the surface area of the fluid flow, A, into or out of the system is considered rather than the surface area of the whole system.

The rate of flow work on the system, $\dot{W}_{flow}$, is the difference between the rate of work done by the fluid at the system inlet(s) and the rate of work done by the fluid at the system outlet(s):

$$\dot{W}_{flow} = \sum_i \dot{W}_i - \sum_j \dot{W}_j \qquad [4.2\text{-}23]$$

where i and j are different inlet and outlet streams, respectively.

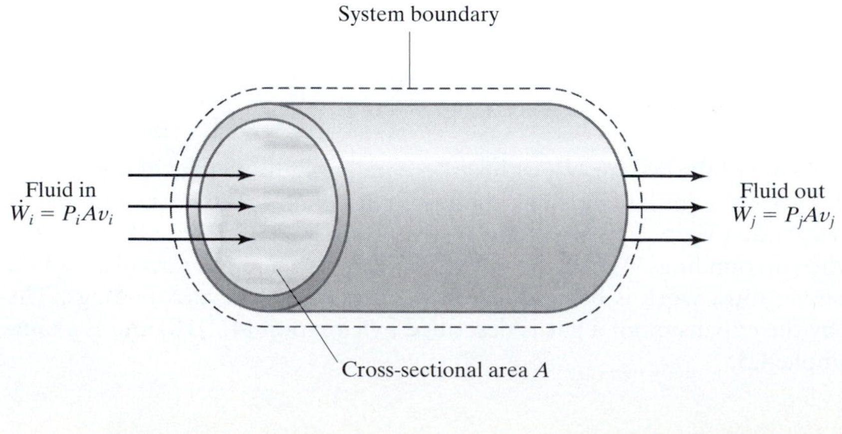

Figure 4.3
Flow work of fluid material entering and leaving a system.

4.2.3 Enthalpy

Enthalpy (H) is a thermodynamic function that is defined as:

$$H = U + PV \qquad [4.2\text{-}24]$$

where U is internal energy, P is pressure, and V is volume. The dimension of enthalpy is that of energy, $[L^2Mt^{-2}]$. Enthalpy is a convenient variable to use in the conservation of energy equation. Specific internal energy, pressure, specific volume, and specific enthalpy are all state functions (see Section 4.5). Specific enthalpy ($\hat{H}$) is defined as:

$$\hat{H} = \hat{U} + P\hat{V} = \hat{U} + \frac{P}{\rho} \qquad [4.2\text{-}25]$$

where $\hat{U}$ is specific internal energy, $\hat{V}$ is specific volume, and ρ is density.

Like internal energy, enthalpy cannot be known in absolute terms and is determined relative to a reference point or state. Calculating the change of enthalpy of a system is usually accomplished by looking at changes in temperature, chemical composition, phase, or pressure. Sections 4.5 and 4.8 are devoted to this task.

The **rate of enthalpy** ($\dot{H}$) is the rate at which enthalpy moves with a fluid or other material:

$$\dot{H} = \dot{U} + P\dot{V} \qquad [4.2\text{-}26]$$

Again, the rate of enthalpy cannot be known in absolute terms, but changes in the rate of enthalpy can be quantified.

EXAMPLE 4.2 **Specific Enthalpy Change in Air**

Problem: Air is sparged into a bioreactor at room temperature (25°C) and heated up to 37°C. The air is allowed to expand during the heating such that the bioreactor maintains a constant pressure of 1.0 atm. Given that the specific internal energy of air increases by approximately 250 J/mol as air is warmed, what is the difference in specific enthalpy of air between these two states? The molecular weight of air is approximately 28.9 g/mol. Assume that air behaves as an ideal gas.

Solution: The ideal gas law ($PV = nRT$) is used to find the specific volume of air at 25°C (298 K) and 37°C (310 K). At room temperature:

$$\hat{V} = \frac{V}{n} = \frac{RT}{P} = \frac{\left(0.08206\dfrac{\text{L}\cdot\text{atm}}{\text{mol}\cdot\text{K}}\right)(298\text{ K})}{1\text{ atm}} = 24.5\ \frac{\text{L}}{\text{mol}}$$

For air at 310 K, $\hat{V}$ is 25.4 L/mol.

The absolute specific enthalpy cannot be found at either temperature. However, the difference in specific enthalpy between the two temperatures is calculated by constructing a difference equation based on equation [4.2-25]:

$$\begin{aligned}\hat{H}_{310\text{K}} - \hat{H}_{298\text{K}} &= \hat{U}_{310\text{K}} - \hat{U}_{298\text{K}} + P(\hat{V}_{310\text{K}} - \hat{V}_{298\text{K}})\\ &= 250\frac{\text{J}}{\text{mol}} + (1\text{ atm})\left(25.4\frac{\text{L}}{\text{mol}} - 24.5\frac{\text{L}}{\text{mol}}\right)\left(\frac{101.3\text{ J}}{\text{L}\cdot\text{atm}}\right) = 341\ \frac{\text{J}}{\text{mol}}\end{aligned}$$

Note that the specific enthalpy change calculated here is on a per-mole basis. ■

4.3 Review of Energy Conservation Statements

The total energy of a system is always conserved. The conservation equation for total energy is a mathematical description of the movement and accumulation of total energy in a system of interest. The law of the conservation of total energy states that total energy can be neither created nor destroyed, but only converted from one form into an equivalent quantity of another form of energy. (An exception to this statement can be made for nuclear reactions.) Although total energy is conserved and remains constant in the universe, specific forms of energy, such as mechanical and electrical energy, are not conserved.

The differential form of the conservation equation is appropriate when rates of energy, heat, work, or a combination of the three are specified. Consider the system shown in Figure 4.4. The rates of mass entering and leaving the system have associated energies rates $\dot{E}_{T,i}$ and $\dot{E}_{T,j}$, respectively, in the form of internal, kinetic, and potential energies. The rate at which flow work, $\dot{W}_{\text{flow}}$, enters the system depends on the movement of mass across the system boundary. The rate at which heat is added to the system is denoted by $\dot{Q}$. The rate at which nonflow work is done on the system by the surroundings is indicated by $\dot{W}_{\text{nonflow}}$.

The generic conservation equation is written to account for the movement of total energy into and out of the system by bulk mass transfer and for the transfer of heat and work across the system boundary. No Generation and Consumption terms are present, since total energy is conserved. The differential form of the conservation of total energy is:

$$\dot{\psi}_{\text{in}} - \dot{\psi}_{\text{out}} = \dot{\psi}_{\text{acc}} = \frac{d\psi}{dt} \qquad [4.3\text{-}1]$$

$$\sum_i \dot{E}_{T,i} - \sum_j \dot{E}_{T,j} + \sum \dot{Q} + \sum \dot{W} = \frac{dE_T^{\text{sys}}}{dt} \qquad [4.3\text{-}2]$$

where $\sum_i \dot{E}_{T,i}$ is the rate at which total energy enters the system by bulk material transfer, $\sum_j \dot{E}_{T,j}$ is the rate at which total energy leaves the system by bulk material transfer, $\sum \dot{Q}$ is the net rate at which the system is heated, $\sum \dot{W}$ is the net rate of flow and nonflow work on the system, and dE_T^{sys}/dt is the rate at which total energy accumulates within the system. Subscripts i and j refer to the inlet and outlet streams, respectively. The Accumulation term is expressed as the instantaneous rate of change of total energy of the system or the rate of accumulation of total energy in the system. When an Accumulation term is present, further information, such as an initial condition, may need to be specified. The dimension of the terms in equation [4.3-2] is $[\text{L}^2\text{Mt}^{-3}]$.

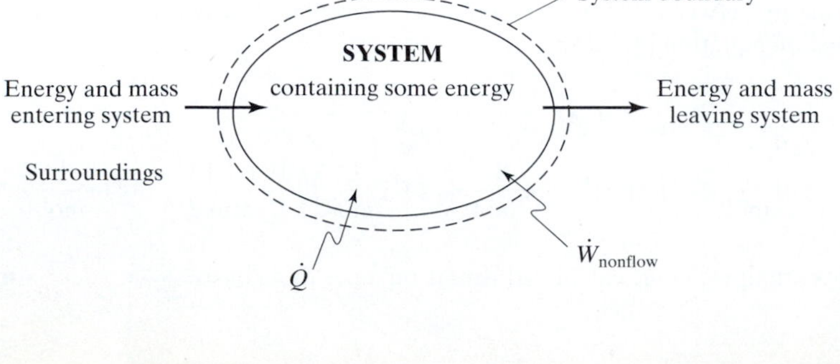

Figure 4.4
System showing modes of rate of energy transfer into and out of a system and rate of energy accumulation in the system.

Since total mass—but not total moles—is conserved in all systems, the development of the conservation of total energy equation is presented using mass and mass rates. Recall equation [4.2-8], which defines the rate of total energy as the sum of potential, kinetic, and internal energies:

$$\dot{E}_T = \dot{E}_P + \dot{E}_K + \dot{U} = \dot{m}\hat{E}_P + \dot{m}\hat{E}_K + \dot{m}\hat{U} \qquad [4.3\text{-}3]$$

Equation [4.3-2] can be rewritten as:

$$\sum_i (\dot{E}_{P,i} + \dot{E}_{K,i} + \dot{U}_i) - \sum_j (\dot{E}_{P,j} + \dot{E}_{K,j} + \dot{U}_j) + \sum \dot{Q} + \sum \dot{W} = \frac{dE_T^{\text{sys}}}{dt} \qquad [4.3\text{-}4]$$

or

$$\sum_i \dot{m}_i(\hat{E}_{P,i} + \hat{E}_{K,i} + \hat{U}_i) - \sum_j \dot{m}_j(\hat{E}_{P,j} + \hat{E}_{K,j} + \hat{U}_j) + \sum \dot{Q} + \sum \dot{W} = \frac{dE_T^{\text{sys}}}{dt} \qquad [4.3\text{-}5]$$

The total rate of work is given as the sum of the flow and nonflow work:

$$\sum \dot{W} = \sum \dot{W}_{\text{flow}} + \sum \dot{W}_{\text{nonflow}} \qquad [4.3\text{-}6]$$

The flow work of any particular flowing stream can be written as:

$$\dot{W}_{\text{flow}} = PAv = \dot{m}\frac{P}{\rho} = \dot{m}P\hat{V} \qquad [4.3\text{-}7]$$

Given that the rate of flow work, $\dot{W}_{\text{flow}}$, is the difference between the rate of work done by the fluid at the system inlet(s), i, and the rate of work done by the fluid at the system outlet(s), j, equation [4.3-6] is rewritten as:

$$\sum \dot{W} = \sum_i \dot{m}_i \frac{P_i}{\rho_i} - \sum_j \dot{m}_j \frac{P_j}{\rho_j} + \sum \dot{W}_{\text{nonflow}} \qquad [4.3\text{-}8]$$

Therefore, equation [4.3-5] can be written as:

$$\sum_i \dot{m}_i\left(\hat{E}_{P,i} + \hat{E}_{K,i} + \hat{U}_i + \frac{P_i}{\rho_i}\right) - \sum_j \dot{m}_j\left(\hat{E}_{P,j} + \hat{E}_{K,j} + \hat{U}_j + \frac{P_j}{\rho_j}\right) + \sum \dot{Q} + \sum \dot{W}_{\text{nonflow}} = \frac{dE_T^{\text{sys}}}{dt} \qquad [4.3\text{-}9]$$

Finally, we can rewrite the equation by substituting specific enthalpy:

$$\sum_i \dot{m}_i(\hat{E}_{P,i} + \hat{E}_{K,i} + \hat{H}_i) - \sum_j \dot{m}_j(\hat{E}_{P,j} + \hat{E}_{K,j} + \hat{H}_j) + \sum \dot{Q} + \sum \dot{W}_{\text{nonflow}} = \frac{dE_T^{\text{sys}}}{dt} \qquad [4.3\text{-}10]$$

Because specific internal energy and specific enthalpy are tabulated for many conditions, equations [4.3-5] and [4.3-10] are the differential forms of the conservation of total energy equation that are used most often.

The integral equation is most appropriate to use when trying to evaluate conditions between two discrete time points. Often the integral form of the equation is used when one or more of the rates of energy, heat, or work are a function of time.

When applying the integral conservation equation, write the differential balance equations [4.3-5] and [4.3-10] and integrate between the initial and final conditions. The integral conservation statements are:

$$\int_{t_0}^{t_f} \sum_i \dot{m}_i(\hat{E}_{P,i} + \hat{E}_{K,i} + \hat{U}_i)\,dt - \int_{t_0}^{t_f} \sum_j \dot{m}_j(\hat{E}_{P,j} + \hat{E}_{K,j} + \hat{U}_j)\,dt + \int_{t_0}^{t_f} \sum \dot{Q}\,dt + \int_{t_0}^{t_f} \sum \dot{W}\,dt = \int_{t_0}^{t} \frac{dE_T^{\text{sys}}}{dt}\,dt \qquad [4.3\text{-}11]$$

$$\int_{t_0}^{t_f} \sum_i \dot{m}_i(\hat{E}_{P,i} + \hat{E}_{K,i} + \hat{H}_i)\,dt - \int_{t_0}^{t_f} \sum_j \dot{m}_j(\hat{E}_{P,j} + \hat{E}_{K,j} + \hat{H}_j)\,dt + \int_{t_0}^{t_f} \sum \dot{Q}\,dt + \int_{t_0}^{t_f} \sum \dot{W}_{\text{nonflow}}\,dt = \int_{t_0}^{t_f} \frac{dE_T^{\text{sys}}}{dt}\,dt \qquad [4.3\text{-}12]$$

where t_0 is the initial time and t_f is the final time. The difference here is that equation [4.3-11] is formulated with internal energy and equation [4.3-12] is formulated with enthalpy. Solving problems with the integral form of the conservation of energy equation can be very difficult and is not undertaken in this textbook.

The algebraic form of the conservation of total energy equation is used in situations with a finite time period and when discrete quantities of matter or energy or both enter or leave the system. The algebraic equation can be derived from the integral equation for a defined, finite time period. The total heat, Q, that acts on a system during the time period is defined as:

$$Q = \int_{t_0}^{t_f} \sum \dot{Q}\,dt \qquad [4.3\text{-}13]$$

and the total work, W, that acts on a system during the time period is defined as:

$$W = \int_{t_0}^{t_f} \sum \dot{W}\,dt \qquad [4.3\text{-}14]$$

With these definitions, the algebraic form of the conservation of energy equation is:

$$\sum_i (E_{P,i} + E_{K,i} + U_i) - \sum_j (E_{P,j} + E_{K,j} + U_j) + Q + W = E_{T,\text{f}}^{\text{sys}} - E_{T,0}^{\text{sys}} \qquad [4.3\text{-}15]$$

where $E_{T,\text{f}}^{\text{sys}}$ and $E_{T,0}^{\text{sys}}$ are the total energies of the system at the final and initial conditions, respectively. The different forms of energy are written in terms of specific energy:

$$\sum_i (m_i(\hat{E}_{P,i} + \hat{E}_{K,i} + \hat{U}_i)) - \sum_j (m_j(\hat{E}_{P,j} + \hat{E}_{K,j} + \hat{U}_j)) + Q + W = E_{T,\text{f}}^{\text{sys}} - E_{T,0}^{\text{sys}} \qquad [4.3\text{-}16]$$

Substitution for the total amount of work and for the definition of enthalpy results in another common form of the algebraic conservation of energy equation:

$$\sum_i (m_i(\hat{E}_{P,i} + \hat{E}_{K,i} + \hat{H}_i)) - \sum_j (m_j(\hat{E}_{P,j} + \hat{E}_{K,j} + \hat{H}_j)) + Q + W_{\text{nonflow}} = E_{T,\text{f}}^{\text{sys}} - E_{T,0}^{\text{sys}} \qquad [4.3\text{-}17]$$

Mass accounting or conservation equations may need to be developed along with the energy conservation equation, depending on the complexity of the system. We use coupled mass and energy equations most often when considering unsteady-state systems.

4.4 Closed and Isolated Systems

In a closed system, no mass crosses the system boundary. Therefore, energy contributions associated with bulk transfer are eliminated from equations [4.3-5] and [4.3-16], leaving the differential and algebraic forms, respectively:

$$\sum \dot{Q} + \sum \dot{W} = \frac{dE_T^{sys}}{dt} \qquad [4.4\text{-}1]$$

$$Q + W = E_{T,f}^{sys} - E_{T,0}^{sys} \qquad [4.4\text{-}2]$$

In both equations, total energy, E_T, is the sum of three types of energy: potential, kinetic, and internal. By definition, the work terms (W and $\dot{W}$) include both nonflow and flow work. In a closed system, because no material moves into or out of the system, there is no flow work. Thus, W and $\dot{W}$ include only nonflow work and can be rewritten with $W_{nonflow}$ and $\dot{W}_{nonflow}$ terms, respectively.

The symbol Δ is often used to signify the difference between two different conditions or points. In this book, Δ is most often used to signify the difference between the outlet stream and inlet stream; however, it is also used to signify the difference in a system between the final condition and initial condition. The conservation of total energy for a closed system is:

$$Q + W = (E_{P,f}^{sys} - E_{P,0}^{sys}) + (E_{K,f}^{sys} - E_{K,0}^{sys}) + (U_f^{sys} - U_0^{sys}) \qquad [4.4\text{-}3]$$

$$Q + W = \Delta E_P^{sys} + \Delta E_K^{sys} + \Delta U^{sys} \qquad [4.4\text{-}4]$$

This equation is also known as the **first law of thermodynamics** for a closed system. It states that the change in energy of a system is equal to the heat and work applied to the system.

Many biologically related thermodynamic problems employ the first law of thermodynamics. One classic nonbiological example is that of a cylinder with a frictionless piston.

EXAMPLE 4.3 Expansion of a Gas

Problem: A container with a gas and a movable piston contains a gas-phase reaction that produces 61.3 J of heat (Figure 4.5). The total number of moles of gas in the system does not change. The gas inside the container has an initial volume of 1.0 L at 298 K and 1 atm. If the temperature rises to 350 K and pressure from the piston remains constant, what will the volume of the gas be? How much work is done on the system when the gas expands? What is the change in internal energy of the gas? Assume that the gas behaves like an ideal gas.

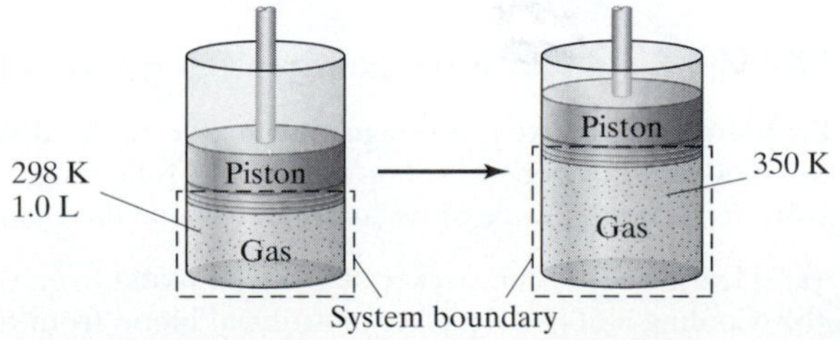

Figure 4.5
Expansion of a gas due to a gas-phase reaction.

Solution: Because the gas is assumed to behave ideally, we can use the ideal gas law (equation [1.5-14]). Because the number of moles is constant, we can rearrange the ideal gas law to solve for the number of moles, n, for both the initial and final conditions and set them equal to each other:

$$n = \frac{PV_0}{RT_0} = \frac{PV_f}{RT_f}$$

We can further simplify this equation by eliminating P and R, since these are constant. We can then rearrange the equation to solve for the final volume:

$$V_f = V_0\left(\frac{T_f}{T_0}\right) = 1.0\text{ L}\left(\frac{350\text{ K}}{298\text{ K}}\right) = -1.17\text{ L}$$

The work done on the system is calculated using equation [4.2-18]:

$$W = -\int_{V_1}^{V_2} P\,dV = -\int_{1.0\text{L}}^{1.17\text{L}} (1\text{ atm})\,dV = -(1\text{ atm})(1.17\text{ L} - 1.0\text{ L}) = -\,0.17\text{ L}\cdot\text{atm}$$

$$W = -0.17\text{ L}\cdot\text{atm}\left(\frac{101.3\text{ J}}{\text{L}\cdot\text{atm}}\right) = -17.2\text{ J}$$

Because the volume of the gas expands ($V_f > V_0$) and the calculated work is negative, work is being done by the system (gas) on the surroundings (piston).

The container with the piston is a closed system. Therefore, the first law of thermodynamics (equation [4.4-3]) is appropriate to use to determine the change in internal energy of the system:

$$Q + W = (E_{P,f}^{sys} - E_{P,0}^{sys}) + (E_{K,f}^{sys} - E_{K,0}^{sys}) + (U_f^{sys} - U_0^{sys})$$

Because the system is stationary and is not changing position in space, changes in kinetic and potential energies are negligible, so

$$U_f^{sys} - U_0^{sys} = Q + W = 61.3\text{ J} - 17.2\text{ J} = 44.1\text{ J}$$

The change in internal energy of the system is 44.1 J. ■

In most closed biological systems, potential energy does not change between the initial and final conditions. In other words, the system does not move up or down in space relative to a fixed reference state. Also, a change in kinetic energy is rare in a closed system, since the velocity of the system rarely changes (i.e., the system does not accelerate). However, when a closed biological system reacts, its internal energy can change. During biological reactions, temperature, chemical composition, phase, and other variables may change; thus, changes in internal energy are often very important to consider. For each closed system of interest, all three forms of energy—potential, kinetic, and internal—should be considered before any terms are discarded.

EXAMPLE 4.4 Internal Energy Changes of a Blood Substitute

Problem: You have just designed an artificial blood substitute and want to test some of its thermodynamic properties. For each of the following cases, state whether nonzero heat and work terms are positive or negative. Determine the change in internal energy for:

(a) Heating a 500-mL packet of artificial blood from room temperature to 37°C.
(b) Cooling a 500-mL packet of artificial blood from room temperature to −70°C.

Solution: The first law of thermodynamics is appropriate to use in both cases, because the systems are closed. Because the packets are stationary and are not changing in space, changes in potential and kinetic energy are negligible. Thus, the first law of thermodynamics is reduced to:

$$Q + W = \Delta U^{sys}$$

(a) No work is done on the system by moving parts. To warm the blood, heat is transferred from the surroundings to the system; thus, Q is positive:

$$Q = \Delta U^{sys}$$

(b) To cool the blood, heat is transferred from the system to the surroundings; thus, Q is negative. Most artificial blood substitutes are composed primarily of water; therefore, we can model the packet of blood substitute as water, which changes phase from a liquid to a solid as it is cooled below 0°C. Water expands when it freezes, so work is done by the system on the surroundings. Therefore, work is also negative:

$$Q + W = \Delta U^{sys} < 0$$ ■

Finally, an isolated system is enclosed by a boundary that does not allow the transfer of any extensive property by any means. In an isolated system, no energy flows by any mechanism into or out of the system. In the case of an isolated system, the heat and work terms are also equal to zero, and equations [4.4-1] and [4.4-2] reduce to:

$$0 = \frac{dE_T^{sys}}{dt} \qquad [4.4\text{-}5]$$

$$0 = E_{T,f}^{sys} - E_{T,0}^{sys} = \Delta E_T^{sys} \qquad [4.4\text{-}6]$$

No energy accumulates in an isolated system; the amount of total energy at the initial condition is equal to that at the final condition. In other words, the total energy in an isolated system is constant. Truly isolated systems are rarely encountered in biological and medical applications.

4.5 Calculation of Enthalpy in Nonreactive Processes

A change in enthalpy can occur as a result of a temperature change, a pressure change, a phase change, mixing, or a reaction. In this section we consider the first four types of changes; changes due to reactions are discussed in Section 4.8. Discussions in other textbooks (e.g., Felder RM and Rousseau RW, *Elementary Principles of Chemical Processes,* 2000) focus on internal energy, which is used more often in the conservation of energy equations for closed systems. However, the conservation of energy equation formulated with enthalpy terms is better suited for solving problems involving open systems commonly found in biomedical applications; therefore, the focus in this section is exclusively on enthalpy.

4.5.1 Enthalpy as a State Function

A **state function** or property is an intensive property that depends only on the current state of the existence of the system and not on the path taken to reach it. Examples of state properties include temperature, pressure, composition, specific enthalpy, and specific volume. Heat and work are not state functions; they are **path functions,** because they depend on the path or method used to deliver the energy.

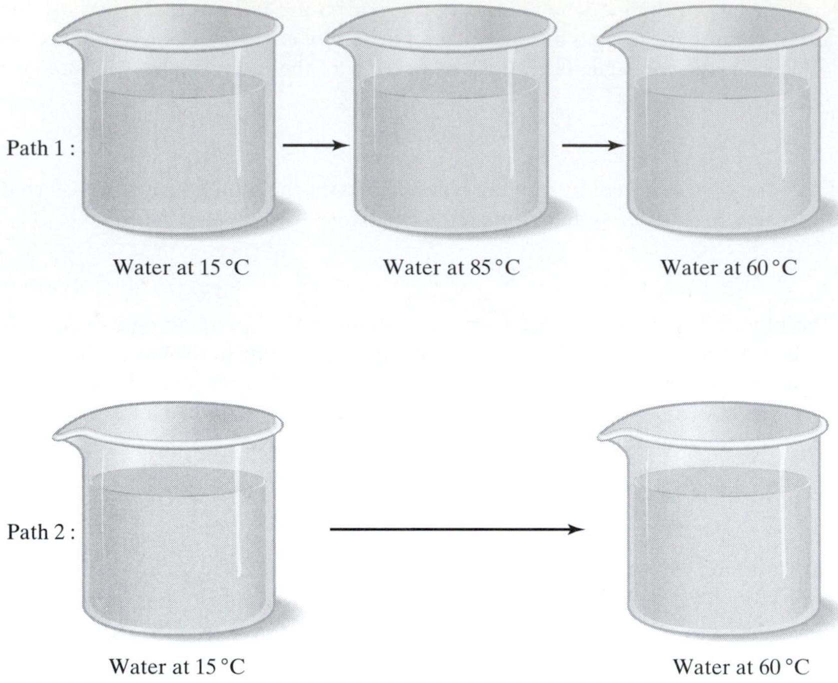

Figure 4.6
Two paths for the heating of water from 15°C to 60°C.

Consider the state property of temperature (Figure 4.6). If you begin with a system of water at 15°C, heat it to 85°C, and then cool it to 60°C (path 1), the temperature of the system is the same as if you begin with the exact same system of water at 15°C and heat it directly to 60°C (path 2). In other words, the temperature of the system does not depend on the path taken to warm the water from 15°C to 60°C; it measures 60°C at the final condition for both scenarios. On the other hand, the amount of heat required for the two paths does differ. More heat is required to warm the water from 15°C to 85°C (path 1) than from 15°C to 60°C (path 2). Since all the heat is not recovered when cooling the water from 85°C to 60°C (path 1), the amount of heat for path 1 is greater than for path 2. Hence, heat is considered a path function.

Specific internal energy, $\hat{U}$, and specific enthalpy, $\hat{H}$, are two important state properties. Like all state functions, $\hat{U}$ and $\hat{H}$ depend on the state of the system, specifically its temperature, phase (gas, liquid, solid, or crystal), and pressure. The fact that $\hat{U}$ and $\hat{H}$ are state functions has an important consequence for the application of the total energy conservation equation. **Even though the absolute values of internal energy and enthalpy of a system can never be known, the difference in internal energy or enthalpy between any two states can be calculated.**

The governing equations in Section 4.3 are constructed in such a way that the difference between the inlet and outlet energy rates or amounts, not the absolute value of the energy rate or amount, is required. For example, recall the differential equation [4.3-10]:

$$\sum_i \dot{m}_i(\hat{E}_{P,i} + \hat{E}_{K,i} + \hat{H}_i) - \sum_j \dot{m}_j(\hat{E}_{P,j} + \hat{E}_{K,j} + \hat{H}_j) + \sum \dot{Q} + \sum \dot{W}_{\text{nonflow}} = \frac{dE_T^{\text{sys}}}{dt} \qquad [4.5\text{-}1]$$

Note that the change in specific enthalpy is given as the difference between the specific enthalpies of the inlet and outlet. The **change in the rate of enthalpy** ($\Delta\dot{H}$) is defined as:

$$\Delta\dot{H} = -\sum_i \dot{m}_i\hat{H}_i + \sum_j \dot{m}_j\hat{H}_j \qquad [4.5\text{-}2]$$

The value of $\Delta\dot{H}$ is the total enthalpy of the output minus the total enthalpy of the input. For calculations involving the differential form of the total energy conservation equation, **the definition for $\Delta\dot{H}$ in equation [4.5-2] is critical because it offers a method to calculate the change in specific enthalpy as the difference between the inlets and outlets of the system.** Sometimes, specific enthalpies are given on a per-mole basis. In this case, the change in the rate of specific enthalpy across a system is:

$$\Delta\dot{H} = -\sum_i \dot{n}_i\hat{H}_i + \sum_j \dot{n}_j\hat{H}_j \qquad [4.5\text{-}3]$$

For a system with a single input, a single output, and no accumulation, the mass conservation equation (equation [3.3-10]) reduces to:

$$\dot{m}_i - \dot{m}_j = 0 \qquad [4.5\text{-}4]$$

Thus, we can reduce equation [4.5-2] for a system with a single input and single output:

$$-\dot{m}_i\hat{H}_i + \dot{m}_j\hat{H}_j = \Delta\dot{H} \qquad [4.5\text{-}5]$$

$$\dot{m}_i(\hat{H}_j - \hat{H}_i) = \dot{m}_i\,\Delta\hat{H} = \Delta\dot{H} \qquad [4.5\text{-}6]$$

where specific enthalpy is given on a per-mass basis. The change in the rate of enthalpy can also be written:

$$\dot{n}_i(\hat{H}_j - \hat{H}_i) = \dot{n}_i\,\Delta\hat{H} = \Delta\dot{H} \qquad [4.5\text{-}7]$$

where specific enthalpy is given on a per-mole basis.

This same discussion can be applied to the algebraic equation [4.3-17]. The change in the enthalpy of a system, ΔH, is defined as:

$$\Delta H = -\sum_i m_i\hat{H}_i + \sum_j m_j\hat{H}_j \qquad [4.5\text{-}8]$$

where specific enthalpy is given on a per-mass basis. For calculations involving the algebraic form of the total energy conservation equation, **the definition for ΔH in equation [4.5-8] is critical because it offers a method to calculate the change in specific enthalpy as the difference between the inputs and outputs of the system.** As with equation [4.5-3], a similar formulation for the change in enthalpy can be made in terms of moles and specific enthalpy on a molar basis.

Since $\hat{H}$ is a state function, the path to convert a system from one state to another can be accomplished using the most convenient pathway or transition. For example, consider the specific enthalpy change from state A to state B:

$$\text{state A} \xrightarrow{\Delta\hat{H}} \text{state B} \qquad [4.5\text{-}9]$$

This process requires the selection of a **reference state**—an arbitrarily chosen phase, temperature, and pressure that is usually assigned a specific enthalpy of zero. (*Note*:

The actual value of the specific enthalpy of the system at the reference state is not zero. Actual values of specific enthalpy can never be known.) Thus, the specific enthalpy at state A is $\hat{H}_A - \hat{H}_{ref}$, where $\hat{H}_{ref}$ is an arbitrarily assigned reference value. The specific enthalpy at state B is $\hat{H}_B - \hat{H}_{ref}$. Since $\hat{H}_{ref}$ is the same for both states, the change in specific enthalpy, $\Delta\hat{H}$, is:

$$\Delta\hat{H} = \hat{H}_B - \hat{H}_A \qquad [4.5\text{-}10]$$

Since $\hat{H}_{ref}$ dropped out of the equation, its absolute value is irrelevant.

Since specific enthalpy changes do not depend on the path, a series of hypothetical steps from one condition to another that are convenient for calculation can be constructed. As discussed in this section and Section 4.8, a path can be constructed to account for changes between the inlet and outlet conditions. Usually each step of the path in a nonreacting system has a change in one of the following: temperature, pressure, or phase. The change in the specific enthalpy across a system (e.g., state A → state B) is the sum of all the steps in the hypothetical path:

$$\Delta\hat{H} = \sum_k \Delta\hat{H}_k \qquad [4.5\text{-}11]$$

where k is the number of steps in the hypothetical path.

To illustrate how to construct a hypothetical path to calculate the enthalpy change, consider the creation of a diamond from graphite in an industrial setting. Diamonds are naturally formed under pressure inside the Earth over millions of years; those found near the Earth's surface are rare and valuable. For industrial purposes, such as cutting tools, lower-quality natural diamonds can be replaced with synthetic ones created in a much shorter period of time under extremely high pressure and temperature (approximately 1 million psi and 1800°C). To calculate the change in enthalpy for the industrial reaction, a hypothetical path needs to be constructed (Figure 4.7). One path could include the

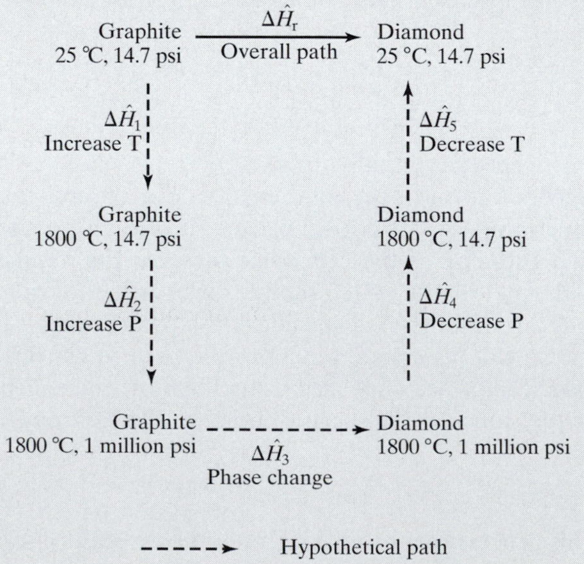

Figure 4.7
Hypothetical path for the enthalpy change associated with the production of industrial diamond.

following five steps: (a) step 1, $\Delta\hat{H}_1$, is the heating of graphite to 1800°C; (b) step 2, $\Delta\hat{H}_2$, represents the increase in pressure; (c) step 3, $\Delta\hat{H}_3$, shows the phase change from graphite to diamond at 1800°C and 1 million psi; (d) step 4, $\Delta\hat{H}_4$, is the decrease in pressure; and (e) step 5, $\Delta\hat{H}_5$, is the cooling of the diamond back to the starting conditions. Note that each step has a change in only one of the following: temperature, pressure, or phase.

EXAMPLE 4.5 **Cooling of Liquid Nitrogen**

Problem: Liquid nitrogen is used in a number of medical applications, such as removing tumors during surgery. Suppose gaseous N_2 at room temperature (298 K) is cooled to liquid N_2 just below its boiling point. N_2 liquefies at 77 K. The specific enthalpy change to cool nitrogen from 298 K to 77 K is −1435 cal/mol. The heat of vaporization, the specific enthalpy change from the liquid to the vapor form, of nitrogen is 1336 cal/mol. (Heat of vaporization is discussed further in Section 4.5.4.) What would be the overall change in specific enthalpy for this process?

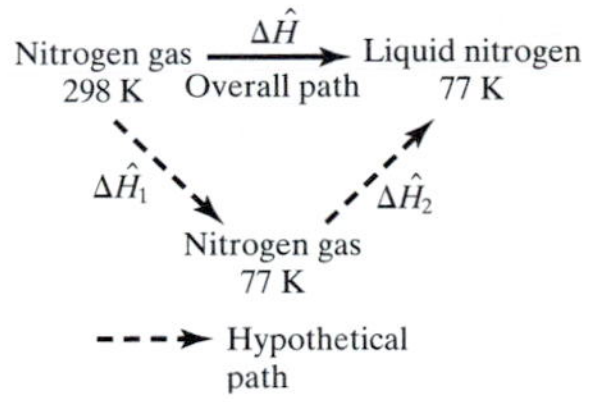

Figure 4.8
Hypothetical path for the enthalpy change associated with the cooling of nitrogen gas to its liquid state.

Solution: Remember that enthalpy is not dependent on path, so we can break up the process into steps. We can set up a hypothetical process involving two steps, since it involves two changes: temperature and phase (Figure 4.8). We can set up step 1 as the cooling of the nitrogen from 298 K to 77 K and step 2 as the liquefaction (vapor phase to liquid phase) of nitrogen. The enthalpy change for the overall process, $\Delta\hat{H}$, is the sum of the two steps:

$$\Delta\hat{H} = \sum_k \Delta\hat{H}_k = \Delta\hat{H}_1 + \Delta\hat{H}_2$$

The specific enthalpy change for step 1 when nitrogen is cooling is $\Delta\hat{H}_1 = -1435$ cal/mol. Liquefaction is the reverse of vaporization, so the specific enthalpy for step 2 is:

$$\Delta\hat{H}_2 = -\Delta\hat{H}_V = -1336\ \frac{\text{cal}}{\text{mol}}$$

The overall change in specific enthalpy is:

$$\Delta\hat{H} = \Delta\hat{H}_1 + \Delta\hat{H}_2 = -1435\frac{\text{cal}}{\text{mol}} - 1336\frac{\text{cal}}{\text{mol}} = -2770\ \frac{\text{cal}}{\text{mol}}$$

Note that the two enthalpy terms are approximately the same order of magnitude. As we discuss later, contributions from phase changes tend to be larger than contributions from temperature changes unless the temperature change is significant (over 200 K in this example). The negative sign for the enthalpy change means that energy must be removed from the system in order to cool and liquefy nitrogen. ■

The total enthalpy change for any process is equal to the sum of the changes in enthalpy in the steps of the hypothetical path. The remainder of the chapter focuses on the methods to calculate enthalpy changes along the different steps of the path. Specifically, we discuss how mixing and changes in temperature, pressure, and phase cause changes in enthalpy. Once the enthalpy change between the inlet and outlet conditions is calculated, the conservation of total energy equation can be applied (Section 4.6).

4.5.2 Change in Temperature

Heat transferred to raise or lower the temperature of a material is called **sensible heat.** Consider an open, steady-state system with no changes in potential or kinetic energy and no nonflow work. The rate at which sensible heat is added or removed is equal to the difference in rate at which enthalpy changes in the system. This system can be mathematically described with the following differential and algebraic equations:

$$\sum_i \dot{m}_i \hat{H}_i - \sum_j \dot{m}_j \hat{H}_j + \sum \dot{Q} = 0 \quad [4.5\text{-}12]$$

$$\sum_i H_i - \sum_j H_j + Q = 0 \quad [4.5\text{-}13]$$

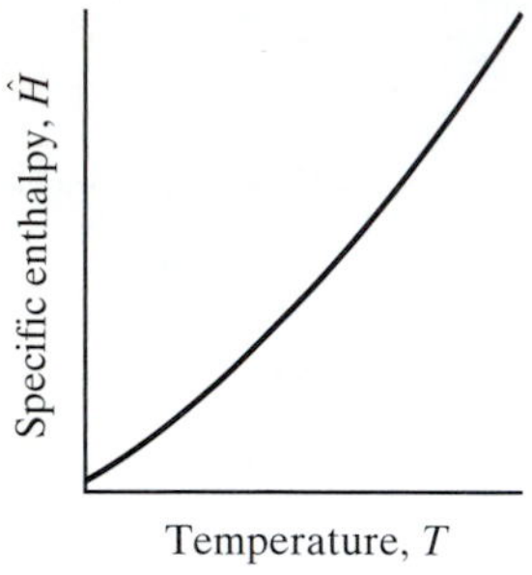

Figure 4.9
Schematic relationship between specific enthalpy and temperature. The slope of the line is the heat capacity at constant pressure.

In this case, the sensible heat is equal to the enthalpy difference between outlet and inlet conditions caused by an increase or decrease in temperature. Using the definitions of $\Delta\dot{H}$ and ΔH from equations [4.5-2] and [4.5-9], the sensible heat is:

$$\sum \dot{Q} = \Delta \dot{H} \quad [4.5\text{-}14]$$

$$Q = \Delta H \quad [4.5\text{-}15]$$

For a system in which the material has a temperature change, the specific enthalpy difference is quantified in $\Delta\dot{H}$ or ΔH.

The specific enthalpy of a substance depends strongly on temperature. Figure 4.9 shows a hypothetical plot of specific enthalpy as a function of temperature for a system under constant pressure. In mathematical terms, a temperature change, ΔT, leads to a change in specific enthalpy, $\Delta\hat{H}$. As ΔT goes to zero, the ratio $\Delta\hat{H}/\Delta T$ approaches the slope of the curve, which is the **heat capacity**:

$$C_p(T) = \lim_{\Delta T \to 0} \frac{\Delta \hat{H}}{\Delta T} \quad [4.5\text{-}16]$$

where C_p is the heat capacity at constant pressure. Note that specific enthalpy increases as temperature increases in a nonlinear fashion; therefore, C_p is given as a function of temperature and is represented by $C_p(T)$. Heat capacity is typically given in units of cal/(mol·°C), J/(mol·°C), or J/(g·°C).

Equation [4.5-16] can be written in its integral form as:

$$\Delta \hat{H} = \int_{T_1}^{T_2} C_p(T)\, dT \quad [4.5\text{-}17]$$

where T_1 is the first temperature and T_2 is the second temperature at constant pressure. Given equation [4.5-15], the integral of the heat capacity across a temperature range is equal to the sensible heat required to warm or cool a material.

Heat capacities for most substances vary with temperature. This means that when calculating the enthalpy change due to changes in temperature, the C_p at that particular temperature must be evaluated. Heat capacities are often tabulated as polynomial functions of temperature, such as:

$$C_p(T) = a + bT + cT^2 + dT^3 \quad [4.5\text{-}18]$$

The coefficients a, b, c, and d used to calculated the C_p for several gases and water at 1 atm are given in Table 4.1. For this table, C_p is calculated using equation [4.5-18] in units of J/(mol·°C), and temperature, T, is in units of degrees Celsius. Heat capacities for other materials are given in Appendices E.1, E.2, and E.3.

TABLE 4.1

Coefficient Heat Capacity Values for Several Gases and Water at 1 atm*†

Species	State	a	$b \times 10^2$	$c \times 10^5$	$d \times 10^9$	Temperature range
Air	gas	28.94	0.4147	0.3191	−1.965	0–1500°C
Carbon dioxide	gas	36.11	4.233	−2.887	7.464	0–1500°C
Hydrogen	gas	28.84	0.00765	0.3288	−0.8698	0–1500°C
Nitrogen	gas	29.00	0.2199	0.5723	−2.871	0–1500°C
Oxygen	gas	29.10	1.158	−0.6076	1.311	0–1500°C
Water	vapor	33.46	0.688	0.7604	−3.593	0–1500°C
Water	liquid	75.4	—	—	—	0–100°C

*Heat capacity is in units of J/mol·°C, and temperatures must be in Celsius.
†Excerpted from Appendix E.1.

For solids and liquids in most biological systems, heat capacities are not a function of temperature. Therefore, the heat capacities for solids and liquids can usually be approximated with just the first term of equation [4.5-18]:

$$C_p = a \qquad [4.5\text{-}19]$$

Since C_p is constant, equation [4.5-17] is integrated as follows:

$$\Delta\hat{H} = C_p(T_2 - T_1) \qquad [4.5\text{-}20]$$

For example, the heat capacity for liquid water (75.4 J/(mol·°C), or 1 cal/(g·°C)) is not a function of temperature in the range of 0–100°C. The first term, a, of equation [4.5-18] is predominant in gases, as well as in the temperature range of most biological systems, as shown in Example 4.6.

Since values of C_p are tabulated on a per-mass and per-mole basis, $\Delta\hat{H}$ may have units of energy per mass or energy per mole. To calculate the absolute change of enthalpy of the system for this step, either amount of mass or moles may be used:

$$\Delta H = m\,\Delta\hat{H} \qquad [4.5\text{-}21]$$

$$\Delta H = n\,\Delta\hat{H} \qquad [4.5\text{-}22]$$

The rate of change of enthalpy of the system for this step may be calculated similarly, using mass flow rate or molar flow rate:

$$\Delta\dot{H} = \dot{m}\,\Delta\hat{H} \qquad [4.5\text{-}23]$$

$$\Delta\dot{H} = \dot{n}\,\Delta\hat{H} \qquad [4.5\text{-}24]$$

For some liquids and gases, such as liquid water in equilibrium with saturated steam, charts have been prepared that give the specific enthalpy as a function of temperature and pressure. Tabulated values for saturated steam are given in Appendices E.5 and E.6, as well as in other textbooks (e.g., Felder RM and Rousseau RW, *Elementary Principles of Chemical Processes*, 2000; Perry RH and Green D, *Perry's Chemical Engineers' Handbook*, 6th ed., 1984).

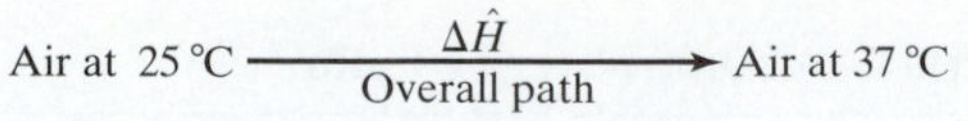

Figure 4.10
Warming of dry air in the lungs.

$$\text{Air at } 25\,^{\circ}\text{C} \xrightarrow[\text{Overall path}]{\Delta\hat{H}} \text{Air at } 37\,^{\circ}\text{C}$$

EXAMPLE 4.6 **Warming of Air During Inhalation**

Problem: The air you breathe is immediately warmed up from the ambient temperature to 37°C before entering your lungs. Calculate the specific enthalpy change when the temperature of air is raised from 20°C to 37°C. Assume the air is bone-dry, which is defined as an environment in which no water is vaporized in the air or in which the humidity is 0%.

Solution: Figure 4.10 shows a path to raise the temperature of air from 20°C to 37°C. The coefficients to calculate C_p as a function of temperature are found in Table 4.1. To calculate the change in specific enthalpy for warming the air, we can use equation [4.5-17]:

$$\begin{aligned}
\Delta\hat{H} &= \int_{T_1}^{T_2} C_p(T)\, dT \\
&= \int_{20^{\circ}\text{C}}^{37^{\circ}\text{C}} (28.94 + 0.4147 \times 10^{-2}T + 0.3191 \times 10^{-5}T^2 - 1.965 \times 10^{-9}T^3)\, dT \frac{\text{J}}{\text{mol} \cdot {}^{\circ}\text{C}} \\
&= 491.98 \frac{\text{J}}{\text{mol}} + 2.01 \frac{\text{J}}{\text{mol}} + 0.045 \frac{\text{J}}{\text{mol}} - 0.00084 \frac{\text{J}}{\text{mol}} \\
&= 494 \frac{\text{J}}{\text{mol}}
\end{aligned}$$

The first term in the heat capacity equation is predominant, which implies that C_p has very little temperature dependence within the specified range. The specific enthalpy change to raise the temperature of the air from 20°C to 37°C is 494 J/mol.

During one breath, the number of moles of gas is constant, while the temperature of the air increases. Therefore, by the ideal gas law, the pressure of air must have increased slightly, assuming that the density of air did not change. In the above calculation, we assume that specific enthalpy is not a function of pressure. This assumption turns out to be a good one, as discussed in Section 4.5.3. ■

4.5.3 Change in Pressure

Changes in enthalpy as a result of changes in pressure are not as important in most biological and medical systems as the other changes considered, but a discussion is warranted for completeness. Recall equation [4.2-25] and consider the difference between the outlet and inlet conditions:

$$\Delta\hat{H} = \Delta\hat{U} + \Delta(P\hat{V}) \qquad [4.5\text{-}25]$$

For solids and liquids, it has been observed experimentally that specific internal energy ($\hat{U}$) and specific volume ($\hat{V}$) are nearly independent of pressure. Therefore, equation [4.5-25] reduces to the following for solids and liquids:

$$\Delta\hat{H} \approx \hat{V}\, \Delta P \qquad [4.5\text{-}26]$$

In biological systems, pressure changes are usually not significant, so $\Delta\hat{H}$ is unaffected.

For ideal gases, specific enthalpy does not depend on pressure; therefore, you can assume that $\Delta\hat{H}$ is zero when considering changes in pressure. This approximation breaks down for ideal gases whose temperatures are below 0°C or whose pressures are well above 1 atm, but these situations rarely occur in biomedical calculations. Consult more advanced textbooks to handle nonideal gases (e.g., Reid RC, Prausnitz JM, and Poling BE, *The Properties of Gases and Liquids*, 1987).

4.5.4 Change in Phase

Phase changes are accompanied by relatively large changes in internal energy and enthalpy as noncovalent bonds between molecules, such as hydrogen bonds, are broken and formed. Consider the conversion of water among its vapor, liquid, and solid phases. In the vapor phase, water molecules move around most freely and have a high specific enthalpy. In the liquid and solid phases, water molecules are more densely packed. In the solid phase they have little rotation or freedom of motion. As shown in Table 4.2, the specific enthalpy of liquid water is lower than that of saturated water vapor at 100°C.

Enthalpy cannot be known in absolute terms and is determined relative to a reference point or state (see Section 4.5.1). In the case of water, the specific enthalpy is defined relative to its triple point, the temperature and pressure at which the liquid, vapor, and solid phases are in equilibrium (0.01°C, 0.00611 bar). The specific enthalpy of water at the triple point is arbitrarily defined to be zero. Remember that values for specific enthalpy can be used only when calculating the difference between two conditions.

The specific enthalpy change associated with the transition of a substance from one phase to another at constant pressure and temperature is known as the **latent heat** of the phase change. As with sensible heat, the term *heat* is used to describe an enthalpy change. A derivation similar to equation [4.5-14] shows that heat is equal to the enthalpy associated with the phase change under specified conditions. Transitions between liquid and vapor, solid and liquid, and solid and vapor are summarized in Table 4.3.

TABLE 4.2

Specific Enthalpy of Water at 100°C and 1 atm

Phase	Specific enthalpy (J/g)
Saturated liquid	419.1
Saturated vapor	2676

TABLE 4.3

Phase Change Processes and Transitions

Process name	Specific enthalpy change*	Initial phase	Final phase
Vaporization or boiling	$\Delta\hat{H}_V$	Liquid	Vapor
Condensation or liquefaction	$-\Delta\hat{H}_V$	Vapor	Liquid
Melting or fusion	$\Delta\hat{H}_M$	Solid	Liquid
Freezing	$-\Delta\hat{H}_M$	Liquid	Solid
Sublimation	$\Delta\hat{H}_S$	Solid	Vapor
Deposition	$-\Delta\hat{H}_S$	Vapor	Solid

*Enthalpy changes are defined in text.

The **latent heat of vaporization,** $\Delta\hat{H}_V$, is the specific enthalpy difference between the liquid and vapor forms of a species at a given temperature and pressure. It describes the specific enthalpy change for the process of evaporation. Evaporation requires the input of energy (e.g., boiling a pot of water). Since condensation is the reverse of vaporization and enthalpy is a state property, the **latent heat of condensation** is the negative of the latent heat of vaporization ($-\Delta\hat{H}_V$). Thus the condensation of a gas to its liquid phase requires removal of energy.

The **latent heat of melting** or fusion, $\Delta\hat{H}_M$, is the specific enthalpy difference between the solid and liquid forms of a species at a given temperature and pressure. Melting requires the input of energy (e.g., melting an ice cube). Since freezing is the reverse of melting and enthalpy is a state property, the **latent heat of freezing** is the negative of the latent heat of melting ($-\Delta\hat{H}_M$). The freezing of a liquid to its solid phase requires removal of energy.

The **latent heat of sublimation,** $\Delta\hat{H}_S$, is the specific enthalpy difference between the solid and vapor forms of a species at a given temperature and pressure. Sublimation requires the input of energy [e.g., sublimation of a block of solid carbon dioxide (i.e., dry ice) to gaseous carbon dioxide]. Since deposition is the reverse of sublimation, the **latent heat of deposition** is the negative of the latent heat of sublimation ($-\Delta\hat{H}_S$). Enthalpy changes associated with phase changes between solids and vapors and between different solid phases are not considered in this textbook.

Strictly speaking, latent heat is a function of both temperature and pressure. In practice, latent heat may vary considerably with temperature but is a very weak function of pressure. Most tabulations of latent heats are given at 1 atm pressure (labeled "standard" latent heats). Latent heats for a few compounds of importance in bioengineering applications are given in Appendix E.4. When using these charts, be sure to look up the latent heat at the temperature of interest.

EXAMPLE 4.7 Vaporization of Water at 37°C

Problem: Calculate the enthalpy change on a per-gram basis for the vaporization of water at 37°C.

Solution: This problem can be solved in two ways. The first way involves using the latent heat of vaporization for water at 37°C, which from Appendix E.5 is 2414 kJ/kg. (*Note*: An interpolation between 36°C and 38°C is required.) The specific heat of vaporization is:

$$\hat{H}_V = 2414\frac{\text{J}}{\text{g}}\left(\frac{1 \text{ cal}}{4.184 \text{ J}}\right) = 577 \frac{\text{cal}}{\text{g}}$$

The second way requires you to consider a hypothetical path. If the heat of vaporization for water is known only at 100°C (its boiling point), you can find it at 37°C by considering the following hypothetical path (Figure 4.11):

$$\Delta\hat{H} = \Delta\hat{H}_1 + \Delta\hat{H}_2 + \Delta\hat{H}_3$$

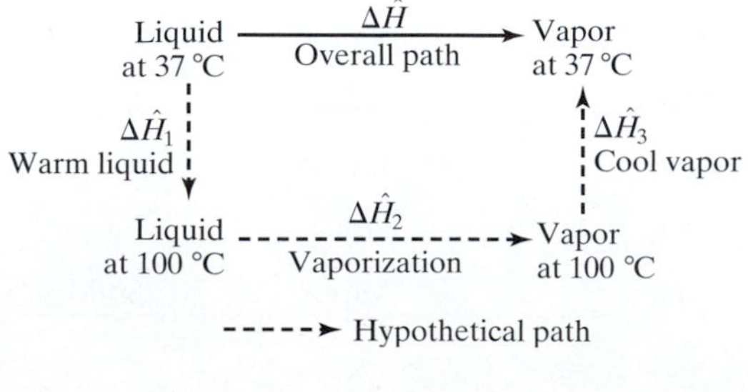

Figure 4.11 Hypothetical path for the enthalpy change associated with the vaporization of water at 37°C.

where $\Delta\hat{H}_1$ is the specific enthalpy change to raise the temperature of water from 37°C to 100°C, $\Delta\hat{H}_2$ is the specific enthalpy change of vaporization of water at 100°C, and $\Delta\hat{H}_3$ is the specific enthalpy change to lower the temperature of water from 100°C to 37°C.

For liquid water, the heat capacity, C_p, is not a function of temperature and is a constant:

$$\Delta\hat{H}_1 = C_p(T_2 - T_1) = \left(75.4 \frac{\text{J}}{\text{mol} \cdot {}^\circ\text{C}}\right)\left(\frac{1 \text{ mol}}{18 \text{ g}}\right)\left(\frac{1 \text{ cal}}{4.184 \text{ J}}\right)(100^\circ\text{C} - 37^\circ\text{C}) = 63.1 \frac{\text{cal}}{\text{g}}$$

The heat of vaporization of water at 100°C is found in Appendix E.5:

$$\Delta\hat{H}_2 = \left(2257 \frac{\text{J}}{\text{g}}\right)\left(\frac{1 \text{ cal}}{4.184 \text{ J}}\right) = 539 \frac{\text{cal}}{\text{g}}$$

The coefficients to calculate C_p as a function of temperature for water vapor are found in Table 4.1. When cooling the water vapor back down, the specific enthalpy change is:

$$\Delta\hat{H}_3 = \int_{100^\circ\text{C}}^{37^\circ\text{C}} C_p(T)\,dT = \int_{100^\circ\text{C}}^{37^\circ\text{C}} (a + bT + cT^2 + dT^3)\,dT$$

$$= \int_{100^\circ\text{C}}^{37^\circ\text{C}} (33.46 + 0.688 \times 10^{-2}T + 0.7604 \times 10^{-5}T^2 - 3.593 \times 10^{-9}T^3)\,dT \frac{\text{J}}{\text{mol}}$$

$$= -2140 \frac{\text{J}}{\text{mol}}\left(\frac{1 \text{ mol}}{18 \text{ g}}\right)\left(\frac{1 \text{ cal}}{4.184 \text{ J}}\right) = -28.4 \frac{\text{cal}}{\text{g}}$$

The overall change in specific enthalpy is the sum of the three steps in the path:

$$\Delta\hat{H} = \Delta\hat{H}_1 + \Delta\hat{H}_2 + \Delta\hat{H}_3 = 63.1\frac{\text{cal}}{\text{g}} + 539\frac{\text{cal}}{\text{g}} - 28.4\frac{\text{cal}}{\text{g}} = 574 \frac{\text{cal}}{\text{g}}$$

Therefore, the latent heat of vaporization at 37°C, calculated through the hypothetical path, is 574 cal/g, which is very close to the first solution (577 cal/g). ■

4.5.5 Mixing Effects

In ideal solutions or ideal mixtures of several compounds, the thermodynamic properties of the mixture are a simple sum of contributions from the individual components. In mixing real solutions, bonds between neighboring molecules in the old solutions are broken, and new bonds between the mixed components are formed. In these solutions, a net absorption or release of energy usually takes place, resulting in a change in the enthalpy of the mixture. For example, energy in the form of heat is released during the dilution of sulfuric acid or hydrochloric acid with water. To account for the change in enthalpy when a solute is added to a liquid, another step along a hypothetical path may need to be added.

The **heat of solution** (ΔH_{sol}) is defined as the change in enthalpy for a process in which one mole of a solute (gas or solid) is dissolved in a specific amount of a liquid solvent at a constant temperature T. As the amount of solvent becomes large, ΔH_{sol} approaches a limiting value known as the heats of solution at infinite dilution. The **heat of mixing** refers to a situation when two liquids are mixed. The calculation for the enthalpy change for the mixing of two liquids is similar to that described for the heat of solution. Values of the heat of solution and mixing are given in various books (e.g., Perry RH and Green D, *Perry's Chemical Engineers' Handbook*, 1984).

In most biological processes, significant changes in enthalpy due to heats of mixing and heats of solution do not occur. Most solutions in vivo and in vitro are dilute aqueous mixtures. For example, your body is over 70% water, in which proteins, sugars, and fats are dissolved at low concentrations. Another example is a bioreactor. Again, most of the nutrient and waste solutions in the aqueous broth are at low concentrations. Therefore, heats of mixing and heats of solution are not considered further in this textbook.

4.6 Open, Steady-State Systems—No Potential or Kinetic Energy Changes

Consider an open nonreacting system with movement of material across the system boundary. If material moves across the boundary, then flow work is also present, and the forms of the conservation of energy equation that include enthalpy can be used. In many biological and biomedical systems, especially those with chemical reactions, high-velocity motion and large changes in height or in position in an electromagnetic field do not generally occur. Thus, we consider a class of problems where changes in potential and kinetic energy are assumed to be negligible. At steady-state, all properties of the system are time invariant. Therefore, the total energy of the system does not change or accumulate. Consider a steady-state, nonreacting system that undergoes no changes in potential and kinetic energy. Formulations for the differential [4.3-10] and algebraic [4.3-17] forms of the conservation of energy equation can be reduced to:

$$\sum_i \dot{m}_i \hat{H}_i - \sum_j \dot{m}_j \hat{H}_j + \sum \dot{Q} + \sum \dot{W}_{\text{nonflow}} = 0 \qquad [4.6\text{-}1]$$

$$\sum_i m_i \hat{H}_i - \sum_j m_j \hat{H}_j + \sum Q + \sum W_{\text{nonflow}} = 0 \qquad [4.6\text{-}2]$$

Variable definitions are given in Section 4.3.

Recall that the symbol Δ is used to represent the difference between the outlet conditions (indexed as j) and the inlet conditions (indexed as i). Given the definitions for the rate of change in enthalpy, $\Delta \dot{H}$ (equation [4.5-2]), and the change in enthalpy, ΔH (equation [4.5-8]), equations [4.6-1] and [4.6-2] reduce to:

$$-\Delta \dot{H} + \sum \dot{Q} + \sum \dot{W}_{\text{nonflow}} = 0 \qquad [4.6\text{-}3]$$

$$-\Delta H + Q + W_{\text{nonflow}} = 0 \qquad [4.6\text{-}4]$$

We avoid needing to know the absolute specific enthalpies given in equations [4-6.1] and [4.6-2] by looking at the difference between the inlet and outlet conditions or streams.

In Section 4.5, we discuss how to calculate the change in specific enthalpy for nonreacting systems with components that changed in temperature, pressure, and phase. **Practically, the strategies in Section 4.5 allow for the calculation of specific enthalpy changes that can then be used in the forms of the conservation of total energy equations given in [4.6-3] and [4.6-4].**

EXAMPLE 4.8 **Heat Loss During Breathing**

Problem: Estimate heat loss during respiration. Assume that a normal person inspires about 6 L/min of bone-dry air at 20°C. Assume that expired air is saturated with water vapor and is at 37°C.

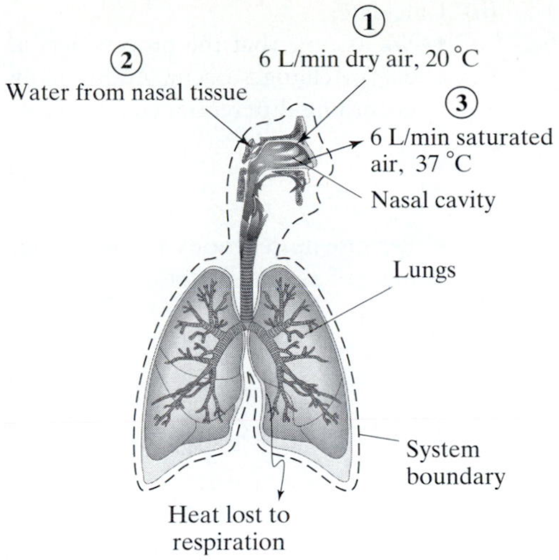

Figure 4.12
Humidification and heating of air during breathing.

Solution:

1. Assemble
   (a) Find: rate of heat loss during respiration.
   (b) Diagram: The respiratory system modeled with the system boundary representing the tissue lining in the respiratory system (Figure 4.12). The incoming bone-dry air at 20°C is represented by stream 1. The outflowing saturated air (i.e., air holding as much water as it can at the given temperature and pressure) at 37°C is represented by stream 3. Stream 2 represents water evaporated from the nasal tissue that enters the air in the system.
2. Analyze
   (a) Assume:
   - Process operates at steady-state.
   - No shaft work.
   - Kinetic and potential energy changes are negligible.
   - No reactions.
   - Air (with and without water vapor) behaves like an ideal gas. This implies enthalpy is not affected by pressure changes.
   - Mass flow rates of inhaled and exhaled air, excluding water, are the same.

   (b) Extra data:
   - Molecular weight of air is 28.84 g/mol.
   - Density of air is 0.0012 g/cm^3.
   - The molal humidity for saturated water vapor at 37°C is 6.7%; therefore, when saturated, there are approximately 0.041 g of water per 1.0 g of dry air.

   (c) Variables, notations, units:
   - Units: °C, cal, g, min, mol.

   (d) Basis: The basis of the inlet stream is calculated using the inlet flow rate of 6 L/min:

$$\dot{m}_1 = \dot{V}\rho = \left(6\frac{\text{L}}{\text{min}}\right)\left(0.0012\frac{\text{g}}{\text{cm}^3}\right)\left(\frac{1000\ \text{cm}^3}{1\ \text{L}}\right) = 7.2\ \frac{\text{g}}{\text{min}}$$

3. Calculate
   (a) Equations: Rates of material flow are given in this system; therefore, the differential equations for the conservation of mass and total energy are needed:

$$\sum_i \dot{m}_i - \sum_j \dot{m}_j = \frac{dm^{\text{sys}}}{dt}$$

$$\sum_i \dot{m}_i(\hat{E}_{P,i} + \hat{E}_{K,i} + \hat{H}_i) - \sum_j \dot{m}_j(\hat{E}_{P,j} + \hat{E}_{K,j} + \hat{H}_j) + \sum \dot{Q} + \sum \dot{W}_{\text{nonflow}} = \frac{dE_T^{\text{sys}}}{dt}$$

(b) Calculate:

- We assume that the process acts at steady-state and that no potential or kinetic energy changes occur. Additionally, no shaft work occurs, so we can reduce the governing differential equation for the conservation of total energy to:

$$-\Delta\dot{H} + \sum\dot{Q} = 0$$

- Because no reactions are occurring, we can write equations counting the moles or mass of air and water:

$$\dot{n}_{1,\text{air}} - \dot{n}_{3,\text{air}} = 0$$
$$\dot{m}_{2,\text{H}_2\text{O}} - \dot{m}_{3,\text{H}_2\text{O}} = 0$$

- The molar flow rates of air into and out of the system are calculated:

$$\dot{n}_{1,\text{air}} = \dot{n}_{3,\text{air}} = \frac{\dot{V}\rho}{M} = \frac{\left(6.0\,\frac{\text{L}}{\text{min}}\right)\left(\frac{0.012\text{ g}}{\text{cm}^3}\right)\left(\frac{1000\text{ cm}^3}{\text{L}}\right)}{28.84\,\frac{\text{g}}{\text{mol}}} = 0.25\,\frac{\text{mol}}{\text{min}}$$

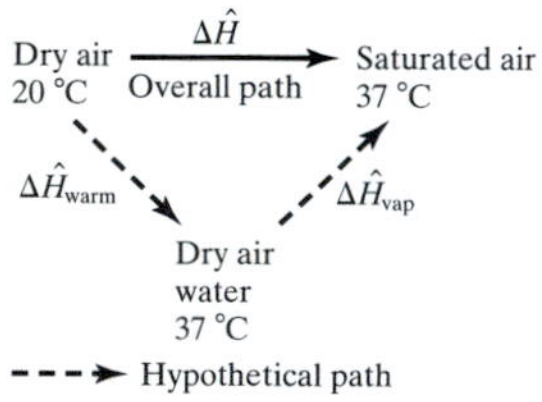

Figure 4.13
Hypothetical path for the enthalpy change associated with the heating of air and vaporization of water at 37°C during breathing.

- Because air saturated with water vapor at 37°C carries 0.041 g of water per 1.0 g of dry air, we can calculate the mass flow rate of water in the outlet stream using the mass flow rate basis:

$$\dot{m}_{3,\text{H}_2\text{O}} = \left(0.041\,\frac{\text{g H}_2\text{O}}{\text{g air}}\right)\left(7.2\,\frac{\text{g air}}{\text{min}}\right) = 0.295\,\frac{\text{g H}_2\text{O}}{\text{min}}$$

- A hypothetical path to model the enthalpy change across the system consists of two steps: (a) heating the dry air from 20°C to 37°C and (b) vaporizing water at 37°C (Figure 4.13). The change in rate of enthalpy is:

$$\Delta\dot{H} = \Delta\dot{H}_{\text{warm}} + \Delta\dot{H}_{\text{vap}}$$

Recall from Example 4.6 that $\Delta\hat{H}_{\text{warm}}$ is equal to 494 J/mol. Therefore:

$$\Delta\dot{H}_{\text{warm}} = \dot{n}_{3,\text{air}}\,\Delta\hat{H}_{\text{warm}} = \left(0.25\frac{\text{mol}}{\text{min}}\right)\left(494\frac{\text{J}}{\text{mol}}\right) = 124\,\frac{\text{J}}{\text{min}}$$

The vaporization of water at 37°C is calculated in Example 4.7. The latent heat of vaporization, $\Delta\hat{H}_{\text{vap}}$, at 37°C is 577 cal/g:

$$\Delta\dot{H}_{\text{vap}} = \dot{m}_{3,\text{H}_2\text{O}}\,\Delta\hat{H}_{\text{vap}} = \left(0.295\frac{\text{g}}{\text{min}}\right)\left(577\frac{\text{cal}}{\text{g}}\right)\left(\frac{4.184\text{ J}}{\text{cal}}\right) = 712\,\frac{\text{J}}{\text{min}}$$

- We can then use the reduced governing equation to determine the energy required by the warming process and vaporization:

$$\sum\dot{Q} = \Delta\dot{H} = \Delta\dot{H}_{\text{warm}} + \Delta\dot{H}_{\text{vap}} = 124\frac{\text{J}}{\text{min}} + 712\frac{\text{J}}{\text{min}} = 836\frac{\text{J}}{\text{min}}$$

4. Finalize

(a) Answer: The rate of heat loss during respiration is 836 J/min. The sensible heat loss is 124 J/min; the heat loss due to the vaporization of water is 712 J/min. Note that the heat loss due to the vaporization of water is about six times as large as that of warming the air.

(b) Check: The energy lost during respiration is compared to published values by consulting a physiology textbook (Guyton and Hall, 2000). The calculated heat loss of 836 J/min is equivalent to approximately 200 cal/min or 288 kcal/day. This value is within range of published values and represents 16–18% of the basal metabolic rate (BMR), the minimum level of energy required to perform chemical reactions in the body and maintain essential activities of the central nervous system, heart, kidney and other organs. ■

EXAMPLE 4.9 Heat Requirement in Warming Blood

Problem: Since blood is refrigerated for storage, it is warmed before contact with a patient to prevent hypothermia. Calculate the rate of heat required to continuously warm 10.0 L/min of blood from 30°C to 37°C using an electric heater, as shown in Figure 4.14. A stirrer adds work to the system at a rate of 0.50 kW. Assume the heat capacity of blood is constant at $1.0 \text{ cal/(g} \cdot {}^\circ\text{C)}$ and the density of blood is 1.0 g/mL. The working volume of the tank is 1.0 L.

Solution:

1. Assemble
   (a) Find: rate of heat required to warm 10.0 L/min of blood from 30°C to 37°C.
   (b) Diagram: Figure 4.14 shows the blood heating device. Blood enters and leaves the heater at a rate of 10.0 L/min. Heat and work are both added to the system.
2. Analyze
   (a) Assume:
   - Tank is well mixed, so conditions inside the tank are the same as the outlet stream (e.g., temperature in the tank and in outlet stream 2 is 37°C).
   - Heat capacity (C_p) of blood does not depend on temperature.
   - Density of blood (ρ) is constant.
   - No evaporation.
   - Heat lost to the surroundings is negligible.
   - System is at steady-state.
   - Potential and kinetic energy changes are negligible.
   - No reactions.

   (b) Extra data: No extra data are needed.
   (c) Variables, notations, units:
   - T_1 = temperature of inlet stream.
   - T_2 = temperature of outlet stream and inside the tank.
   - Units: °C, cal, g, min.

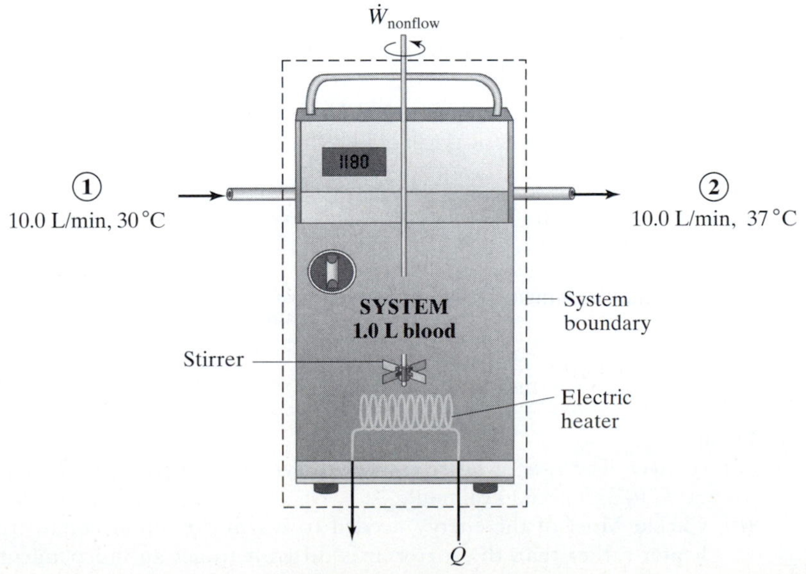

Figure 4.14
Blood heating device (steady-state operation).

(d) Basis: Since we assume the density of blood is 1.0 g/mL, we can use the inlet flow rate of 10.0 L/min of blood in stream 1 to obtain a basis of 10.0 kg/min.

3. Calculate

(a) Equations: Since rates of material flow and work are given, the differential conservation of mass and conservation of total energy equations are most appropriate:

$$\sum_i \dot{m}_i - \sum_j \dot{m}_j = \frac{dm^{sys}}{dt}$$

$$\sum_i \dot{m}_i(\hat{E}_{P,i} + \hat{E}_{K,i} + \hat{H}_i) - \sum_j \dot{m}_j(\hat{E}_{P,j} + \hat{E}_{K,j} + \hat{H}_j) + \sum \dot{Q} + \sum \dot{W}_{\text{nonflow}} = \frac{dE_T^{sys}}{dt}$$

(b) Calculate:

- Because we assume the process has no reactions and the system is at steady-state, the mass flow rates of blood into and out of the system are equal to the basis:

$$\dot{m}_1 = \dot{m}_2 = 10.0 \frac{\text{kg}}{\text{min}}$$

- Since the system is at steady-state and no kinetic or potential energy changes are occurring, the steady-state energy balance equation is applied here. (*Note*: While the inlet and outlet flows do contribute kinetic energy, the change in kinetic energy is zero, since the inlet and outlet flows are the same.) Only one source each of heat and work is identified:

$$\dot{m}_1\hat{H}_1 - \dot{m}_2\hat{H}_2 + \dot{Q} + \dot{W}_{\text{nonflow}} = 0$$

- For the inlet and outlet streams, respectively:

$$\dot{m}_1\hat{H}_1 = \dot{m}_1 C_p(T_1 - T_{\text{ref}})$$
$$\dot{m}_2\hat{H}_2 = \dot{m}_2 C_p(T_2 - T_{\text{ref}})$$

where T_{ref} is an arbitrary reference temperature. Note that $\hat{H}_{\text{ref}}$ is dropped from the above equation since we assume its value is zero. Substituting this into the reduced governing energy equation yields:

$$\dot{m}_1 C_p(T_1 - T_{\text{ref}}) - \dot{m}_2 C_p(T_2 - T_{\text{ref}}) + \dot{Q} + \dot{W}_{\text{nonflow}} = 0$$

- Because $\dot{m}_1 = \dot{m}_2$, we can simplify the above equation to:

$$\dot{m}_1 C_p(T_1 - T_2) + \dot{Q} + \dot{W}_{\text{nonflow}} = 0$$

- Nonflow work (0.5 kW = 500 J/s) is positive because it is added to the system. Substituting numerical values into the above equation gives:

$$10.0\frac{\text{kg}}{\text{min}}\left(1.0\frac{\text{cal}}{\text{g}\cdot{}^\circ\text{C}}\right)\left(1000\,\frac{\text{g}}{\text{kg}}\right)(30^\circ\text{C} - 37^\circ\text{C})$$
$$+ \dot{Q} + 500\frac{\text{J}}{\text{s}}\left(\frac{0.239\ \text{cal}}{\text{J}}\right)\left(60\,\frac{\text{s}}{\text{min}}\right) = 0$$
$$\dot{Q} = 62{,}800\ \frac{\text{cal}}{\text{min}}$$

4. Finalize

(a) Answer: The rate of heat supplied to system to warm 10.0 L/min of blood from 30°C to 37°C is 63 kcal/min.

(b) Check: Most of the energy needed to warm the blood comes from the electric heater rather than the stirrer. It is difficult to get an independent check on this answer. ■

The two examples above were differential equations that contained heat or work terms or both. A special case of the algebraic form of the conservation of energy equation applies for an open, steady-state, nonreacting system that does not have contributions from heat or work. Since no heat is exchanged, the system is considered adiabatic.

Suppose that you have a mass m_1 at temperature T_1. To this is added a mass m_2 of the identical material that is at temperature T_2. The heat capacity of both masses is identical and is given as C_p. To calculate the temperature of the combined masses $(m_1 + m_2)$, imagine that m_1 and m_2 are placed into the system and that the combined mass leaves the system.

The algebraic equation:

$$\sum_i (E_{P,i} + E_{K,i} + H_i) - \sum_j (E_{P,j} + E_{K,j} + H_j) + Q + W_{\text{nonflow}} = E_{T,\text{f}}^{\text{sys}} - E_{T,0}^{\text{sys}} \qquad [4.6\text{-}5]$$

is reduced for a steady-state system with no heat or work and no change in potential or kinetic energy:

$$\sum_i H_i - \sum_j H_j = \Delta H = 0 \qquad [4.6\text{-}6]$$

For the mass m_1, the change in enthalpy is written:

$$\Delta H_1 = m_1 C_p (T_1 - T_{\text{ref}}) \qquad [4.6\text{-}7]$$

where T_{ref} is an arbitrarily selected reference temperature. For the mass m_2, the change in enthalpy is written:

$$\Delta H_2 = m_2 C_p (T_2 - T_{\text{ref}}) \qquad [4.6\text{-}8]$$

The change in enthalpy for the combined mass, $m_1 + m_2$, is:

$$\Delta H_3 = (m_1 + m_2) C_p (T_3 - T_{\text{ref}}) \qquad [4.6\text{-}9]$$

where ΔH_3 and T_3 denote the enthalpy change and temperature of the combined mass, respectively. It is assumed that the heat capacity of the combined mass is equal to that of the original masses. The overall change in enthalpy is:

$$\Delta H = \Delta H_1 + \Delta H_2 - \Delta H_3 = 0 \qquad [4.6\text{-}10]$$

which is reduced to:

$$m_1 T_1 + m_2 T_2 = (m_1 + m_2) T_3 \qquad [4.6\text{-}11]$$

So the temperature T_3 is:

$$T_3 = \frac{m_1 T_1 + m_2 T_2}{m_1 + m_2} \qquad [4.6\text{-}12]$$

It makes sense that T_3 is a linear combination of the temperatures of the masses entering the system in proportion to their respective masses. Absolute temperatures must be used in equations [4.6-11] and [4.6-12].

As an example, consider the addition of 100 g of room temperature water (25°C) and 10 g of ice-cold water (4°C) to a beaker. The resulting system is 110 g of water at a temperature of 23°C. It makes sense that the temperature of the mixture is between that of the two initial substances and is closer to the substance that was a larger contributor to the system's mass.

4.7 Open, Steady-State Systems with Potential or Kinetic Energy Changes

In some engineering scenarios, changes in potential or kinetic energy or both are significant, such as when material has high velocity or the changes in height or position of material in a conservative field are large. At steady-state, no total energy accumulates in the system. Consider the steady-state situation with changes in potential and kinetic energy. The differential (equation [4.3-10]) and algebraic (equation [4.3-17]) forms of the conservation of total energy equation can be reduced:

$$\sum_i \dot{m}_i(\hat{E}_{P,i} + \hat{E}_{K,i} + \hat{H}_i) - \sum_j \dot{m}_j(\hat{E}_{P,j} + \hat{E}_{K,j} + \hat{H}_j) + \sum \dot{Q} + \sum \dot{W}_{\text{nonflow}} = 0 \quad [4.7\text{-}1]$$

$$\sum_i m_i(\hat{E}_{P,i} + \hat{E}_{K,i} + \hat{H}_i) - \sum_j m_j(\hat{E}_{P,j} + \hat{E}_{K,j} + \hat{H}_j) + Q + W_{\text{nonflow}} = 0 \quad [4.7\text{-}2]$$

where:

$$\hat{E}_P = gh \quad [4.7\text{-}3]$$

$$\hat{E}_K = \frac{1}{2}v^2 \quad [4.7\text{-}4]$$

where g is the gravitational acceleration constant, h is height relative to a reference plane, and v is velocity. When changes in enthalpy do not occur due to changes in temperature, pressure, or phase across a system and no chemical reactions occur, equations [4.7-1] and [4.7-2] can reduce to:

$$\sum_i \dot{m}_i(\hat{E}_{P,i} + \hat{E}_{K,i}) - \sum_j \dot{m}_j(\hat{E}_{P,j} + \hat{E}_{K,j}) + \sum \dot{Q} + \sum \dot{W}_{\text{nonflow}} = 0 \quad [4.7\text{-}5]$$

$$\sum_i m_i(\hat{E}_{P,i} + \hat{E}_{K,i}) - \sum_j m_j(\hat{E}_{P,j} + \hat{E}_{K,j}) + Q + W_{\text{nonflow}} = 0 \quad [4.7\text{-}6]$$

These equations also assume no flow work.

If flow work is significant or if changes in the pressure or density between the inlet and outlet conditions are significant, the equations below can be derived starting from equations [4.7-1] and [4.7-2]. Here, it is assumed that no changes in internal energy occur:

$$\sum_i \dot{m}_i\left(\hat{E}_{P,i} + \hat{E}_{K,i} + \frac{P_i}{\rho_i}\right) - \sum_j \dot{m}_j\left(\hat{E}_{P,j} + \hat{E}_{K,j} + \frac{P_j}{\rho_j}\right) + \sum \dot{Q} + \sum \dot{W}_{\text{nonflow}} = 0 \quad [4.7\text{-}7]$$

$$\sum_i\left(E_{P,i} + E_{K,i} + \frac{P_i}{\rho_i}\right) - \sum_j\left(E_{P,j} + E_{K,j} + \frac{P_j}{\rho_j}\right) + Q + W_{\text{nonflow}} = 0 \quad [4.7\text{-}8]$$

EXAMPLE 4.10 Hydroelectric Power

Problem: Hydroelectric power plants convert the energy of moving water into electricity. An advantage of such plants over conventional power plants powered by coal is that emissions linked to acid rain (sulfur dioxide and nitrogen oxides) are substantially lower. Hydroelectric plants may use a dam, a quickly moving river, or a waterfall to produce electricity. A disadvantage of hydroelectric power plants is that they may disrupt aquatic wildlife in the river across which the power plant is built.

Suppose that a river flowing into an hydroelectric power plant usually has a velocity of about 1.0 m/s. After flowing through the plant, the river opens up into a large lake, and its velocity drops to nearly 0 m/s. If water flows through the power plant at 2.8 m^3/s and has a head of 9 m, how much energy can theoretically be produced? (The term *head* is used to describe the vertical distance or elevation of a liquid above a reference plane.)

In this situation, only 190 kW of power is produced. The efficiency of the power plant is estimated as the ratio of the actual energy that is relayed from the plant relative to the ideal or maximum energy generation. What is the efficiency of this hydroelectric power plant? Brainstorm a few reasons why the plant is not 100% efficient.

Solution: Water enters the power plant system traveling at 1.0 m/s with a volumetric flow rate of 2.8 m^3/s. Water leaves the plant and enters the lake. The system is diagrammed in Figure 4.15. We assume the system is at steady-state and has no friction, no internal energy change, and no heat transfer across the system boundary. Our goal is to find the energy (work) that can theoretically be generated by the changes in the potential and kinetic energy of the water, as well as the efficiency of the power plant.

The inlet mass flow rate is:

$$\dot{m}_1 = 2.8\frac{\text{m}^3}{\text{s}}\left(100\,\frac{\text{cm}}{\text{m}}\right)^3\left(1\frac{\text{g}}{\text{cm}^3}\right)\left(\frac{1\text{ kg}}{1000\text{ g}}\right) = 2800\,\frac{\text{kg}}{\text{s}}$$

Given the assumptions above, equation [4.7-5] is reduced:

$$\dot{m}_1(\hat{E}_{P,1} + \hat{E}_{K,1}) - \dot{m}_2(\hat{E}_{P,2} + \hat{E}_{K,2}) + \sum \dot{W}_{\text{nonflow}} = 0$$

The change in gravitational potential energy is:

$$\hat{E}_{P,1} - \hat{E}_{P,2} = g(h_1 - h_2) = \left(9.81\,\frac{\text{m}}{\text{s}^2}\right)(9.0\text{ m} - 0) = 88.3\,\frac{\text{m}^2}{\text{s}^2}$$

Note that the reference height is set to 0. The change in kinetic energy is:

$$\hat{E}_{K,1} - \hat{E}_{K,2} = \frac{1}{2}(v_1^2 - v_2^2) = \frac{1}{2}\left(\left(1.0\,\frac{\text{m}}{\text{s}}\right)^2 - 0\right) = 0.5\,\frac{\text{m}^2}{\text{s}^2}$$

Note that the change in potential energy is a much larger contribution than the change in kinetic energy.

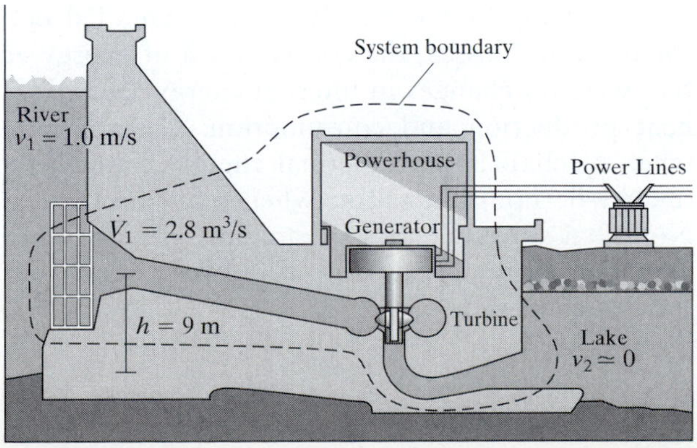

Figure 4.15
Flow of water through a power plant.

Because the system is at steady-state, $\dot{m}_1 = \dot{m}_2$. Solving for nonflow work gives:

$$\dot{W}_{\text{nonflow}} = -\dot{m}_1((\hat{E}_{P,1} - \hat{E}_{P,2}) + (\hat{E}_{K,1} - \hat{E}_{K,2}))$$

$$= -2800\frac{\text{kg}}{\text{s}}\left(88.3\frac{\text{m}^2}{\text{s}^2} + 0.5\frac{\text{m}^2}{\text{s}^2}\right) = -248 \text{ kW}$$

Since the value of the nonflow work is negative, work is being done by the system on the surroundings. This makes sense, since a hydroelectric plant is designed to generate power. If the plant produces 190 kW of energy, then the efficiency η is:

$$\eta = \frac{190 \text{ kW}}{248 \text{ kW}}(100) = 76\%$$

Some reasons for the power plant's lower efficiency could be losses of energy as friction or heat. In addition, some energy may be consumed by electrical equipment, further reducing the amount available to be relayed from the plant. In comparison, power plants that rely on steam power, such as plants that burn coal, have a lower efficiency and lose a significant amount of energy when heating water and converting it to steam.

In summary, the power plant is 76% efficient, producing 190 kW of electricity out of an available 248 kW. Most of the energy comes from the potential energy difference, which is the main driving force for most hydroelectric power plants. ■

Let us compare the differential conservation of energy equation given in [4.7-7] to the extended Bernoulli equation presented in Chapter 6 (equation [6.11-9b]), which is used to describe a system with fluid flow in which shaft (pump) work and frictional losses occur:

$$\dot{m}(\hat{E}_{P,i} - \hat{E}_{P,j}) + \dot{m}(\hat{E}_{K,i} - \hat{E}_{K,j}) + \dot{m}\left(\frac{P_i}{\rho_i} - \frac{P_j}{\rho_j}\right) + \sum \dot{W}_{\text{shaft}} - \sum \dot{f} = 0 \qquad [4.7\text{-}9]$$

Both this equation and equation [4.7-7] describe changes in potential and kinetic energy. Both equations describe flow work and shaft work.

While these equations are very similar, it is important to be careful to pick the right one for the problem at hand. The extended Bernoulli equation is restricted to steady-state systems with one fluid inlet and one fluid outlet, a uniform velocity profile, and an incompressible fluid. In addition, only interconversions between mechanical energy and thermal energy are considered. Although friction does change the thermal energy of a system, friction and heat are not equivalent or interchangeable terms. While the extended Bernoulli equation accounts only for frictional losses, the conservation of energy equation for a steady-state system with no changes in internal energy (equation [4.7-7]) captures all forms of heat production and consumption. Use the conservation of energy equation when mechanical and thermal energy changes occur in the system; use the extended Bernoulli equation when only mechanical energy changes are present. Note that since the above example contained only mechanical energy terms, the extended Bernoulli equation could have been used to solve this problem to yield the same answer.

4.8 Calculation of Enthalpy in Reactive Processes

During chemical reactions, rearranging the bonds between the atoms of reactants and products causes changes in the internal energy of a system. In reactions, energy is required to break the existing bonds of the reactants, and energy is released during bond formation to create the products. The difference between the final and initial energy states of the products and reactants is known as the heat of reaction. We provide a brief overview of heat of reaction and then discuss methods for calculating heat of reaction. **The tools developed in this section allow us to calculate the specific enthalpy change for a reacting system, which can then be used in the total energy conservation equation.**

4.8.1 Heat of Reaction

The **heat of reaction** or enthalpy of reaction ($\Delta\hat{H}_r$) is the enthalpy change for a single process reacting at a specific constant temperature and pressure in which stoichiometric quantities of reactants react completely to form products. The **standard heat of reaction** ($\Delta\hat{H}_r^\circ$) is designated with a degree symbol and is the heat of reaction when both the reactants and products are at a specified reference temperature and pressure, usually 25°C and 1 atm. The heat of reaction, $\Delta\hat{H}_r$, and the standard heat of reaction, $\Delta\hat{H}_r^\circ$, are given on a per-mole basis.

Reactants undergo a chemical reaction at a certain temperature and pressure to form some amount of products. The change in enthalpy for a reacting system, ΔH_r, is equal to the difference in the enthalpy of the products and the reactants:

$$\Delta H_r = \sum_p (n_p\hat{H}_p) - \sum_r (n_r\hat{H}_r) \qquad [4.8\text{-}1]$$

where n is the number of moles actually involved in or produced by the reaction (not necessarily the number of moles in the system), $\hat{H}$ is the specific enthalpy of a species, the subscript p is for product, and the subscript r is for reactant. Because chemical reactions are written on a mole basis, equation [4.8-1] and subsequent equations include the variables of specific enthalpy on a per-mole basis and mole.

Chemical reactions can be classified as either endothermic or exothermic. An **endothermic** process requires more energy to break the bonds of the reactants than is released when the bonds of the products are formed. During an **exothermic** reaction, more energy is released when the bonds of the products are formed than is required to break the bonds of the reactants. That is, an exothermic process generates energy. In an exothermic reaction, the heat of reaction is negative in value. On the other hand, an endothermic reaction consumes energy, and the heat of reaction is positive in value.

When calculating the standard heat of reaction, the phase of the reactants and products must be known. A liquid is denoted (ℓ); solid, (s); gas, (g); and crystal, (c). Consider the balanced liquid (ℓ)-phase reaction:

$$a\text{A}(\ell) + b\text{B}(\ell) \longrightarrow p\text{P}(\ell) + q\text{Q}(\ell) \qquad [4.8\text{-}2]$$

where A and B are reactants; P and Q are products; and a, b, p, and q are the corresponding stoichiometric coefficients. The stoichiometric coefficient of a compound is the number preceding a compound in a balanced reaction. (Refer to Chapter 3 for a more in-depth discussion of reactions and stoichiometric coefficients.) Suppose that a moles of A and b moles of B in a system react completely to form p moles of P and q moles of Q. The heat of reaction in the liquid phase for the balanced equation [4.8-2] is written as $\Delta H_r(\ell)$:

$$\Delta H_r(\ell) = \sum_p (n_p \hat{H}_p) - \sum_r (n_r \hat{H}_r) = p\hat{H}_P + q\hat{H}_Q - a\hat{H}_A - b\hat{H}_B \qquad [4.8\text{-}3]$$

The heat of reaction per mole of reactant A is $\Delta H_r(\ell)$ divided by the stoichiometric coefficient a. The heat of reaction per mole of any reactant or product can be calculated similarly.

The numerical value of the heat of reaction depends on the state of aggregation of the reactants and products. In equation [4.8-2], all the reactants and products are in the liquid phase. If, however, the product Q is in the gaseous state:

$$a\mathrm{A}(\ell) + b\mathrm{B}(\ell) \longrightarrow p\mathrm{P}(\ell) + q\mathrm{Q}(g) \qquad [4.8\text{-}4]$$

then the enthalpy changes for reactions [4.8-4] and [4.8-2] are not equal.

Enthalpy is an extensive property and thus depends on the size of system. Consequently, the stoichiometric equation determines the heat of reaction. The heat of reaction for equation [4.8-2] is $\Delta H_r(\ell)$. The heat of reaction for the following reaction:

$$4a\mathrm{A}(\ell) + 4b\mathrm{B}(\ell) \longrightarrow 4p\mathrm{P}(\ell) + 4q\mathrm{Q}(\ell) \qquad [4.8\text{-}5]$$

is four times that of equation [4.8-2], or $4\Delta H_r(\ell)$. This makes sense, since the number of moles involved in the reaction increases by a factor of four.

It is impossible to compile a complete standard heat of reaction table, since the number of reactions is infinite. **However, using Hess's law and values of the heats of formation or combustion, the heat of reaction for many reacting systems at standard temperature and pressure can be calculated.** According to **Hess's law**, if the original reaction can be written using an algebraic combination of other reactions, then the standard enthalpy of a reaction is the algebraic combination of the enthalpies of the other reactions. Hess's law is a valid method because specific enthalpy is a state function. To calculate the specific enthalpy change across a reacting system at nonstandard temperature or pressure or both, other enthalpy changes may need to be considered as well. Section 4.8.2 considers reacting systems at standard temperature and pressure. Section 4.8.3 considers reacting systems at nonstandard temperature.

4.8.2 Heats of Formation and Combustion

The standard heat of reaction can be calculated using the heat of formation or the heat of combustion. The **standard heat of formation** of a compound ($\Delta\hat{H}_f^\circ$) is the specific enthalpy change associated with the formation of 1 mole of the compound at a reference temperature and pressure (usually 25°C and 1 atm) from its constituent elements. $\Delta\hat{H}_f^\circ$ is given on a per-mole basis. When writing a formation reaction for a compound, use the elements as they naturally occur (e.g., N_2 rather than N). Elemental constituents that are commonly encountered in biochemical reactions are $O_2(g)$, $N_2(g)$, C(s), and $H_2(g)$. The standard heat of formation for these elements and others as they naturally occur is zero.

An example is the formation of urea [$CO(NH_2)_2$]. Its formation equation is written as:

$$\mathrm{C}(s) + 2\mathrm{H_2}(g) + \frac{1}{2}\mathrm{O_2}(g) + \mathrm{N_2}(g) \longrightarrow \mathrm{CO(NH_2)_2}(s) \qquad [4.8\text{-}6]$$

Using the formalism established in equation [4.8-1], the standard heat of formation is the specific enthalpy difference between the product [$CO(NH_2)_2$] and the reactants (C, H_2, O_2 and N_2). The standard heats of formation of C(s), H_2(g), N_2(g) and O_2(g) are zero. The $\Delta\hat{H}_f^\circ$ of $CO(NH_2)_2$, and hence of the formation reaction, is given in Appendix E.7 as −533 kJ/mol. Since $\Delta\hat{H}_f^\circ$ is negative, the reaction is exothermic. Note that the reaction is written such that 1 mole of urea is formed, even though this forces one stoichiometric coefficient to be a fractional value.

The standard heat of reaction is calculated from the standard heats of formation of compounds in the reaction of interest:

$$\Delta\hat{H}_r^\circ = \sum_p(\sigma_p\,\Delta\hat{H}_{f,p}^\circ) - \sum_r(\sigma_r\,\Delta\hat{H}_{f,r}^\circ) \qquad [4.8\text{-}7]$$

where σ is the stoichiometric coefficient, p is product, r is reactant, and $\Delta\hat{H}_f^\circ$ is the standard heat of formation. Using this method, it is necessary to determine the standard heat of formation for each product and reactant.

Recall the hypothetical reaction equation:

$$a\text{A}(\ell) + b\text{B}(\ell) \rightarrow p\text{P}(\ell) + q\text{Q}(\ell) \qquad [4.8\text{-}8]$$

where A and B are reactants; P and Q are products; and a, b, p, and q are the corresponding stoichiometric coefficients. Here, the heat of reaction is calculated from the standard heats of formation of the four different compounds:

$$\Delta\hat{H}_r^\circ = \sum_p(\sigma_p\,\Delta\hat{H}_{f,p}^\circ) - \sum_r(\sigma_r\,\Delta\hat{H}_{f,r}^\circ) = p\,\Delta\hat{H}_{f,\text{P}}^\circ + q\,\Delta\hat{H}_{f,\text{Q}}^\circ - a\,\Delta\hat{H}_{f,\text{A}}^\circ - b\,\Delta\hat{H}_{f,\text{B}}^\circ \qquad [4.8\text{-}9]$$

Lists of standard heats of formation are found in Appendices E.7 and E.8. Standard heats of formation for many compounds are tabulated and listed in chemical engineering textbooks (e.g., Felder RM and Rousseau RW, *Elementary Principles of Chemical Processes*, 2000; Perry RH and Green D, *Perry's Chemical Engineers' Handbook*, 1984). Heats of formation of some compounds can be measured with calorimeters.

EXAMPLE 4.11 **Photosynthesis Reaction**

Problem: The abundance of renewable bioenergy resources is a result of the rapid growth of green plants. Photosynthesis is important to sustaining life on Earth. Photosynthetic organisms convert carbon dioxide and water into glucose and oxygen. Find the standard heat of reaction for photosynthesis:

$$6\,CO_2(g) + 6\,H_2O(\ell) \longrightarrow C_6H_{12}O_6(s) + 6\,O_2(g)$$

Solution: The standard heat of reaction is calculated using the standard heats of formation of the reactants and products:

$$\Delta\hat{H}_r^\circ = \sum_p(\sigma_p\,\Delta\hat{H}_{f,p}^\circ) - \sum_r(\sigma_r\,\Delta\hat{H}_{f,r}^\circ) = 1\hat{H}_{f,C_6H_{12}O_6}^\circ + 6\hat{H}_{f,O_2}^\circ - 6\hat{H}_{f,CO_2}^\circ - 6\hat{H}_{f,H_2O}^\circ$$

The heats of formation for the different species are shown in Table 4.4 and in Appendices E.7 and E.8.

$$\Delta\hat{H}_r^\circ = 1\left(-1274\frac{\text{kJ}}{\text{mol}}\right) + 6(0) - 6\left(-394\frac{\text{kJ}}{\text{mol}}\right) - 6\left(-286\frac{\text{kJ}}{\text{mol}}\right) = 2810\frac{\text{kJ}}{\text{mol}}$$

TABLE 4.4

Heats of Formation for Photosynthetic Species

Species	$\Delta \hat{H}_f^\circ$ (kJ/mol)
Carbon dioxide, $CO_2(g)$	−394
Water, $H_2O(\ell)$	−286
Glucose, $C_6H_{12}O_6(s)$	−1274
Oxygen, $O_2(g)$	0

Thus, the process of photosynthesis is endothermic, which makes sense, since energy from light is required to drive this process. ■

The **standard heat of combustion** ($\Delta \hat{H}_c^\circ$) is the specific enthalpy change associated with the combustion of 1 mole of a substance with oxygen with both reactants and products at a reference temperature and pressure (usually 25°C and 1 atm). Tabulated heat of combustion values assume that all carbon in the reactant is converted to $CO_2(g)$; all hydrogen to $H_2O(\ell)$; all nitrogen to $N_2(g)$; and all sulfur to $SO_2(g)$. Compounds involved in a combustion process most often contain carbon. Also, compounds with elements other than C, N, H, O, and S do not have heat of combustion values. The standard heats of combustion for $O_2(g)$ and the combustion products $CO_2(g)$, $H_2O(\ell)$, $N_2(g)$, and $SO_2(g)$ are zero.

An example is the combustion of caffeine ($C_8H_{10}O_2N_4$):

$$C_8H_{10}O_2N_4(s) + \frac{19}{2}O_2(g) \longrightarrow 8\,CO_2(g) + 5\,H_2O(\ell) + 2\,N_2(g) \quad [4.8\text{-}10]$$

Using the formalism established in equation [4.8-1], the standard heat of combustion is the specific enthalpy difference between the product (CO_2, H_2O, and N_2) and the reactants ($C_8H_{10}O_2N_4$). The standard heats of combustion of $CO_2(g)$, $H_2O(\ell)$, and $N_2(g)$ are zero. The $\Delta \hat{H}_c^\circ$ for one mole of $C_8H_{10}O_2N_4$, and hence of the combustion reaction, is given in Appendix E.9 as −4247 kJ/mol. Since $\Delta \hat{H}_c^\circ$ is negative, the combustion reaction is exothermic. Note that the reaction is written such that 1 mole of caffeine undergoes the combustion reaction, even though this forces the stoichiometric coefficient for oxygen to be a fractional value.

Standard heat of combustion values can be used to calculate the standard heat of reaction, $\Delta \hat{H}_r^\circ$, for reactions involving combustible reactants and combustible products. This process is another application of Hess's law:

$$\Delta \hat{H}_r^\circ = \sum_r (\sigma_r\, \Delta \hat{H}_{c,r}^\circ) - \sum_p (\sigma_p\, \Delta \hat{H}_{c,p}^\circ) \quad [4.8\text{-}11]$$

When calculating the heat of reaction from heats of combustion, we subtract the enthalpy values of products from those of the reactants. This is different from the calculation of heat of reaction from heats of formation, where we subtract the enthalpy values of reactants from those of the products.

Lists of standard heats of combustion are found in Appendices E.7 and E.9. Standard heat of combustion values for many compounds are listed in chemistry and chemical engineering textbooks (e.g., Lide DR, *CRC Handbook of Chemistry and Physics*, 2002; Felder RM and Rousseau RW, *Elementary Principles of Chemical Processes*, 2000; Perry RH and Green D, *Perry's Chemical Engineers' Handbook*, 1984; Doran PM, *Bioprocess Engineering Principles*, 1995).

TABLE 4.5

Standard Heats of Combustion Involved in Biosynthesis of Glycine

Species	$\Delta \hat{H}_c^\circ$ (kJ/mol)
Serine, $C_3H_7O_3N(c)$	−1448
Glycine, $C_2H_5O_2N(c)$	−973
Formaldehyde, $CH_2O(g)$	−571

EXAMPLE 4.12 Biosynthesis of Glycine

Problem: Amino acids are the building blocks of proteins. Biosynthetic pathways in humans have evolved to generate some, but not all, of the amino acids. The conversion of the amino acid serine to the amino acid glycine is catalyzed by the enzyme serine hydroxymethyltransferase. Serine ($C_3H_7O_3N$) is converted to glycine ($C_2H_5O_2N$) and formaldehyde (CH_2O) as follows:

$$C_3H_7O_3N(c) \longrightarrow C_2H_5O_2N(c) + CH_2O(g)$$

Serine and glycine are both in the crystal (c) form for this reaction. Calculate the standard heat of reaction for this catalyzed reaction.

Solution: The heat of reaction is calculated using the standard heats of combustion:

$$\Delta \hat{H}_r^\circ = \sum_r (\sigma_r \, \Delta \hat{H}_{c,r}^\circ) - \sum_p (\sigma_p \, \Delta \hat{H}_{c,p}^\circ)$$

$$\Delta \hat{H}_r^\circ = \Delta \hat{H}_{c,C_3H_7O_3N}^\circ - \Delta \hat{H}_{c,C_2H_5O_2N}^\circ - \Delta \hat{H}_{c,CH_2O}^\circ$$

Given the data in the Table 4.5, the standard heat of reaction is calculated:

$$\Delta \hat{H}_r^\circ = -1448 \frac{\text{kJ}}{\text{mol}} - \left(-973 \frac{\text{kJ}}{\text{mol}}\right) - \left(-571 \frac{\text{kJ}}{\text{mol}}\right) = 96 \frac{\text{kJ}}{\text{mol}}$$

The standard heat of reaction is 96 kJ/mol. Since $\Delta \hat{H}_r^\circ$ is positive, the reaction is endothermic. Enzymes facilitate both endothermic and exothermic reactions. Sometimes endothermic reactions are coupled with the dephosphorylation of adenosine triphosphate (ATP) to adenosine diphosphate (ADP) or another biochemical energy source. Most enzyme-catalyzed reactions also have $\Delta \hat{H}_r^\circ$ values that are fairly small in magnitude (see Section 4.8.3). ■

When reactants are available in stoichiometric amounts, the reaction is at standard temperature and pressure (usually 25°C and 1 atm), and the reaction goes to completion, the **heat of reaction**, ΔH_r, across the system is:

$$\Delta H_r = \Delta H_r^\circ = \frac{n_s}{|\sigma_s|} \Delta \hat{H}_r^\circ \qquad [4.8\text{-}12]$$

where n_s is the number of moles of species s initially placed in the system, σ_s is the stoichiometric coefficient for species s, and $\Delta \hat{H}_r^\circ$ is the standard heat of reaction. For a system with flow rates into and out of the system, the **rate of heat of reaction**, $\Delta \dot{H}_r$, is:

$$\Delta \dot{H}_r = \Delta \dot{H}_r^\circ = \frac{\dot{n}_s}{|\sigma_s|} \Delta \hat{H}_r^\circ \qquad [4.8\text{-}13]$$

where $\dot{n}_s$ is the molar flow rate of species s into the system. The calculated values of ΔH_r and $\Delta \dot{H}_r$ are independent of the species selected for the calculation. Recall that the reaction rate R is a constant for all species and compounds in a reacting system (see Section 3.8). The computed values of ΔH_r and ΔH_r have this same property. Equations [4.8-12] and [4.8-13] are for a system with one reaction; these equations can be generalized for systems with multiple, simultaneous reactions.

Recall Example 4.12 and consider a situation where 10 moles of serine are converted completely to glycine and formaldehyde at 25°C and 1 atm. The heat of reaction, ΔH_r, is calculated:

$$\Delta H_r = \frac{n_s}{|\sigma_s|}\Delta \hat{H}_r^\circ = \frac{10 \text{ mol}}{|-1|}\left(96 \frac{\text{kJ}}{\text{mol}}\right) = 960 \text{ kJ} \qquad [4.8\text{-}14]$$

Systems with nonstandard temperatures or pressures or both, systems with nonstoichiometric amounts of reactants, or systems in which the reaction does not go to completion are addressed in Section 4.8.3.

4.8.3 Heat of Reaction Calculations at Nonstandard Conditions

Biological reactions often do not take place at standard conditions (usually 25°C and 1 atm). Instead, reactions often occur at or near 37°C (physiological temperature). Generally, heats of combustion and formation tables only include enthalpy changes for reactions at 25°C and 1 atm pressure. To calculate the heat of reaction, ΔH_r, for a process at nonstandard temperature or pressure or both, additional calculations must be completed.

Recall that the change in the specific enthalpy in a system is the sum of all the steps in the hypothetical path:

$$\Delta \hat{H} = \sum_k \Delta \hat{H}_k \qquad [4.8\text{-}15]$$

where k is the number of steps in the hypothetical path. Each step along the path should be a chemical reaction at standard temperature and pressure, or a change in pressure, temperature, or phase. In systems with chemical reactions at nonstandard temperatures, several chemical species must be warmed or cooled. Since enthalpy is a state function, the total enthalpy changes across the actual and hypothetical paths are identical.

Consider again the following reaction between compounds A and B to form products P and Q:

$$a\text{A}(\ell) + b\text{B}(\ell) \rightarrow p\text{P}(\ell) + q\text{Q}(\ell) \qquad [4.8\text{-}16]$$

Now assume that this reaction occurs at nonstandard temperature T. The heat of reaction at temperature T, $\Delta H_r(T)$, can be calculated using the hypothetical reaction pathway as shown in Figure 4.16. The first step on the hypothetical path is to either warm or cool the reactants from T°C to 25°C. The second step is the reaction at 25°C at which the standard heat of reaction, ΔH_r°, can be calculated from either heats of formation or heats of combustion data. Finally, the third step is to either cool or warm the products and any excess reactants from 25°C back to T°C. The enthalpy change for a given amount of material at a nonstandard T is calculated from the parts of the hypothetical path:

$$\Delta H_r(T) = \Delta H_1 + \Delta H_2 + \Delta H_3 \qquad [4.8\text{-}17]$$

where ΔH_1 and ΔH_3 are changes in sensible heat and ΔH_2 is equal to ΔH_r°, the heat of reaction at 25°C. Values for ΔH_1 and ΔH_3 are calculated using heat capacities

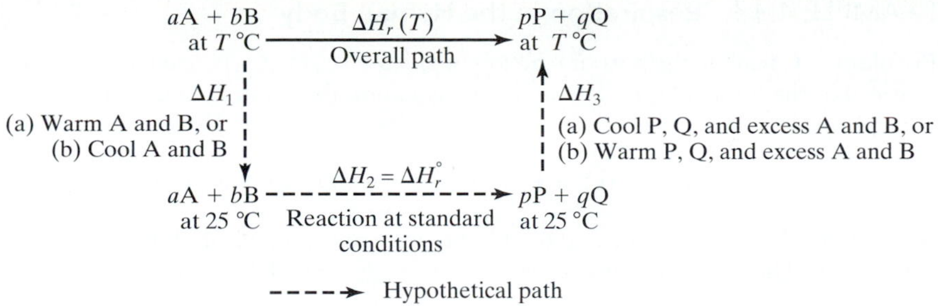

Figure 4.16
Hypothetical reaction pathway for reaction at nonstandard conditions.

and the methods discussed in Section 4.5.2. Similar equations can be written for the rate of change of enthalpy, $\Delta \dot{H}_r(T)$.

A hypothetical path must be constructed for each reaction that occurs at nonstandard temperature or pressure or both. The enthalpy change of the first step, ΔH_1, is usually the change of one nonstandard condition to its standard condition (e.g., change in pressure from 3 atm to 1 atm). When calculating ΔH_1 and other enthalpy changes prior to the reaction, only the reactants need be considered. After the reaction, the products and any excess reactants must be considered in returning the compounds from the standard to the nonstandard conditions. To account for temperature changes, heat capacities are required:

$$\Delta H = \sum_s \left(m_s \int_{T_1}^{T_2} C_{P,s}(T)\, dT \right) \qquad [4.8\text{-}18]$$

or

$$\Delta H = \sum_s \left(n_s \int_{T_1}^{T_2} C_{P,s}(T)\, dT \right) \qquad [4.8\text{-}19]$$

where m_s is the mass of species s, n_s is the moles of species s, $C_p(T)$ is the heat capacity that may be a function of temperature, T_1 is the temperature at which the process began, and T_2 is the temperature at which the process ends. For every species undergoing a temperature change, sensible heat changes are calculated and summed.

Depending on how different the reaction temperature is from the standard temperature, the magnitude of the sensible heat changes may be negligible relative to the magnitude of the heat of reaction. In bioengineering applications, sensible heat changes (e.g., ΔH_1 and ΔH_3 in the above discussion) are usually of the same order of magnitude, but one is a positive value and the other negative. When a reaction occurs at or near 37°C, the sensible heat changes are usually small relative to the heat of reaction. An exception is for reactions that involve biochemical enzymes, where sensible heat changes are often of the same order of magnitude as the heat of reaction. The heat of reaction for single-enzyme reactions is typically small because only small molecular rearrangements occur (e.g., see Example 4.12).

At low and moderate pressures, the heat of reaction is nearly independent of pressure. In most bioengineering applications, reactions occur at or near atmospheric pressure (1 atm). Therefore, for situations in this book, hypothetical paths for pressure changes are not built.

EXAMPLE 4.13 Respiration in the Human Body

Problem: Calculate the heat of reaction of glucose ($C_6H_{12}O_6$) during respiration (i.e., combustion) in the human body. The following equation describes the reactants and products of respiration:

$$C_6H_{12}O_6(s) + 6\,O_2(g) \longrightarrow 6\,CO_2(g) + 6\,H_2O(\ell)$$

Assume that 1 mol of glucose and 6 mol of oxygen are available and that the reaction goes to completion. The relevant heat capacity values are given in Table 4.6.

Solution: The body's temperature, 37°C, is a nonstandard temperature, so we construct a hypothetical path. The steps include (1) cooling the reactants from 37°C to 25°C, (2) reacting and forming the products at 25°C, and (3) warming the products from 25°C to 37°C. The described path is shown in Figure 4.17.

To calculate the sensible heat changes, equation [4.8-19] is required:

$$\Delta H = \sum_s \left(n_s \int_{T_1}^{T_2} C_{P,s}(T)\, dT \right) = \sum_s (n_s C_{P,s}(T_2 - T_1))$$

Glucose and oxygen are cooled along step 1 of the hypothetical path:

$$\Delta H_1 = 1\text{ mol}\left(225.9 \frac{\text{J}}{\text{mol} \cdot {}^\circ\text{C}}\right)(25^\circ\text{C} - 37^\circ\text{C})$$
$$+ 6\text{ mol}\left(29.3 \frac{\text{J}}{\text{mol} \cdot {}^\circ\text{C}}\right)(25^\circ\text{C} - 37^\circ\text{C}) = -4820\text{ J}$$

The products, carbon dioxide and water, are warmed along step 3 of the hypothetical path:

$$\Delta H_3 = 6\text{ mol}\left(36.47 \frac{\text{J}}{\text{mol} \cdot {}^\circ\text{C}}\right)(37^\circ\text{C} - 25^\circ\text{C})$$
$$+ 6\text{ mol}\left(75.4 \frac{\text{J}}{\text{mol} \cdot {}^\circ\text{C}}\right)(37^\circ\text{C} - 25^\circ\text{C}) = 8050\text{ J}$$

TABLE 4.6

Heat Capacities for Compounds Involved in Respiration

Compound	$C_P\left(\frac{\text{J}}{\text{mol} \cdot {}^\circ\text{C}}\right)$
Glucose, $C_6H_{12}O_6(s)$	225.9
Oxygen, $O_2(g)$	29.3
Carbon dioxide, $CO_2(g)$	36.47
Water, $H_2O(\ell)$	75.4

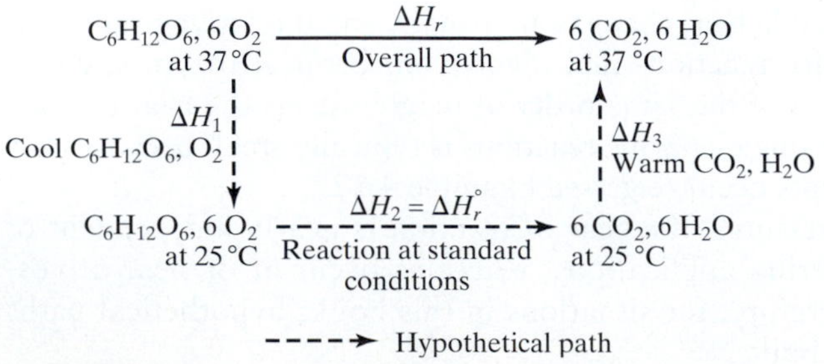

Figure 4.17 Hypothetical reaction pathway for complete combustion of glucose in the human body.

Glucose and oxygen have been completely combusted so they need not be included in the sensible heat change calculation for step 3.

The standard heat of combustion for glucose is found in Appendix E.9:

$$\Delta H_r^\circ = \frac{n_s}{|\sigma_s|}\Delta \hat{H}_r^\circ = \frac{n_{C_6H_{12}O_6}}{|\sigma_{C_6H_{12}O_6}|}\Delta \hat{H}_{c,C_6H_{12}O_6}^\circ = \frac{1\text{ mol}}{|-1|}\left(-2805\,\frac{\text{kJ}}{\text{mol}}\right) = -2805\text{ kJ}$$

Combining the three steps in the path, the heat of reaction can be calculated for the system:

$$\Delta H_r(37°\text{C}) = \Delta H_1 + \Delta H_r^\circ + \Delta H_3 = -4.82\text{ kJ} - 2805\text{ kJ} + 8.05\text{ kJ} = -2800\text{ kJ}$$

Thus, the heat of reaction for 1 mole of glucose with oxygen at 37°C is −2800 kJ. Note that both ΔH_1 and ΔH_3 are much less than ΔH_r°, and as a result, ΔH_r is approximately equal to ΔH_r°. Also, notice that the sensible heat changes captured in ΔH_1 and ΔH_3 are similar in magnitude but opposite in sign. ■

In the above example, the reactants are in stoichiometric proportions, and the reaction goes to completion. When these two constraints are not met, the calculation for $\Delta H_r(T)$ changes. Recall equation [4.8-12], used to calculate the heat of reaction when the fractional conversion for all reactants is one. To account for situations when the fractional conversion is less than one, the heat of reaction is:

$$\Delta H_r^\circ = \frac{f_s n_s}{|\sigma_s|}\Delta \hat{H}_r^\circ \qquad [4.8\text{-}20]$$

where f_s is the fractional conversion of species s (i.e., the proportion of species s that is consumed in the reaction), σ_s is the stoichiometric coefficient of species s, n_s is the number of moles of species s initially placed in the system, and $\Delta \hat{H}_r^\circ$ is the standard heat of reaction. Similarly, for a system with flow rates into and out of the system, the change in the rate of enthalpy is:

$$\Delta \dot{H}_r^\circ = \frac{f_s \dot{n}_s}{|\sigma_s|}\Delta \hat{H}_r^\circ \qquad [4.8\text{-}21]$$

where $\dot{n}_s$ is the molar flow rate of species s into the system. Equations [4.8-20] and [4.8-21] are valid only when the species s is a reactant. Recall that the fractional conversion of a reactant is:

$$f_s = \frac{n_{i,s} - n_{j,s}}{n_{i,s}} \qquad [4.8\text{-}22]$$

or

$$f_s = \frac{\dot{n}_{i,s} - \dot{n}_{j,s}}{\dot{n}_{i,s}} \qquad [4.8\text{-}23]$$

where i and j refer to inlet and outlet, respectively.

Consider the following liquid (ℓ)-phase reaction at standard temperature and pressure:

$$1\text{ A}(\ell) + 3\text{ B}(\ell) \rightarrow 1\text{ P}(\ell) + 2\text{ Q}(\ell) \qquad [4.8\text{-}24]$$

where A and B are the reactants, and P and Q are the products. Suppose that 100 mol of A and 300 moles of B react completely to form 100 moles of P and 200 mol of Q. The standard heat of reaction, ΔH_r°, is known to be 100 kJ/mol. Since A and

B are provided in stoichiometric quantities and the reaction goes to completion, the fractional conversion of A, f_A, and of B, f_B, is one. (Since the fractional conversion is one, equation [4.8-12] could also be used to calculate ΔH_r°.) The overall change in enthalpy from the hypothetical reaction given in equation [4.8-24], ΔH_r°, is calculated for species A and B as:

$$\text{A:} \qquad \Delta H_r^\circ = \frac{f_s n_s}{|\sigma_s|}\Delta \hat{H}_r^\circ = \frac{1.0(100 \text{ mol})}{|-1|} 100\frac{\text{kJ}}{\text{mol}} = 10{,}000 \text{ kJ} \qquad [4.8\text{-}25]$$

$$\text{B:} \qquad \Delta H_r^\circ = \frac{f_s n_s}{|\sigma_s|}\Delta \hat{H}_r^\circ = \frac{1.0(300 \text{ mol})}{|-3|} 100\frac{\text{kJ}}{\text{mol}} = 10{,}000 \text{ kJ} \qquad [4.8\text{-}26]$$

Because ΔH_r° is independent of the species selected for the calculation, the computed ΔH_r° values are the same.

Now suppose that 100 moles of A and 150 moles of B react to form 50 moles of P and 100 moles of Q at standard temperature and pressure using the hypothetical reaction given in equation [4.8-24]. B is the limiting reactant, and 50 moles of A are in excess. The standard heat of reaction, $\Delta \hat{H}_r^\circ$, is known to be 100 kJ/mol. In this case, A and B are not provided in stoichiometric quantities. The fractional conversions of A and B are calculated:

$$\text{A:} \qquad f_A = \frac{n_{i,A} - n_{j,A}}{n_{i,A}} = \frac{100 \text{ mol} - 50 \text{ mol}}{100 \text{ mol}} = 0.5 \qquad [4.8\text{-}27]$$

$$\text{B:} \qquad f_B = \frac{n_{i,B} - n_{j,B}}{n_{i,B}} = \frac{150 \text{ mol} - 0 \text{ mol}}{150 \text{ mol}} = 1.0 \qquad [4.8\text{-}28]$$

The overall change in enthalpy from the reaction, ΔH_r°, is the same regardless of whether the calculation is based on species A or B:

$$\text{A:} \qquad \Delta H_r^\circ = \frac{f_s n_s}{|\sigma_s|}\Delta \hat{H}_r^\circ = \frac{0.5(100 \text{ mol})}{|-1|} 100\frac{\text{kJ}}{\text{mol}} = 5000 \text{ kJ} \qquad [4.8\text{-}29]$$

$$\text{B:} \qquad \Delta H_r^\circ = \frac{f_s n_s}{|\sigma_s|}\Delta \hat{H}_r^\circ = \frac{1.0(150 \text{ mol})}{|-3|} 100\frac{\text{kJ}}{\text{mol}} = 5000 \text{ kJ} \qquad [4.8\text{-}30]$$

Note that the overall change in the enthalpy of reaction for this second case is one-half that of the first case. This makes sense, since only one-half as much product is being formed in the second case as in the first.

When the reactants are not present in stoichiometric proportions or the reaction does not go to completion or both, calculating changes in sensible heat is effected. Care must be taken to make sure that the correct amounts of mass, moles, mass rate, or molar rate of the reactants and products are used in equations [4.8-18] and [4.8-19].

EXAMPLE 4.14 Incomplete Respiration in the Human Body

Problem: Consider Example 4.13 again to calculate the heat of reaction at 37°C during respiration:

$$C_6H_{12}O_6(s) + 6\ O_2(g) \rightarrow 6\ CO_2(g) + 6\ H_2O(\ell)$$

Let us assume instead that 1 mole of glucose and 9 moles of oxygen are provided as reactants and that 0.2 mol of glucose remain after the reaction has ceased. The relevant heat capacity values are given in Table 4.6.

Solution: Because the reaction is incomplete, we first calculate the fractional conversion of glucose:

$$f_{C_6H_{12}O_6} = \frac{n_{i,C_6H_{12}O_6} - n_{j,C_6H_{12}O_6}}{n_{i,C_6H_{12}O_6}} = \frac{1\text{ mol} - 0.2\text{ mol}}{1\text{ mol}} = 0.8$$

To calculate how many moles of each species are consumed during the reaction, the reaction rate, R, needs to be calculated. Using glucose as the species for calculation:

$$R = \frac{n_{i,s}f_s}{-\sigma_s} = \frac{(1\text{ mol})(0.8)}{-(-1)} = 0.8\text{ mol}$$

Thus, the amount of oxygen present after the reaction is:

$$n_{j,O_2} = n_{i,O_2} + \sigma_{O_2}R = 9\text{ mol} + (-6)0.8\text{ mol} = 4.2\text{ mol}$$

For CO_2 and H_2O, 4.8 mol of each are produced during the reaction and leave the system.

The hypothetical path from Example 4.13 is used (Figure 4.18); however, both the products and the excess reactants must be warmed from 25°C to 37°C in step 3 of the path. To calculate the sensible heat changes, equation [4.8-19] is required and simplified to:

$$\Delta H = \sum_s\left(n_s\int_{T_1}^{T_2} C_{P,s}(T)\,dT\right) = \sum_s(n_sC_{P,s}(T_2 - T_1)) = (T_2 - T_1)\sum_s n_sC_{P,s}$$

since the temperature change is the same for all compounds in each step. The change in enthalpy across step 1 of the path includes cooling glucose and oxygen:

$$\Delta H_1 = (25°\text{C} - 37°\text{C})\left[1\text{ mol}\left(225.9\frac{\text{J}}{\text{mol}\cdot°\text{C}}\right) + 9\text{ mol}\left(29.3\frac{\text{J}}{\text{mol}\cdot°\text{C}}\right)\right] = -5.88\text{ kJ}$$

The change in enthalpy across step 3 of the path includes heating the products (carbon dioxide and water) and the excess reactants (glucose and oxygen):

$$\Delta H_3 = (37°\text{C} - 25°\text{C})\left[4.8\text{ mol}\left(36.47\frac{\text{J}}{\text{mol}\cdot°\text{C}}\right) + 4.8\text{ mol}\left(75.4\frac{\text{J}}{\text{mol}\cdot°\text{C}}\right)\right.$$
$$\left. + 0.2\text{ mol}\left(225.9\frac{\text{J}}{\text{mol}\cdot°\text{C}}\right) + 4.2\text{ mol}\left(29.3\frac{\text{J}}{\text{mol}\cdot°\text{C}}\right)\right] = 8.46\text{ kJ}$$

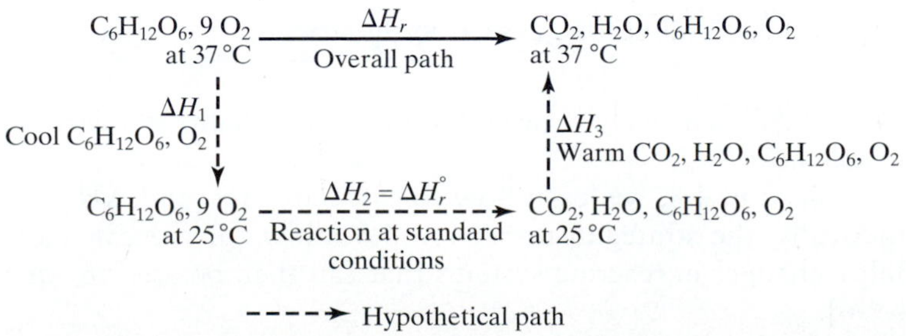

Figure 4.18 Hypothetical reaction pathway for incomplete respiration of glucose in the human body.

To calculate ΔH_2, the standard heat of combustion for glucose is found in Appendix E.9:

$$\Delta H_r^\circ = \frac{f_s n_s}{|\sigma_s|}\Delta \hat{H}_r^\circ = \frac{f_{C_6H_{12}O_6} n_{C_6H_{12}O_6}}{|\sigma_{C_6H_{12}O_6}|}\Delta \hat{H}_{c,C_6H_{12}O_6}^\circ = \frac{0.8(1\text{ mol})}{|-1|}(-2805\text{ kJ}) = -2244\text{ kJ}$$

Combining the three steps in the path:

$$\Delta H_r(37^\circ\text{C}) = \Delta H_1 + \Delta H_r^\circ + \Delta H_3 = -5.88\text{ kJ} - 2244\text{ kJ} + 8.46\text{ kJ} = -2240\text{ kJ}$$

Thus, the heat of reaction for the partial combustion of glucose with oxygen at 37°C is −2240 kJ. Recall that the ΔH_r for complete combustion of glucose is −2800 kJ. As expected, the heat of reaction for partial combustion ($f_{C_6H_{12}O_6} = 0.8$) is less than that for complete combustion ($f_{C_6H_{12}O_6} = 1.0$). Note again that both ΔH_1 and ΔH_3 are much less than H_r°. As a result, ΔH_r is approximately equal toΔH_r°. ■

4.9 Open Systems with Reactions

In Section 4.8, we learn how to calculate the heat of reaction, ΔH_r, for a system containing reacting components. With the ability to calculate the total change in enthalpy across a reacting system, the total energy conservation equation can be applied to reacting systems. For a steady-state system with no change in kinetic or potential energy, the algebraic equation [4.3-17] and the differential equation [4.3-10] reduce to:

$$\sum_i m_i \hat{H}_i - \sum_j m_j \hat{H}_j + Q + W_{\text{nonflow}} = 0 \qquad [4.9\text{-}1]$$

$$\sum_i \dot{m}_i \hat{H}_i - \sum_j \dot{m}_j \hat{H}_j + \sum \dot{Q} + \sum \dot{W}_{\text{nonflow}} = 0 \qquad [4.9\text{-}2]$$

For systems undergoing reactions, the change in enthalpy or rate of change of enthalpy between inlet and outlet is defined:

$$-\Delta H_r = \sum_i m_i \hat{H}_i - \sum_j m_j \hat{H}_j \qquad [4.9\text{-}3]$$

$$-\Delta \dot{H}_r = \sum_i \dot{m}_i \hat{H}_i - \sum_j \dot{m}_j \hat{H}_j \qquad [4.9\text{-}4]$$

Again, we avoid needing to know the absolute specific enthalpies given in equations [4-9.1] and [4.9-2] by looking at the difference between the inlet and outlet amounts or streams. Thus, equations [4.9-1] and [4.9-2] can be rewritten as:

$$-\Delta H_r + Q + W_{\text{nonflow}} = 0 \qquad [4.9\text{-}5]$$

$$-\Delta \dot{H}_r + \sum \dot{Q} + \sum \dot{W}_{\text{nonflow}} = 0 \qquad [4.9\text{-}6]$$

ΔH_r and $\Delta \dot{H}_r$ may include several terms, including the standard heat of reaction and sensible heat.

In Section 4.8, we learn how to calculate ΔH_r and $\Delta \dot{H}_r$ for reacting systems. Practically, the strategies in Section 4.8 allow for the calculation of specific enthalpy changes in reacting systems that can then be used in equations [4.9-5] and [4.9-6].

EXAMPLE 4.15 Photosynthesis in Green Plants

Problem: Photosynthesis is a much more complicated reaction than that described in Example 4.11, in which the plant takes in carbon dioxide and water from the surrounding environment and converts them into glucose and oxygen:

$$6\ CO_2(g) + 6\ H_2O(\ell) \longrightarrow C_6H_{12}O_6(s) + 6\ O_2(g)$$

Photosynthesis consists of two separate reactions: light and dark. The light reactions use light photons to excite electrons in chlorophyll located in the thylakoid membrane of chloroplasts. This generates two energy intermediates: ATP and NADPH. A phosphate group is added to ADP to make ATP, and $NADP^+$ is reduced to make NADPH. In the dark reactions, which take place in the stroma of chloroplasts, energy is released by removing a phosphate group from ATP (i.e., converting ATP back to ADP) and by oxidizing NADPH (i.e., converting NADPH back to $NADP^+$). The ADP and $NADP^+$ from the dark reactions are then returned to the thylakoid membrane (Figure 4.19).

The energy released in the dark reactions is used to attach carbons in glucose synthesis. In all, the production of one glucose molecule requires 18 ATP and 12 NADPH. If 30.5 kJ/mol of energy is liberated in the removal of phosphate groups from ATP, what is the work associated with the oxidation of one mole of NADPH to $NADP^+$?

Assume the plant does not change temperature during the photosynthetic reaction (i.e., the energy in photons is used to excite electrons only; thus, the thermal energy produced from the production of one glucose molecule is negligible). Assume that the reaction takes place at standard conditions (25°C, 1 atm).

Solution: If we model the plant as our system, we can assume that the system is at steady-state; because the system does not move, both potential and kinetic energy do not change.

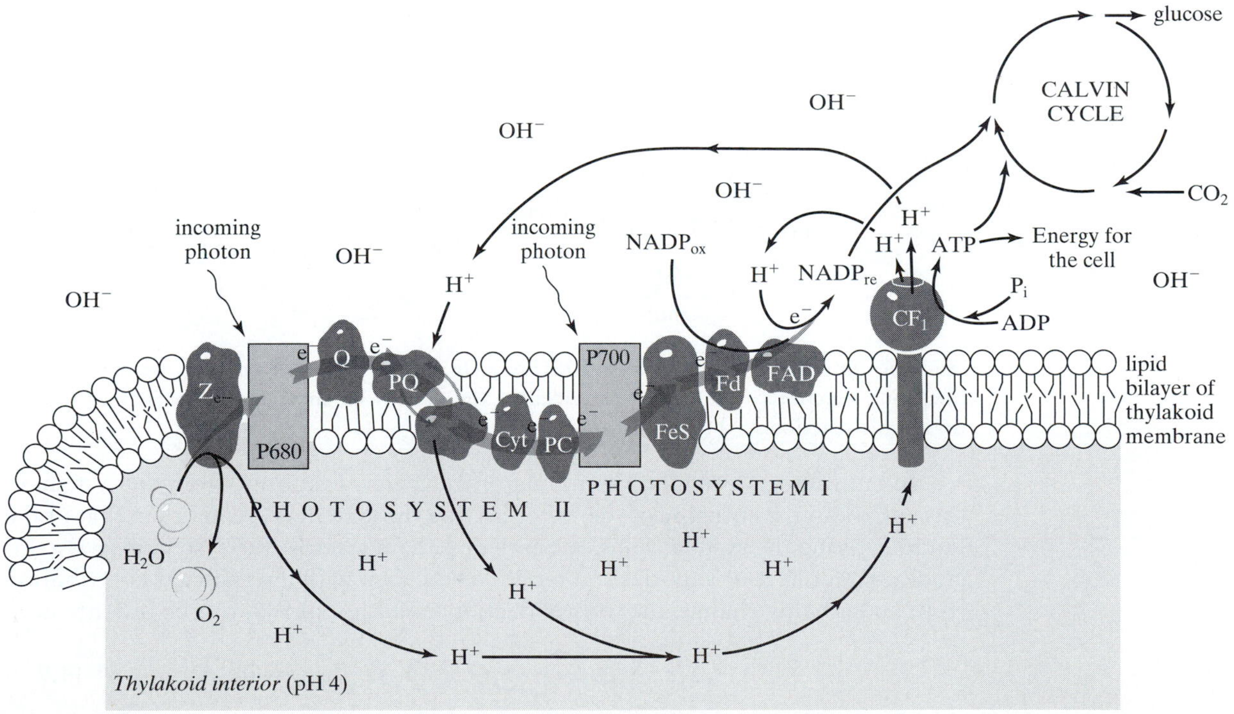

Figure 4.19
Photosynthesis with light and dark reactions. (*Source*: Keeton WT and Gould JL, *Biological Science*, 4th ed. New York: WW Norton, 1986.)

The algebraic conservation of total energy equation can be used to model the system:

$$\sum_i m_i \hat{H}_i - \sum_j m_j \hat{H}_j + Q + W_{\text{nonflow}} = 0$$

The system reacts during photosynthesis, so we can substitute in equation [4.9-3] for the first two terms, which leaves equation [4.9-5]:

$$-\Delta H_r + Q + W_{\text{nonflow}} = 0$$

Since we assume the plant does not exchange heat with the surroundings, this equation is simplified to:

$$-\Delta H_r + W_{\text{nonflow}} = 0$$

Recall from Example 4.11 that the standard heat of reaction of the production of glucose during photosynthesis is calculated as 2810 kJ/mol using the standard heats of formation of the reactants (CO_2 and H_2O) and products (glucose and O_2). Using Avogadro's number, the ΔH_r° of one molecule of glucose at standard conditions is 4.66×10^{-21} kJ.

The release of energy into the system by the removal of a phosphate group from ATP and the oxidation of NADPH is considered nonflow work, because work is energy that flows as a result of a driving force other than temperature. (An alternative approach would be to look at all three reactions—production of glucose, removal of a phosphate from ATP, and oxidation of NADPH—in the ΔH_r term.) Because work is a form of energy, we can calculate the work done by converting 18 ATP to 18 ADP to produce one glucose molecule by using the energy released in converting one mole of ATP (30.5 kJ):

$$W_{\text{ATP}} = \left(30.5 \frac{\text{kJ}}{\text{mol}}\right)\left(\frac{1 \text{ mol}}{6.02 \times 10^{23} \text{ molecules}}\right)(18 \text{ molecules}) = 9.12 \times 10^{-22} \text{ kJ}$$

The work contributed to the system by the conversion of 12 NADPH molecules to 12 $NADP^+$ molecules is calculated by separating the nonflow work term into an ATP term and a NADPH term:

$$\begin{aligned} -\Delta H_r + W_{\text{nonflow}} &= -\Delta H_r + W_{\text{ATP}} + W_{\text{NADPH}} = 0 \\ W_{\text{NADPH}} &= \Delta H_r - W_{\text{ATP}} = 4.66 \times 10^{-21} \text{ kJ} - 9.12 \times 10^{-22} \text{ kJ} \\ &= 3.75 \times 10^{-21} \text{ kJ} \\ W_{\text{NADPH}} &= 3.75 \times 10^{-21} \text{ kJ}\left(\frac{6.02 \times 10^{23} \text{ molecules}}{\text{mol}}\right)\left(\frac{1}{12 \text{ molecules}}\right) \\ &= 188 \frac{\text{kJ}}{\text{mol}} \end{aligned}$$

The conversions of ATP and NADPH to ADP and $NADP^+$, respectively, are both energy-liberating reactions. The energy provided to the system by these reactions enables the endothermic reaction of the production of glucose. ■

In bioreactors, a significant amount of water may evaporate from the system. In this situation, the enthalpy of the vapor leaving the system is greater than that of the liquid entering the system. Thus, the enthalpy change across the system may also include a latent heat of vaporization term. For the case with a significant contribution from an enthalpy change due to vaporization, equation [4.9-6] can be rewritten:

$$-\Delta \dot{H}_r - \Delta \dot{H}_v + \sum \dot{Q} + \sum \dot{W}_{\text{nonflow}} = 0 \qquad [4.9\text{-}7]$$

where $\Delta \dot{H}_v$ is the rate of heat of vaporization and is calculated using either equation:

$$\Delta \dot{H}_v = \dot{m}\, \Delta \hat{H}_v \qquad [4.9\text{-}8]$$

or

$$\Delta \dot{H}_v = \dot{n}\, \Delta \hat{H}_v \qquad [4.9\text{-}9]$$

where $\Delta \hat{H}_v$ is the specific heat of vaporization on a per-mass or per-mole basis.

EXAMPLE 4.16 **Production of Citric Acid**

Problem: A naturally occurring compound in citrus fruits, citric acid ($C_6H_8O_7$) is an important compound in aerobic respiration. As a food additive, the compound is a preservative to prevent discoloration of foods. In industry, citric acid is manufactured continuously using a submerged culture of *Asperigillus niger* in a batch reactor operated at 30°C:

$$C_6H_{12}O_6(s) + aNH_3(g) + bO_2(g) \longrightarrow cCH_{1.79}N_{0.2}O_{0.5}(s)$$
$$+dCO_2(g) + eH_2O(\ell) + fC_6H_8O_7(s)$$

For this reaction, the respiratory quotient is 0.45. The yield of citric acid per mole of glucose consumed is 0.70. The cell mass is given as $CH_{1.79}N_{0.2}O_{0.5}$. The heats of combustion for the compounds in the chemical reaction are given in Table 4.7.

The inlet flow rates of glucose and ammonia are 20 kg/hr and 0.4 kg/hr, respectively. The inlet flow rate of oxygen is 7.5 kg/hr. The fractional conversion of glucose is 0.91. Mechanical agitation of the broth adds 15 kW of power to the system. One-tenth of the water produced by the reaction is evaporated. Estimate the cooling requirements. (Adapted from Doran PM, *Bioprocess Engineering Principles*, 1995.)

Solution:

1. Assemble
   (a) Find: cooling requirement for continuous operation.
   (b) Diagram: The system boundary is the wall of the bioreactor (Figure 4.20).
2. Analyze
   (a) Assume:
   - Tank is well mixed.
   - Sensible heat is negligible.
   - Temperature of the bioreactor is maintained at 30°C.
   - Bone-dry oxygen enters bioreactor.
   - The system is at steady-state.
   - No change in kinetic or potential energy.

   (b) Extra data:
   - The heat of vaporization, $\Delta \hat{H}_v$, of water at 30°C is 2430.7 kJ/kg (Appendix E.5).

   (c) Variables, notations, units:
   - Units: kg, kJ, hr, mol.

   (d) Basis: The inlet flow rate of glucose at 20 kg/hr can be used as a basis:

$$\dot{n}_{\text{in},C_6H_{12}O_6} = \frac{\dot{m}_{\text{in},C_6H_{12}O_6}}{M_{C_6H_{12}O_6}} = \frac{20\dfrac{\text{kg}}{\text{hr}}}{180\dfrac{\text{g}}{\text{mol}}\left(\dfrac{1\text{ kg}}{1000\text{ g}}\right)} = 111\ \frac{\text{mol}}{\text{hr}}$$

TABLE 4.7

Heats of Combustion for Compounds Involved in Production of Citric Acid

Compound	$\Delta \hat{H}_c^\circ$ (kJ/mol)
Glucose, $C_6H_{12}O_6(s)$	−2805
Ammonia, $NH_3(g)$	−382.6
Cell mass, $CH_{1.79}N_{0.2}O_{0.5}(s)$	−552
Citric acid, $C_6H_8O_7(s)$	−1962

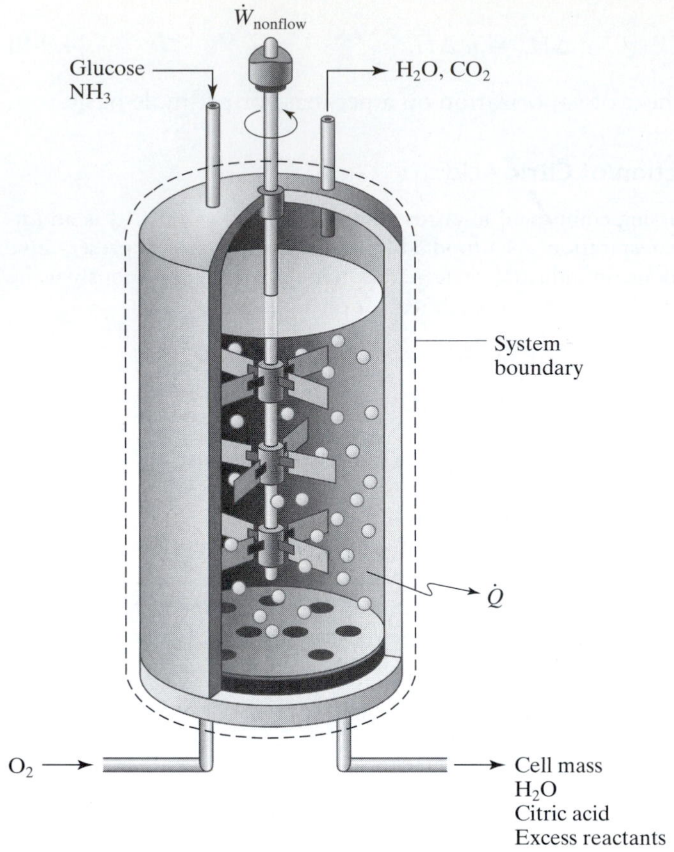

Figure 4.20
System diagram of bioreactor producing citric acid.

(e) Reaction: The reaction is given in the problem statement:

$$C_6H_{12}O_6(s) + aNH_3(g) + bO_2(g) \longrightarrow cCH_{1.79}N_{0.2}O_{0.5}(s) + dCO_2(g) + eH_2O(\ell) + fC_6H_8O_7(s)$$

Element balances need to be constructed to balance the reaction:

C: $-6 + c + d + 6f = 0$
N: $-a + 0.2c = 0$
H: $-12 - 3a + 1.79c + 2e + 8f = 0$
O: $-6 - 2b + 0.50c + 2d + e + 7f = 0$
RQ: $0.45 = \dfrac{d}{b}$
Yield: $0.70 = f$

Since f is already known, we have five equations and five unknowns. Using MATLAB or another computer program, the variables can be determined:

$a = 0.196 \quad b = 1.82 \quad c = 0.979 \quad d = 0.821 \quad e = 2.62$

The balanced equation is written:

$$C_6H_{12}O_6(s) + 0.196\ NH_3(g) + 1.82\ O_2(g) \longrightarrow 0.979\ CH_{1.79}N_{0.2}O_{0.5}(s) + 0.821\ CO_2(g) + 2.62\ H_2O(\ell) + 0.7\ C_6H_8O_7(s)$$

3. Calculate
(a) Equations: Since rates of material flow are given, the differential form of the conservation of energy equation is used:

$$\sum_i \dot{m}_i(\hat{E}_{P,i} + \hat{E}_{K,i} + \hat{H}_i) - \sum_j \dot{m}_j(\hat{E}_{P,j} + \hat{E}_{K,j} + \hat{H}_j) + \sum \dot{Q} + \sum \dot{W}_{nonflow} = \frac{dE_T^{sys}}{dt}$$

(b) Calculate:

- We assume the system is at steady-state, so we can set the Accumulation term to zero. Additionally, no potential or kinetic energy changes occur, so these terms can also be eliminated. However, evaporation does occur, so we can reduce the governing differential equation for the conservation of energy to equation [4.9-7]:

$$-\Delta\dot{H}_r - \Delta\dot{H}_v + \sum\dot{Q} + \sum\dot{W}_{\text{nonflow}} = 0$$

- Because we want to calculate the rate at which the system must have heat removed, we rearrange the equation to solve for heat:

$$\dot{Q} = \Delta\dot{H}_r + \Delta\dot{H}_v - \sum\dot{W}_{\text{nonflow}}$$

- To find $\Delta\dot{H}_r$, we first use equation [4.8-11] to calculate the standard heat of reaction using the heats of combustion for each compound in Table 4.7:

$$\begin{aligned}
\Delta\hat{H}_r^\circ &= \sum_r(\sigma_r\,\Delta\hat{H}_{c,r}^\circ) - \sum_p(\sigma_p\,\Delta\hat{H}_{c,p}^\circ) \\
&= (1)\Delta\hat{H}_{c,C_6H_{12}O_6}^\circ + (0.196)\Delta\hat{H}_{c,NH_3}^\circ - (0.979)\Delta\hat{H}_{c,CH_{1.79}O_{0.50}N_{0.20}}^\circ \\
&\quad - (0.7)\Delta\hat{H}_{c,C_6H_8O_7}^\circ \\
&= (1)\left(-2805\frac{\text{kJ}}{\text{mol}}\right) + (0.196)\left(-382.6\frac{\text{kJ}}{\text{mol}}\right) \\
&\quad - (0.979)\left(-552\frac{\text{kJ}}{\text{mol}}\right) - (0.7)\left(-1962\frac{\text{kJ}}{\text{mol}}\right) \\
&= -966\ \frac{\text{kJ}}{\text{mol}}
\end{aligned}$$

Remember that the standard heats of combustion for O_2 and the combustion products CO_2 and H_2O are zero. Because contributions from sensible heat are negligible compared to the heat of reaction, we ignore those contributions and calculate the rate change in enthalpy, $\Delta\dot{H}_r$, at 30°C by substituting glucose as the compound for the calculation:

$$\Delta\dot{H}_r(30°\text{C}) \cong \Delta\dot{H}_r^\circ = \frac{f_s\dot{n}_s}{|\sigma_s|}\Delta\hat{H}_r^\circ = \frac{f_{C_6H_{12}O_6}\dot{n}_{C_6H_{12}O_6}}{|\sigma_{C_6H_{12}O_6}|}\Delta\hat{H}_r^\circ$$

$$= \frac{0.91\left(111\dfrac{\text{mol}}{\text{hr}}\right)}{|-1|}\left(-966\frac{\text{kJ}}{\text{mol}}\right) = -97{,}600\ \frac{\text{kJ}}{\text{hr}}$$

- To calculate the term for rate of heat of vaporization, $\Delta\dot{H}_v$, we need to find the rate at which water is produced from the reaction and then calculate how much of this water evaporates. Because ammonia and oxygen are supplied in excess, the reaction rate is calculated using glucose:

$$R = \frac{\dot{n}_{\text{in},C_6H_{12}O_6}f_{C_6H_{12}O_6}}{-\sigma_{C_6H_{12}O_6}} = \frac{111\dfrac{\text{mol}}{\text{hr}}(0.91)}{-(-1)} = 101\ \frac{\text{mol}}{\text{hr}}$$

Using the reaction rate, we can now calculate the molar production rate of water:

$$\dot{n}_{\text{out},H_2O} = \dot{n}_{\text{in},H_2O} + \sigma_{H_2O}R = 0 + (2.62)101\frac{\text{mol}}{\text{hr}} = 265\ \frac{\text{mol}}{\text{hr}}$$

One-tenth of the water is evaporated:

$$\begin{aligned}
\Delta\dot{H}_v &= \frac{1}{10}\dot{n}_{\text{out},H_2O}\,\Delta\hat{H}_v = \frac{1}{10}\left(265\frac{\text{mol}}{\text{hr}}\right)\left(2430.7\frac{\text{kJ}}{\text{kg}}\right)\left(18\frac{\text{g}}{\text{mol}}\right)\left(\frac{\text{kg}}{1000\ \text{g}}\right) \\
&= 1160\ \frac{\text{kJ}}{\text{hr}}
\end{aligned}$$

- The power input by mechanical agitation is 15 kW:

$$\dot{W}_{\text{nonflow}} = 15 \text{ kW} = 15\frac{\text{kJ}}{\text{s}}\left(\frac{3600 \text{ s}}{\text{hr}}\right) = 54{,}000 \frac{\text{kJ}}{\text{hr}}$$

- The rate of heat, $\dot{Q}$, is:

$$\dot{Q} = \Delta\dot{H}_r + \Delta\hat{H}_v - \dot{W}_{\text{nonflow}}$$
$$= -97{,}600\frac{\text{kJ}}{\text{hr}} + 1160\frac{\text{kJ}}{\text{hr}} - 54{,}000\frac{\text{kJ}}{\text{hr}} = -150{,}000 \frac{\text{kJ}}{\text{hr}}$$

$\dot{Q}$ is negative, indicating that heat is removed from the system. Note that the energy loss due to the evaporation of water is small compared to the heat of reaction.

4. Finalize
   (a) Answer: Heat must be removed from the batch reactor at a rate of 150,000 kJ/hr to maintain the reaction at 30°C.
   (b) Check: Since $\Delta\dot{H}_r$ for the chemical reaction (i.e., production of citric acid) is calculated to be exothermic and energy is added to the system by mechanical agitation of the broth, it makes sense that heat must be removed from the system. The answer is reasonable because it is the same order of magnitude as the energy input by mechanical agitation and the heat of reaction. ■

4.10 Dynamic Systems

Recall the differential and algebraic conservation of total energy equations:

$$\sum_i \dot{m}_i(\hat{E}_{P,i} + \hat{E}_{K,i} + \hat{H}_i) - \sum_j \dot{m}_j(\hat{E}_{P,j} + \hat{E}_{K,j} + \hat{H}_j) + \sum \dot{Q} + \sum \dot{W}_{\text{nonflow}} = \frac{dE_T^{\text{sys}}}{dt} \quad [4.10\text{-}1]$$

$$\sum_i m_i(\hat{E}_{P,i} + \hat{E}_{K,i} + \hat{H}_i) - \sum_j m_j(\hat{E}_{P,j} + \hat{E}_{K,j} + \hat{H}_j) + Q + W_{\text{nonflow}} = E_{T,\text{f}}^{\text{sys}} - E_{T,0}^{\text{sys}} \quad [4.10\text{-}2]$$

The right-hand side of the equations represents the change in total energy of the system. In dynamic systems, the accumulated energy or rate of accumulation of energy is nonzero.

In this book, several simplifying assumptions are made for unsteady-state systems so that the problems are tractable. The assumptions below are for the differential form of the conservation of total energy equation; similar assumptions can also be made for the algebraic form. First, the system is assumed to have one inlet and one outlet stream, both having the same mass flow rate. Thus, rather than writing $\dot{m}_i$ or $\dot{m}_j$, the term $\dot{m}$ is substituted. Second, changes in the kinetic and potential energy across the system are assumed to be negligible. Third, the system is assumed to be well mixed. The consequence of this assumption is that system variables, such as composition and temperature, are equal to that of the outlet stream. For example, the temperature of the system is equivalent to T_j. In dynamic systems, the system variable of interest, such as the temperature of the system, may change over the course of time.

Other assumptions include no phase changes and no chemical reactions in the system. Also, specific internal energy and enthalpy must not be a function of pressure. Finally, the heat capacities of the contents in the system are assumed to be constant. Recall that the integral of the heat capacity at constant pressure (C_p) across a temperature range is equal to the specific enthalpy, $\Delta\hat{H}$, required to warm or cool a material (equation [4.5-17]). The integral of the heat capacity at constant volume (C_v) across a temperature range is equal to the specific internal energy, $\Delta\hat{U}$:

$$\Delta\hat{U} = \int_{T_1}^{T_2} C_v(T)\,dT \qquad [4.10\text{-}3]$$

where T_1 is the first temperature and T_2 is the second temperature at constant volume. When the $C_v(T)$ is a constant C_v, the specific internal energy $\hat{U}$ is:

$$\hat{U} = C_v(T - T_{\text{ref}}) \qquad [4.10\text{-}4]$$

where T is the temperature of the material of interest and T_{ref} is the reference temperature. Note that the specific internal energy $\hat{U}$ is really the specific internal energy at temperature T relative to the specific internal energy at the reference temperature, which we assume to be zero. With these assumptions, equation [4.10-1] can be reduced to:

$$\dot{m}\hat{H}_i - \dot{m}\hat{H}_j + \sum\dot{Q} + \sum\dot{W}_{\text{nonflow}} = \frac{dE_T^{\text{sys}}}{dt} \qquad [4.10\text{-}5]$$

The difference between the enthalpies of the inlet stream and the outlet stream is determined using only sensible heats, since no reactions or phase changes are present:

$$\dot{m}\hat{H}_i = \dot{m}C_p(T_i - T_{\text{ref}}) \qquad [4.10\text{-}6]$$

$$\dot{m}\hat{H}_j = \dot{m}C_p(T_j - T_{\text{ref}}) = \dot{m}C_p(T - T_{\text{ref}}) \qquad [4.10\text{-}7]$$

where T is the temperature of both the system and the outlet stream. Therefore:

$$\dot{m}\hat{H}_i - \dot{m}\hat{H}_j = \dot{m}C_p(T_i - T) \qquad [4.10\text{-}8]$$

Substituting for internal energy as the product of specific internal energy (equation [4.10-4]) and the mass contained in the system m^{sys}, the time derivative of the total energy of a system is:

$$\frac{dE_T^{\text{sys}}}{dt} = \frac{dU^{\text{sys}}}{dt} = \frac{d}{dt}(m^{\text{sys}}C_v(T - T_{\text{ref}})) = m^{\text{sys}}C_v\frac{dT}{dt} \qquad [4.10\text{-}9]$$

when the mass m^{sys} and C_v are constant with respect to time. (Since T_{ref} is a constant, its derivative is zero.) Substitutions are made for the difference in enthalpy across the system and the rate of change of total energy in the system.

$$\dot{m}C_p(T_i - T) + \sum\dot{Q} + \sum\dot{W}_{\text{nonflow}} = m^{\text{sys}}C_v\frac{dT}{dt} \qquad [4.10\text{-}10]$$

Note that in equation [4.10-10], the temperature T is a function of time. Thus, in an unsteady-state system, the temperature T of the system changes over the time period of interest. With an initial condition, this equation can be integrated.

The algebraic equation for a dynamic system is written similarly:

$$mC_p(T_i - T) + Q + W_{\text{nonflow}} = m^{\text{sys}}C_v(T_f - T_0) \qquad [4.10\text{-}11]$$

where m is the mass transferred across the system boundary. When considering a system for which an algebraic equation is appropriate, the amount of energy that crosses the system boundary, $mC_p(T_i - T)$, is usually equal to zero.

For liquids and solids, C_v is equal to C_p. For gases, $C_v = C_p - R$, where R is the ideal gas constant. Heat capacities are tabulated in Appendices E.1–E.3, E.7, and E.8.

EXAMPLE 4.17 Start-up of a Blood Heating Device

Problem: Consider a start-up scenario for the blood warmer in Example 4.9, in which there is no stirrer to add work to the system (Figure 4.21). Suppose the tank is initially filled with 1.0 L of blood at 30°C. At $t = 0$, a heater begins to warm up the blood at a rate of 70 kcal/min. At the same time the heater is turned on, 10.0 L/min of blood at 30°C starts to flow continuously into and out of the tank. Calculate the time required for the temperature of the blood in the tank and in the outlet stream to reach 37°C.

Solution:

1. Assemble
   (a) Find: time required for the temperature of the blood to reach 37°C.
   (b) Diagram: Figure 4.21 shows the blood heating device. Blood enters and leaves the heater at a rate of 10.0 L/min. Heat is added to the system.
2. Analyze
   (a) Assume:
   - Tank is well mixed.
   - C_p and C_v are equal, constant, and have a numerical value of 1.0 cal/(g·°C).
   - No nonflow work.
   - Density of blood is constant at 1.0 g/cm^3.
   - No evaporation, phase change, or reaction occurs.
   - Heat lost to the surroundings is negligible.
   - No potential or kinetic energy changes.

   (b) Extra data: No extra data are needed.
   (c) Variables, notations, units:
   - T_1 = a constant indicating the temperature of the inlet stream.
   - T = a variable indicating the temperature of the outlet stream, as well as inside the tank.
   - Units: L, min, cal, kg, °C.

   (d) Basis: Since we assume the density of blood is 1.0 g/mL, we can use the inlet flow rate of 10.0 L/min of blood to obtain a basis of 10.0 kg/min.
3. Calculate
   (a) Equations: Since rates of material flow and heat are given, differential conservation equations for mass and energy are appropriate:

$$\sum_i \dot{m}_i - \sum_j \dot{m}_j = \frac{dm^{\text{sys}}}{dt}$$

$$\sum_i \dot{m}_i(\hat{E}_{P,i} + \hat{E}_{K,i} + \hat{H}_i) - \sum_j \dot{m}_j(\hat{E}_{P,j} + \hat{E}_{K,j} + \hat{H}_j) + \sum \dot{Q} + \sum \dot{W}_{\text{nonflow}} = \frac{dE_T^{\text{sys}}}{dt}$$

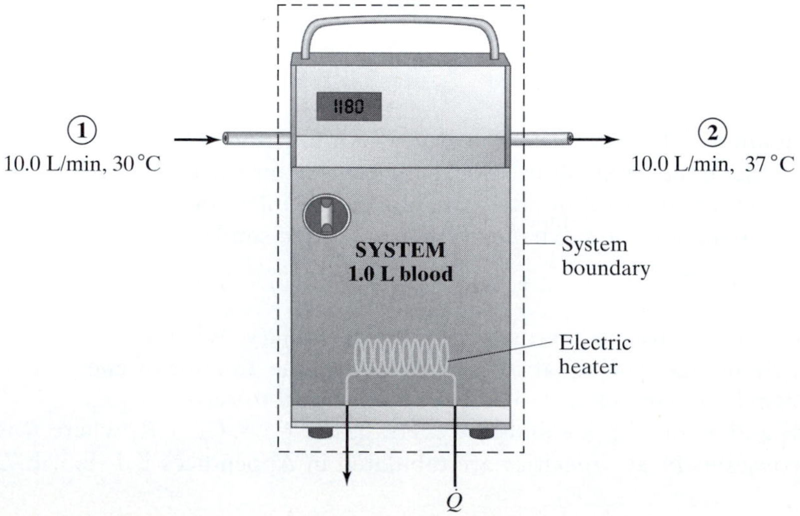

Figure 4.21
Blood heating device (unsteady-state start-up).

(b) Calculate:

- We assume that the process has no reactions and the system with respect to total mass is at steady-state. The mass flow rates of blood into and out of the system are equal to the basis:

$$\dot{m}_1 = \dot{m}_2 = \dot{m} = 10.0 \frac{\text{kg}}{\text{min}}$$

Because the system mass is at steady-state, and the volume inside the tank remains constant, the mass of the blood inside the tank remains constant at 1.0 kg.

- Because potential and kinetic energies do not change, the system does not have reactions, and the system is well mixed, the unsteady-state energy balance equation [4.10-10] is applied. After further reductions, such as nonflow work is reduced to zero, the temperature change as a function of time is given as:

$$\dot{m}C_p(T_1 - T) + \sum \dot{Q} = mC_v \frac{dT}{dt}$$

$$10.0 \frac{\text{kg}}{\text{min}} \left(\frac{1000 \text{ g}}{1 \text{ kg}} \right) 1.0 \frac{\text{cal}}{\text{g} \cdot {}^\circ\text{C}} (30^\circ\text{C} - T) + 70{,}000 \frac{\text{cal}}{\text{min}}$$

$$= 1.0 \text{ kg} \left(\frac{1000 \text{ g}}{1 \text{ kg}} \right) \left(1.0 \frac{\text{cal}}{\text{g} \cdot {}^\circ\text{C}} \right) \frac{dT}{dt}$$

$$370 \frac{\text{kcal}}{\text{min}} - 10T \frac{\text{kcal}}{\text{min} \cdot {}^\circ\text{C}} = \left(1 \frac{\text{kcal}}{{}^\circ\text{C}} \right) \frac{dT}{dt}$$

- To calculate the time needed to reach 37°C, we use the given initial condition (temperature at $t = 0$ is 30°C) and integrate the equation we obtained above to calculate the temperature T as a function of time:

$$\int_{30^\circ\text{C}}^{37^\circ\text{C}} \frac{dT \dfrac{\text{kcal}}{{}^\circ\text{C}}}{370 \dfrac{\text{kcal}}{\text{min}} - 10\, T \dfrac{\text{kcal}}{\text{min} \cdot {}^\circ\text{C}}} = \int_0^t dt$$

$$\frac{-1 \dfrac{\text{kcal}}{{}^\circ\text{C}}}{10 \dfrac{\text{kcal}}{\text{min} \cdot {}^\circ\text{C}}} \ln\left(370 \frac{\text{kcal}}{\text{min}} - 10\, T \frac{\text{kcal}}{\text{min} \cdot {}^\circ\text{C}} \right) \Bigg|_{30^\circ\text{C}}^{37^\circ\text{C}} = t$$

The integral cannot be evaluated because ln(0) cannot be determined. Therefore, the temperature in the tank cannot reach 37°C within a finite period of time. The blood temperature can only approach 37°C (Table 4.8). For practical purposes, the temperature reaches close to 37°C in less than 1 min.

TABLE 4.8

Temperature During Start-up of Blood Heating Device

Time (min)	Temperature (°C)
0.195	36.0
0.264	36.5
0.425	36.9
0.494	36.95
0.724	36.995
0.955	36.9995
1.645	36.9999995
2.267	36.999999999
2.497	36.9999999999

4. Finalize
   (a) Answer: It will take an infinite amount of time for the temperature to reach 37°C. However, the temperature reaches very close to 37°C within 1 min.
   (b) Check: It is hard to get an independent check on this answer. Using the algebraic equation, the time required for the heater to warm 1.0 L of blood from 30°C to 37°C is much less than 1 minute. Since our system is constantly being replenished with cold (30°C) blood, it should actually take longer to warm it to 37°C. ■

Another example of the application of the dynamic energy conservation equation is human metabolism. The metabolism of the body is the sum of all the chemical reactions in all the cells of the body by which energy is provided for vital processes. The metabolic rate is normally expressed in terms of the rate of heat liberated during these chemical reactions. **Basal metabolic rate** (BMR) is the rate at which energy is used in the body during absolute rest but while the person is awake.

Figure 4.22 shows the dependence of BMR on age and gender in kilocalories per square meter of body surface area (this normalizes for size). A typical 30-year-old man (5 ft 8 in, 150 lb_m) has a surface area of about 1.8 m^2 (see Appendix D.2). This implies that he has a BMR of about 67 kcal/hr or 1600 kcal/day.

Rarely do people spend a day at absolute rest. Performing any type of activity other than cellular activity, respiration, and circulation requires energy. The actual metabolic rate depends on the type of activity performed. Some energy expenditure values for some activities are given in Table 4.9.

EXAMPLE 4.18 Metabolism in a Young Man

Problem: Brian, a 19-year-old male (5 ft 8 in, 150 lb_m), has a BMR of 1730 kcal/day. Suppose his recommended daily diet consists of 53 g of protein, 71 g of fat, and 320 g of carbohydrates (Figure 4.23a). The heats of reaction of carbohydrates, fats, and proteins are given in Table 4.10.

Case I: Based on this diet, how much energy can Brian expend each day without dipping into body reserves?

Case II: Assume Brian is very inactive and requires only 20% above BMR to survive. Assume that extra available energy is stored as fat and the conversion of energy to fat is 100% efficient. How much mass does Brian gain per day?

Case III: Suppose Brian is an astronaut wearing a well-insulated space suit during a space walk. The space suit is designed to remove body heat to maintain a constant body temperature. Suppose that the space suit suddenly malfunctions and cannot remove any heat. How much will Brian's body temperature increase in 2 hours due to the heat generated just from basal metabolism? Assume the body has a heat capacity of 0.86 kcal/kg·°C.

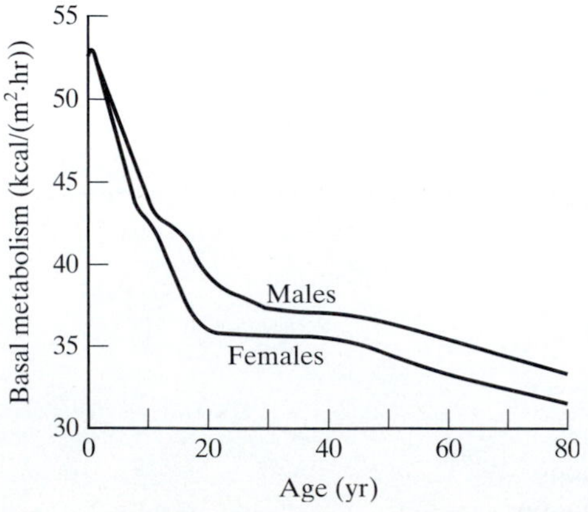

Figure 4.22
Normal BMR at different ages for each sex. (*Source:* Guyton AC and Hall JE, *Textbook of Medical Physiology*. Philadelphia: Saunders, 2000.)

TABLE 4.9

Energy Expenditure During Different Types of Activity for a 70-kg Man*

Form of activity	Energy expenditure (kcal/hr)
Sleep	65
Awake, lying still	77
Sitting at rest	100
Standing relaxed	105
Dressing or undressing	118
Typewriting rapidly	140
Walking slowly (2.6 mph)	200
Swimming	500
Running (5.3 mph)	570
Walking very fast (5.3 mph)	650
Walking up stairs	1100

* Table adapted from Guyton AC and Hall JE, *Textbook of Medical Physiology*. Philadelphia: Saunders, 2000.

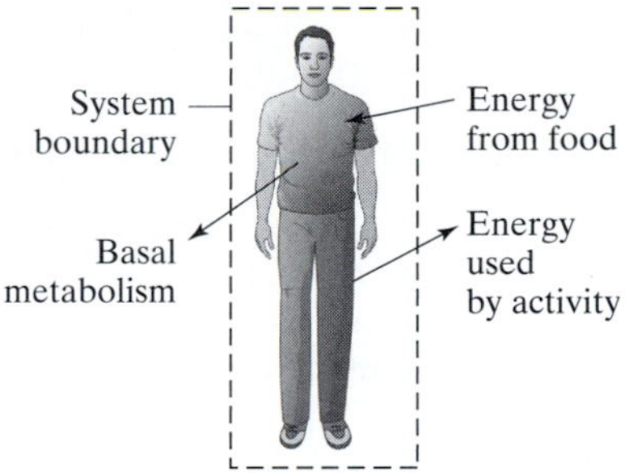

Figure 4.23a
Metabolism in a standard man.

TABLE 4.10

Metabolism of Different Classes of Foods*

Compound	$\Delta\hat{H}_r$, Heat of reaction (kcal/g)
Carbohydrate	4.1
Fat	9.3
Protein	4.5

* Data from Guyton AC and Hall JE, *Textbook of Medical Physiology*. Philadelphia: Saunders, 2000.

Solution:

Case I: The differential form of the conservation of total energy equation is:

$$\sum_i \dot{E}_{T,i} - \sum_j \dot{E}_{T,j} + \sum \dot{Q} + \sum \dot{W} = \frac{dE_T^{sys}}{dt}$$

Since we seek to model this system without dipping into body reserves, we assume that the system is at steady-state. The energy expended as BMR involves energy losses through bulk material transfer (e.g., energy loss during breathing) as well as through non-contact transfer (e.g., heat expended during metabolism). For this problem, we lump all of these expenditures into $\dot{E}_{T,\text{BMR}}$ so the above equation reduces to:

$$\sum_i \dot{E}_{T,i} - \sum_j \dot{E}_{T,j} = 0$$

$$\dot{E}_{T,\text{food}} - \dot{E}_{T,\text{BMR}} - \dot{E}_{T,\text{other}} = 0$$

where $\dot{E}_{T,\text{food}}$ is the rate at which energy enters the system through food, $\dot{E}_{T,\text{BMR}}$ is the rate at which energy is expended by the system through basal metabolism and other mandatory functions, and $\dot{E}_{T,\text{other}}$ is the rate at which energy is expended by the system through physical activities. The rate at which energy enters the system is calculated based on the energy content of the three different types of food (carbohydrate, fat, and protein):

$$\dot{E}_{T,\text{food}} = \dot{m}_{\text{carb}}(\hat{H}_{r,\text{carb}}) + \dot{m}_{\text{fat}}(\hat{H}_{r,\text{fat}}) + \dot{m}_{\text{prot}}(\hat{H}_{r,\text{prot}})$$

$$\dot{E}_{T,\text{food}} = 320\frac{\text{g}}{\text{day}}\left(4.1\frac{\text{kcal}}{\text{g}}\right) + 71\frac{\text{g}}{\text{day}}\left(9.3\frac{\text{kcal}}{\text{g}}\right)$$

$$+ 53\frac{\text{g}}{\text{day}}\left(4.5\frac{\text{kcal}}{\text{g}}\right) = 2210\ \frac{\text{kcal}}{\text{day}}$$

The total energy available is 2210 kcal/day. To find how much energy Brian can expend without dipping into reserves, we rearrange the differential energy conservation equation and substitute the known values:

$$\dot{E}_{T,\text{other}} = \dot{E}_{T,\text{food}} - \dot{E}_{T,\text{BMR}} = 2210\frac{\text{kcal}}{\text{day}} - 1730\frac{\text{kcal}}{\text{day}} = 480\ \frac{\text{kcal}}{\text{day}}$$

The amount of energy available for activity is 480 kcal/day.

Case II: In this system, Brian takes in food and expends energy at 20% above BMR. No nonflow work or heat acts on the system (Figure 4.23b). Since the total mass, and hence the energy of Brian's body, changes as he gains mass, it is appropriate to assume that the system is dynamic. The following differential energy conservation equation is most appropriate:

$$\dot{E}_{T,\text{in}} - \dot{E}_{T,\text{out}} = \frac{dE_T^{\text{sys}}}{dt}$$

The rate at which energy accumulates is the difference between the rate at which energy enters the system through food and the rate at which energy exits the system through basal metabolism and other energy expenditures. We calculated in *Case I* that the system takes in 2210 kcal/day through food. Because Brian's other energy expenditures are 20% above his BMR, we can calculate the rate at which energy leaves the system:

$$\dot{E}_{T,\text{out}} = 1.2\dot{E}_{T,\text{BMR}} = 1.2\left(1730\frac{\text{kcal}}{\text{day}}\right) = 2080\ \frac{\text{kcal}}{\text{day}}$$

Thus, the rate at which energy accumulates in the system can be calculated:

$$\frac{dE_T^{\text{sys}}}{dt} = \dot{E}_{T,\text{in}} - \dot{E}_{T,\text{out}} = 2210\frac{\text{kcal}}{\text{day}} - 2080\frac{\text{kcal}}{\text{day}} = 130\frac{\text{kcal}}{\text{day}}$$

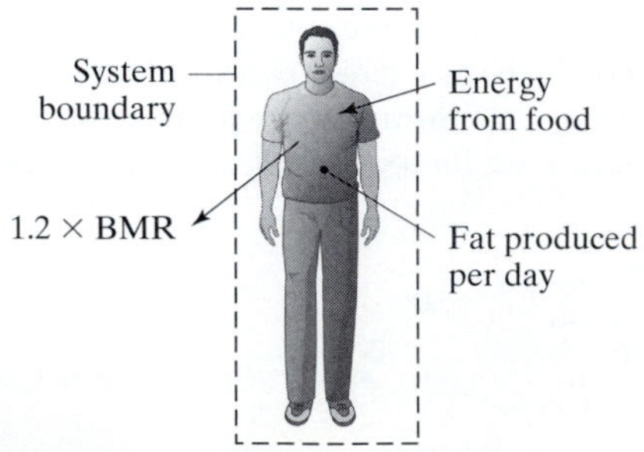

Figure 4.23b
Metabolism of an inactive man.

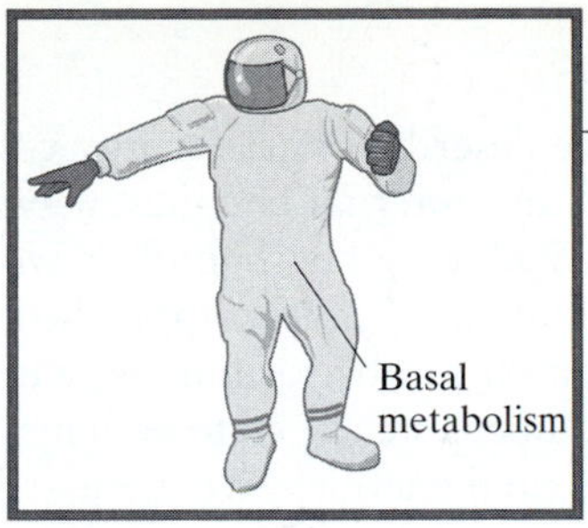

Figure 4.23c
Metabolism of an astronaut wearing an insulated space suit.

The rate at which energy accumulates is 130 kcal/day; this value is assumed to be the rate that energy is stored as fat. Assuming the conversion of energy to fat is 100%, Brian's rate of mass gain is:

$$\frac{\left(130\dfrac{\text{kcal}}{\text{day}}\right)}{9.3\dfrac{\text{kcal}}{\text{g}}} = 14\ \frac{\text{g}}{\text{day}}$$

At this rate, he would gain 0.9 lb_m in a month.

Case III: We assume that when Brian is an astronaut he does not eat any food during the space walk (Figure 4.23c). Because his space suit malfunctions, the energy expended by his body through basal metabolism cannot be released to the environment. With the insulated space suit barrier, the energy released through basal metabolism warms up the body. The unsteady-state total energy conservation equation for this situation is:

$$\dot{E}_{T,\,\text{BMR}} = \frac{dE_T^{\text{sys}}}{dt}$$

Recall from equation [4.10-9] that the change in the system energy is related to the change in the system temperature:

$$\frac{dE_T^{\text{sys}}}{dt} = mC_v\frac{dT}{dt}$$

For a solid, C_v is equal to C_p. The BMR is the rate at which energy accumulates in the system:

$$\frac{dE_T^{\text{sys}}}{dt} = \dot{E}_{T,\text{BMR}} = mC_p\frac{dT}{dt}$$

$$1730\frac{\text{kcal}}{\text{day}}\left(\frac{1\text{ day}}{24\text{ hr}}\right) = 150\text{ lb}_\text{m}\left(\frac{1\text{ kg}}{2.2\text{ lb}_\text{m}}\right)\left(0.86\frac{\text{kcal}}{\text{kg}\cdot{}^\circ\text{C}}\right)\frac{dT}{dt}$$

$$72.1\frac{\text{kcal}}{\text{hr}} = 58.6\frac{\text{kcal}}{{}^\circ\text{C}}\frac{dT}{dt}$$

$$\int_0^{2\text{hr}} 72.1\frac{\text{kcal}}{\text{hr}}dt = \int_{37^\circ\text{C}}^{T} 58.6\frac{\text{kcal}}{{}^\circ\text{C}}dT$$

$$72.1\frac{\text{kcal}}{\text{hr}}(2\text{ hr} - 0) = (58.6T - 2170^\circ\text{C})\frac{\text{kcal}}{{}^\circ\text{C}}$$

$$T = 39.5^\circ\text{C} = 103^\circ\text{F}$$

The astronaut's body temperature after 2 hours is 39.5°C (103°F). ■

Summary

In this chapter, we described basic energy concepts, which included definitions for potential, kinetic, and internal energies, as well as enthalpy. We also discussed energy in transition, including heat, nonflow work, and flow work. We described how the conservation statement can be applied to the extensive property of total energy and how this equation can be formulated with internal energy or enthalpy.

We examined open, steady-state systems, with and without significant changes in kinetic and potential energies. We discussed how to calculate changes in enthalpy with changes in temperature, pressure, phase, and reactions. We then solved problems with systems undergoing changes in enthalpy. Finally, we analyzed how the equations can be used to solve for variables in dynamic systems.

Table 4.11 reinforces that total energy may accumulate in a system because of bulk material transfer across the system boundary or because of direct and nondirect contacts. See the tables concluding other chapters for comparison. In this text, the conservation of total energy is presented as a foundation before the accounting of electrical energy (Chapter 5) and the accounting of mechanical energy (Chapter 6).

Because the focus of this book is strictly on the accounting and conservation equations, many important topics in the field of thermodynamics are not addressed. Other critical concepts, such as entropy, free energy, and the second law of thermodynamics, are addressed in other books (e.g., Kyle BG, *Chemical and Process Thermodynamics*, 3d ed, Upper Saddle River, NJ: Prentice Hall, 1999; Çengel YA and Boles MA, *Thermodynamics: An Engineering Approach*, 3d ed., Boston: McGraw-Hill, 1998.)

TABLE 4.11

Summary of Movement, Generation, Consumption, and Accumulation in the Total Energy Accounting Equation

Accumulation	Input − Output		+ Generation − Consumption	
Extensive property	Bulk material transfer	Direct and non-direct contacts	Chemical reactions	Energy interconversions
Total energy	X	X		

References

1. Zubay G. *Biochemistry*, 2d ed. New York: Macmillan Publishing Co, 1988.
2. Centers for Disease Control and Prevention. "Heat Illnesses and Death." July 23, 2003. www.cdc.gov/communication/tips/heat.htm (accessed January 6, 2005).
3. Bromley LA, Desaussure VA, Clipp JC, and Wright JS. "Heat capacities of sea water solutions at salinities of 1 to 12% and temperatures of 2° to 80°C." *J Chem Eng Data* 1967, 12:202–6.

Problems

4.1 There are numerous methods to estimate the temperature of substances. Some say it is possible to estimate the temperature of air in degrees Fahrenheit by counting the number of chirps a cricket makes in 15 seconds, then adding that number to 37.

Consider a new temperature scale based on the rate that crickets chirp. Assume that the temperature in this scale is equal to the number of times a cricket chirps per minute, and that the temperature on this scale is reported as °X.

(a) Derive an expression that relates temperature in degrees Fahrenheit ($T_{°F}$) to degrees Cricket ($T_{°X}$).

(b) Given that the heat capacity of blood at body temperature is 1.87 J/(g · °F), what is the heat capacity of blood at body temperature in J/(g · °X)?

(c) The heat capacity of blood as a function of temperature is given by:

$$C_p\left[\frac{\text{J}}{\text{g}\cdot{}^\circ\text{F}}\right] = 1.85 + 0.000234T[{}^\circ F]$$

Derive an equation for the heat capacity of blood that uses temperature in °X and gives the heat capacity in J/(g ·°X). In other words, determine what the numerical values are for the constants *a* and *b* in the following equation:

$$C_p\left[\frac{\text{J}}{\text{g}\cdot{}^\circ\text{X}}\right] = a + bT[{}^\circ\text{X}]$$

(d) To check that your answer in part (c) is correct, calculate the heat capacity of blood at 98.6°F using both equations and compare the results. Do the results of your calculations make sense?

4.2 A sample of oxygen is subjected to an absolute pressure of 2.4 atm. If the specific internal energy of the sample at 310 K is 5700 J/mol relative to a known reference state, what is the specific enthalpy of the oxygen relative to that same reference state?

4.3 A weight is added to a piston so that the volume of the gas inside the container is reduced from 2.5 L to 1.0 L at a constant temperature. How much heat do you need to add to the system if you want to increase the volume back to 2.5 L at this new pressure? Assume an ideal gas and an initial pressure of 1.0 atm.

4.4 You have engineered an enzyme that is supposed to break down protein A into polypeptide B. The enzyme has its optimal activity at 37°C, so a process must be designed to maintain this temperature at all times. A batch reactor process is used; i.e., the bioreactor is filled with an initial quantity of enzyme and protein and is permitted to run until the precursor is almost completely exhausted. The ratio of moles of protein A consumed to polypeptide B produced is 1:10. The reaction is irreversible and first order, following the relation:

$$-\frac{dC_A}{dt} = kC_A$$

where k is the rate constant for the reaction ($k = 0.01$ 1/s), C_A is the concentration of protein A, and t is time.

(a) Calculate the time required to consume 99% of the precursor.

(b) Calculate the number of moles of polypeptide B produced during the time period calculated in part (a).

Because the reaction is exothermic, for every mole of B produced, 10 kJ of energy is released. Heat is removed by means of a heat exchanger that dissipates excess heat via a chilled water stream and can be modeled by the equation:

$$\dot{Q} = hA(T_{\text{bioreactor}} - T_{\text{water}})$$

where h is the heat transfer coefficient, A is the surface area of the exchanger, $T_{\text{bioreactor}}$ is the temperature of the bioreactor, and T_{water} is the temperature of the chilled water stream. The heat transfer coefficient h is a function of $\dot{V}_{\text{water}}$, the flow rate of chilled water. The following data are provided: the volume of the bioreactor is 10 L, $C_{A,0}$ is 150 mM, T_{water} is 4°C when the process begins, A is 5 m^2, and h is $\dot{V}_{\text{water}} \times 100$ kJ/(m^5·K).

(c) Calculate the total heat removed from the system in kJ during the calculated time period.

(d) Determine the flow rate of water $\dot{V}_{\text{water}}$ as a function of time in L/min.

4.5 A bomb calorimeter is a device commonly used to measure the internal energy of a substance, especially in combustion reactions (Figure 4.24). A calorimeter is well insulated and designed to maintain a constant volume. For a calorimeter to work properly, its calorimeter constant C must be known. The calorimeter constant is related to the internal energy change as $\Delta U = C\,\Delta T$. For benzoic acid ($C_7H_6O_2$), the heat of combustion ΔH_c is -3226.7 kJ/mol. A 2.53-g sample of benzoic acid is burned in a bomb calorimeter at 25°C, and the temperature increases by 3.72°C. What is the calorimeter constant?

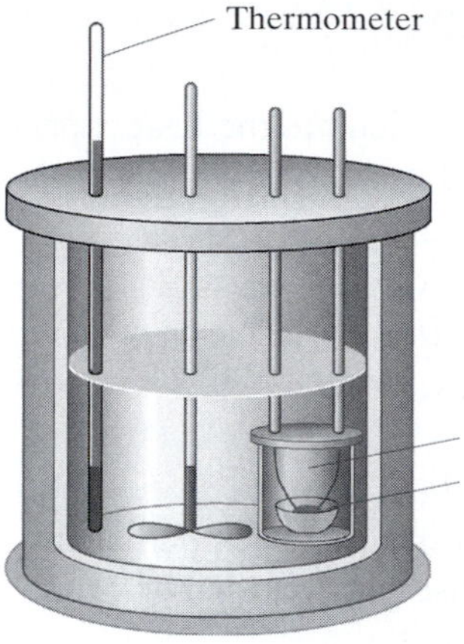

Figure 4.24
Bomb calorimeter.

4.6 In direct calorimetry, a person is placed in a large, water-insulated chamber. The chamber is kept at a constant temperature. While in the chamber, the subject is asked to perform a number of normal activities, such as eating, sleeping, and exercising. The rate of heat released from the subject's body can be measured by the rate of heat gain by the water bath. Would direct calorimetry be a practical way to measure metabolic rate? Why or why not?

A person is placed inside a calorimetric chamber for 24 hours. During this time, the 660-gallon water bath heats up by 3.2°F. What is the subject's metabolic rate during this period? Report your answer in kcal/day. Assume that there is no heat loss from the water to the surroundings.

4.7 Cryogenics has the potential to be useful in a variety of fields, including medicine. Suppose you have engineered a method to successfully deep-freeze and thaw human organs using liquid nitrogen without any freezing damage to the cells and tissue structure. How much heat must be removed from a liver (1.5 kg) to freeze it at 180 K? For liquids and solids, heat capacity at constant pressure, C_p, is approximately equal to heat capacity at constant volume, C_v.

4.8 The heat capacity at constant pressure, C_p, is the slope of the change in specific enthalpy as a function of temperature, as given in equation [4.5-16]:

$$C_p(T) = \lim_{\Delta T \to 0} \frac{\Delta \hat{H}}{\Delta T}$$

The limit as $T \to 0$ becomes

$$C_p(T) = \left(\frac{\partial \hat{H}}{\partial T}\right)_P$$

where the ∂ indicates a partial derivative. Partial derivatives are used when a function ($\hat{H}$ in this case) is dependent on more than one variable (T and P in this case).

There is a similar relationship between the heat capacity at constant volume, C_v, and the partial derivative of specific internal energy, $\hat{U}$, with respect to temperature, as follows:

$$C_v(T) = \left(\frac{\partial \hat{U}}{\partial T}\right)_V$$

From these heat capacity definitions and the definition of enthalpy given in this chapter, derive the relationship between C_p and C_v for an ideal gas in terms of R (ideal gas constant) and other variables that may be necessary.

4.9 Changes in the specific enthalpy of a system were given in the text separately for changes in temperature and pressure.

(a) Write a formula that describes the changes in specific enthalpy for an ideal gas undergoing changes in both temperature and pressure.

(b) Write a formula that describes the change in specific enthalpy for a liquid or solid that is undergoing changes in both temperature and pressure.

4.10 You work for a surgeon who asks you to design a heat-exchanging device to continuously heat 5.0 L/min of blood from 4°C to 37°C by transferring heat from warm water.

(a) Assuming that the specific heat capacity of blood is constant at 1.0 cal/(g·°C) and its density is also constant at 1.0 g/mL, estimate the required rate of heat transfer in cal/min.

(b) A doctor suggested that the blood could be warmed by simply immersing coils of tubing carrying the blood in a large water bath. Using the following assumptions, estimate the necessary volume of the water bath.

- The initial temperature of the water bath is 50°C and no more heat is supplied to the water bath; i.e., it is allowed to cool off during the surgery.
- The surgery lasts 3 hours.
- The final water temperature should not fall below 40°C in order to maintain a proper temperature gradient for heat transfer.
- Heat transfer occurs only between the blood and the water; no heat is exchanged with the environment.

(c) Is the design in part (b) practical? Make recommendations to improve it.

4.11 You need to size a continuous vaporizer for a child's sick room. The device receives liquid water at 20°C and 1 atm and produces steam at a rate of 0.7 g/min. At what rate must energy be supplied if the device is 100% efficient? The standard heat of vaporization of water is 2256.9 kJ/kg and the specific heat capacity of water is 1 cal/ (g·°C).

4.12 Consider a cold winter day with an air temperature of 5°C and relative humidity of about 20% (moisture content 0.001 g H_2O/g dry air). A person waiting at a bus stop inhales at an average rate of 7 g dry air/min and exhales air saturated with water at body temperature (37°C) and 1 atm. The heat capacity of dry air is 1.05 J/ (g·°C). Estimate the rate of heat loss by breathing in kcal/hr.

4.13 A seated person generates 77 kcal/hr of metabolic heat. The same person walking at 5.3 miles per hour generates 650 kcal/hr. How much does this person have to sweat for the evaporation of sweat to remove the difference in energy generated between the states in which the person is sitting and walking? Assume the skin temperature is 33°C and the air temperature is 30°C.

4.14 Ronaldo averages three meals per day, each composed of 25 g protein, 35 g fat, and 80 g carbohydrates.

(a) What is Ronaldo's daily energy intake based on these three meals?

(b) Ronaldo exercises to burn all of the energy gained from the three meals. Figure 4.25 is a graph expressing the rate of caloric depletion in Ronaldo over the course of his exercise time. In order to burn off all the daily calories, for how many hours must the rigorous exercise portion of Ronaldo's workout session last each day? The energy transfer E from the body during warm-up and cool-down phases may be described by:

$$\frac{dE_w}{dt} = -5600(t - 0.25)^2 + 350 \qquad 0 \le t < 0.25 \text{ hr}$$

$$\frac{dE_c}{dt} = -1400(t - x) + 350 \qquad x < t \le x + 0.25 \text{ hr}$$

where t is the time given in hours.

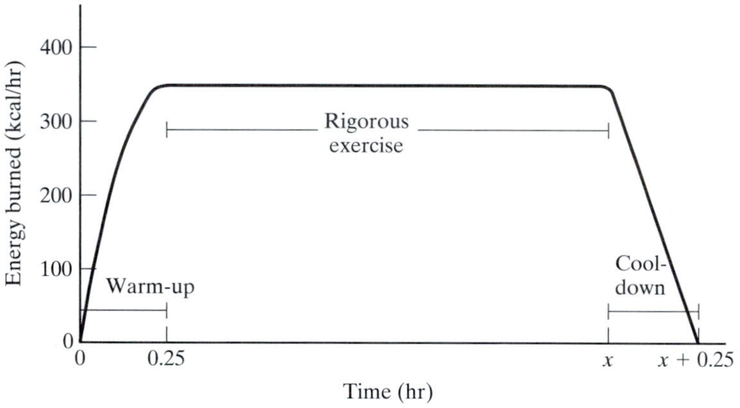

Figure 4.25
Energy burned during exercise.

4.15 Mammalian cells are usually cultured in a carbon dioxide (CO_2) incubator (Figure 4.26), in which both the temperature and carbon dioxide concentration are maintained at a desired level, T_D and $C_{CO_2,D}$. The temperature is controlled by an electrical heating element inside the incubator; the carbon dioxide concentration is maintained by a gas valve. Assume that the carbon dioxide flow is controlled by an on–off controller. In other words, the valve will be turned on to allow gaseous CO_2 to flow into the incubator when the incubator CO_2 concentration is below a certain threshold limit, $C_{CO_2,L}$, and the valve will be

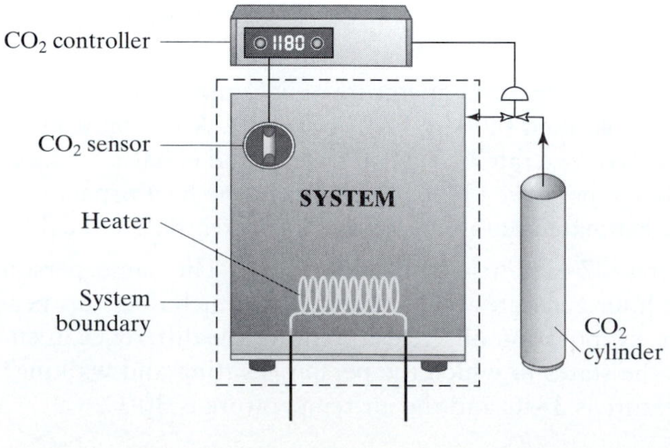

Figure 4.26
Carbon dioxide incubator.

turned off to stop gaseous CO_2 from flowing into the incubator when the incubator CO_2 concentration is above a certain threshold limit, $C_{CO_2,U}$. Although the walls of the incubator are relatively well insulated, heat is lost to the surroundings at a rate estimated to be $\dot{Q}_L$ (kcal/hr).

(a) The system is defined as the incubator. Is the system open, closed, or isolated when the valve is off? Justify your answer.

(b) Is the system open, closed, or isolated when the valve is on? Justify your answer.

(c) Describe how you would estimate the daily heat requirement of the system. Because the system is not perfectly sealed, CO_2 leaks from the incubator even when the door remains shut. In this situation, the average amount of CO_2 flow into the incubator per day is $\dot{m}_{CO_2}$ (g/hr). For this model, assume that nobody opens the incubator during this period. (*Note*: Most of the heat and CO_2 losses occur when the incubator door is ajar.)

(d) Assume that the door is opened ten times. Describe how you would estimate the daily heat requirement of the system.

4.16 A pharmaceutical company has decided to test the feasibility of manufacturing a new drug using biochemical engineering. In this approach, a valuable intermediate, intA, will be produced from raw materials using a genetically engineered bacterial strain. After undergoing a series of chemical steps, this intermediate will then be converted to the final product.

Calculate the heat requirement (Figure 4.27) to convert intA to another more stable intermediate, intB, using a 2-L reactor.

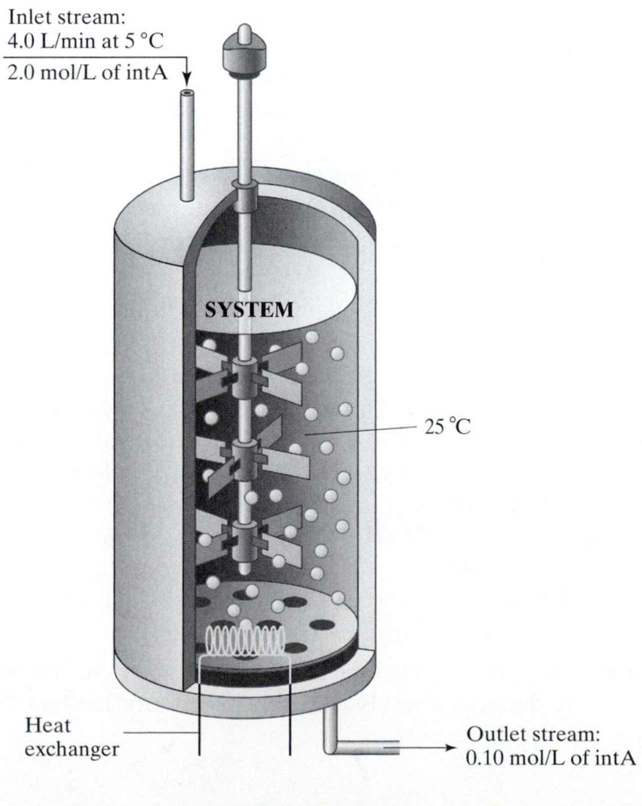

Figure 4.27 Reactor used to manufacture a new drug using a genetically engineered bacterial strain.

The following information is supplied by the technical support group:

- IntA is relatively unstable and has to be maintained at 5°C prior to entering the reactor.
- The flow rate of the inlet stream is 4.0 L/min.
- The reactor operates at 25°C and 1 atm.
- The specific heat capacity of the reactant and product streams is 1 cal/(g·°C). and is a constant.
- The density of the reactant and product streams is 2.0 g/cm^3 and is a constant.
- One mole of intA forms 2 moles of intB with negligible by-product formation:

$$\text{intA} \rightarrow 2\ \text{intB}$$

- The reaction of intA under the given conditions does not go to completion. When 2.0 mol/L of intA flow into the reactor, 0.10 mol/L remains unreacted.
- The standard heat of formation of intA is −2050 kJ/mol.
- The standard heat of formation of intB is −1560 kJ/mol.
- Molecular weights of intA and intB are 1080 g/mol and 540 g/mol, respectively.
- The reactor is well insulated.
- The stirrer does work on the system at a rate of 10 W.

Calculate the rate of heat addition or removal to maintain the reactor at the desired temperature.

4.17 After researching genetically modified (GM) foods, you decide to expand your agricultural business to Illinois.

(a) What are some of the current advantages and drawbacks of GM crop production and foods? In your opinion, do the benefits outweigh the risks? Why?

(b) Once on your new farm, you soon discover that annual rainfall is very low and the nearest stream is too far away to be of any practical use. People tell you that the ground water in the area is 25–35 m below the surface, so you decide to dig a well to provide water for irrigation. If you want to pump the water at 1.0 m/s and your pipe has a cross-sectional area of 0.05 m^2, how much power will you need to provide? Report your answer in units of kW and hp.

(c) Pumps can be purchased only at certain power levels, such as 5 hp, 10 hp, 25 hp, and 50 hp. Which pump would you select?

(d) The depth of ground water varies with the season. With the pump you selected in part (c), what is the maximum depth of ground water at which your pump could still operate to bring water to the surface at the conditions given in part (b)?

4.18 Water parks are popular summertime attractions. Visitors enjoy wading in wave pools and sliding down different types of water slides. A particular water slide has a height of 75 ft and an incline of 70°. At the end of the slide is a 100-ft horizontal segment to slow the slider down.

(a) Assuming that there is no friction or air resistance going down the slide, what is the velocity of a 150-lb$_m$ person at the beginning of the horizontal segment?

(b) How much work is done on the slider in order to slow him down to 5 ft/s at the end of the horizontal segment?

4.19 Upon reentry into the Earth's atmosphere, the bottom of a space shuttle heats up to dangerous levels as the craft slows for landing. If the velocity of the shuttle is

28,500 km/hr at the beginning of reentry and 370 km/hr just prior to landing, how much energy is lost as heat? The shuttle has a mass of 90,000 kg. Assume that the change in potential energy is negligible compared to the change in kinetic energy.

4.20 Adenosine triphosphate (ATP) is a major source of energy for cells in the body. Energy is released when one of the phosphate bonds is broken to form adenosine diphosphate (ADP).

$$\text{ATP} + \text{H}_2\text{O} \rightleftharpoons \text{ADP} + \text{P}_i$$

Given the thermodynamic data in Table 4.12, find the heat of reaction for the forward reaction.

TABLE 4.12

Heats of Formation in Phosphorylation Reactions

Species	$\Delta\hat{H}_f^\circ$ (kJ/mol)*
Adenosine triphosphate (ATP)	−2981.79
H_2O	−286.65
Adenosine diphosphate (ADP)	−2000.19
P_i	−1299.13

*Values from Alberty RA and Goldberg RN, "Standard thermodynamic formation properties for the adenosine 5′-triphosphate series," *Biochemistry* 1992, 31:10610–15.

4.21 Estimate the standard heat of reaction for the synthesis of solid leucylglycine by the following reaction:

$$\text{leucine (s)} + \text{glycine (s)} \longrightarrow \text{leucylglycine (s)} + \text{H}_2\text{O}(\ell)$$

The standard heats of formation are given in Table 4.13.

TABLE 4.13

Standard Heats of Formation for Synthesis of Solid Leucylglycine

Species	$\Delta\hat{H}_f^\circ$ (kcal/mol)
Leucine (s)	−154.16
Glycine (s)	−128.46
Leucylglycine (s)	−205.6
$H_2O(\ell)$	−68.317

4.22 Find the heat of reaction for the combustion of propane:

$$\text{C}_3\text{H}_8(\text{g}) + 5\ \text{O}_2(\text{g}) \longrightarrow 3\ \text{CO}_2(\text{g}) + 4\ \text{H}_2\text{O}(\ell)$$

4.23 What is the heat of combustion for one mole of liquid ethanol (C_2H_6O)? Remember to balance the combustion reaction.

4.24 Upon combusting a small sample of lactic acid, you observe that 2450 kJ energy is released. Calculate the heat of combustion of lactic acid. How many moles of lactic acid were in your sample?

4.25 *Escherichia coli* is a type of bacteria that, depending on the strain, either lives in the human digestive tract without any adverse effect or causes severe illness in humans. A researcher wishes to grow a particular strain of *E. coli* (overall formula $CH_{1.77}O_{0.49}N_{0.24}$) in a bioreactor for protein production. At 5 kg/hr, glycerol ($C_3H_8O_3$) is the limiting reactant. Reaction products are constantly being removed to keep reactor conditions constant.

$$\text{glycerol} + NH_3 + O_2 \longrightarrow E.\ coli + CO_2 + H_2O$$

How much heat needs to be removed from the bioreactor to keep it at a constant temperature? Assume that the heat of combustion of *E. coli* is −22.83 kJ/g and the respiratory quotient is 0.44. Assume that the enthalpy changes required to raise and lower the temperature of compounds are negligible relative to $\Delta \hat{H}_r^\circ$.

4.26 Industrial production of ethanol requires fermentation using *Saccharomyces cerevisiae*, a type of yeast. The elemental formula of *S. cerevisiae* is $CH_{1.83}O_{0.56}N_{0.17}$. Suppose you want to produce ethanol (C_2H_6O) in a batch process at 25°C. You plan to add 0.25 lb_m NH_3 and 5.0 lb_m glucose to the yeast for the following (unbalanced) reaction:

$$C_6H_{12}O_6(s) + NH_3(g) \longrightarrow C_3H_8O_3(\ell) + C_2H_6O(\ell)$$
$$+ CH_{1.83}O_{0.56}N_{0.17}(s) + CO_2(g) + H_2O(\ell)$$

Note that 1 mol glucose produces 0.5 mol glycerol and that the ratio of NH_3 to H_2O is 1:1.

During preparation, you accidentally add too much glucose. However, you decide to let the bioreactor run anyway. After the reaction goes to completion (at least one of the starting reactants is used up), you discover that 1.4 lb_m ethanol had been produced. How much heat had been produced and subsequently removed from the system? How much glucose did you initially add to the reactor? Assume that the heat of combustion of *S. cerevisiae* is −21.2 kJ/g.

4.27 Lactose is broken down into monosaccharides in the stomach through the following reaction:

$$\text{lactose} + H_2O \longrightarrow \text{glucose} + \text{galactose}$$

(a) Find the standard heat of reaction for this conversion.

(b) The above reaction is usually catalyzed by the enzyme lactase. However, lactose-intolerant people are not able to produce lactase and become ill when they consume too much lactose. Suppose that you have developed a new treatment for lactose intolerance. To test its effectiveness, you decide to run a simulation using a bioreactor. Lactose dissolved in water is added to the bioreactor at a rate of 100 g/min. Reaction products, water, and undigested lactose are removed at the same rate. Assume that reactants and products are both at standard conditions (25°C, 1 atm). If 125 J/s heat is removed to keep the bioreactor temperature constant, what percentage of the lactose is broken down into glucose and galactose?

4.28 In the human body, the conversion of a mole of adenosine triphosphate (ATP) to adenosine diphosphate (ADP) liberates about 7.3 kcal of energy. For every mole of glucose the body consumes, 38 moles of ATP are formed. What percent of energy is wasted as heat when the body breaks down glucose to carbon dioxide and water?

4.29 Intravenous feedings are given to hospital patients who cannot eat on their own or cannot tolerate tube feedings. Usually, the feeding solution has a balance of carbohydrates, fats, protein, and vitamins. Using the information from Problem 4.28, if glucose were the only component in the feeding solution, what would be the minimum amount of glucose needed per day to generate an average basal heat production rate of 1650 kcal/day?

4.30 Patients recovering from knee surgery often spend time in a warm pool, performing various rehabilitation exercises. The desired temperature of the pool water, T_{pool}, is 55°C, the tepid water temperature is 18°C, and the volume of the pool is 10,000 L. Three engines may be rented to heat the pool: a 1-MW engine for \$400/day, a 500-kW engine for \$233/day, and a 10-kW engine for \$30/day.

The heat transfer by convection between the pool and the air can be modeled by:

$$\dot{Q} = hA(T_{air} - T_{pool})$$

where h is the heat transfer coefficient and A is the surface area of the pool. These values are 2 W/(m^2·K) and 12 m^2, respectively. Assume that the air temperature T_{air} is constant at 21°C and that the pool walls are well insulated.

(a) How much heat must be transferred to the pool to heat it from 18°C to 55°C?

(b) You have a maximum of two full days to warm the pool before rehabilitation classes begin. Which engine would be the most cost-effective choice? In other words, how much time will each engine take, and how much will each engine cost? Remember that the engines are rented on a per-day basis.

4.31 Hypothermia is a condition in which the core body temperature drops below 35°C. A person suffering from hypothermia is usually treated with warm (approximately 43°C) humidified air and intravenous (IV) fluids.

(a) Basal heat production is 1850 kcal/day at 37°C and increases to three times higher at 33°C. The rate of heat production between these two temperatures can be assumed to be linear. Assuming that the heat capacity of the body is approximately equal to that of water, how long would it take to warm the body without any outside heating?

(b) How much IV fluid would be needed to rewarm the blood in the body if basal heat production was ignored? Considering that the total volume of blood in the body is approximately 5 L, how appropriate is this method as a warming technique?

(c) The breathing rate is approximately 6 L/min. If the heat capacity of air over our range of interest is 29.1 J/(mol·°C), how long would it take to rewarm the body if the only heat source was the warm humidified air?

(d) Your answers should indicate that the warm IV fluid and humidified air do not play a significant role in rewarming a hypothermic person. Instead, they are generally used to prevent further heat loss. For example, the body usually loses heat during breathing. Using the warm humidified air prevents this heat loss. The warm IV and humidified air treatments are especially important in preventing the overcooling of vital organs, such as the brain and heart. Generally they require administration by paramedics or doctors with the proper equipment. Propose a method to warm a hypothermic person if paramedics and hospitals are not available. Explain its advantages and disadvantages.

4.32 Every summer, parents are warned not to leave children or pets in the car with the air conditioner off and windows rolled up. The interior of the car can quickly become much hotter than outside and lead to heat stroke or heat death. In humans, heat stroke occurs when the core body temperature reaches 40°C. Heat death occurs when the core body temperature reaches approximately 41°C. Heat transfer, $\dot{Q}$, from the air to a person can be modeled by:

$$\dot{Q} = hA(T_{air} - T_{body})$$

where h is the heat transfer coefficient and A is the surface area of the person.

(a) Approximate how many minutes it would take to reach heat stroke and heat death in a closed car on a hot, sunny, clear day for a 13-kg toddler given the following assumptions and data:
- The air inside the car heats up to 65°C instantaneously.
- The initial temperature of the person is 37°C.
- Surface area of the child is 0.70 m^2.
- Heat capacity of the human body is approximately 3.6 kJ/(kg·°C).
- The person does not sweat.
- The heat transfer coefficient, h, is 15 W/(m^2·°C).

(b) Repeat the question for an 80-kg adult (surface area of 2.0 m^2).

(c) Times of 10 to 15 minutes are realistic for heat death for a child.[2] How does your calculated answer compare? What are some weaknesses of the above model? How would improvements to the model affect the calculated time?

4.33 Suppose a 75-kg woman wants to lose 5 kg of mass by going to a sauna.

(a) If she decides to go every day for 6 weeks, how long would her sauna sessions have to be to achieve her desired mass loss? Assume:
- The woman's metabolic rate is 2000 kcal/day and that she consumes 2300 kcal/day.
- The metabolism of 9.3 g of fat releases 1 kcal of energy.
- The woman sweats 1 L/hr in intense heat.
- Exposed skin surface area is 1.5 m^2.

(b) Is your answer feasible? In general, saunas recommend a limit of 10 to 15 minutes per session. Why?

(c) Sauna temperatures can reach about 90°C. Calculate what the woman's body temperature would be after the length of time you calculated in part (a).
- Assume a heat transfer coefficient from air to body of 12 W/(m^2·K).
- Assume the body's heat capacity is the same as that of water.

4.34 Suppose you have a vat containing 13 kg of salt dissolved in 100 L of water at 15°C. You want to dilute the solution to 0.030 kg/L, so you constantly add pure water at 5.0 L/min and remove salt solution from the vat at the same rate.

(a) Derive an equation that relates the salt concentration in the outlet stream to the time the dilution process has been operating. How long will it take to achieve the desired salt concentration? Assume that the volume of material and the volume of water in the vat are unchanged during this process.

(b) Heat capacity of a salt solution is dependent on concentration:

$$C_p = 0.996 - 1.17 \times 10^{-3} S$$

where S is 1000 times the weight fraction of the solute.[3] For example, S is 80 if the solution is 8 wt% salt. C_p has units of kcal/(kg·°C). In calculating the salinity, S, do not assume that the mass of salt is negligible in the overall solution.

Suppose the incoming stream is at 25°C and you initially turn on a heater, which supplies 2.5 kW to the vat. Using MATLAB or a similar program, plot the temperature of the solution inside the vat as a function of time. In MATLAB, the functions `diff` and `dsolve` or `ode45` might be useful. From your plot, estimate the temperature of the solution inside the vat when the salt concentration is 0.030 kg/L.

4.35 During heat exhaustion, it is desired to rapidly cool a patient. A doctor in a hospital came up with an idea of cooling a solid metal disk (1 mm thick, 25 cm in diameter) to 0°C in ice water and then cooling the heat-exhausted patient by putting the disk on his chest. The rate of heat exchange can be modeled by the following equation:

$$\dot{Q} = h_e A(T_s - T_c)$$

where T_s is the skin temperature, T_c is the temperature of the copper disk, and A is the contact area. The skin temperature is 30°C and the equivalent heat transfer heat coefficient, h_e, between the disk and the patient skin is 10 W/(m^2·K). Ignore the heat loss from the back and side of the disk. Assume that the temperature of the skin remains constant at 30°C; only the temperature of the disk changes. Properties of the metal disk include a heat capacity of 420 J/(kg·K) and a density of 7800 kg/m^3.

Case 1

(a) Calculate how long it will take the metal disk to reach 27°C.
(b) Estimate the total amount of heat removed.
(c) Plot the temperature of the metal disk as a function of time.
(d) Plot the heat exchange rate as a function of time.

Case 2

(a) If the thickness of the metal disk is increased to 5 mm, calculate how long it will take the metal disk to reach 27°C.
(b) Estimate the total amount of heat removed.
(c) Plot the temperature of the metal disk as a function of time.
(d) Plot the heat exchange rate as a function of time.
(e) Compare the two plots with those of *Case 1* and comment.
(f) From a practical point of view, what might be some potential problem(s) if this idea is implemented?

Case 3

(a) A slightly different design is proposed to replace the solid metal disk with a very thin metal container (with dimensions similar to those of *Case 2*). The container holds 100 g of water. The container with water is chilled to 0°C in ice water before being applied to the chest of the heat-exhausted patient. Calculate how long it will take the container to reach 27°C. Ignore the contribution from the metal container, since the heat capacity of water is much greater than that of metal.
(b) Estimate the total amount of heat removed.
(c) From a practical point of view, what are the potential advantage(s) or disadvantage(s) of this design?

4.36 At the onset of fever the body temperature rises continually with little heat loss from the body. If heat is not removed by such means as cold packs, ice, etc., how long will it take the body to reach a critical temperature of 41°C? Use 0.86 kcal/(kg·°C). as the heat capacity of the human body, 1750 kcal/day as the basal heat production, and 70 kg as the body mass.

4.37 Your friends have just received an ice cream maker. The unit consists of a bowl and a base that rotates the bowl around a stationary mixing paddle. The walls of the bowl contain an unknown mixture that absorbs heat from the ice cream. The bowl must be frozen (−20°C) before use.

Unfortunately, your friends have lost the instructions and can't remember how long they must operate the unit. You tell them not to worry—for you have mastered energy conservation problems and can calculate the amount of time required to freeze the ice cream.

Looking at the box, you find the specifications for the unit:

- Inner surface area of freezing bowl: 600 cm^2
- Heat transfer coefficient for bowl: $h = 0.025$ J/($cm^2 \cdot s \cdot °C$)
- Power required to stir the bowl (100% efficiency): 25 W
- Amount of ice cream mixture added: 1 kg

The rate of heat transfer is:

$$\dot{Q} = hA(T_{\text{bowl}} - T_{\text{milk}})$$

Recalling your freshman chemistry course, you remember that solutes lower the freezing point of water. Assume that the freezing point is lowered to −5°C. Ice cream mixture contains milk, cream, sugar, and vanilla extract. To achieve the consistency of soft-serve ice cream, only half of the water must be frozen. At −5°C, $\Delta\hat{H}_{f,\text{water}} \approx 330$ kJ/kg. Derive an equation, in terms of the above variables, and estimate the total operating time.

4.38 A tank contains 1000 kg water at 24°C. It is planned to heat this water using saturated steam at 130°C in a coil inside the tank. The rate of heat transfer from the steam is given by the equation:

$$\dot{Q} = hA(T_{\text{steam}} - T_{\text{water}})$$

where $\dot{Q}$ is the rate of heat transfer, h is the overall heat transfer coefficient, A is the surface area for heat transfer, and T is the temperature. The heat transfer area provided by the coil is 0.3 m^2; the heat transfer coefficient is 220 kcal/($m^2 \cdot hr \cdot °C$). Condensate leaves the coil saturated. Assume the heat capacity of water is constant, and neglect the heat capacity of the tank walls. (From Doran PM, *Bioprocess Engineering Principles*, 1999.)

(a) The tank has a surface area of 0.9 m^2 exposed to the ambient air. The tank exchanges heat through this exposed surface at a rate given by an equation similar to that above. For heat transfer to or from the surrounding air, the heat transfer coefficient is 25 kcal/($m^2 \cdot hr \cdot °C$). If the air temperature is 20°C, calculate the time required to heat the water to 80°C.

(b) What time is saved if the tank is insulated?

4.39 A heat-sensitive sample stored in a capped test tube is taken from the freezer to be analyzed (Figure 4.28). Soon after the sample tube is immersed in an ice-water bath, the fire alarm goes off. The researcher leaves the room immediately, leaving the sample (still in the ice-water bath) on the bench. Fortunately, the fire alarm is only an unscheduled fire drill.

The rate of heat exchange, $\dot{Q}$, between the ice-water bath and its surrounding air can be modeled by the following equation:

$$\dot{Q} = h_A A(T_A - T_i)$$

where T_A is the air temperature, T_i is the temperature of the ice-water bath, h_A is the overall heat transfer coefficient, and A is the surface area for heat transfer. The

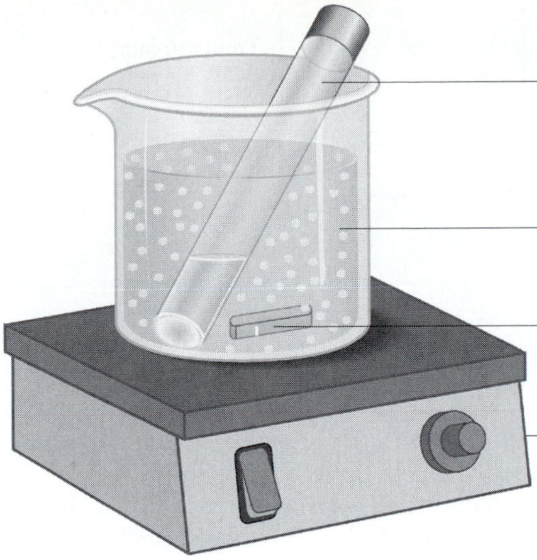

Figure 4.28
Heat-sensitive sample stored in a capped test tube in an ice-water bath.

air temperature is 22°C; the heat transfer area of the ice-water bath is estimated to be 500 cm^2; the heat transfer coefficient h_A is 0.03 cal/(cm^2·min·°C).

Since the ice-water bath is in contact with the magnetic stirrer, heat exchange between these two systems also has to be taken into account. The rate of heat exchange between the ice-water bath and the magnetic stirrer can be similarly modeled by the following equation:

$$\dot{Q} = h_s A(T_s - T_i)$$

where T_s is the stirrer temperature, T_i is the temperature of the ice-water bath, h_s is the overall heat transfer coefficient, and A is the surface area for heat transfer. The magnetic stirrer temperature is 22°C; the heat transfer area is estimated to be 200 cm^2; the heat transfer coefficient h_s is 0.1 cal/(cm^2·min·°C). Assume that:

- The ice-water bath contains 100 g of water and 400 g of ice when the researcher leaves the room.
- The amount of work done by the magnetic stirrer on the system is negligible.
- The total heat capacity of the test tube (with sample) is negligible.

(a) Suppose that the sample is damaged if its temperature rises above 0°C. Estimate the maximum duration of the fire drill for the sample to remain intact. Assume the researcher returns to the laboratory immediately after the fire drill.

(b) Suppose that the sample is damaged if its temperature rises above 5°C. Estimate the maximum duration of the fire drill for the sample to remain intact. Assume the researcher returns to the laboratory immediately after the fire drill.

4.40 An ice bath is used to chill a process stream (Figure 4.29). The process stream is originally at 90°C. Its temperature drops to 10°C at the outlet after passing through a cooling loop inside the ice bath. The ice bath has 100 kg of ice at the beginning. How often does the ice bath need to be replaced? In other words, how long does it take for all the ice to melt? The process stream has a specific heat capacity, C_p, of 1.0 cal/(g·°C) and a density of 1.0 g/mL.

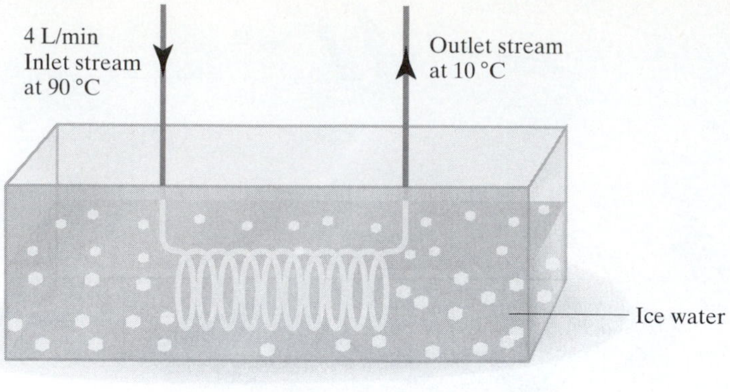

Figure 4.29
Ice bath used to chill a process stream.

4.41 During dive preparation, a diver in a wet suit who has been exposed to the hot sun for a long time might become overheated. This is because the neoprene suit limits the normal emission of heat in the air. Hyperthermia is a term that refers to heat-related sicknesses. The two basic forms of hyperthermia are heat exhaustion and heat stroke. The symptoms include dizziness, disorientation, headache, nausea, weakness, red or pale face, rapid pulse up to 120 beats per minute, frequent breathing, raised body temperature, excessive perspiration, and loss of consciousness. Assuming the heat lost from the body to it surrounding is given by:

$$\dot{Q} = h_e A(T_b - T_s)$$

where T_b is the body temperature, T_s is the surrounding air temperature, A (m^2) is the body surface area, and h_e(kcal/(m^2·hr·°C)) is the overall heat transfer coefficient. Given a diver body mass of m (kg), a metabolic rate of $\dot{M}_R$ (kcal/hr), and a diver heat capacity of C_p (kcal/(kg·°C)), describe how you can estimate the maximum time of exposure by the diver to a certain air temperature of T_s before the body temperature, T_b, reaches a critical temperature of T_c.

4.42 Consider a pulse energy of E_L delivered by a laser during surgery. The energy is absorbed within a tissue volume of 1000 μm^3, which is 80% water. Assume the heat capacity of the tissue (with and without water) is 4.35 kJ/kg K. Describe what happens to the tissue when the pulse energy, E_L, is applied at the different values of 0.1 μJ, 0.5 μJ, and 3.0 μJ.

4.43 A laboratory-scale 10-L glass fermenter used for culture of hybridoma cells contains nutrient medium at 4°C. The fermenter is wrapped in an electrical heating mantle that delivers heat at a rate of 500 W. Before inoculation, the medium and vessel must be at 36°C. The medium is well mixed during heating. Estimate the time required for medium preheating. The glass fermenter vessel has a mass of 13 kg and a C_p of 0.20 cal/(g·°C). The nutrient medium has a mass of 8.0 kg and a C_p of 1.05 cal/(g·°C). (Adapted from Doran PM, *Bioprocess Engineering Principles*, 1999.)

4.44 Consider an ice ball with density ρ (in g/cm^3) and radius r (in cm) immersed in a warm-water bath (Figure 4.30). Assume that the rate of heat transfer, $\dot{Q}$ (in cal/min), from the warm water to the ice ball is given by:

$$\dot{Q} = qA$$

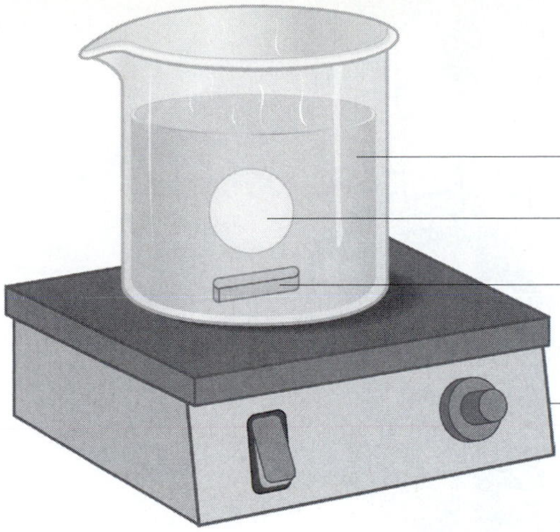

Figure 4.30
Ice ball in a warm water bath.

where $\dot{q}$ is rate of heat transfer per unit area (in cal/(min·cm^2)) and A (in cm^2) is the contact area between the ice ball and water (i.e., the surface area of the ice ball at any instant). Further assume that the heat transfer is uniform, i.e., the ice ball maintains its spherical shape at all times. Let $\Delta\hat{H}_f$ (in cal/g) denotes the heat of fusion of ice.

(a) Perform an analysis on the system and derive an expression for the radius of the ice ball, r, as a function of time in terms of ρ, $\dot{q}$, $\Delta\hat{H}_f$, and any physical constants you think are necessary. Assume $\dot{q}$, the rate of heat transfer per unit area, and ρ, the density of the ice ball, are constants. Assume further that $r = R$ at time zero.

(b) Find the time at which $r = 0.5R$.

4.45 A biotechnology firm has just constructed a new genetically engineered *Escherichia coli* strain that is capable of producing an important recombinant protein. It was found that the production of this recombinant protein is proportional to cell growth. Ammonia is used as a nitrogen source for aerobic respiration of glucose. The recombinant protein has an overall formula of $CH_{1.55}O_{0.31}N_{0.25}$. The yield of biomass ($CH_{1.77}O_{0.49}N_{0.24}$) from glucose is determined to be 0.48 g biomass/g glucose; the yield of recombinant protein from glucose is about 20% that for cells. The following equation can be used to represent the production process:

$$C_6H_{12}O_6(s) + aO_2(g) + bNH_3(g)$$
$$\longrightarrow cCH_{1.77}O_{0.49}N_{0.24}(s) + dCO_2(g) + eH_2O(\ell) + fCH_{1.55}O_{0.31}N_{0.25}(s)$$

where a, b, c, d, e, and f are the stoichiometric coefficients. Since the yield of biomass from glucose is determined to be 0.48 g biomass/g glucose, c is 3.46 mol/mol. Since the yield of recombinant protein from glucose is about 20% that for cells, the yield is calculated as 0.096 g recombinant protein/g glucose.

Suppose a continuous bioreactor with a working volume of 100 L is used to produce the recombinant protein (Figure 4.31). A stream of medium containing essential nutrients including glucose and ammonia flows into the reactor at a rate of 10.0 L/hr. The medium contains 50.0 g/L of glucose and a sufficient quantity of ammonia in water. The exit stream contains the cells that

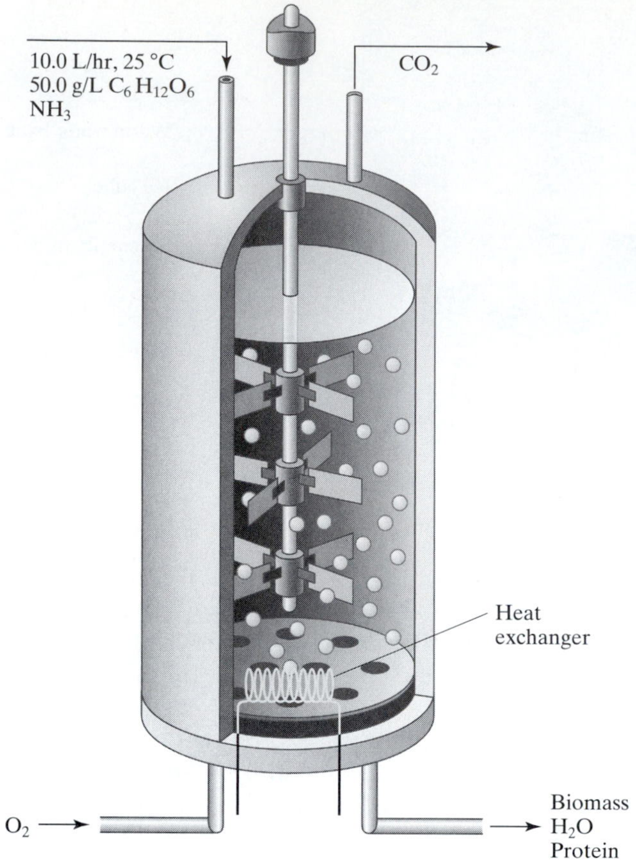

Figure 4.31
Bioreactor used to produce recombinant protein with genetically engineered *Escherichia coli.*

harbor the recombinant protein. Under these conditions, only a negligible quantity of glucose can be detected in the exit stream. The temperatures of the inlet stream and of the bioreactor are set at 25°C. Assume the reactor is well insulated and the amount of shaft work involved due to mixing is negligible. Suppose the reactor has been operated for a while and is at steady-state.

(a) How much ammonia is required?
(b) What is the recombinant protein production rate?
(c) It is proposed that the heat of reaction can be related to the oxygen consumption rate by the following expression:

$$\Delta H_r \approx -460 \text{ kJ/mol of oxygen consumed}$$

How good is this correlation when compared with that calculated using the heats of combustion (Table 4.14)?

TABLE 4.14

Standard Heats of Combustion for Production of a Recombinant Protein

Species	$\Delta\hat{H}_c^\circ$ (kJ/mol)
$C_6H_{12}O_6$ (s)	−2805
NH_3 (g)	−382.6
Biomass (s)	−551
Recombinant protein (s)	−567

(d) What is the heat addition or removal rate in order to maintain the reactor at 25°C?
(e) In the middle of the run, the heat exchanger malfunctions; as a result, no heat can be added or removed from the reactor. Assume that culture behavior remains the same as before within this temperature range. What is the temperature of the reactor one hour after the mishap?

CHAPTER

5 Conservation of Charge

5.1 Instructional Objectives and Motivation

After completing Chapter 5, you should be able to do the following:

- Write and apply the positive and negative charge accounting equations and the net charge conservation equation.
- Derive Kirchhoff's current law (KCL) from the conservation of net charge and apply KCL to a node in a circuit.
- Define electrical energy and specify elements that generate and consume electrical energy.
- Derive Kirchhoff's voltage law (KVL) from the electrical energy accounting equation and apply KVL to a loop in a circuit.
- Explain the relationship between voltage, current, and resistance using Ohm's law.
- Set up and solve circuits involving a variety of circuit elements, including voltage sources, current sources, resistors, capacitors, and inductors.
- Use Einthoven's law to calculate unknown potentials from an electrocardiogram.
- Use the Hodgkins-Huxley equations to model charge flow across a biological membrane.
- Apply charge accounting and conservation equations to reacting systems involving radioactive decay, acid/base reactions, and electrochemical reactions.
- Solve unsteady-state systems using charge and electrical energy accounting equations.

5.1.1 Neural Prostheses

Charge accounting and conservation and electrical energy accounting equations are used in many exciting areas in the field of bioengineering. In previous physics or engineering courses, you have likely encountered Kirchhoff's current law (KCL) and Kirchhoff's voltage law (KVL). These important equations are based on charge and electrical energy accounting equations, respectively. The design of circuit elements to measure biopotentials or to control a biomedical device requires a thorough understanding of KCL, KVL, and other equations such as Ohm's law. Charge accounting and conservation and electrical energy accounting equations are also used to model systems with chemical reactions involving charged species. In this chapter these equations are applied to a wide range of example and homework problems.

In this introductory section we highlight neural prostheses. The development of neural prostheses is an emerging field in bioengineering, where accounting and conservation principles are routinely applied in order to model systems and build new devices. The complex challenge below serves to motivate our discussion of the charge accounting and conservation equations and the electrical energy accounting equation.

Healthy, intact nerves transmit information from the brain to various parts of the body and vice versa by conducting electrochemical signals called action potentials. Nerve fibers have a resting membrane potential of about −90 mV. Action potentials involve rapid changes in membrane potential from negative to positive (depolarization) and back to negative again (repolarization) in a total time period of less than 1 msec. An action potential elicited at any point on an excitable membrane typically induces the excitation of adjacent portions of the membrane, resulting in propagation of the action potential. In this way, the action potential moves along the length of the nerve fiber until it reaches the fiber's end and delivers the signal either to another nerve or to an organ or muscle. Thus, action potentials make possible long-distance communication of signals carrying sensory or motor information in the nervous system.

Input to the nervous system is provided by sensory receptors that detect stimuli such as touch, sound, taste, light, pain, heat, and cold. The type of sensation perceived when a nerve fiber is stimulated depends on the specific point in the central nervous system where the nerve tract terminates. For example, fibers from the retina of the eye terminate in the vision areas of the brain; fibers from the ear terminate in the auditory areas of the brain; and touch fibers terminate in touch areas of the brain. Spinal cord injuries (SCIs), stroke, and neurological disorders such as cerebral palsy may result in the inability of the nerve fibers to successfully transmit these information-bearing electrical signals to the brain. Conversely, the same or other types of injury may render the nerve fibers incapable of carrying motor instructions from the brain to the limbs and organs.

Devices developed in various fields of clinical medicine and biomedical engineering to restore sensory and motor functions of the human body are termed neural prostheses. Neural prostheses use electrical activation of the nervous system to restore function to individuals with neurological impairment. In order to accomplish this goal, product designers must achieve an understanding of the interface between electronics and nerve cells. Neural prosthetic devices function by electrical initiation of action potentials in nerve fibers that carry the signal to another nerve or an organ or muscle. An action potential initiated by a pulse of positive charge from an implanted electronic device is indistinguishable from a naturally occurring action potential.[1] Thus, any organ or muscle under neural control is theoretically a suitable candidate for neural prosthetic control.

Applications for neural prostheses include stimulation in both sensory (e.g., cochlear implants) and motor (e.g., bladder control) systems in order to restore function and provide increased independence to patients. Using an electronic device to stimulate a nerve offers the possibility of restoring hearing for the deaf, sight for the blind, and a variety of motor functions for victims of SCI, stroke, and cerebral palsy. Successful outcomes of motor system neural prostheses to date include restoration of standing and stepping in paraplegic patients, restoration of hand grasping and releasing in quadriplegia, restoration of bladder function (continence and micturition) following SCI, and electrical respiration in high-level quadriplegia.[2] Designers of neural prosthetic systems must know and understand principles of underlying circuits and electronics in addition to biology, and how concepts from the two fields may be combined and applied to produce a device with a specific functional benefit.

Neural prostheses are currently used to treat two principal sensory disorders. Cochlear implants, which are now well established in clinical practice, were developed to restore hearing after loss of the outer hair cells when the inner hair cells and the pathways to the auditory cortex remain intact. The system consists of an extracorporeal microphone, a speech processor that digitizes sounds into coded signals, and a transmitter that sends the code across the skin to the internal implant. The implant converts the code to electrical signals, which are sent to electrodes that are surgically inserted in the inner ear to stimulate the remaining nerve fibers. (Electrodes are conductors used to establish electrical contact with a nerve or other nonmetallic part of a circuit.) The signals received by the electrodes are recognized as sounds by the brain, producing a hearing sensation. The other primary application of neural prostheses for sensory percepts is in the stimulation of retinal neurons to elicit visual sensations in partially or totally blind patients. Though the small volume of the eye and the fragility of the retina pose unique design challenges, the principles underlying neural stimulation are the same.

Patients who suffer SCI, stroke, cerebral palsy, or other neurological disorders may experience a variety of lost and impaired motor functions. For example, depending on the level of the lesion on the spinal cord, SCI can result in partial (paraplegia) or total (quadriplegia) paralysis, loss of bladder and bowel control, sexual dysfunction, muscle atrophy, and chronic pain. Neural prostheses seek to restore these functions. One of the first neural prostheses in clinical practice was the drop foot stimulator, invented for hemiplegic stroke victims who have lost the ability to lift the toes during the swing phase of gait due to paralysis of the ankle dorsiflexor muscle. Drop foot stimulators use surface or implanted electrodes, activated when a contact switch in the shoe detects that the foot has left the ground, to stimulate the peroneal nerve and induce foot flexion.[3] Similar to the drop foot system, the implantable sacral anterior root stimulator activates bladder motor pathways to produce effective voiding.

A third application of neural prostheses for restoration of motor function is the implantable Freehand system (Figure 5.1) by NeuroControl, Inc. (FDA approved in 1997), which aids in the restoration of palmar and lateral grasping function. A position sensor placed on the opposite shoulder translates small shoulder movements

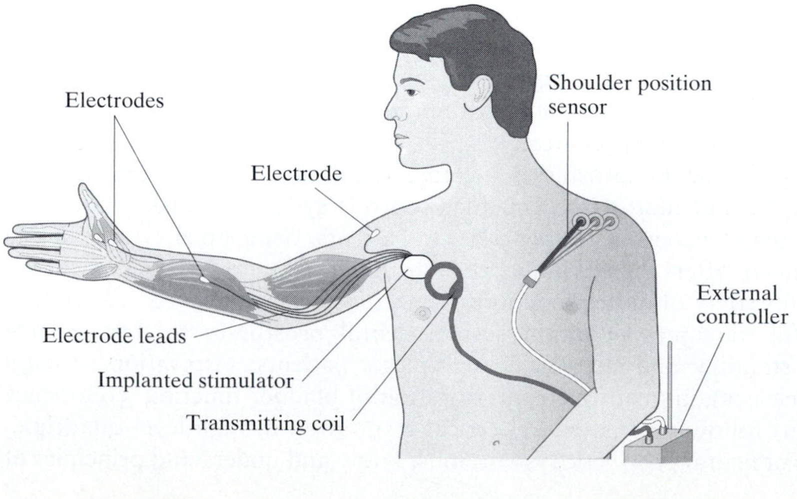

Figure 5.1
Implantable Freehand system from NeuroControl, Inc. (*Source:* Sadowsky CL, "Electrical stimulation in spinal cord injury," *NeuroRehabilitation* 2001, 16:165–9.)

into a control signal. The control signal is sent to an external controller, which processes the information into radio waves to power and control the implant. The stimulator implanted in the chest sends electronic stimuli through wires to eight electrodes placed on the motor points of the hand and forearm muscles to induce the grasping action.[3] This prosthesis is most suitable for SCI patients with injuries at the C5 or C6 cervical vertebra level who have a decreased grasp force and range of motion in the hand but a preserved ability to move the shoulder and elbow joints.[4]

For high cervical SCIs (cervical vertebra C3 or above) that result in the loss of voluntary movement of respiratory muscles, electrical stimulation of the phrenic nerve can cyclically activate the paralyzed diaphragm muscle to produce respiration. A platinum electrode is surgically implanted against the deep surface of the phrenic nerve and connected by a subcutaneous lead to a radio receiver under the skin of the chest. A transmitter outside the body conveys power and control signals to the receiver, allowing adjustment of respiration.[5]

Despite the aforementioned successes, room for improvement remains in the design of neural prostheses for a variety of applications. Many SCI and stroke victims and patients with neurological disorders still do not have access to neural prostheses, and these devices are, in large part, still in the testing stages. The design of neural prostheses requires a thorough understanding of electrical circuits, electronics, and movement of charge in the nervous system. Bioengineers collaborate with neurosurgeons, ophthalmologists, otolaryngologists, orthopedists, physicists, electrical engineers, and materials scientists to develop neural prosthetic systems. Practitioners face many challenges in the design and manufacture of neural prostheses, including:

- *Integration*: The artificial neural prosthesis must be integrated into the natural sensory and motor systems. Ideally, independent implanted devices with an internal power source and integrated sensors would detect commands from the motor cortex and deliver stimulation waveforms to appropriate muscles, bypassing the neural damage altogether and providing natural control in accordance with the user's desired intention.[6] Conversely, sensory information received by the nervous system should be able to be transmitted to the brain without hindrance from neural damage.
- *Individualization*: The optimal neural prosthesis would be designed and adapted to the individual patient in order to ensure the best functional recovery. The role of the neural prosthesis in conjunction with mechanical devices and specific rehabilitation procedures must be considered.
- *Ease of access and use*: Electrodes, control systems, and hardware must be small, inexpensive, and easy to implant in order to make these systems available to the greatest number of patients and to provide patients with the highest quality of life. Long-term maintenance of the system should be minimized.
- *Biocompatibility and protection*: Components of the implants must be biocompatible to avoid inflammatory and hypersensitivity effects in vivo. The implanted device must be hermetically sealed to protect it from corrosion by bodily fluids.[7]
- *Mechanical compatibility*: The implanted prosthesis must be able to withstand forces to which it may be subjected, and the surrounding tissue must correspondingly withstand any forces that might be transmitted to it from the implant.
- *Interference*: As with all electronic devices, neural prostheses are susceptible to electromagnetic interference.

Multidisciplinary teams tackle these research challenges in an attempt to improve the quality of life for affected individuals. Armed with novel laboratory and clinical studies, bioengineers use charge and electrical energy balances to help them understand and model the nervous system and build devices that bridge across its damaged parts. In Examples 5.14 and 5.16, we show how charge and electrical energy balances can be used to model neuron behavior, which must be understood prior to the design of neural prostheses.

This chapter opens with an overview of basic charge and electrical energy concepts, including current and voltage. The accounting and conservation equations are developed, as appropriate, for positive, negative, and net charge as well as electrical, energy. Kirchhoff's current and voltage laws are explored in some depth, with examples focused on circuit analysis and biological systems. Finally, we explore the accounting of charge and electrical energy for both dynamic systems and reacting systems.

5.2 Basic Charge Concepts

To design an electrical system, we must understand how charge moves and how it accumulates. Conceptually, the accounting and conservation of charge are very similar to the those of other extensive properties discussed previously. As with mass, we account for the charge of a system by analyzing how various forms of charge enter and leave the system, are generated and consumed, and are accumulated within the system. Before developing the conservation of charge, we review a few basic definitions.

5.2.1 Charge

Matter is composed of three fundamental particles: the electron, the proton, and the neutron. Atoms of all chemical compounds consist of a nucleus composed of protons and neutrons, with electrons orbiting the nucleus. For example, the element nitrogen, ${}^{14}_{7}N$, has seven protons and seven neutrons in its nucleus. In an uncharged atom like nitrogen, the number of protons in the nucleus equals the number of electrons surrounding the nucleus; therefore, seven electrons orbit the nucleus. The electron and the proton carry **electric charges** of equal magnitude, termed the **elementary charge**. The elementary charge on an electron is -1; that on a proton is $+1$. The neutron carries no charge.

The fundamental unit of electric charge, q, is the coulomb (C). The dimension of charge is [tI]. A coulomb is formally defined as the amount of charge that flows past a reference point in a wire in 1 second when the current in the wire is 1 ampere, but it is more convenient to describe 1 C as being approximately 6.24×10^{18} elementary charges. In other words, 6.24×10^{18} electrons or protons constitute 1 C of charge. Conversely, the charge of a proton is $+1.602 \times 10^{-19}$ C; the charge of an electron is -1.602×10^{-19} C. The mass and charge of each of these fundamental particles is shown in Table 5.1.

TABLE 5.1

Properties of Fundamental Particles

Fundamental particle	Mass (g)	Elementary charge	Charge (C)
Electron	9.11×10^{-28}	-1	-1.602×10^{-19}
Proton	1.67×10^{-24}	$+1$	$+1.602 \times 10^{-19}$
Neutron	1.67×10^{-24}	0	0

5.2.2 Current

Current, i or $\dot{q}$, is the rate of movement or flow of electric charge in a conducting material past a specified point. Conductors are materials such as metals, ionic solutions, and ionized gases, in which individual charges are free to move throughout the material. When an electric field is applied to a conductor, the ordered motion of the individual electric charges, called an **electric current**, results. Current is defined as the time derivative of charge or the rate of charge:

$$i = \frac{dq}{dt} = \dot{q} \qquad [5.2\text{-}1]$$

The dimension of electric current is [I], which is a base physical variable (Table 1.1). Electric current is usually expressed in units of amperes (A), equivalent to C/s. One ampere equals 6.24×10^{18} electrons per second passing a given point in a conducting material.

Electric current is often visualized as the movement of electrons in a conducting material. This model is appropriate when analyzing and designing electrical circuits. However, both positively and negatively charged ions can move in conducting materials, such as aqueous solutions. For example, the flow of positively charged potassium and sodium ions and the flow of negatively charged chloride ions across the cell membrane are essential for the development of membrane potentials in nerve and muscle fibers. During photosynthesis, both positive charges (hydrogen ions) and negative charges (electrons) are transported through a complex cellular milieu in order to synthesize glucose from water and carbon dioxide.

We have noted that electric current is expressed in amperes, defined as the flow of electrons per unit time past a given point. Thus, when electrons or negatively charged ions move in a conducting material, their direction of charge flow is the same as the direction of current. When protons or positively charged ions move in a conducting material, their direction of charge flow is opposite that of the direction of current. In other words, the notion of positive charge flowing in one direction is equivalent to the notion of negative charge flowing in the opposite direction.

5.2.3 Coulomb's Law and Electric Fields

Charges exert electric forces on one another in much the same way masses exert gravitational forces. While all masses are considered positive values, charges fall into two categories—positive and negative. The signs of the charges determine the attractive or repulsive electrostatic force between them: Like charges repel each other while opposite charges attract. To describe the electrostatic force between two point charges, q_1 and q_2, **Coulomb's law** is used:

$$\vec{F}_{12} = \frac{kq_1q_2}{r^2}\vec{r}_{12} \qquad [5.2\text{-}2]$$

where $\vec{r}_{12}$ is the unit vector that points from q_1 to q_2, r is the distance between the two charges, and the constant k is $9.0 \times 10^9\ (\text{N} \cdot \text{m}^2)/\text{C}^2$. By convention, repulsive forces are positive, and attractive forces are negative.

An **electric field** is a region associated with a distribution of electric charge. The location of an electric charge within an electric field allows the charge to experience a force. In general, this chapter will not deal specifically with the forces acting on individual charges. We will, however, be concerned with the consequences of this law. Since charge is related to force and force is related to work and energy, charge is also related to work and energy. The energy associated with electric charge forms part of the foundation for the rest of this chapter.

5.2.4 Electrical Energy

Electrical energy refers to the energy associated with the flow of electric current and with electromagnetic energy. Electromagnetic energy is associated with electric and magnetic fields and includes the energy of radio waves, gamma waves, microwaves, x-rays, infrared light, visible light, and ultraviolet light. These forms of electrical energy are not addressed in this text. We explore only electric potential energy, which we refer to simply as electrical energy.

Charged particles placed in an electric field possess potential energy, analogous to mass possessing gravitational energy in a gravitational field. The potential energy per unit charge, or specific potential energy, is simply called **electric potential.** Note that previously (in Chapters 1, 3, and 4), the term *specific* referred to a physical variable reported on a per-mass or per-mole basis. Throughout Chapter 5, this term refers to a variable on a per-charge basis.

Voltage (v) refers to a difference in electric potential between two specified points or the potential energy change per unit charge in moving from one point to another. The term voltage is often used interchangeably with electric potential and potential difference. The dimension of voltage is $[L^2Mt^{-3}I^{-1}]$. The most common unit of potential difference and voltage is the volt (V), defined as joule/C.

Recall that measurements of gravitational potential energy are defined with respect to a reference height. Gravitational potential energy is most relevant when examining a difference in height, where movement of the object causes the potential energy to be converted to some other form of energy. This concept is analogous to that of electrical potential energy. Voltage is usually given as a definite number, where the number implies the difference of electric potentials between two points. Usually, the ground state serves as a point of reference, analogous to a height of zero for gravitational potential. When an electric potential difference exists between two points, the voltage can be used to do work, such as when a battery powers an electric or mechanical device.

When a charge moves from one electric potential to a different one, electric potential energy is either generated or consumed. The **electric potential energy** $\left(E_E\right)$ of a single charge is:

$$E_E = qv \quad [5.2\text{-}3]$$

The dimension of electrical energy is $[L^2Mt^{-2}]$. Common units of electrical energy are joule (J) and kW · hr. Note that the voltage in equation [5.2-3] is not an absolute voltage but a voltage difference measured relative to a reference state, which may be the ground.

EXAMPLE 5.1 Excitation of an Electron During Photosynthesis

Problem: Many reactions that occur during photosynthesis are driven by the electron transfer chain. When a photon of light is absorbed by a chlorophyll molecule, the photon's energy causes an electron to jump to a higher energy level. The energy from excited electrons is harvested in photophosphorylation, also called the light reactions of photosynthesis. Suppose that during the course of photosynthesis, an electron is excited from a state of +0.5 V to −1.0 V. What energy is imparted to this electron?

Solution: Equation [5.2-3] can be applied to calculate the electrical energy. The charge on an electron is negative. The electric potential difference is the difference between the excited state (ex) and unexcited state (un) of the electron.

$$E_E = qv = q(v_{ex} - v_{un})$$

$$E_E = (-1.602 \times 10^{-19}\text{ C})(-1.0\text{ V} - 0.5\text{ V}) = 2.403 \times 10^{-19}\text{ J}$$

■

Electrical potential energy can also move into or out of a system at a charge flow rate, i. The **rate of electrical energy** $(\dot{E}_E)$ is defined as the product of the current and the specific potential energy (voltage) for that current:

$$\dot{E}_E = iv \qquad [5.2\text{-}4]$$

The dimension of rate of electrical energy is $[L^2Mt^{-3}]$. Rate of energy is known as power; the most common metric unit of power in circuit analysis is the watt (W).

EXAMPLE 5.2 Hair Dryer

Problem: How much charge passes through a 1200-W, 120-V hair dryer in 5 minutes?

Solution: The wattage is the power or rate at which electrical energy is used by the hair dryer. Equation [5.2-4] is rearranged to solve for electric current:

$$i = \frac{\dot{E}_E}{v} = \frac{1200\text{ W}}{120\text{ V}} = 10\text{ A} = 10\ \frac{\text{C}}{\text{s}}$$

Thus, the hair dryer pulls 10 A of electric current. In 5 min (300 s), 3000 C of charge flows through the circuitry of the hair dryer. ■

Consider an analogy between charge and mass (Figure 5.2). Imagine a water molecule of mass m. A water molecule moving through a hose has a mass flow rate of $\dot{m}$. The opening of the hose is at a height h above the ground, so using the gravitational acceleration constant, we can find the potential energy per unit mass of water $(\hat{E}_P = g\,\Delta h)$. The rate of potential energy change $(\dot{E}_P)$ can be measured as the water molecule changes position in the gravitational field.

Similarly, we can apply this analogy to an electron, where a single electron is like a single water molecule. An electron moving though a wire has a flow rate or current, i. Just as the water molecule travels from one position in the hose to another, an electron can travel from a point in a wire to another, moving from one electric potential to another. The difference in potential between these two points is an electric potential difference or voltage (v). Using the voltage, we can find the electric potential energy per unit charge $(\hat{E}_E = v)$. Like the water molecule, the rate of electric potential energy $(\dot{E}_E)$ can be measured as the electron changes positions in an electric field.

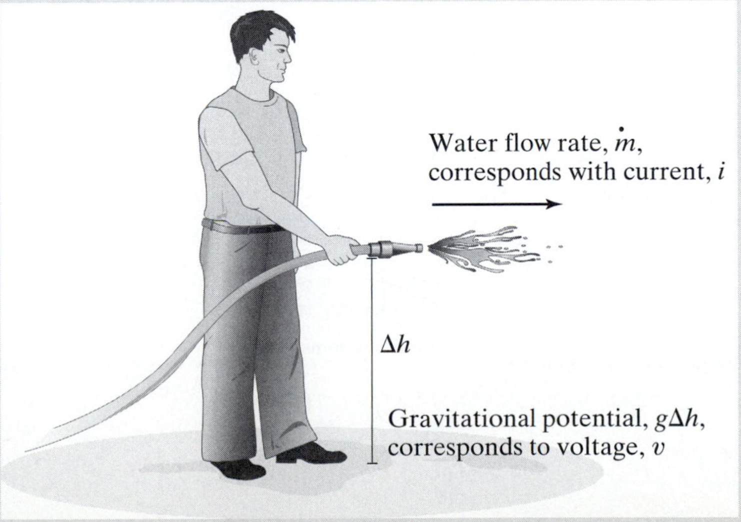

Figure 5.2 Analogy between mass and charge.

5.3 Review of Charge Accounting and Conservation Statements

Like mass, charge is an inherent property of matter. The number of negatively charged electrons and positively charged protons present in a species determines the charge of that species. The charges themselves, present in the electron or proton, can be neither created nor destroyed by most reactions within the scope of this text. However, by transferring electrons from one molecule to another, charged species, such as a positively charged sodium ion, can be created.

Net charge is always conserved in a system. Thus, net charge is neither created nor destroyed in a system or in the universe. **On the other hand, positive and negative charges are not conserved and can be created or consumed in a system and in the universe.** However, to preserve the conservation of net charge, when a positive charge is created, a negative charge must also be created. The same is true when a negative charge is consumed: A positive charge must also be consumed. That is, in all cases, equal amounts of positive and negative charges must be created or consumed together in the universe.

Accounting and conservation equations are frequently used to account for the number of charged particles present in a system. In the context of an accounting equation, positive and negative charges refer to the species present that bear a positive or negative charge.

A schematic representation of a system is shown in Figure 5.3. Charges enter and leave the system across the system boundary. Generation and consumption of positive and negative charges occur within the system. Charges may also accumulate in the system.

5.3.1 Accounting Equations for Positive and Negative Charge

Recall the generic algebraic accounting statement from equation [2.4-2]:

$$\psi_{\text{in}} - \psi_{\text{out}} + \psi_{\text{gen}} - \psi_{\text{cons}} = \psi_{\text{acc}} \qquad [5.3\text{-}1]$$

Algebraic equations are most appropriate to use when discrete quantities or chunks of charge are specified. Positive charge, q_+, and negative charge, q_-, are counted in separate equations:

$$\text{Positive:} \quad \sum_k q_{+,k} - \sum_j q_{+,j} + q_{+,\text{gen}} - q_{+,\text{cons}} = q_{+,\text{f}}^{\text{sys}} - q_{+,0}^{\text{sys}} \qquad [5.3\text{-}2]$$

$$\text{Negative:} \quad \sum_k q_{-,k} - \sum_j q_{-,j} + q_{-,\text{gen}} - q_{-,\text{cons}} = q_{-,\text{f}}^{\text{sys}} - q_{-,0}^{\text{sys}} \qquad [5.3\text{-}3]$$

where $\sum_k q_{\pm,k}$ is the quantity of positive or negative charge entering the system by bulk transfer during the time period, $\sum_j q_{\pm,j}$ is the quantity of positive or negative

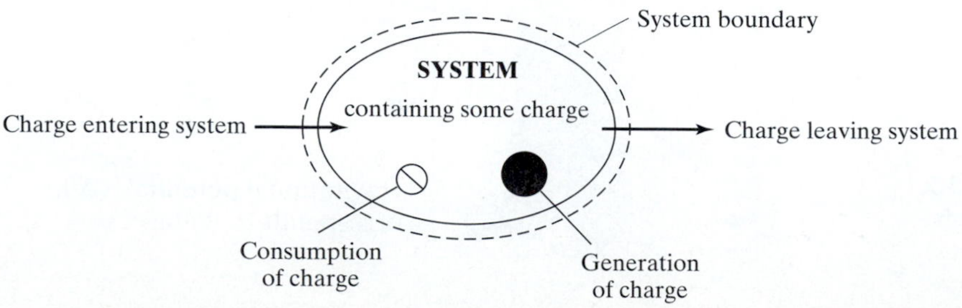

Figure 5.3
Schematic drawing of the movement, generation, consumption, and accumulation of charges in a system.

charge leaving the system by bulk transfer, $q_{\pm,\text{gen}}$ is the quantity of positive or negative charge generated in the system, $q_{\pm,\text{cons}}$ is the quantity of positive or negative charge consumed by the system, $q^{\text{sys}}_{\pm,\text{f}}$ is the quantity of positive or negative charge contained within the system at the end of the time period, and $q^{\text{sys}}_{\pm,0}$ is the quantity of positive or negative charge contained within the system at the beginning of the time period. The subscripts k and j refer to the inlet and outlet, respectively. Generation and consumption of charge typically occurs when chemical reactions happen in a system. The dimension of the terms in equations [5.3-2] and [5.3-3] is [tI].

The differential forms of the charge accounting equations are appropriate when rates of charge are present. Recall that flow of charge into or out of a system corresponds with current, i, which can also be written as $\dot{q}$:

$$\text{Positive:} \quad \sum_k \dot{q}_{+,k} - \sum_j \dot{q}_{+,j} + \dot{q}_{+,\text{gen}} - \dot{q}_{+,\text{cons}} = \frac{dq^{\text{sys}}_{+}}{dt} \qquad [5.3\text{-}4]$$

$$\text{Negative:} \quad \sum_k \dot{q}_{-,k} - \sum_j \dot{q}_{-,j} + \dot{q}_{-,\text{gen}} - \dot{q}_{-,\text{cons}} = \frac{dq^{\text{sys}}_{-}}{dt} \qquad [5.3\text{-}5]$$

where $\sum_k \dot{q}_{\pm,k}$ is the rate of positive or negative charge (i.e., current) entering the system by bulk transfer, $\sum_j \dot{q}_{\pm,j}$ is the rate of positive or negative charge (i.e., current) leaving the system by bulk transfer, $\dot{q}_{\pm,\text{gen}}$ is the rate at which positive or negative charge is generated in the system, $\dot{q}_{\pm,\text{cons}}$ is the rate at which positive or negative charge is consumed by the system, and $dq^{\text{sys}}_{\pm}/dt$ is the rate at which positive or negative charge accumulates within the system. While current, defined explicitly as the rate at which charge flows in a conductor, is appropriate strictly for the movement of material, it is not appropriate to use to describe Generation and Consumption of charge. The Generation and Consumption terms describe charge reactions, not movement, so $\dot{q}$ is retained. The Accumulation term expresses the instantaneous rate of change of positive or negative charge in the system, or the rate at which charge accumulates in the system. When an Accumulation term is present, other information, such as an initial condition, may need to be specified before a system can be solved. The dimension of the terms in equations [5.3-4] and [5.3-5] is [I].

The integral form of the charge accounting equation is most appropriate when trying to evaluate conditions between two discrete time points. When applying the integral equation, write the differential balance equation and integrate between the initial and final states:

$$\text{Positive:} \quad \int_{t_0}^{t_f} \sum_k \dot{q}_{+,k}\, dt - \int_{t_0}^{t_f} \sum_j \dot{q}_{+,j}\, dt + \int_{t_0}^{t_f} \dot{q}_{+,\text{gen}}\, dt - \int_{t_0}^{t_f} \dot{q}_{+,\text{cons}}\, dt = \int_{t_0}^{t_f} \frac{dq^{\text{sys}}_{+}}{dt} dt \qquad [5.3\text{-}6]$$

$$\text{Negative:} \quad \int_{t_0}^{t_f} \sum_k \dot{q}_{-,k}\, dt - \int_{t_0}^{t_f} \sum_j \dot{q}_{-,j}\, dt + \int_{t_0}^{t_f} \dot{q}_{-,\text{gen}}\, dt - \int_{t_0}^{t_f} \dot{q}_{-,\text{cons}}\, dt = \int_{t_0}^{t_f} \frac{dq^{\text{sys}}_{-}}{dt} dt \qquad [5.3\text{-}7]$$

where t_0 is the initial time and t_f is the final time. The dimension of the terms in equations [5.3-6] and [5.3-7] is [tI].

5.3.2 Conservation Equation for Net Charge

Net charge, q, is defined as the amount of positive charge minus the amount of negative charge:

$$q = q_+ - q_- \qquad [5.3\text{-}8]$$

Thus, we can write the following algebraic and differential equations, respectively, for net charge:

$$\text{Net:} \quad \sum_k q_k - \sum_j q_j + q_{\text{gen}} - q_{\text{cons}} = q_f^{\text{sys}} - q_0^{\text{sys}} \qquad [5.3\text{-}9]$$

$$\text{Net:} \quad \sum_k i_k - \sum_j i_j + \dot{q}_{\text{gen}} - \dot{q}_{\text{cons}} = \frac{dq^{\text{sys}}}{dt} \qquad [5.3\text{-}10]$$

Net charge is a conserved extensive property, so it is conserved in the system and in the universe. A single positive charge or a single negative charge has not been observed to create or to consume itself. **Instead, it has been observed that a pair of electric charges, one negative and one positive, is created or consumed simultaneously within systems.** Thus, when a pair of electric charges is created or destroyed, the net charge of the system is unchanged.

> Consider an analogy between the conservation of charge and the conservation of mass. In a reacting system, the mass associated with specific chemical species may change, and accounting equations for those chemical species may contain Generation or Consumption terms. However, the total mass of the system is constant.
>
> Likewise, neutral species can chemically dissociate or react to create charged species. The accounting equations for positive and negative charges may contain Generation or Consumption terms. However, the net charge of the system is constant, so a conservation equation can be used to describe net charge.

Since net charge has been observed to be neither generated nor consumed, an extra constraint can be placed on equation [5.3-9]. Because the positive and negative charges generated in a system must be equal, the net charge generated in the system is zero:

$$q_{\text{gen}} = q_{+,\text{gen}} - q_{-,\text{gen}} = 0 \qquad [5.3\text{-}11]$$

Because the positive and negative charges consumed in a system must be equal, the net charge consumed in the system is zero:

$$q_{\text{cons}} = q_{+,\text{cons}} - q_{-,\text{cons}} = 0 \qquad [5.3\text{-}12]$$

Thus, equation [5.3-9] is rewritten:

$$\text{Net:} \quad \sum_k q_k - \sum_j q_j = q_f^{\text{sys}} - q_0^{\text{sys}} \qquad [5.3\text{-}13]$$

where k is the index of the inlet and j is the outlet. Equation [5.3-13] states the **conservation of net charge**.

Just as net charge is conserved, the rate of net charge is also conserved. Therefore, the rate at which positive charge is generated is equal to the rate at which negative charge is generated. The same is true for the rates at which positive and negative charges must be consumed:

$$\dot{q}_{gen} = \dot{q}_{+,gen} - \dot{q}_{-,gen} = 0 \quad [5.3\text{-}14]$$

$$\dot{q}_{cons} = \dot{q}_{+,cons} - \dot{q}_{-,cons} = 0 \quad [5.3\text{-}15]$$

Therefore, the differential equation for conservation of charge is written:

$$\sum_k i_k - \sum_j i_j = \frac{dq^{sys}}{dt} \quad [5.3\text{-}16]$$

The integrated form of equation [5.3-16] is:

$$\int_{t_0}^{t_f} \sum_k i_k \, dt - \int_{t_0}^{t_f} \sum_j i_j \, dt = \int_{t_0}^{t_f} \frac{dq^{sys}}{dt} dt \quad [5.3\text{-}17]$$

Equations [5.3-16] and [5.3-17] are used when a current or a rate of charge is given. Since net charge is neither generated nor consumed, the accumulation of net charge is limited to the difference between the charges moving into and out of the system.

5.4 Review of Electrical Energy Accounting Statement

Many different types of energy can be measured, such as mechanical, electrical, and thermal energy. Magnetic and electric fields interact with electric current and vice versa; the energy associated with the flow of electric current is known as electrical energy. In Chapters 4 and 6, conservation and accounting equations describing total energy and mechanical energy, respectively, are developed. In this section, an accounting equation is developed for electrical energy. Information about the number of charges flowing in a circuit is usually given as current, so the algebraic form of the electrical energy accounting equation is not commonly used for solving this class of problems; thus, we do not present it here.

The total energy of a system must be conserved; however, electrical energy is not always conserved. Thus, an accounting equation must be used to mathematically describe how electrical energy moves into and out of a system, how it is generated and consumed in a system, and how it accumulates in a system.

Consider the system shown in Figure 5.4. The rates of charges entering and leaving the system are represented by i_k and i_j, respectively. Electrical energy can move into or out of the system when bulk material that is charged flows across the

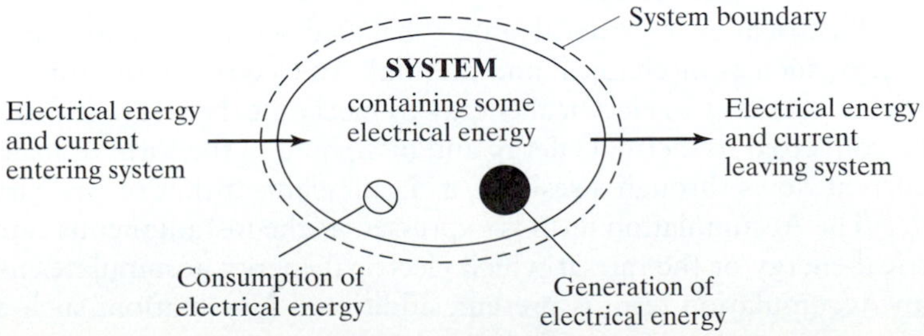

Figure 5.4
Schematic drawing of the rate at which charge (i.e., current) is moved, generated, consumed, and accumulated in a system.

system boundary. Usually, electrical energy is generated or consumed in the system when it is converted from or to another type of energy. Either of these processes can result in accumulation of electrical energy within the system.

The generic accounting equation tracks the movement of electrical energy and the generation, consumption, and accumulation of electrical energy in the system. The differential form of the accounting equation is appropriate when rates of electrical energy are specified:

$$\dot{\psi}_{\text{in}} - \dot{\psi}_{\text{out}} + \dot{\psi}_{\text{gen}} - \dot{\psi}_{\text{cons}} = \frac{d\psi}{dt} \qquad [5.4\text{-}1]$$

$$\sum_k \dot{E}_{E,k} - \sum_j \dot{E}_{E,j} + \sum \dot{G}_{\text{elec}} - \sum \dot{W}_{\text{elec}} = \frac{dE_E^{\text{sys}}}{dt} \qquad [5.4\text{-}2]$$

where $\sum_k \dot{E}_{E,k}$ is the rate at which electrical energy enters the system by bulk charge transfer, $\sum_j \dot{E}_{E,j}$ is the rate at which electrical energy leaves the system by bulk charge transfer, $\sum \dot{G}_{\text{elec}}$ is the rate at which electrical energy is generated in the system, $\sum \dot{W}_{\text{elec}}$ is the rate at which electrical energy is consumed by the system, and dE_E^{sys}/dt is the rate at which electrical energy accumulates within the system. The indices k and j refer to inlet and outlet, respectively. The dimension of the terms in equation [5.4-2] is $[\text{L}^2\text{Mt}^{-3}]$, which is the same dimension as that of power.

Electrical energy enters and leaves the system across the boundary as a current. (Contributions to electrical energy from electric and magnetic fields are neglected in this book.) The rate of electrical energy, $\dot{E}_E$, is defined as the product of the current and the specific potential energy of that current (equation [5.2-4]), so equation [5.4-2] can be written as:

$$\sum_k i_k v_k - \sum_j i_j v_j + \sum \dot{G}_{\text{elec}} - \sum \dot{W}_{\text{elec}} = \frac{dE_E^{\text{sys}}}{dt} \qquad [5.4\text{-}3]$$

This is the differential form of the governing electrical energy accounting equation.

The primary source of production and consumption of electrical energy is the conversion of one form of energy to another form. Many devices generate electrical energy by energy conversion; for example, a battery converts chemical energy to electrical energy. Another example is an electric power station. By heating up water to steam, which turns a turbine, thermal energy is converted to mechanical energy. The turbine is connected to a generator, turning mechanical energy into electrical energy.

A thermocouple is a device that uses the conversion of thermal energy to electrical energy to take a temperature measurement. A pair of dissimilar metal wires (e.g., copper and iron) are fused together to make a thermocouple. By keeping one junction of the wires at a known reference temperature, the other junction can be heated such that an electric potential difference results. This potential difference causes current to flow. To measure this difference in potential, an instrument such as a voltmeter can be connected.

Electrical energy can also be consumed when it is converted to other forms of energy, such as mechanical and thermal. An electric motor, for example, can convert chemical energy to electrical energy to mechanical energy. Electrical energy can also be converted to thermal energy and dissipated in the form of heat when an electrical current flows through a **resistor**, a circuit element that resists current flow.

The Accumulation term is expressed as the instantaneous rate of change of electrical energy or the rate at which electrical energy accumulates in the system. When an Accumulation term is present, additional information, such as an initial condition, may need to be specified to solve the problem.

Electrical energy can be stored in electronic devices known as capacitors and inductors. A **capacitor** stores energy in an electric field, while an **inductor** stores energy in a magnetic field. In systems describing electric circuits, the magnitude of electrical energy of the system (E_E) is the sum of the energies stored in capacitors ($E_{E,C}$) and inductors ($E_{E,L}$). Further discussion about the nature and functions of these devices is presented in Section 5.8.

When trying to account for the movement, generation, consumption, and accumulation of electrical energy in a system between two specific time points, the integral form of the electrical energy accounting equation is most appropriate to use:

$$\int_{t_0}^{t_f} \sum_k i_k v_k \, dt - \int_{t_0}^{t_f} \sum_j i_j v_j \, dt + \int_{t_0}^{t_f} \sum \dot{G}_{\text{elec}} \, dt - \int_{t_0}^{t_f} \sum \dot{W}_{\text{elec}} \, dt = \int_{t_0}^{t_f} \frac{dE_E^{\text{sys}}}{dt} dt \quad [5.4\text{-}4]$$

where t_0 is the beginning and t_f is the end of the time period of interest. The dimension of the terms in equation [5.4-4] is $[L^2Mt^{-2}]$.

5.5 Kirchhoff's Current Law (KCL)

The most common application of the differential conservation of net charge equation is in circuit analysis. If a system is at steady-state, equation [5.3-16] reduces to:

$$\text{Net:} \quad \sum_k i_k - \sum_j i_j = 0 \quad [5.5\text{-}1]$$

where subscript k refers to inlet currents and subscript j refers to outlet currents. Equation [5.5-1] is known as **Kirchhoff's current law** (KCL), which states that at any junction point, the sum of all currents entering the junction must equal the sum of all currents leaving the junction. Electrons do not accumulate at any point in a conducting material, so KCL can be applied to electrical networks made of conductors.

To apply KCL, the system boundary is defined around a **node**, a point in the circuit where two or more circuit elements meet. A circuit element can be any number of electrical devices, such as a wire, a battery, a resistor, a capacitor, or an inductor. In KCL, currents entering a node are considered Input terms, whereas currents leaving a node are considered Output terms; the algebraic sum of all the currents entering and leaving any node in a circuit equals zero. In other words, **KCL states that the sum of the current flowing toward any point in a circuit is equal to the sum of the currents flowing away from that point.** KCL is one of the most useful and widely used equations in circuit analysis and design.

A node at which three or more circuit elements meet can analogously be compared to the junction of three or more fluid streams. Consider a circuit with one inlet wire and two outlet wires (Figure 5.5a); the inlet current has two possible paths to leave the node. This is analogous to the steady-state flow of blood in a single vessel that splits into two vessels (Figure 5.5b). The sum of the current in the two outlet wires (i_B and i_C) must equal the current in the inlet wire (i_A), just as the sum of the mass flow rates of the two outlet streams ($\dot{m}_E$ and $\dot{m}_F$) must equal the mass flow rate of the inlet ($\dot{m}_D$).

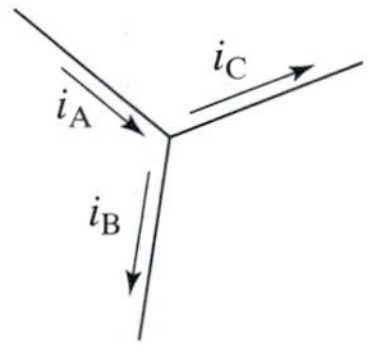

Figure 5.5a
Circuit with one inlet and two outlet wires.

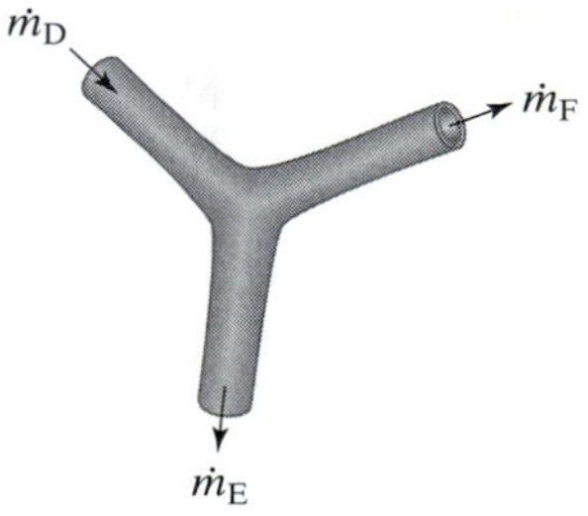

Figure 5.5b
One vessel bifurcating into two.

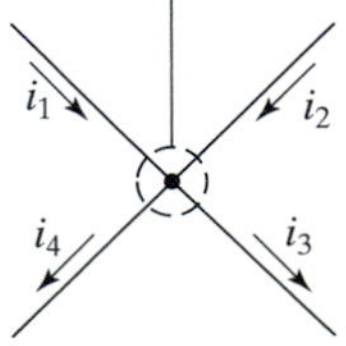

Figure 5.6
Four wires connected at a node.

EXAMPLE 5.3 Kirchhoff's Current Law Applied to a Simple Circuit

Problem: Figure 5.6 illustrates a system with four wires connected at a node. Use KCL to develop an equation describing the current flow at the node.

Solution: The system boundary surrounds the node where the four circuit elements (i.e., wires) meet. Currents i_1 and i_2 enter the node; currents i_3 and i_4 leave the node. Applying KCL gives:

$$\sum_k i_k - \sum_j i_j = 0$$
$$i_1 + i_2 - i_3 - i_4 = 0$$

Note that the inlet currents are positive and the outlet currents are negative. ■

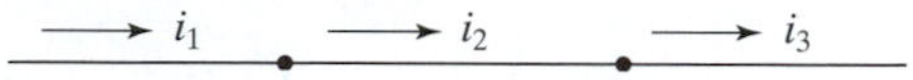

Figure 5.7
Three wires (1, 2, 3) connected in series.

Circuit elements can be connected in two different ways: in series or in parallel. Two components connected in **series** are attached such that the end of one circuit element meets the end of another circuit element. Thus, if you move through one component to its end, the only place you can move is into the next component. When two elements connect at a single node, the elements are always in series. Figure 5.7 illustrates three wires in series. Applying KCL to wires connected in series indicates that the magnitude of the current is the same in each wire. Thus, the circuit in Figure 5.7 has $i_1 = i_2 = i_3$.

Circuit elements connected in **parallel** are attached such that the current is split and enters into multiple circuit elements, recombining when the element branches meet again. In Figure 5.8, wires 2, 3, and 4 are connected in parallel. Note that each of these three wires is joined with the other two at both ends. Applying KCL to wires in parallel indicates that current splits at nodes where wires are connected in parallel. The circuit in Figure 5.8 has $i_1 = i_2 + i_3 + i_4$ for node A and $i_2 + i_3 + i_4 = i_5$ for node B. Thus, the magnitude of the current in wires 2, 3, and 4 is less than those in wires 1 and 5. If the current is split equally among wires 2, 3, and 4, then the current in wires 2, 3, and 4 is one-third that in wire 1.

In addition to how they can be connected, circuits can be described in terms of the continuity of the circuit elements. An **open circuit** has a gap or break in continuity so that current cannot flow. Open circuits can be used in making measurements, such as temperature. Filling the gap with a conductor closes the circuit. In a **closed circuit,** current can flow freely.

For a circuit with n nodes, applying KCL yields n current equations. Of these n equations, only $n - 1$ are linearly independent. In circuit analysis, most system boundaries are specified as nodes, but other system boundaries are certainly possible. In some situations, current carried in a wire may enter a cluster of components. If the system boundary encompasses that cluster, an overall balance may be written on the inlet and outlet current to the system. The configuration shown in Figure 5.9 is an example of a supernode.

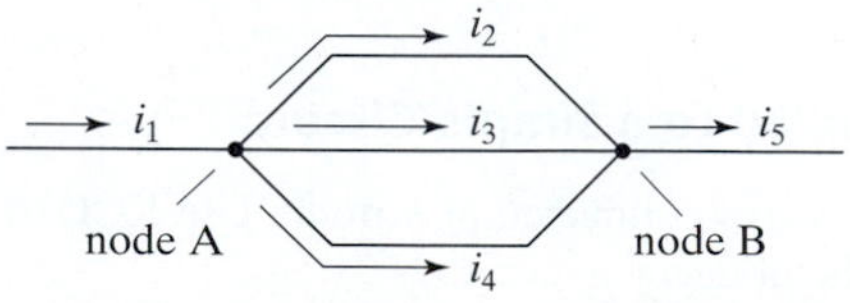

Figure 5.8
Three wires (2, 3, 4) connected in parallel.

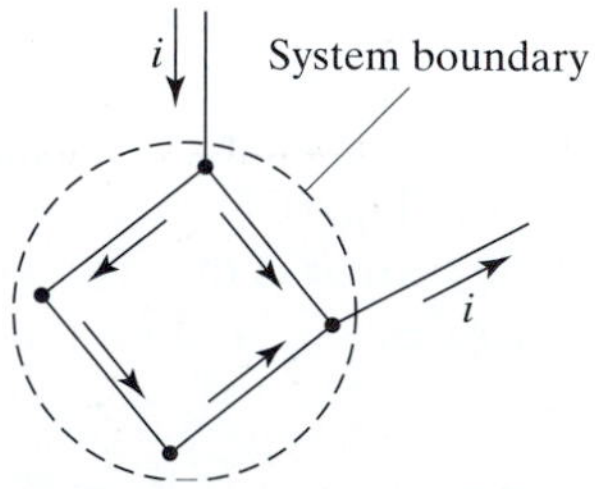

Figure 5.9
System boundary encompassing a supernode.

Biomedical instrumentation is made of circuit hardware that contains both simple and complex configurations. The examples below are simple configurations that could be found in designs for various types of electronic sensing or biomedical measurement devices.

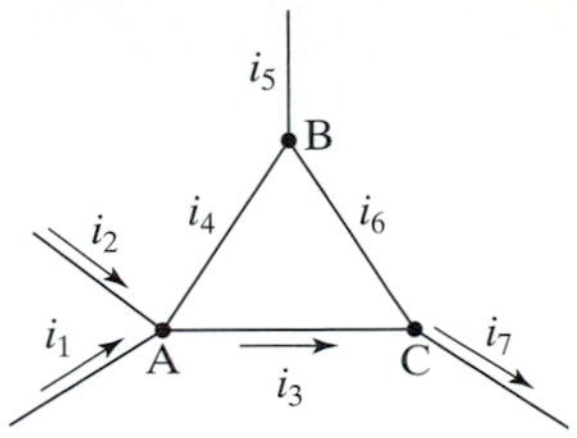

Figure 5.10a
Closed-circuit system with seven wires and three nodes.

EXAMPLE 5.4 Kirchhoff's Current Law Applied to a Closed Circuit

Problem: Figure 5.10a shows a closed circuit with seven wires and three nodes (A, B, and C). The following currents are known: $i_1 = 6.0$ A, $i_2 = 2.5$ A, $i_7 = 1.0$ A. For the closed circuit, the relationship $i_4 = 16i_3$ is also known. Determine the direction and magnitude of all unknown currents.

Solution: The system is first defined as the overall cluster of elements (Figure 5.10b). The differential conservation equation for net charge (KCL) is written for wires 1, 2, 5, and 7. Since the direction of current in wire 5 is unknown, we arbitrarily assume the current travels out of node B. The overall equation for Figure 5.10b is:

$$i_1 + i_2 - i_5 - i_7 = 0$$

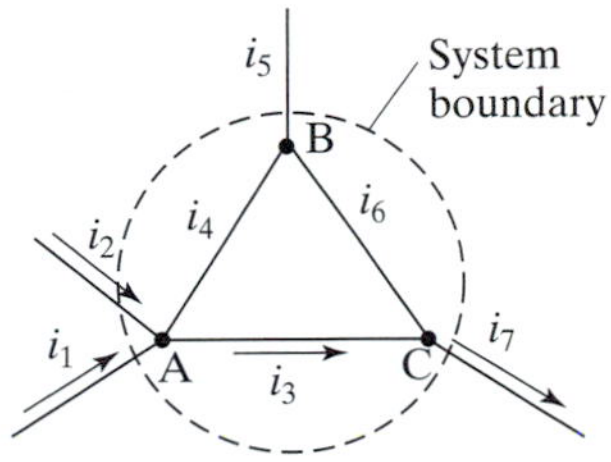

Figure 5.10b
System boundary defined for a cluster of circuit elements.

KCL gives one current equation each for the three nodes. A system boundary is drawn around each node (Figures 5.10c–e). Since the directions of the currents in wires 4 and 6 are unknown, we assume arbitrary directions for each. We assume the current in wire 4 flows out of node A, and the current in wire 6 flows out of node C. Applying KCL to the three nodes gives:

$$\begin{aligned} \text{A:} \quad & i_1 + i_2 - i_3 - i_4 = 0 \\ \text{B:} \quad & i_4 + i_6 - i_5 = 0 \\ \text{C:} \quad & i_3 - i_6 - i_7 = 0 \end{aligned}$$

Thus, with the three nodal equations and one overall equation, we have four equations total. However, only three of these are linearly independent. Any one of the equations may be derived by using the other three.

Using the equation derived for node A, the known magnitudes of the currents i_1 and i_2, and the relationship $i_4 = 16i_3$, we can solve for i_3:

$$\begin{aligned} \text{A:} \quad i_1 + i_2 - i_3 - i_4 &= 6.0\text{ A} + 2.5\text{ A} - i_3 - 16i_3 = 0 \\ 17i_3 &= 8.5\text{ A} \\ i_3 &= 0.5\text{ A} \end{aligned}$$

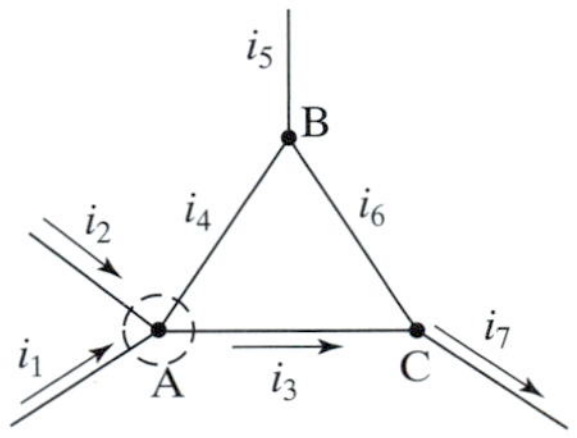

Figure 5.10c
System boundary draw around node A.

Having solved for i_3, we also know that:

$$i_4 = 16i_3 = 8.0\text{ A}$$

Because the magnitude of i_4 is a positive value, we know that we correctly assumed the direction of current in wire 4, which is out of node A and into node B. If we had originally assumed that the current in wire 4 flows in the opposite direction, we would have calculated $i_4 = -8.0$ A.

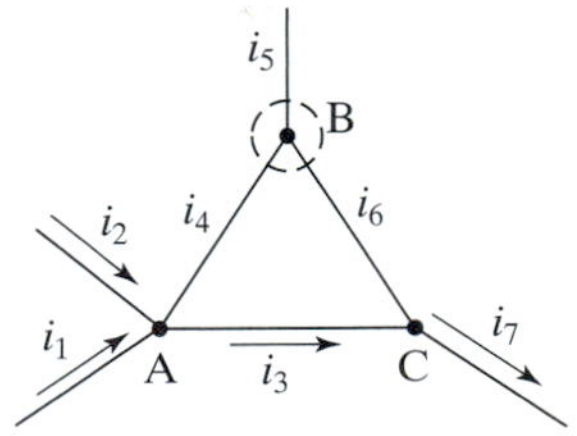

Figure 5.10d
System boundary drawn around node B.

Node C is analyzed next, since it has fewer unknown values (i_6) than node B (i_5 and i_6). We can use the KCL equation developed for node C and substitute the values we just found to solve for the remaining unknown value:

$$\begin{aligned} \text{C:} \quad i_3 - i_6 - i_7 &= 0.5\text{ A} - i_6 - 1.0\text{ A} = 0 \\ i_6 &= -0.5\text{ A} \end{aligned}$$

Because the magnitude of the current in i_6 is negative, this means that the current flows through wire 6 in the opposite direction of that assumed. Instead, 0.5 A of current flows out of node B and into node C.

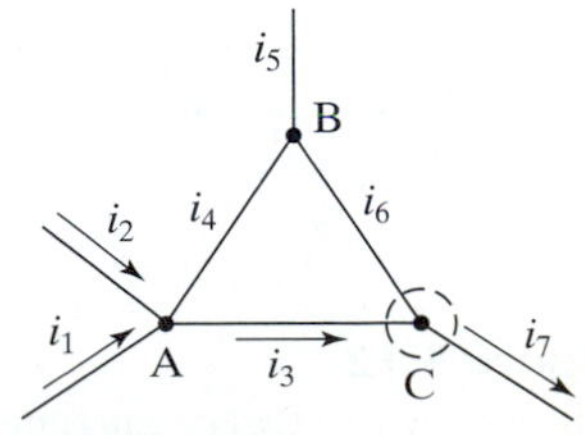

Figure 5.10e
System boundary drawn around node C.

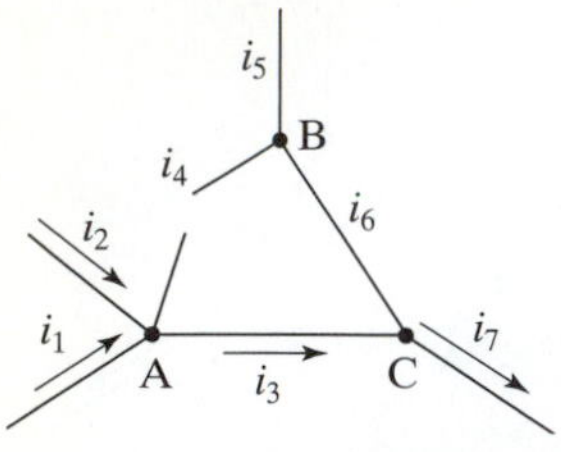

Figure 5.11a
Open-circuit system with seven wires and three nodes.

Now we can use either the nodal equation for B or the overall KCL equation to solve for the current in wire 5. Substituting the known values into the nodal equation for B, we can find i_5:

$$\text{B:} \quad i_4 + i_6 - i_5 = 8.0\text{ A} - 0.5\text{ A} - i_5 = 0$$
$$i_5 = 7.5\text{ A}$$

Thus, 7.5 A of current flows out of node B. ∎

EXAMPLE 5.5 Kirchhoff's Current Law Applied to an Open Circuit

Problem: Consider Example 5.4, which examined a closed circuit with seven elements (wires) and three nodes (A, B, and C). Figure 5.11a shows the same system with wire 4 cut to create an open circuit. The following currents are known: $i_1 = 6.0$ A, $i_2 = 2.5$ A, $i_7 = 1.0$ A. Determine the direction and magnitude of all unknown currents.

Solution: The gap in an open circuit blocks current flow, so no current flows through wire 4. Current flows through the circuit as though wire 4 were not present, so Figure 5.11b is drawn to indicate this change.

As in Example 5.4, we can draw a boundary around each node to apply KCL. We assume the same directions of any unknown currents as in Example 5.4:

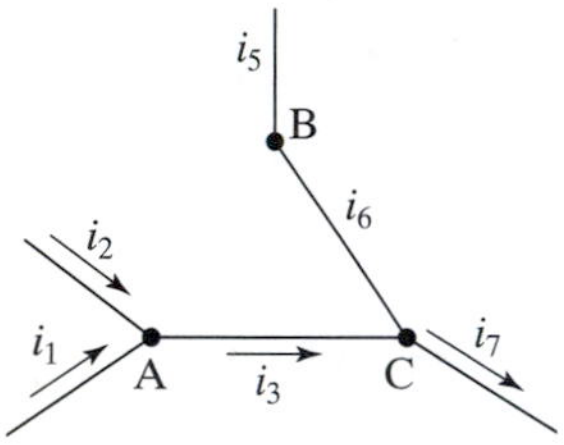

Figure 5.11b
Open-circuit system with six wires and three nodes.

$$\text{A:} \quad i_1 + i_2 - i_3 = 0$$
$$\text{B:} \quad i_6 - i_5 = 0$$
$$\text{C:} \quad i_3 - i_6 - i_7 = 0$$

We can substitute the known values into the nodal equation for A to yield:

$$\text{A:} \quad i_1 + i_2 - i_3 = 6.0\text{ A} + 2.5\text{ A} - i_3 = 0$$
$$i_3 = 8.5\text{ A}$$

We can then substitute this value into the nodal equation for C to get:

$$\text{C:} \quad i_3 - i_6 - i_7 = 8.5\text{ A} - i_6 - 1.0\text{ A} = 0$$
$$i_6 = 7.5\text{ A}$$

Wires 5 and 6 are in series, so the current that flows through them must be of the same magnitude and direction. Thus, $i_5 = 7.5$ A, and the direction of current flow in wire 5 is out of node B. ∎

EXAMPLE 5.6 Kirchhoff's Current Law Applied to a Complex Circuit

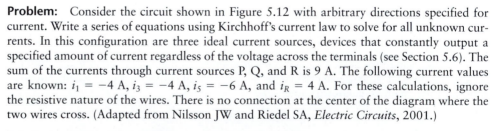

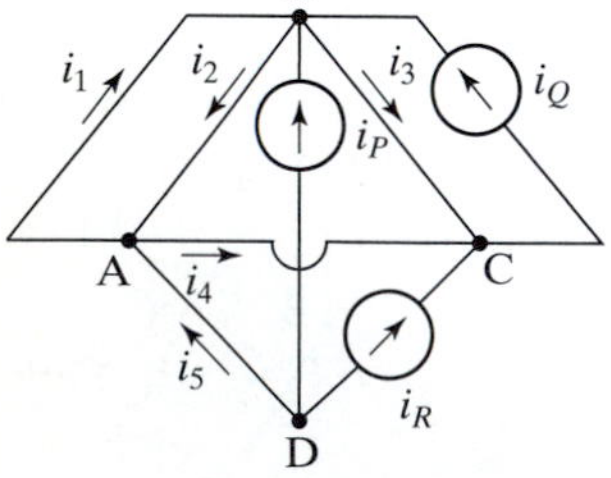

Figure 5.12
Circuit with three current sources.

Problem: Consider the circuit shown in Figure 5.12 with arbitrary directions specified for current. Write a series of equations using Kirchhoff's current law to solve for all unknown currents. In this configuration are three ideal current sources, devices that constantly output a specified amount of current regardless of the voltage across the terminals (see Section 5.6). The sum of the currents through current sources P, Q, and R is 9 A. The following current values are known: $i_1 = -4$ A, $i_3 = -4$ A, $i_5 = -6$ A, and $i_R = 4$ A. For these calculations, ignore the resistive nature of the wires. There is no connection at the center of the diagram where the two wires cross. (Adapted from Nilsson JW and Riedel SA, *Electric Circuits*, 2001.)

Solution: KCL can be applied to nodes A, B, C, and D:

$$\text{A:} \quad i_2 + i_5 - i_1 - i_4 = 0$$
$$\text{B:} \quad i_1 - i_2 - i_3 + i_P + i_Q = 0$$
$$\text{C:} \quad i_3 + i_4 + i_R - i_Q = 0$$
$$\text{D:} \quad -i_R - i_P - i_5 = 0$$

Only three of these four equations are linearly independent. Another equation can be written based on the sum of the currents through the current sources:

$$\text{Source:} \quad i_P + i_Q + i_R = 9 \text{ A}$$

Using the current source equation, equations for nodes A, B, and C, and the given values, the above equations are simplified:

$$\begin{aligned} \text{Source:} \quad & i_P + i_Q = 5 \text{ A} \\ \text{A:} \quad & i_2 - i_4 = 2 \text{ A} \\ \text{B:} \quad & -i_2 + i_P + i_Q = 0 \\ \text{C:} \quad & i_4 - i_Q = 0 \end{aligned}$$

With four equations and four unknowns, we can now solve for the currents. This problem is suited for solving in MATLAB by setting up a matrix with the four equations. These scalar equations can be represented by the following matrix equation in the form $A\vec{x} = \vec{y}$:

$$\begin{bmatrix} 1 & 1 & 0 & 0 \\ 0 & 0 & 1 & -1 \\ 1 & 1 & -1 & 0 \\ 0 & -1 & 0 & 1 \end{bmatrix} \begin{bmatrix} i_P \\ i_Q \\ i_2 \\ i_4 \end{bmatrix} = \begin{bmatrix} 5 \\ 2 \\ 0 \\ 0 \end{bmatrix}$$

We can solve for the solution vector $\vec{x}$ in MATLAB using the following set of commands:

```
>> A = [1 1 0 0; 0 0 1 -1; 1 1 -1 0; 0 -1 0 1];
>> y = [5; 2; 0; 0];
>> x = A\y
```

The solution is: $i_P = 2$ A, $i_Q = 3$ A, $i_2 = 5$ A, and $i_4 = 3$ A. The solution can be checked through back-substitution. ■

5.6 Kirchhoff's Voltage Law (KVL)

A second law commonly used in circuit analysis is Kirchhoff's voltage law (KVL). In contrast to KCL, the derivation of KVL uses the electrical energy accounting equation as its starting point:

$$\sum_k \dot{E}_{E,k} - \sum_j \dot{E}_{E,j} + \sum \dot{G}_{\text{elec}} - \sum \dot{W}_{\text{elec}} = \frac{dE_E^{\text{sys}}}{dt} \qquad [5.6\text{-}1]$$

Consider a simple closed circuit with only one loop (Figure 5.13). A **loop** is a path that traces a set of circuit elements in series; the location at which the loop path starts is the same location at which it stops in the circuit. This path also does not trace any circuit element more than once. The system is then defined around the circuit so that no current crosses the system boundary. For a steady-state system with no inlets or outlets of electrical energy, we can reduce equation [5.6-1] to:

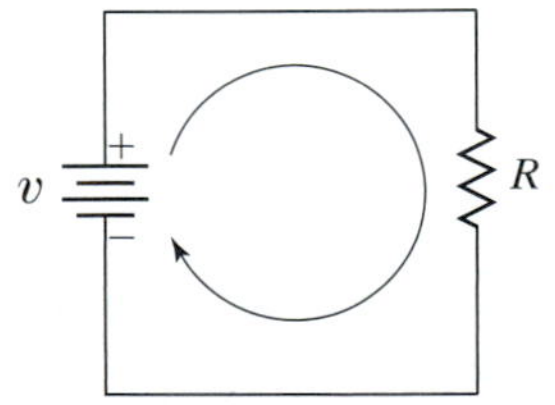

Figure 5.13
Simple closed circuit with one loop including a voltage source and a battery.

$$\sum \dot{G}_{\text{elec}} - \sum \dot{W}_{\text{elec}} = 0 \qquad [5.6\text{-}2]$$

This equation states that the total rate of electrical energy generated within the system is equal to the total rate of electrical energy consumption. KVL is derived from equation [5.6-2].

In Section 5.6.3, we first develop KVL for a simple circuit with one loop. We then demonstrate that KVL equations can be similarly developed and are appropriate for steady-state systems with inlet and outlet currents.

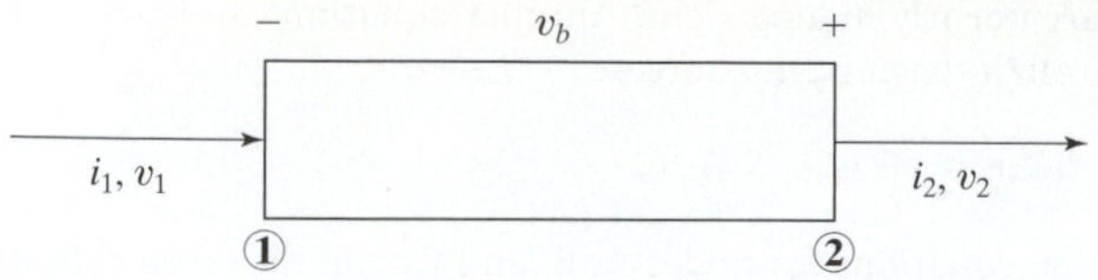

Figure 5.14
Circuit element in which current travels from lower to higher voltage.

5.6.1 Elements that Generate Electrical Energy

Recall that voltage is a difference in electrical potential. By definition, if a positive change in potential energy is accomplished when moving a test charge from position A to B, then the electric potential at point B is higher than at point A, and the voltage $(v_B - v_A)$ is positive.

When specifying a voltage, it is necessary to designate a reference or ground state, since voltage is a measure of difference in potential. For example, the voltage across each circuit element is a measure of the difference in electrical potential across that element. However, it is common to refer to a circuit element as having a particular voltage. It is important to remember that the stated voltage represents the difference in potential between the element and the ground state.

Consider a voltage change across the element in Figure 5.14. In electronics, a negative sign is used to signify the lower-voltage end and a positive sign to signify the higher-voltage end. When current is moved from the negative terminal (labeled 1) to the positive terminal (labeled 2) of this element, the rate of potential energy of the element is increased. This is an example of the rate of generation of electrical energy, $\dot{G}_{\text{elec}}$, in the element:

$$\sum \dot{G}_{\text{elec}} = i_2 v_2 - i_1 v_1 \qquad [5.6\text{-}3]$$

where i_1 and i_2 are the inlet and outlet currents, and v_1 and v_2 are the voltages at the inlet and outlet, respectively. Remember that v_1 and v_2 are shorthand references for potential differences between that point in the circuit and the ground state.

Using KCL, we know that i_1 is equal to i_2, so we can reduce equation [5.6-3] to:

$$\sum \dot{G}_{\text{elec}} = i_1(v_2 - v_1) = i_1 v_b \qquad [5.6\text{-}4]$$

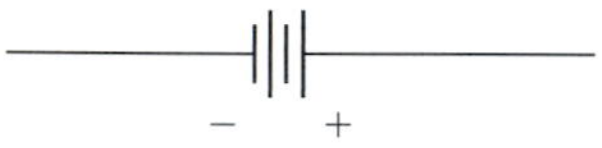

Figure 5.15a
Circuit symbol for a voltage source.

where v_b is the voltage gain across the element. When the current through this element is positive, the element generates electrical energy at a particular rate. Examples of voltage sources include batteries, piezoelectric disks, and generators.

An **ideal voltage source** is a circuit element that maintains a prescribed voltage across its terminals regardless of the current flowing in those terminals. The symbol for a voltage source in a circuit is shown in Figure 5.15a. A **battery** is often modeled as an ideal voltage source that provides a steady, constant, specified voltage to a circuit.

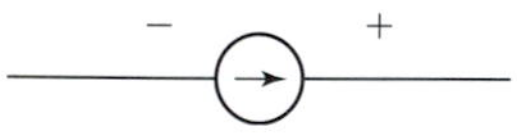

Figure 5.15b
Circuit symbol for a current source.

An **ideal current source** is a device that constantly outputs a specified amount of current regardless of the voltage across those terminals. Although very difficult to find in nature, ideal current sources can be simulated with a collection of several electronic components. A current source generates as much or as little voltage across its terminals as necessary to produce a particular amount of current. A current source generates electrical energy at a rate equal to the current multiplied by the voltage across the terminals. The symbol for a current source in a circuit is shown in Figure 5.15b.

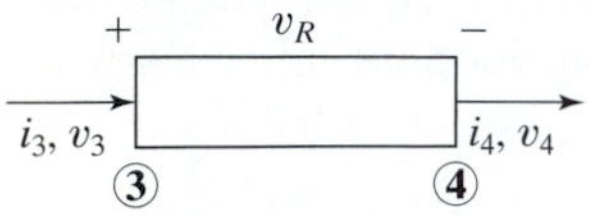

Figure 5.16
Circuit element in which current travels from higher to lower voltage.

5.6.2 Resistors: Elements that Consume Electrical Energy

Consider a voltage change across the element in Figure 5.16. When current is moved from the positive end (labeled 3) to the negative end (labeled 4) of this element, the rate of potential energy of the element is decreased. This is an example of the rate of consumption of electrical energy, $\dot{W}_{elec}$, in the element:

$$-\sum \dot{W}_{elec} = i_4 v_4 - i_3 v_3 \quad [5.6\text{-}5]$$

Using KCL, we know that i_3 is equal to i_4, so we can reduce equation [5.6-5] to:

$$+\sum \dot{W}_{elec} = i_3(v_3 - v_4) = i_3 v_R \quad [5.6\text{-}6]$$

where v_R is the voltage drop across the element. When the current through this element is positive, the element consumes electrical energy at a particular rate. In electronics the most common example of an element that consumes electrical energy is a **resistor**.

Most materials exhibit measurable **resistance** (R) to the flow of electric current. When current passes through a material resisting the flow of electrons, such as the electrical component of a resistor, the voltage of the current drops and electrical energy is consumed. Usually when electrical energy is consumed, it is dissipated as thermal energy. Charge flows from high potential (+) to low potential (−) in a resistor since it is a passive device. The symbol for a resistor, labeled R in a circuit diagram, is shown in Figure 5.17. The SI unit of resistance is the ohm (Ω), which has the unit of volt/ampere and a dimension of $[L^2Mt^{-3}I^{-2}]$.

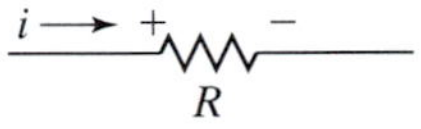

Figure 5.17
Circuit symbol for a resistor.

The resistance of a specific piece of a material is proportional to the material's **resistivity** (ρ) and the ratio of the material's length to its cross-sectional area. While resistivity is a property of a material, resistance is a property of a particular piece of the material. In symbolic form, this law states:

$$R = \frac{\rho l}{A} \quad [5.6\text{-}7]$$

where ρ is the resistivity constant of the resistive material, l is the length of the resistive material, and A is the cross-sectional area of the resistive material. This law was first observed for metal wires, but it can be applied to other systems. Typically, the numerical value of resistance in resistors in electronic systems is specified.

Two or more resistors connected in series (Figure 5.18) behave like an equivalent resistor that has a resistance equal to the sum of the individual resistances in series:

$$R_{eq} = R_1 + R_2 + \cdots + R_n \quad [5.6\text{-}8]$$

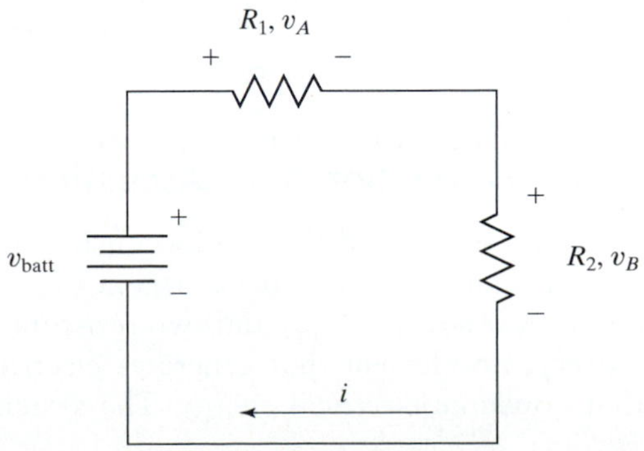

Figure 5.18
Two resistors connected in series with a battery.

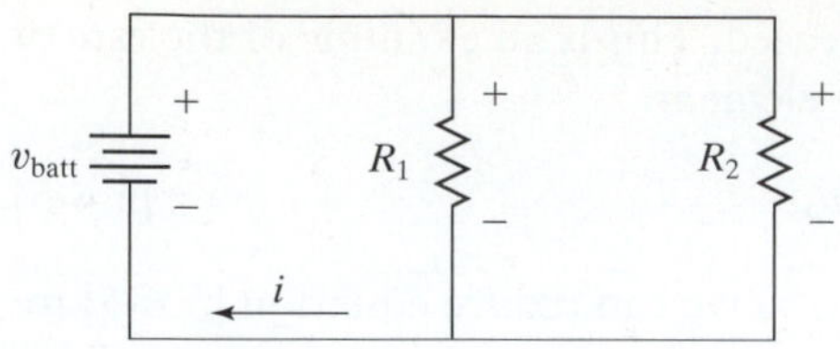

Figure 5.19
Two resistors connected in parallel with a battery.

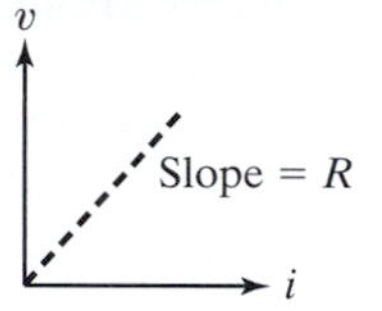

Figure 5.20
Illustration of Ohm's law, showing the linear relationship between voltage and current.

where n is the number of resistors connected in series. An **equivalent resistor** (R_{eq}) or an **effective resistor** (R_{eff}) has an effect on the circuit equivalent to that of the resistors it is replacing. For resistors in series, the total resistance is greater than the individual resistance of any particular resistor. Adding extra resistors in series has a similar effect as increasing the length of a piece of resistive material.

Two or more resistors connected in parallel (Figure 5.19) combine to give an effective resistance according to the relationship:

$$\frac{1}{R_{eq}} = \frac{1}{R_1} + \frac{1}{R_2} + \cdots + \frac{1}{R_n} \qquad [5.6\text{-}9]$$

Because a parallel combination offers multiple paths through which current can travel, the net resistance is always lower than the lowest individual resistance. Adding extra resistors in parallel combination has a similar effect as increasing the cross-sectional area of a piece of resistive material. One very powerful tool in circuit analysis is to simplify complex configurations of resistors by reducing series and parallel configurations to equivalent resistances.

For an ideal resistor, the relation between the applied voltage and current is shown in Figure 5.20. The linear relationship between voltage and current is known as **Ohm's law**:

$$v = iR \qquad [5.6\text{-}10]$$

where v is the voltage drop across the resistor, i is the current through the resistor, and R is the resistance of the resistor. Ohm's law is often used in conjunction with KCL and KVL to solve circuit problems.

5.6.3 Derivation and Discussion of KVL

KVL can be derived for any loop using the electrical energy accounting equation and the conservation of net charge. Consider the circuit shown in Figure 5.18 with one power source (v_{batt}) and two resistors (v_A, v_B) in series. The system contains a battery, an element that generates electrical energy, and two resistors, elements that consume electrical energy. The system boundary is defined to surround the entire circuit so that no current crosses the system boundary. Because the system is at steady-state, the electrical energy accounting equation [5.4-2] reduces to:

$$\sum \dot{G}_{\text{elec}} - \sum \dot{W}_{\text{elec}} = 0 \qquad [5.6\text{-}11]$$

$$iv_{\text{batt}} - iv_A - iv_B = 0 \qquad [5.6\text{-}12]$$

In addition, since no current sources or elements affect the current, the current is constant throughout the whole system, which reduces equation [5.6-12] to:

$$v_{\text{batt}} - v_A - v_B = 0 \qquad [5.6\text{-}13]$$

This equation is an illustration of **Kirchhoff's voltage law**, which states that the algebraic sum of voltage drops taken around any closed loop in a circuit is equal to zero. Generally, KVL is stated:

$$\sum_{\text{loop}} v_{\text{elements}} = 0 \qquad [5.6\text{-}14]$$

where a loop is a closed path in a circuit. For a circuit with n loops, KVL yields n voltage equations. Of these, only $n - 1$ equations are linearly independent.

By convention, the high-voltage end of an element is labeled with a positive sign, and the low-voltage end with a negative sign. To assign whether voltage should be a drop or gain in the KVL equation for a loop, find the sign of an element corresponding to the side immediately following the element and transfer that to the equation. For example, if the current in the loop moves across an element from positive to negative, such as across a resistor, then the voltage contribution of that element should be subtracted. Conversely, if the current in the loop moves across the element from negative to positive, such as across a battery, that element adds to the voltage of the loop. When applying KVL, elements that generate electrical energy have a positive sign and elements that consume electrical energy have a negative sign.

EXAMPLE 5.7 Kirchhoff's Voltage Law in a Simple Series Circuit

Problem: Consider the circuit in Figure 5.21 that contains one power source and three resistors. The following information is known: $v_B = 120$ V, $R_1 = 20\ \Omega$, $R_3 = 10\ \Omega$, $i = 3$ A. Use KVL to find R_2.

Solution: To apply KVL, we arbitrarily designate that the current travels in a clockwise direction around the path. The KVL equation for the circuit is:

$$\sum_{\text{loop}} v_{\text{elements}} = v_B - v_{R_1} + v_{R_2} - v_{R_3} = 0$$

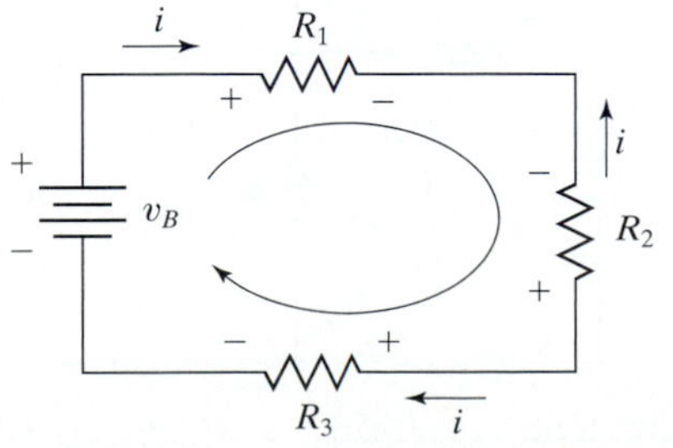

Figure 5.21
Circuit with one battery and three resistors connected in series.

Note that when using a clockwise loop, a positive sign immediately follows the elements v_B and R_2, so these voltage gains are added in the applied KVL equation. Voltages across R_1 and R_3 are negative.

Using KCL, we know that the magnitude of the current is constant throughout the circuit. We can use Ohm's law to substitute for the unknown voltage gains and drops across each resistor and substitute the known values of currents and resistances to solve for R_2.

$$v_B - iR_1 + iR_2 - iR_3 = 120\text{ V} - (3\text{ A})(20\ \Omega) + (3\text{ A})R_2 - (3\text{ A})(10\ \Omega) = 0$$
$$R_2 = -10\ \Omega$$

The magnitude of R_2 is 10 Ω. Since resistors are passive devices, they can only consume electrical energy and cannot generate it. The term iR_2 is calculated as -30 V, and thus R_2 consumes energy. If the polarities for R_2 in Figure 5.21 were reversed, then the calculated value for R_2 would be $+10\ \Omega$. The reading of a negative resistance means that the positive and negative leads of a voltmeter, which read potential difference, are reversed when placed across R_2.

One major source of errors in the application of KCL, KVL, and Ohm's law is mistakes from sign conventions. You may encounter situations where the polarities are not assigned. Labeling the high-voltage and low-voltage ends of elements and stating the direction of current flow before applying KVL may help you minimize mistakes. ■

We have demonstrated that the electrical energy accounting equation reduces to KVL for a circuit with just a single loop. For circuit configurations that contain elements in parallel, several possible loops may be drawn (e.g., Figure 5.22a). When applying the electrical energy accounting equation to each of these loops, Input and Output terms may be present. However, the governing equation always reduces to KVL when the system is at steady-state.

For example, consider loop 1 in Figure 5.22a. The system boundary is drawn to include only the part of the circuit involved in loop 1 (Figure 5.22b). The differential form of the governing electrical energy accounting equation (equation [5.4-3]) is:

$$\sum_k i_k v_k - \sum_j i_j v_j + \sum \dot{G}_{\text{elec}} - \sum \dot{W}_{\text{elec}} = \frac{dE_E^{\text{sys}}}{dt} \qquad [5.6\text{-}15]$$

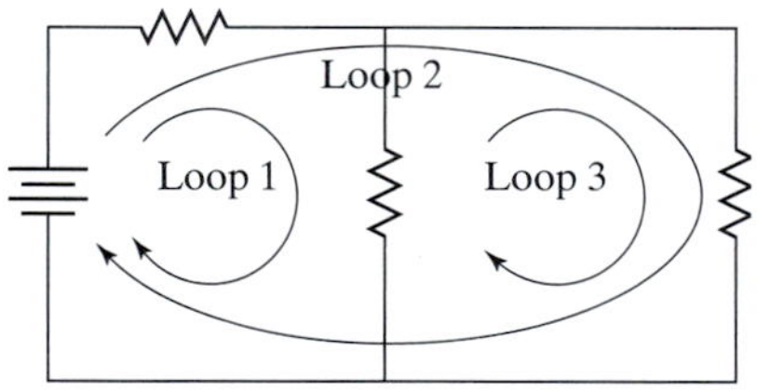

Figure 5.22a
Possible loop configurations for a parallel circuit.

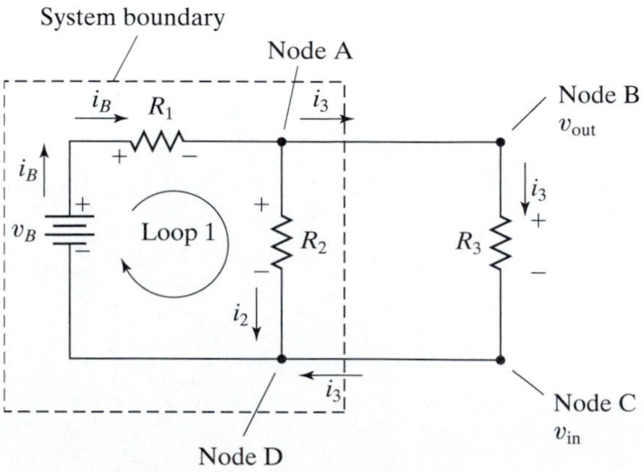

Figure 5.22b
System designation for loop 1 in parallel circuit.

In this circuit, the current i_B splits at node A into i_2, which travels through resistor R_2 in the system, and i_3, which travels through resistor R_3 outside of the system. The battery generates electrical energy in the system at a rate of $i_B v_B$. Resistors R_1 and R_2 consume electrical energy at rates of $i_B v_1$ and $i_2 v_2$, respectively, where v_1 and v_2 are the voltage drops across resistors R_1 and R_2.

Electrical energy leaves the system at node A and enters the system at node D. Let v_{out} be the potential at node B before current i_3 passes through resistor R_3 and v_{in} be the potential at node C after i_3 passes through R_3. Then electrical energy leaves the system shown at a rate of $i_3 v_{out}$ and enters the system at a rate of $i_3 v_{in}$. The system is at steady-state, so no electrical energy accumulates in the loop. Substituting these values into equation [5.6-15] gives:

$$i_3 v_{in} - i_3 v_{out} + i_B v_B - i_B v_1 - i_2 v_2 = 0 \quad [5.6\text{-}16]$$

$$-i_3(v_{out} - v_{in}) + i_B v_B - i_B v_1 - i_2 v_2 = 0 \quad [5.6\text{-}17]$$

where $v_{out} - v_{in}$ is the voltage drop across resistor R_3. Since resistors R_2 and R_3 are in parallel, the voltage drop across the two resistors is the same:

$$v_{out} - v_{in} = v_2 \quad [5.6\text{-}18]$$

(See Example 5.8 for proof.) Thus, equation [5.6-17] becomes:

$$i_B v_B - i_B v_1 - (i_2 + i_3)v_2 = 0 \quad [5.6\text{-}19]$$

For node A, we can apply KCL:

$$i_B - i_2 - i_3 = 0 \quad [5.6\text{-}20]$$

Substituting equation [5.6-20] into equation [5.6-19] gives:

$$i_B v_B - i_B v_1 - i_B v_2 = 0 \quad [5.6\text{-}21]$$

Since the current i_B is constant, we can reduce the equation such that i_B is cancelled:

$$v_B - v_1 - v_2 = 0 \quad [5.6\text{-}22]$$

which is an illustration of KVL for loop 1. The same principles can be applied to loops 2 and 3 in Figure 5.22a. Thus, the differential electrical energy accounting equation reduces to KVL for any loop when the system is at steady-state.

EXAMPLE 5.8 Current-Divider Circuit

Problem: The configuration in Figure 5.23a is called a current-divider circuit. It consists of two resistors connected in parallel across a voltage source or current source. The purpose of a current divider is to divide the current across two or more elements. Calculate the voltage across each resistor in the circuit. Relate the current across the battery to the currents across the two resistive elements.

Solution: The system encloses the circuit such that no current crosses the boundary. We can draw two loops for the circuit, as shown in Figure 5.23b, and apply KVL to the steady-state system:

Loop 1: $v_B - v_{R_1} = 0$

Loop 2: $v_B - v_{R_2} = 0$

where v_B is the voltage across the battery and R_1 and R_2 are the resistive elements. From these equations, we see that $v_B = v_{R_1} = v_{R_2}$. Thus, the voltage drop across two resistors in parallel is the same regardless of the magnitude of the resistances.

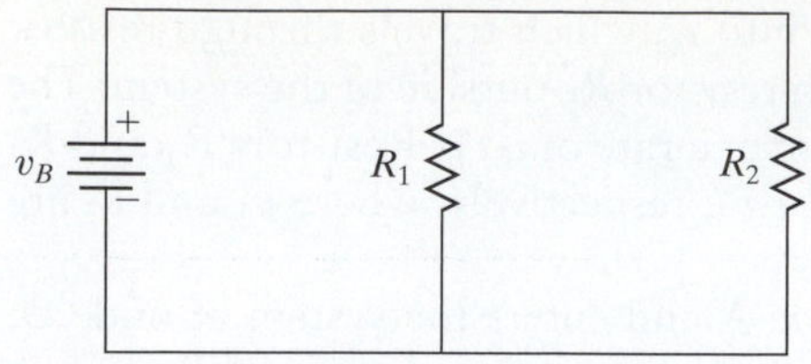

Figure 5.23a
Current-divider circuit.

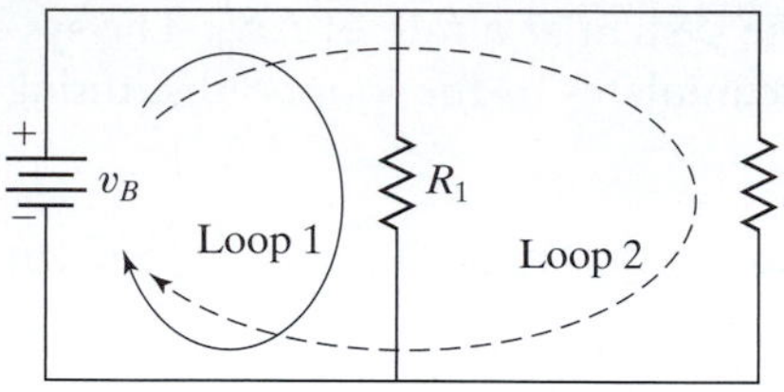

Figure 5.23b
Two loops drawn in arbitrary directions to indicate assumed direction of current flow.

We can use Ohm's law to calculate the voltages of the two resistors:

$$v_B = i_1 R_1 = i_2 R_2$$

A single equivalent resistor R_{eq} can be substituted for the two resistors (Figure 5.23c). Because the resistors are in parallel, we can use equation [5.6-9] to determine the equivalent resistance:

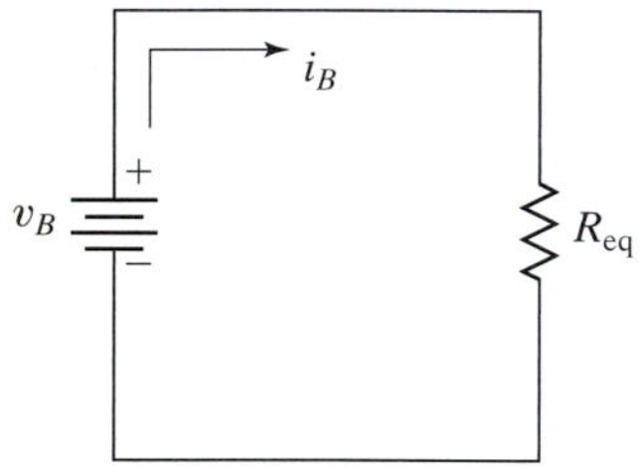

Figure 5.23c
Replacement of two resistors from Figure 5.23a with an equivalent resistor.

$$\frac{1}{R_{eq}} = \frac{1}{R_1} + \frac{1}{R_2} = \frac{R_1 + R_2}{R_1 R_2}$$

$$R_{eq} = \frac{R_1 R_2}{R_1 + R_2}$$

Because of KCL, we know that the same current that flows across the battery, i_B, flows through the equivalent resistance system. So we can assume the current flows clockwise in the loop with the equivalent resistor and apply KVL:

$$\sum_{\text{loop}} v_{\text{elements}} = v_B - v_R = 0$$

$$v_B = v_R = i_B R_{eq} = \frac{R_1 R_2}{R_1 + R_2} i_B$$

Using Ohm's law and the fact that the voltages across each resistor are equal to the voltage across the battery, we can now calculate the currents through the two resistive elements:

$$i_1 = \frac{v_B}{R_1} = \frac{R_2}{R_1 + R_2} i_B$$

$$i_2 = \frac{v_B}{R_2} = \frac{R_1}{R_1 + R_2} i_B$$

Since both $R_2/(R_1 + R_2)$ and $R_1/(R_1 + R_2)$ are always less than 1, the current through each branch is less than the current through the battery. Thus, this example illustrates how the circuit configuration divides the current. By selecting appropriate resistive elements for a current divider, you can design a circuit to meet a specified need. ■

KVL can be used together with KCL to solve more complex circuits. Often, neither law by itself provides enough equations to solve for the unknown variables in circuits. For example, in a given circuit, there will be more unknown currents than loops that provide linearly independent equations using KVL. To relate some of these currents to one another, KCL may be used. Ohm's law may also provide additional linearly independent equations.

EXAMPLE 5.9 Simultaneous Application of KCL and KVL

Problem: For the circuit in Figure 5.24a, calculate the voltage drop and current across each resistor.

Solution:

1. Assemble
   (a) Find: The voltage drop and current across each resistor.
   (b) Diagram: Figure 5.24b shows the circuit with three voltage loops that have arbitrarily assigned directions of current flow. System boundaries enclose the circuit elements inclusive of that loop.
2. Analyze
   (a) Assume:
   - The circuit is at steady-state.

   (b) Extra data: No extra data are needed.
   (c) Variables, notations, units:
   - Use A, V, Ω.
3. Calculate
   (a) Equations: In this circuit, there are elements that generate and elements that consume electrical energy. Depending on where the system boundary is drawn, the system may include Input and Output terms for the flow of electrical energy, so the differential form of the electrical energy accounting equation [5.4-3] may be used. However, we demonstrated that for any system at steady-state, the governing equation reduces to KVL:

$$\sum_{\text{loop}} v_{\text{elements}} = 0$$

   Since we assume the circuit is at steady-state, all the nodes in the circuit are at steady-state, so we can reduce the differential conservation of charge equation [5.3-18] to KCL:

$$\sum_k i_k - \sum_j i_j = 0$$

   To relate voltage and current, we use Ohm's law:

$$v = iR$$

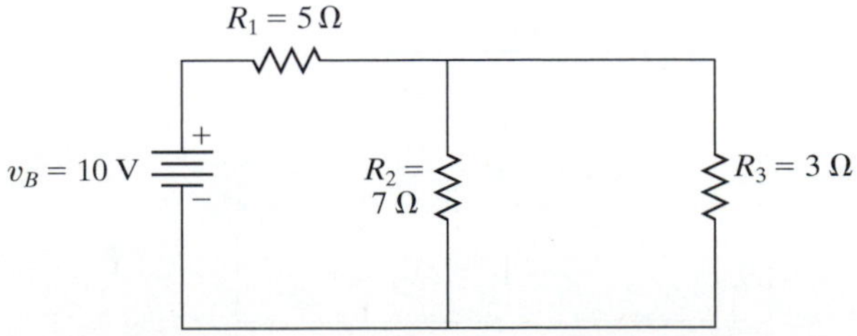

Figure 5.24a
Circuit with two parallel resistors connected in series with a third resistor.

(b) Calculate:

- Because each loop in the circuit is a steady-state system, we can write one KVL equation for each loop:

$$\begin{aligned} &\text{Loop 1:} \quad v_B - v_1 - v_2 = 0 \\ &\text{Loop 2:} \quad v_B - v_1 - v_3 = 0 \\ &\text{Loop 3:} \quad v_2 - v_3 = 0 \end{aligned}$$

  At first glance, it seems like there are three equations and three unknowns, perfect for solving for the three voltages. However, the equation for loop 3 may be obtained from the equations for loops 1 and 2, so only two of these three equations are linearly independent; so, the problem is currently underspecified. We can use KCL and Ohm's law to obtain more equations to solve for the unknown variables.

- We use the arbitrarily defined directions of current flow in Figure 5.24b and apply KCL to node A (Figure 5.24c):

$$i_1 - i_2 - i_3 = 0$$

- Now there are six unknowns and three linearly independent equations. The remaining three equations are obtained by applying Ohm's law across each of the resistors:

$$\begin{aligned} v_1 &= i_1 R_1 \\ v_2 &= i_2 R_2 \\ v_3 &= i_3 R_3 \end{aligned}$$

- With six equations and six unknowns, we can now solve for the voltage and current across each resistor. This problem is suited for solving in MATLAB by setting up a matrix with the six equations. Each equation must be rewritten such that all unknown values are located together on one side of the equal sign. The equations are rewritten with known values substituted in as follows:

$$\begin{aligned} v_B - v_1 - v_2 = 0 &\quad\Rightarrow\quad v_1 + v_2 = v_B = 10 \text{ V} \\ v_B - v_1 - v_3 = 0 &\quad\Rightarrow\quad v_1 + v_3 = v_B = 10 \text{ V} \\ i_1 = i_2 + i_3 &\quad\Rightarrow\quad i_1 - i_2 + i_3 = 0 \\ v_1 = i_1 R_1 &\quad\Rightarrow\quad v_1 - i_1 R_1 = v_1 - i_1(5\ \Omega) = 0 \\ v_2 = i_2 R_2 &\quad\Rightarrow\quad v_2 - i_2 R_2 = v_2 - i_2(7\ \Omega) = 0 \\ v_3 = i_3 R_3 &\quad\Rightarrow\quad v_3 - i_3 R_3 = v_3 - i_3(3\ \Omega) = 0 \end{aligned}$$

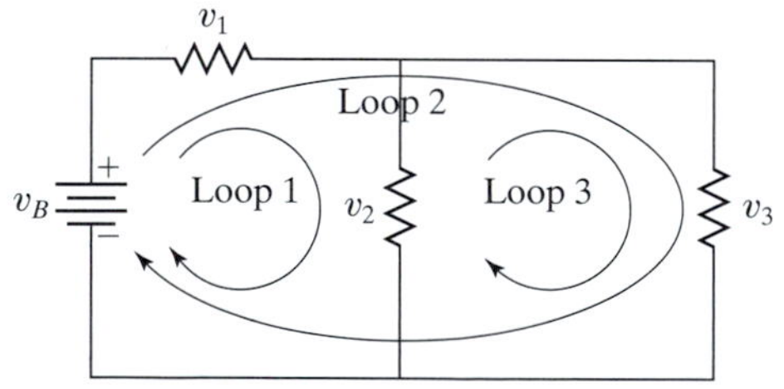

Figure 5.24b
Three possible loop configurations with arbitrarily defined directions of current flow.

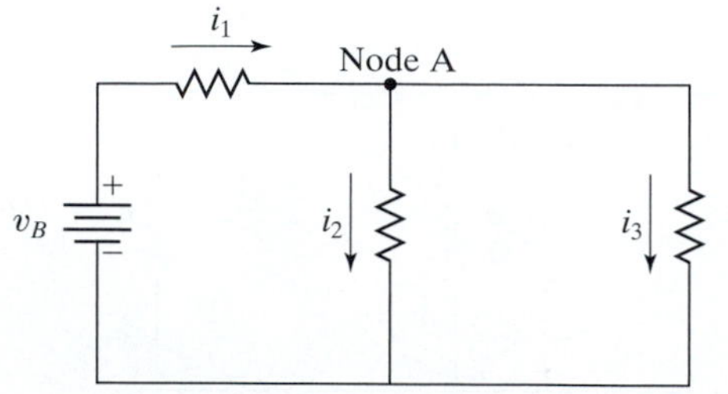

Figure 5.24c
Arbitrarily assigned directions of current flow.

These scalar equations can be represented by the following matrix equation in the form $A\vec{x} = \vec{y}$:

$$\begin{bmatrix} 1 & 1 & 0 & 0 & 0 & 0 \\ 1 & 0 & 1 & 0 & 0 & 0 \\ 0 & 0 & 0 & 1 & -1 & -1 \\ 1 & 0 & 0 & -5 & 0 & 0 \\ 0 & 1 & 0 & 0 & -7 & 0 \\ 0 & 0 & 1 & 0 & 0 & -3 \end{bmatrix} \begin{bmatrix} v_1 \\ v_2 \\ v_3 \\ i_1 \\ i_2 \\ i_3 \end{bmatrix} = \begin{bmatrix} 10 \\ 10 \\ 0 \\ 0 \\ 0 \\ 0 \end{bmatrix}$$

We can solve for the solution vector $\vec{x}$ in MATLAB using the following set of commands:

```
>> A = [1 1 0 0 0 0; 1 0 1 0 0 0; 0 0 0 1 -1 -1; 1 0 0 -5 0 0; 0 1 0 0 -7 0;
        0 0 1 0 0 -3];
>> y = [10; 10; 0; 0; 0; 0];
>> x = A\y
```

The solution is displayed as:

```
x =

    7.04
    2.96
    2.96
    1.41
    0.42
    0.99
```

4. Finalize
   (a) Answer: The voltage drops and currents across each resistor are: $v_1 = 7.04$ V, $v_2 = 2.96$ V, $i_1 = 1.41$ A, $i_2 = 0.42$ A, and $i_3 = 0.99$ A.
   (b) Check: Using back-substitution into the original six linearly independent equations confirms that these results are correct.

An alternate method to solve Example 5.9 is to reduce the parallel and series resistors to an equivalent resistor. R_2 and R_3 are in parallel; their effective resistance is calculated using equation [5.6-9] as:

$$\frac{1}{R_{eq23}} = \frac{1}{R_2} + \frac{1}{R_3} = \frac{1}{7\ \Omega} + \frac{1}{3\ \Omega}$$

$$R_{eq23} = 2.1\ \Omega$$

R_1 is in series with R_{eq23}. Using equation [5.6-8], the overall equivalent resistance is:

$$R_{eq} = R_1 + R_{eq23} = 5\ \Omega + 2.1\ \Omega = 7.1\ \Omega$$

Using Ohm's law, the current through the battery is:

$$i_B = \frac{v_B}{R_{eq}} = \frac{10\ \text{V}}{7.1\ \Omega} = 1.41\ \text{A}$$

Because of KCL, the value of i_1, the current through R_1, is also 1.41 A; this value is consistent with the previous solution.

The voltage drop across R_1 is:

$$v_1 = i_1 R_1 = (1.41\text{ A})(5\,\Omega) = 7.04\text{ V}$$

which is consistent with the previous solution. Using KVL around loops 1 and 3, the voltage drops across R_2 and R_3 can be calculated. Then, the current through these resistive elements can be determined using Ohm's law.

■

5.6.4 Einthoven's Law

The heart's contractions are stimulated by electrical impulses. When stimulated by an impulse, the current also spreads into the tissues adjacent to the heart, and a small proportion of the current reaches the body's surface. Electrodes can be placed on the skin on the limbs and chest to record the electrical potentials generated by the current. This recording of the heart's electrical activity, which traces voltages as a function of time, is called an **electrocardiogram** (ECG or EKG).

Monitoring the electrical activity of the heart can help in the diagnosis of cardiac disease and other disorders. ECGs provide information for the diagnosis of a variety of cardiac problems, including heart enlargement, congenital heart defects, arrhythmias, coronary occlusions (blockage of the heart's arteries), abnormal positions of the heart, heart inflammation (pericarditis or myocarditis), cardiac arrest, electrical conduction disturbances, and imbalances in the electrolytes that control heart activity.

The standard arrangement of the electrodes for an ECG is a configuration known as **Einthoven's triangle** (Figure 5.25a), which involves electrodes positioned on each of three limbs (right arm, left arm, left leg) so that an electric potential difference can be

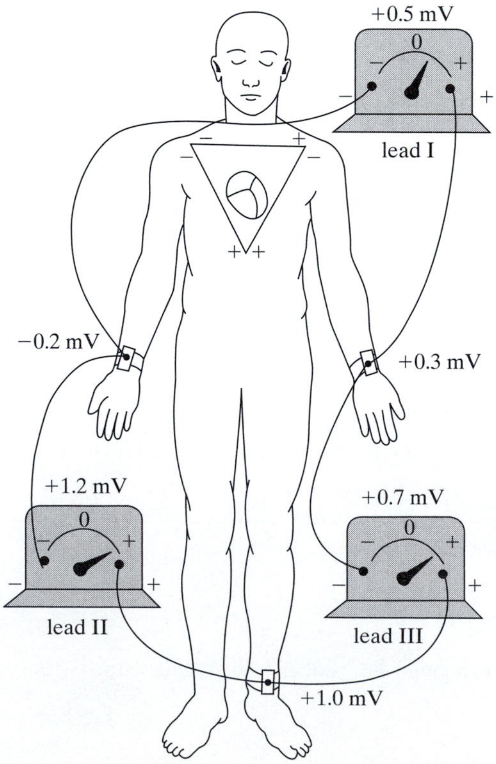

Figure 5.25a
Einthoven's triangle.
(*Source:* Guyton AC and Hall JE, *Textbook of medical Physiology*. Philadelphia Sounders, 2000.)

measured between two electrodes. The apices of a triangle drawn surrounding the heart represent the points at which the right arm, left arm, and left leg connect electrically with the fluids surrounding the heart (Figure 5.25b). Each electrode is connected by a wire to an electrocardiograph, which records the ECG signal. Each pair of electrodes forms a closed circuit with the electrocardiograph. For example, lead I is a scalar value equal to the electric potential difference between the left arm and the right arm.

Although this is not a traditional system of wires and resistors, the concepts previously developed can be applied. The electrodes forming the closed loop of Einthoven's triangle (right arm → left arm → left leg → right arm) have measurable voltage differences, so KVL can be applied. When tracing around the loop clockwise, leads I and III are voltage gains, whereas lead II is a voltage loss:

$$\sum_{\text{loop}} v_{\text{elements}} = \text{lead I} + \text{lead III} - \text{lead II} = 0 \qquad [5.6\text{-}23]$$

Einthoven's law follows from KVL and states that at any given time, if the potential of any two leads is known, then the third potential can be calculated. Equation [5.6-23] is usually rearranged to express lead II in terms of the others:

$$\text{lead I} + \text{lead III} = \text{lead II} \qquad [5.6\text{-}24]$$

Using the ECG, we can reconstruct the **cardiac vector**, which represents the mean depolarization of the heart at a given instant, to give us a three-dimensional view of the heart's action. The cardiac vector (Figure 5.26) can be trigonometrically calculated from the voltage data of any two of the three leads. As a vector, it has both a magnitude, which is usually measured in millivolts, and a direction. The magnitude of the vector plotted as a function of time can be used to identify stages of the cardiac cycle—the depolarization, contraction, and repolarization of the

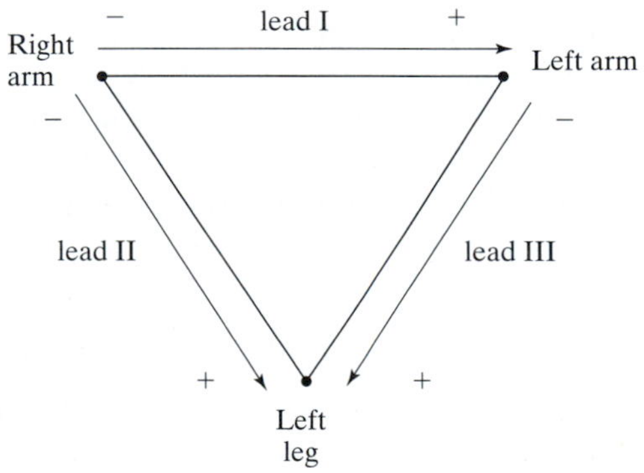

Figure 5.25b
Configuration of electrodes for recording electrocardiographs.

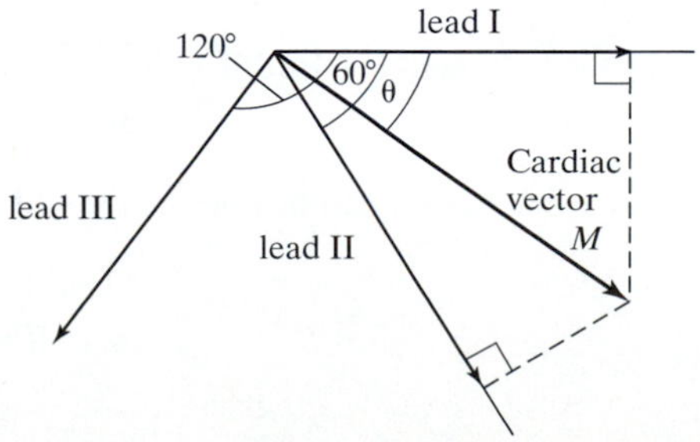

Figure 5.26
Calculating the magnitude and direction of the cardiac vector.

atria and ventricles. The direction can reveal several facts, such as the orientation of the heart and the relative strength of the left and right sides of the heart. Using this in conjunction with the ECG, a doctor can tell a lot about how the heart is acting in a patient.

The cardiac vector can be determined at any point in the cardiac cycle if the voltages of any two leads are known, but leads I and II are most commonly used. By convention, lead I lies on a horizontal (0°) axis, lead II falls 60° clockwise of lead I, and lead III falls 120° clockwise of lead I. The lengths of the projections of the leads are equal to the measured voltages. To draw the cardiac vector, lines are drawn perpendicularly to the leads at the heads of the projections on the leads. The cardiac vector starts at the point of origin of the three lead axes and ends at the point of intersection of the perpendicular lines (Figure 5.26). The angle of deflection of the cardiac vector clockwise from lead I is θ, which determines the direction that approximates the mean electrical axis of the heart, and the magnitude of the cardiac vector (M) is the length of this vector, which approximates the mean potential of the heart.

We can also use trigonometry to relate the potentials of the leads to the cardiac vector:

$$\text{lead I} = M\cos(\theta) \qquad [5.6\text{-}25]$$

$$\text{lead II} = M\cos(60^\circ - \theta) \qquad [5.6\text{-}26]$$

$$\text{lead III} = -M\cos(60^\circ + \theta) \qquad [5.6\text{-}27]$$

These equations relate the magnitude and the direction of the cardiac vector to the measured voltages of the leads.

It is important to highlight the distinction between vector and scalar quantities in these operations. Recall that a vector has both magnitude and direction. When measurements are taken across the leads, scalar values of voltages are obtained. Einthoven's law relates these scalar values; it does not involve vector addition. The lengths of the perpendicular vectors drawn from the leads to determine the cardiac vector are obtained using the magnitudes from the ECG. When translating these scalar quantities onto the lead formation (Figure 5.26) used to determine the cardiac vector, the predetermined angles between the leads allow you to write trigonometric equations to solve for the direction of the cardiac vector.

EXAMPLE 5.10 Application of Einthoven's Laws

Problem: At a particular point in time, lead I is measured as 0.82 mV and lead II is measured as 0.91 mV. Calculate the value of lead III, the magnitude of the cardiac vector, and the angle of deflection.

Solution: Lead III can be obtained by applying Einthoven's law:

$$\text{lead I} - \text{lead II} + \text{lead III} = 0.82\text{ mV} - 0.91\text{ mV} + \text{lead III} = 0$$
$$\text{lead III} = 0.09\text{ mV}$$

The cardiac vector can be obtained by solving two equations simultaneously:

$$\text{lead I} = M\cos(\theta)$$
$$\text{lead II} = M\cos(60^\circ - \theta)$$

$$0.82\ \text{mV} = M\cos(\theta)$$
$$0.91\ \text{mV} = M\cos(60^\circ - \theta)$$

You can either solve this system of equations by hand or use a computer program. MATLAB has a built-in solver called *solve* for systems of equations. The program performs calculations in radians, however, so the angle must be converted:

```
>> [M, θ] = solve('0.82= M *cos(θ)','0.91= M *cos(pi/3- θ)')

M =

[ -1.00281]
[ 1.00281]

θ =

[ -2.528]
[ 0.613]
```

MATLAB appears to return two different solutions to this system of equations. With scrutiny, however, these two solutions can be shown to be the same. The magnitude of the vector, M, is 1 mV, and the angle of deflection, θ, is 35°.

An alternate method to solve for M and θ is to do a graphical analysis using the template shown in Figure 5.26. ■

5.6.5 Hodgkin-Huxley Model

Another biological example of electrical phenomena is the relationship of ion flow, resistance, and potential in the **Hodgkin-Huxley model**. This model is based on Ohm's law and mathematically states that the flow of an ion y is proportional to the difference between the membrane potential and the equilibrium potential and inversely proportional to the membrane resistance:

$$i_y = \frac{v_m - v_{e,y}}{R_y} \qquad [5.6\text{-}28]$$

where i_y is the current of the ionic species y, v_m is the membrane potential, $v_{e,y}$ is the equilibrium potential of the ionic species y, and R_y is the resistance of the membrane to the flow of ionic species y. The potential difference, in this case, $v_m - v_{e,y}$, is the driving force for the movement of charged particles. The value of $v_{e,y}$, the equilibrium potential, can be calculated using the Nernst equation. This form of the Hodgkin-Huxley model is often used to describe the flow of sodium, potassium, chloride, and other ions across the cellular membrane during an action potential.

Membrane **conductance** (g) is the inverse of the membrane resistance. The Hodgkin-Huxley model can also be written as:

$$i_y = g_y(v_m - v_{e,y}) \qquad [5.6\text{-}29]$$

where g_y is the membrane conductance of the ionic species y. If the quantity $(v_m - v_{e,y})$ is greater than zero, the direction of transport of the ion will be out of the cell; if the quantity is less than zero, the ion is transported into the cell. Thus, the flow of a specific ion into or out of a cell across the membrane generates a current that can be related through the Hodgkin-Huxley model.

When a membrane is at steady-state (i.e., not undergoing depolarization or repolarization), several charged species exhibit a concentration gradient across the membrane. Because the species are charged, this gradient gives rise to a potential difference across the membrane. To calculate the membrane potential, $v_{e,y}$, from a particular concentration gradient for a specific species, the **Nernst equation** can be used:

$$v_{e,y} = \frac{RT}{FZ_y}\ln\left(\frac{[y_o]}{[y_i]}\right) \qquad [5.6\text{-}30]$$

where $v_{e,y}$ is the equilibrium potential for y, a charged species; R is the gas constant; T is the absolute temperature; F is Faraday's constant (96,485 coulombs/mole); Z_y is the valence of y; $[y_o]$ is the concentration of y outside the cell; and $[y_i]$ is the concentration of y inside the cell.

The potential acting across the cell membrane on each ion can be analyzed. For example, chloride ions are present in higher concentrations in the extracellular fluid than in the cell interior, and they tend to diffuse along this concentration gradient into the cell. However, the interior of the cell is negative relative to the exterior, and chloride ions are pushed out of the cell along this electrical gradient. Equilibrium is reached when the influx and efflux of chloride ions are equal. For example, the Nernst equation for the chloride ion is:

$$v_{e,\mathrm{Cl}^-} = \frac{RT}{FZ_{\mathrm{Cl}^-}}\ln\left(\frac{[\mathrm{Cl}_o^-]}{[\mathrm{Cl}_i^-]}\right) \qquad [5.6\text{-}31]$$

Converting from the natural log to the base 10 log and replacing some of the constants with numerical values, the equation becomes:

$$v_{e,\mathrm{Cl}^-} = 61.5\log\left(\frac{[\mathrm{Cl}_i^-]}{[\mathrm{Cl}_o^-]}\right)\ \text{mV at } 37^\circ\text{C} \qquad [5.6\text{-}32]$$

By making these substitutions, we obtain an expression for the equilibrium membrane potential in millivolts. Note that in converting to the simplified expression, the concentration ratio is reversed, because the -1 valence of Cl^- has been removed from the expression. Table 5.2 shows typical intracellular and extracellular concentrations for a few key ions. Since chloride ions are present at a higher concentration in the extracellular fluid than in the cell interior, the equilibrium membrane potential is negative and has a calculated value of approximately -70 mV. Similar expressions and calculations can be developed for the potassium and sodium ions.

EXAMPLE 5.11 Flow of Sodium Ions During Depolarization

Problem: Using the Hodgkin-Huxley model, calculate the flow of sodium ions through voltage-gated sodium channels in a membrane at the onset of depolarization. Assume a membrane

TABLE 5.2

Intra- and Extracellular Ion Concentrations of Mammalian Spinal Motor Neurons*

	Concentration (mmol/L H_2O)		Approximate
Ion	Inside cell	Outside cell	equilibrium potential (mV)
Na^+	15.0	150.0	+60
K^+	150.0	5.5	−90
Cl^-	9.0	125.0	−70

*Data from Ross G, ed. *Essentials of Human Physiology*. Chicago: Year Book Med Pub, 1978.

in the human body has a surface area of 1μm^2 and 75 sodium channels. The threshold membrane potential for sodium channels is -65 mV. The membrane resistance of one sodium channel is 250 GΩ.

Solution: We can use the Hodgkin-Huxley model given in Equation [5.6-28] to solve for the current generated by the flow of sodium ions. At the start of depolarization, the membrane potential, v_m, is given as -65 mV. We use the Nernst equation to find the membrane equilibrium potential of sodium. Values for the intracellular and extracellular concentrations of sodium are given in Table 5.2. Assuming a body temperature of 37°C, the membrane equilibrium potential of sodium is calculated:

$$v_{e,Na^+} = 61.5 \log\left(\frac{[Na_o^+]}{[Na_i^+]}\right) \text{mV} = 61.5 \log\left(\frac{150 \text{ mM}}{15 \text{ mM}}\right) \text{mV} = 61.5 \text{ mV}$$

Substituting all values into the Hodgkin-Huxley model gives:

$$i_{Na^+} = \frac{v_m - v_{e,Na^+}}{R_{Na^+}} = \frac{-65 \text{ mV} - 61.5 \text{ mV}}{\dfrac{250 \text{ G}\Omega}{\text{channel}}} = -5.1 \times 10^{-10} \frac{\text{mA}}{\text{channel}}$$

To find the total ion flow for the surface area of the membrane (1 μm^2) at the onset of depolarization:

$$i_{Na^+} = -5.1 \times 10^{-10} \frac{\text{mA}}{\text{channel}} \, 75 \text{ channels} = -3.8 \times 10^{-8} \text{ mA}\left(\frac{10^{12} \text{ pA}}{10^3 \text{ mA}}\right) = -38 \text{ pA}$$

The flow of sodium ions through 75 voltage-gated sodium channels in a membrane with a surface area of 1 μm^2 at the start of depolarization is -38 pA. Since the difference in potentials, and hence the current, is less than zero, the ions are transported into the cell at the start of depolarization. ■

The cell membrane can be modeled as a circuit. The behavior of key ions involved in generating an action potential, as well as other features such as charge storage, can be included in the model, depending on the desired complexity. Here, we model the charge flow of sodium, potassium, and chloride ions across a cell membrane that is in equilibrium.

The flow of each ionic species is modeled with a series combination of a resistor and an electric potential equal to the ion's Nernst potential. Since the ions flow in parallel through the cell membrane, it makes sense that the elements in the circuit model are also in parallel (Figure 5.27). At equilibrium, there is no net flow of charged ions (i.e., current) across the membrane. Applying KCL to node A gives:

$$i_{K^+} + i_{Na^+} + i_{Cl^-} = 0 \qquad [5.6\text{-}33]$$

Thus, the net current across the membrane is equal to zero.

At equilibrium, there is a potential across the membrane, v_m. Given the equilibrium potential of an ion y, $v_{e,y}$, and the Hodgkin-Huxley model, we can substitute for each current given in equation [5.6-33] using equation [5.6-28] for each ion y:

$$\frac{v_m - v_{e,K^+}}{R_{K^+}} + \frac{v_m - v_{e,Na^+}}{R_{Na^+}} + \frac{v_m - v_{e,Cl^-}}{R_{Cl^-}} = 0 \qquad [5.6\text{-}34]$$

We can also use equation [5.6-29] to substitute for terms in equation [5.6-33]:

$$g_{K^+}(v_m - v_{e,K^+}) + g_{Na^+}(v_m - v_{e,Na^+}) + g_{Cl^-}(v_m - v_{e,Cl^-}) = 0 \qquad [5.6\text{-}35]$$

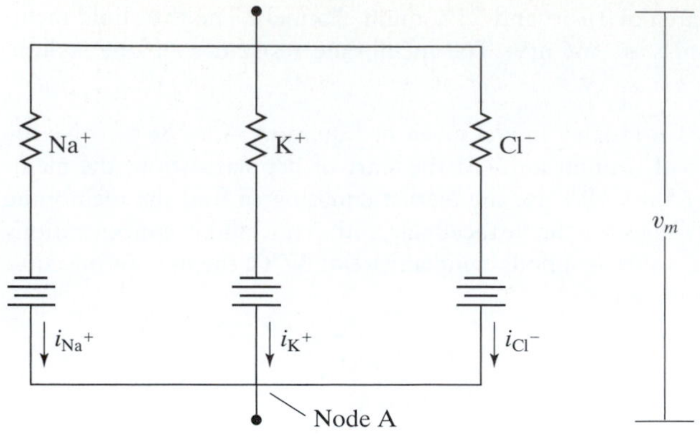

Figure 5.27
Circuit model of the flow of sodium, potassium, and chloride ions through the cell membrane.

when substituting the membrane conductance for the inverse of the membrane resistance. Solving for the membrane potential gives:

$$v_m = \frac{\sum_y v_{e,y} g_y}{\sum_y g_y} \qquad [5.6\text{-}36]$$

Thus, the membrane potential can be calculated using the Nernst equilibrium potentials for the various ionic species and their respective conductances. The SI unit of conductance is the siemens (S), which is the inverse of resistance in the unit of ohms.

EXAMPLE 5.12 **Equilibrium Membrane Potential**

Problem: Estimate the membrane potential at equilibrium for a membrane containing sodium, potassium, and chloride ions. Use the intracelluar and extracellular ion concentrations from Table 5.2. The following conductances are given: $g_{Na^+} = 1$ pS, $g_{K^+} = 33$ pS, and $g_{Cl^-} = 3$ pS.

Solution: We use the Nernst equation [5.6-30] to calculate the equilibrium Nernst potenial for the three ions. For potassium:

$$v_{e,K^+} = \frac{RT}{FZ_{K^+}} \ln\left(\frac{[K_o^+]}{[K_i^+]}\right) = 61.5 \log\left(\frac{5.5\text{ mM}}{150\text{ mM}}\right) = -88\text{ mV}$$

In a similar manner, v_{e,Na^+} is calculated as 61.5 mV, and v_{e,Cl^-} as −70.3 mV. Using equation [5.6-36], the membrane potential is calculated as:

$$v_m = \frac{\sum_y v_{e,y} g_y}{\sum_y g_y} = \frac{v_{e,Na^+} g_{Na^+} + v_{e,K^+} g_{K^+} + v_{e,Cl^-} g_{Cl^-}}{g_{Na^+} + g_{K^+} + g_{Cl^-}}$$

$$v_m = \frac{61.5\text{ mV}(1\text{ pS}) - 88\text{ mV}(33\text{ pS}) - 70.3\text{ mV}(3\text{ pS})}{1\text{ pS} + 33\text{ pS} + 3\text{ pS}} = -82.5\text{ mV}$$

This calculated value is close to the known resting potential of a motor neuron. Note that value of the membrane potential is determined largely by the equilibrium potential of the potassium since its conductance is an order of magnitude higher than the other two ions. ■

While this model is helpful for elucidating simple behavior, it does not capture the time-dependent nature of the depolarization and repolarization of a cell membrane during an action potential. In addition, the conductance of ions through the membrane changes as a function of time. Circuit models of increased complexity that include capacitors and other time-dependent elements have been developed to more accurately capture the dynamic nature of an action potential. In Example 5.16, we consider a slightly more complicated model.

5.7 Dynamic Systems—Focus on Charge

In a dynamic or unsteady-state system, charge accumulates in the system, so the initial and final conditions are not the same. Recall the differential forms of the charge accounting equations, which are appropriate to use when rates of charge are given:

$$\text{Positive:} \quad \sum_k \dot{q}_{+,k} - \sum_j \dot{q}_{+,j} + \dot{q}_{+,\text{gen}} - \dot{q}_{+,\text{cons}} = \frac{dq_+^{\text{sys}}}{dt} \qquad [5.7\text{-}1]$$

$$\text{Negative:} \quad \sum_k \dot{q}_{-,k} - \sum_j \dot{q}_{-,j} + \dot{q}_{-,\text{gen}} - \dot{q}_{-,\text{cons}} = \frac{dq_-^{\text{sys}}}{dt} \qquad [5.7\text{-}2]$$

$$\text{Net:} \quad \sum_k i_k - \sum_j i_j = \frac{dq^{\text{sys}}}{dt} \qquad [5.7\text{-}3]$$

In a dynamic system, the rate of charge (positive, negative, or net) stored in or depleted from the system is nonzero. Thus, the term on the right-hand side of the equation is nonzero.

The integral charge accounting equation may be appropriate for dynamic systems when the condition of the system is considered between two discrete time points. Recall the integral forms of the charge accounting equations:

$$\text{Positive:} \quad \int_{t_0}^{t_f} \sum_k \dot{q}_{+,k}\, dt - \int_{t_0}^{t_f} \sum_j \dot{q}_{+,j}\, dt + \int_{t_0}^{t_f} \dot{q}_{+,\text{gen}}\, dt - \int_{t_0}^{t_f} \dot{q}_{+,\text{cons}}\, dt = \int_{t_0}^{t_f} \frac{dq_+^{\text{sys}}}{dt} dt \qquad [5.7\text{-}4]$$

$$\text{Negative:} \quad \int_{t_0}^{t_f} \sum_k \dot{q}_{-,k}\, dt - \int_{t_0}^{t_f} \sum_j \dot{q}_{-,j}\, dt + \int_{t_0}^{t_f} \dot{q}_{-,\text{gen}}\, dt - \int_{t_0}^{t_f} \dot{q}_{-,\text{cons}}\, dt = \int_{t_0}^{t_f} \frac{dq_-^{\text{sys}}}{dt} dt \qquad [5.7\text{-}5]$$

$$\text{Net:} \quad \int_{t_0}^{t_f} \sum_k i_k\, dt - \int_{t_0}^{t_f} \sum_j i_j\, dt = \int_{t_0}^{t_f} \frac{dq^{\text{sys}}}{dt} dt \qquad [5.7\text{-}6]$$

Recall that since net charge is never generated or consumed, the Generation and Consumption terms are eliminated.

Capacitors are electrical elements that use a pair of oppositely charged conductors to store charge. A standard capacitor consists of a pair of parallel metal plates, such as aluminum or copper, separated by a small distance that is filled with a nonconducting material, such as air. Capacitors are commonly found in dynamic or unsteady-state electrical systems. The circuit symbol for a capacitor is shown in Figure 5.28a.

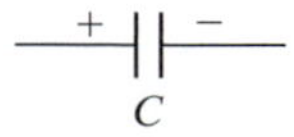

Figure 5.28a
Circuit symbol for a capacitor.

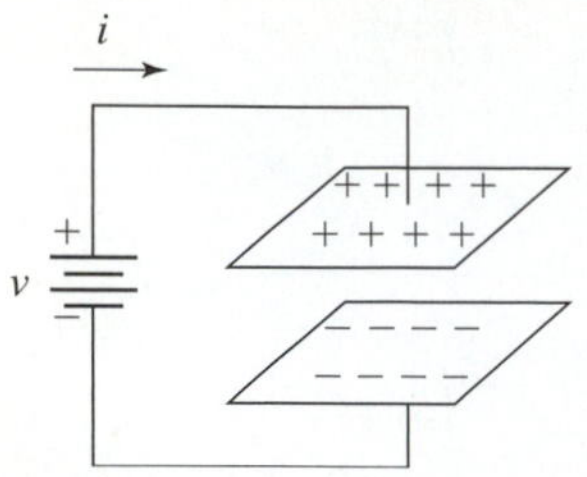

Figure 5.28b
Capacitor plates with separated positive and negative charges.

If a battery or other source of electrical energy provides a flow of charges to a capacitor, the capacitor quickly becomes charged. The voltage source provides the work to transfer charge (usually electrons) from one conductor to the other. When fully charged, positive charge, q_+, has accumulated on one plate, and an equal amount of negative charge, q_-, has accumulated on the other plate. Note that the net charge of a capacitor is always zero. Figure 5.28b shows capacitor plates with positive and negative charge.

The separation of positive and negative charges in a capacitor creates an electric field. Because the distance between plates for a given capacitor is constant, the electric field between the plates is proportional both to the voltage, v, and the charge, q, that has been transferred across the plates. In an ideal capacitor, the v across the capacitor is directly proportional to the magnitude of the charge q on the capacitor:

$$q = Cv_c \qquad [5.7\text{-}7]$$

where C is the capacitance and v_c is the voltage, or potential difference, across the capacitor element. **Capacitance** is a characteristic of a given capacitor that depends on the structure and dimensions of the capacitor. The unit of capacitance (C) is the farad (F), which is equivalent to coulomb/volt. The dimension of capacitance is $[L^{-2}M^{-1}t^4I^2]$.

For a parallel-plate capacitor whose plates have area A and are separated by distance d, the capacitance, C, is given by:

$$C = \frac{\varepsilon_0 A}{d} \qquad [5.7\text{-}8]$$

where ε_0 is the permittivity constant. The constant ε_0 is equal to 8.85×10^{-12} $C^2/(N \cdot m^2)$ (or F/m) for capacitors with vacuum between the plates.

Most capacitors contain an insulating sheet called a dielectric between the plates. Common materials used as dielectrics in capacitors include air, glass, paper, polyethylene, polystyrene, Teflon, and water. The dielectric replaces the constant ε_0 in equation [5.7-8] with a material property, affording more flexibility in designing the capacitance of a capacitor. Furthermore, a dielectric allows the plates to be placed closer together without touching, and the decreased distance between the plates results in an increased capacitance.

EXAMPLE 5.13 Charging of a Capacitor

Problem: Current enters a capacitor plate at a rate of $i = \alpha e^{-\beta t}$, where α is 5.0 A and β is 25 1/s. If no net charge is initially on the plate, how much net positive charge is deposited onto the plate after 50 ms?

Solution: We assume that charging a capacitor does not cause any reactions. We also assume current does not leave the system, which is defined as the capacitor plate (Figure 5.29). Because we are given the current and a specific time interval, we can use the integral form of the conservation of net charge equation [5.7-6]:

$$\int_{t_0}^{t_f} \sum_k i_k \, dt - \int_{t_0}^{t_f} \sum_j i_j \, dt = \int_{t_0}^{t_f} \frac{dq^{sys}}{dt} dt$$

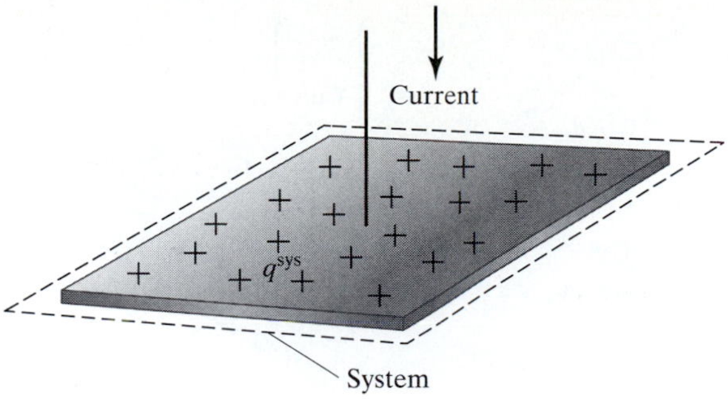

Figure 5.29
Charging of the positive plate of a capacitor.

No current flows out of the system, and only one current flows into the system, so we can reduce the governing equation to:

$$\int_{t_0}^{t_f} i_k \, dt - \int_{t_0}^{t_f} \frac{dq^{sys}}{dt} dt = \int_{q_0^{sys}}^{q_f^{sys}} dq^{sys}$$

Substituting the given values into the reduced equation to solve for the charge, q_f^{sys}, at $t = 50$ ms gives:

$$\int_0^{q_f^{sys}} dq^{sys} = \int_{t_0}^{t_f} i_k \, dt = \int_0^{0.05 \text{ s}} (5e^{-(25 \text{ 1/s})t} \text{ A}) \, dt = \int_0^{0.05 \text{ s}} \left(5e^{-(25 \text{ 1/s})t} \frac{\text{C}}{\text{s}} \right) dt$$

$$q_f^{sys} = (-0.2e^{-(25 \text{ 1/s})t} \text{ C})\Big|_0^{0.05 \text{ s}} = -0.057 \text{ C} - (-0.2 \text{ C}) = 0.14 \text{ C}$$

The charge on the plate after 50 ms is 0.14 C. ■

EXAMPLE 5.14 Discharge of a Defibrillator

Problem: Fibrillation is a coronary event in which rapid, uncoordinated twitching of small muscle fibers in the heart replaces the normal rhythmic contraction, causing the heart to stop pumping blood. Without rapid intervention, fibrillation may result in brain damage or even cardiac arrest, also known as a heart attack. During fibrillation, 10% of the ability to restart the heart is lost every minute.

A defibrillator is an electronic device that delivers an electric shock to the fibrillating heart to restore the normal rhythm. A common type is the capacitive-discharge defibrillator, which uses a capacitor to store and quickly deliver charge to a patient. The charge delivered to the heart as an electric shock can sometimes restore the heart's normal electrical activity and beating.

A fully charged capacitor is switched on at $t = 0$ so that the current out of the defibrillator is:

$$i = 40e^{-(500 \text{ 1/s})t} \text{ A}$$

Assuming that the capacitor cannot be charged while it is being discharged, how long will it take for the capacitor to be 99% discharged? Assume that the capacitor has 0.080 C of charge at $t = 0$.

Solution:

1. Assemble
   (a) Find: Amount of time needed for the capacitor to be 99% discharged.
   (b) Diagram: The system is drawn in Figure 5.30.

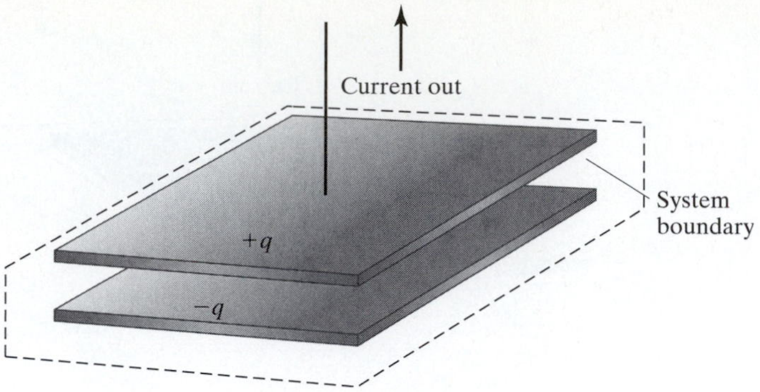

Figure 5.30
Current leaving system during defibrillator discharge.

2. Analyze
   (a) Assume:
   - Current does not enter the system.
   - Discharging the defibrillator does not cause any reactions.

   (b) Extra data: No extra data are needed.
   (c) Variables, notations, units:
   - Use C, s.
3. Calculate
   (a) Equations: Since current is given, as well as a discrete time period, the integral form of the conservation of net charge equation [5.7-6] can be used:

$$\int_{t_0}^{t_f} \sum_k i_k \, dt - \int_{t_0}^{t_f} \sum_j i_j \, dt = \int_{t_0}^{t_f} \frac{dq^{\text{sys}}}{dt} dt = \int_{q_0^{\text{sys}}}^{q_f^{\text{sys}}} dq^{\text{sys}}$$

   (b) Calculate:
   - Because we assume no current can flow into the system drawn in Figure 5.30, we can reduce the governing equation to:

$$-\int_{t_0}^{t_f} i_j \, dt = \int_{q_0^{\text{sys}}}^{q_f^{\text{sys}}} dq^{\text{sys}}$$

   - Substituting the given values into the reduced equation to solve for the time needed for the capacitor to discharge gives:

$$\int_{q_0^{\text{sys}}}^{q_f^{\text{sys}}} dq^{\text{sys}} = -\int_{t_0}^{t_f} i_j \, dt = \int_0^t -40e^{-(500\ 1/\text{s})t} \frac{\text{C}}{\text{s}} dt$$

$$q_f^{\text{sys}} - q_0^{\text{sys}} = \left(0.080e^{-(500\ 1/\text{s})t}\ \text{C}\right)\Big|_0^t = 0.080e^{-(500\ 1/\text{s})t}\ \text{C} - 0.080\ \text{C}$$

   - Initially, the capacitor has a charge of 0.080 C. At the time of interest, the capacitor is 99% discharged, so it holds only 1% of its initial charge and thus $q_f^{\text{sys}} = 0.01q_0^{\text{sys}}$. We can now substitute q_f^{sys} into the integrated equation in which we previously solved for the time needed for the capacitor to discharge:

$$q_f^{\text{sys}} - q_0^{\text{sys}} = 0.080e^{-(500\ 1/\text{s})t}\ \text{C} - 0.080\ \text{C}$$

$$0.01q_0^{\text{sys}} - q_0^{\text{sys}} = 0.080e^{-(500\ 1/\text{s})t}\ \text{C} - 0.080\ \text{C}$$

$$-0.99q_0^{\text{sys}} = -0.99\ (0.080\ \text{C}) = 0.080e^{-(500\ 1/\text{s})t}\ \text{C} - 0.080\ \text{C}$$

$$-0.0792\ \text{C} = 0.080e^{-(500\ 1/\text{s})t}\ \text{C} - 0.080\ \text{C}$$

$$0.01 = e^{-(500\ 1/\text{s})t}$$

$$t = 0.0092\ \text{s}$$

4. Finalize
   (a) Answer: It takes 9.2 ms for the capacitor to become 99% discharged.
   (b) Check: According to literature (Webster, *Medical Instrumentation: Application and Design*, 1998), the capacitor in a capacitive-discharge defibrillator should discharge in approximately 10 ms. Our value of 9.2 ms matches well with this estimate, so it is a reasonable answer. ■

Essentially all cells of the body have electrical potentials across their membranes. Furthermore, some cells, such as nerve and muscle cells, are "excitable"—that is, capable of self-generation of electrochemical impulses at their membranes. In many ways, cellular membranes act like capacitors when it comes to storing charge on membrane surfaces and transporting charge through the membranes.

Healthy, intact nerves transmit information from the brain to various parts of the body and vice versa by conducting electrochemical signals called **action potentials**. Nerve cells have a resting membrane potential ranging from −70 mV to −90 mV relative to the outside of the cell. Action potentials involve rapid (~1 ms) changes in membrane potential from negative to positive (depolarization) and back to negative again (repolarization) (Figure 5.31). As the signal passes through each region of the axon, sodium channels in the membrane open, and the interior is flooded with sodium ions. In this stage of the action potential, also known as **depolarization**, the potential rises to +35 mV across the membrane. A few ten-thousandths of a second later, the sodium channels begin to close and potassium channels open. The efflux of potassium ions causes **repolarization** and lowers the potential of the cell membrane to −110 mV. Eventually, the ion gradients and resting potential are reestablished at −90 mV so that the neuron can fire again.

An action potential elicited at any point on an excitable membrane typically induces the excitation of adjacent portions of the membrane and later adjacent cells, resulting in propagation of the action potential. In this way, the action potential moves along the length of the nerve fiber to deliver a signal either to another nerve or to an organ or muscle. Thus, action potentials make possible long-distance communication of signals carrying sensory or motor information in the nervous system. During an action potential, the cell is often modeled as a dynamic system, since the concentration gradients of species are changing.

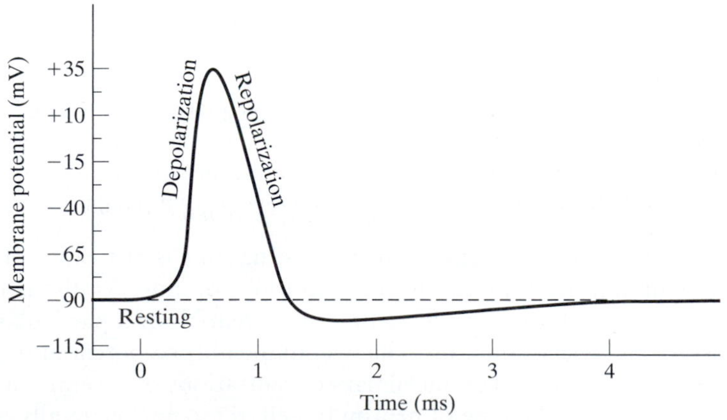

Figure 5.31
Membrane potential during an action potential. (*Source:* Modified from Guyton AC and Hall JE, *Textbook of Medical Physiology*. Philadelphia: Saunders, 2000.)

EXAMPLE 5.15 Charge Accumulation During an Action Potential

Problem: Consider a cell during the depolarization and repolarization phases. A system is defined such that it includes a patch of the membrane measuring 1 μm^2 and the interior of the cell directly below this patch. During the depolarization phase, which lasts 0.1 ms, sodium ions are estimated to flow into the neuron at a rate of 7.8×10^{15} ions/($cm^2 \cdot s$). During the repolarization phase, which lasts 0.2 ms, potassium ions are estimated to flow out of the neuron at a rate of 4.5×10^{15} ions/($cm^2 \cdot s$). After both phases, what is the accumulation of positive charge inside the cell?

Solution: The given inlet and outlet rates are dependent on the area over which the ions flow, so it is necessary to calculate the rates for our system:

$$\text{Na}^+: \quad \left(7.8 \times 10^{15} \frac{\text{ions}}{\text{cm}^2 \cdot \text{s}}\right)(1\ \mu\text{m}^2)\left(\frac{1\ \text{cm}^2}{10^8\ \mu\text{m}^2}\right) = 7.8 \times 10^7\ \frac{\text{ions}}{\text{s}}$$

Both sodium and potassium ions have a charge of +1 elementary charge or $+1.6 \times 10^{-19}$ C. We convert the ion flux to current:

$$\text{Na}^+: \quad \left(7.8 \times 10^{7} \frac{\text{ions}}{\text{s}}\right)\left(1.6 \times 10^{-19} \frac{\text{C}}{\text{ion}}\right) = 1.25 \times 10^{-11}\ \frac{\text{C}}{\text{s}}$$

Similar calculations can be performed to determine the potassium ion flux, which gives 7.2×10^{-12} C/s. This conversion gives us rates of charge flow in our system.

Since a discrete time period is also specified, we can use the integral accounting equation for positive charge [5.7-4]:

$$\int_{t_0}^{t_f} \sum_k \dot{q}_{+,k}\, dt - \int_{t_0}^{t_f} \sum_j \dot{q}_{+,j}\, dt + \int_{t_0}^{t_f} \dot{q}_{+,\text{gen}}\, dt - \int_{t_0}^{t_f} \dot{q}_{+,\text{cons}}\, dt = \int_{t_0}^{t_f} \frac{dq_+^{\text{sys}}}{dt} dt = \int_{q_{+,0}^{\text{sys}}}^{q_{+,f}^{\text{sys}}} dq_+^{\text{sys}}$$

During an action potential, charges only move through the membrane; they are neither generated nor consumed. Sodium ions flow in, and potassium ions flow out. Consequently, we can reduce equation [5.7-4] and substitute our known variables to solve for the final charge of the system:

$$\int_{t_0}^{t_f} \sum_k \dot{q}_{+,k}\, dt - \int_{t_0}^{t_f} \sum_j \dot{q}_{+,j}\, dt = \int_{q_{+,0}^{\text{sys}}}^{q_{+,f}^{\text{sys}}} dq_+^{\text{sys}}$$

$$\int_0^{0.0001\ \text{s}} (1.25 \times 10^{-11}\ \text{A})\, dt - \int_{0.0001\ \text{s}}^{0.0003\ \text{s}} (7.2 \times 10^{-12}\ \text{A})\, dt = q_{+,f}^{\text{sys}} - q_{+,0}^{\text{sys}} = q_{+,\text{acc}}^{\text{sys}}$$

$$q_{+,\text{acc}}^{\text{sys}} = (1.25 \times 10^{-11}\ \text{A})(0.0001\ \text{s}) - (7.2 \times 10^{-12}\ \text{A})(0.0003\ \text{s} - 0.0001\ \text{s})$$

$$q_{+,\text{acc}}^{\text{sys}} = -1.9 \times 10^{-16}\ \text{C}$$

During the course of depolarization and repolarization, 1.9×10^{-16} C of positive charge exits a 1-μm^2 neuronal membrane patch. To transmit another signal, the cell must return its resting potential to −90 mV. During this time, the net accumulation of positive charge will reach zero through the use of active sodium-potassium pumps. ■

A membrane system during the resting state is at steady-state, as ion pumps help to maintain the appropriate concentrations of ions. As discussed in Section 5.6, the steady-state or equilibrium membrane potential for a particular species is calculated using the Nernst equation. The overall membrane potential of a cell is a function of the intracellular and extracellular concentrations of several ions (see Example 5.12). Depending on the type of neuron or cell, these ions generally include sodium, chlorine, potassium, and calcium.

TABLE 5.3

Permeability Coefficients for Cell Membrane of Frog Skeletal Muscle*

Ion	Permeability Coefficient (cm/s)
A^{-}†	~0
Na^{+}	2×10^{-8}
K^{+}	2×10^{-6}
Cl^{-}	4×10^{-6}

*Values are equivalents diffusing through 1 cm^2 under specified conditions. For the purposes of comparison, P_{K^+} in water is 10.

†A^{-} is a generic anion. (Data from Hodgkin AL and Horowicz P, "The influences of potassium and chloride ions on the membrane potential of single muscle fibers," *J Physiol* 1959, 148:127–60.)

Cell membranes are selectively impermeable to most intracellular proteins and organic anions, which make up most of the intracellular anions. However, the membranes are moderately permeable to sodium ions and freely permeable to chloride and potassium ions. The permeability to potassium ions is fifty to a hundred times greater than that to sodium ions. This permeability allows for action potentials and for specific changes in the membrane potential that are used for cellular signaling.

During the firing of an action potential, the membrane becomes more permeable to certain ions. As ions flow through the membrane, the system becomes dynamic, and some charges accumulate on each side of the membrane to change the potential. It should be noted, however, that very few ions need to cross the membrane to change the membrane potential. The effect is extremely localized, and therefore, the overall concentrations of ions inside and outside the cell remain roughly constant. The variation of the cell membrane permeability for different ions allows for a specific response during the action potential. Thus, during an action potential, the value of v_m changes as a function of time. Table 5.3 for the cell membrane of frog skeletal muscle is representative of ion permeabilities. It should be noted that permeability of these ions in the membrane, although appreciable, is a fraction of their permeability in water.

5.8 Dynamic Systems—Focus on Electrical Energy

In addition to charge, capacitors also store electrical energy. As charge builds on the plates, the separation of charge generates an electric field. The field results in an electric force that opposes the accumulation of more charge. Consequently, work must be done to move additional charge onto the plates. As current flows into the capacitor, the electric field becomes stronger. If the voltage source is removed from the circuit, current rapidly flows through the circuit from the positively charged capacitor plate to the negatively charged plate, thereby discharging the capacitor. This discharging occurs because the voltage source no longer provides the work required to maintain the charge separation on the capacitor plates. The electric field disappears, and the electrical energy stored in the electric field is dissipated, typically as heat in the resistors of the circuit.

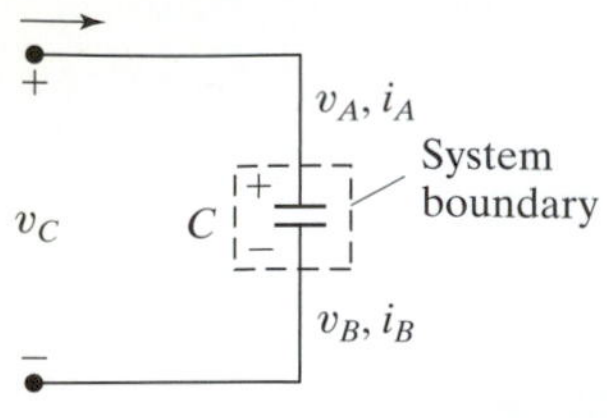

Figure 5.32
Circuit with a charging capacitance element.

Consider a capacitance element that stores energy in a system (Figure 5.32). Since the net charge of the two plates at all times is always zero for the system, no net charge accumulates in the system. If we consider the capacitor as a node and apply KCL, we have:

$$i_A - i_B = 0 \qquad [5.8\text{-}1]$$

$$i_A = i_B = i \qquad [5.8\text{-}2]$$

Recall the electrical energy accounting equation for a system without generation or consumption:

$$\sum_k i_k v_k - \sum_j i_j v_j = \frac{dE_E^{sys}}{dt} \qquad [5.8\text{-}3]$$

Substituting the equation for the difference in potential of a capacitor gives:

$$i(v_A - v_B) = \frac{dE_{E,C}^{sys}}{dt} \qquad [5.8\text{-}4]$$

Recall equation [5.7-7], in which the voltage drop across a capacitor is equal to the charge divided by the capacitance. We can then simplify the unsteady-state equation describing a system with a capacitor such that:

$$i(v_A - v_B) = iv_c = \frac{iq}{C} = \frac{dE_{E,C}^{sys}}{dt} \qquad [5.8\text{-}5]$$

where v_c is the voltage difference across the capacitor.

Consider only the positively charged top plate. If the capacitor is being charged, the current is the rate of change of charge:

$$i = \frac{dq}{dt} = \frac{d}{dt}(v_c C) \qquad [5.8\text{-}6]$$

For a system with constant capacitance, the current across the capacitor is:

$$i = C\frac{dv_c}{dt} \qquad [5.8\text{-}7]$$

Consequently, we can substitute the relationship between current and charge to get the rate of change of electrical energy:

$$\frac{dE_{E,C}^{sys}}{dt} = \frac{q}{C}i = \frac{q}{C}\left(\frac{dq}{dt}\right) \qquad [5.8\text{-}8]$$

$$\frac{dE_{E,C}^{sys}}{dt} = Cv_c\frac{dv_c}{dt} = \frac{d}{dt}\left(\frac{1}{2}Cv_c^2\right) \qquad [5.8\text{-}9]$$

This is a more practical formula to describe the accumulation of electrical energy in a capacitor and is useful in describing the charging and discharging of a capacitor in an unsteady-state system. Recall from physics that the electrical energy stored in a capacitor is given as $E_{E,C} = \frac{1}{2}Cv_c^2$. This is the algebraic form of equation [5.8-9].

EXAMPLE 5.16 **Natural Response of an *RC* Circuit**

Problem: Consider the electrical circuit shown in Figure 5.33. The capacitor is connected to a three-position switch.

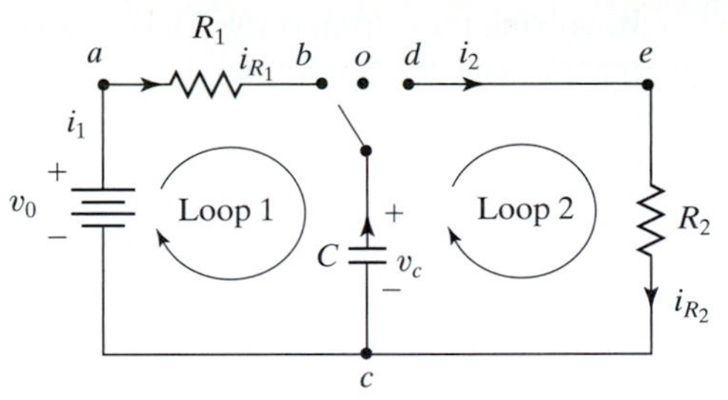

Figure 5.33
Capacitor in a circuit connected to a three-position switch.

Case 1: The switch is originally at position *o*, which indicates the circuit is open and the capacitor is discharged ($v_c = 0$). At $t = 0$, the switch is moved to position *b*. Find the steady-state voltage v_c across the capacitor.

Case 2: The switch is originally at position *b*, and the system is at steady-state. At $t = 0$, the switch is moved to position *d*. (a) Derive an equation for the voltage across the capacitor and resistor as a function of time. (b) Derive an equation for the current through the capacitor and resistor as a function of time. (c) Use the electrical energy accounting equation to explain the energy conversion in the circuit.

Solution: *Case 1*: Using KVL and Ohm's law around loop 1, we can write the equation:

$$\text{loop 1:} \qquad \sum_{\text{loop}} v_{\text{elements}} = v_0 - i_{R_1}R_1 - v_c = 0$$

When the switch is closed, the capacitor begins to charge, and the voltage across the capacitor, v_c, increases. Because the resistor and capacitor are in series, the total voltage across the series combination is the battery voltage, v_0. Thus, as v_c increases and approaches v_0, the voltage across the resistor decreases to zero. Since the current through the resistor (i_{R_1}) is proportional to the voltage drop across the resistor, i_{R_1} also decreases as the capacitor charges. At steady-state, the capacitor is charged to the full battery voltage, and $i_{R_1} = 0$. Therefore we have:

$$v_0 - v_c = 0$$
$$v_c = v_0$$

The direction of the current during the charging of the capacitor is in the direction of loop 1.

Case 2:

(a) At $t = 0$, the switch is moved from position *b* to *d*. For the arbitrarily drawn loop 2 in the circuit in Figure 5.33, net charge cannot accumulate in the resistor or the capacitor. At $t > 0$, the conservation of net charge implies that the current is constant around the loop ($i_2 = i_{R_2}$).

The current through the capacitor is given by:

$$i_2 = C\left(\frac{dv_c}{dt}\right)$$

We can apply KVL and Ohm's law around loop 2 to obtain the equation:

$$\text{loop 2:} \qquad \sum_{\text{loop}} v_{\text{elements}} = i_2 R_2 + v_c = 0$$

$$i_2 = \frac{-v_c}{R_2}$$

By solving for i_2 from this KVL equation, we can set it equal to the equation for the current through the capacitor:

$$i_2 = C\left(\frac{dv_c}{dt}\right) = \frac{-v_c}{R_2}$$

We can rearrange this equation to this form:

$$\left(\frac{dv_c}{dt}\right) = -\left(\frac{1}{R_2C}\right)v_c$$

Integrating with the initial condition $v_{c,0}$ for the voltage on the capacitor at $t = 0$ gives:

$$\int_{v_{c,0}}^{v_c}\frac{dv_c}{v_c} = -\int_0^t\left(\frac{1}{R_2C}\right)dt$$

$$\ln\left(\frac{v_c}{v_{c,0}}\right) = -\left(\frac{1}{R_2C}\right)(t-0) = -\frac{t}{R_2C}$$

$$v_c = v_{c,0}e^{-\frac{t}{R_2C}}$$

As expected, at $t = 0$, $v_c = v_{c,0}$. As time goes to infinity, v_c approaches 0, meaning the capacitor becomes fully discharged. According to KVL, the voltage across the resistor is equal to that of the capacitor (i.e., $v_{R_2} = v_c$); therefore, the voltage across the resistor can also be described by the expression obtained for the voltage on the capacitor. Thus, the voltage across the resistor also decays exponentially.

(b) The current through the resistor can be calculated by applying Ohm's law and considering the magnitude of the voltage drop:

$$i_{R_2} = \frac{v_{R_2}}{R_2} = \frac{v_{c,0}e^{-\frac{t}{R_2C}}}{R_2}$$

The current through the capacitor can be calculated by:

$$i_2 = C\frac{dv_c}{dt} = C\frac{d}{dt}\left(v_{c,0}e^{-\frac{t}{R_2C}}\right) = C\left(-\frac{1}{R_2C}\right)v_{c,0}e^{-\frac{t}{R_2C}} = -\left(\frac{v_{c,0}}{R_2}\right)e^{-\frac{t}{R_2C}}$$

Because we calculated the current to be negative, this indicates that the direction of current flow for i_2 through the capacitor is the opposite of the direction of current flow through the capacitor in Case 1. The direction of current through the capacitor during discharge is opposite the direction of current during charging. (*Note*: the current in loop 2 flows in the same direction as shown in Figure 5.33.)

(c) If we define the system outside of loop 2, no current flows into or out of the system. Thus, we can reduce the differential accounting equation for electrical energy (equation [5.4-2]) to:

$$\sum\dot{G}_{\text{elec}} - \sum\dot{W}_{\text{elec}} = \frac{dE_E^{\text{sys}}}{dt}$$

We can reduce the equation further, since no electrical energy is generated in the system, so the $\sum\dot{G}_{\text{elec}}$ term is eliminated. However, electrical energy is consumed by the resistor and converted to thermal energy. For the resistor, the power consumption, $\dot{W}_{\text{elec}}$, is equal to the product of the voltage and current across the resistor:

$$\dot{W}_{\text{elec}} = v_{R_2}i_{R_2} = v_{c,0}e^{-\frac{t}{R_2C}}\left(\frac{v_{c,0}e^{-\frac{t}{R_2C}}}{R_2}\right) = \frac{v_{c,0}^2}{R_2}e^{\left(-\frac{2t}{R_2C}\right)}$$

The electrical energy stored by the capacitor is calculated using equation [5.8-9]:

$$\frac{dE^{sys}_{E,C}}{dt} = \frac{d}{dt}\left(\frac{1}{2}Cv_c^2\right) = \frac{d}{dt}\left(\frac{1}{2}Cv_{c,0}^2 e^{\left(-\frac{2t}{R_2C}\right)}\right) = -\frac{v_{c,0}^2}{R_2}e^{\left(-\frac{2t}{R_2C}\right)}$$

The power consumed by the resistor is equal and opposite to the change in electrical energy stored by the capacitor, which verifies that our reduced governing equation is true. As the capacitor discharges, electrical energy stored in its electric field is converted to thermal energy in the resistor. ■

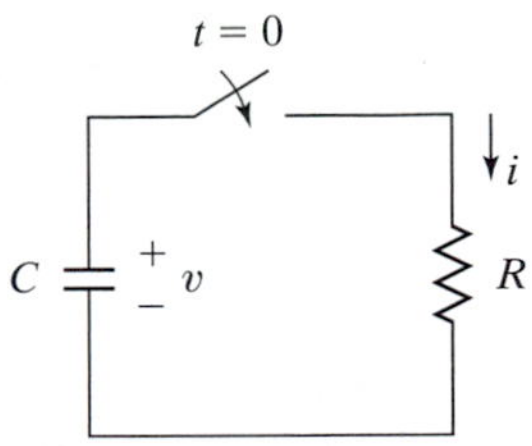

Figure 5.34
Generalized RC circuit.

The above problem calculated the voltage, current, and power consumption of a specific RC circuit. These equations can be generalized for a natural response of an RC circuit as shown in Figure 5.34. Initially, the voltage across the capacitor is v_0. At time equal to zero, the circuit is closed and the capacitor begins to discharge.

The time constant τ for an RC circuit equals the product of the resistance, R, and the capacitance, C, and is defined:

$$\tau = RC \qquad [5.8\text{-}10]$$

The voltage (v), current (i), and power consumption ($\dot{W}_{elec}$) for the RC circuit are written in terms of the time constant, τ, respectively, as:

$$v = v_0 e^{-\frac{t}{\tau}} \qquad [5.8\text{-}11]$$

$$i = -\left(\frac{v_o}{R}\right)e^{-\frac{t}{\tau}} \qquad [5.8\text{-}12]$$

$$\dot{W}_{elec} = \frac{v_0^2}{R}e^{\left(-\frac{2t}{\tau}\right)} \qquad [5.8\text{-}13]$$

where v is voltage, v_0 is the initial voltage, t is time, i is current, R is resistance, and τ is the time constant for an RC circuit.

Equations for the charging of a capacitor can also be written with the time constant τ. The magnitude of the time constant reveals characteristics of the system during these dynamic periods.

EXAMPLE 5.17 Modeling of a Neuron

Problem: In designing circuits for neural prostheses, one design objective is to mimic the behavior of neurons so that remaining intact neurons can be stimulated. The neuron membrane can be modeled as a simple circuit containing three voltage sources, three resistors, and a capacitor, as shown in Figure 5.35a. The three voltage sources represent the neuron resting potential, v_r; tonic current driving potential, v_t; and synaptic driving potential, v_s. The voltage across the membrane, which is modeled as a capacitor, is designated as v_m. Using KCL, develop a time-dependent mathematical model that relates the voltage sources, the known resistances, the capacitance of the membrane, and the voltage across the membrane.

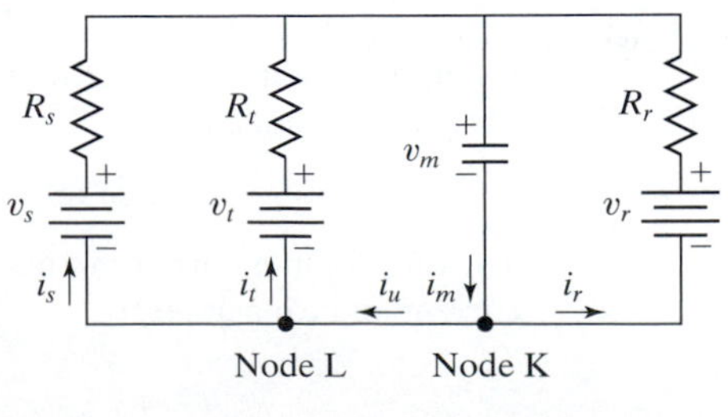

Figure 5.35a
Neuron membrane modeled as a simple circuit containing three voltage sources, three resistors, and a capacitor. (*Source:* Jung R, Brauer EJ, and Abbas JJ, "Real-time interaction between a neuromorphic electronic circuit and the spinal cord," *IEEE Trans Neural Syst Rehabil Eng* 2001, 9:319–26.)

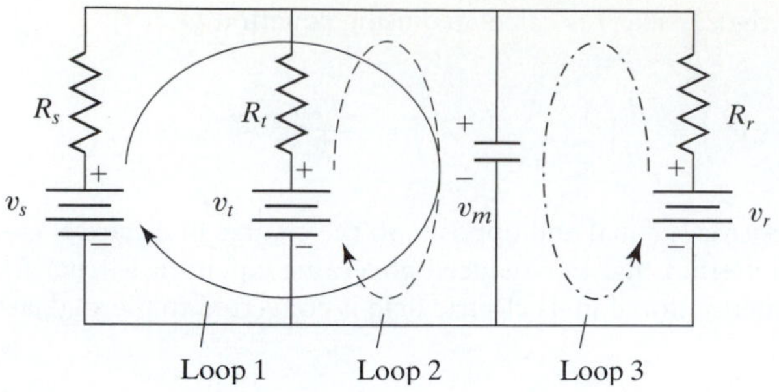

Figure 5.35b
Three possible loop configurations for the neuron membrane model.

Solution:

1. Assemble
   (a) Find: A time-dependent mathematical model that relates the voltage sources, resistances, membrane capacitance, and voltage across the membrane.
   (b) Diagram: The diagram of the circuit is shown in Figure 5.35a. Figure 5.35b shows three loops drawn in arbitrary directions under consideration.
2. Analyze
   (a) Assume:
   - The model drawn in Figure 5.35a is a reasonable representation of the cell membrane.
   - The capacitance of the cell membrane is constant.

   (b) Extra data: No extra data are needed.
   (c) Variables, notations, units:
   - m = membrane
   - r = resting
   - t = tonic
   - s = synaptic
3. Calculate
   (a) Equations: The problem statement asks us to develop a model using KCL:

$$\sum_k i_k - \sum_j i_j = 0$$

Because resistors are present in the model, we can also use KVL and Ohm's law:

$$\sum_{\text{loop}} v_{\text{elements}} = 0$$

$$v = iR$$

Because we assume the capacitance of the cell membrane is constant, we can use equation [5.8-7] to relate the capacitor in the model to the current:

$$i = C\frac{dv_c}{dt}$$

   (b) Calculate:
   - Using KCL, we can obtain two equations for the two nodes depicted in Figure 5.35a:

$$\text{node K:} \quad i_m = i_u + i_r$$

$$\text{node L:} \quad i_u = i_s + i_t$$

Using the equation obtained at node L, we can substitute for i_u in the equation obtained for node K:

$$\text{node K:} \quad i_m = i_u + i_r = i_s + i_t + i_r$$

   - To solve for the current across the membrane, we use the relationship between current and capacitance:

$$i_m = C_m\frac{dv_m}{dt}$$

- To find relationships for the voltage sources and the voltage across the membrane, we can use KVL to develop an equation for each loop drawn in Figure 5.35b:

$$\begin{aligned} &\text{loop 1:} \quad v_s - v_{R_s} - v_m = 0 \quad \text{or} \quad v_{R_s} = v_s - v_m \\ &\text{loop 2:} \quad v_t - v_{R_t} - v_m = 0 \quad \text{or} \quad v_{R_t} = v_t - v_m \\ &\text{loop 3:} \quad v_r - v_{R_r} - v_m = 0 \quad \text{or} \quad v_{R_r} = v_r - v_m \end{aligned}$$

Using Ohm's law, we can determine the current in each of these loops by defining them in terms of the voltage drops across the resistors:

$$i_s = \frac{v_{R_s}}{R_s} = \frac{v_s - v_m}{R_s}$$

$$i_t = \frac{v_{R_t}}{R_t} = \frac{v_t - v_m}{R_t}$$

$$i_r = \frac{v_{R_r}}{R_r} = \frac{v_r - v_m}{R_r}$$

Note the similarity between these equations and the Hodgkins-Huxley model. In both cases, a potential difference is the driving force that generates a current.

- We can now substitute the current values determined into the simplified KCL equation written for node K and set the KCL equation equal to the relationship between current and capacitance:

$$i_m = i_r + i_s + i_t = \frac{v_s - v_m}{R_s} + \frac{v_t - v_m}{R_t} + \frac{v_r - v_m}{R_r} = C_m \frac{dv_m}{dt}$$

4. Finalize

(a) Answer: A simple model of the cell membrane that relates the voltage sources, resistances, membrane capacitance, and voltage across the membrane is:

$$C_m \frac{dv_m}{dt} = \frac{v_s - v_m}{R_s} + \frac{v_t - v_m}{R_t} + \frac{v_r - v_m}{R_r}$$

(b) Check: It is difficult to verify the reasonableness of our model because we have performed a theoretical analysis of a neuronal membrane involving no numerical values. However, we have taken all of the neuron potentials and resistances into account and have also accounted for the membrane capacitance. The form of this solution is similar to equation [5.6-34] except that this model includes a capacitance term. ■

Just as capacitors store electrical energy in an electric field, **inductors** are devices that store electrical energy in a magnetic field. An inductor is a coil of wire that carries a current. When current moves along the coiled wire, it induces a magnetic field that runs along the axis of the coil. If the current changes, the resulting magnetic field changes, and an electric potential difference is created. Note that for a potential difference to be present, there must be a change in the current through the coil; if the current is constant, no voltage is generated. Inductors function as a type of inertial element, opposing changes in current (recall Lenz's law from physics).

Mathematically, the voltage drop, v_L, across an inductor is given by:

$$v_L = L\frac{di_L}{dt} \tag{5.8-14}$$

where i_L is the current through the inductor and L is the inductance, a constant that depends on the physical properties of the inductor. Inductance has the dimension of $[L^2Mt^{-2}I^{-2}]$ and is measured in henrys (H), which is $(V \cdot s)/A$.

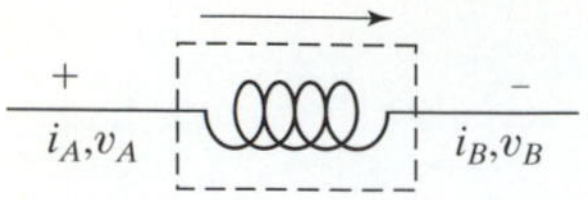

Figure 5.36
Current flow through an inductor.

Using the accounting equation for electrical energy (equation [5.4-3]) and the conservation equation for net charge (equation [5.7-3]), we can calculate the storage of electrical energy for the inductor in Figure 5.36:

$$\frac{dE_{E,L}^{sys}}{dt} = i_A v_A - i_B v_B \quad [5.8\text{-}15]$$

Because charge does not accumulate in an inductor, the differential conservation of net charge equation is simplified to:

$$i_A - i_B = 0 \quad [5.8\text{-}16]$$

$$i_A = i_B = i_L \quad [5.8\text{-}17]$$

Therefore, equation [5.8-15] is reduced to:

$$\frac{dE_{E,L}^{sys}}{dt} = i_L(v_A - v_B) = i_L v_L \quad [5.8\text{-}18]$$

where v_L is the voltage difference across the inductor.

Given equation [5.8-14], the rate of change of electrical energy in an inductor is written as:

$$\frac{dE_{E,L}^{sys}}{dt} = L\left(i_L \frac{di_L}{dt}\right) \quad [5.8\text{-}19]$$

Recall from physics that the electrical energy stored in an inductor is given as $E_{E,L} = \frac{1}{2}Li^2$. This is the algebraic form of equation [5.8-19].

Note the similarities and differences between equations [5.8-7] and [5.8-14] and between equations [5.8-9] and [5.8-19].

EXAMPLE 5.18 **Energy Storage in an Inductor**

Problem: At $t = 0$, a switch is closed in the circuit shown in Figure 5.37. Based on readings from an ammeter, the current through the circuit is determined to be:

$$i = 2.0t\frac{\text{A}}{\text{s}}$$

The inductor has an inductance of 50 mH. How much energy is stored in the inductor after 10 ms?

Solution: From equation [5.8-19], we know that:

$$\frac{dE_{E,L}^{sys}}{dt} = L\left(i_L \frac{di_L}{dt}\right)$$

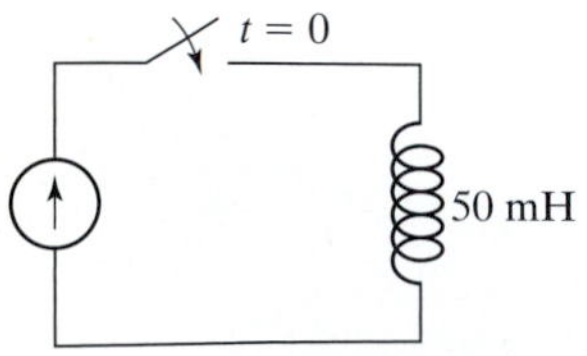

Figure 5.37
Circuit with a current source, inductor, and open switch.

Before the switch is closed at $t = 0$, no current flows through the inductor, so no electrical energy is stored in the inductor. Thus, we can use the algebraic form of the equation that gives energy stored in an inductor:

$$E_{E,L}^{sys} = \frac{1}{2}Li_L^2$$

Substituting in the given numerical values gives:

$$E_{E,L}^{sys} = \frac{1}{2}Li_L^2 = \frac{1}{2}(50\text{ mH})\left(\frac{1\text{ H}}{1000\text{ mH}}\right)\left(2.0t\frac{\text{A}}{\text{s}}\right)^2 = 0.1t^2\frac{\text{J}}{\text{s}^2}$$

At $t = 10$ ms:

$$E^{sys}_{E,L} = 0.1(10 \times 10^{-3}\ \text{s})^2 \frac{\text{J}}{\text{s}^2} = 1.0 \times 10^{-5}\ \text{J}$$

The inductor has stored 1.0×10^{-5} J of energy after 10 ms. ■

Consider the natural response of an RL circuit as shown in Figure 5.38. Assume that the inductor has stored energy, and at time equal to zero the circuit is closed. Using KVL gives:

$$v_L + v_R = 0 \qquad [5.8\text{-}20]$$

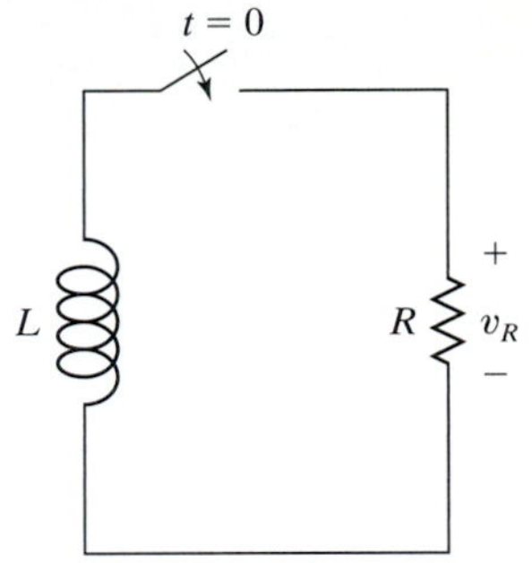

Figure 5.38
Generalized RL circuit.

Substituting the voltage drop across an inductor from equation [5.8-14] gives:

$$L\frac{di}{dt} + iR = 0 \qquad [5.8\text{-}21]$$

Since L and R are constants, equation [5.8-21] can be integrated to:

$$i = i_0 e^{-\frac{R}{L}t} \qquad [5.8\text{-}22]$$

where i is current, i_0 is the initial current, t is time, R is resistance, and L is inductance.

Similar to an RC circuit, the time constant τ for an RL circuit is defined as:

$$\tau = \frac{L}{R} \qquad [5.8\text{-}23]$$

so that equation [5.8-22] can be written as:

$$i = i_0 e^{-\frac{t}{\tau}} \qquad [5.8\text{-}24]$$

The magnitude of the time constant reveals characteristics of the system during this dynamic period. The voltage across the resistor, v, is then calculated as:

$$v = iR = i_0\,\text{Re}^{-\frac{t}{\tau}} \qquad [5.8\text{-}25]$$

The power consumption, $\dot{W}_{elec}$, in the resistor is equivalent to the electrical energy stored by the inductor:

$$\dot{W}_{elec} = i_0^2\,\text{Re}^{-\frac{2t}{\tau}} \qquad [5.8\text{-}26]$$

Equations that describe the time-dependent process of storing electrical energy in an inductor can also be written with the time constant τ. Note the similarity and differences between the equations describing an RC circuit and an RL circuit.

5.9 Systems with Generation or Consumption Terms—Focus on Charge

In Chapters 3 (Conservation of Mass) and 4 (Conservation of Energy), we focus on reactions in which atoms in chemical compounds are rearranged to form new compounds. In this chapter, the definition of reaction is expanded to include the rearrangement of electrons or protons within or between chemical species. In this section, we examine electrochemical reactions and equilibrium dissociation reactions, in which the exchange of charged species occurs.

Recall that positive and negative charges can be generated simultaneously in a reacting system. For example, the algebraic accounting equations for positive and negative charge are:

$$\text{Positive:} \quad \sum_k q_{+,k} - \sum_j q_{+,j} + q_{+,\text{gen}} - q_{+,\text{cons}} = q_{+,\text{f}}^{\text{sys}} - q_{+,0}^{\text{sys}} \qquad [5.9\text{-}1]$$

$$\text{Negative:} \quad \sum_k q_{-,k} - \sum_j q_{-,j} + q_{-,\text{gen}} - q_{-,\text{cons}} = q_{-,\text{f}}^{\text{sys}} - q_{-,0}^{\text{sys}} \qquad [5.9\text{-}2]$$

When considering positive and negative charge, Generation or Consumption terms or both terms may be present. However, equal amounts of positive and negative charges are always created or consumed during a reaction. Therefore, the algebraic form of the net accounting equation always reduces to the conservation of charge equation:

$$\text{Net:} \quad \sum_k q_k - \sum_j q_j = q_{\text{f}}^{\text{sys}} - q_0^{\text{sys}} \qquad [5.9\text{-}3]$$

The positive and negative accounting equations and the net charge conservation equation can also be written in the differential and integral forms.

Many introductory circuit analysis textbooks do not discuss reacting systems. However, because chemical reactions commonly occur at the interface between medical equipment and the human body, we discuss the application of charge accounting and conservation equations to biological and medical systems. These equations can be used to track and account for charged species in many medically relevant reactions.

5.9.1 Radioactive Decay

In **radioactive decay**, a chemical element decays or decomposes to a distinctly different chemical element with fewer protons or neutrons. Usually, electrons are ejected as well. Radioactive elements are used as tracers in many biomedical applications, such as in the diagnoses and treatments of thyroid disorders, heart disease, brain disorders, and cancer. One major clinical application for radioactive tracers is radioguided surgery (RGS), a technique in which the surgeon identifies tissue marked by a radionuclide before surgery.

Radioactive isotope compounds, such as ^3H, ^{14}C, ^{125}I, and ^{131}I, are widely used as markers or tracers in laboratory biomedical research. The isotopic marker behaves chemically the same as other atoms in the compound, but the differing number of neutrons allows it to be detected separately from other atoms of the same element. Isotopic markers are the basis of nuclear magnetic resonance (NMR), which is used to investigate the mechanisms of chemical reactions. NMR also forms the underlying principle of magnetic resonance imaging (MRI), a technology used to generate images of the inner spaces of opaque organs to visualize pathological or physiological alterations of living tissues.

The balanced chemical equations written to describe radioactive decay are an illustration of the conservation of net charge. A chemical element decays or decomposes to a distinctly different chemical element with fewer protons or neutrons. A particle is ejected from the original atom, removing mass or charge or both. A list of decay components is given in Table 5.4. One example is alpha decay, in which a helium atom, which has two neutrons and two protons, is ejected from the nucleus. When an electrically neutral isotope undergoes alpha decay, its mass and atomic element are reduced and the resulting atom carries a −2 charge. The ejected helium atom carries a +2 charge. Thus, net charge is preserved in the universe.

Beta decay works in a similar way, with an electron or positron being ejected. No mass is lost except for the electron or positron (which typically does not affect the atomic weight), so the total element mass stays the same in the decaying atom. When an electron is ejected, the decaying atom has changed electrically to have one more positive charge. This is balanced by the electron becoming a separate entity with a single negative charge. The positive and negative charges cancel one another, and net charge is preserved in the universe. The same is true for the loss of a positron, since a positron is a positive charge that has the same mass and magnitude of charge as an electron, but the charges of the atoms and the positron are the reverse of those in the situation with the electron.

TABLE 5.4

Constituents of Radioactive Decay

Symbol	Name	Charge
${}^{0}_{1}\beta$	Positron	+1
${}^{0}_{-1}\beta$	Electron	−1
ν	Neutrino	0
$\tilde{\nu}$	Antineutrino	0
γ	Gamma ray	0

A positron is a particle that has the same mass and magnitude of charge as an electron but carries a positive charge. When an emitted positron and an electron combine and are annihilated, a gamma ray is produced. In positron emission tomography (PET), a patient is injected with a radionuclide that decays by positron emission and is then scanned.

In the diagnosis of cancer and Alzheimer's disease, glucose metabolism is measured using fluoro-2-deoxy-D-glucose (FDG), which is labeled with fluorine-18, a radionuclide that decays according to the following reaction:

$$ {}^{18}_{9}F \longrightarrow {}^{18}_{8}O + {}^{0}_{1}\beta + {}^{0}_{-1}\beta + \nu \longrightarrow {}^{18}_{8}O + \gamma + \nu $$

where ${}^{0}_{1}\beta$ is a positron, ${}^{0}_{-1}\beta$ is an electron, ν is a neutrino with no charge, and γ is gamma radiation (Table 5.4). ${}^{18}_{8}O$ is a stable, electrically neutral, naturally occurring oxygen isotope that does not have any adverse effect on humans.

Net charge is conserved in both steps of this reaction (Figure 5.39). Positive and negative charges are generated or consumed simultaneously in pairs. In the first reaction, ${}^{18}_{9}F$ is consumed, and one ${}^{18}_{8}O$, one positron, one electron, and one neutrino are generated. ${}^{18}_{9}F$ and ${}^{18}_{8}O$ are electrically neutral. One positron carrying a +1 charge and one free electron carrying a −1 charge are generated simultaneously. Because q_{gen} and q_{cons} are both equal to zero, we have shown that net charge is conserved.

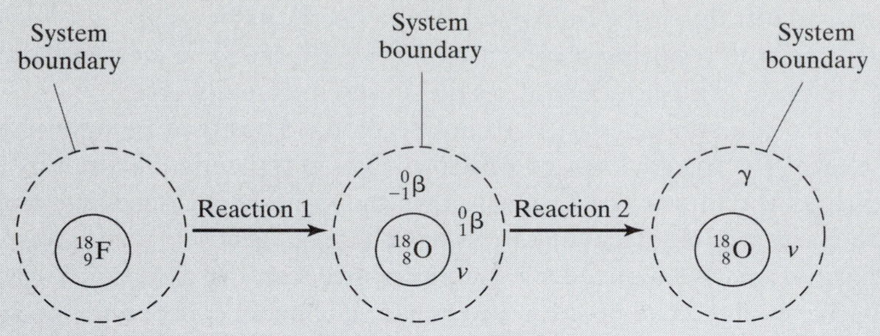

Figure 5.39 Fluorine-18 radioactive decay.

In the second reaction, we see that the neutrino and the oxygen do not react any further, and two species ($^{0}_{1}\beta$, $^{0}_{-1}\beta$) are consumed. The negatively charged electron and the positively charged positron combine to generate electrically neutral gamma radiation. For the second reaction, q_{gen} and q_{cons} are again equal to zero, and net charge is conserved.

5.9.2 Acids and Bases

Many compounds are composed of two or more charged chemical constituents. Examples include hydrochloric acid (HCl) and sodium hydroxide (NaOH). When placed in water, these two compounds dissociate: HCl to H^+ and Cl^-; NaOH to Na^+ and OH^-. An **acid** is defined as a proton (H^+) donor; a **base** is defined as a proton acceptor. **Strong acids** and **bases** almost completely dissociate in water.

Weak acids and **bases** only partially dissociate in water. Therefore, the contribution of a weak acid, such as acetic acid (CH_3COOH), carbonic acid (H_2CO_3), or lactic acid ($CH_3CH(OH)COOH$), to the hydrogen ion concentration is much less than the total concentration of added acid. Weak acids and bases are often used as biological buffers, which can reversibly bind hydrogen ions and serve to help maintain a relatively stable pH. An example is phosphate buffered saline (PBS), which includes the salts, NaCl and KCl, as well as Na_2HPO_4 and NaH_2PO_4, in H_2O.

Solutions are frequently described by their **pH**, a unitless value that indicates the concentration of H^+ in the solution:

$$pH = -\log [H^+] \qquad [5.9\text{-}4]$$

where $[H^+]$ is the concentration of hydrogen ions in the solution in units of mol/L (M). The logarithmic scale compensates for the tremendous range of potential H^+ concentrations that may be present in aqueous solutions. A change in pH by one unit signifies a change in the concentration of H^+ by a factor of ten (one order of magnitude). The concentrations of H^+ and OH^- in aqueous solution balance such that:

$$[H^+][OH^-] = 10^{-14}\ M^2 \qquad [5.9\text{-}5]$$

where $[OH^-]$ is the concentration of hydroxide ions in the solution in units of mol/L (M). Note that this equation holds strictly only at room temperature (25°C).

The pH values of typical solutions range from 0 to 14. A pH value lower than 7 indicates an acidic solution; a pH over 7 indicates a basic solution. A pH of 7 indicates a neutral solution, such as pure water, which has equal amounts of H^+ and OH^-. In pure water, equal H^+ and OH^- concentrations of 10^{-7} M are expected.

Because strong acids and bases are assumed to dissociate completely in solution, the contribution of H^+ ions to solution by a strong acid is equal to the total concentration of the acid. For example, a 0.01-M solution of HCl has:

$$[H^+] = 0.01 \text{ M} \qquad [5.9\text{-}6]$$

$$\text{pH} = -\log(0.01) = 2 \qquad [5.9\text{-}7]$$

The pH of a strong base can be calculated similarly. A 0.01-M solution of NaOH has:

$$[OH^-] = 0.01 \text{ M} \qquad [5.9\text{-}8]$$

$$[H^+] = \frac{10^{-14}\text{ M}^2}{[OH^-]} = \frac{10^{-14}\text{ M}^2}{10^{-2}\text{ M}} = 10^{-12}\text{ M} \qquad [5.9\text{-}9]$$

$$\text{pH} = -\log(10^{-12}) = 12 \qquad [5.9\text{-}10]$$

The dissociation of a generic acid (HA) in aqueous solution is given by:

$$HA \rightleftharpoons H^+ + A^- \qquad [5.9\text{-}11]$$

where A^- is the conjugate base of HA, or the base formed when HA donates a hydrogen ion. Note that net charge is conserved in this dissociation reaction.

The **equilibrium constant** (K) relates concentrations of the products and reactants of a chemical reaction in equilibrium. For the chemical reaction given by equation [5.9-11], the **acid dissociation equilibrium constant** (K_a) is:

$$K_a = \frac{[H^+][A^-]}{[HA]} \qquad [5.9\text{-}12]$$

The acid dissociation constant is a measure of the strength of an acid. In a weak acid, the products of the dissociation reaction will have low concentrations, resulting in a low K_a. In a strong acid, the dissociation reaction goes almost to completion, leaving a very small concentration of HA and a large K_a. Like pH, because of the large range of magnitude of values for K_a, we can describe K_a using a logarithmic scale:

$$\text{p}K_a = -\log K_a \qquad [5.9\text{-}13]$$

where K_a is determined from concentrations specified in units of mol/L (M). A strong acid has a large K_a and a very small pK_a; a weak acid has a relatively small K_a and large pK_a. It can be shown that for the dissociation of an acid HA, the pH and pK_a of the acid are related by the **Henderson-Hasselbach equation:**

$$\text{pH} = \text{p}K_a + \log\frac{[A^-]}{[HA]} \qquad [5.9\text{-}14]$$

EXAMPLE 5.19 Effect of Aspirin on Blood Acidity

Problem: Acetylsalicylic acid ($C_9H_8O_4$), commonly known as aspirin, has been used for over 100 years as an effective pain reliever. Aspirin works by suppressing the production of prostaglandins, chemicals that enhance sensitivity to pain. Bayer®, a leading producer of aspirin, recommends a dosage of one or two 325-mg tablets every 4 hours to relieve pain. If no buffers are in the blood, what will the pH of blood be after two aspirin tablets are consumed and completely absorbed into the bloodstream? Assume that no hydrogen ions are initially present in the blood and that the body contains 5.0 L of blood.

Solution:

1. Assemble
   (a) Find: pH of blood after ingesting two aspirin tablets, assuming no buffer system.
   (b) Diagram: The system is the body's entire blood volume.
2. Analyze
   (a) Assume:
   - Extent of dissociation of acetylsalicylic acid is the same regardless of presence or absence of buffer.
   - No other source of hydrogen ions in bloodstream other than the given reaction.
   - No movement of charged species across the system boundary.

   (b) Extra data:
   - The pK_a of acetylsalicylic acid is 3.5.

   (c) Variables, notations, units:
   - Units: mol, L.

   (d) Basis: The initial amount of aspirin (HA) in the blood is:

$$n_{\text{HA},0}^{\text{sys}} = 2 \text{ tablets}\left(\frac{325 \text{ mg}}{\text{tablet}}\right)\left(\frac{1 \text{ g}}{1000 \text{ mg}}\right)\left(\frac{1 \text{ mol}}{180.2 \text{ g}}\right) = 3.607 \times 10^{-3} \text{ mol}$$

   (e) Reactions: When no buffer system is present, only the chemical dissociation of the aspirin tablet must be considered. Positive (H^+) and negative ($C_9H_7O_4^-$, labeled as A^-) charges are generated when acetylsalicylic acid ($C_9H_8O_4$, labeled as HA) dissociates:

$$\text{HA} \rightleftharpoons \text{H}^+ + \text{A}^-$$

3. Calculate
   (a) Equations: No charges move across the system boundary, and no charges are consumed within the system. Because we assume that no buffers are in the blood, the initial number of moles of H^+ and A^- is equal to zero, so we can reduce the accounting equations for positive and negative charge to:

$$\text{Positive } (\text{H}^+)\text{:} \quad n_{\text{gen}} = n_{\text{H}^+,\text{f}}^{\text{sys}} - n_{\text{H}^+,0}^{\text{sys}} = n_{\text{H}^+,\text{f}}^{\text{sys}}$$

$$\text{Negative } (\text{A}^-)\text{:} \quad n_{\text{gen}} = n_{\text{A}^-,\text{f}}^{\text{sys}} - n_{\text{A}^-,0}^{\text{sys}} = n_{\text{A}^-,\text{f}}^{\text{sys}}$$

$$\text{Neutral (HA):} \quad n_{\text{cons}} = n_{\text{HA},\text{f}}^{\text{sys}} - n_{\text{HA},0}^{\text{sys}}$$

   where n_{gen} is defined as the number of moles of charged species (H^+ or A^-) generated upon dissociation and n_{cons} is defined as the number of moles of HA consumed upon dissociation. Note that n_{gen} is equal to n_{cons}, the number of moles of acetylsalicylic acid that dissociate.

   (b) Calculate:
   - The K_a of acetylsalicylic acid, calculated from the pK_a and adjusted to a mole basis (K_a'), is:

$$pK_a = -\log(K_a)$$

$$K_a = 10^{-pK_a} = 10^{-(3.5)} = 3.16 \times 10^{-4} \frac{\text{mol}}{\text{L}}$$

$$K_a' = 3.16 \times 10^{-4} \frac{\text{mol}}{\text{L}}(5.0 \text{ L}) = 1.58 \times 10^{-3} \text{ mol}$$

   - Assuming that the volume of blood is constant, the concentrations of reactants and products in equation [5.9-12] may be replaced with the number of moles of these species. Substituting in equilibrium amounts from the charge accounting equations above:

$$K_a' = \frac{(n_{\text{H}^+,\text{f}}^{\text{sys}})(n_{\text{A}^-,\text{f}}^{\text{sys}})}{n_{\text{HA},\text{f}}^{\text{sys}}} = \frac{(n_{\text{gen}})(n_{\text{gen}})}{(3.607 \times 10^{-3} \text{ mol} - n_{\text{gen}})} = 1.58 \times 10^{-3} \text{ mol}$$

$$n_{\text{gen}} = 1.72505 \times 10^{-3} \text{ mol}$$

The value of n_{gen} is equal to the number of moles of H^+ and A^- generated and also present at the final condition.

- To calculate the pH of a solution, the molar concentration of H^+, not the amount, must be used:

$$[H^+]_f^{sys} = \frac{n_{H^+,f}^{sys}}{V_{blood}} = \frac{1.73 \times 10^{-3}\text{ mol}}{5.0\text{ L}} = 3.45 \times 10^{-4}\ \frac{\text{mol}}{\text{L}}$$

Thus, after taking aspirin, the pH of blood without any buffer would be:

$$\text{pH} = -\log[H^+]_f^{sys} = -\log[3.45 \times 10^{-4}] = 3.46$$

4. Finalize
   (a) Answer: If no buffer system is present, the pH of blood after ingesting two aspirin tablets will be 3.5.
   (b) Check: This pH value of 3.5 is considerably lower than the normal pH of blood, which is 7.4. If buffers were not present, taking 2 aspirins would considerably alter blood pH, well past the point of death. The assumption that no buffer system is present is not valid. A more realistic scenario is discussed in the following example.

■

EXAMPLE 5.20 Effect of Aspirin on Blood Acidity in the Presence of Buffers

Problem: When individuals take acetylsalicylic acid ($C_9H_8O_4$), commonly known as aspirin, buffers in the blood help dampen changes in blood pH. The main buffer is the bicarbonate system:

$$H_2CO_3 \rightleftharpoons HCO_3^- + H^+$$

The hydrogen ions released in the dissociation of acetylsalicylic acid bind to HCO_3^- ions in the bicarbonate buffer system to make H_2CO_3. Sources of HCO_3^- ions in the blood include H_2CO_3 and $NaHCO_3$.

What will the pH of blood be after two aspirin tablets are consumed with this buffer system in place? The pK_a of H_2CO_3 is 6.1 at body temperature. Assume that the blood initially contains 2.66×10^{-2} mol/L of completely dissociated sodium bicarbonate ($NaHCO_3$) and 1.4×10^{-3} mol/L of undissociated carbonic acid (H_2CO_3). Use the equilibrium amount of H^+ calculated in Example 5.19 for the initial amount of H^+. Assume that the body contains 5.0 L of blood.

Solution:

1. Assemble
   (a) Find: pH of blood with bicarbonate buffer system after ingesting two aspirin tablets.
   (b) Diagram: The system is the body's entire blood volume.
2. Analyze
   (a) Assume:
      - The presence of the buffer does not shift the dissociation equilibrium of acetylsalicylic acid.
      - No other source of hydrogen ions in bloodstream other than the given reactions.
      - Complete dissociation of $NaHCO_3$.
      - No movement of charged species across the system boundary.

   (b) Extra data:
   The pK_a of acetylsalicylic acid is 3.5.
   (c) Variables, notations, units:
      - Units: mol, L.

   (d) Basis: Initial amount of H^+ calculated as 1.72505×10^{-3} mol in Example 5.19.

(e) Reactions: In the presence of the bicarbonate buffer system, two additional chemical dissociations must also be considered:

$$H_2CO_3 \rightleftharpoons HCO_3^- + H^+$$
$$NaHCO_3 \longrightarrow Na^+ + HCO_3^-$$

in addition to the dissociation of acetylsalicylic acid:

$$HA \rightleftharpoons H^+ + A^-$$

where HA is $C_9H_8O_4$ and A^- is $C_9H_7O_4^-$.

3. Calculate
(a) Equations:
- The charge balance for the buffered system is more complicated than that for the unbuffered system (Example 5.19). Specifically, there are two weak acids, HA and H_2CO_3. Acetylsalicylic acid is a strong acid relative to H_2CO_3 (pK_a of 3.5 compared to 6.1). Therefore we make the simplifying assumption that the presence of the buffer does not shift the dissociation equilibrium of acetylsalicylic acid.
- This problem is set up so that the H^+ ions from the already dissociated acetylsalicylic acid bind with the HCO_3^- ions in solution (from the already dissociated $NaHCO_3$) to form carbonic acid (H_2CO_3). Essentially, we let the acetylsalicylic acid dissociate in the absence of the buffer and then apply the buffer to bind with free protons. We neglect the more intricate interactions between the compounds.
- To simplify the charge accounting equations, no charges move across the system boundary. Given the problem setup, no H^+ ions should be generated. The accounting equation for moles of positive charge of H^+ is:

$$-n_{H^+,cons} = n_{H^+,f}^{sys} - n_{H^+,0}^{sys}$$

No moles of HCO_3^-, a negatively charged species, are generated; however, they are consumed:

$$-n_{HCO_3^-,cons} = n_{HCO_3^-,f}^{sys} - n_{HCO_3^-,0}^{sys}$$

The accounting equation for moles of H_2CO_3 is:

$$n_{H_2CO_3,gen} = n_{H_2CO_3,f}^{sys} - n_{H_2CO_3,0}^{sys}$$

Notice that the number of moles of HCO_3^- consumed by reaction with H+ from the acetylsalicylic acid ($n_{HCO_3^-,cons}$) is equal to the number of moles of H_2CO_3 generated ($n_{H_2CO_3,gen}$).

(b) Calculate:
- The initial amounts of HCO_3^- and H_2CO_3 are given on a molar basis. Multiplying the concentration of the species by the blood volume gives:

$$n_{HCO_3^-,0}^{sys} = 0.133 \text{ mol}$$

from the already dissociated $NaHCO_3$ and

$$n_{HCO_3^-,0}^{sys} = 7.0 \times 10^{-3} \text{ mol}$$

- The final amount of H^+ from Example 5.19 is used as the initial amount for this system:

$$n_{H^+,f}^{sys} = n_{H^+,0}^{sys} - n_{cons} = 1.72505 \times 10^{-3} \text{ mol} - n_{cons}$$

For bicarbonate:

$$n_{HCO_3^-,f}^{sys} = n_{HCO_3^-,0}^{sys} - n_{cons} = 0.133 \text{ mol} - n_{cons}$$

and carbonic acid:

$$n_{H_2CO_3,f}^{sys} = n_{H_2CO_3,0}^{sys} + n_{gen} = 7.0 \times 10^{-3} \text{ mol} + n_{cons}$$

Note that the species-specific consumption and generation terms are equal and therefore have been simplified in the above equations.

- To find the equilibrium constant of H_2CO_3 dissociation:

$$pK_a = -\log K_a = 6.1$$
$$K_a = 7.94 \times 10^{-7}\ \text{M}$$

Once again, this needs to be converted to moles:

$$K_a' = 7.94 \times 10^{-7}\frac{\text{mol}}{\text{L}}(5.0\ \text{L}) = 3.97 \times 10^{-6}\ \text{mol}$$

- Recall that we may substitute equilibrium amounts in moles for concentrations if the volume of the blood is constant. The equilibrium of carbonic acid is:

$$K_a' = \frac{(n_{H^+,f}^{sys})(n_{HCO_3^-,f}^{sys})}{n_{H_2CO_3,f}^{sys}} = \frac{(1.72505 \times 10^{-3}\ \text{mol} - n_{cons})(0.133\ \text{mol} - n_{cons})}{(7.0 \times 10^{-3}\ \text{mol} + n_{cons})}$$
$$K_a' = 3.97 \times 10^{-6}\ \text{mol}$$
$$n_{cons} = 1.72478 \times 10^{-3}\ \text{mol}$$

- Using the balance equation for H^+:

$$n_{H^+,f}^{sys} = n_{H^+,0}^{sys} - n_{cons} = 1.72505 \times 10^{-3}\ \text{mol} - 1.72478 \times 10^{-3}\ \text{mol}$$
$$= 2.7 \times 10^{-7}\ \text{mol}$$

Significant figures were carried for the calculation in order to maintain accuracy of the final answer. For this problem, rounding intermediate numbers to two or three significant figures may result in an inaccurate final answer.

- The pH of the solution is calculated using the molar concentration of H^+:

$$[H^+]_f^{sys} = \frac{n_{H^+,f}^{sys}}{V_{blood}} = \frac{2.7 \times 10^{-7}\ \text{mol}}{5.0\ \text{L}} = 5.4 \times 10^{-8}\ \frac{\text{mol}}{\text{L}}$$

Thus, the pH of blood with buffer would be:

$$\text{pH} = -\log[H^+]_f^{sys} = 7.3$$

4. Finalize
   (a) Answer: In the presence of the simplified bicarbonate buffer system, the pH of blood after ingesting two aspirin tablets is 7.3.
   (b) Check: The pH of blood is still slightly lower than normal, but the bicarbonate system buffered the decrease significantly. There are approximately 10^4 times fewer H^+ ions in the blood than there were in the system without buffers, because the hydrogen ions released from the acetylsalicylic acid are taken out of circulation by binding to HCO_3^-.

   Remember that we made the major simplifying assumption that the equilibrium dissociation of aspirin is the same in the presence of the buffering system as in its absence. In reality, both the aspirin and the H_2CO_3 would dissociate slightly different amounts, such that they would both be in equilibrium and balance each other. It is also important to recognize that the blood is not simply a passive buffering system with just one buffering reaction. In addition to other compounds that dissociate near a pH of 7, active metabolic steps are taken to ensure that blood pH remains in a healthy range. ■

5.9.3 Electrochemical Reactions

Electrochemical reactions involve the oxidation and reduction of materials. **Oxidation** is a reaction in which a chemical species (usually a metal) loses one or more electrons and typically forms a cation. **Reduction** is the corresponding reaction in which a chemical species (usually a nonmetal) gains one or more electrons and typically forms an anion. Rusting metal, tarnished silver, and copper plating are visible products of electrochemical reactions. For example, iron-containing metals rust when the iron in the metal reacts with oxygen in the air in the presence of water. The metallic iron (Fe) molecules are oxidized to Fe^{3+} while the oxygen molecules (O_2) are reduced to O^{2-}. The result is hydrated iron oxide (Fe_2O_3), more commonly known as rust.

A **battery** is a device that uses electrochemical reactions to generate electric potential energy. It does so by converting chemical energy into electrical energy by raising the potential energy of charged particles. A battery has a positive electrode (cathode), at which a material is reduced, and a negative electrode (anode), at which a material is oxidized. Electrochemical reactions that cause a buildup of electrons at the anode generate and maintain an electric potential difference between the positive and negative terminals. This potential difference can be used to run a circuit or other electric or mechanical device, since electrons want to move to the cathode to eliminate the potential difference.

Many types of batteries and fuel cells are produced today, from tiny lithium watch batteries, to lead oxide car batteries, to hydrogen fuel cells used in the space shuttle. Batteries are also used in hospital equipment and biomedical devices, such as pacemakers, drug pumps, neurostimulators, cardiac defibrillators, and left ventricular assist devices (LVADs). Among this wide range of uses for batteries, the theory behind them remains the same.

EXAMPLE 5.21 **Charge Produced by a Lithium-Iodide Battery**

Problem: Lithium-iodide batteries are commonly used as power sources for implantable pacemakers (Figure 5.40a). Since pacemakers are implanted in the body, replacing the battery requires a surgical procedure. Hence, a battery with an extended lifetime is a key design criterion.

A lithium-iodide battery has an overall reaction of:

$$2\ \mathrm{Li} + \mathrm{I_2} \longrightarrow 2\ \mathrm{LiI}$$

The reduction half-reaction that takes place at the cathode is:

$$\mathrm{I_2} + 2\ e^- \longrightarrow 2\ \mathrm{I}^-$$

and the oxidation half-reaction that takes place at the anode is:

$$\mathrm{Li} \longrightarrow \mathrm{Li}^+ + e^-$$

If the lithium-iodide battery contains 0.5 g of lithium, how much charge can flow from the anode if the battery is completely discharged?

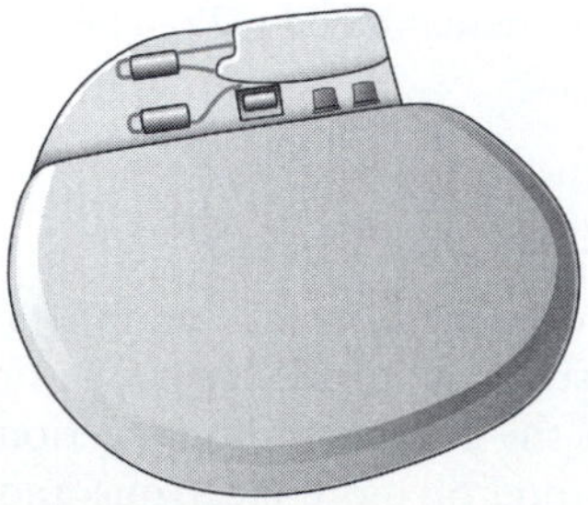

Figure 5.40a
Lithium-iodide battery model.

Solution: The system is the lithium anode (Figure 5.40b). To calculate the amount of charge that can flow from the anode, we can use the accounting equation for negative charge [5.9-2]:

$$\sum_k q_{-,k} - \sum_j q_{-,j} + q_{-,\text{gen}} - q_{-,\text{cons}} = q_{-,\text{acc}}^{\text{sys}}$$

We assume that no charges accumulate within the system. We also assume no negative charges flow into the system, and no negative charges are consumed within the system. This simplifies our governing equation to:

$$-q_{-,\text{out}} + q_{-,\text{gen}} = 0$$

$$q_{-,\text{out}} = q_{-,\text{gen}}$$

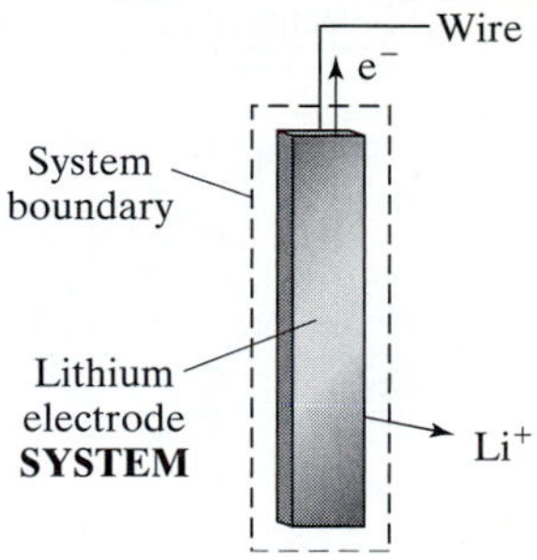

Figure 5.40b
Lithium anode system.

Thus, the negative charge flowing out of the system is equal to the negative charge generated at the lithium electrode, assuming that lithium is completely dissociated to Li^+ and electrons.

Because lithium is completely oxidized (i.e., lithium is consumed) to generate the electrons in a 1:1 ratio, we can use the given initial mass of 0.5 g of lithium and the molecular weight of lithium to calculate a molar basis for calculating the amount of charge flowing out of the anode:

$$n_{\text{Li}} = \frac{m_{\text{Li}}}{M_{\text{Li}}} = \frac{0.5\text{ g}}{6.941\dfrac{\text{g}}{\text{mol}}} = 0.072\text{ mol Li}$$

Thus, the negative charge $(-)$ leaving the system is equal to the amount of charge generated by the oxidation of lithium:

$$q_{-,\text{out}} = q_{-,\text{gen}} = 0.072\text{ mol Li} = 0.072\text{ mol}(-)$$

Using Faraday's constant, we can convert the molar amount of charge to coulombs:

$$q_{-,\text{out}} = (0.072\text{ mol}(-))\left(\frac{96{,}485\text{ C}}{\text{mol}(-)}\right) = 6950\text{ C}$$

Thus, approximately 7000 C of charge is available when the lithium-iodide battery is completely discharged. A typical pacemaker battery has a capacity of approximately 6000 C to 8000 C, so this is a reasonable answer. ■

5.10 Systems with Generation or Consumption Terms—Focus on Electrical Energy

In a reacting system, electrical energy can be either generated or consumed or both. When rates of electrical energy are given, it is appropriate to use the differential form of the electrical energy accounting equation [5.4-2]:

$$\sum_k \dot{E}_{E,k} - \sum_j \dot{E}_{E,j} + \sum \dot{G}_{\text{elec}} - \sum \dot{W}_{\text{elec}} = \frac{dE_E^{\text{sys}}}{dt} \quad [5.10\text{-}1]$$

The integral equation is formulated in equation [5.4-4].

In Section 5.6, we discuss elements that generate and consume electrical energy, such as batteries and resistors, respectively. As shown in equation [5.6-4], the rate at which elements generate electrical energy is:

$$\sum \dot{G}_{\text{elec}} = iv_b \quad [5.10\text{-}2]$$

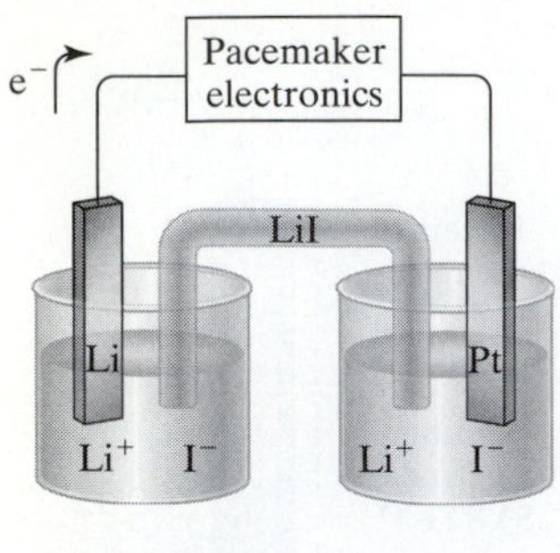

Figure 5.41a
Pacemaker attached to a lithium-iodide battery.

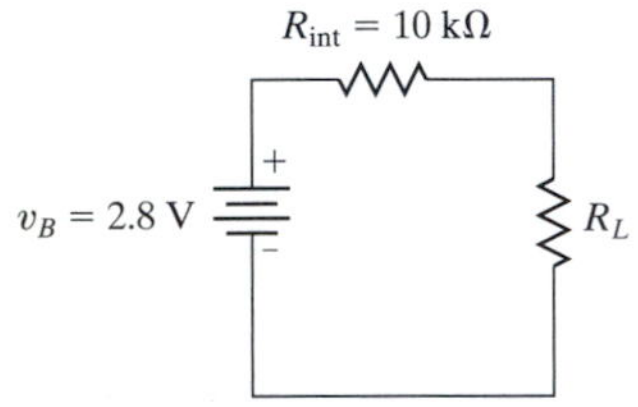

Figure 5.41b
Model of lithium-iodide electrochemical cells and the pacemaker as a power source with two resistances.

where i is current through the element and v_b is the voltage gain across the element. As shown in equation [5.6-6], the rate at which elements consume electrical energy is:

$$\sum \dot{W}_{elec} = iv_R \qquad [5.10\text{-}3]$$

where v_R is the voltage drop across the element. The terms $\sum \dot{G}_{elec}$ and $\sum \dot{W}_{elec}$ are both representations of power. Thus, the presented derivations and explanations are consistent with the formula from physics:

$$P = iv \qquad [5.10\text{-}4]$$

where P is power, i is current, and v is voltage.

The derivation and application of KVL relies on these expressions for the generation and consumption of electrical energy. The power consumption during the release of energy from a capacitor or an inductor is included in Section 5.8. In this section, we consider a few more applications involving the Generation and Consumption terms of the governing electrical energy accounting equation.

EXAMPLE 5.22 **Lithium-Iodide Battery in Pacemakers**

Problem: A pacemaker is attached to a lithium-iodide battery similar to the one in Example 5.21. The electrochemical half-cell setup for a lithium iodide battery is shown in Figure 5.41a. We can model the electrochemical cells and the pacemaker as a power source with two resistances (Figure 5.41b). The open-circuit voltage (v_B), measured across the terminals with no load attached, is 2.8 V. (An open-circuit voltage assumes no internal resistance from the battery or wiring itself. Batteries typically have some inherent resistance, which is often ignored in solving theoretical problems.) Suppose the reactions of the electrochemical cells in the battery have an internal resistance (R_{int}) of 10 kΩ. If the battery contains 0.60 g of lithium metal at the anode, what is the average resistance of the pacemaker, R_L, at the end of a battery's life (8 to 10 years)? What is the power generated by the battery?

Solution:

1. Assemble
   (a) Find: The average resistance of and the power generated by the pacemaker at 8 years and at 10 years.
   (b) Diagram: Two systems must be defined to solve this problem. The first system is the same as in Example 5.21, where the system is defined as the lithium anode (Figure 5.40b). The second system is defined as the entire circuit model with the pacemaker and battery (Figure 5.41b).
2. Analyze
   (a) Assume:
   - The battery solution is well mixed.
   - The battery and the pacemaker do not leak charge.
   - The systems are at steady-state.

   (b) Extra data: No extra data are needed.
   (c) Variables, notations, units:
   - Units: mol, C, V.

   (d) Basis: The battery initially contains 0.60 g of lithium, so we calculate a basis:

$$n_{Li} = \frac{m_{Li}}{M_{Li}} = \frac{0.60\ g}{6.941 \frac{g}{mol}} = 0.086\ mol\ Li$$

   (e) Reactions:

$$\text{Cathode: } I_2 + 2\ e^- \longrightarrow 2\ I^-$$
$$\text{Anode: } Li \longrightarrow Li^+ + e^-$$

3. Calculate
   (a) Equations: As in Example 5.21, we can use the governing equation for negative charge [5.9-2] to calculate the total amount of charge produced by the battery:

$$\sum_k q_{-,k} - \sum_j q_{-,j} + q_{-,\text{gen}} - q_{-,\text{cons}} = q_{-,\text{acc}}^{\text{sys}}$$

We use the differential form of the electrical energy accounting equation, since reactions are involved:

$$\sum_k \dot{E}_{E,k} - \sum_j \dot{E}_{E,j} + \sum \dot{G}_{\text{elec}} - \sum \dot{W}_{\text{elec}} = \frac{dE_E^{\text{sys}}}{dt}$$

For the steady-state circuit, we can use KCL, KVL, and Ohm's law to determine the average resistances of the battery:

$$\sum_k i_k - \sum_j i_j = 0$$

$$\sum_{\text{loop}} v_{\text{elements}} = 0$$

$$v = iR$$

   (b) Calculate:
   - As in Example 5.21, the first system has no negative charges flowing in, being consumed, or being accumulated in the system, so we can reduce the governing equation for negative charge to:

$$-q_{-,\text{out}} + q_{-,\text{gen}} = 0$$

From the accounting equation for negative charge, we know that the negative charge flowing out of the system is equal to the negative charge generated at the lithium electrode:

$$q_{-,\text{out}} = q_{-,\text{gen}} = 0.086 \text{ mol Li}\left(\frac{1 \text{ mol}(-)}{1 \text{ mol Li}}\right)\left(\frac{96{,}485 \text{ C}}{\text{mol}(-)}\right) = 8298 \text{ C}$$

Thus, approximately 8300 C of charge flows out of the battery and into the rest of the circuit over a period of 8 to 10 years.
   - For the second system, we know the system is at steady-state and the voltages across the battery and each resistor can be calculated using KVL:

$$\sum_{\text{loop}} v_{\text{elements}} = v_B - v_{\text{int}} - v_L = 0$$

$$v_B = v_{\text{int}} + v_L$$

where v_B is the open-circuit voltage, v_{int} is the voltage across the resistor that represents the internal resistance of the battery, and v_L is the voltage across the pacemaker. Substituting in Ohm's law gives:

$$v_B = i_{\text{int}} R_{\text{int}} + i_L R_L$$

   - Because net charge does not accumulate anywhere in the system, we can apply KCL to the developed KVL equation. Therefore, as we discuss in Sections 5.5 and 5.6, the current along any point in the circuit is constant. This simplifies our KVL equation to:

$$v_B = i(R_{\text{int}} + R_L)$$

where i is the current through the circuit.
   - Using the definition of current, we can calculate the amount of current that flows through the circuit at 8 and 10 years:

$$\text{8 years:} \qquad i = \frac{dq}{dt} = \frac{8298\ \text{C}}{8\ \text{years}}\left(\frac{1\ \text{year}}{365\ \text{days}}\right)\left(\frac{1\ \text{day}}{24\ \text{hr}}\right)\left(\frac{1\ \text{hr}}{3600\ \text{s}}\right) = 33\ \mu\text{A}$$

Using this formula to calculate the current at 10 years gives a value of 26 μA.

- Using the calculated currents at 8 and 10 years, we can rearrange our simplified KVL equation to calculate R_L:

$$\text{8 years:} \qquad R_L = \frac{v_B}{i} - R_{\text{int}} = \frac{2.8\ \text{V}}{33\ \mu\text{A}} - 10\ \text{k}\Omega = 75\ \text{k}\Omega$$

Performing a similar calculation for the pacemaker at 10 years gives an average resistance of 98 kΩ.

- The first system has no electrical energy flowing in (as current), being consumed, or being accumulated in the system. So, we can reduce the governing equation for electrical energy to:

$$-\sum_j \dot{E}_{E,j} + \sum \dot{G}_{\text{elec}} = 0$$

Using the values at 8 years, the power generated by the battery is:

$$\sum \dot{G}_{\text{elec}} = \sum_j \dot{E}_{E,j} = iv = (33\ \mu\text{A})(2.8\ \text{V}) = 9.24 \times 10^{-5}\ \frac{\text{J}}{\text{s}}$$

Using a time of 10 years, the generated power is 7.28×10^{-5} J/s. Some of that power is dissipated to run the pacemaker; some is lost in the battery itself.

4. Finalize
   (a) Answer: Assuming a battery lasts from 8 to 10 years, the range of load resistance the pacemaker sees is from 75 kΩ to 98 kΩ. The power generated by the battery ranges from 7.28×10^{-5} J/s to 9.24×10^{-5} J/s.
   (b) Check: It is hard to get an independent check on these values. Checking with a pacemaker manufacturer might help you decide whether the values are reasonable. ■

EXAMPLE 5.23 Power of a Thermocouple

Problem: A thermocouple is a device that uses the conversion of thermal energy to electrical energy to take a temperature measurement. A thermocouple is composed of a pair of dissimilar metal wires (e.g., copper and iron) fused together. Keeping one junction of the wires at a known reference temperature, the other junction of the thermocouple can be heated such that an electric potential difference results, causing current to flow. A potential difference of 10 mV is produced by a certain thermocouple. The resulting current through the connected circuit is 1000 μA. At what rate is heat converted into electric power?

Solution: The system encompasses part of the thermocouple (Figure 5.42). Thermocouples do not accumulate charge or energy, so the system is steady-state. The temperature gradient induces

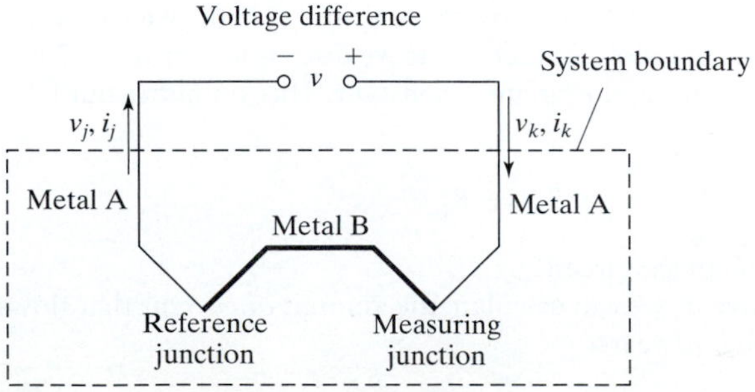

Figure 5.42
Thermocouple system. (*Source:* Cogdell JR, *Foundations of Electrical Engineering*, 2d ed. Upper Saddle River, NJ: Prentice Hall, 1996.)

an energy generation term from within the thermocouple. Energy at a specific rate enters and leaves the system with this current. The differential energy accounting equation is simplified to:

$$\sum_k \dot{E}_{E,k} - \sum_j \dot{E}_{E,j} + \sum \dot{G}_{\text{elec}} = 0$$

Based on KCL, a constant current (i) flows in and out of the system. The rate of energy generation is written in terms of the current and voltage difference:

$$\sum \dot{G}_{\text{elec}} = \sum_j \dot{E}_{E,j} - \sum_k \dot{E}_{E,k} = i_j v_j - i_k v_k = i(v_j - v_k)$$

Substituting known values of the voltage difference and current into the equation gives:

$$\sum \dot{G}_{\text{elec}} = (1000\ \mu\text{A})(10\ \text{mV}) = 1 \times 10^{-5}\ \frac{\text{J}}{\text{s}}$$

This thermocouple produces 10 μW of electrical power because of the temperature gradient it experiences between its junctions. In order to satisfy the conservation of total energy, this temperature gradient must be maintained by some external thermal power source. ■

Summary

In this chapter, we described basic charge and electrical energy concepts, which included definitions for current and voltage. We formulated the accounting statement for the extensive properties of positive charge, negative charge, and electrical energy. We described why the conservation statement can be applied to net charge.

We explored common circuit elements including resistors, capacitors, batteries, and inductors. We derived Kirchhoff's current law and Kirchhoff's voltage law from the appropriate charge and electrical energy accounting equations. Together with Ohm's law, KCL and KVL were applied to various circuits and to model biological membranes. We analyzed how the equations can be used to solve for variables in dynamic systems. We posed and solved a range of problems involving reacting systems.

Table 5.5 reinforces that electrical energy may accumulate in a system because of bulk material transfer across the system boundary, direct and nondirect contacts, or energy interconversions. Positive and negative charge may accumulate because of bulk material transfer or chemical reactions; net charge accumulates only because of bulk material transfer. See the tables concluding other chapters for comparison.

TABLE 5.5

Summary of Movement, Generation, Consumption, and Accumulation in the Electrical Energy and Charge Accounting Equations

Accumulation	Input − Output		+ Generation − Consumption	
Extensive property	Bulk material transfer	Direct and non-direct contacts	Chemical reactions	Energy interconversions
Electrical energy	X	X		X
Net charge	X			
Positive charge	X		X	
Negative charge	X		X	

References

1. Jaeger RJ. "Principles underlying functional electrical stimulation techniques." *J Spin Cord Med* 1996, and 19:93–6.
2. Grill WM and Kirsch RF. "Neuroprosthetic applications of electrical stimulation." *Assist Technol* 2000, 12:6–20.
3. Stieglitz T, Schuettler M, and Koch KP. "Neural prostheses in clinical applications—trends from precision mechanics towards biomedical microsystems in neurological rehabilitation." *Biomed Tech (Berl)* 2004, 49:72–7.
4. Sadowski CL. "Electrical stimulation in spinal cord injury." *NeuroRehabilitation* 2001, 16:165–9.
5. Peckham PH and Creasey GH. "Neural prostheses: Clinical applications of functional electrical stimulation in spinal cord injury." *Paraplegia* 1992, 30:96–101.
6. Bhadra N, Kilgore KL, and Peckham PH. "Implanted stimulators for restoration of function in spinal cord injury." *Med Eng Phys* 2001, 23:19–28.
7. Craelius W. "The bionic man: Restoring mobility." *Science* 2002, 295:1018–21.
8. Jung R, Brauer EJ, and Abbas JJ. "Real-time interaction between a neuromorphic electronic circuit and the spinal cord." *IEEE Trans Neural Syst Rehabil Eng* 2001, 9:319–26.
9. Cobbold RSC. *Transducers for Biomedical Measurements: Principles and Applications.* New York: John Wiley & Sons, 1974.
10. Dekker C and Ratner M. "Electronic properties of DNA." *Physics World* 2001. http://physicsweb.org/articles/world/14/8/8 (accessed January 8, 2005).
11. National Nanofabrication Users Network. "The Research Experience for Undergraduates Program: Research Accomplishments 2000." http://www.nnin.org/doc/2000NNUNreuRA.pdf (accessed January 24, 2006).
12. Guyton AC and Hall JE. *Textbook of Medical Physiology.* Philadelphia: Saunders, 2000.

Problems

5.1 In cells, ions often flow through conduits in the membrane known as ion channels. These channels allow the passage of various kinds of charged particles through the nonpolar membrane. A particular channel exchanges positively-charged hydrogen ions with negatively-charged carbonate ions. Assume that during the time the channel is open, 4.9×10^9 H^+ ions flow into the cell and the same amount of CO_3^- ions flow out. Given that the length of a typical ion channel is 16 Å and that the current produced is 6.2×10^{-12}A, what is the average velocity of the ions in cm/s? Assume that all the ions fit entirely within the channel at one time.

5.2 Cellular membranes create voltage potentials by separating the charged ions. A membrane with a 70 mV absolute potential difference has 1×10^4 Na^+ ions flowing through it. Assuming the potential is constant, what change of electrical potential energy is experienced by the Na^+ ions?

5.3 Suppose the voltage of a battery is rated at 6 V. The current that is produced is 3 A. What is the power output of the battery?

5.4 Apply the following questions to both the circuit shown in Figure 5.43(a) (resistors in series) and the circuit shown in Figure 5.43(b) (resistors in parallel).

(a) Use Ohm's law with Kirchhoff's current and voltage laws to derive equations for currents i_1, i_2, and i (through resistor R_1, resistor R_2, and the voltage source, respectively) in terms of R_1, R_2, and v.

(b) R_1 and R_2 can be interchanged with an equivalent resistor with resistance R without changing the values of v and i. Show the derivation of an equation for the equivalent resistance R in terms of R_1 and R_2.

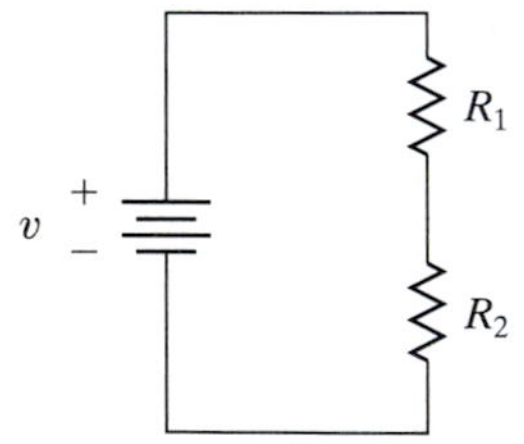

(a)

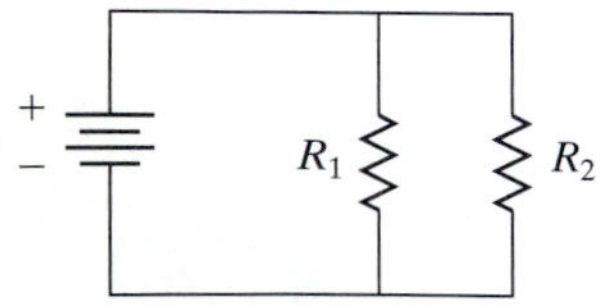

(b)

Figure 5.43
Circuit diagrams for Problem 5.4.

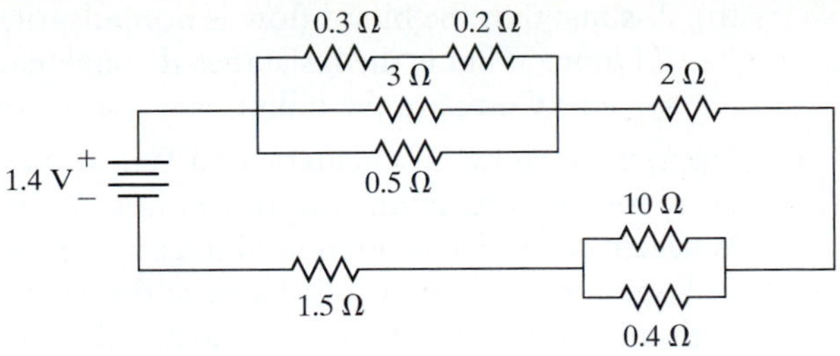

Figure 5.44
Circuit diagram for Problem 5.5.

5.5 For the circuit shown in Figure 5.44, use Ohm's law with Kirchhoff's current and voltage laws to determine the following:
(a) The currents through each of the resistors and the voltage source.
(b) The equivalent resistance of all of the resistors. (*Note:* It might be useful to use a program such as MATLAB to solve the system of equations.)

5.6 Often systems of mass or material flow can be described using circuit analogs, because of similarities between mass flow and current. Just as electric charge can be driven in a current by potential difference, mass can be driven by differences in pressure between two points. Current flowing through resistors results in a voltage drop. Likewise, as mass flows, it also experiences a decrease in pressure as it moves through frictional (resistive) elements.

A model of blood flow through systemic and pulmonary circulation is shown in Figure 5.45. Between each two components of the circulatory system (modeled as circuit elements), an approximate blood pressure is given.
(a) Derive an equation relating mass flow to pressure drop. Verify that an analog of Kirchhoff's voltage law applies to the system shown. Based on the derivation of KVL, what can you say about the system (i.e., is the system at steady-state)?

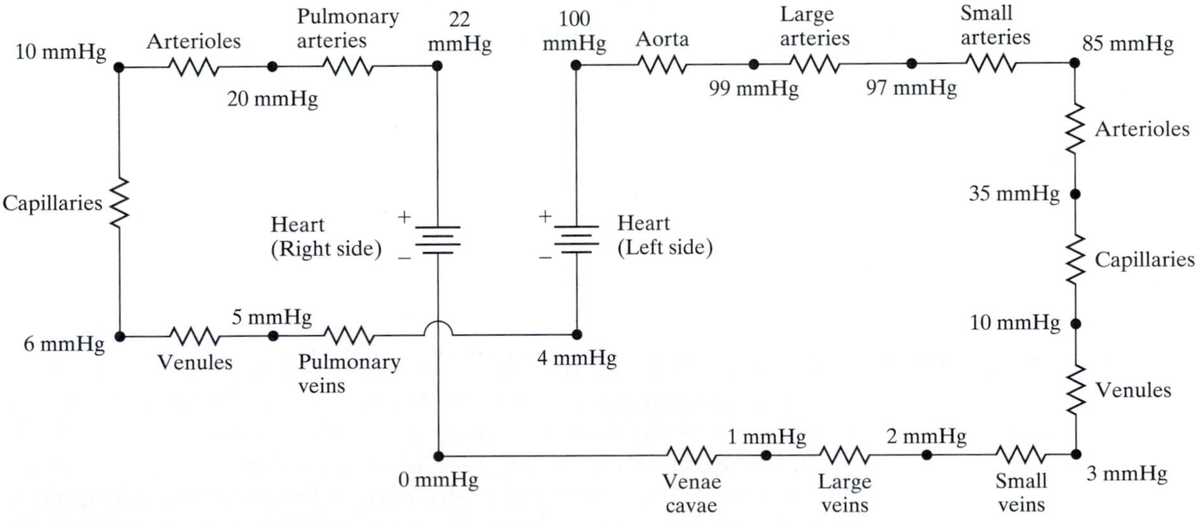

Figure 5.45
A model of blood flow through systemic and pulmonary circulations.

(b) Assume that the blood flow is nonpulsatile and the volumetric flow rate is 5 L/min. What is the resistance through each component of the circulatory system? Compare the pulmonary and systemic resistances.

5.7 Figure 5.46 shows a schematic of a Wheatstone bridge, a circuit configuration used to measure unknown resistances. In bioengineering applications, the Wheatstone bridge is often used in gauges that evaluate mechanical properties of bones, muscles, and cells because the resistances of those materials change with mechanical deformation. The circuit element denoted G represents a galvanometer, a device that measures small amounts of current. Resistances R_1 and R_2 are fixed and known. To determine the unknown resistance R_x, R_3 is varied so that the current through the galvanometer is zero. Using Ohm's law, KCL, and KVL, determine the unknown resistance (R_x) in terms of the known resistances.

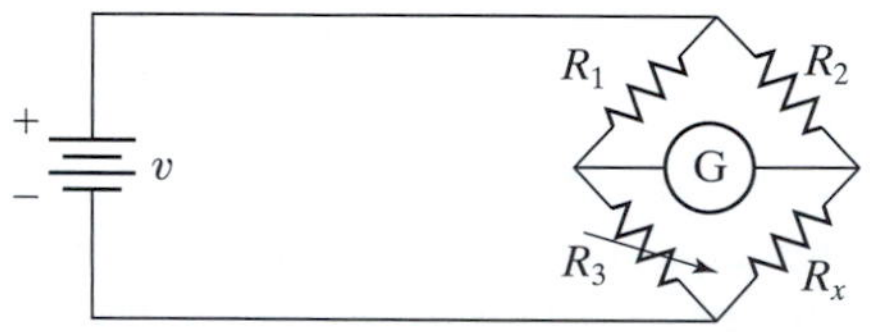

Figure 5.46
Wheatstone bridge with a galvanometer.

5.8 A thermistor is a device whose resistance decreases as temperature increases. The resistance of the thermistor (R_t) is related to its absolute temperature (T) in the following equation:

$$\frac{dR_t}{dT} = -\frac{(\beta \times R_t)}{T^2}$$

where β is the material constant for the thermistor and T is the temperature in degrees Kelvin. To determine the temperature of a premature infant, you attach a thermistor to the infant's abdomen. The thermistor is part of the Wheatstone bridge shown in Figure 5.47, with $R_1 = R_2 = 4500\ \Omega$. The thermistor is known to have a resistance of 5000 Ω at 25°C and it has a material constant of 4000 K. If the current through the galvanometer is zero when R_3 is 3100 Ω, what is the temperature of the baby?

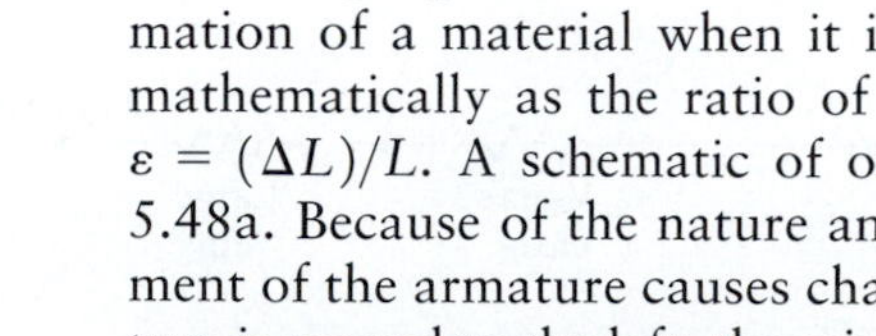

Figure 5.47
Wheatstone bridge with a thermistor.

5.9 A strain gauge is a device that uses resistance to measure strain, the deformation of a material when it is subjected to a force. Strain (ε) is defined mathematically as the ratio of change in length to original length, L, as: $\varepsilon = (\Delta L)/L$. A schematic of one type of strain gauge is shown in Figure 5.48a. Because of the nature and positioning of the resistance wires, movement of the armature causes changes in resistance. For example, if the armature is moved to the left, the wires of R_1 and R_4 are stretched identically. The resulting increase in length and decrease in cross-sectional area of the wires causes an increase in resistance of those wires. In addition, deformation of

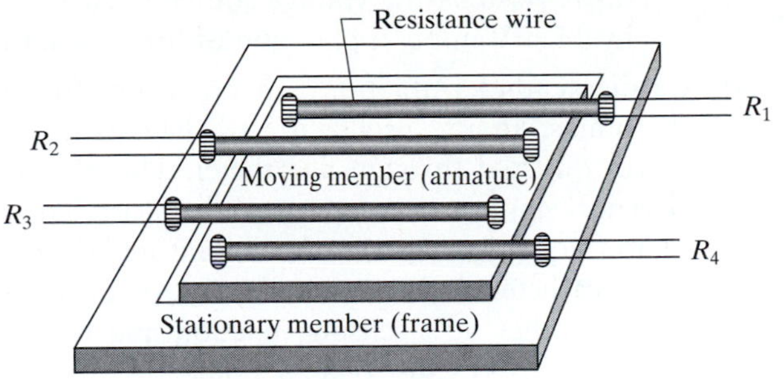

Figure 5.48a
A schematic of a type of strain gauge. (*Source:* Cobbold RSC, *Transducers for Biomedical Measurements: Principles and Applications*. New York: John Wiley & Sons, 1974, p. 121. Figure originally from Bartholomew D., *Electrical Measurements and Instrumentation*. Boston: Allyn and Bacon, 1963.)

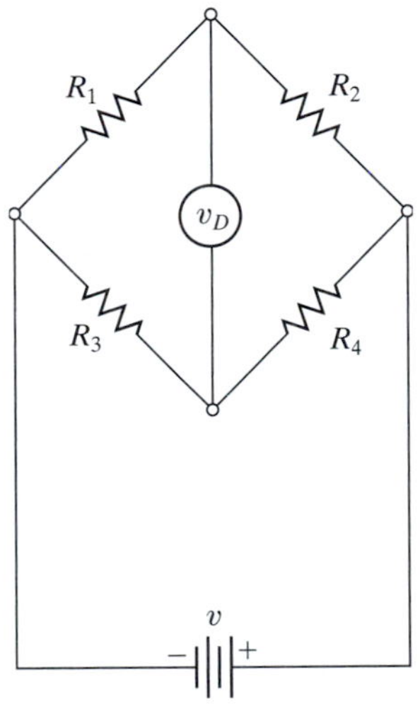

Figure 5.48b
Wheatstone bridge with a voltmeter.

the wires may also change their resistivity, ρ. At the same time, the slight tension release in resistance wires R_2 and R_3 causes a decrease in length, an increase in cross-sectional area, and therefore a decrease in resistance. These parameters can be combined into a gauge factor, G:

$$G = \frac{\Delta R/R}{\Delta L/L}$$

The changes in resistance are measured using the Wheatstone bridge circuit shown in Figure 5.48b. Note that the circuit is similar to the one used in Problem 5.7, except it uses a voltmeter instead of a galvanometer. The voltmeter measures the potential difference across the elements in parallel with it. Voltmeters have such a high resistance that current through them can be considered negligible.

(a) Use KCL, KVL, and Ohm's law to derive the following equation:

$$\Delta v_0 = v \cdot \frac{\Delta R}{R}$$

where $R_1 = R_4 = R + \Delta R$, $R_2 = R_3 = R - \Delta R$, v is the potential difference across the voltage source, and v_0 is the potential difference measured by the voltmeter.

(b) The resistance wires are made of manganin, which has a gauge factor of 0.47. Assume the voltage source is 10 V. If the voltmeter reading changes by 15 mV upon application of force, what is the strain on the material?

5.10 Cardiac catheter ablation is a procedure frequently used to correct heart arrhythmias. In the procedure, cardiac tissue is heated using radio-frequency waves directed through a catheter. This produces scarring, which blocks the electrical signal to some parts of the heart. The temperature rise induced by the radio-frequency waves is often monitored using a thermocouple that is attached to the catheter. A thermocouple consists of two different metals that are welded to each other at a sensing junction and to a resistor at a reference junction (Figure 5.49). Because of a phenomenon called the Seebeck effect, when the temperature at the sensing junction is greater than the temperature at the reference junction, current flows through the circuit. A voltmeter across the resistor shows a voltage related to the sensing junction temperature by the sensitivity:

$$S = \frac{dv}{dT_S}$$

where the sensitivity S is dependent on the metals in the thermocouple, the reference temperature, T_S, is the temperature at the sensing junction in °C, and v is given in μV. The sensitivity of copper and constantan ($Cu_{57}Ni_{43}$) is 45 μV/°C when the reference temperature, T_R, is 20°C.[9] If the temperature of the reference junction in the catheter thermocouple is 20°C and the voltmeter in the catheter thermocouple reads 3.8 mV, what is the temperature of the cardiac tissue?

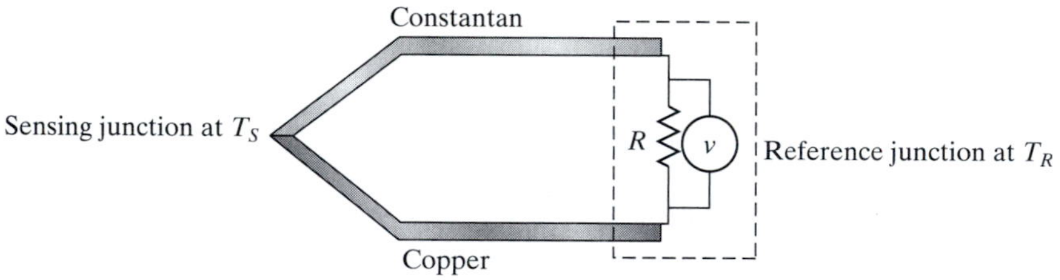

Figure 5.49
Thermocouple consisting of constantan and copper.

5.11 Bob is building a phonocardiograph, a device that records heart sounds, for his bioengineering class. When he is finished, he plugs the device into an electrical outlet, which delivers a maximum voltage of 120 V. Unfortunately, the plug breaks. Assume the body follows Ohm's law. (*Note:* Given resistances are approximate and should NOT be tested.)

(a) Bob puts the palm of his hand over the electrical outlet, so that the alternating current is being delivered to his palm. A dry human palm has a resistance of about 5 kΩ. Analysis of several accidents has shown that a person feels pain when exposed to 3 mA. Will Bob feel pain?

(b) It has also been shown that tissue is burned if exposed to more than 5 A. Will the tissue in the palm of Bob's hand be burned?

(c) Bob's friend Edna decides to help Bob out by removing the plug pieces from the electrical outlet. She takes one plug piece in each hand, causing the current to flow through her arms and chest. Assume that the resistance of each arm is 750 Ω, the resistance of the chest is 500 Ω, and resistance is uniform across the chest. It has been shown that the heart stops if exposed to 4 A. Will Edna's heart stop?

(d) If the heart is exposed to an outside current of 75 mA, it will fibrillate (flutter in a way that does not efficiently pump blood). Will Edna's heart fibrillate?

(e) Bob's friend Doris decides to remove the plug pieces in the same manner as Edna, but using rubber gloves with a resistance of 20 MΩ. What will be the current through Doris's hands, arms, and chest? How will that current affect her body?

5.12 Researchers have shown that DNA appears to be capable of transporting charge. Although the exact mechanisms of the charge transport are unknown, DNA could be used in molecular electronics, which is defined as the area of science and technology that studies electronics and sensors based on molecular organization.

One of the first theories for the transport mechanism was that DNA is like a conductor, a so-called "molecular wire." Researchers looked at the relationship between current flowing through strands of DNA and the potential difference and found DNA to be nearly ohmic.[10]

(a) Using the chart in Figure 5.50,[11] find the resistance of the ohmic region and calculate the power dissipated by the strands of DNA if current is 50 pA.

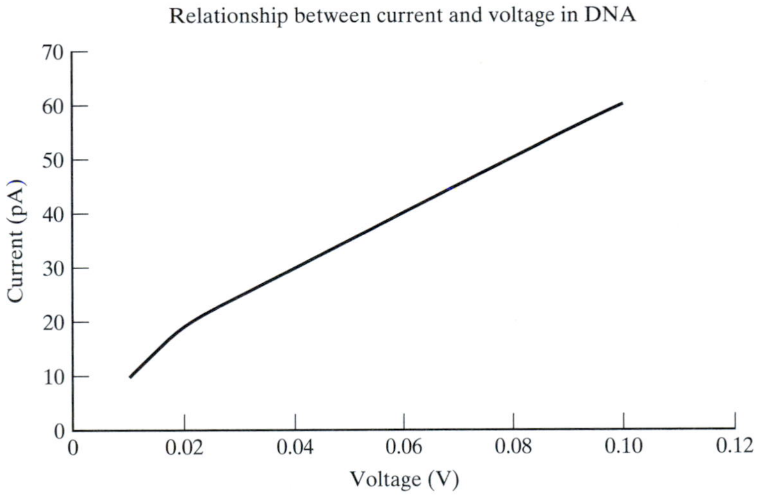

Figure 5.50 Relationship between current and voltage in DNA. (*Source:* Modified from Douglas E, "Electrical conductivity in oriented DNA." National Nanofabrication Users Network, The Research Experience for Undergraduates Program: Research Accomplishments 2000.

(b) Experiments by other researchers have shown that earlier experiments may have been contaminated by residue of other conductors and that the molecular wire idea may be incorrect. If DNA proves to be more of an insulator than a conductor, would you expect the dissipated power to be higher? Why?

5.13 (a) A voltage divider is a circuit used to split the voltage between two resistors in series. Find the voltage of each resistor and the current going through them in Figure 5.51a.

(b) A current divider is a circuit used to split the current between two resistors. Find the current to each resistor and the voltage drop associated with them in the circuit in Figure 5.51b.

(c) Compare and contrast the results from (a) and (b).

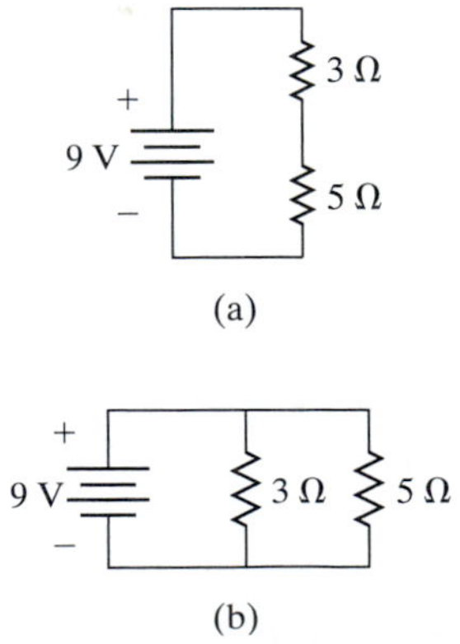

Figure 5.51 Voltage and current divider circuits for Problem 5.13.

5.14 Combining a current and voltage divider, both functions can be found in the circuits in Figure 5.52(a) and Figure 5.52(b). Find the voltage and current across each resistor.

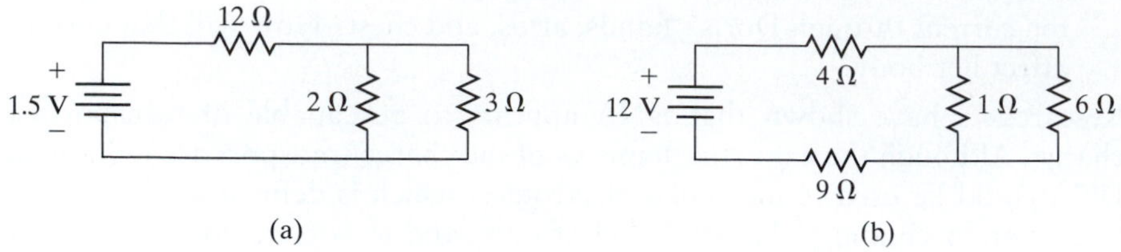

Figure 5.52
Circuit diagrams for Problem 5.14.

5.15 Voltage-divider circuits return a voltage that is a linear function of the input voltage. This function is dependent upon two resistances. For the voltage divider in Figure 5.53, what percentage of the input voltage (v_{in}) is the output voltage (v_{out}), in terms of the two resistances, R_1 and R_2?

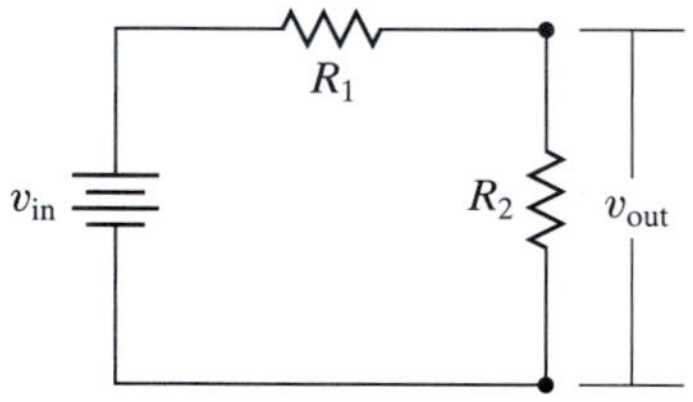

Figure 5.53
Circuit diagram for Problem 5.15.

5.16 For the circuits in Figures 5.54a and 5.54b, the resistances and voltage are: $R_1 = 5\text{ k}\Omega$, $R_2 = 100\text{ k}\Omega$, $R_3 = 200\text{ k}\Omega$, $R_4 = 150\text{ k}\Omega$, $R_5 = 250\text{ k}\Omega$, and $v_1 = 100\text{ V}$.

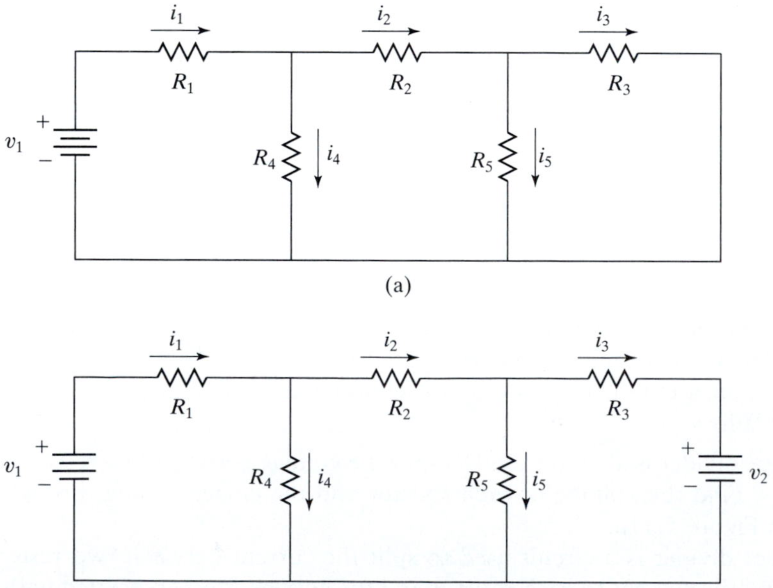

Figure 5.54
Circuit diagrams for Problem 5.16.

(a) Solve for the values of the currents i_1, i_2, i_3, i_4, and i_5 in Figure 5.54a.
(b) Voltage source v_2 is added to the circuit in Figure 5.54b. Suppose that each resistor is rated to carry a current of no more than 1 mA. Determine the allowable range of positive values for the voltage v_2. You might find the function *syms* in MATLAB to be helpful.

5.17 An ECG takes the potential of three limbs relative to the average electric potential of the body. The right arm has a potential of −0.15 mV, the left arm +0.55 mV, and the left leg +0.93 mV. What are the magnitude and the angle of deflection of the cardiac vector at this moment?

5.18 Calculate lead II from lead I and lead III based on the data in Table 5.6. Graph the ECG trace from lead II.

TABLE 5.6

ECG Leads I and III

Time (s)	Voltage (mV) lead I	Voltage (mV) lead III	Time (s)	Voltage (mV) lead I	Voltage (mV) lead III
0.01	0.00915	−0.06866	0.36	0.12222	−0.15640
0.02	0.01464	−0.06989	0.37	0.12619	−0.15610
0.03	0.00900	−0.06882	0.38	0.14129	−0.15808
0.04	0.01968	−0.05890	0.39	0.15594	−0.15640
0.05	0.02914	−0.07126	0.40	0.17135	−0.16541
0.06	0.04943	−0.07568	0.41	0.18493	−0.16617
0.07	0.05645	−0.05997	0.42	0.21148	−0.16815
0.08	0.07415	−0.05905	0.43	0.23178	−0.17365
0.09	0.06851	−0.03662	0.44	0.25390	−0.16678
0.10	0.06134	−0.07156	0.45	0.28457	−0.17624
0.11	0.04791	−0.08484	0.46	0.31921	−0.17365
0.12	0.04821	−0.07782	0.47	0.35430	−0.15762
0.13	0.03814	−0.07431	0.48	0.39062	−0.14923
0.14	0.01068	−0.07004	0.49	0.41793	−0.13672
0.15	0.00915	−0.06836	0.50	0.44006	−0.12512
0.16	0.00900	−0.06470	0.51	0.42480	−0.09750
0.17	0.00091	−0.05905	0.52	0.36834	−0.06790
0.18	0.00061	−0.06058	0.53	0.28503	−0.04883
0.19	−0.00061	−0.05890	0.54	0.18493	−0.03189
0.20	−0.00427	−0.05829	0.55	0.10940	−0.03464
0.21	−0.04044	−0.01968	0.56	0.05630	−0.03937
0.22	0.26550	−0.13580	0.57	0.02822	−0.04105
0.23	0.57754	0.36712	0.58	0.01861	−0.04944
0.24	0.66955	0.98327	0.59	0.00915	−0.04227
0.25	−0.10773	0.71655	0.60	0.01037	−0.04990
0.26	−0.17227	−0.20477	0.61	0.01876	−0.05615
0.27	−0.19013	−0.12802	0.62	0.02121	−0.05859
0.28	−0.03616	−0.10864	0.63	0.02639	−0.06027
0.29	0.05874	−0.12726	0.64	0.02777	−0.05890
0.30	0.06271	−0.13702	0.65	0.02868	−0.06866
0.31	0.06805	−0.13458	0.66	0.02822	−0.06943
0.32	0.07965	−0.13672	0.67	0.02883	−0.06866
0.33	0.08773	−0.14069	0.68	0.02853	−0.06882
0.34	0.09857	−0.14618	0.69	0.02868	−0.05859
0.35	0.11627	−0.15686	0.70	0.02563	−0.06119

(Continued)

TABLE 5.6 (*Continued*)

ECG Leads I and III

Time (s)	Voltage (mV) lead I	Voltage (mV) lead III	Time (s)	Voltage (mV) lead I	Voltage (mV) lead III
0.71	0.01861	−0.06165	1.02	0.05432	−0.05295
0.72	0.01892	−0.05890	1.03	0.07095	−0.04807
0.73	0.01892	−0.05890	1.04	0.06607	−0.01801
0.74	0.01236	−0.05096	1.05	0.04806	−0.05524
0.75	0.01785	−0.05920	1.06	0.03921	−0.06958
0.76	0.01419	−0.04883	1.07	0.03677	−0.06287
0.77	0.00793	−0.04288	1.08	0.02838	−0.04929
0.78	0.01022	−0.04730	1.09	−0.00793	−0.04883
0.79	0.00762	−0.04929	1.10	0.00061	−0.05890
0.80	0.00915	−0.04913	1.11	−0.00458	−0.04669
0.81	0.00366	−0.04913	1.12	−0.01114	−0.04868
0.82	0.00137	−0.05630	1.13	−0.01984	−0.04288
0.83	0.00351	−0.05249	1.14	−0.01465	−0.04028
0.84	−0.00031	−0.04883	1.15	−0.02029	−0.04593
0.85	0.00640	−0.05096	1.16	−0.05325	0.00808
0.86	−0.00122	−0.05142	1.17	0.21759	−0.10498
0.87	−0.00076	−0.04868	1.18	0.56915	0.31143
0.88	−0.00031	−0.04929	1.19	0.70739	0.95703
0.89	0.00671	−0.05417	1.20	−0.08911	0.81192
0.90	0.00534	−0.05280	1.21	−0.18814	−0.10407
0.91	0.00320	−0.05936	1.22	−0.20508	−0.11398
0.92	0.00885	−0.05051	1.23	−0.07141	−0.09186
0.93	0.00305	−0.05890	1.24	0.03707	−0.09857
0.94	0.00305	−0.05768	1.25	0.04287	−0.10223
0.95	0.00061	−0.05325	1.26	0.04898	−0.10742
0.96	−0.00015	−0.05905	1.27	0.06835	−0.10834
0.97	−0.00122	−0.04868	1.28	0.07537	−0.11780
0.98	−0.00351	−0.05493	1.29	0.07888	−0.11612
0.99	0.00793	−0.03983	1.30	0.09246	−0.12589
1.00	0.01953	−0.06012	1.31	0.09765	−0.12680
1.01	0.03387	−0.05371	1.32	0.11413	−0.12253

5.19 Use KVL, Ohm's law, the definition of current, and the definition of capacitance to find an equation for the amount of charge stored in the capacitor for the circuit in Figure 5.55. Your equation should be in terms of capacitance, voltage, resistance, and time. Assume that the battery is attached at time $t = 0$, when the charge on the capacitor is zero.

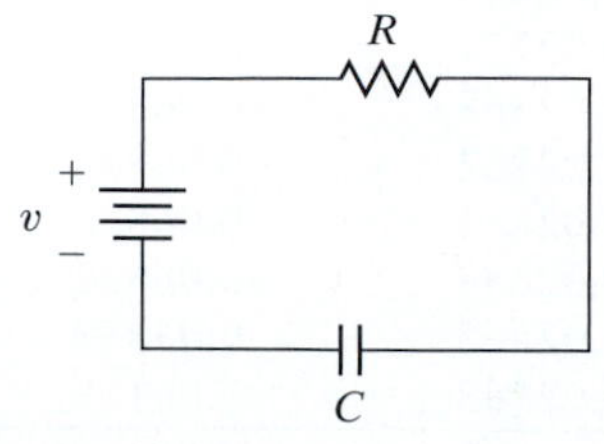

Figure 5.55
Circuit diagram for Problem 5.19.

5.20 Apply the following questions to circuit (a) (capacitors in series) and circuit (b) (capacitors in parallel), shown in Figure 5.56.

(a) Use the definition of capacitance with KVL to derive equations for the amounts of charge (q_1 and q_2) stored by each capacitor and the total charge (q) stored by both capacitors. Your answers should be in terms of C_1, C_2, and v.

(b) C_1 and C_2 can be interchanged with an equivalent capacitor with capacitance C that stores the same amount of charge q for a certain voltage v. Derive an equation for the equivalent capacitance C in terms of C_1 and C_2.

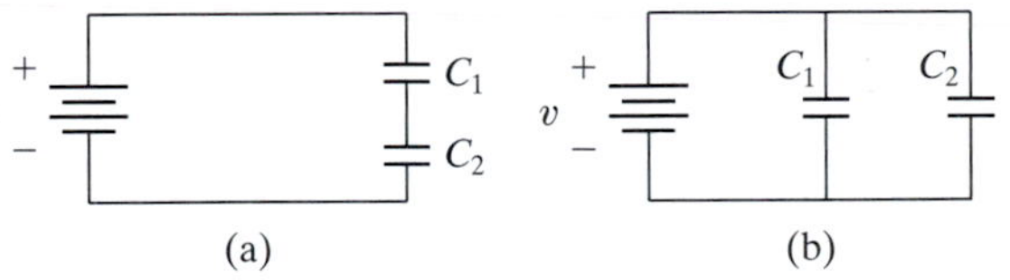

Figure 5.56
Circuit diagrams for Problem 5.20.

5.21 For each of the circuits (a), (b), and (c) in Figure 5.57, use the definition of capacitance along with KVL and KCL to determine the charge stored by each of the capacitors, the total charge stored, and the equivalent capacitance for the circuits.

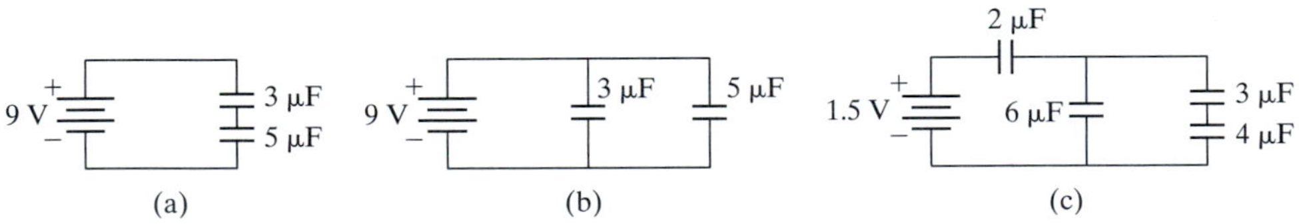

Figure 5.57
Circuit diagrams for Problem 5.21.

5.22 Over time, a capacitor that is hooked up to a voltage source will charge until its voltage matches the source voltage. The time this takes is dependent on the capacitance and the resistance of the circuit. Answer the following questions for circuits (a) and (b) in Figure 5.58.

(a) If the battery is connected to the circuit at time $t = 0$, what is the charge on the capacitor as a function of time as the capacitor is charging?

(b) What is the current through the resistor as a function of time as the capacitor is charging?

(c) At what time when the capacitor is charging will the current through the resistor be 1 μA?

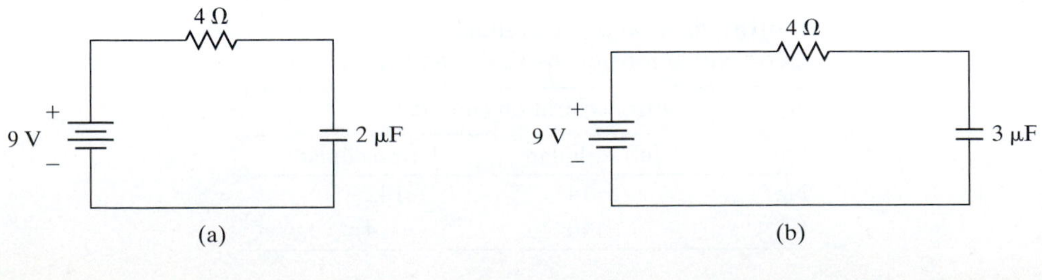

Figure 5.58
Circuit diagrams for Problem 5.22.

5.23 The circuits (a) and (b) in Figure 5.59 contain capacitors that can be discharged. Answer the following questions for both circuits.

(a) When the capacitor is fully charged, the battery is removed from the circuit. What is the charge on the capacitor as a function of time as the capacitor is discharging? Assume that the battery is removed at time $t = 0$.

(b) What is the current through the resistor as a function of time as the capacitor is discharging?

(c) At what time when the capacitor is discharging will the current through the resistor be 1 μA?

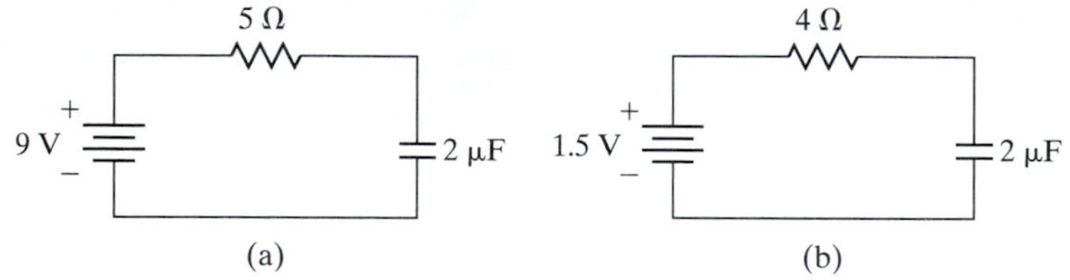

Figure 5.59
Circuit diagrams for Problem 5.23.

5.24 The cell membrane can be modeled as a plate capacitor, with the lipid membrane as the insulator and the intracellular and extracellular fluid as conducting plates. It has been experimentally shown that biological membranes typically have a capacitance of 1 μF per square centimeter of membrane. Use the definition of capacitance with Faraday's constant to determine the charge and the number of moles of excess ions stored on 1 cm^2 of each of the following membranes:

(a) Smooth muscle cell, with a resting intracellular potential of −50 mV to −60 mV.

(b) Large nerve fiber, with a resting intracellular potential of −90 mV.

(c) Large nerve fiber, with an "overshoot" action potential of +35 mV.

5.25 An action potential is propagated down a neuron by means of several sodium and potassium channels and pumps. Normally, each section of neuron has a resting membrane potential of −90 mV. This potential is created by a Na^+/K^+ pump, which pumps three Na^+ ions out of the cell for every two K^+ ions pumped into the cell, and by K^+/Na^+ leak channels, which are 100 times more permeable to K^+ than to Na^+. The potential difference change during an action potential is shown in Figure 5.31.

(a) If the concentrations of Na^+ and K^+ for the equilibrium state of the cell are shown in Table 5.7, what is the contribution of the ions to the resting membrane potential of the nerve? Does this answer differ from the expected resting potential? Explain.

(b) As mentioned in Problem 5.24, the cell membrane can be modeled as a capacitor. The equation for the definition of capacitance can be differentiated with respect to time to determine the current across the capacitor. Use

TABLE 5.7

Intracellular and Extracellular Ion Concentrations of the Cell Membrane

	Concentration (mEq/L)	
Ion	Intracellular	Extracellular
Na^+	14	142
K^+	140	4

Figure 5.31 to determine the current through 1 cm^2 of the neuron membrane during depolarization and repolarization. What is the rate of sodium and potassium ions passing through 1 cm^2 of the membrane during depolarization and repolarization? Recall that the charge of a proton is 1.6×10^{-19} C and that biological membranes typically have a capacitance of 1 μF per square centimeter of membrane.

5.26 Before digital timers were employed, analog timers based on *RC* circuits were employed to keep track of time. In the circuit in Figure 5.60, time was measured by how fast the capacitor charged up, thus giving a constant current.

(a) If the battery is connected to the circuit at time $t = 0$, find the charge on the capacitor as a function of time as the capacitor is charging.

(b) What is the current through each resistor as a function of time as the capacitor is charging?

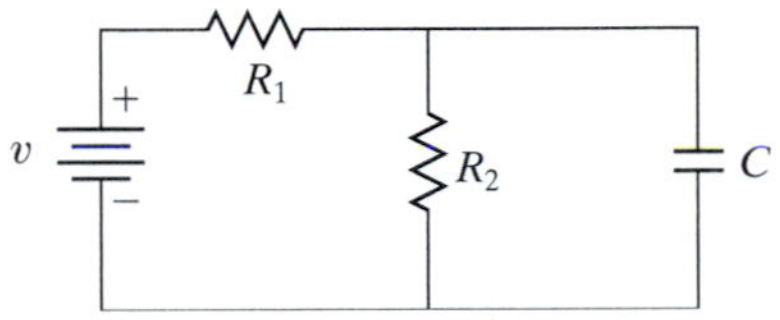

Figure 5.60
Circuit diagram for Problem 5.26.

5.27 Analog timers found use as the countdown mechanism in various triggering devices, such as for a time bomb. The idea was that when the current flow stopped after the discharging of a capacitor, the triggering mechanism would be activated.

(a) Consider the circuit in Figure 5.60. When the capacitor is fully charged, the battery is removed, creating a short circuit. What is the charge on the capacitor as a function of time as the capacitor is discharging?

(b) What is the current through each resistor as a function of time as the capacitor is discharging?

5.28 Ventricular fibrillation is a serious cardiac arrhythmia that can quickly become fatal. Fibrillation happens when contraction of individual cardiac muscle cells is not synchronized. A short, strong current can be applied to defibrillate the heart and throw the cells back into sync.

(a) For patients at high risk of tachycardia and fibrillation, implantable cardioverter-defibrillators (ICDs) can be put inside the body to provide quick treatment if needed. ICDs often contain capacitors. An ICD contains two identical capacitors, each with a capacitance of 200 μF and a maximum energy storage capacity of 75 J. What is the maximum voltage that can be applied to one of the capacitors?

(b) The energy stored in the capacitors may not all be delivered to a patient because the system is not ideal (e.g., there are losses along the circuit and at the electrodes). If you need only 750 V to shock the heart back into synchronization, what would the minimum efficiency of the ICD in part (a) have to be?

5.29 An electrical capacitor consists of two conductors separated by an insulator. In terms of this definition, the cell membrane can be modeled as a capacitor, with the intracellular fluid and extracellular fluid being the two conductors and the membrane the insulating layer. The cell membrane is more complicated than a simple capacitor, however, because there are ion channels that allow ion flow, and thus current is generated. One model of a cell membrane is shown in Figure 5.61.

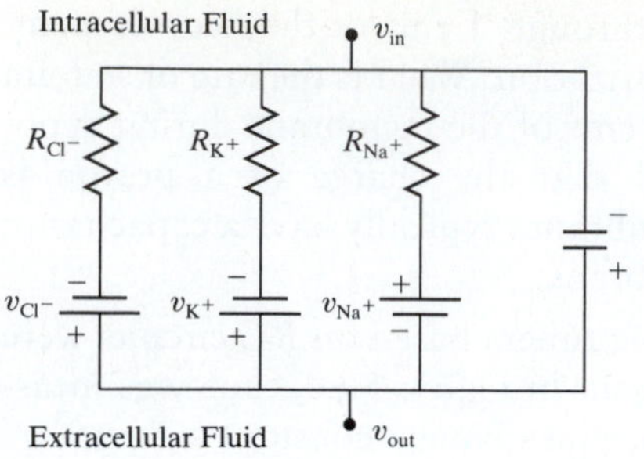

Figure 5.61
Circuit model of cell membrane ion channels.

The resistors represent the resistance through the ion channels to ion flow. The voltage sources (batteries) represent the potential difference across the membrane caused by concentration gradients of each type of ion. Given this model of the cell membrane, derive an equation for the current across the cell membrane, i_m, in terms of the capacitance, the potential differences and resistances of the ions in the model, and the overall potential difference across the membrane.

5.30 An artificial pacemaker provides an electrical stimulus to bring cardiac muscle cells to threshold and initiate action potentials when the pacemaker cells in the heart are not working properly. Suppose that the pacemaker provides a current applied as a square-wave pulse (consider it to be a step input, as you are only interested in the sudden increase in current). The cell membrane can be modeled as a resistor and capacitor in parallel, as shown in Figure 5.62.

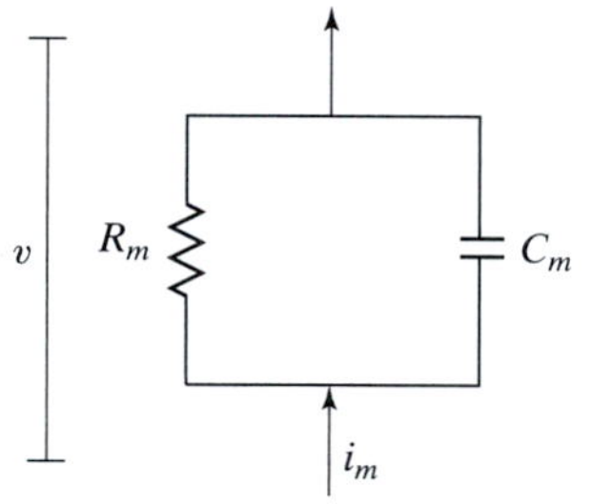

Figure 5.62
Cell membrane modeled as a resistor and capacitor in parallel.

You want to bring the membrane potential up from -90 mV to the threshold value of -55 mV (after which the Na^+ channels open and the action potential begins), so you need to apply a voltage of 35 mV. The value of the membrane resistance is 3300 Ω and the membrane capacitance is 1.5 μF. If you want the increase in voltage to occur within 5 ms, what should be the value of the applied current (i_m)?

5.31 We model a voltage source, v_s, in a defibrillator such that when it is switched on at time $t = 0$, the voltage is:

$$v_s(t) = 4000e^{-(5500\ 1/\text{s})t}\ \text{V}$$

Assuming that the human torso has a resistance of 100 Ω and that the circuit contains an inductor with an inductance of 50 mH, what is the current across the patient's torso as a function of time?

5.32 In a positron emission tomography (PET) scan, a radionuclide that decays by positron emission is injected into a patient. The PET scan works by detecting gamma rays emitted in two directions when the emitted positron combines with an electron and both are annihilated.

(a) Oxygen-15 and ^{15}O-labeled water are often used with PET scans to study oxygen metabolism. For example, PET scans can be used to determine the viability of heart tissue to determine the effectiveness of heart surgery. Oxygen-15, which has a half-life of 2.03 minutes, decays according to the following reactions:

$$^{15}_{8}\text{O} \longrightarrow {}^{15}_{7}\text{N} + {}^{0}_{+1}\beta + {}^{0}_{-1}\beta + \nu_e \longrightarrow {}^{15}_{7}\text{N} + \gamma + \nu_e$$

Nitrogen-15 is a stable, naturally occurring isotope. Show that net charge is conserved during these reactions.

(b) Carfentanil labeled with carbon-11 has been used to study opiate receptors in monkey and human brains. Carbon-11, which has a half-life of 20.4 minutes, decays according to the following reactions:

$$^{11}_{6}\text{C} \longrightarrow {}^{11}_{5}\text{B} + {}^{0}_{+1}\beta + {}^{0}_{-1}\beta + \nu_e \longrightarrow {}^{11}_{5}\text{B} + \gamma + \nu_e$$

$^{11}_{5}$B is a stable, naturally occurring isotope. Show that net charge is conserved during these reactions.

5.33 Strontium-89 has been shown to relieve the pain of bone metastases in patients with certain kinds of cancer. Strontium-89, with a half-life of 50.5 days, is administered intravenously and is incorporated into bone with a preference for metastatic regions. When it decays, the isotope becomes more stable by transforming one of its neutrons into a proton, then emitting a beta particle (electron) and an antineutrino:

$$^{89}_{38}\text{Sr} \longrightarrow {}^{89}_{39}\text{Y}^{+} + {}^{0}_{-1}\beta + \tilde{\nu}$$

where $^{0}_{-1}\beta$ is an electron and $\tilde{\nu}$ is an antineutrino, a particle with no charge. The electrons can destroy some metastases and may also deaden some nerve endings. How much charge is accumulated during this reaction?

5.34 Iodine-131, a radioactive isotope of iodine, is used to test thyroid function and treat thyroid disorders, such as hyperthyroidism and cancer.

(a) The decay of ^{131}I results in release of a beta particle and gamma radiation as well as a stable element. What is this stable element? Write out the decay reaction of ^{131}I.

(b) Given that the half-life of ^{131}I is approximately 8 days, how much negative charge does 25 g of iodine lose as beta particles in 15 days as it decays? A decay reaction may be modeled by the equation:

$$[\text{A}] = [\text{A}]_0 e^{-kt}$$

where k is the rate constant, t is time, [A] is the quantity of interest of substance A, and $[\text{A}]_0$ is the initial quantity of substance A.

5.35 In medical imaging, iodine-123 is often used to detect abnormalities in the thyroid because of its affinity toward that organ. The imaging device detects the neutrinos emitted during electron capture by iodine-123:

$$^{123}_{53}\text{I}^{+} + {}^{0}_{-1}\beta \longrightarrow {}^{123}_{52}\text{Te} + \nu_e$$

(a) How much charge is accumulated during this reaction?

(b) How many neutrinos are emitted if 2 mg of iodine-123 is given to a patient?

5.36 You are working in a candy factory that makes both cherry and lemon candy. One afternoon after lunch, you get sleepy and accidentally put the ingredients for cherry candy into the lemon candy vat. You remember from your bioengineering class that sour taste receptors detect hydrogen ion concentration to signal a sour taste, and that sweet taste receptors detect organic substances. You decide to fix your problem by simply adjusting the pH of the lemon vat. The pH of an uncontaminated vat of liquid lemon candy is 2.85. The pH of the contaminated lemon candy liquid is 3.4, and the vat contains 50 gallons of liquid. Assume that the contaminated lemon candy liquid behaves as a strong acid.

(**a**) You first think about adjusting the pH using 0.25 M hydrochloric acid. What volume of hydrochloric acid should you add in order to correct the mistake?

(**b**) You then decide that HCl probably tastes bad and probably shouldn't be ingested, so you change your acid to 1.0 M ascorbic acid (vitamin C), which has a pK_a of 4.17. (For this problem, assume only one dissociation.) How much ascorbic acid should you add in order to correct your mistake?

(**c**) How would the calculated volume change if the ascorbic acid were 2.0 M?

5.37 One of the main functions of saliva is to buffer against acid from food and plaque, which contributes significantly to the formation of cavities. While there are several buffers in the saliva, carbonic acid (H_2CO_3) has the highest concentration and has the greatest effect on pH.

(**a**) While the salivary concentration of carbonic acid stays at a fairly constant 1.3 mM, the level of bicarbonate (HCO_3^-) can vary with the rate that saliva flows from salivary glands. For low flow rates, the bicarbonate concentration is around 2 mM; for medium flow rates, it is 30 mM; and for high flow rates, around 60 mM. The pK_a of carbonic acid at body temperature is 6.1. Assuming that the pH of saliva is determined primarily by carbonic acid and bicarbonate, determine the pH of saliva for each of the three flow rates. The normal pH of saliva is about 6.3.[12]

(**b**) The most prevalent bacterium in the mouth, *Streptococcus mutans*, breaks down sugar and releases lactic acid ($pK_a = 3.86$). If *S. mutans* has produced 10^{-8} mole of lactic acid since your last swallow, what is the pH of your saliva? What would the pH be without the bicarbonate buffer? Assume that your mouth contains about 1 mL of saliva and that your saliva is flowing at a low rate.

(**c**) You take a drink of orange juice, and after you swallow, 0.5 mL remains in your mouth. What is the pH of your saliva if your mouth contains 1 mL of pure saliva, and if you model orange juice as 1.0 mM citric acid ($pK_a = 3.13$; assume only one dissociation)?

(**d**) Why do you think some toothpastes contain baking soda (sodium bicarbonate)?

5.38 Acetylsalicylic acid ($C_9H_8O_4$), commonly known as aspirin, has been used for over a hundred years as an effective pain reliever. Aspirin works by suppressing the production of prostaglandins, chemicals that enhance sensitivity to pain. A person is considered to have acidosis if the blood pH falls below the normal value of 7.4. The lower limit of blood pH is approximately 6.8, below which a person could go into shock or die. How much aspirin must be consumed for blood pH to fall below this lower limit? State your assumptions.

5.39 A typical pH meter consists of two adjacent electrodes, which are placed in a solution of unknown pH. One electrode is often made of calomel, which is protected from voltage effects of any chemicals in the measured solution by a salt bridge. Another electrode, often made of Ag/AgCl, is placed inside the glass bulb filled with a solution of known pH, typically HCl. The Nernst equation is applied to the glass electrode to compare the potential difference across the glass electrode membrane with the pH of the solution to be analyzed. Answer the following questions, assuming that the solution inside the glass electrode is 1.0 M HCl.

(a) Use the Nernst equation to develop an equation for pH in terms of voltage (mV) across the glass membrane if the pH meter is at room temperature (25°C).

(b) Repeat part (a) for 0°C and 37°C. Is the relationship between pH and voltage dependent on temperature?

(c) Determine the pH of a solution of 0.5 M H_3PO_4, a common buffer in the human body, in two different ways. First, calculate the pH from the voltage from the pH meter; then, calculate the pH from the pK_a of H_3PO_4. The voltage across the glass membrane in the pH meter is −71 mV. The pK_a of H_3PO_4 at 25°C is 2.12 (assume only one dissociation).

(d) You are working for a clinic in a developing country with very limited medical resources. In fact, the only diagnostic tools to which you have access are your five senses, some sterile cups, and a broken pH meter that shows voltage across the glass electrode instead of pH. (You have calibrated the broken pH meter using a pH indicator strip that you had and determined that the relationship between pH and voltage from part (a) is still valid.) A 30-year-old man comes in with severe pain in his side and back. He also reports that he has been feeling the urge to urinate more frequently than usual. You suspect that he suffers from a kidney stone. You know that patients with renal calculi (a type of kidney stone) often have urine that is slightly more alkaline than normal.[12] You also know that the normal pH of urine is between 4.6 and 8. The broken pH meter shows a potential difference of −510 mV across the glass membrane. Is the man likely to be suffering from renal calculi?

5.40 Kidneys are the body's natural defense mechanism against acidosis. The proximal tubules produce NH_3 to counter the effects of hydrogen ions by removing them from the blood stream to form NH_4^+. Assuming that the blood pH is at 7.2, how many of moles of NH_3 must be produced to raise the blood pH back to a normal 7.4? The K_a of NH_4^+ is 5.6×10^{-10}M.

CHAPTER 6

Conservation of Momentum

6.1 Instructional Objectives and Motivation

After completing Chapter 6, you should be able to do the following:

- Explain the concepts behind and the applications of the conservation of linear and angular momentum.
- Identify various methods of momentum transfer, specifically material transfer and the application of forces on a system.
- Distinguish between situations requiring a differential or integral momentum conservation equation.
- Set up and solve systems involving rigid-body statics and fluid statics.
- Make appropriate simplifications to the conservation of linear momentum for isolated, steady-state systems.
- Apply the concepts of kinetic energy and the coefficient of restitution to systems with collisions.
- Apply the steady-state conservation of linear momentum equation to systems with mass flow.
- Relate conservation of linear momentum for unsteady-state systems to Newton's second law of motion.
- Determine the Reynolds number for fluid flow in closed conduits, and describe the meaning and significance of laminar flow and turbulent flow.
- Apply the steady-state mechanical energy accounting equation for systems with pump work, frictional losses, or both.
- Recognize systems in which the Bernoulli equation is applicable, and use the equation to analyze systems with flowing liquids.

6.1.1 Bicycle Kinematics

Linear and angular momentum conservation equations are used widely in the field of bioengineering. When considering the forces acting on rigid-body or fluid static systems, the basic momentum conservation equations are helpful. The momentum conservation equation and the mechanical energy accounting equation are often used to solve problems in systems involving fluid flow, such as blood and air flow in the human body and flow through industrial piping. The conservation of momentum can also be applied to model systems involving collisions of cells and other biologically relevant material. In this chapter, the conservation of momentum is applied to a wide range of example and homework problems.

In this introductory section of Chapter 6, we highlight the application of the conservation of momentum to kinematics, with a particular emphasis on cycling. The complex challenges below serve to motivate our discussion of the linear and angular momentum conservation equations.

Athletic activities demand from our bodies a wide range of motion. Bioengineers study body kinematics to develop models of the intricate mechanical motions of the human body. One form of exercise that helps scientists study body kinematics is cycling. Since most of the movement and propulsive force involved in cycling occurs in the legs, biomechanical study of cycling focuses on lower-extremity kinematics. Studying how bones, muscles, tendons, and ligaments affect leg motion in cycling and how these parts can be injured prepares bioengineers for designing equipment to enhance the performance and increase the safety of cyclists, and for developing new methods to treat and to prevent cycling injuries.

Because the knee is anatomically complex and subject to large and repetitive stresses, knee injuries are very common in cycling. Repeated cycles of tension in connective tissues (tendons and ligaments) can cause microstructural tearing of fibers, manifested in tendonitis. Frequent leg injuries include degeneration of patellar cartilage, patellar tendonitis, and quadriceps tendonitis. Injuries in the neck, back, and shoulders are also common among cyclists.

Bioengineers may attempt to optimize the bicycle–rider system for peak performance by understanding the complex relationship between bicycle geometry and rider kinematics. For example, research has demonstrated that cycling involves internal and external rotation of the tibia about its long axis, translation of the knee toward and away from the bicycle, and motion of the leg away from the plane of the bike. Altering the seat height changes the amount the muscle lengthens, which affects the muscle's ability to produce the forces needed to power a bicycle. Such findings and knowledge have allowed engineers to prevent injury by developing better models and advising appropriate training.

Biomechanical analysis requires an understanding of how forces, reaction forces, and torques affect the interaction between the rider and the bicycle (Figure 6.1). To develop a universal model of how these interactions apply to cyclists, bioengineers often make assumptions to simplify their calculations, such as modeling the thigh,

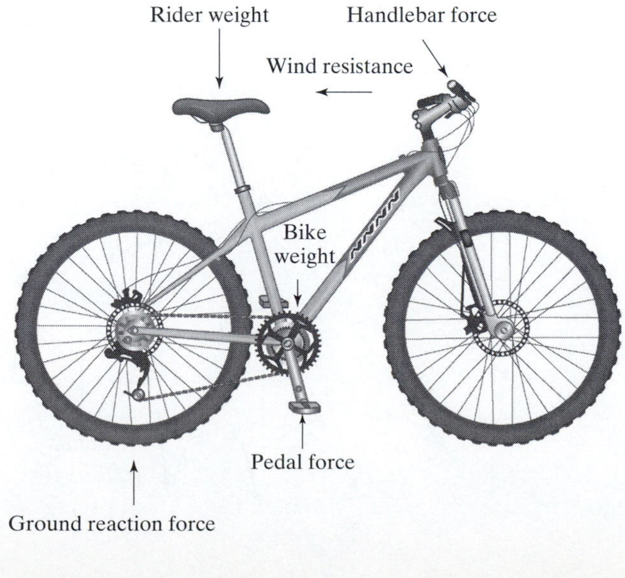

Figure 6.1
External forces acting on a bicycle.

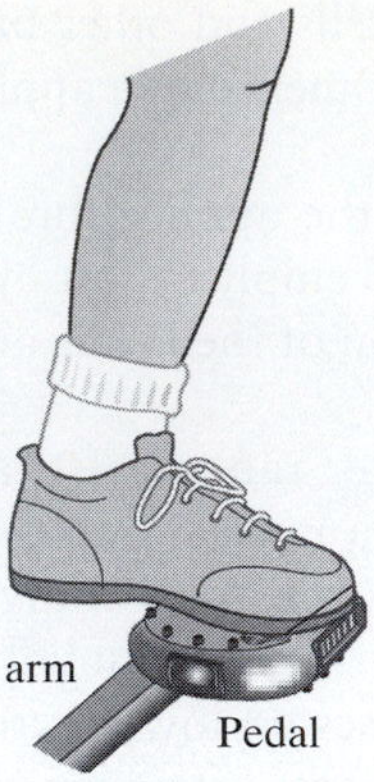

Figure 6.2
Pedal force components.

lower leg, and foot as a system of linked rigid bodies that work together to impart power to the crank (Figure 6.2). The pressure distribution over the surface of the pedal has also been an area of study, as the forces involved in energizing the bicycle cannot realistically be modeled as uniform (Figure 6.3).

Computer software is currently available to allow real-time collection and display of three-dimensional motion data. Using advanced computer algorithms, engineers can determine and analyze a variety of kinematical parameters, including the angular displacement of the hip and knee; the patterns of muscle length change; the force profile; the pressure distribution on the sole of the shoe; and torque patterns for the ankle, knee, and hip.[1]

Although discoveries in kinematics research and rapidly advancing technology have improved safety, bioengineers will continue to model body kinematics for the purpose of designing equipment and techniques to enhance performance while minimizing injuries and without compromising safety. Many challenges face engineers in the study of body kinematics. Some areas specific to cycling include:

- *Equipment advances*: Competitive athletes in all sports are constantly seeking ways to increase speed, performance, and comfort. Biomechanical research

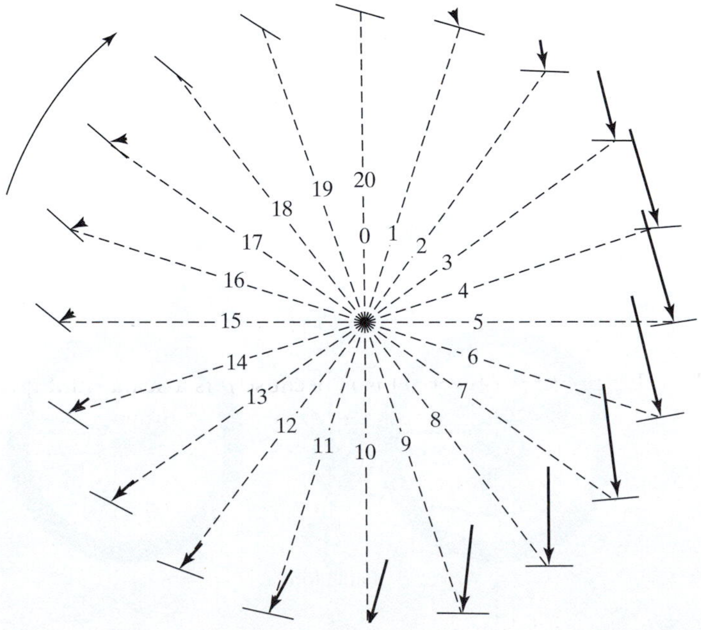

Figure 6.3
Relative forces exerted by a foot during one complete pedal rotation. Arrows show the direction and relative magnitude of the force at 20 locations as the foot/pedal system undergoes one rotation.

provides insight on how equipment can maximize performance. For example, engineers and cyclists are investigating ways to reduce aerodynamic drag. Performance-enhancing equipment, such as shoes, cycling suits, and helmets, is constantly being redesigned based on development of new materials and better models.

- *Injury treatment*: A more complete understanding of the function and activity of each body part may lead to novel ideas about treating or replacing injured parts.
- *Injury prevention*: Although treatment may alleviate the pain from injury, bioengineers must advise how to prevent an injury from happening. Biomechanical studies may suggest alternate equipment design and riding techniques that cause fewer injuries. In cycling, evaluating injury potential includes understanding the relationships between skeletal variations (e.g., leg length) and bicycle geometry (e.g., seat height).[2]

Multidisciplinary teams around the world tackle these research challenges in sports medicine facilities, industry, and academia. Along with intricate measurements of human motion and sophisticated computational tools, bioengineers use momentum balances to help them model different aspects of body kinematics. In Examples 6.1, 6.5, and 6.11, we tie this material into the chapter by examining how the conservation of momentum plays a part in the study of lower-extremity kinematics and cycling. Remember that kinematics is only one of many exciting areas where linear and angular momentum conservation equations can be applied to bioengineering and its related fields.

This chapter first discusses the types of momentum that can affect a system and how the governing equations are written when modeling linear momentum and angular momentum. Certain assumptions about a system, such as whether it is steady-state or static, can determine the form and use of the governing equations to solve for momentum. How bulk material transfer and external forces change the system's momentum is also explored. Finally, we look at how the mechanical energy accounting equation and Bernoulli equation can be used with conservation of momentum to solve systems with fluid flow.

6.2 Basic Momentum Concepts

Any moving object possesses both linear and angular momentum. **Linear momentum** ($\vec{p}$ [LMt^{-1}]) is an extensive property that quantifies the motion of a particle or a system proportional to the mass. **Angular momentum** ($\vec{L}$ [L^2Mt^{-1}]) is an extensive property, proportional to the mass of the system, which applies to any object undergoing motion, particularly rotational motion about a point. Angular momentum is used to describe the torque on bodies in static and dynamic analyses of structures.

Both types of momentum are described by three-dimensional vector quantities. Linear momentum, obtained by multiplying the mass of an object by its velocity, has the same direction as the velocity of the object. This is because $\vec{p}$ is a scalar multiple of the velocity vector, where mass is the scalar multiplier. Angular momentum of a particle or body is the cross product of the particle's position vector and its linear momentum. The mathematics for describing systems with angular momentum can be complex and is outside the scope of this text. A more thorough analysis of angular momentum can be found in other engineering textbooks (e.g., Glover C, Lunsford KM, and Fleming JA, *Conservation Principles and the Structure of Engineering*, 1994).

Figure 6.4
Girl jumping off a boat on a lake. (*Source*: Bedford A and Fowler W, *Engineering Mechanics: Statics and Dynamics*. Upper Saddle River, NJ: Prentice Hall, 2002.)

6.2.1 Newton's Third Law

Newton's third law of motion states that forces always exist by the interaction of two or more bodies and that the force on one body is equal and opposite to the force acting on the other. When a **force** is applied to a free-body object, the object accelerates in the direction of the applied force; thus, force is a vector quantity. **Forces may affect the momentum of a system of interest, but the net momentum in the universe is unchanged, because the force affecting the system has an equal and opposite force that acts outside the system.** This can be illustrated by a girl jumping off a boat (Figure 6.4). Imagine the girl as the system and the boat as the surroundings. Initially, the girl stands still in the boat, which is not moving on the lake. At this point, neither the boat nor the girl has any momentum, since neither has any velocity. When she jumps off, her feet push against the bottom of the boat, and the boat exerts an equal and opposite force. The force of the boat adds momentum to the girl (the system), but the force from the girl adds the same amount of momentum *in the opposite direction* to the boat (the surroundings), causing it to drift off away from the girl. Therefore, the net momentum in the universe is unchanged, since momentum is neither created nor destroyed in the universe—that is, momentum in the universe is conserved.

Momentum can transfer across the system boundary by two major modes: (1) by mass and (2) by forces. First, any moving object has momentum, and this momentum can be transferred into or out of a system by bulk material transfer. Second, momentum can be added or removed from a system when forces in its surroundings act upon the system. Recall that Input and Output terms can describe the exchange or transfer of an extensive property. Both bulk material transfer that crosses system boundaries and forces that act on the system to carry momentum into or out of a system are represented in the Input and Output terms of the conservation equation.

6.2.2 Transfer of Linear Momentum Possessed by Mass

Any moving object possesses linear momentum ($\vec{p}$). When a mass crosses a system boundary, linear momentum enters or leaves the system with it. Mass crosses the system boundary at a particular linear velocity ($\vec{v}$) which has both a magnitude and direction. The amount of linear momentum that crosses the system boundary is described as the product of the mass m and the velocity $\vec{v}$ of that mass at the system boundary:

$$\vec{p} = m\vec{v} \qquad [6.2\text{-}1]$$

Common units for linear momentum are kg·m/s, g·cm/s, and lb_m·ft/s. Linear momentum can enter and leave the system through many different objects, each with a different velocity.

EXAMPLE 6.1 Linear Momentum of a Bicycle

Problem: A cyclist who has a mass of 70 kg races on a bicycle that has a mass of 9 kg. Calculate the linear momentum of the system, composed of the rider and bicycle, when the cyclist is traveling 10 mph.

Solution: The linear momentum $\vec{p}$ is calculated using equation [6.2-1]. We assume the rider travels forward, so we define the direction in which he travels as $\vec{i}$:

$$\vec{p} = m\vec{v} = (70 \text{ kg} + 9 \text{ kg})\left(10\vec{i}\ \frac{\text{mi}}{\text{hr}}\right)\left(\frac{1 \text{ hr}}{3600 \text{ s}}\right)\left(\frac{1 \text{ m}}{0.0006214 \text{ mi}}\right) = 353\vec{i}\ \frac{\text{kg}\cdot\text{m}}{\text{s}}$$

The rider and bicycle have a linear momentum of about $350\vec{i}$ kg·m/s. Remember that the directions of velocity and linear momentum are the same. ■

The **center of mass** (CM) of a system is the point in space that describes the average position of all the mass in the system. For a simple symmetric object with a constant density, the center of mass is located at its geometric center. The concept implies two important applications. First, although the components of a system may not move with the same velocity, the total $\vec{p}$ can be calculated by applying equation [6.2-1] to each component separately and summing the individual momentums. However, you can also calculate the total $\vec{p}$ by multiplying the total mass of the system by the velocity of the CM, which is often considerably easier. The second application of CM involves gravitational forces. In a uniform gravitational field, which is the only type this text considers, the weight of an object acts at its CM.

Mass flow rate ($\dot{m}$) is a scalar quantity that measures the rate at which mass moves or flows. Thus, since flowing mass can carry momentum across a system boundary, the **rate of linear momentum** ($\dot{\vec{p}}$) crossing the system boundary by bulk mass transfer can be described by the product of $\dot{m}$ and the velocity $\vec{v}$ as follows:

$$\dot{\vec{p}} = \dot{m}\vec{v} \qquad [6.2\text{-}2]$$

The dimension of ($\dot{\vec{p}}$) is [LMt^{-2}]. Common units of rate of linear momentum are newton (N, also kg·m/s^2), dyne (g·cm/s^2), and pound-force (lb_f). The velocity term can also be thought of as linear momentum possessed per unit mass. (Note that both velocity and linear momentum per mass have dimensions of [Lt^{-1}].)

6.2.3 Transfer of Linear Momentum Contributed by Forces

Linear momentum within a system can change when surrounding external forces ($\vec{F}$) act on the system. The dimension of $\vec{F}$ is [LMt^{-2}], which is the same as the rate of linear momentum. Free-body diagrams are often constructed to help identify and label the various forces in a system. There are two major classes of forces: (1) surface or contact forces and (2) body forces.

Surface or **contact forces** act on the system at the system boundary. Examples include solid-solid contacts, pressure exerted on the system boundary, and drag or frictional forces. An example of a solid-solid contact force is a suspension cable that holds a bridge (the system) in place. Another example is the attachment between the Achilles tendon and the calcaneus bone.

When different pressures are exerted on different surfaces or parts of the system, the **pressure force** ($\vec{F}_p$) needs to be considered:

$$\vec{F}_p = -\iint_A P\vec{n}\, dA \qquad [6.2\text{-}3]$$

where P is the pressure exerted by the surroundings on the system, $\vec{n}$ is a unit vector that is normal to the system's plane of interest and points out of the system, and A is the area over which the pressure acts. In the scope of this book, the direction of the unit vector and the pressure exerted on the area will always be constant (i.e., not a function of position), so it is appropriate to rewrite equation [6.2-3] as:

$$\vec{F}_p = -P\vec{n}\iint_A dA \qquad [6.2\text{-}4]$$

Solving an equation such as this requires knowledge of surface integrals, a concept that is described in detail in multivariable calculus textbooks and is not used in this text. Instead, here the integral of dA will always simplify to the area of a uniform section or object over which the pressure acts. Thus, equation [6.2-4] is always used in this text as:

$$\vec{F}_p = -P\vec{n}A \qquad [6.2\text{-}5]$$

where A is the area over which the pressure acts, and A is often a cross-sectional area. It is important to remember that equation [6.2-3] is the fundamental equation to apply when pressure forces are involved in any system.

When a constant pressure is applied over the entire surface of the system, $\vec{F}_p$ does not need to be considered. When an entire surface has a uniformly applied pressure, each force has a corresponding force that negates it, since the two forces are of equal magnitude but opposite direction. **$\vec{F}_p$ needs to be considered in cases such as when inlet and outlet streams are at different pressures or when there are different pressures acting across the system boundaries.**

EXAMPLE 6.2 Air Cylinder

Problem: Air cylinders are used to generate specific forces or motions at precise points in many mechanical devices. In certain biomedical applications, air cylinders are used to test various forces on the spine, such as testing deformations under certain loading conditions. To do mechanical work, small cylinders can be hooked up to a pneumatic system. As air expands inside the pressurized interior, a corresponding force pushes on a plunger connected to a crank or shaft that does work.

Suppose an air cylinder with a 1-inch diameter can be pressurized up to 200 psig. Assuming that it operates at atmospheric pressure, what force is the cylinder capable of generating?

Solution: Because we are interested in the forces the cylinder can exert, we need to draw the system boundary such that it shows unbalanced forces acting on it. The system is the plunger, since it has atmospheric pressure on top and compressed air inside acting on the bottom (Figure 6.5). The wall of the cylinder exerts an equal pressure all around the plunger, so it does not need to be considered (i.e., the forces acting in the r-direction cancel each other out).

The plunger can be modeled as a disk with a 1-inch diameter. On the interior face, the pressure is 200 psig or 214.7 psia. On the exterior face, the pressure is 1.0 atm or 14.7 psia. The unit vectors point out of the plunger system from each face. Given the coordinate system, the unit vectors on the two faces of the plunger disk are defined as $\vec{n}_{\text{bottom}} = -1$ and $\vec{n}_{\text{top}} = 1$.

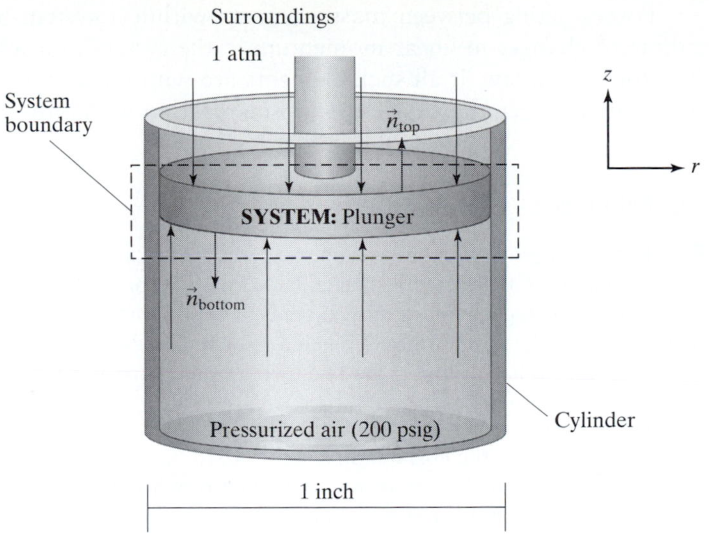

Figure 6.5
Pressurized air cylinder.

Because all P and $\vec{n}$ values are constant, and because the cross-sectional area of the disk can be calculated, equation [6.2-5] can be used. Equation [6.2-5] is applied to each face, and the individual contributions of the forces on each side are then added to determine the total force on the plunger:

$$\sum \vec{F}_{plunger} = \vec{F}_{bottom} - \vec{F}_{top} = -P_{bottom}\vec{n}_{bottom}A - P_{top}\vec{n}_{top}A = -P_{bottom}(-1)A - P_{top}(1)A$$

$$= (P_{bottom} - P_{top})A = \left(214.7\frac{\text{lb}_f}{\text{in}^2} - 14.7\frac{\text{lb}_f}{\text{in}^2}\right)\pi(0.500\ \text{in})^2 = 157\ \text{lb}_f$$

Thus, the cylinder pressure applied to the plunger is capable of exerting up to 157 lb_f, which pushes the plunger up out of the cylinder in the positive z-direction. ■

The other type of force that can contribute to linear momentum is a **body force**, which acts on the total mass (m) in the system. Examples include gravitational and electromagnetic fields. Forces acting on the system because of electrical fields are considered in Chapter 5. The most common body forces considered in problems involving momentum are forces acting on the system because of gravity ($\vec{F}_g$):

$$\vec{F}_g = m\vec{g} \qquad [6.2\text{-}6]$$

where $\vec{g}$ is the gravitational constant. The direction of the gravitational constant is dependent on the coordinate system you define for a problem. For an arbitrary mass of 1 lb_m, the force of gravity is calculated to be:

$$\vec{F}_g = m\vec{g} = (1\ \text{lb}_m)\left(32.2\frac{\text{ft}}{\text{s}^2}\right)\left(\frac{1\ \text{lb}_f \cdot \text{s}^2}{32.2\ \text{lb}_m \cdot \text{ft}}\right) = 1\ \text{lb}_f \qquad [6.2\text{-}7]$$

Do not forget to use the conversion factor g_c when converting force units in the British system! A common mistake is to see the calculation of the gravitation force acting on 1 lb_m to equal 1 lb_f and then conclude that 1 lb_m equals 1 lb_f. This conclusion is totally wrong! The force (or weight) of a mass of 1 lb_m under Earth's gravitational pull is 1 lb_f. The units of lb_m characterize mass, and the units of lb_f characterize force; force and mass are not the same.

Forces acting between mass elements within a system boundary do not contribute to changes in linear momentum in the system as a whole. This is related to Newton's third law. If all such elements are within the system boundary, any force operating between them has an opposite reaction within the system as well.

EXAMPLE 6.3 Hospital Table

Problem: A person who has sustained serious spinal or neck damage in an accident must be placed on an immobile spine board i.e.,back board before transport to the hospital. The spine board and head support allow critical areas to be supported e.g.,c-collar and help prevent further trauma to the spine or neck. When a patient arrives in the emergency room, the patient, spine board, and head support are laid directly on a gurney, a table with wheels.

(a) Maria suffers a serious neck injury in a head-on car collision. Emergency personnel transport her to the hospital on a spine board (i.e., backboard) with a head support (e g., c-collar)and transfer her to a gurney. Before wheeling her to the emergency room, the team prepares her for surgery, placing an oxygen bottle and supply bag on the gurney (Figure 6.6). Estimate the total force the gurney legs must exert to balance Maria and the objects on the table. The mass of each item and the distance from the end of the table to the center of mass (CM) of each item are given in Table 6.1.

(b) Maria's body has lost a lot of blood from the accident, so the doctor starts her on an intravenous (IV) drip, which is wheeled alongside the gurney. The IV bag drips at a rate of 45 mL/min with a linear velocity of 0.5 ft/s. Estimate the rate of momentum contributed to the system from the IV.

Solution:

(a) Because we are trying to find the total force the gurney legs must exert to keep Maria and the tabletop contents in equilibrium, the system should be modeled to include Maria, the spine board, the gurney tabletop, and the equipment placed on the tabletop (Figure 6.6).

By convention, we define the body forces due to gravity on each object in the system to point in the −y-direction. Since the legs are in contact with the tabletop, these forces are surface forces. The forces between Maria and the spine board and between the spine board and the tabletop do not need to be considered, since they are between elements in a system and do not act across the system boundary. Drawing a free-body diagram can help determine which forces need to be included in the force balance. Recall from equation

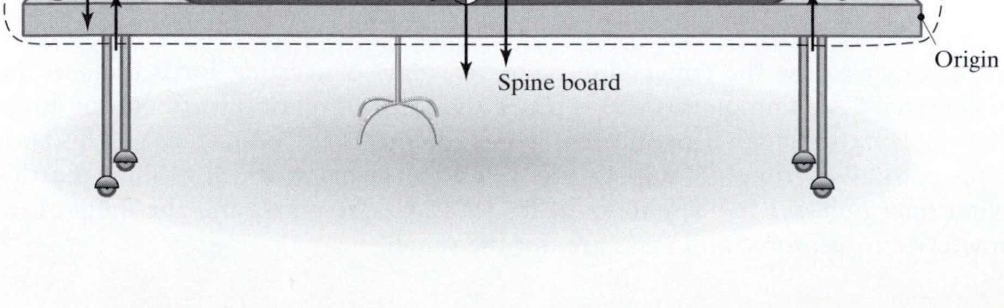

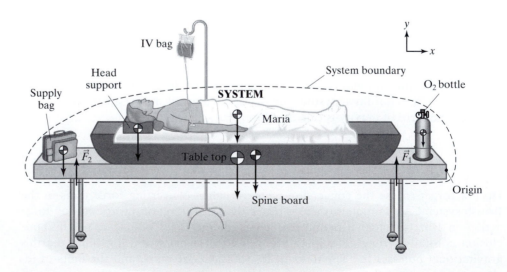

Figure 6.6
Hospital gurney table. Diagram not to scale.

TABLE 6.1

Mass and Positions of Objects on Gurney

Item	Mass (lb_m)	Distance of CM from origin (the end of the table) (cm)
O_2 bottle	3	10
Table legs		30
Spine board	15	90
Maria	120	100
Tabletop	10	110
Head support	3	180
Table legs		190
Supply bag	15	210

[6.2-7] that a 1-lb_m object has a weight of 1 lb_f in Earth's gravitation field. Consider the known forces on the system:

$$-\vec{F}_{O_2} - \vec{F}_{board} - \vec{F}_{Maria} - \vec{F}_{tabletop} - \vec{F}_{head} - \vec{F}_{bag} + \vec{F}_{legs} = 0$$
$$-3\ lb_f - 15\ lb_f - 120\ lb_f - 10\ lb_f - 3\ lb_f - 15\ lb_f + \vec{F}_{legs} = 0$$
$$\vec{F}_{legs} = 166\ lb_f$$

Thus, the four gurney legs exert a total upward force of 166 lb_f to keep Maria and the tabletop contents from collapsing.

(b) The system remains the same, since the IV bag is outside the boundary and we want to track the rate at which momentum moves into the system. IV fluid enters the system by bulk transfer of mass, which carries momentum. We make the assumption that the density of the IV solution is about 1.0 g/mL, since the solution is used to replace the blood Maria's body lost. The rate of linear momentum entering the system is calculated using equation [6.2-2]:

$$\dot{\vec{p}} = \dot{m}\vec{v} = \rho\dot{V}\vec{v}$$
$$= \left(1.0\frac{g}{mL}\right)\left(45\frac{mL}{min}\right)\left(0.5\frac{ft}{s}\right)\left(\frac{1\ min}{60\ s}\right)\left(\frac{1\ lb_m}{453.6\ g}\right)\left(\frac{1\ lb_f \cdot s^2}{32.17\ lb_m \cdot ft}\right)$$
$$= 2.57 \times 10^{-5}\ lb_f$$

Notice that this contribution is negligible relative to the forces (i.e., weight) from Maria and the equipment. Section 6.3 shows how rate of momentum and force terms are included in the conservation of linear momentum equation. ■

6.2.4 Transfer of Angular Momentum Possessed by Mass

Any moving object, including a rotating object, possesses **angular momentum** ($\vec{L}$). If a mass traveling in a linear direction crosses the system boundary, angular momentum crosses the boundary as well. Examples include a car wheel rolling into the system of a garage or an ice hockey puck sliding into the system of a goal.

A discrete quantity of mass (m) crossing the boundary contributes angular momentum to the system. The amount of $\vec{L}$ that crosses the system boundary is determined by the cross product of the position vector with the linear momentum ($\vec{p}$) of the object as follows:

$$\vec{L} = \vec{r} \times \vec{p} = \vec{r} \times (m\vec{v}) \qquad [6.2\text{-}8]$$

where $\vec{r}$ is the position vector and $\vec{v}$ is the velocity of the mass crossing the system boundary.

As with $\vec{p}$, it is important to note that $\vec{L}$ involves a position vector that must be defined with respect to a particular coordinate system. The particular magnitude and direction of the angular momentum of the system depend on the choice of a reference point. For many rotating objects, it is convenient to select the axis of rotation to be the reference point.

It is easy to think—mistakenly—that angular momentum can be applied only when an object rotates. However, *all* moving objects have angular momentum—how much is always relative to a specified reference point, so that a position vector can be defined. Angular momentum has little noticeable effect on systems if it acts in the line of the axis in which the reference point is defined.

How angular momentum affects a system can best be seen in circular motion. Imagine a playground merry-go-round as a system (Figure 6.7). The merry-go-round is initially not in motion (i.e., there is no momentum). Miriam runs at a velocity $\vec{v}$ in a straight line tangential to the edge of the merry-go-round; when she jumps on, she causes the merry-go-round to turn. A moment later, Ben also runs at a velocity $\vec{v}$ straight toward the merry-go-round's center and jumps on. Ben has little effect on making the merry-go-round turn faster or slower. If we take the position vectors of both Miriam and Ben to find their angular momentum with respect to the center of the merry-go-round, we would find that Ben's angular momentum is zero and Miriam's is nonzero. Thus, Ben does not carry any angular momentum into the system, but Miriam does.

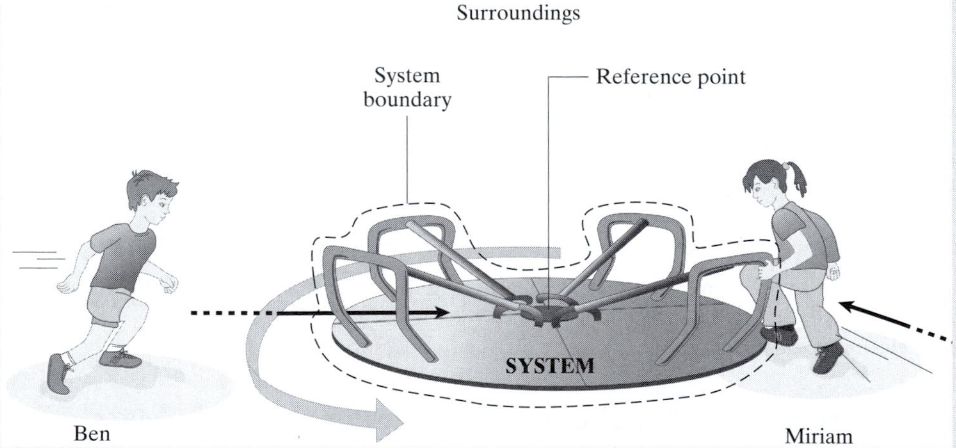

Figure 6.7
Angular momentum in a merry-go-round.

EXAMPLE 6.4 Satellite

Problem: A geosynchronous satellite traveling at a constant speed rotates in synchrony with the Earth once every 24 hours. Suppose a 200-kg satellite has an orbit 35,786 km above the Earth's surface. What is the angular momentum possessed by this satellite about the center of the Earth?

Solution: Recall that the direction of the linear momentum of an object is the same as the direction of its velocity. Assuming the orbit is a uniform circle, the position vector points radially outward from the center (i.e., the radius of the satellite orbit), and the direction of the velocity vector is always perpendicular to the position vector relative to the center of the Earth. To find the magnitude of the position vector, we add the Earth's radius, 6370 km, to the height the satellite orbits above Earth, so the total radius is 42,156 km. With this information, we calculate that, based on a 24-hour orbital period, the satellite's speed is 11,040 km/hr. Because the direction of velocity constantly changes along the path of a circle, we calculate the magnitude of the linear momentum using equation [6.2-1] and define an arbitrary direction $\vec{j}$ to find angular momentum:

$$\vec{p} = m\vec{v} = (200\ \text{kg})\left(11{,}040\vec{j}\,\frac{\text{km}}{\text{hr}}\right)\left(\frac{1\ \text{hr}}{3600\ \text{s}}\right)\left(\frac{1000\ \text{m}}{1\ \text{km}}\right) = 6.13 \times 10^5\,\vec{j}\,\frac{\text{kg}\cdot\text{m}}{\text{s}}$$

We define the direction such that $\vec{r} = 42{,}160\vec{i}$ km and $\vec{p} = 6.13 \times 10^5\,\vec{j}$ (kg · m/s), since the position and velocity vectors are perpendicular to each other. The angular momentum of the system is calculated using equation [6.2-8]:

$$\vec{L} = \vec{r} \times \vec{p} = (42{,}160\vec{i}\ \text{km}) \times \left(6.13 \times 10^5\,\vec{j}\,\frac{\text{kg}\cdot\text{m}}{\text{s}}\right)\left(\frac{1000\ \text{m}}{1\ \text{km}}\right) = 2.58 \times 10^{13}\,\vec{k}\,\frac{\text{kg}\cdot\text{m}^2}{\text{s}}$$

The satellite orbits the Earth at an angular momentum of $2.58 \times 10^{13}\,\vec{k}$ kg · m²/s. Notice the direction of the angular momentum is perpendicular to both the position and linear momentum vectors. If the satellite was orbiting the equator, the direction of the angular momentum would lie along the Earth's longitudinal axis. ■

The **rate of angular momentum** ($\dot{\vec{L}}$) crossing the system boundary is the cross product of the position vector with the rate of linear momentum:

$$\dot{\vec{L}} = \vec{r} \times \dot{\vec{p}} = \vec{r} \times (\dot{m}\vec{v}) \qquad [6.2\text{-}9]$$

where $\dot{m}$ is the mass flow rate across the system boundary. The dimension of ($\dot{\vec{L}}$) is [L^2Mt^{-2}]. The cross product of $\vec{r}$ and $\vec{v}$ can be thought of as angular momentum per unit mass. (Note that both $\vec{r} \times \vec{v}$ and angular momentum per unit mass have the dimension of [L^2t^{-1}].)

6.2.5 Transfer of Angular Momentum Contributed by Forces

When a force acts on a system, it may produce a **torque** ($\vec{\tau}$), which is a measure of how a force changes the rotational motion of an object. Torque has both magnitude and direction and is calculated as the cross product of the position vector ($\vec{r}$) and the applied external force ($\vec{F}$):

$$\vec{\tau} = \vec{r} \times \vec{F} \qquad [6.2\text{-}10]$$

where $\vec{r}$ measures the position vector from the point of application of the force to the reference point. Torque is a vector that is perpendicular to the plane formed by $\vec{r}$ and $\vec{F}$, and it is around the axis along which the vector lies that any rotation would occur (recall the right-hand rule from physics). The dimension of $\vec{\tau}$ is [L^2Mt^{-2}], which is the same as the rate of angular momentum.

The angular momentum of a system can change when external forces act upon the system to create a torque. Torque contributes angular momentum by the same two classes of forces: (1) surface or contact forces and (2) body forces. Recall that surface or contact forces act at the system boundary and can include solid-solid contacts, pressure exerted on the system boundary, and drag or frictional forces. One example of a solid-solid contact that contributes torque is bone-to-cartilage contact, such as the hip or knee joint. Frictional forces in joints can also cause torque, although many biological materials have extremely low frictional coefficients. Torque can also result from body forces such as gravitational, electrical, and magnetic forces that act on the total mass in the system. Torque due to gravitational force ($\vec{\tau}_g$) is given as follows:

$$\vec{\tau}_g = \vec{r} \times \vec{F}_g = \vec{r} \times (m\vec{g}) \qquad [6.2\text{-}11]$$

where $\vec{g}$ is the gravitational constant. The direction of the gravitational constant is dependent on the coordinate system you define for a problem. Forces acting between mass elements within the system boundary do not contribute to angular momentum changes of the system as a whole.

EXAMPLE 6.3 **Hospital Table (continued)**

Problem: Recall the problem statement in Example 6.3 and figure given in Figure 6.6. Determine the torque that the legs must balance to maintain equilibrium. Do not consider the IV bag. Assume that the force acts upward at two locations (where each location has two legs), denoted by $\vec{F}_1$ at 45 cm and $\vec{F}_2$ at 175 cm from the table end (the origin).

Solution: Since we are given the center of mass for each item on the table, we calculate the torque of each and add them together to find the total torque. Because the system does not rotate, we assume the sum of the torques equals zero. (The development of this equation is given in Section 6.5.) Table 6.1 gives the center of mass (CM) of each item.

$$\sum\vec{\tau} = \sum\vec{r} \times \vec{F} = 0$$

$$(10\vec{i}\text{ cm} \times -3\vec{j}\text{lb}_f) + (30\vec{i}\text{ cm} \times \text{F}_1\vec{j}\,) + (90\vec{i}\text{ cm} \times -15\vec{j}\text{ lb}_f) + (100\vec{i}\text{ cm} \times -120\vec{j}\text{ lb}_f)$$
$$+ (110\vec{i}\text{ cm} \times -10\vec{j}\text{ lb}_f) + (190\vec{i}\text{ cm} \times F_2\vec{j}) + (180\vec{i}\text{ cm} \times -3\vec{j}\text{ lb}_f)$$
$$+ (210\vec{i}\text{ cm} \times -15\vec{j}\text{ lb}_f) = 0$$

$$(30\text{ cm})F_1\vec{k} + (190\text{ cm})F_2\vec{k} = 18{,}170\vec{k}\text{ cm}\cdot\text{lb}_f$$

From Example 6.3, we know that:

$$\vec{F}_{\text{legs}} = \vec{F}_1 + \vec{F}_2 = 166\text{ lb}_f$$

Solving these two equations simultaneously yields $\vec{F}_1 = 83.6\text{ lb}_f$ and $\vec{F}_2 = 82.4\text{ lb}_f$. Note that these forces are very similar. This makes sense, since the main mass (Maria) is in the middle, and her weight is evenly distributed across the system, while the smaller objects on the end balance out. ■

6.2.6 Definitions of Particles, Rigid Bodies, and Fluids

The conservation of linear and angular momentum can be applied to systems containing particles, rigid bodies, and fluids. Examples of each are shown throughout this chapter.

Particles and rigid bodies are treated differently in many areas of mechanics and biomechanics. A **particle** is an idealized point mass having finite mass and no volume. When considering a particle, assume it occupies only a point in space. A consequence of this definition is that all the contact and body forces (e.g., gravitational) act at the point in space that the particle occupies. Examples of systems containing objects treated like particles include colliding cells (Example 6.10) and an item suspended from a cable (Figure 6.8a).

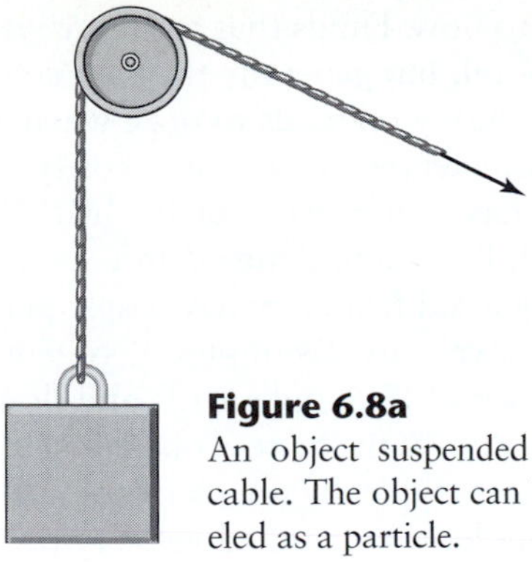

Figure 6.8a
An object suspended from a cable. The object can be modeled as a particle.

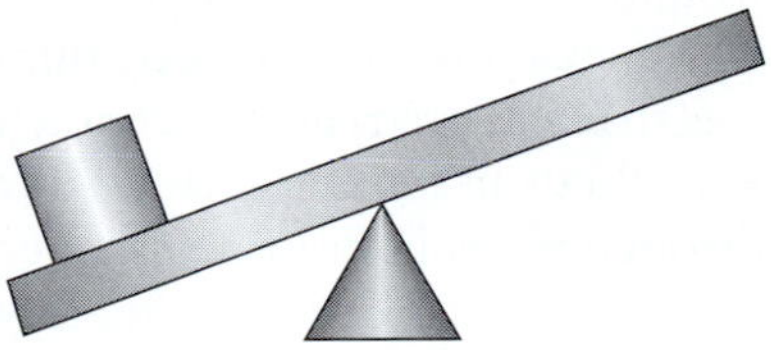

Figure 6.8b
Seesaw system, which can be modeled as a rigid body.

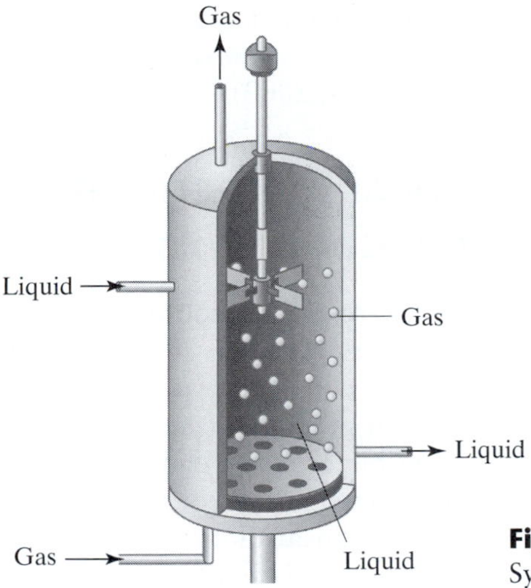

Figure 6.8c
System with fluids in motion.

In contrast, a **rigid body** has a specific finite mass and size, and its components are fixed within the body. In other words, no change in relative position of any two elements in the body can occur, and the body cannot be deformed. In addition, no mass enters or leaves the body. However, contact and body forces may act differently on different parts of the body. For example, since the body has finite mass and volume, different points on it may experience different pressure forces, such as in Example 6.3 with Maria lying on the spine board. Other solid elements outside it may come in contact with discrete points on the rigid body, but do not necessarily cover the whole body. Since the body is rigid, all its components at the same position relative to the reference point travel at the same angular velocity. In the absence of any rotation, the body also moves at the same linear velocity. Examples of systems containing items that are treated like rigid bodies include the arm in a static hold (Example 6.6), a lever, and a seesaw (Figure 6.8b).

A **fluid** is a substance that tends to flow under applied forces or that conforms to the outline of its container. Depending on their density and viscosity, gases and liquids may be fluids. Although they are both considered fluids, there are several important differences. The most obvious is that the density of gas is typically much less than that of liquid. Since molecules in gases are considerably more active and further apart, they are able to be compressed more readily than liquids, and a particular amount of gas can expand or compress to fill a wider range of volumes than the same amount of liquid.

Viscosity (μ) is a measure of a fluid's resistance to flow. Fluids that are more viscous seem thicker. There are many different kinds of oil, but generally they are considerably more viscous than water. Consider the way heavy oil tends to ooze when it moves, whereas water flows much more easily. At body temperature, the viscosity of blood is about three times that of water. Viscosity has a dimension of $[L^{-1}Mt^{-1}]$. Common units of viscosity are poise (P, also g/cm · s), Pa · s, and dyne · s/cm^2.

In examples in this text, fluids in motion are regarded from a macroscopic perspective and are typically characterized by an average velocity. Examples of systems with fluids in motion include flow through a blood vessel (Example 6.13) and flow through pipes in bioprocessing equipment (Figure 6.8c). With the tools in this text, the velocity profile of a liquid flowing through a vessel is typically not characterized. In contrast, texts that focus on transport phenomena look at detailed velocity profiles in flowing fluids from a microscopic perspective. The study of transport phenomena can be found in other texts (e.g., Truskey GA, Yuan F, and Katz DF, *Transport Phenomena in Biological Systems*, 2004; Bird RB, Stewart WE, and Lightfoot EN, *Transport Phenomena*, 2002).

6.3 Review of Linear Momentum Conservation Statements

Linear and angular momentum are always conserved in the universe (see Section 6.2.1). Thus, momentum cannot be produced or destroyed in a system or in the universe. Recall that the Generation and Consumption terms describe the production and destruction of an extensive property in a system. In problems involving momentum, Generation and Consumption terms are eliminated from the accounting equation, which then reduces to the conservation equation.

The conservation equation for linear momentum is a mathematical description of the movement of momentum into and out of the system by bulk material transfer, net external forces acting on the system, and the accumulation of momentum. The Input and Output terms account for momentum that transfers across the system boundary by net external forces and by the movement of mass. The Accumulation term accounts for any change in the amount of linear momentum in the system in the time period of interest.

Recall the definition of conservation of linear momentum from physics, which says that when the net external force on a system is zero, the total momentum of the system is constant. This definition is not used in this book! The concept of conservation of linear momentum is the same in physics and in bioengineering; however, how physicists and bioengineers define systems is different. This changes how bioengineers characterize the system, which affects how the conservation equation is applied.

When describing and solving problems involving momentum, differential and integral balances are much more common than algebraic balances because they can account for the time-dependent nature of momentum. The differential form of the conservation equation [2.4-11] is appropriate to use when rates of linear momentum are specified:

$$\dot{\psi}_{\text{in}} - \dot{\psi}_{\text{out}} = \frac{d\psi}{dt} \qquad [6.3\text{-}1]$$

The conservation equation is written to account for the movement of linear momentum into and out of the system with bulk material transfer, such as with mass flow ($\dot{m}$), and the application of external forces to the system (Figure 6.9):

$$\sum_i \dot{\vec{p}}_i - \sum_j \dot{\vec{p}}_j + \sum \vec{F} = \frac{d\vec{p}^{\,sys}}{dt} \qquad [6.3\text{-}2]$$

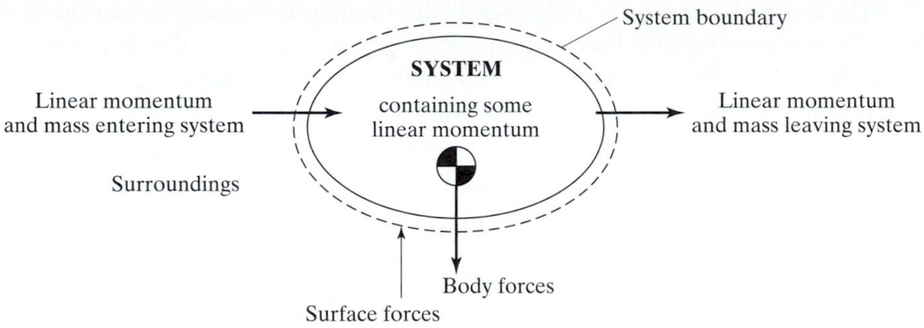

Figure 6.9
Graphical representation of linear momentum conservation equation.

$$\sum_i \dot{m}_i\vec{v}_i - \sum_j \dot{m}_j\vec{v}_j + \sum \vec{F} = \frac{d\vec{p}^{\,sys}}{dt} \qquad [6.3\text{-}3]$$

where $\sum_i \dot{\vec{p}}_i$ and $\sum_i \dot{m}_i\vec{v}_i$ are the sums of all rates of linear momentum entering the system by bulk material transfer, $\sum_j \dot{\vec{p}}_j$ and $\sum_j \dot{m}_j\vec{v}_j$ are the sums of all rates of linear momentum leaving the system by bulk maerial transfer, $\sum \vec{F}$ is the sum of the external forces acting on the system, and $d\,\vec{p}^{\,sys}/dt$ is the rate of accumulation of linear momentum contained within the system. The indices i and j refer to the numbered inlets and outlets, respectively.

The Accumulation term is expressed as the instantaneous rate of change of linear momentum of the system. When an Accumulation term is present, further information such as an initial condition or the system's acceleration may need to be specified. The dimension of the terms in equations [6.3-2] and [6.3-3] is [LMt^{-2}].

The momentum of the system $\vec{p}^{\,sys}$ can be calculated using a strategy appropriate for the complexity of the system. When the system is a particle, it is assumed that the particle travels at one velocity, and $\vec{p}^{\,sys}$ is calculated as the product of the mass of the system and its velocity.

When the system is more complex, it can sometimes be reduced to a number of segments or compartments. Each segment is assumed to have a constant velocity, even if the segments are rotating or translating relative to each other. For n segments in the system, the linear momentum of a system $\vec{p}^{\,sys}$ is described as:

$$\vec{p}^{\,sys} = \sum_k m_k\vec{v}_k \qquad [6.3\text{-}4]$$

where m is the mass of an individual segment, $\vec{v}_k$ is the velocity of that segment, and k is the index of the segments.

Another approach is to identify the density (ρ) of the mass in the system. The momentum of the system can be calculated by integrating over the volume of the system, V:

$$\vec{p}^{\,sys} = \iiint_V \rho\vec{v}\,dV \qquad [6.3\text{-}5]$$

This representation of momentum is particularly helpful when describing fluids, and it is often used in the derivation and application of transport equations describing fluid flow. In this text, we limit the description to the momentum of simple rigid bodies and particles. Applications involving accumulation of momentum in systems with multiple segments or fluids can be found in other texts.

The integral equation is most useful when trying to evaluate conditions between two discrete time points. The integral conservation equation is developed by writing the differential balances (equations [6.3-2] and [6.3.3]) integrating between the initial and final times. The integral conservation statements are as follows:

$$\int_{t_0}^{t_f} \sum_i \dot{\vec{p}}_i \, dt - \int_{t_0}^{t_f} \sum_j \dot{\vec{p}}_j \, dt + \int_{t_0}^{t_f} \sum \vec{F} \, dt = \int_{t_0}^{t_f} \frac{d\vec{p}^{\text{sys}}}{dt} dt \qquad [6.3\text{-}6]$$

$$\int_{t_0}^{t_f} \sum_i \dot{m}_i \vec{v}_i \, dt - \int_{t_0}^{t_f} \sum_j \dot{m}_j \vec{v}_j \, dt + \int_{t_0}^{t_f} \sum \vec{F} \, dt = \int_{t_0}^{t_f} \frac{d\vec{p}^{\text{sys}}}{dt} dt \qquad [6.3\text{-}7]$$

where $\int_{t_0}^{t_f} \sum_i \dot{\vec{p}}_i \, dt$ and $\int_{t_0}^{t_f} \sum_i \dot{m}_i \vec{v}_i \, dt$ are the sums of all linear momentums entering by bulk material transfer between times t_0 andt_f, $\int_{t_0}^{t_f} \sum_i \dot{\vec{p}}_i \, dt$ and $\int_{t_0}^{t_f} \sum_j \dot{m}_j \vec{v}_j \, dt$ are the sums of all linear momentums leaving by bulk material transfer between times t_0 and t_f, $\int_{t_0}^{t_f} \sum \vec{F} \, dt$ is the total linear momentum contributed by all external forces acting on the system between times t_0 and t_f, and $\int_{t_0}^{t_f} (d\vec{p}^{\,\text{sys}}/dt)$ is the total linear momentum accumulating in the system between times t_0 and t_f.

The indices i and j again refer to the numbered inlets and outlets, respectively. The dimension of the terms in equations [6.3-6] and [6.3-7] is [LMt^{-1}]. Information on the conditions of the system at t_0 and t_f may be needed to solve problems using the integral equation.

The individual terms in equations [6.3-6] and [6.3-7] may or may not be functions of time. In either case, the inlet and outlet terms describing the rate of momentum and the term for the linear momentum of the system may be integrated over the specified time interval as follows:

$$\sum_i \vec{p}_i - \sum_j \vec{p}_j + \int_{t_0}^{t_f} \sum \vec{F} \, dt = \vec{p}_{\text{f}}^{\text{sys}} - \vec{p}_0^{\text{sys}} \qquad [6.3\text{-}8]$$

$$\sum_i m_i \vec{v}_i - \sum_j m_j \vec{v}_j + \int_{t_0}^{t_f} \sum \vec{F} \, dt = \vec{p}_{\text{f}}^{\text{sys}} - \vec{p}_0^{\text{sys}} \qquad [6.3\text{-}9]$$

where $\sum_i \vec{p}_i$ and $\sum_i m_i \vec{v}_i$ are the sums of the linear momentums entering the system by bulk material transfer between t_0 and t_f, $\sum_j \vec{p}_j$ and $\sum_j m_j \vec{v}_j$ are the sums of the linear momentums leaving the system by bulk material transfer between t_0 and t_f, $\int_{t_0}^{t_f} \Sigma \vec{F} \, dt$ is the total linear momentum contributed by all external forces acting on the system between times t_0 and t_f, $\vec{p}_{\text{f}}^{\text{sys}}$ is the linear momentum of the system at the final condition t_f, and $\vec{p}_0^{\text{sys}}$ is the linear momentum of the system at the initial condition t_0. Again, i and j are the indices of the inlets and outlets, respectively.

The formulation of the integral equation [6.3-8] may look like an algebraic equation, particularly the Inlet, Outlet, and Accumulation terms. However, the integral force term should remind you that this is a special case of the integral equation.

6.4 Review of Angular Momentum Conservation Statements

As with linear momentum, the angular momentum of a system is always conserved. The conservation equation for angular momentum is a mathematical description of the movement of angular momentum into and out of a system, the torques acting on a system, and the accumulation of angular momentum in a system. Again, differential and integral balances are more common than algebraic balances when discussing angular momentum.

The differential form of the conservation equation is appropriate to use when torques or rates of angular momentum are specified. The conservation equation is written to account for the movement of angular momentum into and out of the system with bulk mass transfer and the application of external forces to the system as follows:

$$\sum_i \dot{\vec{L}}_i - \sum_j \dot{\vec{L}}_j + \sum (\vec{r} \times \vec{F}) = \frac{d\vec{L}^{\text{sys}}}{dt} \qquad [6.4\text{-}1]$$

where $\sum_i \dot{\vec{L}}_i$ is the sum of all of the rates of angular momentum entering the system by bulk material transfer, $\sum_j \dot{\vec{L}}_j$ is the sum of all of the rates of angular momentum leaving the system by bulk material transfer, $\sum (\vec{r} \times \vec{F})$ is the sum of the external torques acting on the system by the surroundings, and $d\vec{L}^{\text{sys}}/dt$ is the rate of accumulation of angular momentum contained within the system. Again, i and j are the indices of the inlet and outlet streams, respectively. Recall that $\vec{r}$ is the position vector.

The Accumulation term is expressed as the instantaneous rate of change of angular momentum of the system. When an Accumulation term is present, further information, such as an initial condition or the system's angular acceleration, may need to be specified. The dimension of the terms in equation [6.4-1] is $[L^2Mt^{-2}]$.

Substituting the definition of the rate of angular momentum in equation [6.2-9] gives:

$$\sum_i (\vec{r}_i \times \dot{\vec{p}}_i) - \sum_j (\vec{r}_j \times \dot{\vec{p}}_j) + \sum (\vec{r} \times \vec{F}) = \frac{d\vec{L}^{\text{sys}}}{dt} \qquad [6.4\text{-}2]$$

$$\sum_i (\vec{r}_i \times (\dot{m}_i\vec{v}_i)) - \sum_j (\vec{r}_j \times (\dot{m}_j\vec{v}_j)) + \sum (\vec{r} \times \vec{F}) = \frac{d\vec{L}^{\text{sys}}}{dt} \qquad [6.4\text{-}3]$$

The angular momentum of the system $\vec{L}^{\text{sys}}$ can be calculated using a strategy appropriate for the complexity of the system. The methods and procedures for calculating the angular momentum and moments of inertia of particles, systems of particles, rigid bodies, and fluids can be found in physics and engineering textbooks (e.g., Glover C, Lunsford KM, and Flemin JA, *Conservation Principles and the Structure of Engineering*, 1994). Problems in this book requiring the application of the conservation of angular momentum equations are limited to steady-state systems, so calculating $\vec{L}^{\text{sys}}$ is not necessary.

The integral equation is most useful when trying to evaluate conditions between two discrete time points. The integral conservation equation uses the differential balance equation [6.4-2] and integrates it between the initial and final times. The integral angular momentum conservation statement is:

$$\int_{t_0}^{t_f} \sum_i (\vec{r}_i \times \dot{\vec{p}}_i)\, dt - \int_{t_0}^{t_f} \sum_j (\vec{r}_j \times \dot{\vec{p}}_j)\, dt + \int_{t_0}^{t_f} \Sigma(\vec{r} \times \vec{F})\, dt = \int_{t_0}^{t_f} \frac{d\vec{L}^{\text{sys}}}{dt} dt \qquad [6.4\text{-}4]$$

where $\int_{t_0}^{t_f} \sum_i (\vec{r}_i \times \dot{\vec{p}}_i)\, dt$ is the sum of all angular momentums entering by bulk material transfer between times t_0 and t_f, $\int_{t_0}^{t_f} \sum_j (\vec{r}_j \times \dot{\vec{p}}_j)\, dt$ is the sum of all angular momentums leaving by bulk material transfer between times t_0 and t_f, $\int_{t_0}^{t_f} \Sigma(\vec{r} \times \vec{F})\, dt$ is the total angular momentum contributed by all external torques on the system between times t_0 and t_f, and $\int_{t_0}^{t_f} d\vec{L}^{\text{sys}}/dt$ is the total angular momentum accumulating in the system between times t_0 and t_f.

Again, i and j are the indices of the inlets and outlets, respectively. The dimension of the terms in equation [6.4-4] is $[\text{L}^2\text{Mt}^{-1}]$. Information on the conditions of the system at t_0 and t_f may be needed to solve problems using the integral equation.

The individual terms in equation [6.4-4] may or may not be functions of time. In either case, the inlet and outlet terms describing the flow rates of angular momentum and the terms for the angular momentum of the system may be integrated over the specified time interval as follows:

$$\sum_i (\vec{r}_i \times \vec{p}_i) - \sum_j (\vec{r}_j \times \vec{p}_j) + \int_{t_0}^{t_f} \Sigma(\vec{r} \times \vec{F})\, dt = \vec{L}_{\text{f}}^{\text{sys}} - \vec{L}_0^{\text{sys}} \qquad [6.4\text{-}5]$$

where $\sum_i (\vec{r}_i \times \vec{p}_i)$ is the sum of the angular momentums entering by bulk mass transfer between t_0 and t_f, $\sum_j (\vec{r}_j \times \vec{p}_j)$ is the sum of the angular momentums leaving by bulk mass transfer between t_0 and t_f, $\int_{t_0}^{t_f} \Sigma(\vec{r} \times \vec{F})\, dt$ is the total angular momentum contributed by all external torques acting on the system between t_0 and t_f, $\vec{L}_{\text{f}}^{\text{sys}}$ is the total angular momentum in the system at the final condition, and $\vec{L}_0^{\text{sys}}$ is the total angular momentum in the system at the initial condition.

6.5 Rigid-Body Statics

One common class of engineering problems involves the application of the conservation of momentum to closed, steady-state systems. In a closed (but not isolated) system, movement of momentum through bulk material transport across the system boundary does not occur. However, external forces may act on the system. For this situation, the differential form of the statement of the conservation of linear momentum is often used and is reduced from equation [6.3-2] for the closed, steady-state situation:

$$\sum \vec{F} = 0 \qquad [6.5\text{-}1a]$$

Breaking the vector into its components, scalar equations in each dimension can be written for a system in rectangular coordinates as:

$$\sum F_x = 0, \qquad \sum F_y = 0, \qquad \sum F_z = 0 \qquad [6.5\text{-}1b]$$

Since the systems described by these equations do not move and no mass moves into or out of them, engineers often call these systems **static**. Other engineers, such as mechanical engineers, solve problems involving static systems that include particles, bodies, and structures (e.g., frames and trusses).

Introductory physics courses often cover particle statics. An example is the calculation of the forces on a mass suspended by a cable (e.g., Figure 6.8a). The human body and other biomedical devices and systems are not commonly modeled as particles. Therefore, we do not consider the application of equation [6.5-1] to particles. Static systems composed of rigid bodies and structures are very common in bioengineering applications. This text considers rigid-body statics as an introduction to more complex applications further explored in biomechanics textbooks.

The mass of a rigid body is finite and does not change with time. Contact and body forces may act at different locations on a body, but sometimes it can be assumed that the contact and body forces act at one point. For example, the force of gravity is often assumed to act at the center of mass.

The differential form of the statement of the conservation of angular momentum is reduced from equation [6.4-1] for a closed, steady-state system to just the sum of the torques as follows:

$$\sum(\vec{r} \times \vec{F}) = 0 \qquad [6.5\text{-}2a]$$

As before, equations in each dimension can be written for a system in rectangular coordinates as:

$$\sum(\vec{r} \times \vec{F})_x = 0, \qquad \sum(\vec{r} \times \vec{F})_y = 0, \qquad \sum(\vec{r} \times \vec{F})_z = 0 \qquad [6.5\text{-}2b]$$

Using equations [6.5-1a] and [6.5-2a] simultaneously is sufficient to solve many rigid-body statics problems.

EXAMPLE 6.5 Forces While Bicycling

Problem: Figure 6.10a models a leg pedaling a bicycle. Point a is the ankle, point p is where the pedal connects to the crank, and point b is the gear about which the crank rotates. Between a and p is the distance between the ankle and the pedal, and between p and b is the crank, the bar that rotates when you move the pedal. The gear (point b) is fixed in space relative to the leg and to the ground.

Treat each segment as a rigid body with no mass. Suppose the bicyclist is not moving and remains stationary. The lower leg exerts 289 N (at point a) in the $-y$-direction on the ankle. Calculate the force in the x-direction at the ankle (a), as well as the forces in the x- and y-directions at the pedal (p). The connection between the ankle and the pedal is 14 cm long and makes a 50° angle with the horizontal.

Solution: Figure 6.10b shows the system that includes the ankle and pedal and the known forces. To find the forces acting on the pedal and the ankle, a number of assumptions are required, including:

- The segment between the ankle and the pedal is a rigid body with no mass.
- The bicycle and the foot are stationary.

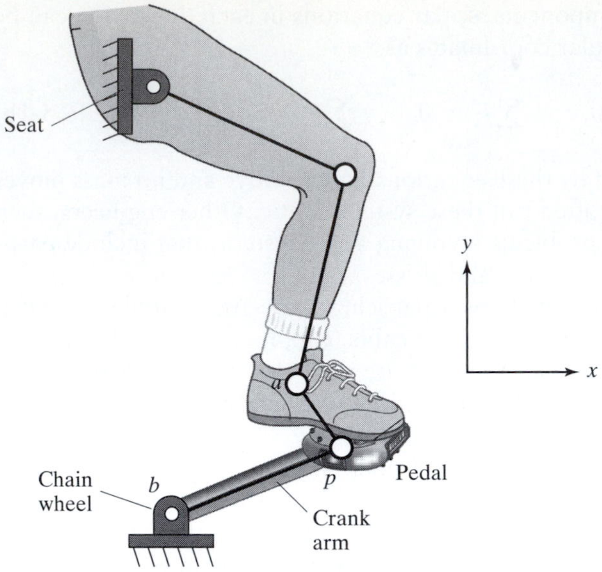

Figure 6.10a
Leg pedaling on a bicycle. (Adapted from: Burke ER, ed., *High-Tech Cycling*. Champaign, IL: Human Kinetics Publishers, 1995.)

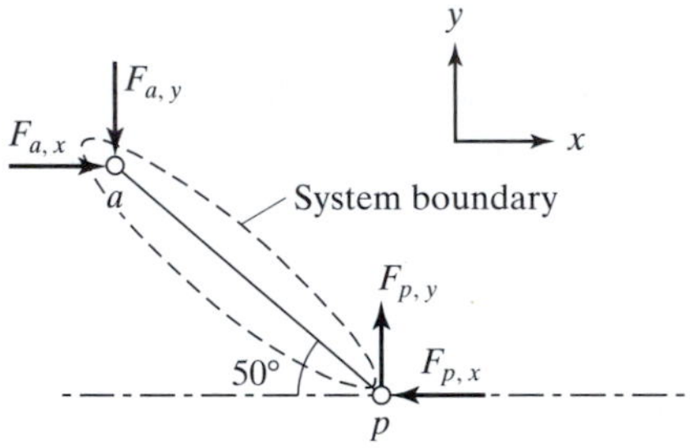

Figure 6.10b
Ankle and pedal system.

- The system is at steady-state.
- Forces are constant.
- No forces act in the lateral direction (z-axis).

The differential conservation of linear momentum equation (equation [6.3-2]) is reduced for a closed, steady-state system with no mass flow:

$$\sum \vec{F} = 0$$

Similarly, the differential conservation of angular momentum (equation [6.4-1]) is reduced for a steady-state system with no mass flow:

$$\sum (\vec{r} \times \vec{F}) = 0$$

For the system containing the ankle and the pedal, equations in the x- and y-directions are written:

$$x: \quad F_{a,x} - F_{p,x} = 0$$
$$y: \quad -F_{a,y} + F_{p,y} = -289 \text{ N} + F_{p,y} = 0$$

To determine the external torques, a point of origin must be selected so that values of $\vec{r}$ can be established. We set the point of origin at point p so that $\vec{r}_{p,x} = 0$ and $\vec{r}_{p,y} = 0$. The conservation of angular momentum for a closed, steady-state system is then applied:

$$(\vec{r}_{a,y} \times \text{F}_{a,x}\vec{i}) + (\vec{r}_{a,x} \times F_{a,y}(\vec{j})) = 0$$
$$((0.14 \text{ m})(\sin 50°\vec{j}) \times F_{a,x}\vec{i}) + ((0.14 \text{ m})(\cos 50°(-\vec{i})) \times 289(-\vec{j})) = 0$$

Thus, the magnitude of $F_{a,x}$ is calculated as 242 N. All the magnitudes of the forces for the ankle-pedal system are obtained:

$$F_{a,x} = 242\vec{i}\ \text{N}, \qquad F_{a,y} = -289\vec{j}\ \text{N}, \qquad F_{p,x} = -242\vec{i}\ \text{N}, \qquad F_{p,y} = 289\vec{j}\ \text{N}$$

In this instance, the force in the vertical direction is about 20% more than the horizontal force, though this ratio changes dramatically depending on the angle the ankle-pedal link makes with the horizontal. When the leg is fully extended and the pedal is at the lowest point in the cycle, the only force is in the y-direction. As the knee bends and the leg contracts, more force is exerted in the horizontal direction. In this way, each of the joint forces in the leg and the ankle and at the pedal depends on the location of the pedal in the pedaling cycle.

Two major simplifying assumptions—the stationary cyclist and the ankle and pedal of no mass—limit the use of this model. In a more realistic model, the forces would not be balanced; the force differential would be converted to forward motion. The magnitudes of the forces in the system change with time, and the rider-bicycle system would not be static. Furthermore, each segment (e.g., foot) has a certain mass that exerts an external force, causing variation in the forces from the calculated values. For our analysis to have any engineering value (e.g., for study of injury prevention), the model must be modified to take these factors into account. ■

EXAMPLE 6.6 **Forces on the Forearm**

Problem: The *biceps brachii*, a muscle in the arm, connects the radius, a bone in the forearm, to the scapula in the shoulder (Figure 6.11a). The muscle attaches at two places on the scapula (thus the name *biceps*) but at only one on the radius. To move or hold the arm in place, the biceps muscle balances the weight of the arm and the force at the elbow joint. Assume that the center of mass of the arm is 15 cm from the elbow joint. Also assume that the diameters of the upper arm and lower arm are each 6 cm and that the muscles are attached at the locations shown in Figure 6.11b. The horizontal force of the elbow joint ($F_{E,x}$) on the forearm is 6.5 N when the forearm is held parallel to the ground. Assume that the biceps supports the entire weight of the forearm. Calculate the necessary force from each branch of the biceps to hold the forearm parallel to the ground.

Solution:

1. Assemble
   (a) Find: force of each branch in the biceps.
   (b) Diagram: The system diagram is shown in Figure 6.11b. The unknown biceps forces are F_A and F_B.
2. Analyze
   (a) Assume:
   - The center of mass, and thus the point of action of weight, is 15 cm from the elbow (point E).
   - The biceps are the only muscles holding up the forearm.
   - Arm is stationary.
   - Forces act only in the xy-plane and are constant.
   - The system is at steady-state.

   (b) Extra data:
   - Mass of the average person is 150 lb_m.
   - One lower forearm is 2.3% of the body mass of the average person.

   (c) Variables, notations, units:
   - W = weight
   - F_A, F_B = force of the two branches of the biceps
   - F_E = force of elbow
   - Use cm, N.

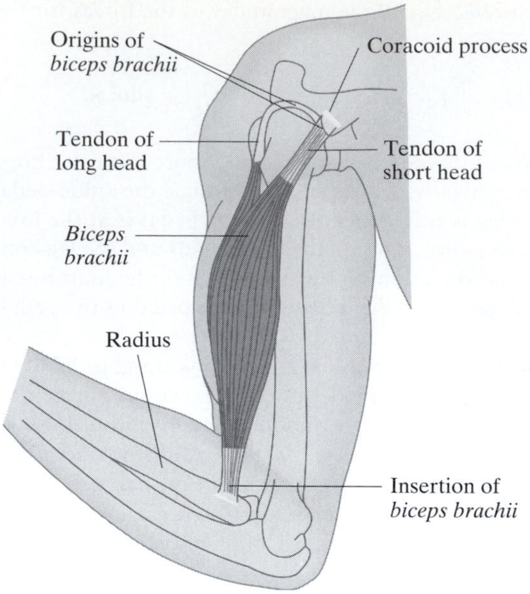

Figure 6.11a
Attachment of *biceps brachii*. (*Source*: Shier D, Hole JW, Butler J, and Lewis R, *Hole's Human Anatomy & Physiology*. Columbus: McGraw-Hill, 2002.)

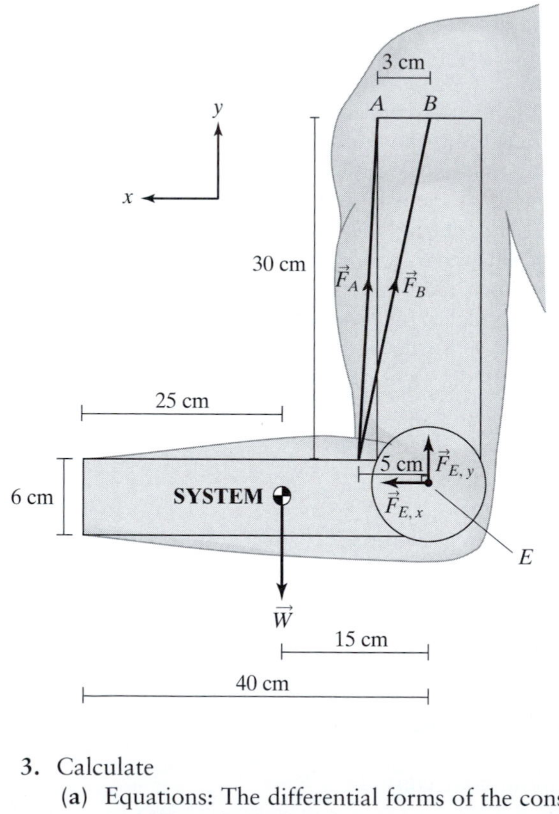

Figure 6.11b
Forces on the forearm. Not drawn to Scale.

3. Calculate
   (a) Equations: The differential forms of the conservation of linear momentum (equation [6.2-2]) and angular momentum (equation [6.4-1]) are needed. Because no mass flows in the system, rates of linear momentum entering and leaving the system through bulk material transfer are zero. Because the forearm is a rigid-body, static system at steady-state, the Accumulation term is zero. This simplifies these equations to:

$$\sum \vec{F} = 0$$

$$\sum (\vec{r} \times \vec{F}) = 0$$

(b) Calculate:

- We arbitrarily define a coordinate system such that point E at the elbow is the origin. Summing the forces in the x- and y-directions gives:

$$x: \quad \sum F_x = F_{A,x} + F_{B,x} - F_{E,x} = 0$$
$$y: \quad \sum F_y = F_{A,y} + F_{B,y} + F_{E,y} - W = 0$$

- Only components of forces acting perpendicular to an axis contribute a torque. Summing the torques around point E gives:

$$(\vec{r}_{F_A} \times \vec{F}_A) + (\vec{r}_{F_B} \times \vec{F}_B) + (\vec{r}_W \times \vec{W}) = 0$$

- When the forearm is held parallel to the ground, the horizontal (x-direction) force of the elbow joint on the forearm, $F_{E,x}$, is 6.5 N. The weight of the forearm in the y-direction is calculated:

$$\vec{W} = m\vec{g} = 0.023(150\ \text{lb}_\text{m})\left(32.2\frac{\text{ft}}{\text{s}^2}\right)\left(\frac{\text{s}^2 \cdot \text{lb}_\text{f}}{32.2\ \text{lb}_\text{m} \cdot \text{ft}}\right)\left(\frac{1\ \text{N}}{0.225\ \text{lb}_\text{f}}\right) = 15.3\ \text{N}$$

- The x- and y-components of the forces $\vec{F}_A$ and $\vec{F}_B$ are determined by figuring out the angles θ_A and θ_B using trigonometry (Figure 6.11c), which are $\theta_A = 86.2°$ and $\theta_B = 80.5°$. Substituting these values into the forces in the x- and y-directions gives:

$$x: \quad F_A(\cos 86.2°) + F_B(\cos 80.5°) - 6.5\ \text{N} = 0$$
$$0.06627F_A + 0.1650F_B = 6.5\ \text{N}$$
$$y: \quad F_A(\sin 86.2°) + F_B(\sin 80.5°) + F_{E,y} - 15.3\ \text{N} = 0$$
$$0.9978F_A + 0.9863F_B + F_{E,y} = 15.3\ \text{N}$$

- The position vector $\vec{r}$ is determined by figuring out the x- and y-components from the tail of the force vector to the origin independent of the given coordinate system, so $\vec{r}_{F_A} = \vec{r}_{F_B} = (5\vec{i}, 3\vec{j})$ and $\vec{r}_W = (15\vec{i})$. Substituting in values for the torques around point E gives:

$$\sum(\vec{r} \times \vec{F})_E = (\vec{r}_{F_A} \times \vec{F}_A) + (\vec{r}_{F_B} \times \vec{F}_B) + (\vec{r}_W \times \vec{W}) = 0$$
$$\{(5\vec{i}\ \text{cm}) \times [F_A(\sin 86.2°)\vec{j}] + (3\vec{j}\ \text{cm}) \times [-F_A(\cos 86.2°)\vec{i}]\}$$
$$+ \{(5\vec{i}\ \text{cm}) \times [F_B(\sin 80.5°)\vec{j}] + (3\vec{j}\ \text{cm}) \times [-F_B(\cos 80.5°)\vec{i}]\}$$
$$+ [(15\vec{i}\ \text{cm}) \times (-15.3\vec{j}\ \text{N})] = 0$$

$$5.188F_A\vec{k}\ \text{cm} + 5.426F_B\vec{k}\ \text{cm} - 230.0\vec{k}\ \text{N} \cdot \text{cm} = 0$$
$$5.188F_A + 5.426F_B = 230.0\ \text{N}$$

- Now there are three equations for the three unknown variables F_A, F_B, and $F_{E,y}$. These three equations can be solved algebraically or by using MATLAB:

$$0.06627F_A + 0.1650F_B = 6.5\ \text{N}$$
$$0.9978F_A + 0.9863F_B + F_{E,y} = 15.3\ \text{N}$$
$$5.188F_A + 5.426F_B = 230.0\ \text{N}$$

The magnitudes of F_A, F_B, and $F_{E,y}$ are determined by defining the matrix and vector that represents this set of equations. Arranging these scalar equations into the matrix equation form $Ax = y$ gives:

$$\begin{bmatrix} 0.06627 & 0.1650 & 0 \\ 0.9978 & 0.9863 & 1 \\ 5.188 & 5.426 & 0 \end{bmatrix} \begin{bmatrix} F_A \\ F_B \\ F_{E,y} \end{bmatrix} = \begin{bmatrix} 6.5 \\ 15.3 \\ 230.0 \end{bmatrix}$$

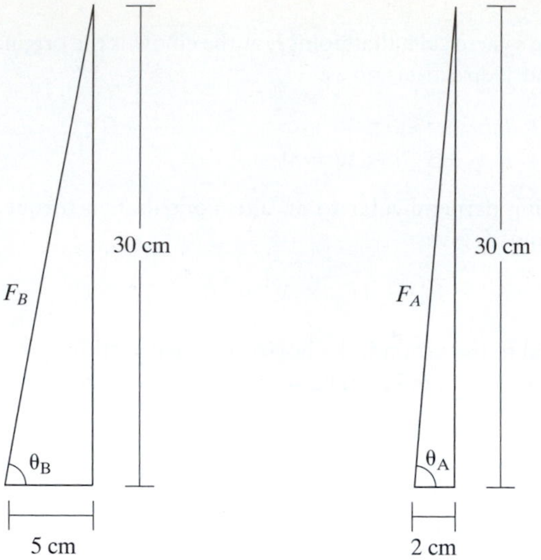

Figure 6.11c
Forces broken down into trigonometric components.

Substituting this matrix equation into MATLAB yields:

$$x = \begin{bmatrix} F_A \\ F_B \\ F_{E,y} \end{bmatrix} = \begin{bmatrix} 5.40 \\ 37.2 \\ -26.8 \end{bmatrix}$$

4. Finalize
   (a) Answer: The magnitudes of the forces of the two biceps branches and the horizontal force at the elbow, respectively, are $F_A = 5.40$ N, $F_B = 37.2$ N, and $F_{E,y} = -26.8$ N. The directions of these forces are shown in Figure 6.11d.
   (b) Check: Note that F_B is much greater than F_A. $F_{E,y}$ acts in the direction opposite to what was initially assumed. This makes sense, since $F_{E,y}$ is the force of the elbow on the forearm. The force of the biceps in the upward direction results in a compressive force in the elbow joint. The compressive force on the forearm at the elbow must act in the downward direction, because the forearm is the lower bone of the joint. ∎

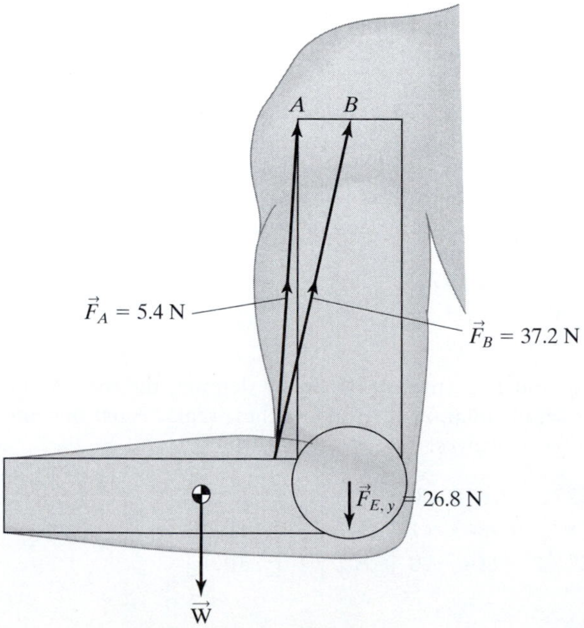

Figure 6.11d
Directions of forces acting on the forearm.

EXAMPLE 6.7 Attachment of a Fibroblast Cell

Problem: Human dermal fibroblast (HDF) cells are often used in tissue culture laboratories when connective tissue cells are needed for research. Frequently, they are grown in small clear flasks (Figure 6.12a) and covered with a thin, nutrient-filled layer of media. HDF cells are anchorage dependent, meaning that they must form attachments with a surface to proliferate. While cells are generally not rigid bodies, some of the forces can be examined with this assumption.

The flask is set such that the wall to which the cells are attached is perpendicular to the ground (Figure 6.12b). Consider the diagram of the HDF cell attached to the flask at two points, as shown in Figure 6.12c. (In reality the cell would be attached at many more points through protein–protein bonds.) The $+y$-axis opposes the direction of gravity; the x-axis lies in the plane of the wall with attached cells and is parallel to the ground. The cell is modeled as a flat shape that lies in the plane of the wall. The distances between the attachment points and the center of mass are $\vec{r}_1$ and $\vec{r}_2$, which are known. Assuming the weight of an HDF cell to be W, what can you say about the forces at the attachment points?

Solution: As in the previous examples, the system is steady-state and static, so the sum of the forces and the sum of the torques about the center of mass are both equal to zero:

$$\begin{aligned}
&x: \quad \Sigma F_x = F_{2,x} - F_{1,x} = 0 \\
&y: \quad \Sigma F_y = F_{1,y} + F_{2,y} - W = 0 \\
&\text{torque:} \quad (\vec{r}_1 \times \vec{F}_1) + (\vec{r}_2 \times \vec{F}_2) = 0 \\
&-(r_{1,y}F_{1,x})_z - (r_{1,x}F_{1,y})_z + (r_{2,y}F_{2,x})_z + (r_{2,x}F_{2,y})_z = 0
\end{aligned}$$

where $F_{1,x}$ and $F_{2,x}$ are the components of the attachment forces in the x-direction, $F_{1,y}$ and $F_{2,y}$ are the components in the y-direction, $r_{1,x}$ and $r_{2,x}$ are the position vectors in the x-direction, and $r_{1,y}$ and $r_{2,y}$ are the position vectors in the y-direction. All four values of the position vectors are known.

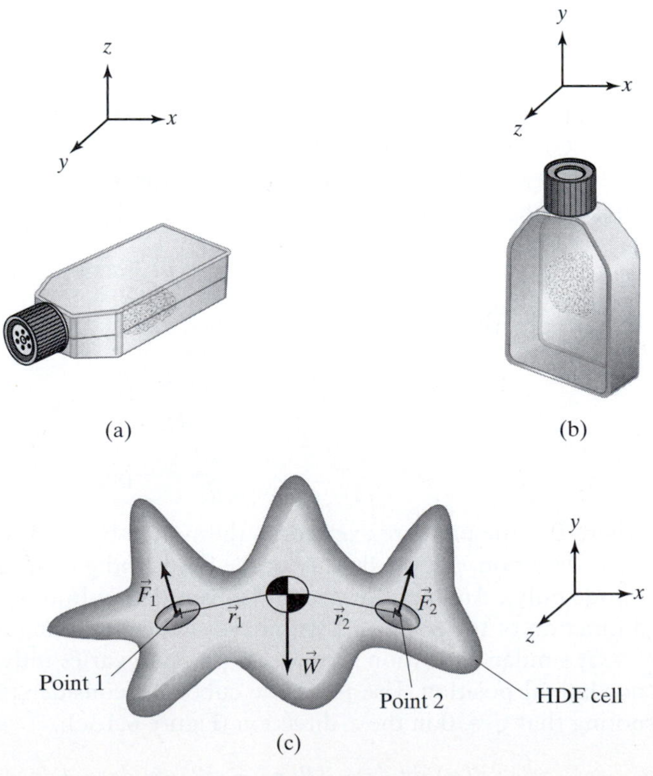

Figure 6.12
(a) Flasks growing cells in culture. (b) Attached cells shown in vertical flask. (c) Attached HDF cell with two attachment points, labelled 1 and 2. Cell is in the x-y plane; z is out of the plane.

The directions of the resulting cross products are obtained by the right-hand rule. Crossing a moment arm in the $+x$-direction with a force component in the $+y$-direction gives a resulting vector in the $+z$-direction (i.e., out of the page). In all the equations above that do not include vector components, the variables are given as magnitudes. Thus, $-(r_{1,y}F_{1,x})_z$ is interpreted such that $r_{1,y}F_{1,x}$ is the magnitude, and the vector points in the $-z$-direction, as denoted by the negative sign.

Note that there are four unknowns ($F_{1,x}$, $F_{1,y}$, $F_{2,x}$, $F_{2,y}$) but only three equations. A system having more unknowns than equations is called **underspecified** or, in mechanics, **statically indeterminate**. There is not enough information to determine the forces using conservation of momentum equations. Often, some geometric property can be used to obtain a relationship between two or more of the unknowns. If enough of these can be found so that the number of equations is equal to the number of unknowns, a unique solution can be determined. ■

6.6 Fluid Statics

Linear momentum can enter and leave a system because of surface forces or body forces or both. Consider a steady-state system with no bulk mass transfer. Equation [6.3-2] reduces to:

$$\sum \vec{F} = 0 \qquad [6.6\text{-}1]$$

In Section 6.5, we apply the above equation to rigid bodies. Another class of problems using equation [6.6-1] involves fluids that do not move, or **static fluids**. A static fluid's viscosity does not affect its behavior, since it is not flowing.

Consider a cube with sides dx, dy, and dz that exists within a still fluid of density ρ (Figure 6.13a). The cube edges define the system boundary. Gravity exerts a body force on the mass of the fluid. Each face of the cube is subject to a pressure force. Equation [6.6-1] is rewritten for the z-direction:

$$\sum \vec{F}_z = \vec{F}_p + \vec{F}_g = 0 \qquad [6.6\text{-}2]$$

where $\vec{F}_P$ is the pressure force and $\vec{F}_g$ is the gravitational force, which are defined in Section 6.1. In the z-direction, the pressure force balances the gravitational force. Surface pressure forces act on two faces, at location z and $z + \Delta z$; the differential pressure, dP, captures any difference in fluid pressure between the two opposing faces. Gravity acts on the entire cube (Figure 6.13b). $\vec{F}_P$ is calculated using equation [6.2-4]; remember that the normal vector points out of the face of the cube. Substituting these values into equation [6.6-2] gives:

$$z: \quad P(z)\,dx\,dy - (P(z) + dP)\,dx\,dy - (dx\,dy\,dz)\rho g = 0 \qquad [6.6\text{-}3]$$

Dividing by the volume ($dx\,dy\,dz$) gives:

$$\frac{P(z)}{dz} - \frac{P(z) + dP}{dz} - \rho g = 0 \qquad [6.6\text{-}4]$$

$$\frac{dP}{dz} = -\rho g \qquad [6.6\text{-}5]$$

where P is the pressure exerted on the system by the surroundings, z is the height in the z-direction, ρ is the density of the fluid, and g is the gravitational constant (magnitude only). An important result from equation [6.6-5] is that the pressure varies as a function of the position z within a system containing a static fluid.

A similar derivation shows that pressure varies only as a function of the height, not lateral position. Use the same cube and consider the forces in the x-direction, noting that $g = 0$ in the x-direction (Figure 6.13c):

$$P(x)\,dy\,dz - (P(x) + dP)\,dy\,dz - (dx\,dy\,dz)\rho(0) = 0 \qquad [6.6\text{-}6]$$

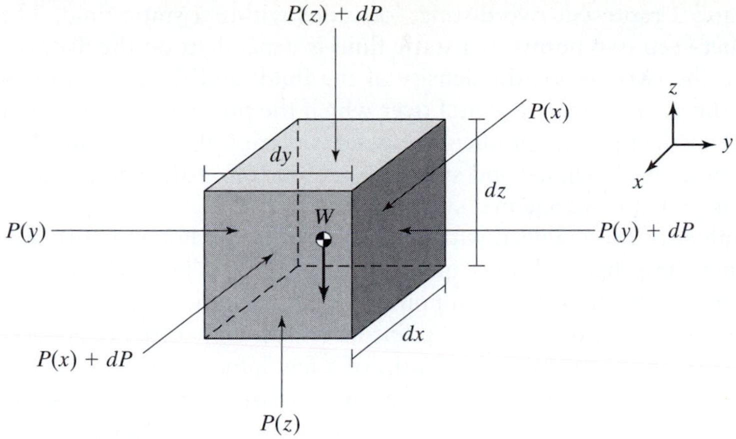

Figure 6.13a
Surface pressure and gravitational forces acting on the cube.

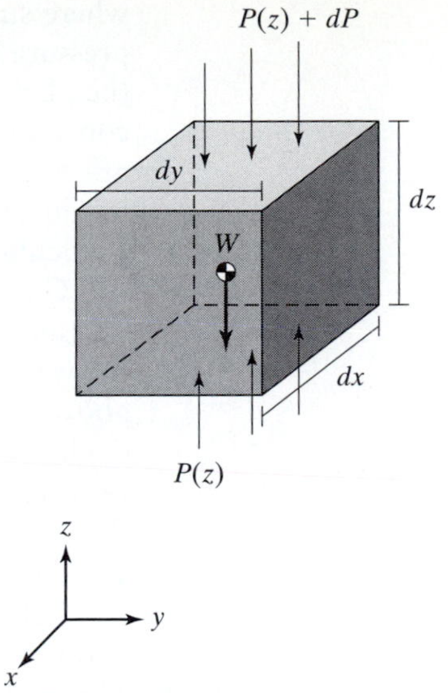

Figure 6.13b
Surface pressure and gravitational forces in the z-direction.

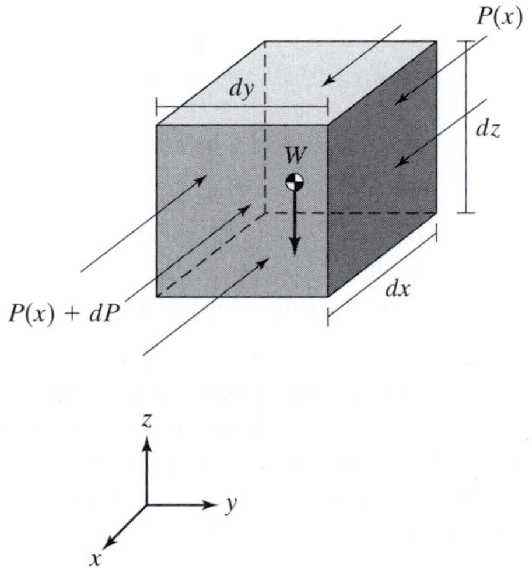

Figure 6.13c
Surface pressure forces in the x-direction.

Dividing by the volume $(dx\ dy\ dz)$ gives:

$$\frac{dP}{dx} = 0 \qquad [6.6\text{-}7]$$

Therefore, the pressure does not change as a function of the x-position. An identical process can be applied to show that pressure is independent of the y-position as well. Thus, pressure forces act only in the z-direction on static fluids.

Consider a fluid with a constant density ρ. Equation [6.6-5] is integrated as:

$$\int_{P_1}^{P_2} dP = -\rho g \int_{z_1}^{z_2} dz \qquad [6.6\text{-}8]$$

$$P_2 - P_1 = -\rho g(z_2 - z_1) \qquad [6.6\text{-}9]$$

$$\Delta P = -\rho g\ \Delta z \qquad [6.6\text{-}10]$$

where subscripts 1 and 2 represent two distinct locations within a static fluid. **The pressure difference between two points in a static fluid is dependent on the distance (i.e., height) between the two points, the density of the fluid, and the gravitational constant.** Note that the area (in the *xy*-plane) over which the pressure acts does not affect the calculation of the pressure gradient. Engineers often talk about a **head** or height of fluid (captured in the right-hand side of equation [6.6-10]) that creates a particular pressure or pressure change in a system.

Consider a container of fluid labeled with points *A*, *B*, and *C* (Figure 6.14). The pressure at *A* is higher than that at *B* by a quantity $\rho g(z_B - z_A)$. This difference is the weight per unit area of the fluid between points *A* and *B*. As the height of liquid above a point increases, the pressure at that point increases. Consider underwater exploration. People can safely scuba dive to depths of a few hundred feet. However, for deep underwater exploration, submersibles and robots must be built to withstand the hydrostatic pressure of water. For example, at a depth of 0.5 miles, the hydrostatic water pressure of 78 atm would crush a human body.

Referring back to Figure 6.14, there is no change in pressure between points *A* and *C*, since they are at the same vertical distance beneath the surface of the liquid. This means that two points that have the same height of fluid with the same density above them have the same hydrostatic pressure exerted upon them. The hydrostatic pressure you feel on your body submerged ten feet in a swimming pool, for example, is the same as if you were submerged ten feet in a freshwater lake. The size or area of the system does not affect the hydrostatic pressure.

In summary, the pressure in a static fluid is dependent on the height of, and the position within, the fluid. As long as there is a continuous, static path through the fluid, these conclusions are valid for all types of bodies or vessels.

EXAMPLE 6.8 **Hydrostatic Pressure Difference Between Shoulder and Ankle**

Problem: Estimate the difference in hydrostatic pressure from the weight of fluid in the body between the shoulder and the ankle. Does the weight of the person enter the calculations? Justify your answer. Assume the fluid is static and the density of blood is 1.056 g/cm^3.

Solution: Because the person is the system and the fluid is static, no fluid moves within, into, or out of the system, making it a steady-state system. No bulk mass crosses the boundary. The system involves only forces due to hydrostatic pressure and gravity.

Since the density of the fluid is constant, equation [6.6-10] can be applied, where location 2 is the shoulder and location 1 is the ankle. We estimate that the difference in length between the shoulder and the ankle is 5 feet (Figure 6.15). The pressure difference ΔP is assumed to be positive:

$$\Delta P = P_1 - P_2 = \rho g(z_2 - z_1)$$

$$= \left(1.056\frac{\text{g}}{\text{cm}^3}\right)\left(9.81\frac{\text{m}}{\text{s}^2}\right)(5\text{ ft} - 0\text{ ft})\left(\frac{0.305\text{ m}}{\text{ft}}\right)\left(\frac{100\text{ cm}}{\text{m}}\right)^3\left(\frac{1\text{ kg}}{1000\text{ g}}\right)\left(\frac{760\text{ mm Hg}}{101{,}300\dfrac{\text{kg}}{\text{m}\cdot\text{s}^2}}\right)$$

$$= 119\text{ mmHg}$$

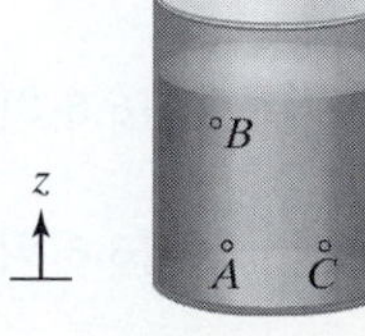

Figure 6.14
Container of fluid with points *A*, *B*, and *C*.

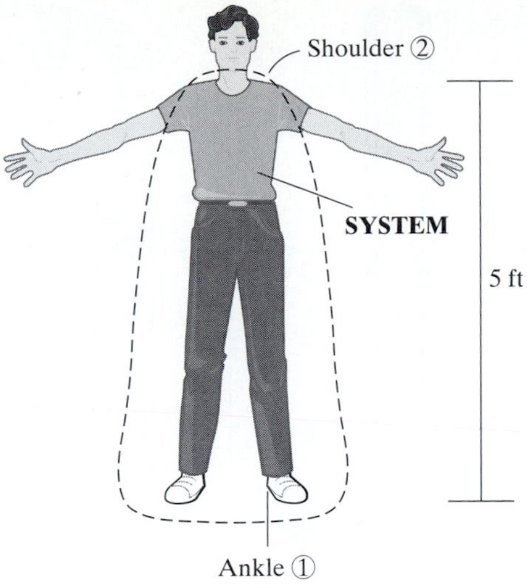

Figure 6.15
Difference in hydrostatic pressure between shoulder and ankle.

Thus, the hydrostatic pressure in the ankle is 119 mmHg greater than the pressure in the shoulder. This makes sense, because the ankle has a much greater hydrostatic head than the shoulder has.

The pressure difference between the ankle and the shoulder is not dependent on the weight of the person. As noted earlier, the pressure across the x- and y-planes does not vary, and the area over which the pressure acts is not captured in the equations describing hydrostatic fluids. In other words, the circumference of the person (i.e., whether thin or fat) does not matter. In contrast, the absolute pressure at the feet is dependent on the person's weight and dimensions of the body and feet. ■

EXAMPLE 6.9 Force Due to Hydrostatic Pressure in Two Containers

Problem: Consider two containers, R and S, filled with water (Figure 6.16). Neglecting the effect of the annulus at height y for container R, calculate the force on the base caused by the hydrostatic pressure. Assume the pressure of air above the fluid is negligible, so the pressure at the top of each container is zero.

Solution: The total height of fluid $(x + y)$ in each container is the same. Consequently, using equation [6.6-10], the change in pressure between the top and the base of each container is also the same:

$$\Delta P = \rho g(x + y)$$

where ρ is the density of water. The pressure at the base of each container is also the same, since the pressure at the top is zero.

$$P_{\text{base}} = P_{\text{top}} + \Delta P = \rho g(x + y)$$

Because the cross-sectional area at the base of each container is the same, we can conclude that the force of hydrostatic pressure $\vec{F}_P$ on the base of each container is identical and equal to:

$$\vec{F}_P = AP_{\text{base}} = A\rho g(x + y)$$

where A is the cross-sectional area of the base.

Thus, the force caused by hydrostatic pressure on the base of each container is the same. This result may seem counterintuitive at first, since the volume and weight of the water in

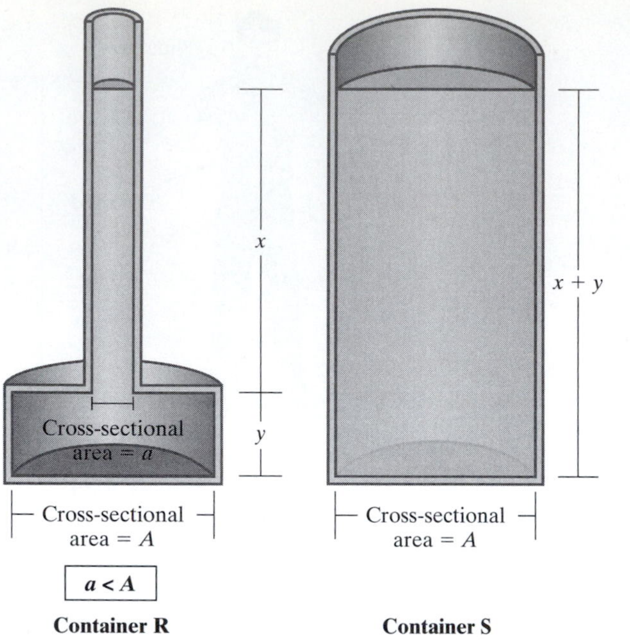

Figure 6.16
Hydrostatic pressure forces on the base of two containers with the same base cross-sectional area but different dimensions.

container S is greater. A table supporting container R with less water should bear less weight than a table supporting container S with more water. Actually, this intuition is consistent with the given solution.

We can use Newton's third law to explain how container R exerts less weight on the table than container S but has the same pressure force on the base of the container. The total force the table exerts on a container must balance the force the container exerts on the table. We also know that the total force the container exerts must be equal to its weight. You may notice that the pressure force on the base in container R exceeds the weight of the water. Therefore, there must be an additional force present such that the sum of the two forces is equal to the weight. This force acts through the container walls and is directly related to the pressure present at the annulus at height y above the base. How both forces work together is the focus of Problem 6.18. ■

6.7 Isolated, Steady-State Systems

Consider a system with no bulk mass transfer across the system boundary and no external forces. Such a system is isolated, and no momentum accumulates in it. Under these conditions, the integral formulation of the equation is commonly used. In these situations, equation [6.3-8] is reduced to the following:

$$0 = \vec{p}_{\mathrm{f}}^{\,\mathrm{sys}} - \vec{p}_{0}^{\,\mathrm{sys}} \qquad [6.7\text{-}1]$$

where $\vec{p}_{\mathrm{f}}^{\,\mathrm{sys}}$ is the total linear momentum of the system at the final condition t_f, and $\vec{p}_{0}^{\,\mathrm{sys}}$ is the total linear momentum of the system at the initial condition t_0.

In a steady-state system, the linear momentum of the system at the final condition is equal to that at the initial condition. It is important to note that the term steady-state describes the total momentum of the system. A steady-state system in this section should not be confused with the steady-state, static systems discussed in Sections 6.5 and 6.6. Components in a steady-state system in this section can move relative to each other within the confines of the system boundary, whereas they do not move in static, steady-state systems including rigid bodies or fluids. The calculation of the term $\vec{p}^{sys}$ describes the momentum contained within or of the system.

Where a system boundary is can influence whether or not a system is considered steady-state. Although every action has an opposing reaction, the distinction between a steady-state and an unsteady-state system is that the former has actions and reactions that act entirely within the system boundary and any external forces acting on the system are balanced. Thus, these forces have no effect on changing the momentum of a steady-state system.

The same statement that the linear momentum of the system at the initial and final conditions is equal can be obtained from classical mechanics. Recall Newton's third law of motion: **If body A exerts a force on body B, then body B must exert an equal but opposite force on body A. It is impossible for any force to occur without its corresponding reactive force—forces always occur in pairs.** This means that all internal forces of components in a system add up to zero, where force can be defined as the rate of change of momentum with respect to time. Therefore, the net derivative of the system's momentum is zero, and the total momentum of the system is constant. This statement is consistent with equation [6.7-1] above.

EXAMPLE 6.10 Platelet Adhesion

Problem: You devise an experiment to investigate if epinephrine, which has been demonstrated to increase platelet adhesion in vivo, induces platelet binding. To mimic the platelets' natural environment but isolate them from other proteins in the blood, you fill a flow chamber with a prepared saline solution containing epinephrine. Two platelets are injected into the solution in the directions and velocities shown in Figure 6.17a. If epinephrine can independently induce platelet adhesion, the adhered platelets should travel together with a new velocity. Assuming the platelets adhere, what will be the magnitude and direction of the velocity? Neglect water resistance and gravity. Each platelet weighs 22 picograms.

Solution:

1. Assemble
   (a) Find: Magnitude and direction of adhered platelets' velocity.
   (b) Diagram: See Figures 6.17a and b. The system is drawn outside of the platelets so that the platelets do not cross the system boundary.

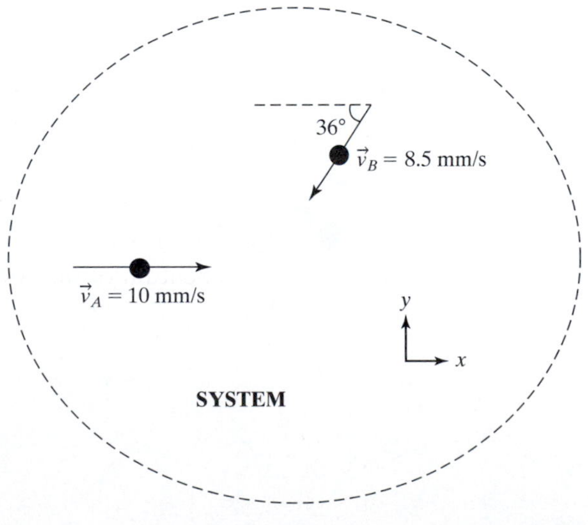

Figure 6.17a
Initial directions and velocities of two platelets.

2. Analyze
   (a) Assume:
      - Neglect the effects of gravity, water resistance, friction, and other external forces.
      - Platelets adhere completely after the collision; i.e., the collision is perfectly inelastic.
      - Steady-state system.
      - System is the area occupied by the platelets before and after they collide.
   (b) Extra data: No extra data are needed.
   (c) Variables, notations, units:
      - Subscript 0 identifies the period before collision; subscript f identifies the period after collision.
      - *A* identifies platelet A; *B* identifies platelet B.
      - Use mm, s, degrees, pg.

3. Calculate
   (a) Equations: The integral equation [6.3-6] is selected since we are evaluating the system between two discrete time points—before collision and after collision. We assume that no external forces, such as gravity, act on the system. In addition, no momentum crosses the system boundary. Therefore, the system is isolated and at steady-state (equation [6.7-1]):

$$0 = \vec{p}_{\text{f}}^{\text{sys}} - \vec{p}_0^{\text{sys}}$$

   The system momentum is the product of the system mass and the system velocity. Because there are multiple particles, momentum at the initial and final conditions must be summed over all particles:

$$0 = \sum m_{\text{f}}\vec{v}_{\text{f}}^{\text{sys}} - \sum m_0\vec{v}_0^{\text{sys}}$$

   The governing equation states that the initial momentum of the system is equal to the final momentum of the system.
   (b) Calculate:
      - First, calculate the velocity vectors of platelets A and B at t_0:

$$\vec{v}_{A,0} = 10\vec{i}\ \frac{\text{mm}}{\text{s}}$$

$$\vec{v}_{B,0} = (-8.5(\cos 36°)\vec{i} - 8.5(\sin 36°)\vec{j})\ \frac{\text{mm}}{\text{s}}$$

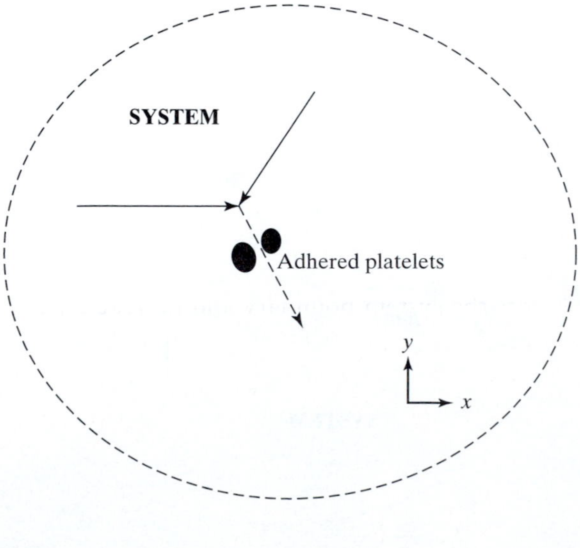

Figure 6.17b
Final direction of two adhered platelets.

- To find the velocity of the adhered cells from the momentum of the system, we need to know that at t_f, the total mass (m_{tot}) of the final system is 44 pg. All values can then be substituted into the governing equation written in both the x- and y-directions:

$$x: \quad 0 = m_{tot}v_{x,f} - (m_A v_{Ax,0} + m_B v_{Bx,0})$$

$$0 = (44 \text{ pg})v_{x,f} - (22 \text{ pg})\left(10\frac{\text{mm}}{\text{s}}\right) - (22 \text{ pg})\left(-8.5 \cos 36°\frac{\text{mm}}{\text{s}}\right)$$

$$v_{x,f} = 1.6 \ \frac{\text{mm}}{\text{s}}$$

$$y: \quad 0 = m_{tot}v_{y,f} - (m_A v_{Ay,0} + m_B v_{By,0}) = m_{tot}v_{y,f} - m_B v_{By,0}$$

$$0 = (44 \text{ pg})v_{y,f} - (22 \text{ pg})\left(-8.5 \sin 36°\frac{\text{mm}}{\text{s}}\right)$$

$$v_{y,f} = -2.5 \ \frac{\text{mm}}{\text{s}}$$

4. Finalize
   (a) Answer: The velocity of the adhered platelets is $\vec{v}_f = (1.6\vec{i} - 2.5\vec{j})$ mm/s.
   (b) Check: The direction makes sense, since the magnitude of the x-component of the momentum of platelet A is greater than that of platelet B. Because the momentum of platelet A does not have a y-component, the adhered platelets would move in the same y-direction as the momentum of platelet B. ■

The platelet adhesion example concerned particles that stuck together and moved uniformly as one body. This is an illustration of a **perfectly inelastic** or **plastic collision**. The two equations formed from the conservation of linear momentum (x- and y-directions) are sufficient to solve for the velocity of the center of mass of the adhered particles after their collision. However, when particle collision is **elastic**, two particles colliding together will result in each particle having its own unique direction and speed after the collision. In such cases, applying the conservation of momentum equation where only the initial conditions are known results in an underspecified system, since there are more unknown variables than equations. Additional information or equations or both must be provided to solve for the velocities of the particles after collision.

In an elastic system, two equations commonly provide more relationships to help solve a problem. In a **perfectly elastic collision**, the kinetic energy of a system of two colliding particles is constant. In physics, this special case is often termed the *conservation of kinetic energy*. While the kinetic energy in a system can be constant, it is important to remember that this definition of conservation is not used in this text. Recall that kinetic energy, E_K, is given as:

$$E_K = \frac{1}{2}mv^2 \qquad [6.7\text{-}2]$$

In the special case of a perfectly elastic collision within a steady-state system with no movement of mass or energy across the system boundary and no reactions, the initial kinetic energy of the system is equal to the final kinetic energy of the system:

$$0 = E_{K,f}^{sys} - E_{K,0}^{sys} \qquad [6.7\text{-}3]$$

where $E_{K,f}^{sys}$ is the total kinetic energy of the system at final condition t_f and $E_{K,0}^{sys}$ is the total kinetic energy of system at initial condition t_0. Note that kinetic energy is scalar and does not have a direction, so equating the initial and final kinetic energies of a system contributes only one equation to solving a system involving a perfectly elastic collision.

Kinetic energy can be lost when colliding objects transform some of their energy to other forms, such as when an object deforms, heat is given off, or sound is generated. A relationship describing the elasticity of colliding objects, the **coefficient of restitution** *(e)*, can be used to develop an additional equation to solve a problem. The coefficient of restitution is a ratio of linear impulses (forces) that the objects exert upon one another during a collision. For a more complete discussion of this relationship, see statics and dynamics books such as Bedford and Fowler's *Engineering Mechanics: Statics and Dynamics* (2002). The coefficient of restitution can be measured by the ratio of the relative velocities before and after impact:

$$e = \frac{v_{\text{separation}}}{v_{\text{approach}}} \qquad [6.7\text{-}4]$$

where $v_{\text{separation}}$ is the difference in velocities after collision and v_{approach} is the difference in velocities before collision.

The value of e is usually determined experimentally and depends on the properties of the objects (such as their material) and the velocities and orientations of the objects as they collide. In most applications, the value of e ranges between 0 for a perfectly inelastic (plastic) collision and 1 for a perfectly elastic collision, when friction and kinetic energy losses are negligible.

The coefficient of restitution must be evaluated for each coordinate direction. Recall from physics that changing the acceleration of an object (and thus its velocity) in one direction will not change its acceleration in the other directions. That is, if an external force were applied to an object in the x-direction, it would not change the velocity of the object in the y-direction. The effects of forces on accelerations in one direction are independent from those in other directions. When evaluating e, remember to find the differences in velocities in one direction only.

In this text, the coefficient of restitution is applied to two different types of impacts: direct central impacts and oblique central impacts. In a **direct central impact**, the centers of mass of particles A and B approach each other along a straight line (i.e., along one direction, such as the x-direction) and separate along the same line after impact (Figure 6.18). The coefficient of restitution for a direct central impact is:

$$e = \frac{v_{\text{B,f}} - v_{\text{A,f}}}{v_{\text{A,0}} - v_{\text{B,0}}} \qquad [6.7\text{-}5a]$$

where $v_{\text{B,f}}$ is the velocity of particle B after collision, $v_{\text{A,f}}$ is the velocity of particle A after collision, $v_{\text{B,0}}$ is the velocity of particle B before collision, and $v_{\text{A,0}}$ is the velocity of particle A before collision.

In the other type of impact, **oblique central impact**, particles A and B collide at an angle (Figure 6.19). The analysis is like that for direct central impact but is applied only to a single component of the velocity vectors. Suppose that the centers of mass of particles A and B approach with arbitrary velocities $\vec{v}_{\text{A,0}}$ and $\vec{v}_{\text{B,0}}$, respectively. Let us assume that the forces they exert on each other at impact point toward their centers of mass and act only parallel to the x-axis; thus, no forces are exerted in the y- or z-direction and their velocities in these

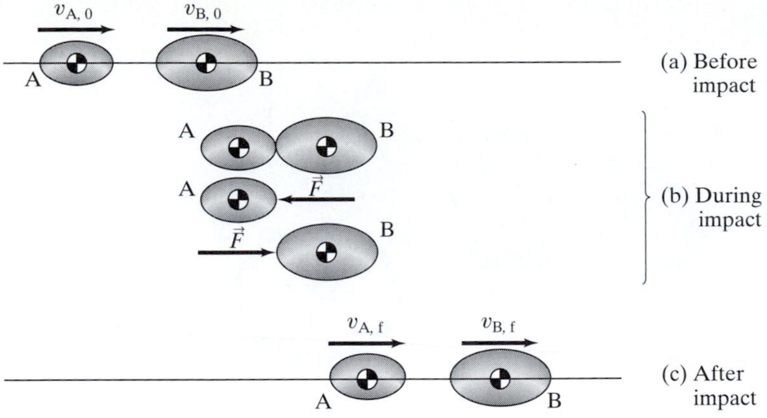

Figure 6.18
Direct central impact of objects A and B. The force exerted on the objects upon impact is given as $\vec{F}$. (*Source*: Bedford A and Fowler W, *Engineering Mechanics: Statics and Dynamics*. Upper Saddle River, NJ: Prentice Hall, 2002.)

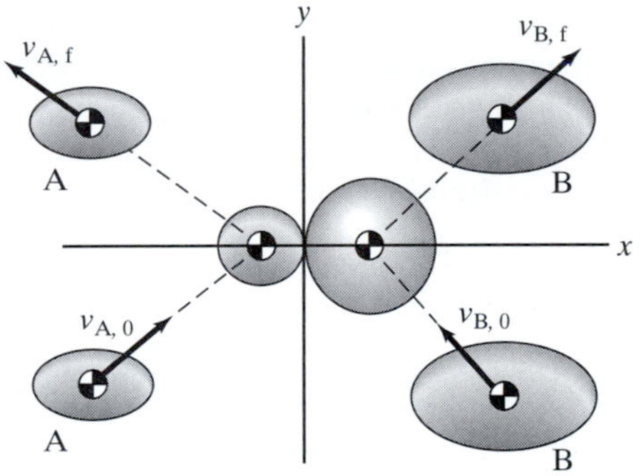

Figure 6.19
Oblique central impact of objects A and B. (*Source*: Bedford A and Fowler W, *Engineering Mechanics: Statics and Dynamics*. Upper Saddle River, NJ: Prentice Hall, 2002.)

directions are unchanged. The coefficient of restitution for this system is defined as:

$$e = \frac{v_{Bx,f} - v_{Ax,f}}{v_{Ax,0} - v_{Bx,0}} \qquad [6.7\text{-}5b]$$

where $v_{Bx,f}$ is the velocity of particle B in the x-direction after collision, $v_{Ax,f}$ is the velocity of particle A in the x-direction after collision, $v_{Bx,0}$ is the velocity of particle B in the x-direction before collision, and $v_{Ax,0}$ is the velocity of particle A in the x-direction before collision. For an oblique central impact, the coefficient of restitution in the y-direction is not applicable, since the directions of the velocities in the y-direction do not change.

Together, the conservation of linear momentum, the kinetic energy equation, and information on the coefficient of restitution should be adequate to form a solution for a system describing the elastic collision of two particles.

EXAMPLE 6.11 Helmet Crash

Problem: Protecting the head with a helmet is a major concern for professional and recreational cyclists. Fortunately you are wearing a helmet when you lose your balance standing on your bike, causing your head to hit the ground at 6.3 m/s (Figure 6.20a). Your helmet fits loosely such that when you fall, two separate collisions occur. First, your helmet crashes into the ground. Then, as your helmet rebounds from the ground, your head collides with the helmet.

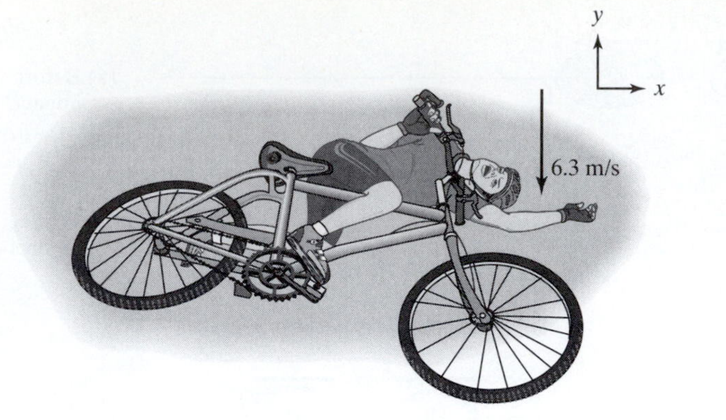

Figure 6.20a
Cyclist falls from his bicycle and his head hits the ground.

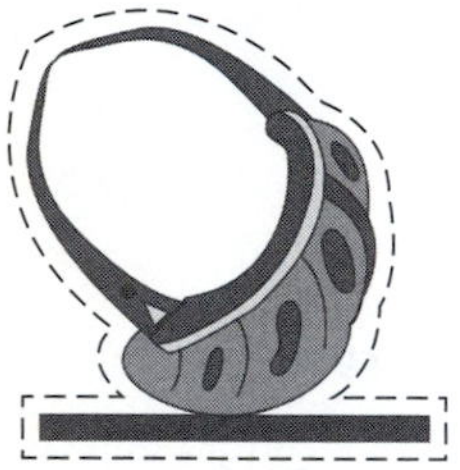

Figure 6.20b
System 1: helmet and ground.

Calculate the velocities of your helmet and your head after the two collisions. Estimate that your head has a mass of 5 kg and your helmet a mass of 330 g. The coefficient of restitution is 0.82 for the impact between the helmet and the ground and 0.17 for the impact between the helmet and your head. (The first coefficient is more elastic, while the second is more plastic. This makes sense, since the outside of your helmet is hard and bounces when hitting the ground; the inside of your helmet is soft and cushions your head.) Assume the effects of gravity, air resistance, and friction are negligible.

Solution: Two systems are drawn (Figures 6.20b and c). A number of assumptions need to be made, including:

- The systems are at steady-state.
- No material crosses the system boundaries.
- No external forces, such as the forces exerted by the neck as a reaction to the fall, act on the systems.
- Both collisions are direct central impact collisions.
- All movements (i.e., velocities) and coefficient of restitution values are along the y-axis.
- Gravity, air resistance, and friction effects are negligible.

We use H to denote helmet, G for ground, and D for head; subscript 0 indicates before the collision and subscript f indicates after the collision.

System 1: For the system in Figure 6.20b, we examine the collision between your helmet and the ground. We assume $v_{G,0} = v_{G,\text{f}} = 0$. Applying the coefficient of restitution equation [6.7-5a] to this system gives:

$$e = 0.82 = \frac{v_{H,\text{f}} - v_{G,\text{f}}}{v_{G,0} - v_{H,0}} = \frac{v_{H,\text{f}}}{-v_{H,0}}$$

$$v_{H,\text{f}} = 0.82\left[0 - \left(-6.3\ \frac{\text{m}}{\text{s}}\right)\right] = 5.2\ \frac{\text{m}}{\text{s}}$$

Thus, the velocity of your helmet after the initial collision is 5.2 m/s in the positive y-direction.

The assumption that the ground is stationary may seem to violate conservation of momentum. In reality, the momentum of the Earth does change to balance the change in momentum experienced by the helmet. However, because of the Earth's large mass, the change in velocity is negligible; thus, we may safely assume the Earth remains stationary without contradicting the premises of this chapter.

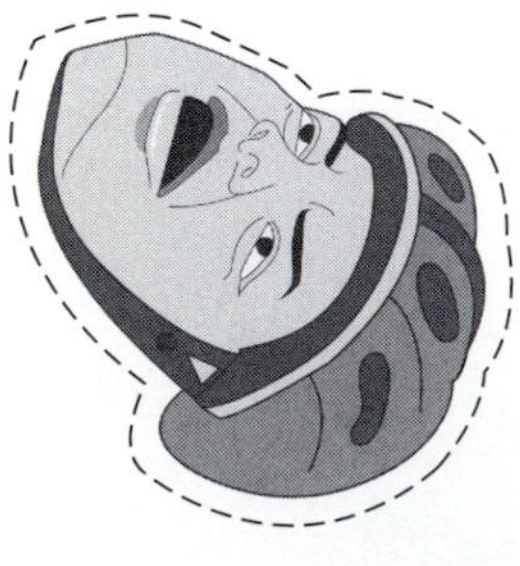

Figure 6.20c
System 2: helmet and head.

System 2: In the collision in Figure 6.20c, we analyze the system that includes your head and your helmet. We notice that the initial velocity of your helmet before colliding with your head is the same as the final velocity of your helmet after colliding with the ground ($\vec{v}_{H,0} = 5.2$ m/s away from the ground). The velocity of your head before collision with the helmet is the same as the velocity of the helmet before it hits the ground ($\vec{v}_{D,0} = 6.3$ m/s toward the ground). Applying the coefficient of restitution equation [6.7-5a] to the second system gives:

$$e = 0.17 = \frac{v_{H,f} - v_{D,f}}{v_{D,0} - v_{H,0}} = \frac{v_{H,f} - v_{D,f}}{-6.3\ \frac{\text{m}}{\text{s}} - 5.2\ \frac{\text{m}}{\text{s}}}$$

$$-\ 2.0\ \frac{\text{m}}{\text{s}} = v_{H,f} - v_{D,f}$$

where all velocities are along the y-axis.

To find both values, the conservation of momentum equation is simplified to equation [6.7-1], because the system is at steady-state, has no mass flow, and has no external forces acting on the overall system:

$$0 = \vec{p}_f^{\,sys} - \vec{p}_0^{\,sys}$$

$$0 = m_H\vec{v}_{H,f} + m_D\vec{v}_{D,f} - (m_H\vec{v}_{H,0} + m_D\vec{v}_{D,0})$$

$$0 = (0.330\text{ kg})(\vec{v}_{H,f}) + (5\text{ kg})(\vec{v}_{D,f}) - \left((0.330\text{ kg})\left(5.2\vec{j}\ \frac{\text{m}}{\text{s}}\right) + (5\text{ kg})\left(-6.3\vec{j}\ \frac{\text{m}}{\text{s}}\right)\right)$$

$$-\ 29.8\vec{j}\,\frac{\text{kg}\cdot\text{m}}{\text{s}} = (0.330\text{ kg})(\vec{v}_{H,f}) + (5\text{ kg})(\vec{v}_{D,f})$$

This gives us a second equation to solve for the two unknown variables $\vec{v}_{H,f}$ and $\vec{v}_{D,f}$. Reformulating the first equation such that $\vec{v}_{H,f} = \vec{v}_{D,f} - 2.0\vec{j}$ m/s and substituting it into the second equation gives:

$$-29.8\vec{j}\,\frac{\text{kg}\cdot\text{m}}{\text{s}} = (0.330\text{ kg})\left(\vec{v}_{D,f} - 2.0\vec{j}\ \frac{\text{m}}{\text{s}}\right) + (5\text{ kg})(\vec{v}_{D,f})$$

$$\vec{v}_{D,f} = \frac{-29.8\vec{j}\,\frac{\text{kg}\cdot\text{m}}{\text{s}} + 0.66\vec{j}\,\frac{\text{kg}\cdot\text{m}}{\text{s}}}{(5\text{ kg} + 0.330\text{ kg})} = -5.5\vec{j}\ \frac{\text{m}}{\text{s}}$$

$$\vec{v}_{H,f} = \vec{v}_{D,f} - 2.0\vec{j}\ \frac{\text{m}}{\text{s}} = -5.5\vec{j}\ \frac{\text{m}}{\text{s}} - 2.0\vec{j}\ \frac{\text{m}}{\text{s}} = -7.5\vec{j}\ \frac{\text{m}}{\text{s}}$$

Following its collision with the helmet, your head travels toward the ground at 5.5 m/s, and the helmet travels 7.5 m/s toward the ground. If you had not been wearing a helmet, your head would have hit the ground while traveling at 6.26 m/s, the velocity at which the helmet impacted the ground. Thus, the helmet succeeds in reducing the impact to the head during a fall. It diffuses the collision over two smaller impacts. ■

6.8 Steady-State Systems with Movement of Mass Across System Boundaries

Movement of mass across a system boundary is very common when considering systems involving the human body and biomedical equipment designed to assist the functioning of the human body. In addition, bioprocessing equipment is often operated under conditions where mass flows into and out of units. **Whenever mass crosses a system boundary by fluid flow, momentum is transferred to or from the system.** Examples of systems with movement of momentum by a fluid include blood flow through the heart or through an assist device, such as a left ventricular assist device (LVAD). Momentum can also enter or leave a system when a discrete quantity of mass crosses the system boundary. Refer to Chapter 3 for an extensive discussion of the movement of mass across the system boundary.

For systems with fluids (liquid or gas) crossing the system boundary at specified rates, it is most appropriate to use the differential form of the conservation of linear momentum equation. Recall equation [6.3-3] in Section 6.3 for multiple inlets and outlets:

$$\sum_i \dot{m}_i\vec{v}_i - \sum_j \dot{m}_j\vec{v}_j + \sum \vec{F} = \frac{d\vec{p}^{\,\text{sys}}}{dt} \qquad [6.8\text{-}1]$$

For steady-state systems, the Accumulation term equals zero, so equation [6.8-1] reduces to:

$$\sum_i \dot{m}_i\vec{v}_i - \sum_j \dot{m}_j\vec{v}_j + \sum \vec{F} = 0 \qquad [6.8\text{-}2]$$

This equation is most often used when mass enters or leaves a system as a flowing fluid specified as a rate.

Momentum can also enter or leave a system as a discrete quantity or chunk. Often, mass enters or leaves a system with a particular velocity; hence, using the integral form of the conservation of linear momentum equation is typically appropriate to calculate the momentum contributed to the system. For example, when a baseball strikes a baseball glove, the incoming ball has a specific mass and velocity and contributes momentum to the system of the baseball glove when it is caught. Discrete quantities of mass can also move into and out of a system carried in a fluid. A biological example is thromboembolism, in which a particle has broken away from a blood clot and circulates in the bloodstream until it blocks another part of the vessel. The particle broken from the clot flows into another part of the vessel (the system) at a particular velocity and contributes momentum to the system.

Recall the integral form of the conservation of linear momentum equation [6.3-6] for multiple inlets and outlets:

$$\int_{t_0}^{t_f} \dot{\vec{p}}_i\,dt - \int_{t_0}^{t_f} \dot{\vec{p}}_j\,dt + \int_{t_0}^{t_f} \sum \vec{F}\,dt = \int_{t_0}^{t_f} \frac{d\vec{p}^{\,\text{sys}}}{dt}\,dt \qquad [6.8\text{-}3]$$

The rate of momentum terms in equation [6.8-3] can be integrated:

$$\sum_i \vec{p}_i - \sum_j \vec{p}_j + \int_{t_0}^{t_f} \sum \vec{F}\,dt = \vec{p}_{\text{f}}^{\,\text{sys}} - \vec{p}_0^{\,\text{sys}} \qquad [6.8\text{-}4]$$

$$\sum_i m_i\vec{v}_i - \sum_j m_j\vec{v}_j + \int_{t_0}^{t_f} \sum \vec{F}\,dt = \vec{p}_{\text{f}}^{\,\text{sys}} - \vec{p}_0^{\,\text{sys}} \qquad [6.8\text{-}5]$$

For steady-state systems, equation [6.8-5] reduces to:

$$\sum_i m_i\vec{v}_i - \sum_j m_j\vec{v}_j + \int_{t_0}^{t_f} \sum \vec{F}\,dt = 0 \qquad [6.8\text{-}6]$$

This equation is most often used when mass enters or leaves a system as a discrete quantity.

Many external forces can act on a system of interest. Surface forces, such as pressure forces, and body forces, such as gravitational forces, may contribute to the term describing external forces. For flowing fluids, another force known as the resultant force may be important.

When fluid flows through a tube or a pipe or against an object, such as a platform, a force from the surroundings (e.g., the material supporting the tube or platform) may be required to hold the system with the flowing fluid in place. This force is often termed the **resultant force** (F_R). When a fluid flowing through a pipe changes direction, a resultant force in the surroundings must be exerted on both the fluid and pipe to keep the pipe from moving. Changes in fluid pressure between an inlet and outlet can also contribute to a resultant force. **The resultant force is described as one of the forces in the $\sum \vec{F}$ term.**

Typically, resultant forces may need to be considered for systems involving the flow of liquids. In contrast, they are often ignored for systems involving the flow of gas because they are not significant in magnitude. For example, in the lungs, because of the low density of air and the minimal pressure drop across the lungs from the trachea to alveoli, the resultant forces that act to hold the vessels in place are very small.

Consider the U-bend in Figure 6.21. Here, fluid in a pipe enters the system with a velocity $\vec{v}_1$ in a tube with cross-sectional area A_1 and leaves at a velocity $\vec{v}_2$ in a tube with cross-sectional area A_2. Neglecting the effects of gravitational forces, two forces are considered: the fluid pressure force ($\vec{F}_P$) and the resultant force ($\vec{F}_R$) required to hold the pipe in place. If we assume that the pressures at the inlet and outlet are P_1 and P_2, respectively, the following reduction to equation [6.8-2] is appropriate for the x-direction:

$$x: \quad \dot{m}_1 v_{1,x} - \dot{m}_2(-v_{2,x}) + \sum(\vec{F}_p + \vec{F}_{R,x}) = 0 \qquad [6.8\text{-}7]$$

Recall equation [6.2-5] to describe $\vec{F}_P$. We define the normal vector $\vec{n}$ to point out of the system, which is in the $-x$ direction for both the inlet and outlet streams. Substituting terms into equation [6.2-5] gives:

$$\sum \vec{F}_P = -P_1(-\vec{i})A_1 - P_2(-\vec{i})A_2 \qquad [6.8\text{-}8]$$

Assuming the fluid has a constant density ρ, we can substitute for the specific mass flow rate and for the pressure force and then solve for the resultant force:

$$\dot{m}_1 v_{1,x} + \dot{m}_2 v_{2,x} + P_1A_1 + P_2A_2 + F_{R,x} = 0 \qquad [6.8\text{-}9]$$

$$\rho A_1 v_1^2 + \rho A_2 v_2^2 + P_1A_1 + P_2A_2 + F_{R,x} = 0 \qquad [6.8\text{-}10]$$

$$F_{R,x} = -\rho A_1 v_1^2 - \rho A_2 v_2^2 - P_1A_1 - P_2A_2 \qquad [6.8\text{-}11]$$

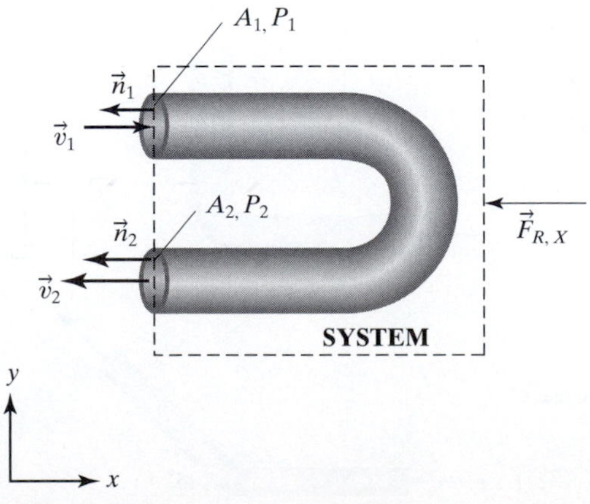

Figure 6.21
A resultant force is needed to hold the U-bend in place.

$F_{R,x}$ describes the solid force that must be exerted on the walls of the U-bend pipe to hold it in place. Note that the directions of the momentum and pressure terms are opposite to that of the resultant force. Because no pressure difference or material transfer occurs across the system boundary in the y-direction, no resultant forces are needed to hold the pipe in place in the y-direction.

The method used above to derive an equation describing the resultant force can be used for other systems. Often, the external forces acting in both the x- and y-directions need to be considered. In some situations, only contributions from fluid flow need to be included, because gravitational forces or pressure forces do not need to be considered, such as when pressure forces do not change along the length of the pipe.

EXAMPLE 6.12 **Resultant Forces Across Pipe Bends**

Problem: Consider the three pipes shown in Figure 6.22. For each case, the mass flow rate for the fluid in the inlet stream is $\dot{m}$, and the linear velocity of the inlet stream is $v\vec{i}$. What is the resultant force needed to stabilize each pipe, and how do these forces compare? Assume each system is at steady-state and the diameter is constant throughout the pipe. Ignore the effects of pressure and gravity.

Solution: Each system we consider is bounded by the natural material boundary of the pipe depicted in the picture. The systems are at steady-state, so we apply equation [6.8-2]. Pressure and gravitational forces are neglected, so the only force of interest is the resultant force. Considering each case separately yields:

Case A:
$$\dot{m}(v\vec{i}) - \dot{m}(v(-\vec{i})) + \vec{F}_R = 0$$
$$\vec{F}_R = \dot{m}v(-2\vec{i})$$

Case B:
$$\dot{m}(v\vec{i}) - \dot{m}(v\vec{j}) + \vec{F}_R = 0$$
$$\vec{F}_R = \dot{m}v(-\vec{i} + \vec{j})$$

Case C:
$$\dot{m}(v\vec{i}) - \dot{m}\left(v\left(\frac{\sqrt{2}}{2}\vec{i} + \frac{\sqrt{2}}{2}\vec{j}\right)\right) + \vec{F}_R = 0$$
$$\vec{F}_R = \dot{m}v\left(\left(\frac{\sqrt{2}}{2} - 1\right)\vec{i} + \frac{\sqrt{2}}{2}\vec{j}\right)$$

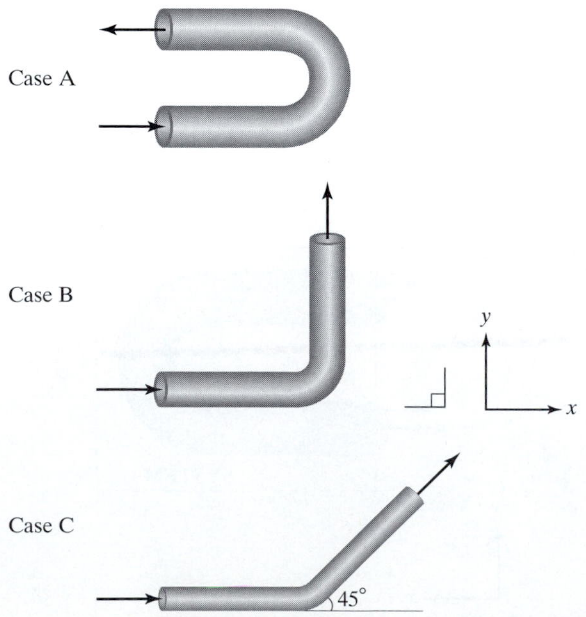

Figure 6.22
Directions of fluid flows in three pipes. Arrows indicate direction of fluid flow.

Notice that in each case, the total resultant force necessary to keep each pipe stationary is the sum of two force components (x- and y-directions). One force component is equal and opposite to the momentum rate of the inlet stream. The second is equal to the momentum rate of the outlet stream. This is consistent with Newton's third law, since the mass flow from the fluid produces a force on the wall of the pipe, so the pipe exerts a resultant force on the fluid to keep the pipe stationary.

The extent to which the force components superimpose on each other is dependent on the angle of the bend. In Case A, the inlet and outlet are in opposite directions, so the two force components are in the same direction and their magnitudes are added. If we were to consider a straight section of pipe with all the same assumptions made in the three cases above, the force components would cancel each other completely and the resultant force would be zero. Cases B and C are between these two extremes; the magnitudes of their net resultant forces lie between zero and $2\dot{m}v$. ■

EXAMPLE 6.13 Flow Around Bend in Total Artificial Heart

Problem: The AbioCor™ total artificial heart (TAH) is an implant designed to replace the anatomical heart when it prematurely loses function. ABIOMED (Danvers, MA) produces a two-chambered heart that can pump 5 L/min at 80 beats per minute. Hypothetical directions of blood flowing through the four vessels that connect to a schematic model of the TAH are shown in Figure 6.23. The cross-sectional area of each of the four vessels is given in Table 6.2. Determine the force the TAH must be able to withstand from the change in movement of blood in both the pulmonic system (from the venae cavae to pulmonary artery) and the systemic system (from the pulmonary vein to the aorta). Also determine the magnitude of these forces. Additional forces, such as those resulting from pumping the blood should not be considered for this calculation.

Solution:

1. Assemble
   (a) Find: magnitude and direction of force exerted by the body to withstand the change in direction of blood flow.
   (b) Diagram: See Figure 6.23. The system boundary encompasses the TAH. The four vessels cross the system boundary. The resultant forces act from the body against the TAH.

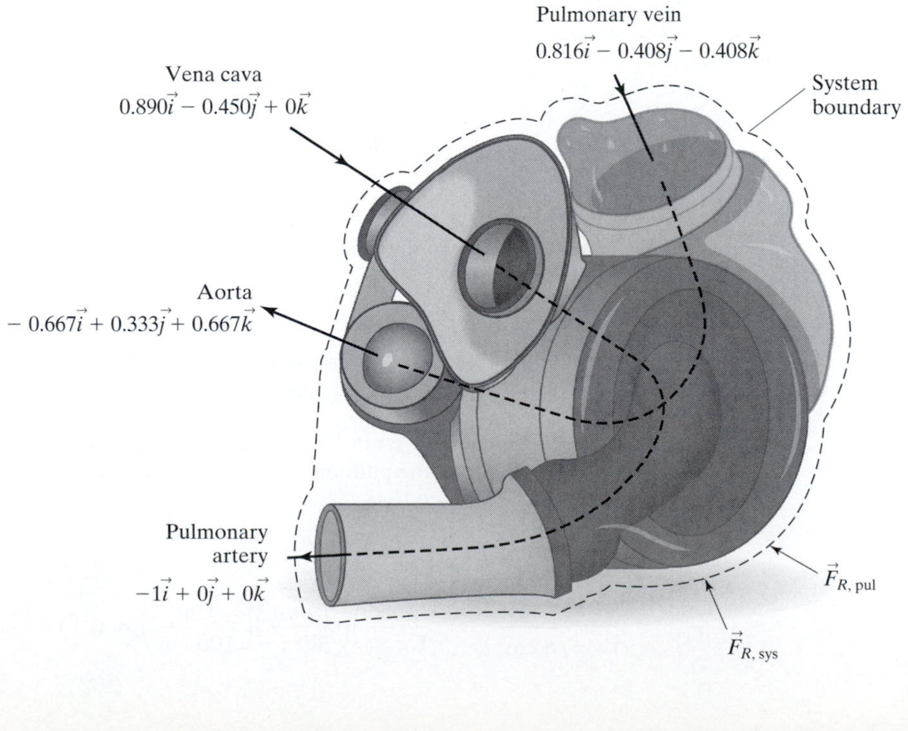

Figure 6.23
Directions of blood flows in a model total artificial heart.

TABLE 6.2

Cross-Sectional Areas of Heart Vessels

System	Vessel	Cross-sectional area (cm^2)
Systemic	Pulmonary vein	6.0
	Aorta	2.5
Pulmonic	Vena cava	8.0
	Pulmonary artery	4.0

2. Analyze
   (a) Assume:
   - The system is at steady-state (i.e., no blood accumulates in the TAH).
   - The blood flow rates and velocities are constant (i.e., nonpulsatile).
   - No leaks occur in the TAH.
   - There are no reactions in the blood or at the blood-wall interface.
   - No frictional losses.
   - Neglect all other force effects, such as gravity and pressure changes, and forces from the contraction of the TAH.

   (b) Extra data:
   - The density of blood is 1.056 g/cm^3.

   (c) Variables, notations, and units:
   - pv = pulmonary vein
   - ao = aorta
   - vc = vena cava
   - pa = pulmonary artery
   - sys = systemic system
   - pul = pulmonic system
   - Use kg, m, s, N.

3. Calculate
   (a) Equations: The differential conservation of linear momentum equation [6.8-1] can be simplified for one inlet and one outlet. Because the system is nonreacting, we can find the mass flow rates by using the differential mass conservation equation [3.3-10]. Because the system is at steady-state, the Accumulation terms for both equations are zero:

$$\dot{m}_i\vec{v}_i - \dot{m}_j\vec{v}_j + \sum\vec{F} = 0$$
$$\dot{m}_i - \dot{m}_j = 0$$

   (b) Calculate:
   - First, we find the mass flow rate into and out of each chamber of the TAH. Since we assume each stroke of the TAH pumps all the blood out, the mass flow in must equal the mass flow out for both halves of the TAH:

$$\dot{m}_i - \dot{m}_j = \rho\dot{V}_i - \rho\dot{V}_j = 0$$
$$\dot{m}_i = \dot{m}_j = \left(1.056\frac{\text{kg}}{\text{L}}\right)\left(5\frac{\text{L}}{\text{min}}\right)\left(\frac{1\text{ min}}{60\text{ s}}\right) = 0.088\ \frac{\text{kg}}{\text{s}}$$

   - The direction of the fluid flow is given in Figure 6.23. We calculate the absolute magnitude of the fluid velocity in the pulmonary vein by equation [3.2-4]:

$$|v| = \frac{\dot{V}}{A}$$
$$|v_{\text{pv}}| = \frac{\dot{V}_{\text{pv}}}{A_{\text{pv}}} = \frac{5\frac{\text{L}}{\text{min}}}{6\text{ cm}^2}\left(\frac{1000\text{ cm}^3}{\text{L}}\right)\left(\frac{1\text{ min}}{60\text{ s}}\right)\left(\frac{1\text{ m}}{100\text{ cm}}\right) = 0.139\ \frac{\text{m}}{\text{s}}$$

$$\vec{v}_{pv} = 0.139\frac{m}{s}(0.816\vec{i} - 0.408\vec{j} - 0.408\vec{k})$$

Velocities are calculated similarly for the other three vessels:

$$\vec{v}_{ao} = 0.333(-0.667\vec{i} + 0.333\vec{j} + 0.667\vec{k})\frac{m}{s}$$

$$\vec{v}_{vc} = 0.104(0.89\vec{i} - 0.45\vec{j} + 0.0\vec{k})\frac{m}{s}$$

$$\vec{v}_{pa} = 0.208(-1.0\vec{i} + 0.0\vec{j} + 0.0\vec{k})\frac{m}{s}$$

Resultant forces for the systemic and pulmonary systems are calculated separately using momentum balances in the x-, y-, and z-directions. For the systemic system, solving the resultant force in the x-direction yields:

$$\dot{m}_i\vec{v}_i - \dot{m}_j\vec{v}_j + \sum\vec{F}_R = 0$$

$$x: \quad \left(0.088\frac{kg}{s}\right)\left(0.139\frac{m}{s}\right)(0.816\vec{i})$$

$$-\left(0.088\frac{kg}{s}\right)\left(0.333\frac{m}{s}\right)(-0.667\vec{i}) + \sum\vec{F}_{R,x} = 0$$

$$\sum\vec{F}_{R,x} = -0.0295 \text{ N}$$

The resultant forces in the y- and z-directions are solved similarly. The total resultant force in the systemic system is $\vec{F}_{R,sys} = (-0.0295\vec{i} + 0.0147\vec{j} + 0.0245\vec{k})$ N. Solving for the resultant force in the pulmonic system in the same manner yields $\vec{F}_{R,pul} = (-0.0264\vec{i} + 0.00412\vec{j})$ N.

- To find the magnitude of the forces, take the square root of the sum of the squared values in the force vectors:

$$F_{R,sys} = \sqrt{(-0.0295\vec{i})^2 + (0.0147\vec{j})^2 + (0.0245\vec{k})^2} \text{ N} = 0.0411 \text{ N}$$

Similarly, $F_{R,pul} = 0.0267$ N.

4. Finalize
   (a) Answer: The net forces exerted by the body on the TAH to support changes in blood flow are $\vec{F}_{R,sys} = (-0.0295\vec{i} + 0.0147\vec{j} + 0.0245\vec{k})$ N and $\vec{F}_{R,pul} = (-0.0264\vec{i} + 0.00412\vec{j})$ N. The magnitude of the resultant force of the systemic system is 0.0411 N and that for the pulmonic system is 0.0267 N.
   (b) Check: The directions of the resultant force vectors make sense, since they oppose the force created by the momentum of the blood. Both resultant forces push up and against the bends in the TAH to keep the tubes in place. The magnitude of the force to hold the vessels in place is higher on the systemic side of the heart (which pumps blood to the whole body) than on the pulmonary side (which pumps blood only to the lungs). The net magnitude of the resultant forces is smaller (by two or three orders of magnitude) than the forces encountered when pressure differences are taken into account. ■

EXAMPLE 6.14 Split of Water Main

Problem: Consider a water main that provides drinking water for a residential area. The main line splits at a T intersection to route water to two separate neighborhoods, as shown in Figure 6.24. Assume the water is evenly divided between two outlet pipes of the same diameter. If the inlet flow is continuous, what are the required resultant forces to support the water flow?

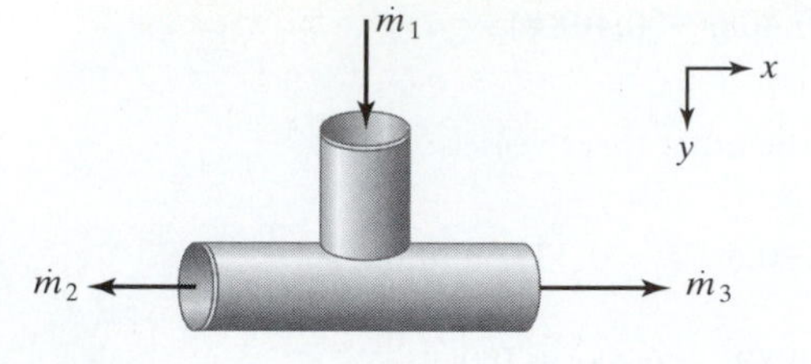

Figure 6.24
Directions of mass flow rates in the splitting of a water main.

Solution: The system includes one inlet ($\dot{m}_1$) and two outlets ($\dot{m}_2$, $\dot{m}_3$). We define a coordinate system such that the fluid enters the inlet flowing in the $+y$-direction. Remember, the water is divided evenly among two outlet pipes that have the same diameter. Using the conservation of mass, this allows us to write the following relationships:

$$\dot{m}_2 = \dot{m}_3, \qquad \vec{v}_2 = -\vec{v}_3, \qquad A_2 = A_3$$

Because the water flows continuously and does not accumulate, the system is at steady-state. Substituting these variables into the differential conservation of linear momentum equation at steady-state [6.8-2] yields equations describing the rate of momentum contributed by fluid flow, pressure forces, and resultant forces in the x- and y-directions:

$$x: \quad -\dot{m}_2(-\vec{v}_2) - \dot{m}_3\vec{v}_3 - P_2(-1)A_2 - P_3(1)A_3 + F_{R,x} = 0$$
$$y: \quad \dot{m}_1\vec{v}_1 - P_1(-1)A_1 + F_{R,y} = 0$$

Since the magnitudes of the outlet mass flow rates, velocities, and cross-sectional areas are equal, the x-direction equation is rewritten as:

$$x: \quad (P_2 - P_3)A_2 + F_{R,x} = 0$$

The resultant force in the y-direction opposes the direction of the water flow. Its magnitude is the sum of the rate of momentum contributed by the water flow as well as the fluid pressure force. The magnitude of the resultant force in the x-direction describes the difference between the fluid pressure forces in the x-direction, but not any rate of momentum terms contributed by water flow. The direction of the resultant force can be either positive or negative depending on the relative magnitude of the two outlet pressures. In the case where the two outlet pressures are identical, there is no resultant force in the x-direction. ■

6.9 Unsteady-State Systems

In an unsteady-state system, at least one variable describing the system (e.g., pressure, flow rate) changes with time. Unsteady-state systems gain or lose momentum with bulk material transfer or when external forces act on them; thus, the Accumulation term is always nonzero. Typically, the differential or integral form of the conservation equation (equation [6.3-3] or [6.3-6]) is required. Depending on the problem statement, an initial or final condition may need to be specified.

For systems with no bulk transfer of material across the system boundary, the differential form of the conservation of linear momentum equation [6.3-3] reduces to the following:

$$\sum\vec{F} = \frac{d}{dt}(\vec{p}^{\,sys}) = \frac{d}{dt}(m^{sys}\,\vec{v}^{\,sys}) \qquad [6.9\text{-}1]$$

where $\sum\vec{F}$ is the sum of all the external forces acting on the system, m^{sys} is the mass of the system, and $\vec{v}^{sys}$ is the velocity of the system. This equation states that the change of momentum of a system with respect to time is equal to the sum of the external forces on the system.

Equation [6.9-1] can be rewritten for the momentum of the system:

$$\sum \vec{F} = m^{\text{sys}} \frac{d\vec{v}^{\text{sys}}}{dt} + \vec{v}^{\text{sys}} \frac{dm^{\text{sys}}}{dt} \qquad [6.9\text{-}2]$$

Recall that the change in velocity with respect to time ($d\vec{v}^{\text{sys}}/dt$) is equal to the acceleration of a system ($\vec{a}^{\text{sys}}$). With no bulk transfer of material, the mass of the system is constant, and dm^{sys}/dt is equal to zero. Therefore, equation [6.9-2] is rewritten:

$$\sum \vec{F} = m^{\text{sys}} \vec{a}^{\,\text{sys}} \qquad [6.9\text{-}3]$$

Equation [6.9-3] mathematically states **Newton's second law of motion,** in which the acceleration of a body is inversely proportional to its mass and directly proportional to the resultant external force acting on it.

EXAMPLE 6.15 **Force on an Astronaut During Takeoff**

Problem: An astronaut sits in a spacecraft (Figure 6.25a). The total mass of the astronaut and her spacesuit is 120 kg. During takeoff, the spacecraft accelerates upward at a constant $6g$ (i.e., six times the normal acceleration of gravity on Earth).

If the system consists of the astronaut and her spacesuit (Figure 6.25b), what force does the chair in the spacecraft exert on the system during takeoff? In reality, astronauts are not positioned as shown in Figure 6.25b, but instead are in an almost horizontal position. Based on your solution, explain the reasoning behind horizontal positioning.

Solution: For this problem, we consider forces and acceleration only in the y-direction. We define the direction of acceleration of the spacecraft to be positive. The astronaut is assumed to accelerate at the same rate as the spacecraft.

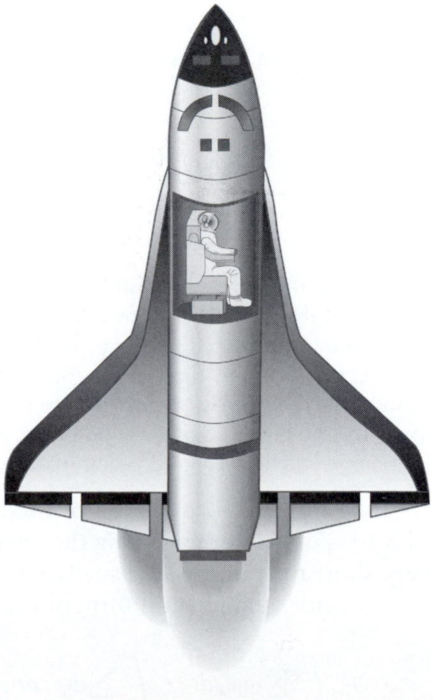

Figure 6.25a
Astronaut sitting in a spacecraft. Picture not drawn to scale.

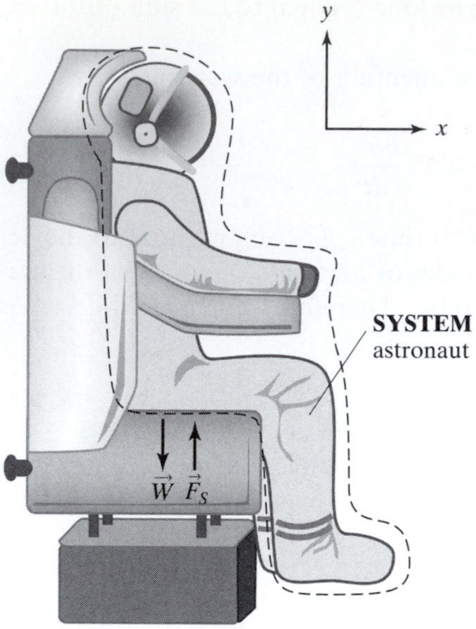

Figure 6.25b
Forces between the astronaut and her chair.

Because no time period is given, the governing differential conservation of linear momentum equation [6.3-3] can be used. Since no mass flows across the system boundaries, the governing equation reduces to the equation for an unsteady-state system with no bulk material transfer (equation [6.9-1]).

The forces acting on the astronaut and her spacesuit (the system) include gravity and the force of the chair ($\vec{F}_s$). The gravitational force is calculated using equation [6.2-6] and is equivalent to the weight (W) of the astronaut and her spacesuit on Earth. Since the mass of the system (m^{sys}) does not change, we can simplify the momentum balance to Newton's second law (equation [6.9-3]). The forces and acceleration act only in the y-direction, so we can consider scalar values:

$$\sum F_y = -W + F_s = m^{sys}a_y^{sys}$$

The acceleration of the system is six times that of the gravitational constant g. Substituting values into the equation above yields:

$$-(120\text{ kg})\left(9.81\frac{\text{m}}{\text{s}^2}\right) + F_s = (120\text{ kg})\left(6\left(9.81\frac{\text{m}}{\text{s}^2}\right)\right)$$

$$F_s = 8230\text{ N}$$

The chair in the spacecraft exerts a force of 8320 N on the astronaut and spacesuit. The portion of this force that is experienced by the astronaut is a lot for a person to withstand. If astronauts were placed in this position for takeoff, their blood would pool in their legs and feet, leaving little in their heads. Since the external force of gravity during takeoff directs blood flow downward, astronauts are placed in an almost horizontal orientation to minimize areas for the blood to pool. In addition, astronauts often wear antigravity suits, which apply pressure to the legs to keep blood from pooling there. ■

The integral formulation of the conservation of linear momentum in equation [6.3-6] is useful for analyzing the effects of **impulse forces**, which are applied during very short time intervals (usually much less than one second).

Consider the integral form of the conservation of linear momentum:

$$\int_{t_0}^{t_f} \sum_i \dot{m}_i \vec{v}_i \, dt - \int_{t_0}^{t_f} \sum_j \dot{m}_j \vec{v}_j \, dt + \int_{t_0}^{t_f} \sum \vec{F} \, dt = \int_{t_0}^{t_f} \frac{d\vec{p}^{sys}}{dt} dt \quad [6.9\text{-}4]$$

For a system with no bulk material transfer:

$$\int_{t_0}^{t_f} \sum \vec{F}\, dt = \int_{t_0}^{t_f} \frac{d\vec{p}^{\,sys}}{dt}\, dt \qquad [6.9\text{-}5]$$

If there is only a single, constant force, integrating yields:

$$\vec{F}(t_f - t_0) = \vec{p}_f^{\,sys} - \vec{p}_0^{\,sys} \qquad [6.9\text{-}6]$$

$$\vec{F}\,\Delta t = \Delta\vec{p}^{\,sys} \qquad [6.9\text{-}7]$$

where Δt is the time over which the impulse of force acts, and $\Delta\vec{p}^{\,sys}$ is the change in the overall momentum of the system. This equation is known as the **impulse-momentum theorem.**

The formulas describing impulse are useful when the momentum of a system changes very rapidly when a force is applied, such as during a collision. Often, for a system with an impulse force, the Accumulation term or change in momentum of the system, $\Delta\vec{p}^{\,sys}$, is calculated using equation [6.9-7].

EXAMPLE 6.16 Force Platform

Problem: One way to measure the impulsive forces involved in walking, running, jumping, and other ambulatory activities is to use a force platform (Figure 6.26a). The platform records the force exerted on its upper surface and outputs force as a function of time.

Before testing a new prosthetic device, you collect data describing normal jumping. The electronic recording from a normal jump is shown in Figure 6.26b. When the person stands still on the platform, the force scale is calibrated initially to read 0 kN, so the effects of gravity can be ignored. Calculate the person's mass and his change in momentum when he pushes down for takeoff. Calculate his vertical velocity as he pushes off. (Adapted from Özkaya N and Nordin M, *Fundamentals of Biomechanics*, 1999.)

Solution: The system is defined as the person. The following assumptions are necessary:

- Force platform calibrated, so effects of gravity do not need to be considered.
- No external body or surface forces act on the system.
- All movement and force measurements are in the y-direction.

The mass of the person can be calculated during his "hang time" (i.e., when he is completely off the platform). Recall that the scale is normalized to zero when the person is on the platform; therefore, when he is off the platform, the absence of this force is a measure of his

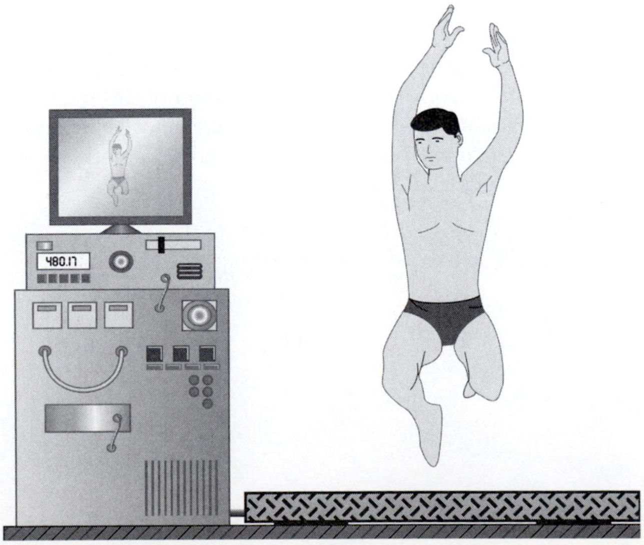

Figure 6.26a
Force platform used to measure impulsive forces.
(*Source*: Özkaya N and Nordin M, *Fundamentals of Biomechanics*. New York: Springer, 1999.)

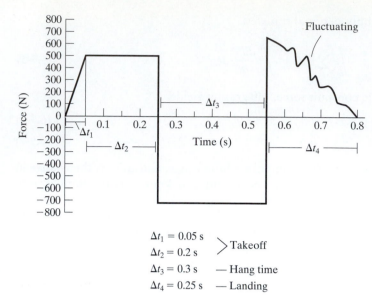

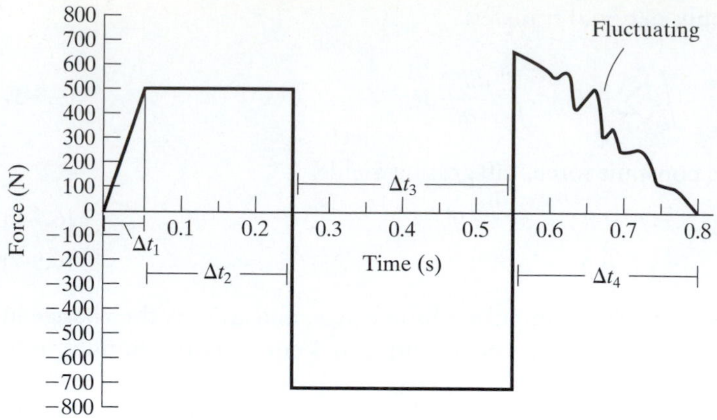

Figure 6.26b
Electronic recording from a normal jump.

weight. According to Figure 6.26b, the recorded force when the person is in the air is −700 N. This weight is used to solve for the person's mass:

$$W = -700 \text{ N} = mg = m(-9.81 \frac{\text{m}}{\text{s}^2})$$

$$m = 71.4 \text{ kg}$$

Because the force output is a function of time, the change in momentum of the system during takeoff can be calculated using the impulse-momentum theorem (equation [6.9-5]). The force is calculated by finding the area under the curve for the takeoff time period in Figure 6.26b. Since the impulse forces are different in Δt_1 and Δt_2, they are calculated individually and added together:

$$\int_{t_0}^{t_f} \vec{F}\, dt = \int_0^{0.25 \text{ s}} \vec{F} dt = \vec{F}_1 \,\Delta t_1 + \vec{F}_2 \,\Delta t_2$$

$$= \frac{1}{2} 500 \text{ N } (0.05 \text{ s}) + 500 \text{ N } (0.2 \text{ s}) = 112.5 \text{ N} \cdot \text{s}.$$

At the beginning of takeoff, the velocity of the person is zero, so $\vec{p}_0^{\text{sys}}$ is equal to zero. At the end of the takeoff period, the momentum of the system is the product of the mass of the person (the system) and his velocity, where velocity is in the $+y$-direction. The force of the platform on the person is in the $+y$-direction, so plugging in the values for the force and the mass gives a calculated velocity of:

$$\int_{t_0}^{t_f} \vec{F}\, dt = \vec{p}_f^{\text{sys}} = \vec{m}_f^{\text{sys}} \vec{v}_f^{\text{sys}}$$

$$112.5 \text{ N} \cdot \text{s} = (71.4 \text{ kg}) \vec{v}_f^{\text{sys}}$$

$$\vec{v}_f^{\text{sys}} = 1.58 \vec{j} \,\frac{\text{m}}{\text{s}}$$

Thus, the velocity when he initially jumps up is 1.58 m/s. ■

Currently, there are 1.3 million amputees in the United States alone.[3] The design and production of artificial limbs is a challenging task for bioengineers that requires the integration of many areas, including mechanics, electronics, and biomaterials. Designing an artificial leg is particularly complex if the knee joint must be included in the prosthetic device. Important aspects include understanding how force is transmitted between the natural limb and the prosthetic, how well the prosthetic mimics the natural limb, and how the prosthetic functions in regular ambulatory activities. Using a force platform can help bioengineers analyze the forces involved in moving the natural limb and help them translate their knowledge to produce a fully functional prosthetic device.

The above text and example problems consider unsteady-state systems with external forces but no mass flow. However, dynamic systems can also have material flow across their system boundary. We consider how to use the primary governing equations to solve for unsteady-state systems with material flow but no external forces. These systems gain or lose momentum as momentum is added to or removed from the system, respectively, as a result of bulk material transfer. One classic example of an unsteady-state system with bulk mass flow is the thrust of a rocket as it leaves Earth's orbit. A biological example involves the propulsion of a squid underwater (Problem 6.34).

In dynamic systems with no external forces acting on them, the differential form of the conservation of linear momentum equation [6.3-2] reduces to:

$$\sum_i \dot{\vec{p}}_i - \sum_j \dot{\vec{p}}_j = \frac{d\vec{p}^{\text{sys}}}{dt} \qquad [6.9\text{-}8]$$

The integral conservation of linear momentum equation [6.3-6] is reduced for dynamic systems with no external forces acting on them:

$$\int_{t_0}^{t_f} \sum_i \dot{\vec{p}}_i \, dt - \int_{t_0}^{t_f} \sum_j \dot{\vec{p}}_j \, dt = \int_{t_0}^{t_f} \frac{d\vec{p}^{\text{sys}}}{dt} dt \qquad [6.9\text{-}9]$$

EXAMPLE 6.17 Acceleration of a Rocket in Space

Problem: Imagine a stationary rocket in outer space. Initially, the rocket and all the fuel currently on board have a combined mass of 1000 kg. For a period of 5 s, the rocket ignites and starts to move forward. It discharges spent fuel at a rate of 5 kg/s, and the vapor leaves the nozzle at a constant speed of 500 m/s. At the end of the ignition burst, how fast is the rocket moving? Ignore the effects of any gravitational fields.

Solution: Consider the rocket's casing as the system boundary. At the initial condition, the system consists of the fuel, the rocket, and all its internal components. At the end of the ignition burst (final condition), the system has lost some fuel. Because this lost mass changes the momentum of the system and because no other forces act on the system, this is an unsteady-state system with no external forces. Since a time interval is given, the integral form of the conservation of linear momentum equation is most appropriate to use, which can be simplified to equation [6.9-9]. There are no inlets and only one outlet, so equation [6.9-9] becomes:

$$-\int_{t_0}^{t_f} \dot{\vec{p}}_j \, dt = \int_{t_0}^{t_f} \frac{d\vec{p}^{\text{sys}}}{dt} dt$$

The momentum rate of the expelled fuel is constant, so the left-hand side is the product of the momentum rate and the time interval. No coordinate system was given, so we define an arbitrary coordinate system such that the vapor leaving is in the $+x$-direction, so $\dot{\vec{p}}_j$ can be calculated:

$$-\dot{\vec{p}}_j(t_f - t_0) = \vec{p}_{\text{f}}^{\text{sys}} - \vec{p}_0^{\text{sys}}$$

$$-\dot{m}_j\vec{v}_j(t_f - t_0) = m_{\text{f}}^{\text{sys}}\vec{v}_{\text{f}}^{\text{sys}} - m_0^{\text{sys}}\vec{v}_0^{\text{sys}}$$

$$-\left(5\,\frac{\text{kg}}{\text{s}}\right)\left(500\vec{i}\,\frac{\text{m}}{\text{s}}\right)(5\,\text{s} - 0\,\text{s}) = m_{\text{f}}^{\text{sys}}\vec{v}_{\text{f}}^{\text{sys}} - (1000\,\text{kg})\left(0\vec{i}\,\frac{\text{m}}{\text{s}}\right)$$

Since the system starts at rest, the final momentum equals the change in momentum exactly. The mass of the system at the final condition is found by subtracting the mass of the fuel expelled during the time interval from the initial mass. Given this, we calculate the final velocity of the rocket:

$$-12{,}500\vec{i}\,\frac{\text{kg}\cdot\text{m}}{\text{s}} = \left(1000\,\text{kg} - \left(5\,\frac{\text{kg}}{\text{s}}\right)(5\,\text{s})\right)\vec{v}_{\text{f}}^{\text{sys}}$$

$$\vec{v}_{\text{f}}^{\text{sys}} = -12.8\vec{i}\,\frac{\text{m}}{\text{s}}$$

Thus, the final velocity of the rocket is 12.8 m/s in the opposite direction of the discharged fuel. ■

Although Newton's laws of motion have been widely known for several hundred years, only recently were they correctly understood in the context of rocketry. Robert Goddard, who is considered a founder of modern rocketry, did much of his work in the early 1900s. At this time many people believed that a rocket could not operate in space, citing Newton's third law as evidence. For a rocket to have a forward acceleration, they reasoned, there must exist some external matter against which it can push off. In the atmosphere, air suffices. However, in the vacuum of space, they argued, there is no medium for this necessary reaction.

As Goddard demonstrated, no external matter is required, because a rocket that expels its own fuel is capable of creating both the action and corresponding reaction to satisfy Newton's laws. As long as the spent fuel leaves the rocket (the system), the rocket is capable of acceleration. Although it took years for some to accept this fact, we can demonstrate it quickly using conservation equations, as in Example 6.17.

6.10 Reynolds Number

Up to this point, the formulas that include velocity have assumed that the fluid can be described by an average velocity. However, the velocity profiles of flow in pipes and other closed conduits vary under different conditions. When applying the mechanical energy accounting equation (Section 6.11), it is important to identify the flow profile, which can be characterized by the Reynolds number.

The **Reynolds number** (Re) is a way to mathematically predict the type of fluid flow and consequently the shape of the velocity profile. For fluid in a circular pipe:

$$\mathrm{Re} = \frac{\rho v D}{\mu} \qquad [6.10\text{-}1]$$

where ρ is the fluid density, v is the average velocity of the fluid, D is the diameter of the pipe through which the fluid flows, and μ is the fluid viscosity. Note that Re is dimensionless. Re is a ratio of the inertial forces to the viscous forces in a flowing fluid. The Reynolds number is present in more complex equations, such as the Navier-Stokes equation for Newtonian fluids used in transport calculations. This book is concerned only with using Re to identify the two main classes of fluid flow in cylindrical vessels: laminar and turbulent.

How a fluid moves through a pipe can be described by the velocity profile, which can reveal certain characteristics about the fluid. A **laminar** velocity profile for a Newtonian fluid is one where the velocity varies as a function of radial position in a parabolic manner (Figure 6.27a). All gases and most simple liquids may be modeled as Newtonian fluids. A mathematically rigorous definition of a Newtonian fluid can be found elsewhere (e.g., Bird RB, Stewart WE, and Lightfoot EN, *Transport Phenomena*, 2002; Truskey GA, Yuan F, and Katz DF, *Transport Phenomena in Biological Systems*, 2004).

Consider a fluid flowing through a stationary cylindrical vessel. Since a thin layer of fluid sticks to the wall, the velocity of the fluid at the wall is zero. Fluid velocity is usually given as an average; thus, a region of the fluid must have a velocity greater than the average. The fluid at the center or midline of the pipe travels at the highest velocity, and the velocity decreases as it approaches the walls. Each layer of fluid travels at a slightly different velocity than its neighboring layer, such that the layers slip along each other very smoothly. Together, the fluid moves along in the direction of the pipe or conduit in a very orderly, smooth way. When $\mathrm{Re} < 2100$ for flow in a cylindrical pipe, the fluid flow is considered laminar. In most areas of the body, the fluid flow is laminar.

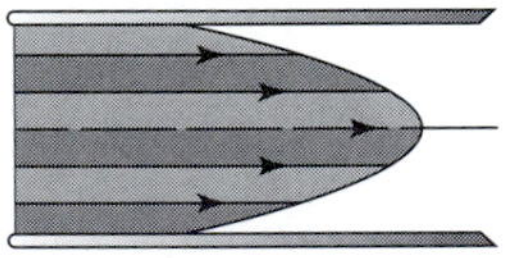

Figure 6.27a
Laminar velocity profile of a homogeneous fluid. Shading shows that the layers of fluid slip smoothly over one another.

A **turbulent** velocity profile is one where the velocity profile is nearly flat, with most regions of the flow traveling at the same velocity down the pipe (Figure 6.27b). Turbulent flow is often described as having a uniform velocity profile. The fluid mixes locally in the tube, creating eddies as it moves along the pipe. Turbulent flow is often called **plug-flow**, since the fluid moves down the tube as a plug of fluid. When $\mathrm{Re} > 4000$ for flow in a cylindrical pipe, the fluid flow is considered turbulent. Turbulent flow is common in industrial applications.

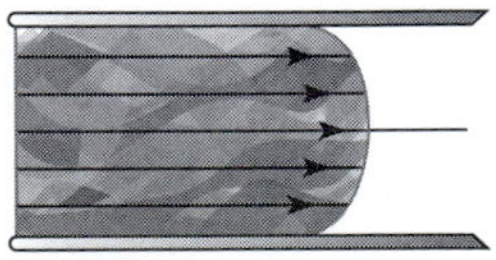

Figure 6.27b
Turbulent velocity profile of a homogeneous fluid. At the microscopic level, mixing and eddies are seen.

In between Reynolds numbers of 2100 and 4000, the flow is considered in **transition,** showing characteristics of both types of flow. These numerical cutoffs for laminar and turbulent flow were determined by looking at empirical data.

Consider a fluid with a fixed density and viscosity flowing through a pipe with a fixed diameter. At a low velocity, the flow patterns are smooth and orderly; in this region, the fluid flow is laminar. As the velocity increases, the fluid becomes more chaotic, disorganized, and mixed; in this region, the fluid flow is turbulent. You can manipulate ρ, D, and μ in equation [6.10-1] to see how each term affects Re.

The assumption that fluid flows at an average velocity may be too simplistic in some complex fluid dynamics systems. Details about velocity as a function of spatial position may need to be known, such as in evaluations of prosthetic heart valve designs. Also, the flow patterns of non-Newtonian fluids may be complex. In these situations, more complex fluid dynamics equations based on mass and momentum conservation equations are required (e.g., Bird RB, Stewart WE, and Lightfoot EN,

Transport Phenomena, 2002; Truskey GA, Yuan F, and Katz DF, *Transport Phenomena in Biological Systems*, 2004; Fournier RL, *Basic Transport Phenomena in Biomedical Engineering*, 1998).

EXAMPLE 6.18 **Air Flow in Trachea**

Problem: For airflow in the trachea, approximate the Reynolds number during inhalation.

Solution: To solve this problem, we must assume the following:

- System is at steady-state.
- The trachea is cylindrical.
- Velocity and properties of air are constant throughout the trachea.
- Flow rate of air into and out of the trachea is 12 L/min.
- The diameter of the trachea is 1.8 cm.

To calculate the Reynolds number, the volumetric flow rate must be converted to a linear velocity:

$$v = \frac{\dot{V}}{A} = \frac{\left(12\dfrac{\text{L}}{\text{min}}\right)\left(\dfrac{1\text{ min}}{60\text{ s}}\right)\left(\dfrac{1000\text{ cm}^3}{\text{L}}\right)}{\pi(0.9\text{ cm})^2} = 78.6\ \frac{\text{cm}}{\text{s}}$$

The Reynolds number can be calculated with this value and known values for the density and viscosity of air:

$$\text{Re} = \frac{\rho v D}{\mu} = \frac{\left(1.225 \times 10^{-6}\dfrac{\text{kg}}{\text{cm}^3}\right)\left(78.6\dfrac{\text{cm}}{\text{s}}\right)(1.8\text{ cm})}{\left(1.79 \times 10^{-7}\dfrac{\text{kg}}{\text{cm}\cdot\text{s}}\right)} = 968$$

A Reynolds number of 970 is laminar, which is consistent with what we know about the body. Beyond this, it is difficult to say if the number is reasonable. It will vary significantly based on the individual, his or her level of activity, and the type of breathing. ■

6.11 Mechanical Energy and Bernoulli Equations

The mechanical energy equation is a powerful one that applies to many systems with fluid flow. While total energy is a conserved property, mechanical energy is not. Hence, mechanical energy must be described with an accounting equation. The Bernoulli equation is one form of the mechanical energy equation for a particular set of conditions. These equations are used to describe and characterize systems with fluid flow under the conditions described.

6.11.1 Mechanical Energy Accounting Equation

Mechanical energy involves the motion and displacement of fluids and bodies, as well as the forces that can change motion and displacement. **Mechanical energy** is the summation of the kinetic energy, potential energy, and work of a system. Other types of energy include thermal energy, which is the sum of internal energy and heat (see Chapter 4), and electrical energy (see Chapter 5).

Energy can be converted from one form to another. For example, when you rub your hands together, they begin to feel warmer. The work you do rubbing your hands converts mechanical energy to thermal energy because of friction. **The mechanical energy equation accounts only for mechanical energy and its conversion to and from other types of energy.**

Like other extensive properties, mechanical energy can enter or leave, be generated or consumed, and accumulate in the system. Since mechanical energy is not conserved, the accounting equation is appropriate:

$$\psi_{in} - \psi_{out} + \psi_{gen} - \psi_{cons} = \psi_{acc} \qquad [6.11\text{-}1]$$

Movement of mass flowing as a fluid transfers mechanical energy into and out of a system in the forms of kinetic energy, potential energy, and flow work. When a fluid is in motion at a given velocity, the fluid possesses **kinetic energy**. The **potential energy** a fluid possesses results from its position in a gravitational field. **Flow work** is the energy required to push the fluid into and out of a system.

Mechanical energy can be generated from another type of energy; conversely, mechanical energy can be consumed or converted to another type of energy. A common energy interconversion in flowing systems is of mechanical energy to thermal energy through frictional losses, or through expansion or compression of a fluid. In flowing fluids, **frictional losses** are the irreversible conversion from mechanical energy to thermal energy. Recall that positive work is work done on the system by the surroundings. Thus, frictional losses are considered negative work (i.e., a loss of mechanical energy from the system) and are described in the Consumption term of the mechanical energy accounting equation.

Shaft work is work done on or by the system using a compressor, pump, turbine, or other device. Shaft work can be either positive or negative, depending upon whether the work is done on or by the system. In approaching the mechanical energy equation, shaft work is considered as a Generation or Consumption term. Recall that these terms are reserved for contributions that change the net amount of that property in the universe (Chapter 2). Since the net mechanical energy of the universe changes when shaft work is applied, shaft work is accounted for as a Generation or Consumption term in the mechanical energy accounting equation. This differs from the approach taken in Chapter 4. In the total energy conservation equation, shaft work is treated as an Input or Output term; an energy gain through shaft work is balanced by a loss of another form of energy, keeping the total energy of the universe constant. Therefore, shaft work is considered an Input or Output term when looking at the total energy of the system, but as a Generation or Consumption term when looking at mechanical energy.

The macroscopic mechanical energy accounting equation is derived from the conservation of momentum (see Bird RB, Stewart WE, and Lightfoot EN, *Transport Phenomena*, 2002, for a full derivation). Its derivation requires knowledge of reasonably complicated mathematics outside the scope of this text. For this book, you need only know that the process of deriving the mechanical energy accounting equation results in an equation that is independent of the linear momentum conservation law. Thus, the two equations are not redundant and can often be used together to solve problems. For these reasons the mechanical energy accounting equation is included here rather than in Chapter 4 with other energy equations.

Because of the complexity of the derivation, we focus here on explaining the equation. The steady-state **mechanical energy accounting equation** is:

$$\dot{m}(\hat{E}_{P,i} - \hat{E}_{P,j}) + \dot{m}(\hat{E}_{K,i} - \hat{E}_{K,j}) + \dot{m}\left(\frac{P_i}{\rho_i} - \frac{P_j}{\rho_j}\right) - \dot{m}\int P\,d\hat{V} + \sum \dot{W}_{shaft} - \sum \dot{f} = 0 \qquad [6.11\text{-}2]$$

where $\dot{m}$ is the mass flow rate, $\hat{E}_P$ is the specific potential energy (potential energy per unit mass), $\hat{E}_K$ is the specific kinetic energy (kinetic energy per unit mass), P_i and P_j are the pressures at the system boundary where mass flow enters and leaves, ρ is the density of the fluid, P is the pressure of the system, $\hat{V}$ is the specific volume (volume per unit mass), $\sum \dot{W}_{\text{shaft}}$ is the total nonflow and nonexpansion work (i.e., shaft work), and $\sum \dot{f}$ is the total frictional losses. Again, i and j indicate the indices for the inlet and outlet, respectively. The dimension of equation [6.11-2] is $[L^2Mt^{-3}]$, which is standard for the rate of energy.

Equation [6.11-2] is widely regarded as the mechanical energy accounting equation. This equation is limited to fluid flow systems that meet the following criteria:

- Steady-state
- One inlet and one outlet
- Interconversions only between mechanical and thermal energies
- No chemical reactions

Because of the limitation to a system at steady-state with a single inlet and a single outlet, the conservation of mass requires that the inlet mass flow rate is equal to the outlet mass flow rate. The mass flow rate is constant throughout the system and is designated $\dot{m}$ in equation [6.11-2].

The first three terms of equation [6.11-2] represent the changes in potential energy, kinetic energy, and flow work from the inlet condition to the outlet condition. The integral term captures the reversible conversion between internal energy and mechanical energy of the fluid due to its expansion or compression as it flows. The last two terms describe the shaft work done on and frictional losses of the system. Since the system is at steady-state, no accumulation of mechanical energy occurs.

Note the similarities between the mechanical energy accounting equation (equation [6.11-2]) and the steady-state differential conservation of total energy equation (equation [4.7-7]) rewritten here:

$$\sum_i \dot{m}_i\left(\hat{E}_{P,i} + \hat{E}_{K,i} + \frac{P_i}{\rho_i}\right) - \sum_j \dot{m}_j\left(\hat{E}_{P,j} + \hat{E}_{K,j} + \frac{P_j}{\rho_j}\right) + \sum \dot{Q} + \sum \dot{W}_{\text{nonflow}} = 0 \qquad [6.11\text{-}3]$$

This formulation of the conservation of total energy equation describes an open, steady-state system with potential and kinetic energy changes but no internal energy changes. Recall that this equation is derived from the steady-state differential conservation of total energy equation when significant flow work or changes in pressure or density between the inlet and outlet conditions are present.

Comparing this modified conservation of total energy equation with the mechanical energy accounting equation shows that they both capture changes in potential and kinetic energy differences, as well as flow work and shaft work. While friction does change the thermal energy of a system, friction is not equivalent to heat, and the two terms are not interchangeable. The mechanical energy accounting equation [6.11-2], accounts only for friction losses; the conservation of total energy equation [6.11-3], which is for a steady-state system with no changes in internal energy, captures all forms of heat transfer. While the two equations are very similar, it is important to pick the appropriate one for each problem encountered. **Use the conservation of energy equation when mechanical and thermal energy changes occur in the system; use the mechanical energy accounting equation when considering only mechanical energy changes and mechanical energy conversions.** Remember that the mechanical energy accounting equation requires a number of conditions and restrictions to be met. Note that since Example 4.10 in Chapter 4 contains only mechanical energy terms, this problem could have been solved with the mechanical energy accounting equation, and the answer would have been the same.

An **incompressible fluid** has a density that is constant over a range of pressures. Assuming that a liquid is incompressible is almost always valid for biological and biomedical systems. In contrast, the density of gases does change as a function of pressure. (For analyzing systems with flowing gases, refer to other books such as Batchelor GK and Batchelor GK, *An Introduction to Fluid Dynamics,* 2000 or Landau LD and Lifshitz EM, *Fluid Dynamics*, 1987.) For an incompressible fluid, the term $\dot{m}\int Pd\hat{V}$ is zero because the fluid specific volume does not change as the fluid flows, reducing equation [6.11-2] to:

$$\dot{m}(\hat{E}_{P,i} - \hat{E}_{P,j}) + \dot{m}(\hat{E}_{K,i} - \hat{E}_{K,j}) + \dot{m}\left(\frac{P_i}{\rho_i} - \frac{P_j}{\rho_j}\right) + \sum \dot{W}_{\text{shaft}} - \sum \dot{f} = 0 \quad [6.11\text{-}4]$$

It is common to see equation [6.11-4] modified by the division of $\dot{m}$ throughout:

$$(\hat{E}_{P,i} - \hat{E}_{P,j}) + (\hat{E}_{K,i} - \hat{E}_{K,j}) + \left(\frac{P_i}{\rho_i} - \frac{P_j}{\rho_j}\right) + \sum \frac{\dot{W}_{\text{shaft}}}{\dot{m}} - \sum \frac{\dot{f}}{\dot{m}} = 0 \quad [6.11\text{-}5]$$

The dimension of equation [6.11-4] is $[L^2Mt^{-3}]$, and the dimension of equation [6.11-5] is $[L^2t^{-2}]$.

Further definitions and explanations of potential energy and kinetic energy are given in Chapter 4. Specific potential energy is:

$$\hat{E}_P = gh \quad [6.11\text{-}6]$$

For systems with a uniform velocity profile, specific kinetic energy is:

$$\hat{E}_K = \frac{1}{2}v^2 \quad [6.11\text{-}7]$$

A uniform velocity profile is typically a good assumption for flow in cylindrical pipes in turbulent regimes. In some cases, a uniform velocity profile can be an acceptable approximation for flow in laminar regimes.

Substituting these into equations [6.11-4] and [6.11-5] yields:

$$\dot{m}(gh_i - gh_j) + \dot{m}\left(\frac{1}{2}v_i^2 - \frac{1}{2}v_j^2\right) + \frac{\dot{m}}{\rho}(P_i - P_j) + \sum \dot{W}_{\text{shaft}} - \sum \dot{f} = 0 \quad [6.11\text{-}8]$$

and

$$(gh_i - gh_j) + \left(\frac{1}{2}v_i^2 - \frac{1}{2}v_j^2\right) + \frac{1}{\rho}(P_i - P_j) + \sum \frac{\dot{W}_{\text{shaft}}}{\dot{m}} - \sum \frac{\dot{f}}{\dot{m}} = 0 \quad [6.11\text{-}9a]$$

which is rewritten as:

$$\left(gh_i + \frac{1}{2}v_i^2 + \frac{P_i}{\rho}\right) - \left(gh_j + \frac{1}{2}v_j^2 + \frac{P_j}{\rho}\right) + \sum \frac{\dot{W}_{\text{shaft}}}{\dot{m}} - \sum \frac{\dot{f}}{\dot{m}} = 0 \quad [6.11\text{-}9b]$$

This is known as the **extended Bernoulli equation.** Equations [6.11-8] and [6.11-9] are used to describe steady-state systems with fluid flow with a uniform velocity profile through one inlet and one outlet in which shaft work and frictional losses occur.

6.11.2 Bernoulli Equation

The **Bernoulli equation** relates the velocity, pressure, and elevation of two points along the path of a fluid in steady-state flow. It can be derived directly from the conservation of linear momentum equation or as a reduction of the mechanical energy accounting equation. The Bernoulli equation is applicable to systems meeting restrictions appropriate for the mechanical energy accounting equation and that also have no frictional loss or work done on the system. Thus, in addition to the list following equation [6.11-2], application of this equation also requires a system to meet the following criteria:

- Inviscid flow (i.e., no viscous energy losses due to friction)
- Incompressible flow
- No shaft work

Equations [6.11-5] and [6.11-9a] reduce as follows:

$$(\hat{E}_{P,i} - \hat{E}_{P,j}) + (\hat{E}_{K,i} - \hat{E}_{K,j}) + \left(\frac{P_i}{\rho_i} - \frac{P_j}{\rho_j}\right) = 0 \qquad [6.11\text{-}10]$$

$$(gh_i - gh_j) + \left(\frac{1}{2}v_i^2 - \frac{1}{2}v_j^2\right) + \frac{1}{\rho}(P_i - P_j) = 0 \qquad [6.11\text{-}11]$$

Note that equation [6.11-11] requires the additional criteria that the velocity profile is uniform. Equation [6.11-11] is often written with different notation:

$$g\,\Delta h + \frac{1}{2}\Delta v^2 + \frac{\Delta P}{\rho} = 0 \qquad [6.11\text{-}12]$$

where Δh is the change in height between the inlet and outlet fluid streams, Δv^2 is the change in the squares of the velocities of the inlet and outlet, and ΔP is the change in pressure between the inlet and outlet. Note that Δv^2 is not the difference of the inlet and outlet velocities squared (i.e., $(v_i - v_j)^2$), but is the difference of the squares of the inlet and outlet velocities (i.e., $(v_i^2 - v_j^2)$). Equations [6.11-11] and [6.11-12] are commonly known as the Bernoulli equation.

Consider a fluid flowing along a pipe with a constant diameter D. The fluid and the system are subject to all the limitations listed above, so the Bernoulli equation can be applied. Let us also assume that the pipe does not have any elevation changes as it flows, reducing equation [6.11-11] to:

$$\left(\frac{1}{2}v_i^2 - \frac{1}{2}v_j^2\right) + \frac{1}{\rho}(P_i - P_j) = 0 \qquad [6.11\text{-}13]$$

Because total mass flow is conserved and the pipe is of constant diameter, the conservation of mass equation [3.3-9] is reduced to:

$$v_i = v_j \qquad [6.11\text{-}14]$$

Substituting this into equation [6.11-13], the net result is:

$$P_i = P_j \qquad [6.11\text{-}15]$$

Across short segments of pipe or an ideal system, this result may be reasonable. However, over long segments of pipe or nonideal systems, this result is counterintuitive.

Typically, fluid converts mechanical energy to thermal energy because of viscous frictional losses as the fluid flows. To account for such losses, the mechanical energy accounting equation for a system with no potential energy change and no pump work is appropriate:

$$\left(\frac{1}{2}v_i^2 - \frac{1}{2}v_j^2\right) + \frac{1}{\rho}(P_i - P_j) - \sum\frac{\dot{f}}{\dot{m}} = 0 \qquad [6.11\text{-}16]$$

Since $v_i = v_j$, the pressure drop is equal to the frictional losses:

$$\frac{1}{\rho}(P_i - P_j) - \sum\frac{\dot{f}}{\dot{m}} = 0 \qquad [6.11\text{-}17]$$

This states the expected result that the **pressure in a pipe drops along the length of the pipe as mechanical energy is lost to friction.**

To minimize the pressure drop along a pipe, pumps are added to increase the mechanical energy of the system and to counter the frictional losses. If we add a pump, which contributes shaft work to the system, equation [6.11-17] becomes:

$$\frac{1}{\rho}(P_i - P_j) + \sum\frac{\dot{W}_{\text{shaft}}}{\dot{m}} - \sum\frac{\dot{f}}{\dot{m}} = 0 \qquad [6.11\text{-}18]$$

A pipe with this configuration can more easily overcome frictional losses and maintain pressure along the pipe.

EXAMPLE 6.19 **Pressures in a Stenotic Vessel**

Problem: Stenotic blood vessels are narrowed or constricted blood vessels, such as those that have been blocked by the buildup of fat or cholesterol (e.g., atherosclerosis) or blood clots (e.g., thrombosis). Imagine three points of a stenosed vessel, where the diameters at either side of the stenosis are $D_1 = D_3$, and the diameter of the narrowest stenotic site is D_2, which is one-tenth of D_1 (Figure 6.28a). The velocity of the blood at the first point is v_1. The density of the blood is ρ, and the viscosity is μ. Neglect frictional losses.

(a) Calculate the Reynolds number for each of these three points. What do these numbers tell you about the flow at these three points?

(b) Use the Bernoulli equation to find the pressure differences between the first and second points and between the first and third points as a function of ρ and the velocity at the first point. To apply the Bernoulli equation to a problem, we have to make the assumption that the velocity profile is uniform. This approximation uses an average velocity to capture the behavior of the fluid.

Solution:

(a) *Reynolds numbers*: Although blood vessels in the body can constrict and dilate under different conditions, they are nearly circular in cross-section, so we assume the vessel is

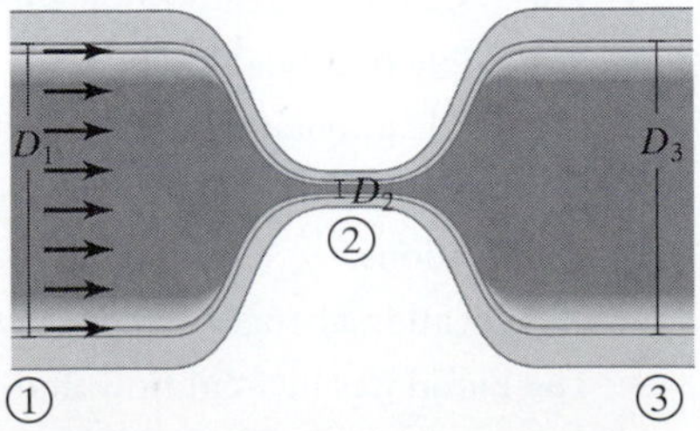

Figure 6.28a
A stenotic vessel.

cylindrical to find the Reynolds number. Because the vessel system is at steady-state and has only one inlet and one outlet, the total mass flow rate in must equal the total mass flow rate out by the conservation of mass [3.4-3].

Looking at Figure 6.28b at points 1 and 2 in the vessel:

$$\dot{m}_1 - \dot{m}_2 = 0$$

$$\rho v_1 \pi \left(\frac{D_1}{2}\right)^2 - \rho v_2 \pi \left(\frac{D_2}{2}\right)^2 = 0$$

$$v_1 D_1^2 - v_2 D_2^2 = 0$$

Knowing that the relationship between the diameters at the first and second points is $D_1 = 0.1D_2$, we solve for the velocity at the second point:

$$v_2 = \frac{v_1 D_1^2}{(0.1D_1)^2}$$

$$v_2 = 100v_1$$

Since the mass flow in must equal the mass flow out, it makes sense that the velocity must increase when the blood flow is forced through a smaller cross-sectional area. Substituting these values for velocity and diameter at point 2 as a function of point 1 into equation [6.10-1], the Reynolds number at the second point in the vessel is:

$$\text{Re}_2 = \frac{\rho v_2 D_2}{\mu} = \frac{\rho(100v_1)(0.1D_1)}{\mu} = \frac{10\rho v_1 D_1}{\mu}$$

The diameters and velocities at the first and third points are the same. In terms of the variables at the first point, this makes the Reynolds numbers for both points:

$$\text{Re}_1 = \text{Re}_3 = \frac{\rho v_1 D_1}{\mu}$$

The Reynolds number is ten times larger at the second point than at the first and third points. If we evaluate the Reynolds numbers using realistic data, the flow at the first and third points is likely laminar, as it is in most vessels, and the flow at the second point likely exhibits turbulent characteristics. This result challenges our assumption that the velocity profile is uniform throughout the system. However, while the solution is approximate, it aptly describes the changes known to exist in a stenotic vessel system. In most regions of the body, prolonged flow with turbulent characteristics can lead to deleterious physiological effects, such as thromboembolism.

(b) *Pressure differences*: Two systems must be established to calculate the two different pressure drops. The first system encircles the first and second points (Figure 6.28b). The second system encircles the first through the third (Figure 6.28c).

We assume the following:

- The flow in the vessel has a uniform velocity profile.
- The vessel is cylindrical.
- The system is at steady-state with one inlet and one outlet.
- Negligible frictional losses.
- Negligible gravitational effects.
- No pump work.
- No reactions.
- No elevation changes across the vessel.
- The blood has inviscid flow and is incompressible.

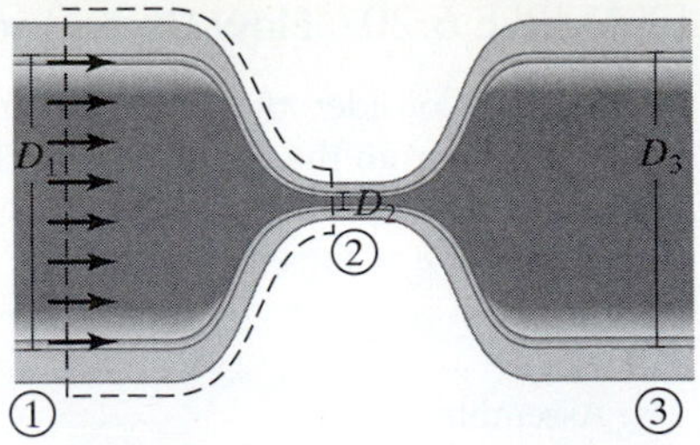

Figure 6.28b
System enclosing the first two points in the stenotic vessel.

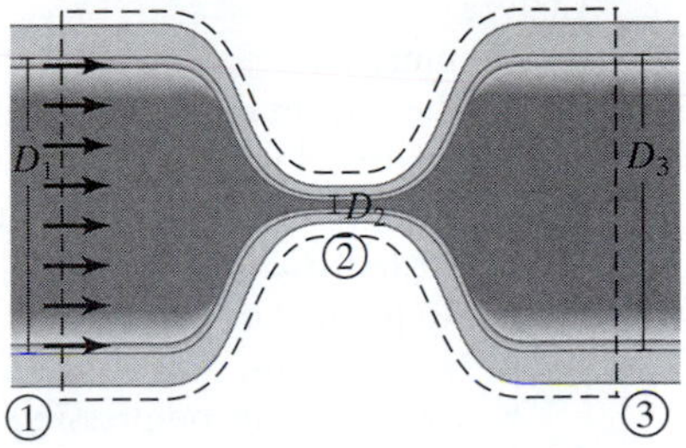

Figure 6.28c
System enclosing the first and third points in the stenotic vessel.

Using the Bernoulli equation is appropriate, since no shaft work or frictional losses occur. Because we assume the points all have the same height in the vessel with respect to the gravitational plane (i.e., $h_i = h_j$), the Bernoulli equation reduces to:

$$\left(\frac{1}{2}v_i^2 - \frac{1}{2}v_j^2\right) + \frac{1}{\rho}(P_i - P_j) = 0$$

Substituting the values in for the pressure drop and velocities at the first and second points gives:

$$\frac{1}{2}(v_1^2 - v_2^2) + \frac{P_1 - P_2}{\rho} = 0$$

$$P_2 - P_1 = \frac{\rho}{2}(v_1^2 - (100\, v_1)^2) = -\frac{9999}{2}\rho v_1^2$$

The velocity of the blood at the first and third points is the same. The pressure drop across the first and third points is:

$$P_3 - P_1 = \frac{\rho}{2}(v_1^2 - v_3^2) = 0$$

Thus, the pressure difference across the entire system is zero.

For liquid to flow through any pipe, the pressure must drop in the direction of the flow, which certainly happens between the points where blood flow enters the narrowed vessel site. The result that the pressure between the points where blood flow enters and exits the vessel is equal is true only because we neglect frictional losses. ■

6.11.3 Additional Applications Using the Mechanical Energy and Bernoulli Equations

The Bernoulli equation and the extended Bernoulli equation are very powerful tools for analyzing systems with flowing liquids. If some information about changes in height, flow rate, or pressure is known, these equations can often be used to finish describing the system.

EXAMPLE 6.20 Flow Up an Inclined Pipe

Problem: Consider the vertical transition in the pipe transporting water (Figure 6.29). Water travels from the opening at the base (0.05 m radius) to the opening at the top (0.03 m radius). The height of the transition is 1 m. If the pressure at the top is 1 atm, what is the required pressure at the base so that the velocity at the base is 1.5 m/s?

Solution:

1. Assemble
   (a) Find: Pressure required at base for fluid velocity at base to be 1.5 m/s.
   (b) Diagram: System shown in Figure 6.29. The system boundary is the wall of the pipe.
2. Analyze
   (a) Assume:
   - The flow in the vessel has a uniform velocity profile (Re $\cong$ 150,000).
   - The vessel is cylindrical.
   - The system is at steady-state with one inlet and one outlet.
   - Neglect frictional losses.
   - No pump work.
   - No reactions.
   - The fluid is incompressible.

   (b) Extra data:
   - $\rho_{\text{water}} = 1000 \dfrac{\text{kg}}{\text{m}^3}$

   (c) Variables, notations, units:
   - Subscripts *base* and *top* refer to the respective heights of the pipe.
   - Use kg, m, s, atm, Pa.
3. Calculate
   (a) Equations: Using the Bernoulli equation [6.11-12] is appropriate, since no shaft work or frictional losses occur.

$$g\,\Delta h + \frac{1}{2}\Delta v^2 + \frac{\Delta P}{\rho} = 0$$

   (b) Calculate:
   - Based on our steady-state assumption, we calculate the outlet velocity by the conservation of mass using the radii of the two ends:

$$\dot{m}_{\text{base}} - \dot{m}_{\text{top}} = 0$$

$$\rho v_{\text{base}} \pi r_{\text{base}}^2 - \rho v_{\text{top}} \pi r_{\text{top}}^2 = 0$$

$$v_{\text{top}} = \frac{v_{\text{base}} r_{\text{base}}^2}{r_{\text{top}}^2} = \frac{\left(1.5\,\frac{\text{m}}{\text{s}}\right)(0.05\text{ m})^2}{(0.03\text{ m})^2} = 4.17\,\frac{\text{m}}{\text{s}}$$

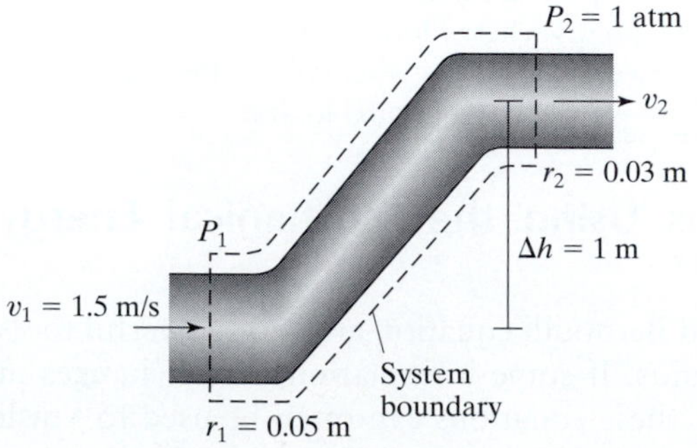

Figure 6.29
Vertical transition of water in a pipe. Figure not drawn to scale.

- The difference in the squared velocities, Δv^2, is $-15.1\ \text{m}^2/\text{s}^2$.
- Substituting this value into the rearranged Bernoulli equation gives:

$$\Delta P = -\rho\left(g\,\Delta h + \frac{1}{2}\Delta v^2\right)$$

$$\Delta P = -1000\frac{\text{kg}}{\text{m}^3}\left(\left(9.81\frac{\text{m}}{\text{s}^2}\right)(-1\ \text{m}) + \frac{1}{2}\left(-15.1\frac{\text{m}^2}{\text{s}^2}\right)\right)$$

$$\Delta P = 1.74 \times 10^4\ \text{Pa}$$

- Comparing the pressure at the base to the pressure at the top gives:

$$\Delta P = P_{\text{base}} - P_{\text{top}}$$

$$P_{\text{base}} = P_{\text{top}} + \Delta P = 1\ \text{atm} + (1.74 \times 10^4\ \text{Pa})\left(\frac{1\ \text{atm}}{1.013 \times 10^5\ \text{Pa}}\right)$$

$$P_{\text{base}} = 1.17\ \text{atm}$$

4. Finalize
   (a) Answer: For the base to have a fluid velocity of 1.5 m/s with 1 atm present at the top, the pressure at the base of the pipe must be 1.17 atm.
   (b) Check: The base pressure is larger than the pressure at the top, which is consistent with intuition, because the fluid gains kinetic and potential energy as it flows up into a smaller pipe. The order of magnitude is about the same, so the answer is realistic. ■

Work can be done on a system by a pump or other device. In the human body, the heart is the pump for the circulatory system. Two examples below illustrate the work the heart does to keep blood in circulation and how that energy is dissipated during circulation.

EXAMPLE 6.21 Work Done by the Heart

Problem: Estimate the work done by the heart to keep the blood circulating. (Adapted from Cooney DO, *Biomedical Engineering Principles: An Introduction to Fluid, Heat, and Mass Transport Processes*, 1976.)

Solution:

1. Assemble
   (a) Find: The work done by the heart to keep the blood circulating.
   (b) Diagram: Shown in Figure 6.30.
2. Analyze
   (a) Assume:
   - The fluid that flows through the heart has a uniform velocity profile.
   - The system can be modeled as steady-state with one inlet and one outlet.
   - Neglect frictional losses in the pumping heart.
   - No reactions occur.
   - Negligible elevation changes in the heart (i.e., the points all lie at the same height in the heart).
   - The blood is inviscid and incompressible.

   (b) Extra data:
   - $\dot{V}_{\text{blood}} = 5.0$ L/min.
   - $\rho_{\text{blood}} = 1.056$ kg/L.
   - The blood flow velocities calculated using the cross-sectional areas of the heart vessels are given in Example 6.13.

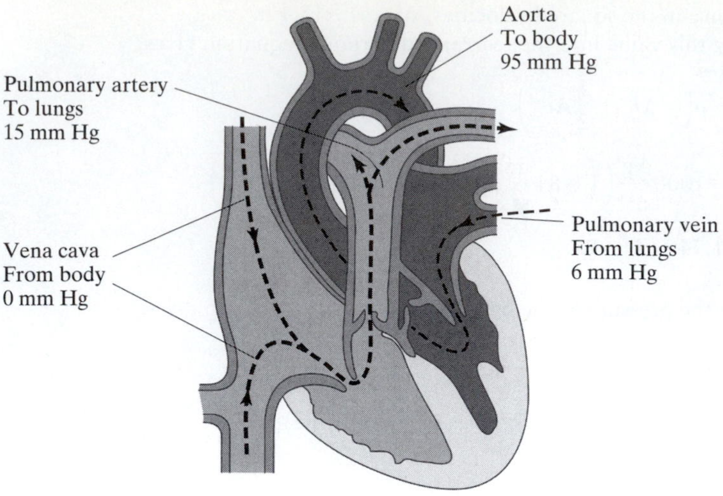

Figure 6.30
Pressures and flow direction in the heart. Systemic side is shaded dark; pulmonic side is shaded light. (*Source*: Cooney, DO. *Boimedical Engineering Principles*: *An introduction to Fluid, Heat and Mass Transport Processes*. New York: Marcel Dekker, 1976.)

- The approximate pressures in the vessels are:
  - pulmonary vein: 6 mmHg
  - aorta: 95 mmHg
  - venae cavae: 0 mmHg
  - pulmonary artery: 15 mmHg

(c) Variables, notations, units:
- pv = pulmonary vein
- ao = aorta
- vc = venae cavae
- pa = pulmonary artery
- Use kg, cm, s, mmHg, L, hp.

3. Calculate

(a) Equations: Using the extended Bernoulli equation [6.11-9a] is most appropriate, since we must account for pressure and shaft work:

$$(gh_i - gh_j) + \left(\frac{1}{2}(v_i)^2 - \frac{1}{2}(v_j)^2\right) + \frac{1}{\rho}(P_i - P_j) + \sum \frac{\dot{W}_{\text{shaft}}}{\dot{m}} - \sum \frac{\dot{f}}{\dot{m}} = 0$$

(b) Calculate:
- Because we assume no elevation changes and no friction losses in the heart, we reduce the extended Bernoulli equation to:

$$\left(\frac{1}{2}(v_i)^2 - \frac{1}{2}(v_j)^2\right) + \frac{1}{\rho}(P_i - P_j) + \sum \frac{\dot{W}_{\text{shaft}}}{\dot{m}} = 0$$

- Pressures are given as gauge pressures. However, since we want a pressure difference, we do not need to convert them to absolute pressures.
- Solving for the systemic system yields:

$$\left(\frac{1}{2}(v_{\text{pv}})^2 - \frac{1}{2}(v_{\text{ao}})^2\right) + \frac{1}{\rho}(P_{\text{pv}} - P_{\text{ao}}) + \sum \frac{\dot{W}_{\text{shaft}}}{\dot{m}} = 0$$

The kinetic energy term is:

$$\left(\frac{1}{2}(v_{\text{pv}})^2 - \frac{1}{2}(v_{\text{ao}})^2\right) = \left(\frac{1}{2}\left(13.9\frac{\text{cm}}{\text{s}}\right)^2 - \frac{1}{2}\left(33.3\frac{\text{cm}}{\text{s}}\right)^2\right) = -457.8\ \frac{\text{cm}^2}{\text{s}^2}$$

The flow work term is:

$$\frac{1}{\rho}(P_{pv} - P_{ao}) = \frac{1}{1.056\frac{\text{g}}{\text{cm}^3}}(6\text{ mmHg} - 95\text{ mmHg})\left(\frac{1.01329 \times 10^6\frac{\text{dynes}}{\text{cm}^2}}{760\text{ mmHg}}\right)$$

$$= -1.12 \times 10^5\frac{\text{dynes}\cdot\text{cm}}{\text{g}} = -1.12 \times 10^5\frac{\text{cm}^2}{\text{s}^2}$$

Thus, the shaft work is:

$$\sum\frac{\dot{W}_{shaft}}{\dot{m}} = -\left(-457.8\frac{\text{cm}^2}{\text{s}^2} - 1.12 \times 10^5\frac{\text{cm}^2}{\text{s}^2}\right) = 1.12 \times 10^5\ \frac{\text{cm}^2}{\text{s}^2}$$

$$\dot{W}_{shaft} = \dot{V}\rho\left(1.12 \times 10^5\frac{\text{cm}^2}{\text{s}^2}\right)$$

$$\dot{W}_{shaft} = \left(5\frac{\text{L}}{\text{min}}\right)\left(1.056\frac{\text{kg}}{\text{L}}\right)\left(1.12 \times 10^5\frac{\text{cm}^2}{\text{s}^2}\right)\left(\frac{1\text{ min}}{60\text{ s}}\right)\left(\frac{1\text{ m}^2}{10{,}000\text{ cm}^2}\right)$$

$$\dot{W}_{shaft} = 0.986\frac{\text{J}}{\text{s}} = 0.00132\text{ hp}$$

Note that the pressure difference term is three orders of magnitude larger than the kinetic energy term. Therefore, we can ignore the kinetic energy in calculating the shaft work in the pulmonic system. The shaft work for the pulmonic system is $\dot{W}_{shaft} = 0.166$ J/s = 0.000223 hp.

4. Finalize
   (a) Answer: The total work done by the heart is the sum of the work done by the systemic and pulmonic systems, which is 1.15 J/s or 0.00154 hp.
   (b) Check: Checking the value against those given in literature shows that the order of magnitude is the same; thus, our answer is reasonable. As an aside, note that a typical lawnmower can have a 5-hp engine. This has a power equivalent to about 3000 human hearts. However, the lawnmower does not run continuously for 80 or more years without taking a break! ■

EXAMPLE 6.22 **Friction Losses in Circulation**

Problem: Calculate the frictional losses in the vessels in the entire circulatory system. (Adapted from Cooney DO, *Biomedical Engineering Principles: An Introduction to Fluid, Heat, and Mass Transport Processes*, 1976.)

Solution: To use the extended Bernoulli equation [6.11-8] for this problem, we must have only one inlet and one outlet stream, so two arbitrary locations adjacent to each other in the circulatory system (denoted by 1 and 2) are selected (Figure 6.31). Since the inlet (1) and outlet (2) locations are in approximately the same location, several important simplifying assumptions can be made:

- No height change between locations 1 and 2. Therefore, the potential change energy $(\Delta\hat{E}_P)$ is zero.
- No velocity change between locations 1 and 2. Therefore, the kinetic energy $(\Delta\hat{E}_K)$ change is zero.
- No pressure change between locations 1 and 2. Therefore, the flow work $(\Delta P/\rho)$ is zero.
- The system is at steady-state.

Thus, the extended Bernoulli equation is simplified to:

$$\sum\dot{W}_{shaft} - \sum\dot{f} = 0$$

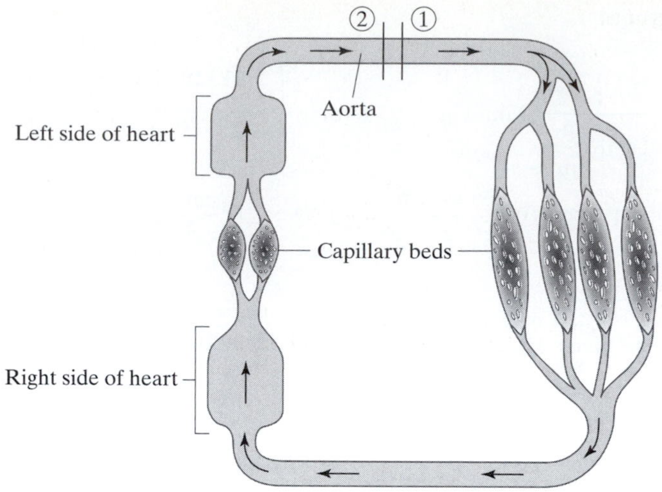

Figure 6.31
Schematic diagram of side of circulation.(*Source*: Cooney, DO. *Boimedical Engineering Principles: An introduction to Fluid, Heat and Mass Transport Processes*. New York: Marcel Dekker, 1976.)

Using the values we calculated in Example 6.21, we find that the frictional losses in the circulatory system are:

$$\sum \dot{f} = \sum \dot{W}_{\text{shaft}} = 0.00154 \text{ hp}$$

The frictional losses in the circulatory system equal the work done by the heart. This is a significant conclusion, because it demonstrates that the heart must work continuously to overcome the frictional energy losses from fluid flow, bifurcations, bends, and other occurrences during blood circulation. ■

In industry, the use of pumps, fans, blowers, and compressors adds energy to a system by increasing the pressure of the fluid. Pumps are used for systems containing liquids; the other three are applied to systems with gases. **Recall that frictional losses occur in flowing fluids over a long distance, so adding a pump can increase the distance over which the fluid can flow for a given outlet pressure.** Pumps are sized to meet design criteria, including the pressure of the liquid outlet stream.

Fans, blowers, and compressors in gas systems function much as pumps do in liquid systems. However, the three pieces of equipment have different methods of increasing pressure in the outlet stream. The primary function of a fan is to move gas, so it is only capable of producing small pressure changes. When more compression of the gas is desired, a blower can be used. Blowers function in a similar manner to fans, but can increase the gas pressure by about 1 atm. To increase the pressure of a gas system to a greater extent, compressors are typically used.

Machines also exist to perform the opposite function—to remove mechanical energy from a system. Turbines are used to convert the mechanical energy of a gas or liquid into another form of energy. For example, a fluid may obtain thermal energy from sources such as a heater in a power plant. This thermal energy serves to heat up the water to make steam, such that the steam moves through the turbine and causes it to turn. The rotating turbine is connected to a generator, which can convert the mechanical energywater to electrical energy. In this example, the steam is the fluid flow energy that serves as an intermediate between thermal and electrical energy. Combining the use of a pump and a turbine at opposite ends of a process allows mechanical energy to be transferred through a system by a fluid.

Bioprocessing units and manufacturing plants require many of these pieces of equipment. Almost any bioreactor that requires a continuous inlet stream also requires a pump. Centrifugal pumps, which operate with low pressure changes, are used in heart–lung bypass machines. Constant-displacement pumps are used for the delivery of drugs, such as insulin. The total artificial heart is a pump that has been tailored to match as closely as possible the specifications of the human heart, which is itself a pump.

EXAMPLE 6.23 Pump-and-Treat Remediation System

Problem: Some regions of groundwater, located in reservoirs below the ground, have been contaminated by methyl *tertiary*-butyl ether (MTBE). A soluble, but not readily biodegradable contaminant under normal conditions, the octane-enhancing fuel oxygenate was added to gasoline from 1979 until 2000, when many state governments limited its use or banned it from further use. One proposal to purify drinking-water sources taken from groundwater contaminated with MTBE is to use a method called pump-and-treat, where the polluted water is extracted using a pump and treated in a remediation system above the ground.

Take the extraction well and surrounding water as your system (Figure 6.32). Find the work performed by the pump in terms of the other variables to bring the groundwater to the surface for treatment. Then find the work performed by the pump to raise groundwater from a reservoir 150 ft below ground through a 6-inch-diameter pipe to the surface at a flow rate of 80 gpm.

Solution:

1. Assemble
   (a) Find: Work done by pump to bring ground water to the surface.
   (b) Diagram: The system is shown in Figure 6.32.
2. Analyze
   (a) Assume:
   - The Earth's surface is the reference height set at zero ($h_{\text{out}} = 0$).
   - Near the base of the well, groundwater flows very slowly ($v_{\text{in}} \cong 0$).
   - The pressure difference between the Earth's surface and the groundwater source is negligible.
   - The velocity profile in the well is uniform.
   - The extraction well is cylindrical.
   - The system is at steady-state with one inlet and one outlet.
   - Negligible frictional losses.

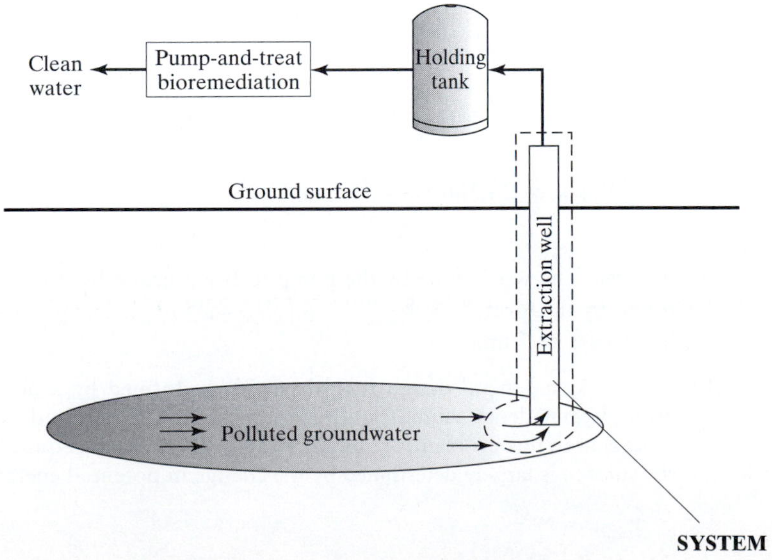

Figure 6.32 Pumping groundwater through an extraction well. *Adapted from* epa.gov.

- No reactions occur.
- The fluid is incompressible.
- The density of the fluid flowing to the well is equal to that flowing out ($\rho_{in} = \rho_{out} = 1.0$ kg/L).

(b) Extra data: No extra data are needed.

(c) Variables, notations, units:

- The subscript *in* refers to the point in the polluted groundwater pool near where the water enters the extraction well, and the subscript *out* to the point at the ground surface where the groundwater transfers from the extraction well to the holding tank.
- Use hp, L, s, kg.

3. Calculate

(a) Equations: Using the extended Bernoulli equation [6.11-8] is most appropriate, since we must account for shaft work:

$$\dot{m}(gh_i - gh_j) + \dot{m}\left(\frac{1}{2}v_i^2 - \frac{1}{2}v_j^2\right) + \frac{\dot{m}}{\rho}(P_i - P_j) + \sum \dot{W}_{\text{shaft}} - \sum \dot{f} = 0$$

(b) Calculate:

- Because we assume the reference height and the groundwater flow entering the extraction well to be zero, we can simplify the specific potential and kinetic energy terms. Since we also assume the pressure difference between the Earth's surface and the groundwater is negligible and that no frictional losses occur, these terms can be eliminated. This simplifies the extended Bernoulli equation. Rearranging the equation to solve for the shaft work gives:

$$\dot{m}gh_{\text{in}} - \frac{1}{2}\dot{m}v_{\text{out}}^2 + \sum \dot{W}_{\text{shaft}} = 0$$

$$\sum \dot{W}_{\text{shaft}} = -\dot{m}\left(gh_{\text{in}} - \frac{1}{2}v_{\text{out}}^2\right)$$

- To solve for the work done by the pump to extract groundwater, the mass flow rate is calculated from the volumetric flow rate to be 5.05 kg/s. To calculate the exit velocity, we model the conduit as a cylinder to find the cross-sectional area and use equation [3.2-4]. The outlet velocity v_{out} is 0.277 m/s.
- The groundwater must be raised 150 feet from a reservoir. Note that the elevation is negative ($h_{\text{in}} = -150$ ft), since the Earth's surface is the reference point. The work done by the pump is:

$$\sum \dot{W}_{\text{shaft}} = -\dot{m}\left(gh_{\text{in}} - \frac{1}{2}v_{\text{out}}^2\right)$$

$$\sum \dot{W}_{\text{shaft}} = -\left(5.05\frac{\text{kg}}{\text{s}}\right)\left(\left(9.81\frac{\text{m}}{\text{s}^2}\right)(-150\text{ ft})\left(\frac{3.2808\text{ m}}{1\text{ ft}}\right) - \frac{1}{2}\left(0.277\frac{\text{m}}{\text{s}}\right)^2\right)$$

$$\sum \dot{W}_{\text{shaft}} = 24{,}380\text{ W} = 32.7\text{ hp}$$

4. Finalize

(a) Answer: The work done by the pump to bring groundwater to the surface can be found by the formula $\sum \dot{W}_{\text{shaft}} = -\dot{m}\left(gh_{\text{in}} - 1/2\, v_{\text{out}}^2\right)$. The pump size would need to be at least 32.7 hp.

(b) Check: As a general design rule, the work performed by a pump depends on the groundwater depth, liquid flow rate, any friction losses, and extraction well size. Note that in this problem, the mechanical energy of work required to pump water to the surface is largely determined by the change in potential energy. ■

Frictional losses arise when fluid flows through pipes and other conduits. As the fluid shears and resists flow, mechanical energy is converted to thermal energy. Frictional energy losses can occur as fluid flows in straight pipes, around bends, through expansions or contractions, through fittings, or in a variety of other configurations.

Empirical values for friction losses can be estimated using formulas developed by application of theory and experimental measurements. Formulas have been developed that estimate the **friction factor** (f) as a function of a number of system characteristics, which may include fluid density, fluid viscosity, pipe diameter, pipe surface roughness, average fluid velocity, length of the pipe, and geometric factors. Geometric factors typically arise when estimating friction losses across contractions, enlargements, fittings, valves, or bends (e.g., elbows).

In laminar flow in a straight pipe, f is dependent only on the Reynolds number:

$$f = \frac{16}{\text{Re}} \qquad [6.11\text{-}19]$$

In turbulent and transitional flow ($2100 < \text{Re} < 100{,}000$) in a straight pipe that is hydraulically smooth, the f is approximately:

$$f = \frac{0.0791}{(\text{Re})^{1/4}} \qquad [6.11\text{-}20]$$

In turbulent flow in a straight pipe with rough surfaces, the friction factor is determined using charts requiring knowledge of the Reynolds number and the relative pipe surface roughness. Consult fluid dynamics textbooks (e.g., Bird RB, Stewart WE, and Lightfoot EN, *Transport Phenomena*, 2002) for guidance in calculating friction factors for turbulent flow in rough pipes.

The friction factor f can then be related to the frictional energy losses, $\dot{f}$. To calculate friction loss along a straight pipe with turbulent flow, the following relationship is used:

$$\frac{\dot{f}}{\dot{m}} = \frac{1}{2}\langle v\rangle^2 \frac{4L}{D} f \qquad [6.11\text{-}21]$$

where $\langle v\rangle^2$ is the square of the average velocity, L is the length of the pipe, D is the diameter of the pipe, and f is the friction factor. It should make sense that the magnitude of friction loss is proportional to the length of the pipe. The estimate for the frictional energy losses $\dot{f}$ can then be used in the mechanical energy accounting and extended Bernoulli equations.

EXAMPLE 6.24 Frictional Losses in the Trans-Alaska Pipeline

Problem: Stretching from Prudhoe Bay on the northern coast of Alaska to the port town of Valdez on the southern coast, the Trans-Alaska Pipeline uses pumps to transport crude oil 800 miles across the Alaskan wilderness. The original design featured a dozen pumping stations, each equipped with four pumps to provide the necessary shaft work to overcome frictional losses. The number of pumps needed varies with pipeline operations, but most are still used today.

The Trans-Alaska Pipeline requires shaft work to maintain fluid flow over long distances. Although the path of the pipeline does have some changes in elevation, the primary reason for the pumps is to offset frictional losses during fluid flow. Estimate the frictional losses in transporting crude oil through the Trans-Alaska Pipeline.

Solution:

1. Assemble
   (a) Find: Frictional losses from crude oil transport through the Trans-Alaska Pipeline.
   (b) Diagram: A map with the pipeline is shown in Figure 6.33.
2. Analyze
   (a) Assume:
   - Pipe is cylindrical, straight, and hydraulically smooth.
   - Flow is a continuous stream through the pipeline. Daily volumetric flow rate corresponds to 24 hours of oil flow at a constant velocity.
   - The flow in the vessel is turbulent and has a uniform velocity profile.
   - The system is at steady-state with one inlet and one outlet.
   - No reactions occur.
   - The fluid is incompressible.

   (b) Extra data:
   - Pipe length: 800 miles or 4.2×10^6 ft
   - Pipe diameter: 4 ft
   - Volumetric flow rate: 1.3 million barrels (petroleum products) per day
   - Approximate oil viscosity: $0.5\ \text{lb}_m/(\text{ft} \cdot \text{s})$
   - Approximate oil density: $51\ \text{lb}_m/\text{ft}^3$

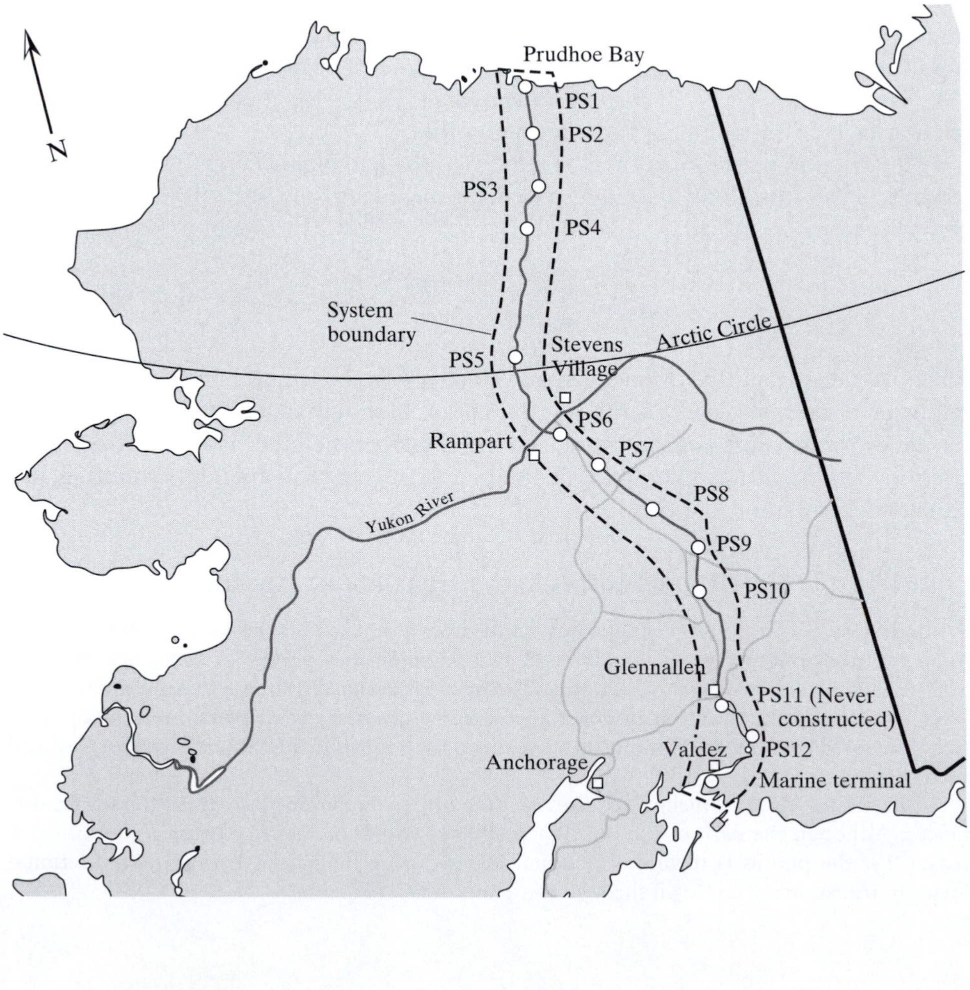

Figure 6.33
Trans-Alaska Pipeline system. PS is a pump station (*Adapted from* wikipedia.org.)

(c) Variables, notations, units:
- Use hp, gal, min, ft, lb_m.

3. Calculate
(a) Equations: We use the Reynolds number equation [6.10-1] to determine if the fluid is turbulent. Because we model a straight, hydraulically smooth pipe with turbulent flow, we use equations [6.11-20] and [6.11-21] to find the friction factor and the frictional loss along the pipeline:

$$\text{Re} = \frac{\rho v D}{\mu}$$

$$f = \frac{0.0791}{(\text{Re})^{1/4}}$$

$$\frac{\dot{f}}{\dot{m}} = \frac{1}{2}\langle v\rangle^2 \frac{4L}{D} f$$

(b) Calculate:
- To find the Reynolds number, we first calculate the volumetric flow rate so we can find the linear velocity:

$$\dot{V} = \left(1.3 \times 10^6 \frac{\text{barrel}}{\text{day}}\right)\left(\frac{42 \text{ gal}}{\text{barrel}}\right)\left(\frac{\text{ft}^3}{7.48 \text{ gal}}\right)\left(\frac{\text{day}}{86{,}400 \text{ s}}\right) = 84.5 \frac{\text{ft}^3}{\text{s}}$$

$$v = \langle v\rangle = \frac{\dot{V}}{A} = \frac{84.5 \frac{\text{ft}^3}{\text{s}}}{\pi(2 \text{ ft})^2} = 6.72 \frac{\text{ft}}{\text{s}}$$

$$\text{Re} = \frac{\rho v D}{\mu} = \frac{\left(51 \frac{\text{lb}_m}{\text{ft}^3}\right)\left(6.72 \frac{\text{ft}}{\text{s}}\right)(4 \text{ ft})}{\left(0.5 \frac{\text{lb}_m}{\text{ft}\cdot\text{s}}\right)} = 2742$$

- This Reynolds number value for the flow is transitional, but the same equation for f is applicable for both turbulent and transitional flow:

$$f = \frac{0.0791}{(\text{Re})^{1/4}} = \frac{0.0791}{(2742)^{1/4}} = 0.0109$$

(Note that modeling it as laminar flow returns a similar value for f.)
- Substituting this value into equation [6.11-21] gives:

$$\frac{\dot{f}}{\dot{m}} = \frac{1}{2}\langle v\rangle^2 \frac{4L}{D} f = \frac{1}{2}\left(6.72 \frac{\text{ft}}{\text{s}}\right)^2 \frac{4(4.2 \times 10^6 \text{ ft})}{(4 \text{ ft})}(0.0109) = 1.03 \times 10^6 \frac{\text{ft}^2}{\text{s}^2}$$

To find the frictional losses, first calculate the mass flow rate:

$$\dot{m} = \rho\dot{V} = \left(51 \frac{\text{lb}_m}{\text{ft}^3}\right)\left(84.5 \frac{\text{ft}^3}{\text{s}}\right) = 4310 \frac{\text{lb}_m}{\text{s}}$$

The frictional losses along the entire system are:

$$\dot{f} = \left(1.03 \times 10^6 \frac{\text{ft}^2}{\text{s}^2}\right)\dot{m}$$

$$\dot{f} = \left(1.03 \times 10^6 \frac{\text{ft}^2}{\text{s}^2}\right)\left(4310 \frac{\text{lb}_m}{\text{s}}\right)\left(\frac{\text{lb}_f \cdot \text{s}^2}{32.2 \text{ lb}_m \cdot \text{ft}}\right)\left(\frac{1.34 \times 10^{-3} \text{ hp}}{0.738 \frac{\text{lb}_f \cdot \text{ft}}{\text{s}}}\right)$$

$$\dot{f} = 2.5 \times 10^5 \text{ hp}$$

4. Finalize
   (a) Answer: The frictional losses in the Trans-Alaska Pipeline total approximately 250,000 hp.
   (b) Check: There are about ten pumping stations along the route, generally equipped with several pumps. The pumps each run on 18,000 horsepower, so the total power along the pipeline can be up to 500,000 hp.[4] Therefore, the pipeline system has sufficient power to overcome frictional losses of the magnitude we approximated. ■

Accounting or conservation equations for mass, momentum, and mechanical energy are often used together to analyze systems. Using equations in conjunction with one another is especially useful when solving systems with multiple unknowns requiring several independent equations.

EXAMPLE 6.25 Flow Constrictor

Problem: Differential manometers are used to measure the change in pressure across sections of pipe found associated with bioreactors and other applications. Consider the flow constrictor depicted in Figure 6.34. The diameter of the pipe decreases from 0.5 m to 0.3 m. The density of the process fluid, ρ_f, is 1.0 g/mL, while the manometer fluid density, ρ_m, is 1.3 g/mL. Given that the manometer head is 0.3 m, calculate the horizontal force that must be applied to the flow constrictor to keep it stationary.

Solution:

1. Assemble
   (a) Find: Horizontal force necessary to hold the flow constrictor stationary.
   (b) Diagram: A flow constrictor is shown in Figure 6.34.
2. Analyze
   (a) Assume:
      - No friction losses through the constrictor.
      - Fluid flows continuously through the constrictor.
      - The flowing fluid in the vessel has a uniform velocity profile.
      - No shaft work occurs.
      - The system is at steady-state with one inlet and one outlet.
      - The height of the flowing fluid does not change.
      - No reactions occur.
      - The flowing fluid is incompressible.
      - Manometer fluid and process fluid do not mix.

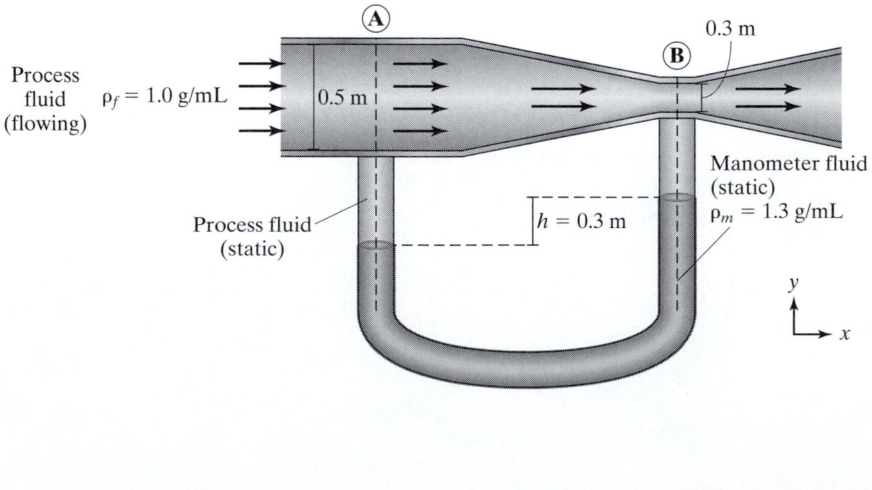

Figure 6.34
Flow constrictor.

(b) Extra data: No extra data are needed.
(c) Variables, notations, units:
- Subscripts A and B refer to two locations in the system.
- Use kg, m, s, N.

3. Calculate
(a) Equations: Because we are interested in the horizontal force needed to stabilize the flow constrictor, we use conservation of linear momentum equation [6.3-3] to find the resultant forces. In conjunction, we also use the Bernoulli equation [6.11-11] to characterize the flowing fluid and the equation for static fluids to model the manometer:

$$\sum_i \dot{m}_i \vec{v}_i - \sum_j \dot{m}_j \vec{v}_j + \sum \vec{F} = \frac{d\vec{p}^{sys}}{dt}$$

$$(gh_i - gh_j) + \left(\frac{1}{2}(v_i)^2 - \frac{1}{2}(v_j)^2\right) + \frac{1}{\rho}(P_i - P_j) = 0$$

$$P_2 - P_1 = -\rho g(h_2 - h_1)$$

(b) Calculate:
- Because we only want the horizontal force and the system is at steady-state, applying the conservation of linear momentum in the x-direction gives:

$$x: \quad \dot{m}_A v_A - \dot{m}_B v_B + \sum F_{R,x} = 0$$

Conservation of total mass requires that $\dot{m}_A = \dot{m}_B$:

$$\sum F_{R,x} = \dot{m}_A(v_B - v_A)$$

- To find the linear velocity of the fluid at A, the conservation of total mass can be used:

$$\dot{m}_A - \dot{m}_B = \rho_f v_A A_A - \rho_f v_B A_B = 0$$

Since density is a constant across the system:

$$\pi(0.25m)^2 v_A - \pi(0.15m)^2 v_B = 0$$
$$v_A = 0.36 v_B$$

- Considering the manometer, the equation for static fluids (equation [6.6-9]) is used to determine the difference between the pressures at points A and B:

$$P_B - P_A = -\rho_m g(h_B - h_A) = -\left(1300\frac{\text{kg}}{\text{m}^3}\right)\left(9.81\frac{\text{m}}{\text{s}^2}\right)(0.3 \text{ m}) = -3826\frac{\text{kg}}{\text{m} \cdot \text{s}^2}$$

- The process fluid is analyzed using the Bernoulli equation. The change in the height of the flowing fluid is zero, so the equation is simplified. Substituting the values gives:

$$\frac{1}{2}(v_A^2 - v_B^2) + \left(\frac{P_A - P_B}{\rho_f}\right) = \frac{1}{2}(v_A^2 - v_B^2) + \left(\frac{3826\frac{\text{kg}}{\text{m} \cdot \text{s}^2}}{1000\frac{\text{kg}}{\text{m}^3}}\right) = 0$$

$$v_B^2 - v_A^2 = 7.65\frac{\text{m}^2}{\text{s}^2}$$

Substituting $v_A = 0.36 v_B$ gives:

$$v_B^2 - (0.36 v_B)^2 = 7.65\ \frac{\text{m}^2}{\text{s}^2}$$

$$v_B = 2.96\ \frac{\text{m}}{\text{s}} \qquad v_A = 1.07\ \frac{\text{m}}{\text{s}}$$

- Either velocity can be used to solve for the mass flow rate:

$$\dot{m}_A = \rho_f v_A A_A = \left(1000\frac{\text{kg}}{\text{m}^3}\right)\left(1.07\frac{\text{m}}{\text{s}}\right)(\pi(0.25\text{ m})^2) = 210\frac{\text{kg}}{\text{s}}$$

- The resultant force is obtained by substituting in all values:

$$\sum F_{R,x} = \dot{m}_A(v_B - v_A) = 210\frac{\text{kg}}{\text{s}}\left(2.96\frac{\text{m}}{\text{s}} - 1.07\frac{\text{m}}{\text{s}}\right) = 398\text{ N}$$

4. Finalize
   (a) Answer: The horizontal force required to offset the change in momentum is 398 N and acts in the direction of the outlet stream.
   (b) Check: As the fluid moves into a region of smaller diameter, it must move faster to maintain the same mass flow rate. Because the same amount of material flows out of the system faster than it flows in, an external force must be applied in the direction of the flow, so the direction of force makes sense. It is harder to establish an intuitive sense of the magnitude; however, 398 N (90 lb_f) could easily be applied by a few strong bolts. ■

Summary

This chapter opened with the types of momentum that can affect a system, including bulk material flow and the application of forces. The conservation of linear momentum and the conservation of angular momentum were formulated in the differential and integral forms. Two applications of the conservation equations for steady-state, static systems—rigid bodies and static fluids—were discussed. How bulk material transfer and external forces change the momentum of a system was explored in steady-state and unsteady-state systems. Discussions of resultant forces, elastic and inelastic collisions, and the coefficient of restitution were included. Newton's second and third laws were derived from the conservation of linear momentum for special cases. The significance of laminar flow and turbulent flow was explained in the context of the definition of Reynolds number. Finally, we looked at how the mechanical energy accounting equation and the Bernoulli equation can be used with the conservation of momentum to model many systems with fluid flow.

Table 6.3 reinforces that linear and angular momentum may accumulate in a system because of bulk material transfer across the system boundary. Mechanical energy may accumulate in a system because of bulk mass transfer across the system boundary and energy interconversions. See the tables concluding other chapters for comparisons.

TABLE 6.3

Summary of Movement, Generation, Consumption, and Accumulation in the Momentum and Mechanical Energy Accounting Equations

Accumulation	Input −	Output	+ Generation	− Consumption
Extensive property	Bulk material transfer	Direct and non-direct contacts	Chemical reactions	Energy interconversions
Mechanical energy	X	X		X
Linear momentum	X	X		
Angular momentum	X	X		

References

1. Gregor RJ and Conconi F. *Road Cycling*. Boston: Blackwell Publishing, 2000.
2. Burke ER, ed. *High-Tech Cycling*. Champaign, IL: Human Kinetics Publishers, 1995.
3. Grose TK. "Smart parts." *ASEE Prism* 2002, 11:16–21.
4. Armistead TF. "Alyeska system upgrade set to kick off in early March." McGraw-Hill's enr.com. January 12, 2004. http://enr.construction.com/news/powerindus/archives/040112.asp (accessed January 11, 2005).
5. Anderson EJ and DeMont ME. "The mechanics of locomotion in the squid *Loligo pealei*: Locomotory function and unsteady hydrodynamics of the jet and intramantle pressure." *J Exp Biol* 2000, 203 Pt 18:2851–63.

Problems

6.1 The College World Series national champions are practicing for a game. The pitcher can throw the baseball at 90 mph. A baseball has a mass of 145 g. What is the linear momentum of the baseball? If the ball leaves the bat at 110 mph, what is its linear momentum after it is hit?

6.2 A pebble of mass 0.50 g is stuck on the wheel of a bicycle, marking a spot on the tire. When the pebble reaches the apex of its path, it is moving 10 mph relative to the axle. The radius of the tire is 8 in. What are the linear momentum and angular momentum about the wheel axle for the pebble?

6.3 Glaucoma is one of the more common causes of blindness and results from an elevated intraocular pressure. Normal gauge pressure values are 13–17 mmHg. Gauge pressures above 20 mmHg are potentially dangerous, so optometrists regularly screen their patients with a technique called tonometry. There are several varieties of tonometry, but their common feature involves applying a small force to the eye. The intraocular pressure is measured as a function of the displacement of the cornea.

The Goldmann tonometer is a specific instrument used for this procedure. It includes a piece that has a diameter of about 3.0 mm that physically comes into contact with the eye. How much force must be applied to this part of the instrument in order to balance the intraocular pressure of a healthy eye over the contact area?

6.4 Many women complain about lower back pain during pregnancy. As their physician and a trained bioengineer, you decide to estimate the force on the lower back during pregnancy for one of your favorite patients.

You focus your attention on the third lumbar vertebra in the lower spine (Figure 6.35). You know that the extensor muscles that run along the back of the spine operate to balance the weight of the body in the chest and gut region. To solve this problem, you make a number of simplifying assumptions:

- Before pregnancy, the woman weighs 130 lb_f.
- The weight of the body above the third lumbar vertebra, $\vec{W}_B$, is 55% of the total weight of the body. (The third lumbar vertebra does not support the weight of the whole body.)
- The forces exerted by the extensor muscles, $\vec{F}_m$, act 2 inches posterior to the center of the vertebral body.
- The weight of the body above the third lumbar vertebra, $\vec{W}_B$, acts 2 inches anterior to the center of the vertebral body.
- Compressive forces, $\vec{F}_C$, act on the center of the vertebral body.

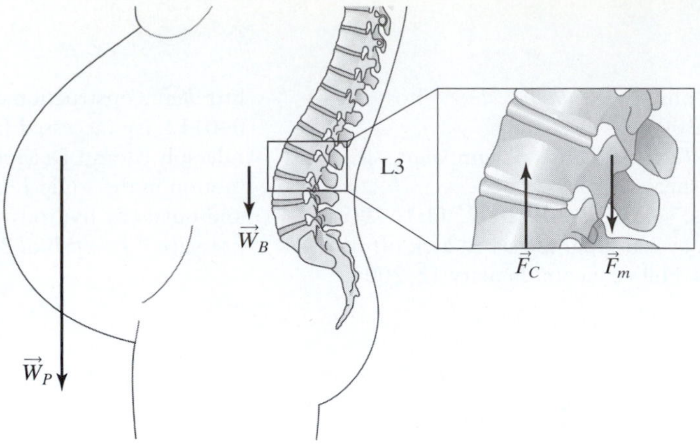

Figure 6.35
Forces on the spine of a pregnant woman.

- During pregnancy, the weight of the abdomen, $\vec{W}_P$, (including baby, placenta, amniotic fluid, etc.), increases by 20 lb_f.
- The center of mass of the additional abdomen weight is 10 inches anterior to the center of the vertebral body.

(a) Calculate the forces exerted by the extensor muscles, $\vec{F}_m$, on the third lumbar vertebra before and during pregnancy.

(b) Calculate the compressive forces, $\vec{F}_C$, which the vertebral body feels before and during pregnancy.

6.5 You are separating DNA strands in a tube containing a sugar/agarose gel using ultracentrifugation (Figure 6.36). The sugar/agarose gel has been prepared with a density gradient as follows:

$$\rho_{gel} = 1.1 + 0.004d^2$$

where d is the distance into the gel in the tube. (ρ_{gel} has units of g/cm^3 when d is specified in units of cm.) The mass of the DNA is 3.2×10^{-12} g, and the volume of the DNA is 2.56×10^{-12} cm^3. The distance between the center of the ultracentrifuge and the top of the gel is 5 cm. The ultracentrifuge is operating at 12,000 rpm.

Note that the acceleration of the DNA in the radial direction is the sum of the centripetal acceleration and linear acceleration toward the end of the tube. Ignore drag so that the only force acting on the DNA is the buoyant force from the gel. When examining the buoyant force, only consider the centripetal acceleration for the gel.

At what depth does the DNA stop? In other words, at what point does the buoyant force result in exactly the centripetal acceleration?

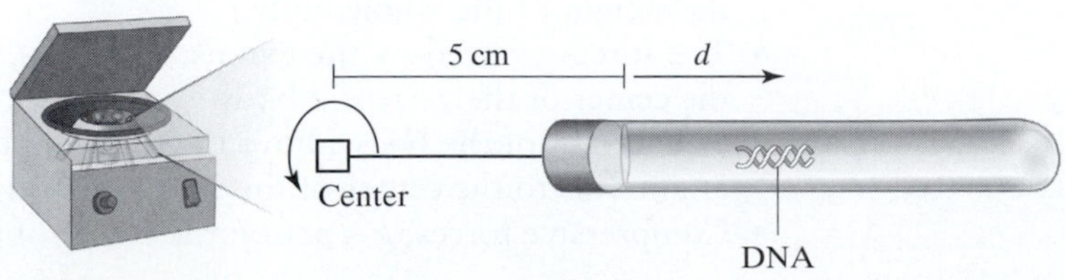

Figure 6.36
Separation of DNA strands using ultracentrifugation. Not drawn to scale.

Figure 6.37
Holding a bowling ball.

6.6 You are helping your little brother hold his 8-lb_m bowling ball as you wait for your lane. You are holding the bowling ball as shown in Figure 6.37, with the force of your hands at a 45° angle from horizontal ($\theta = 45°$). What forces are you and your brother exerting on the ball if it is stationary? How are the forces different if you are instead holding the ball with the force directed at 60° from horizontal?

6.7 A zoologist estimates that the jaw of a lion is subjected to a force $\vec{P}$ as large as 800 N (Figure 6.38). What forces $\vec{T}$ and $\vec{M}$ must be exerted by the temporalis and masseter muscles, respectively, to support this value of $\vec{P}$? (From Bedford A and Fowler W, *Engineering Mechanics: Statics and Dynamics*, Upper Saddle River, NJ: Prentice Hall, 2002.)

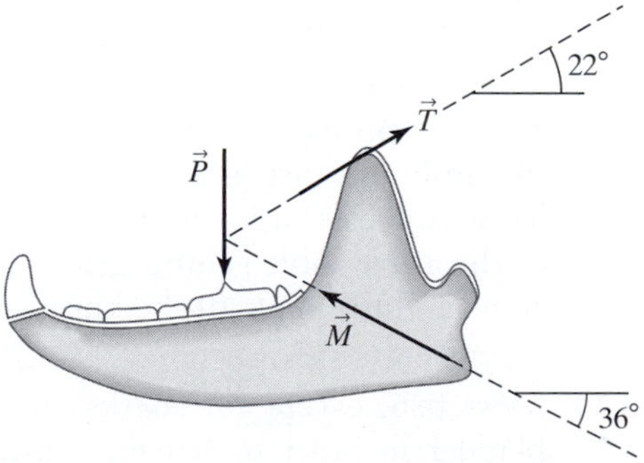

Figure 6.38
Muscle forces needed to support the jaw. (*Source*: Bedford A and Fowler W, *Engineering Mechanics: Statics and Dynamics*. Upper Saddle River, NJ: Prentice Hall, 2002.)

6.8 The sport of gymnastics requires both impressive physical strength and extensive training for balance. The iron cross is an exercise performed on two suspended rings, which the gymnast grips with his hands. Suppose that a male gymnast wishes to execute an iron cross during a gymnastics session (Figure 6.39). The total mass of the gymnast is 125 lb_m. Each ring supports half of the gymnast's weight. Assume that the weight of one of his arms is 5% of his total body weight. The distance from his shoulder joint to where his hand holds the ring is 56 cm. The distance from his hand to the center of mass of his arm is 38 cm. The horizontal distance from his shoulder to the middle of his chest, directly above the center of mass of his body, is 22 cm. If the gymnast is still, how much force and torque are at one of his shoulder joints?

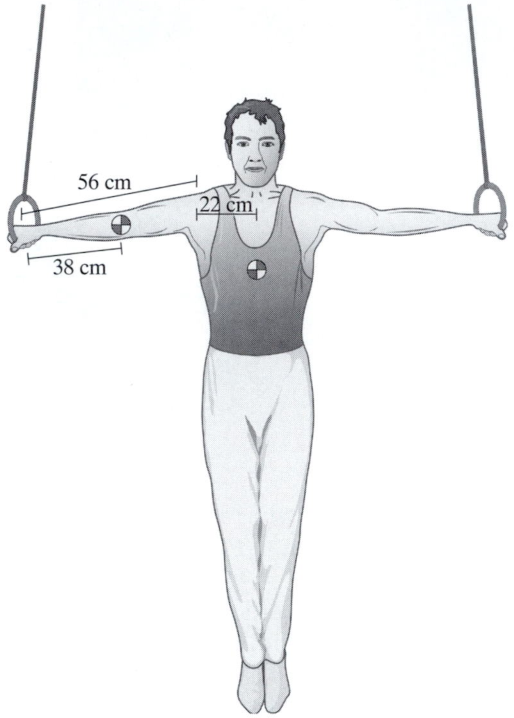

Figure 6.39
Gymnast performing the iron cross.

6.9 Reexamine Example 6.6 in the text concerning the muscles in the forearm. Find the same forces if there were a 5.0-lb_m weight held in the hand. The center of mass of the weight held in the hand is 45 cm from the elbow. The x-component of the force in the elbow is 24 N.

6.10 When bones are healing, it is critical that they are held in a fixed position. This is a primary purpose of casts. It is also sometimes necessary to suspend the limb and cast in a fixed position. Consider the two cables suspending a leg with a cast in Figure 6.40. If the leg and cast weigh 150 N and the angle of the ankle cable is 60 degrees above the horizontal, what are the cable tensions and the angle of the knee cable from the horizontal such that the leg is supported?

6.11 Most fish, except for sharks and their relatives, use an organ called an air bladder in order to remain neutrally buoyant. Calculate the net force on a 5-lb_m fish in salt water ($\rho = 1.024$ g/cm^3), given that the volume of the fish is 135.1 in^3. Now suppose there is a problem with the air bladder, causing it to deflate and the fish to lose 4% of its volume. Calculate the net force on the fish under these conditions. The mass of the fish can be considered constant.

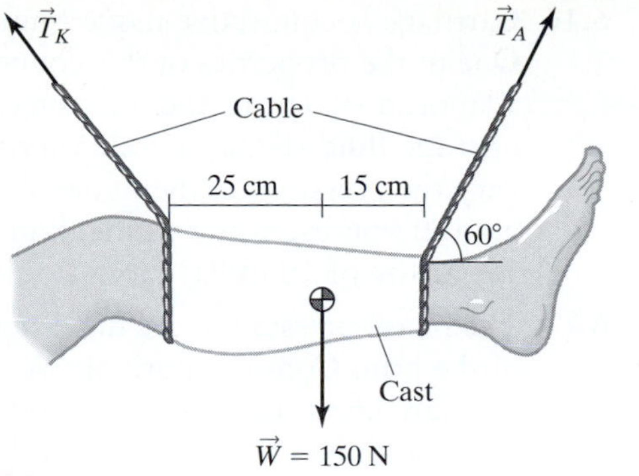

Figure 6.40
Two cables suspending a leg cast.

6.12 Consider a cylinder with a static fluid of density ρ (Figure 6.41). Using the cylindrical coordinate system, derive the change in pressure with position in the cylinder as a function of r, z, and θ. Follow the same type of derivation as outlined in Section 6.6.

6.13 The bacteria found living near hydrothermal vents are notable because they exist at high temperatures and pressures. Their enzymes are useful in PCR (polymerase chain reaction) machines because the operating conditions inside the PCR machines are similar to the bacteria's original habitat. Assume that the ocean is a static fluid. The vents are found 2000 to 2500 ft below sea level. At approximately what pressures do these bacteria live?

6.14 Water and other nutrients move upward through tree trunks to reach branches and leaves. Botanists have pondered the question, "How does water flow upward in the trunks of some of the largest trees?" It can be shown that capillary action and root pressure are not enough to transport water up a 120-m-tall tree. Assume that the pressure at the top of the column of water is almost zero. What pressure is required at the base of the tree in order to maintain a column of water at that height? (In reality, the force is greater, because the water is not only maintained at that height, but is also forced upward.) Do further research into cohesion theory to understand how this phenomenon is possible.

6.15 The deep-sea diving vessel, *Alvin*, can dive up to 12,800 feet under water. The vessel's purpose is to explore depths where the hydrostatic pressure is too great for human divers. What is the maximum pressure that the submersible can withstand? List some ideas about how you might design such a vessel.

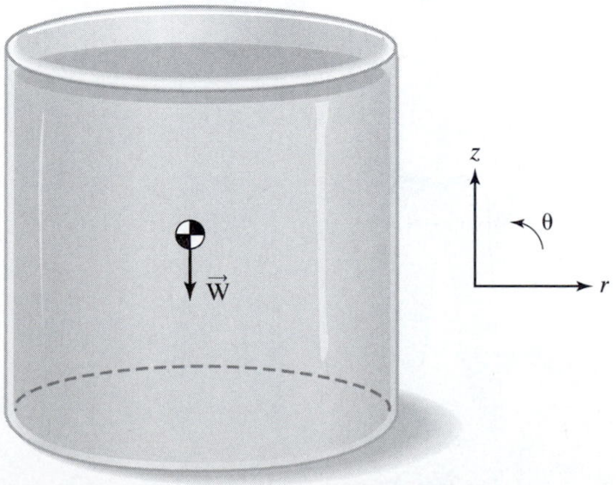

Figure 6.41
Cylinder with static fluid.

6.16 Cartilage is connective tissue found in the human body between some bones. One of the properties of this connective tissue is its ability to support force. In a laboratory setting, the force on cartilage tissue can be simulated by a column of static fluid (Figure 6.42). Varying the amount of fluid above the tissue can vary the pressure on the tissue. Approximately 0.5 m of fluid can fill the column. If you use water as the fluid, how high would you fill the column to get a pressure of 105 kPa? How would you simulate a pressure of 110 kPa?

6.17 The blood pressure in the fingertips can be different depending on the position of the arm. Consider positions A and B, as shown in Figure 6.43. Assume the arms are lifted slowly so the blood pressure has time to equilibrate, the blood is static, and the blood pressure in the head remains constant. What is the difference in blood pressure in the capillaries of the fingertips for positions A and B? (*Hint*: To determine arm length, measure your roommate or your study partner.)

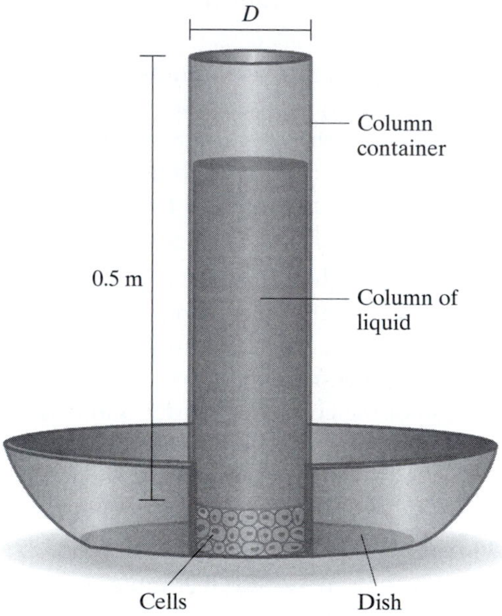

Figure 6.42
Column of static fluid simulating force on cartilage cells. Not drawn to scale.

Figure 6.43
Two different positions of the arms.

6.18 Example 6.9 shows that the force exerted on the bottom area of the two containers depicted in Figure 6.16 is equal. However, you notice that this value exceeds the weight of the water in container R. This apparent discrepancy is reconciled by forces that act through the walls of the vessel and originate from the top. Show that the sum of the forces due to pressures that act on all horizontal surfaces is exactly equal to the weight of the water in the vessel. As a comparison, perform the same procedure on container S.

6.19 A water tower is filled with water to a height of 100 ft (Figure 6.44). What is the water pressure for a faucet 3 ft above the ground?

6.20 A tank containing distilled water to be used in a bioreactor is pressurized such that the pressure at the outlet valve is 25 psig (Figure 6.45). What is the pressure at the top of the tank, which is 4 ft above the outlet valve?

6.21 What is the pressure difference between the shoulder and the ankle when a person reclines on a horizontal surface?

6.22 During mating season, male reindeer attract the attention of female reindeer by fighting each other with their horns and hooves. Consider a scenario in which Dasher, whose mass is 300 lb_m, is charging directly at Dancer, whose mass is 400 lb_m. Just before the collision, Dasher is traveling at 25 mph, and Dancer is traveling at 20 mph.

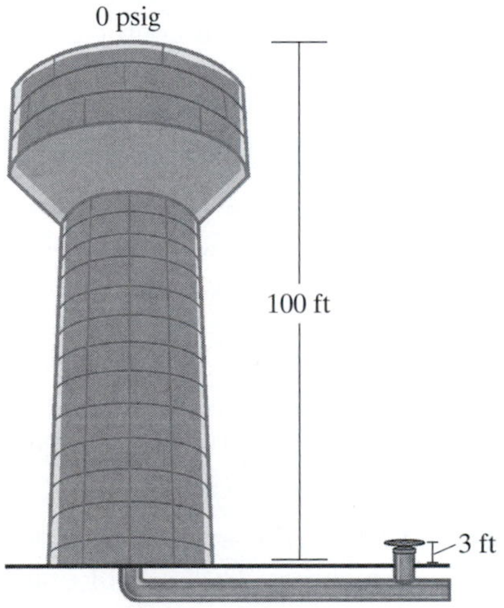

Figure 6.44
Water tower.

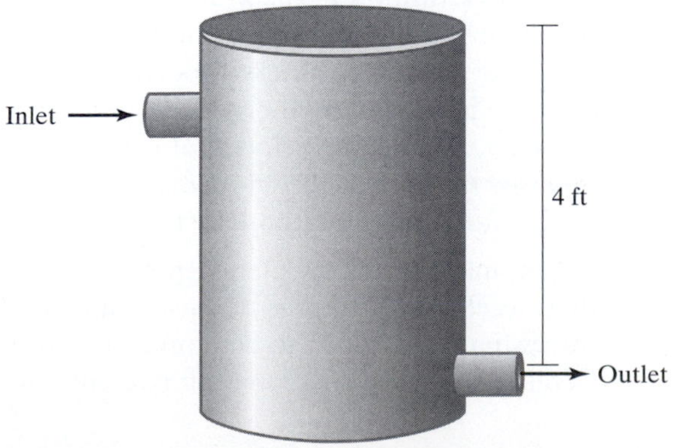

Figure 6.45
Distilled water in a pressurized tank.

(a) Assume that Dasher and Dancer collide and bounce off each other with no energy lost to heat or deformation (i.e., a perfectly elastic collision). What are their speeds after the collision?
(b) Occasionally, reindeer antlers entangle during fights. If Dasher and Dancer's antlers entangle, what is their velocity just after the collision?
(c) Assuming again that their antlers entangle, at what velocity should Dancer be traveling before the collision if their velocity just after the collision is zero?

6.23 An increased concern for automobile safety has led to changes in the design of vehicles. For example, cars are now designed to crumple and absorb energy rather than having the full force of a car crash affect the driver.

Unfamiliar with the city layout, Deborah turns onto a one-way street heading the wrong way. She collides head-on with Charles, who is moving at a velocity of 20 mph at the time of impact. Deborah's initial velocity was 12 mph. Charles' vehicle has a mass of 1500 kg and Deborah's vehicle has a mass of 2100 kg. The coefficient of restitution for the collision is 0.4. What are the velocities of the cars immediately after the collision?

6.24 Daniel is riding a 20-lb_m bicycle at 10 mph heading directly north. Victoria is riding a 30-lb_m bicycle at 7 mph heading 30° east of due north. Daniel has a mass of 150 lb_m, and Victoria has a mass of 100 lb_m. Through some unfortunate mishap, Daniel and Victoria collide. Treat their collision as completely inelastic (i.e., the coefficient of restitution is practically 0). What will be their initial velocity after the collision? If the coefficient of restitution for the collision in the x-direction is 0.2 instead of 0, calculate the velocity of each rider. Assume a smooth and oblique collision. Ignore any external forces such as gravity.

6.25 In Example 6.10, the purpose of the experiment was to determine whether or not epinephrine induces platelet adhesion. The example problem calculated the velocity of the platelets when the epinephrine was successful in causing the platelets to stick together. If the platelets do not stick together, their velocities and directions after the collision are different. Calculate the velocity of each platelet if the epinephrine does not cause platelet adhesion. Assume that the platelets have a completely elastic, oblique collision. Neglect water resistance and gravity.

6.26 Two erythrocytes (red blood cells) collide in a venule after exiting separate capillaries. The wall of one capillary is 135° from the wall of the venule, and the wall of the other capillary is 150° from the wall of the venule, as shown in Figure 6.46. Assume that the erythrocytes can be modeled as hard and smooth cylinders, and that they collide obliquely on their curved faces with a coefficient of restitution in the x-direction of 0.8. Assume that the erythrocytes are traveling parallel to the walls of their respective capillaries just before they collide. The density of an erythrocyte is 1.093 g/mL, and the volume of an erythrocyte is 86 μm^3. If the velocity of each erythrocyte is 0.05 cm/s just before collision, what is the velocity of each after collision? Neglect any effects from the flow of plasma. (From Altman PL and Dittmer DS, eds., *Blood and Other Body Fluids*, Washington DC: FASEB, 1961, pp. 110–111.)

6.27 When an erythrocyte enters a junction of blood vessels, it experiences forces that accelerate it further downstream. The magnitude and direction of these forces may be difficult to determine due to the complex flow patterns of blood. Given a vessel intersection with two entering streams of radius r and velocities

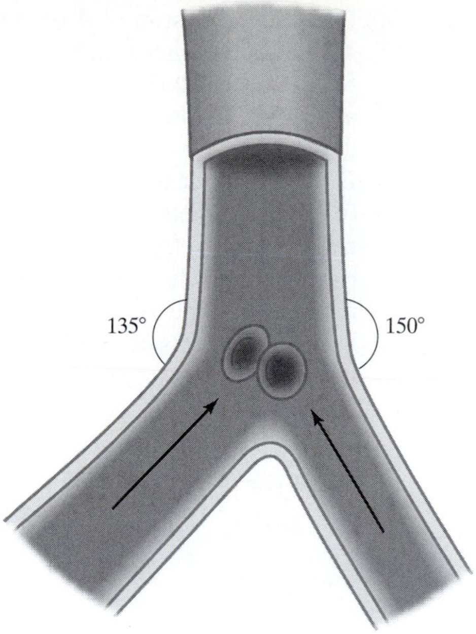

Figure 6.46
Two erythrocytes colliding in a venule after exiting separate capillaries.

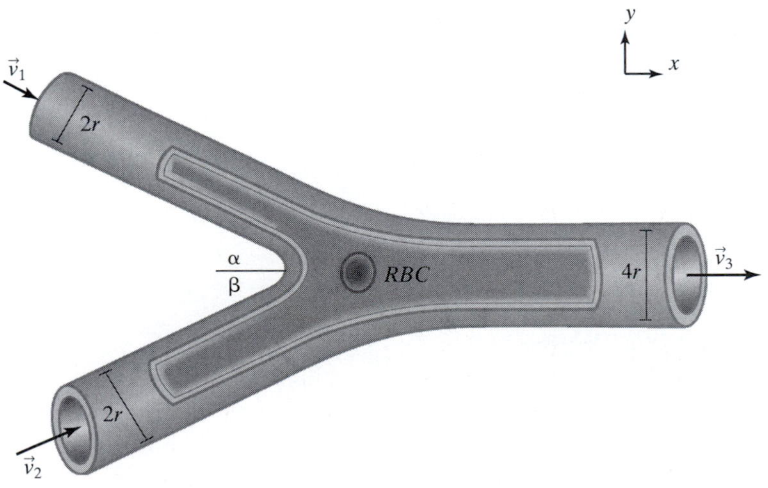

Figure 6.47
Erythrocyte at blood vessel junction.

$\vec{v}_1$ and $\vec{v}_2$, respectively, and one outgoing stream with radius $2r$ and velocity $\vec{v}_3$ (Figure 6.47), determine the magnitude of the force of the fluid surrounding the cell at the intersection of the two streams in terms of the magnitudes of the velocities (v_1, v_2, v_3), r, the resultant forces on the vessel ($F_{R,x}$, $F_{R,y}$), and ρ. The angle α is 30° and β is 60°.

6.28 Two blood vessels join to form a larger vessel. The first inlet vessel has a diameter of 0.5 cm and a blood velocity of 100 cm/s. The second inlet vessel has a diameter of 0.75 cm and a blood velocity of 100 cm/s. The outlet vessel has a diameter of 1 cm. Assume that the density of blood is 1.0 g/cm^3. Assume that the system is at steady-state.

(a) Determine the sum of the external forces $\left(\Sigma\vec{F}\right)$ in the x- and y-directions acting on the system in Figure 6.48a. (The force term that you are solving for describes all the external forces, including pressure forces, gravitational forces, resultant forces, and others.) Report your answer in units of $\text{g}\cdot\text{cm/s}^2$.

(b) During a surgical procedure, your needle accidentally pokes a hole with a 0.75-cm diameter in the blood vessel in Figure 6.48b. You measure the outlet velocity from the hole at 30 cm/s. Assume that the pressure of the two inlet vessels is 800 mmHg. Assume that the pressure at the hole is atmospheric. The resultant force in the x-direction is 340,000 $\text{g}\cdot\text{cm/s}^2$. Ignore the effects of gravity. Determine the pressure of the outlet vessel (labeled as stream 4 or *Out*).

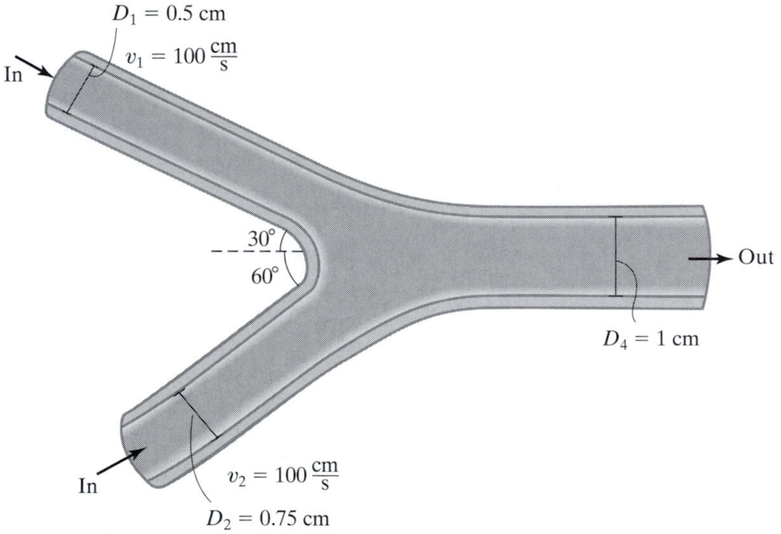

Figure 6.48a
Two blood vessels joining to form larger vessel.

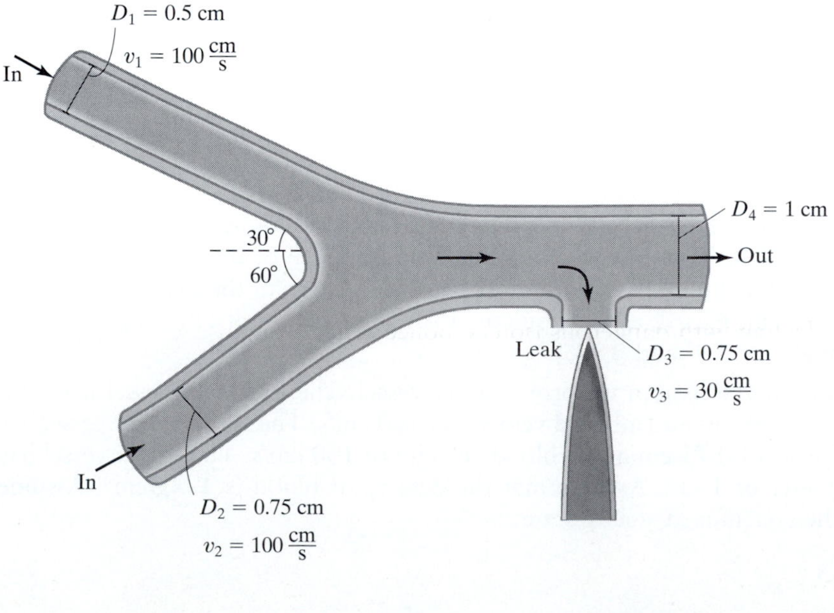

Figure 6.48b
Vessel configuration after needle puncture.

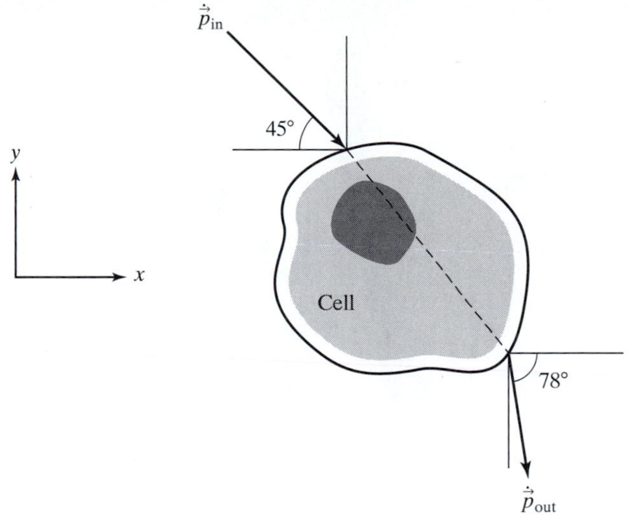

Figure 6.49
Forces exerted on a cell by a laser beam.

6.29 Although small, forces and mass flow are associated with coughing. Momentum is transferred out of the body as a stream of air is expelled from the lungs. Consider the system of the body for the duration of the cough. Is this system open, closed, or isolated? What forces are present? Is the system steady-state or dynamic? Write a general momentum accounting equation for this occurrence, and discuss the relative magnitudes of the terms.

6.30 Optical tweezers are a tool that use focused laser light to manipulate microscopic objects. Because cells have an optical density different than water, light bends as it passes through them. This results in a momentum change and an applied force. Unlike other manipulation methods, there is no risk of contamination, because the tool merely consists of photons carrying momentum. Determine the forces exerted by a typical laser beam (Figure 6.49) with a diameter of 1 μm, power of 500 mW, and a wavelength of 1060 nm. The beam enters at a 45° angle from the horizontal and exits at a 78° angle from the horizontal.

(a) Using the equation $p_{\text{photon}} = h/\lambda$, determine p_{photon} for a single photon passing through a cell. (Planck's constant, h, is 6.626×10^{-34} J · s.)

(b) Determine the number of photons passing through the optical tweezers beam every second, N, given the equation:

$$P = Nhf$$

where P is the power of the beam, h is Planck's constant, and f is frequency.

(c) Calculate the constant force exerted by the laser on the cell. (*Hint:* The force exerted on the cell is opposite to the force that would be required to hold the cell in place.)

6.31 When dealing with dangerous rioters, police officers might use a very powerful water hose to control them. Suppose that a riot hose is mounted on a vehicle and is being fired at a rioter who has obtained a riot shield (Figure 6.50). Water exits the hose at 150 gal/min. The speed of the water stream is 100 ft /s. The rioter holds his riot shield against the water stream so that water is deflected off the shield at a 90° angle in all directions, in equal quantity and speed. How much force is required by the rioter to keep the shield in place? Ignore the weight of the shield.

Figure 6.50
Water from a hose fired at a rioter holding a riot shield perpendicular to the stream of water.

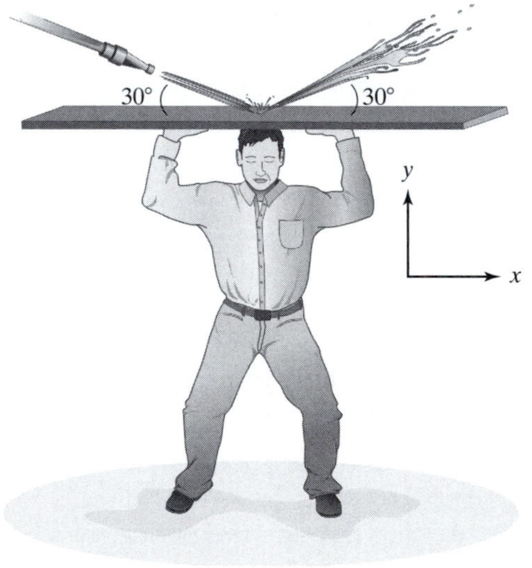

Figure 6.51
Water from a hose fired at a rioter holding a riot shield parallel to the ground above his head.

6.32 Refer to Problem 6.31 in which a riot hose fires a stream of water at a riot shield. Suppose now that the rioter holds the riot shield parallel to the ground above his head (Figure 6.51). The riot shield has a mass of 5 kg. The hose fires its water stream, at the same mass flow rate and velocity given in Problem 6.31, toward the shield at a 30° angle relative to the ground. Assume that all of the water bounces off the shield and leaves at an angle of 150° relative to the ground. (Remember that the angle of incidence is equal to the angle of reflection.) What force must the rioter use to keep the shield in place? Compare the answer to that of Problem 6.31.

6.33 The renal tubules in your kidney are not straight. Rather, they are convoluted and bend around one another. This design allows for liquid to be filtered across a great linear distance of tubule that is packed into a small volume

(Figure 6.52a). Consider the 1-mm-long segment of distal tubule modeled in Figure 6.52b. The diameter of the distal tubule is 20 μm. The velocity of filtrate entering the segment is 420 cm/min. The specific gravity of the filtrate is assumed to be 1.02. Assuming steady-state conditions, what force must the body exert on this segment of distal tubule to keep it stationary? Report your answer in units of dynes.

6.34 Squids are aquatic creatures that use jet propulsion for motion.[5] Taking up water in their main cavity, they eject it out of a siphon, which can be oriented in many different directions. A 0.20-kg squid has a cavity that holds 68 mL of water. When the squid needs to move, the cavity volume decreases by 40% and the expelled water allows the squid to attain a maximum velocity of 1.25 m/s. Assume that the squid shoots water out of its siphon at 6.7 g/s at a velocity of 3.8 m/s. What is the magnitude of the external force that would be required to keep the squid from accelerating?

6.35 Modern fighter jets can easily travel at twice the speed of sound. Thus, the dangers of high acceleration rates are applicable to pilots as well as astronauts (see Example 6.15). When coming out of a dive, fighter pilots can experience up to 9*g* of force pushing upward on them. Suppose the fighter pilot has a mass of 200 lb_m. What vertical force does the plane exert on him if he experiences 9*g*?

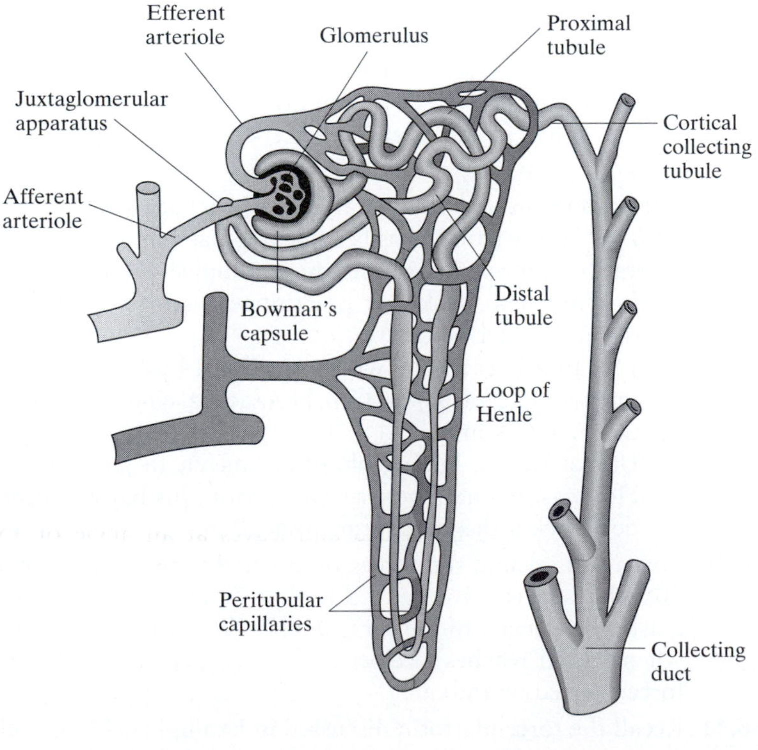

Figure 6.52a
Convolutions of renal tubules in the kidney.
(*Source*: Guyton AC and Hall JE, *Textbook of Medical Physiology*. Philadelphia: Saunders, 2000.)

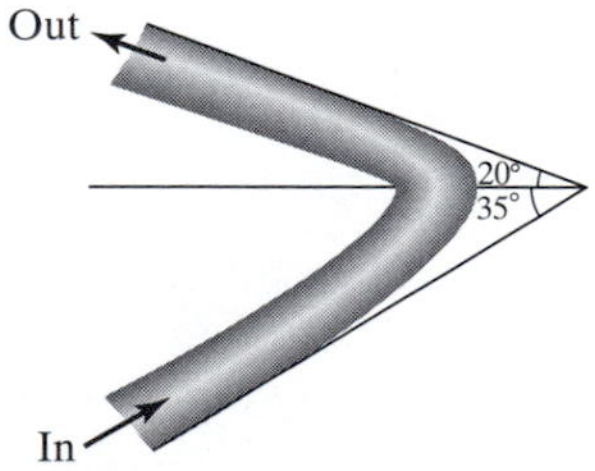

Figure 6.52b
Model of fluid flow in distal tubule.

6.36 Air resistance (the resistance to motion in air) is often ignored when modeling systems on Earth. However, when objects are falling at high velocities through the atmosphere, air resistance cannot be ignored. You convince your 60-kg friend to jump out of a plane and take some measurements on herself.

(a) When she first jumps out of the plane, her velocity is so small that air resistance is negligible. At this point, she falls with an acceleration of 9.81 m/s^2. Sketch a graph of force from air resistance versus acceleration that includes the time period from when your friend leaves the plane until she reaches a constant terminal velocity.

(b) After about 12 s, she reached a constant terminal velocity. (She can tell this because her accelerometer now reads zero.) What is the force from air resistance at this time?

6.37 Consider the astronaut in Example 6.15. In reality, the acceleration of the spacecraft is not constant. Acceleration during firing of the first booster, which is ignited for approximately 1.5 s, can be approximated using the equation $a = 3.1t^2 + 2$, where t is time in seconds and a is acceleration in g's ($1g = 9.8$ m/s^2), directed upward. Find the force that the craft exerts on the astronaut as a function of time.

6.38 The Helios Gene Gun is designed to insert DNA plasmids into cells for gene therapy applications. The DNA plasmids form a coating layer around a gold particle, which is then fired into a cell, pushed by a burst of helium gas. In Example 1.12, a gold particle covered with DNA with a mass of 8.09×10^{-11} g moving at 1100 mph is calculated to have a linear momentum of 3.98×10^{-11} kg·m/s. The helium burst can accelerate the particle from rest to its final speed in 3.4 μs. What is the force that the helium burst puts on a gold particle? Once the particle leaves the gene gun and enters the body, it is no longer accelerated by the helium. Instead, material between the surface and the target cell slows the particle down by collisions and frictional forces. Assume that the velocity of the gold particle when it reaches the target cell is zero. The time that it takes to travel to the target cell is 4.1 μs. What is the average force experienced by the particle inside the body?

6.39 The eastern martial art of Karate-do (a Japanese word, which literally translates to "empty hand") has become popular across the world. Karate uses concentration, power, and physics to accomplish seemingly superhuman feats. A world-renowned black belt practitioner uses karate to crack thick wooden boards and cinder blocks.

(a) He can deliver a chop with a speed of 14 m/s with his 1-kg hand during an impact time of 5 ms. Can he crack a cinder block that can withstand 2.5×10^6 N/m^2?

(b) Unlike karate, boxing blows are meant to jar, not break, the opponent. How fast would a boxer have to move his hand in order to break the cinder block if the impact time was 20 ms?

6.40 You are modeling the forces on a car during a crash. In your simulation, a 2000-lb$_m$ car is driven into a wall at 20 mph. Assume that at the end of the crash, the velocity of the car is 0 mph. From the moment of impact to the time when the car reaches a velocity of zero, a period of 0.37 s elapses. What is the force exerted on the car?

6.41 Recall the force platform discussed in Example 6.16. The electronic recording from a normal jump is shown in Figure 6.26b. In this test, the person pushes off the platform (takeoff period of 0.25 s), is airborne for a short time (hangtime period of 0.3 s), and then lands on the platform in a crouched position (landing period of 0.25 s). Calculate the landing velocity of the person.

6.42 Rockets rely on conservation of momentum to propel them with great force through space. The largest rocket ever constructed was the Saturn V that had a mass of 3×10^6 kg and propelled the first manned mission to the moon. During liftoff, when mission control ignited the engines, fuel was ignited at a rate of 13.84×10^3 kg/s and expelled with an exhaust velocity of 4300 m/s relative to the craft. This exerted tremendous force on the craft and astronauts. If 46% of the Saturn V was payload and the rest was booster fuel weight, what acceleration did the craft and astronauts experience at burnout? Assume at burnout the velocity of the craft was 6700 km/hr.

6.43 You are drinking water with a 20-cm straw. The last 3 cm of the straw is bent over to reach your mouth and therefore do not contribute to the height of the straw. The water flows into your mouth at approximately 0.050 m/s. Consider the velocity of the water at the submerged end liquid in place in a continer that of the straw to be zero. What is the pressure difference between the ends of the straws?

6.44 One way to remove fluid from a container in the absence of a pump is to use a siphon. A siphon is simply a tube positioned such that one end is submerged in liquid the container and the other end is placed in a container that is lower than the first. Once a flow is established, the fluid is drawn continuously through the siphon and the container empties.

A container is filled with water to a height of 0.75 m. The base of the container is also elevated 1 m above the ground. A siphon with a diameter of 1 cm is used to remove the water from the container. Assuming the free end of the tube lies on the ground, use the Bernoulli equation to calculate the volumetric flow rate through the siphon.

6.45 The flow of urine from the bladder, through the urethra, and out of the body, is induced by increased pressure in the bladder resulting from muscle contractions around the bladder with simultaneous relaxation of the muscles in the urethra. The mean pressure in the bladder can be estimated using the velocity of urine as it exits the body. Assume that the bladder is about 5 cm above the external urethral orifice. (This height is different for males and females.) The flow rate of urine from the bladder can be approximately described with the following equations, where t is time in seconds, and $\dot{V}$ is flow rate in mL/s:

$$\dot{V} = -0.306 \times (t - 7)^2 + 15 \qquad 0 \leq t \leq 12$$
$$\dot{V} = -3 \times \sqrt[3]{t - 12} + 7.35 \qquad 12 \leq t \leq 26.7$$

(a) How much total urine volume is excreted during this time period?
(b) Develop equations for the velocity of urine as it exits the body. Assume that the urethra is 5.6 mm in diameter.
(c) Develop an equation for the mean pressure in the bladder as a function of time. (*Note*: In reality, the muscles around the bladder rapidly contract and relax, causing quick variations in bladder pressure.)

6.46 Physicians often use IV lines to administer fluids and drugs rapidly. The rate of application of the IV fluid is determined to some extent by the height of the IV bag above the patient. Assume that the IV line enters the patient through her venous system. The gauge pressure in that vein is 80 mmHg, which is equivalent to an absolute pressure of 112 kPa. Assume that the flow in the line has a uniform velocity profile.

(a) If an IV bag contains 0.5 L and drains through a 0.2-cm diameter line, what is the correct positioning of the bag such that the entire bag is emptied in 10 min?

(b) Design an IV system with a rack to hold an IV bag, a line connecting the IV bag and the patient, and a patient. The system must be able to deliver a 0.5 L IV bag over a time range of 10 min to 8 hr. It is practical to have tubing with two or three different diameters at your disposal (but no more). (*Hint*: Make sure that you understand the impact of the kinetic energy term in the Bernoulli equation.) Your solution must be realistic and able to be implemented.

6.47 Venturi meters are used to measure flow rates of turbulent fluids in small pipes. The design of a venturi meter is such that the fluid velocity increases and the pressure decreases across the constriction when a pipe is constricted. By measuring the height difference h of the manometer fluid between positions 1 and 2 in Figure 6.53, the velocity of the process fluid before the constriction can be determined.

The sections of pipe before and at the constriction are connected with a U-tube manometer filled with liquid of density ρ_M. The process fluid flowing through the pipe has density ρ_F. Before constriction, the cross-sectional area of the pipe is A_1 and the velocity of the fluid is v_1. After the constriction, the cross-sectional area of the pipe is A_2, and the velocity of the fluid is v_2.

Write an equation for the velocity, v_1, as a function of ρ_M, ρ_F, h, g (gravitational constant), A_1, and A_2. Assume that the pressure along position 1 is P_1 and that the pressure along position 2 is P_2. Assume no other forces act on the system. The manometer fluid is static.

6.48 The functional unit of the kidney is the nephron; there are approximately one million nephrons in a kidney. Blood is filtered in the glomerulus and then travels through a system comprised of several tubules and ducts.

Calculate the Reynolds number for each structure given in Table 6.4. The density of the filtrate is approximately that of plasma, which is 1.02 g/mL.

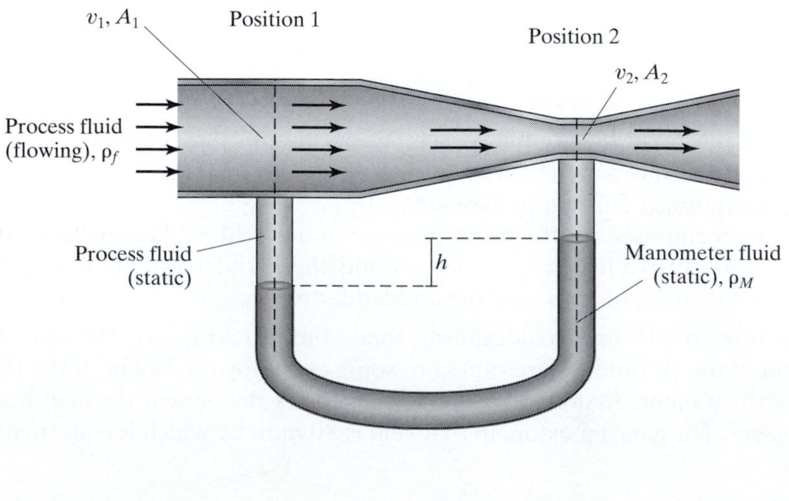

Figure 6.53
Venturi meter with manometer fluid.

TABLE 6.4

Diameters and Flow Rates for Structures in the Kidneys*

Structure	Diameter (μm)	Range of total flow rates (mL/min)
Proximal tubule	30	24–125
Loop of Henle	12	17–24
Distal tubule	20	7–17
Collecting duct	100	1–7

*Data from Cooney DO, *Biomedical Engineering Principles: An Introduction to Fluid, Heat, and Mass Transport Processes*. New York: Marcel Dekker, 1976.

TABLE 6.5

Diameters and Blood Velocities of Vessels in the Human Circulatory System *

Structure	Diameter (cm)	Blood velocity (cm/s)
Aorta	2.0	45
Main arterial branches	0.3	23
Arterioles	0.002	0.3
Capillaries	0.0008	0.07
Venules	0.003	0.1
Main venous branches	0.5	15
Venae cavae	2.0	11

*Data from Cooney DO, *Biomedical Engineering Principles: An Introduction to Fluid, Heat, and Mass Transport Processes*, New York: Marcel Dekker, 1976; and Guyton AC and Hall JE, *Textbook of Medical Physiology*, Philadelphia: Saunders, 2000.

Assume that the viscosity of the filtrate is approximately equivalent to water (1.793×10^{-3} kg/(m · s)). Determine whether the flow through each structure is laminar or turbulent.

6.49 In the human circulatory system, large vessels split into two (bifurcate) or more smaller vessels in progression from the aorta to the arterioles and finally the capillaries. In returning blood to the heart, the capillaries join to form venules and then finally the venae cavae. The diameter and blood velocity are given in Table 6.5 for each type of blood vessel. The viscosity of blood is 0.035 poise. Is the flow through each of these vessels laminar, turbulent, or in transition?

6.50 The water pumped into Puget Sound from the metropolitan area of Seattle must first be cleansed from impurities accumulated during its time in human service. One water treatment plant processes about 133 million gallons of wastewater per day. The steps of wastewater treatment are described in Example 3.13.

(a) Two pipes bring the local wastewater to the water treatment plant. One has a diameter of 144 in, the other a diameter of 88 in. Assume the water is equally distributed between these two pipes. What is the Reynolds number for the flow in each pipe?

(b) Assume that the water treatment facility is 2 miles away from the final discharge location of Puget Sound. To calculate the friction loss per mass flow rate in a smooth pipe, use the following equation:

$$\frac{\dot{f}}{\dot{m}} = 0.005v^2\frac{L}{r}$$

where v is the velocity of the fluid in the pipe, L is the length of the pipe, and r is the radius of the pipe. (From Bird RB, Stewart WE, and Lightfoot EN, *Transport Phenomena*. New York: John Wiley, 2002.) The diameter of the discharge pipe is 144 in. Calculate the friction loss in the pipe.

(c) The pipeline from the wastewater treatment plant to Puget Sound contains a 200-hp pump. Assume that the water treatment facility is at sea level. What is the maximum height that the pipeline can rise above sea level?

Case Studies

CHAPTER 7

Three case studies designed to bridge and integrate the different conserved properties of mass, momentum, charge, and energy are presented in this chapter. Case Study A focuses on modeling the human lungs and the design of an artificial heart-lung machine. Case Study B focuses on the human heart and the design of a total artificial heart. Case Study C focuses on modeling the human kidneys and the design of a dialysis machine. These examples have physical phenomena at the cellular, tissue, and whole-body levels.

Each case study presents some background physiological information about the system. Two or three worked examples are also presented in the text for each case study. Many open-ended, modeling, and design problems are provided at the end of each case study. Problems are identified as requiring knowledge in the areas of mass (M), energy (E), charge (C), momentum (P), or general (G). The case studies synthesize and reinforce material presented in Chapters 1–6 and provide more complete real-world examples of engineering in biology and medicine.

Case Study A

Breathe Easy: The Human Lungs

The major function of the lungs is to continually exchange gases between the blood from the body and the air outside the body during respiration. In the lungs, blood gains oxygen for transport to other tissues, which require oxygen. Aerobic metabolism requires oxygen to break down food to obtain energy for cellular activities like protein synthesis, muscle contraction, and DNA replication. In addition, the lungs remove carbon dioxide, the waste product of cellular metabolism and a chemical component important for maintaining the acid-base balance in the blood.

Each breath begins with movement of the diaphragm (Figure 7A.1). During inspiration, the diaphragm contracts, causing the downward movement of the lower surfaces of the lungs. This increase in volume in the lungs causes a drop in pressure, creating a pressure difference between the interior of the lungs and the surrounding air. This pressure difference pulls air down into the lungs to equalize the pressure gradient, a process known as negative-pressure breathing. Air enters the pulmonary system through either the nose or mouth or both and then proceeds to the trachea. From here, the trachea bifurcates (splits in two) to form a pair of bronchi. The bronchi continue to branch into smaller vessels known as bronchioles. Each time a vessel splits, it forms a new generation. Each bronchiole continues to split into two

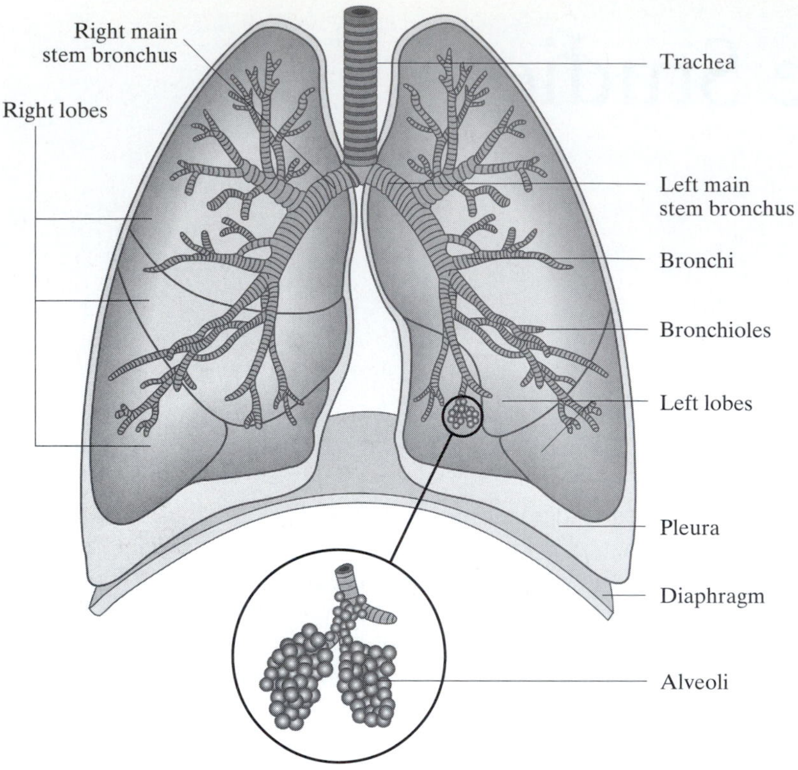

Figure 7A.1
The human lungs and diaphragm.

or three vessels, forming a branching bronchial tree, before terminating at the alveoli. In these millions of alveolar sacs, oxygen and carbon dioxide are exchanged between the lungs and the blood. This repeated bifurcation of vessels creates a large surface area ($\sim 70\ m^2$) over which gas exchange occurs.

During expiration, the diaphragm relaxes, and aided by the chest wall and abdominal structures, the elastic recoil of the lungs compresses the lungs, decreasing the volume and increasing the pressure to force air out. Under resting conditions, a normal breath usually contains 500 mL of air, known as the tidal volume. The breathing rate for healthy adults is usually 12 breaths per minute.

Small bronchial arteries originating from the aorta in the systemic circulatory system supply oxygenated blood for respiration to the lung tissues, including the connective tissue, septa, and large and small bronchi. The pulmonary artery, which carries oxygen-poor blood, travels from the heart to the lungs and then branches into smaller and smaller vessels until the blood reaches the pulmonary capillaries, where the exchange of oxygen and carbon dioxide occurs. Carbon dioxide diffuses from the pulmonary blood across the capillary and alveolar walls and into the alveolar air sacs. Oxygen diffuses from the alveoli into the capillaries to reoxygenate the oxygen-depleted pulmonary blood. The reoxygenated blood returns to the heart through the left atrium for distribution to the body organs and tissues.

EXAMPLE 7A.1 **Friction Losses in the Lungs**

Problem: Blood enters the lungs from the pulmonary artery with a mean pressure of 15 mmHg. After passage through the lungs, the blood returns to the left atrium through the pulmonary vein with a mean pressure of 2 mmHg. Estimate the total frictional loss ($\dot{f}$) in the lungs. Give examples of events in pulmonary circulation that can contribute to frictional loss. The extended Bernoulli equation can be modified for an incompressible fluid with no potential energy change or nonflow work:

$$\frac{1}{2\alpha}\dot{m}(v_1^2 - v_2^2) + \dot{m}\left(\frac{P_1 - P_2}{\rho}\right) - \dot{f} = 0$$

where α equals 0.5 for laminar flow.

Solution:

1. Assemble
   (a) Find: total frictional loss in the lungs.
   (b) Diagram: Figure 7A.2 shows the system.
2. Analyze
   (a) Assume:
   - No change in height of vessels occurs (i.e., no potential energy change).
   - No nonflow work is done on the system.
   - All vessels can be modeled as cylinderical pipes.
   - Blood is an incompressible fluid.

   (b) Extra data:
   - Cardiac output is 5 L/min.
   - The diameters of the pulmonary artery and pulmonary vein are 2.5 cm and 3.0 cm, respectively.

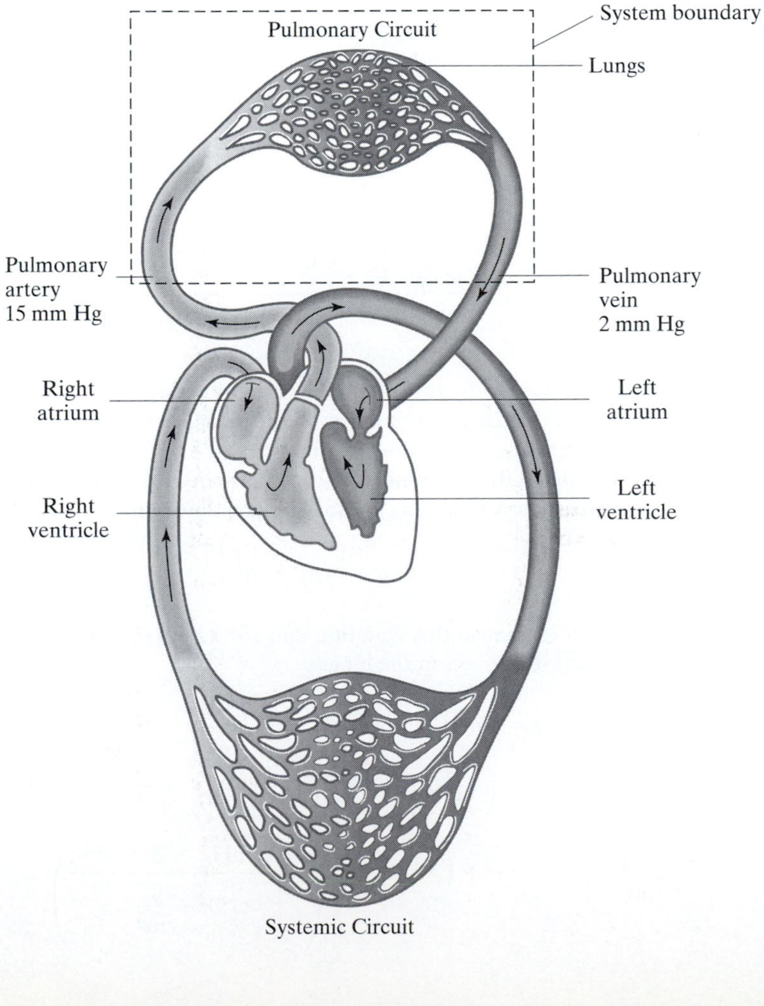

Figure 7A.2
Pressures and direction of blood flow in the circulatory system.

- The density of whole blood is 1.056 g/cm^3.
- The viscosity of whole blood is 3.0 cP = 0.03 g/(cm · s).

(c) Variables, notations, units:
- PA = pulmonary artery
- PV = pulmonary vein
- Use L, min, cm, g, J, mmHg.

(d) Basis: Using the density of blood and cardiac output, we can find the mass flow rate of blood, which we use as a basis:

$$\dot{m} = \rho\dot{V} = \left(1.056\,\frac{\text{g}}{\text{cm}^3}\right)\left(5\,\frac{\text{L}}{\text{min}}\right)\left(1000\,\frac{\text{cm}^3}{1\text{ L}}\right) = 5280\,\frac{\text{g}}{\text{min}}$$

3. Calculate

(a) Equations: We can use the given Bernoulli equation to calculate the frictional losses for an incompressible fluid with laminar flow. To verify that the blood flows in the pulmonary vein and artery are both laminar, we can use the Reynolds number equation [6.10-1]:

$$\text{Re} = \frac{\rho v D}{\mu}$$

We are not given the velocities of blood flow in these vessels, but we can find them using equation [3.2-4]:

$$\dot{V} = Av = \frac{\dot{m}}{\rho}$$

(b) Calculate:
- To be able to use the given Bernoulli equation, we first need to verify that the blood flows in the pulmonary vessels are laminar. For the pulmonary vein:

$$v_{\text{PV}} = \frac{\dot{V}}{A_{\text{PV}}} = \frac{\dot{V}}{\frac{\pi}{4}D^2} = \frac{5\frac{\text{L}}{\text{min}}}{\frac{\pi}{4}(3.0\text{ cm})^2}\left(\frac{1000\text{ cm}^3}{1\text{ L}}\right) = 707\,\frac{\text{cm}}{\text{min}}$$

$$\text{Re}_{\text{PV}} = \frac{\rho v_{\text{PV}} D_{\text{PV}}}{\mu} = \frac{\left(1.056\frac{\text{g}}{\text{cm}^3}\right)\left(707\frac{\text{cm}}{\text{min}}\right)(3.0\text{ cm})}{\left(0.03\frac{\text{g}}{\text{cm}\cdot\text{s}}\right)\left(\frac{60\text{ s}}{1\text{ min}}\right)} = 1244$$

Calculating the Reynolds number for the pulmonary artery in the same manner gives a value of 1490. Both of these Reynolds numbers fall within the range for laminar flow.

- Since blood flow is laminar, we can simplify the given extended Bernoulli equation by substituting α as 0.5 for laminar flow, which yields:

$$\dot{m}(v_1^2 - v_2^2) + \dot{m}\left(\frac{P_1 - P_2}{\rho}\right) - \dot{f} = 0$$

We can rearrange this equation and substitute the numerical values to solve for the frictional losses in the lungs:

$$\dot{f} = \dot{m}(v_1^2 - v_2^2) + \dot{m}\left(\frac{P_1 - P_2}{\rho}\right)$$

$$\dot{f} = \left(5280\frac{\text{g}}{\text{min}}\right)\left(\left(1018\frac{\text{cm}}{\text{min}}\right)^2 - \left(707\frac{\text{cm}}{\text{min}}\right)^2\right)$$

$$+ \left(5280\frac{\text{g}}{\text{min}}\right)\left(\frac{15\text{ mmHg} - 2\text{ mmHg}}{1.056\frac{\text{g}}{\text{cm}^3}}\right)$$

$$\times \left(\frac{101{,}325 \text{ Pa}}{760 \text{ mmHg}}\right)\left(\frac{\frac{\text{kg}}{\text{m}\cdot\text{s}^2}}{\text{Pa}}\right)\left(\frac{1000 \text{ g}}{\text{kg}}\right)\left(\frac{1 \text{ m}}{100 \text{ cm}}\right)\left(\frac{60 \text{ s}}{\text{min}}\right)^2$$

$$\dot{f} = \left(3.09 \times 10^{11}\frac{\text{g}\cdot\text{cm}^2}{\text{min}^3}\right)\left(\frac{1 \text{ kg}}{1000 \text{ g}}\right)\left(\frac{1 \text{ m}}{100 \text{ cm}}\right)^2\left(\frac{1 \text{ min}}{60 \text{ s}}\right)^3$$

$$\dot{f} = 0.143 \frac{\text{J}}{\text{S}}$$

4. Finalize
   (a) Answer: The blood vessels undergo contractions, expansions, bends, and branches that contribute to frictional losses in the lungs. The total frictional loss over the lungs is estimated as 0.14 J/s.
   (b) Check: In Examples 6.21 and 6.22, the total frictional loss for the entire circulatory system is estimated as 1.15 J/s. It is reasonable that the friction loss through the lungs is lower than this value. ■

The air we breathe is approximately 79% nitrogen and 21% oxygen, with trace amounts of carbon dioxide and water vapor, which are all transported throughout the entire respiratory system. The partial pressure of the gases in the respiratory system is directly proportional to the molar concentration of the gas molecules. The partial pressures of nitrogen, oxygen, and carbon dioxide in normal air at 23°C are 597 mmHg, 159 mmHg, and 0.3 mmHg, respectively. Saturated water vapor pressure at 23°C is 21.1 mmHg. Saturated water vapor pressure at 37°C is 47 mmHg.

Oxygen and carbon dioxide diffuse across the alveolar membranes during respiration (Figure 7A.3). Just as a pressure differential drives air into the lungs, a partial pressure gradient across the alveolar membranes separating the alveolar space from the blood drives gas exchange. For diffusion of oxygen to occur, the oxygen pressure gradient across the alveolar membranes must exceed a threshold of 34 mmHg. On the

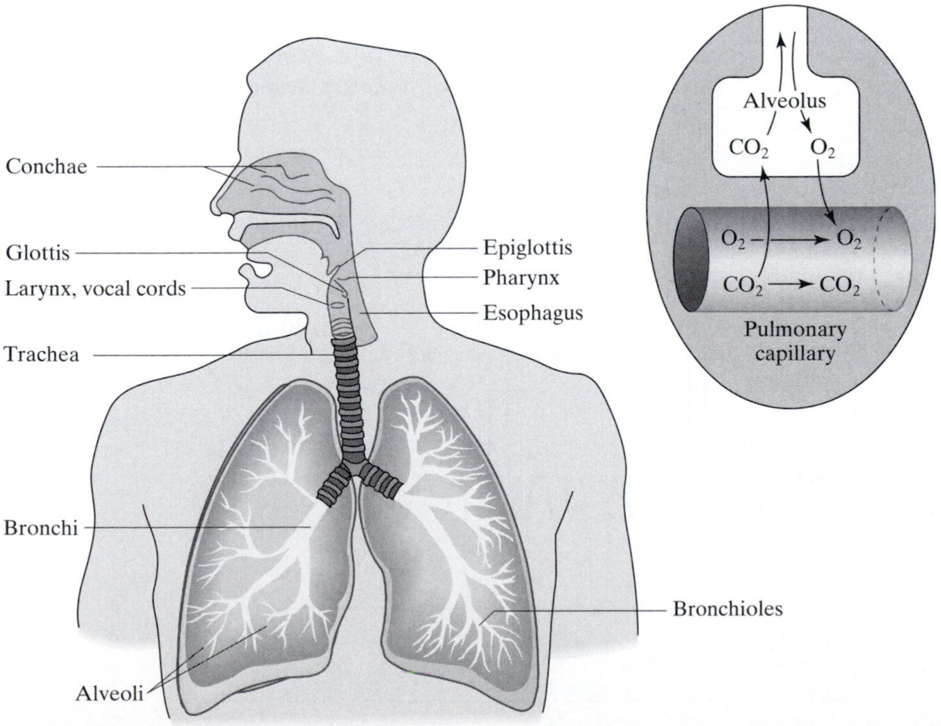

Figure 7A.3
Oxygen and carbon dioxide exchange across the alveolar membranes during respiration. (Modified from: Guyton AC and Hall JE, *Textbook of Medical Physiology*, 10th ed. Philadelphia; WB Saunders, 2000.)

other hand, a partial pressure difference of only 1.0 mmHg is sufficient to initiate the diffusion of carbon dioxide across the alveolar membranes. Diffusion of oxygen into the blood at biological temperature and pressure (BTP), which is at 37°C and 1 atm, occurs at an average rate of 284 mL O_2/min. Carbon dioxide is removed from the blood at a rate of 227 mL CO_2/min at BTP.

The human lungs can be modeled with varying levels of complexity. Part II in the Problems at the end of this section involves developing a multicompartment model of the human lungs.

Gas exchange across the alveolar membranes is a continuous process. A volume of gas (approximately 2300 mL), similar in composition to atmospheric air and known as the functional residual capacity, remains in the terminal airways after each normal breath to ensure constant exposure of the alveolar membranes to oxygen-rich air. If the breathing rate changes suddenly (e.g., during periods of increased physical activity), this residual air provides a constant source of air in the alveoli. Total lung capacity, the maximum volume the lungs can hold, is about 5.8 L (Figure 7A.4).

During open-chest surgeries (e.g., coronary bypass), the surgical field must be still and bloodless, so sometimes the heart must be stopped or clamped down to keep it from beating. Additionally, the gas exchange in the alveoli in the lungs is often bypassed during cardiac surgeries, so a gas exchanger must mimic their function. In these cases, a cardiopulmonary bypass (CPB) machine, also known as a heart-lung machine, may be used to replace the heart's function and to exchange gas in place of the lungs (Figure 7A.5). A mechanical pump creates a steady blood flow rate equal to or slightly less than the patient's cardiac output, sending oxygen-poor blood from the venae cavae to the oxygenating device and oxygen-rich blood back to the aorta. A small amount of blood or isotonic solution is required to initially prime the machine to prevent any disruption of blood flow.

The first successful heart-lung machine was created in the mid-1950s by John Gibbon. Since then, different gas-exchanger designs have been developed (Figure 7A.6). The first design, the bubble oxygenator, directly bubbled oxygen through the oxygen-depleted blood. However, it required blood to be filtered to satisfactorily remove any gas bubbles before returning the blood to the patient. Gas bubbles in a blood vessel can cause an embolus (an obstruction to normal blood circulation), which can lead to thrombosis (blood clot formation). Another model was the film

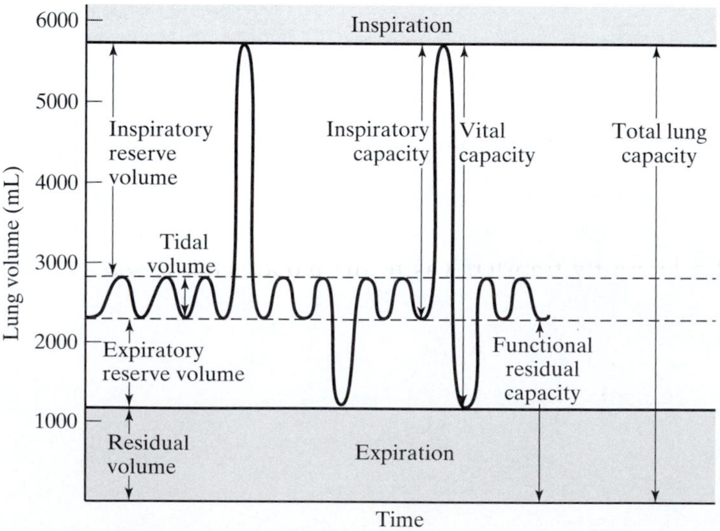

Figure 7A.4
Lung capacities during normal breathing, maximal inspiration, and maximal expiration. (*Source*: Guyton AC and Hall JE, *Textbook of Medical Physiology*. Philadelphia: Saunders, 2000.)

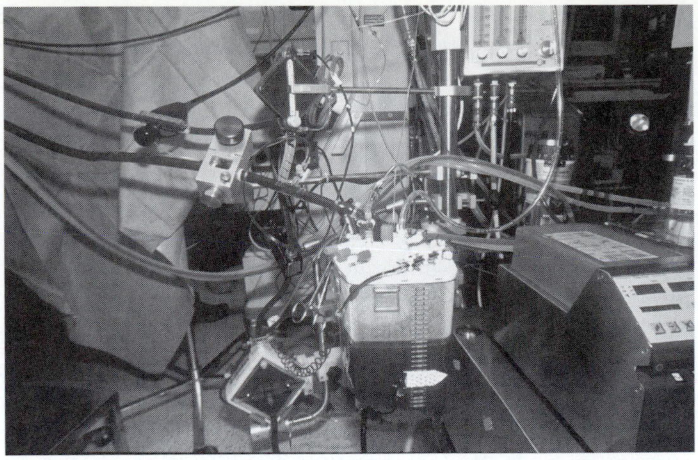

Figure 7A.5
Cardiopulmonary bypass machine.

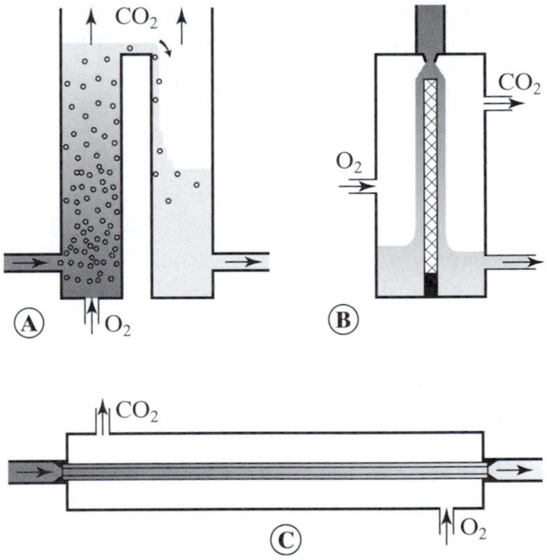

Figure 7A.6
Three basic classes of artificial lungs: (a) bubble, (b) film, and (c) membrane. (*Source*: Cooney DO, *Biomedical Engineering Principles: An Introduction to Fluid, Heat, and Mass Transport Processes*. New York: Marcel Dekker, 1976. Figure originally from Galletti PM and Brecher GA, *Heart-Lung Bypass*. New York: Grune and Stratton, 1962.)

oxygenator, which exposed a thin film of blood to an oxygen-rich atmosphere. However, a film that was not thin enough could not properly equilibrate with the gas phase surrounding it. Rotating the blood thinned out the film and increased gastransfer efficiency, but was prone to damaging the blood cells.

Today, the membrane oxygenator is the most widely used model of gas exchangers. Just as the alveolar membrane separates the blood from oxygen, the membrane oxygenator also indirectly transfers gas to the blood, since it was found in earlier designs that blood in direct contact with gas caused blood trauma (e.g., protein denaturation, hemolysis, bubble formation, fibrin deposition). The gas exchanger of a membrane oxygenator imitates the alveoli in two important ways to allow high rates of gas diffusion: (a) large surface area and (b) permeable membrane wall. With a larger surface area, more gas can be exchanged. Compared with the 70 m^2 available in the natural lungs, membrane oxygenators only have surface areas ranging from 2 m^2 to 10 m^2. However, the heart-lung machine compensates for the lower exchange area with a larger oxygen gradient and a longer blood transit time.

The permeable membrane wall allows gas transfer to occur with lower chances of hemolysis. Since the synthetic membrane is much thicker than that in the lungs, and therefore less permeable than the lungs, the gas exchanger operates at a higher partial pressure of oxygen for diffusion into the blood. Synthetic membranes favor carbon dioxide diffusion over oxygen, so which synthetic membrane to use typically depends on the clinically necessary transfer rate of carbon dioxide.

The main task of the heart-lung machine is to keep oxygen-rich blood flowing through the patient's body after the heart has been stopped. To further reduce risk of thrombosis, a patient undergoing cardiac surgery may receive heparin, a powerful anticoagulant that thins the blood. The patient is then hooked up to the heart-lung machine. To stop the heart from pumping, the heart is either clamped down or treated with a cardioplegia solution, usually filled with potassium to stop electrical activity. By cooling the blood through the CPB, which effectively lowers the body temperature, the induced hypothermia decreases demand for oxygen from other organs in the body. Reducing such a demand on the heart muscles preserves the myocardial cells for up to six hours, allowing ample time for surgery.[1] When the surgical procedure is complete, the heart is restarted, and a heat exchanger in the machine increases blood temperature, restoring normal body temperature before the patient is weaned from the CPB. Part IV in the Problems at the end of this section focuses on the design criteria of a CPB machine.

EXAMPLE 7A.2 **Energy Losses in Cooling Blood**

Problem: To induce hypothermic conditions, most artificial heart-lung machines use a cooler to help lower and maintain a target blood temperature. Blood initially enters the cooler at body temperature (37°C) and then cycles through the cooler at the rate of cardiac output (5 L/min) until the blood reaches the target temperature. For an average adult, the cooler removes heat from the blood until it reaches about 30°C. If the temperature difference between the blood entering and exiting the cooler is initially 5°C, calculate the initial rate of heat removal from the blood. Assume the average-sized adult has a blood volume of about 5 L. Blood density is 1.056 g/cm^3, and the heat capacity of whole blood is 3.740 J/(g · °C).

Solution:

1. Assemble
   (a) Find: initial rate of heat removal to cool a patient's blood.
   (b) Diagram: The temperature of the blood flowing in is T_1, and the temperature of the blood flowing out is T_2 (Figure 7A.7). The heat removed from the system is $\dot{Q}$.
2. Analyze
   (a) Assume:
   - The system is at steady-state (i.e., no energy accumulates in the cooler).
   - No heat is lost in the tubing or other parts of the bypass machine.
   - No changes in potential or kinetic energy in system.
   - No nonflow work occurs.
   - The heat capacity of blood equals the heat capacity of water. (This is not a good assumption, but is necessary for simplification!)

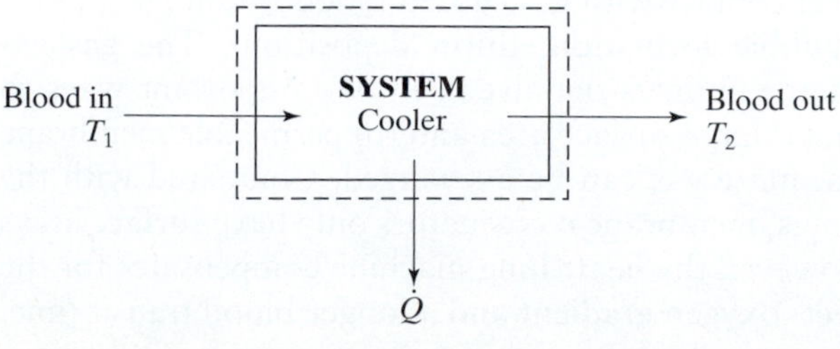

Figure 7A.7
Schematic diagram of temperature differences and heat removal in a blood cooling system.

(b) Extra data: No extra data are needed.
(c) Variables, notations, units:
- Use L, min, °C, g, J.

(d) Basis: The mass flow rate of blood through the cooler is equal to cardiac output. Using the density of blood, cardiac output is equivalent to a mass flow rate of:

$$\dot{m} = \rho\dot{V} = \left(1.056\ \frac{\text{g}}{\text{cm}^3}\right)\left(5\ \frac{\text{L}}{\text{min}}\right)\left(1000\ \frac{\text{cm}^3}{\text{L}}\right) = 5280\ \frac{\text{g}}{\text{min}}$$

which can be used as the basis.

3. Calculate
(a) Equations: Because we are solving for heat, which is a type of energy, we need to use the conservation equation for total energy. We are given values in terms of rates, so we can use the differential form [4.3-10]:

$$\sum_i \dot{m}_i(\hat{E}_{P,i} + \hat{E}_{K,i} + \hat{H}_i) - \sum_j \dot{m}_j(\hat{E}_{P,j} + \hat{E}_{K,j} + \hat{H}_j) + \sum \dot{Q} + \sum \dot{W}_{\text{nonflow}} = \frac{dE_T^{\text{sys}}}{dt}$$

We can find the change in enthalpy of blood using equation [4.5-20]:

$$\Delta\hat{H} = C_p\ (T_2 - T_1)$$

(b) Calculate:
- Since we assume no changes in potential or kinetic energy in the system and no nonflow work, these terms become zero. Also, the steady-state assumption means no energy accumulates in the cooler, so dE_T^{sys}/dt also becomes zero. This simplifies the differential energy conservation equation to:

$$\sum_i \dot{m}_i\hat{H}_i - \sum_j \dot{m}_j\hat{H}_j + \sum \dot{Q} = 0$$

$\Delta\hat{H}$ is equal to the difference between the specific enthalpies of the inlet and outlet streams:

$$\dot{m}C_p(T_1 - T_2) + \dot{Q} = 0$$

- Rearranging the equation and substituting the basis mass flow rate, we can calculate the initial rate of heat removal. We know the initial temperature difference is 5°C, so:

$$\dot{Q} = -\dot{m}C_p(T_1 - T_2) = -\left(\frac{5280\text{ g}}{\text{min}}\right)\left(\frac{3.740\text{ J}}{\text{g}\cdot{}^\circ\text{C}}\right)(5^\circ\text{C})\left(\frac{1\text{ kJ}}{1000\text{ J}}\right) = -98.7\ \frac{\text{kJ}}{\text{min}}$$

4. Finalize
(a) Answer: The initial rate of heating is −98.7 kJ/min. Thus, heat removal is 98.7 kJ/min.
(b) Check: Since the blood decreases in temperature as it is cooled, it makes sense that the rate of heat is negative (i.e., a loss). This model is a gross approximation of the actual cooling machine, which does not have a constant rate of heat removal or a constant temperature difference between the entering and exiting blood streams. As soon as the blood is cooled and cycled from the machine back into the body, the cooled blood mixes and begins to cool the surrounding blood in the body before the blood again reaches the cooling machine. Thus, the entering blood temperature (T_1) starts at 37°C and gradually decreases to 30°C, while the exiting blood temperature (T_2) drops from 32°C to 30°C. Since this is the case, the temperature difference ($T_1 - T_2$) is not a constant 5°C. ■

Even with technically sound devices, health and safety concerns associated with heart-lung machines must also be considered during their design and use. The

Food and Drug Administration, the Department of Health and Human Services, and various other agencies regulate how individual parts are made and how the machine is constructed. Minimizing leaks, foreign particles, blood clots, and air bubbles anywhere in the machine or tubing is necessary, as such contaminants can cause the patient to lose blood, go into shock, or suffer a stroke. To maintain proper lung function, the gas exchanger must meet strict requirements. Furthermore, the pump must maintain a regular flow rate, as well as pressure comparable to that of the patient's bloodstream.

An ideal heart-lung bypass machine should meet certain design criteria. It should provide a highly permeable surface to facilitate gas exchange. Since open-chest surgery can last anywhere between 20 minutes and several hours, the machine must maintain high levels of hemoglobin-saturated blood (95–100%) to carry oxygen to the body at regular cardiac output (5 L/min). The device must simultaneously remove enough carbon dioxide (an approximate P_{CO_2} of about 40 mmHg) to maintain a proper physiological pH. Gentle handling of the blood is necessary to avoid lysing blood cells or denaturing proteins, problems that can lead to clotting. Materials in the pump and gas exchanger must be biocompatible to reduce the probability of a negative immune reaction. Sterilization is crucial to patient safety, so sterile tubing is used once and then thrown away. Such safety concerns are just as important as the technological parameters involved in the design and implementation of a heart-lung machine.

References

1. Baldwin JC, Elefteriades JA, and Kopf GS. "Heart Surgery." BL Zaret, M Moser, and LS Cohen, eds. *Yale University School of Medicine Heart Book*. New York: Hearst Books, 1992.

Problems

Part I—Pulmonary Air Flow

7A.1 (M) Sketch a diagram of the trachea and lungs. Do a balance on the inhaled air in generations 0–3 (0 represents the trachea) of the lungs to determine the linear velocity of the air in generation 3. Assume there are no external forces acting to change the direction of the air. Estimate the linear velocity in the bronchioles and alveoli using values in Table 7A.1.

7A.2 (M, P) Determine the Reynolds number in the listed generations in Table 7A.1. Compare the order-of-magnitude changes in diameter, velocity, and Reynolds number across the generations.

TABLE 7A.1

Path Dimensions Through Generations in the Lungs

Generation	Name	Diameter (cm)	Number in lungs
0	Trachea	1.8	1
1		1.2	2
2		0.8	4
3		0.6	8
6		0.32	115
12	Bronchioles	0.08	8,000
18		0.05	500,000
24	Alveoli	0.02	300,000,000

7A.3 (E) The respiratory muscles must perform work on the lungs in order for the breathing process to occur. The work of inspiration can be divided into three categories: (1) compliance work, or the work to expand the lungs against the lung and chest elastic forces; (2) tissue resistance work, or the work required to overcome the viscosity of the lungs and chest wall structures; and (3) airway resistance work, or the work needed to overcome the resistance of movement of air into the lungs. Using the pressure-volume graph in Figure 7A.8, calculate the different types of work required to inspire 0.5 L of air at atmospheric pressure.

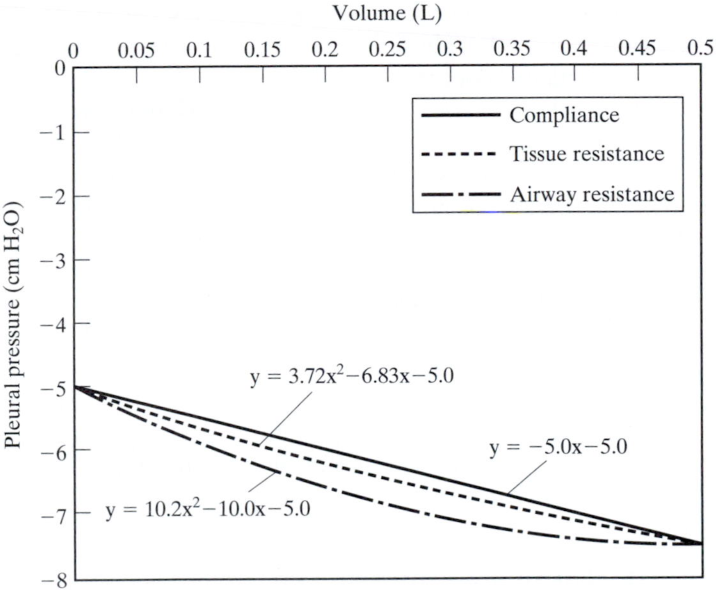

Figure 7A.8
Pressure versus volume for three different categories of the work of inspiration.

7A.4 (G) What is a surfactant? What is the role of surfactant in the lungs in regulating the surface tension of water in the alveolar space? If surfactant is not present or is present but in less than normal quantities, then medical problems (such as collapse of the alveoli) may occur. Explain how the absence of, or reduced levels of, surfactant affect surface tension and pressure in the alveoli.

7A.5 (E) The surface tension of normal fluids with normal amounts of surfactant that line the alveoli is 5–30 dynes/cm. The surface tension of normal fluids without surfactant that line the alveoli is 50 dynes/cm. Surface tension is related to the pressure, P, as follows:

$$P = \frac{2\sigma}{r}$$

where σ is the surface tension and r is the radius of the alveoli. Premature babies usually have alveoli with radii one-quarter that of normal adults. In addition, since surfactant does not usually begin to be secreted in the alveoli until the sixth month of gestation, premature babies usually do not have surfactant. Calculate the work required to inflate a premature baby's lungs by modifying Figure 7A.8. Assume that the surface tension is 25 dynes/cm with normal levels of surfactant and the capacity of the baby's lungs is 8 mL.

7A.6 (M) Consider gas exchange at the alveolar level. Draw a model of the alveoli that shows the interface between the alveoli and the capillary. Assume that alveolar air is 15.4 mol% oxygen. Determine how much oxygen is available for exchange to the capillary in units of moles per minute. It can be assumed that since there is such a large volume of air in the alveoli, the concentration of each gas in the alveolar air is constant. When a body is at rest, venous blood contains 14.4 mL O_2 bound/dL blood with a P_{O_2} of 40 mmHg. The exiting arterial blood contains 19.4 mL O_2 bound/dL blood with a P_{O_2} of 100 mmHg. Calculate the amount of oxygen a body needs at rest in units of moles per minute. Compare the amount of oxygen available for exchange and the amount needed by the body. Why is there a difference?

7A.7 (C) Oxygen concentration is relatively unimportant with respect to direct control of the respiratory center. In fact, carbon dioxide levels in the blood play a major role in controlling respiratory rates.

(a) Explain this physiological process of respiratory control. Where does the regulation occur?

(b) Carbon dioxide in the blood reacts with water to form carbonic acid (H_2CO_3). The carbonic acid then dissociates into hydrogen and bicarbonate ions. Thus, when the blood carbon dioxide concentration increases, more hydrogen ions are released.

$$CO_2 + H_2O \rightleftharpoons H_2CO_3 \rightleftharpoons H^+ + HCO_3^-$$

The K_a for the second reaction is 4.3×10^{-7} M. Henry's coefficient for CO_2 is 17.575 mmHg/μM. Assume that the first reaction goes to completion. Also assume that all of the H^+ and HCO_3^- in the system are solely from the dissociation of carbonic acid. If the blood initially has a P_{CO_2} of 40 mmHg, what is the pH? If the pH is then reduced by 0.1 units, what is the final P_{CO_2}? The pH of blood is normally 7.4. Why are the pH values calculated here different from the pH of blood?

Part II—Modeling the Lungs

7A.8 (M) Design and draw a multicompartment model of the human lungs. Your model must include at least three compartments and may include more. Describe the compartments. Justify your choice of compartments and related assumptions using physiological arguments. What is the volume of each compartment?

7A.9 (M) Your goal is to develop a physiologically realistic model of the gas transport and exchange in the human lungs. Because of the complexity of this task, the computational strategy suggested below is to develop a simple model and then to relax key assumptions. For each step, complete mass balances on the important gases in the air in each compartment of the lungs. Report the partial pressure, volume, and percent composition of each gas in each compartment for one full respiratory cycle (inhalation and exhalation). Comment on the extent of mixing within each compartment.

All Cases

Assume the following characteristics for the ambient air: relative humidity, 17.6%; temperature, 23°C; pressure, 1.0 atm; partial pressures of N_2, 597 mmHg; partial pressure of O_2, 159 mmHg; partial pressure of CO_2, 0.3 mmHg. Reaction/exchange rates of CO_2 and O_2 in the alveoli must be included.

Case I

Assume that all inspired air is transferred to the alveolar space (i.e., there is no dead space). Assume that the lungs are always at atmospheric pressure (during inhalation and exhalation). Assume the system is steady-state.

Case II

Relax the assumption that the lungs are always at atmospheric pressure. Instead, model the lungs with realistic inhalation and exhalation pressures. How do these changes affect the composition of alveolar and exhaled air? Assume the system is steady-state.

Case III

Take into account the dead space in the trachea and lungs. Why are the compositions of alveolar air and exhaled air different? How much of an effect does the dead space air have on the composition of the exhaled air? Assume the system is steady-state. Use *Case II* (i.e., realistic pressures) as a starting point for this calculation.

Case IV

Breathing is really a continuous process, with a constant flow of air into and out of the lungs. Given that inhalation and exhalation times are different, calculate the rate of flow of important gases during inhalation and exhalation. Use *Case III* (i.e., realistic pressures and accounting for dead space) as a starting point for this calculation.

7A.10 (M) During breathing, how constant or steady are the partial pressures of oxygen and carbon dioxide in the alveoli? Does your model reflect this? Explain.

7A.11 (M) How could your model be improved? In other words, what changes could be made to create a model that is more accurate or applicable to a wider range of situations?

Part III—Diseases of the Lungs

During the normal breathing cycle, a great deal of air remains in the lungs, called the residual volume. The residual and tidal volumes are important to consider in Problems 7A.12–7A.14.

7A.12 (M) Assume an allergen is in an aerosol form (1.0 g/(L air)) and can be inspired and expired at the same rate as normal air. After a sufficiently long exposure time, the equilibrium concentration of allergen in the lungs is 1.0 g/(L air). Write equations describing the loss of allergen from the lung for a normal lung, assuming no further exposure. How long will it take to decrease the allergen concentration in the alveoli to less than 1.0% of its inspired concentration (i.e., less than 0.010 g/(L air))? Develop a graph of the concentration of allergen in the lung as a function of time. Use an integral mass balance to check your answer.

7A.13 (M) Assume the allergen from Problem 7A.12 causes asthma that constricts the airways by 50%. Assume an allergen is in an aerosol form (1.0 (g/L air)) and can be inspired and expired at same rate as normal air and reaches an equilibrium concentration of allergen in the lungs of 1.0 g/(L air). Write equations describing the loss of allergen from the asthmatic lung, assuming no further exposure. (*Hint*: What happens to the tidal volume and the residual volume during an asthmatic attack?) How long will it take to decrease the allergen concentration in the alveoli to less than 1.0% of its inspired concentration (i.e., less than 0.010 g/(L air)? Develop a graph of the concentration of allergen in the lung as a function of time. Use an integral mass balance to check your answer.

7A.14 (M) A clean lung is exposed to the allergen at 1.0 g/(L air). Assuming no asthma is induced, write an equation describing the mass uptake of allergen into the lung. Determine the equilibrium concentration of the allergen in the lung and total mass of allergen in the lung. Calculate the time required to reach 50% and 99% of the equilibrium value. Use an integral mass balance to check your answer.

7A.15 (M) Discuss the similarities and differences between the equations developed for Problems 7A.12–7A.14.

7A.16 (G) What is the ventilation/perfusion ratio? Is it dependent on location in the lung? Why or why not?

7A.17 (G) Describe three pathologies of the lung. For each, explain the physiologic causes for reduced gas exchange. How is the ventilation/perfusion ratio affected by these pathologies?

7A.18 (G) Discuss two different lung imaging technologies and how they differ with respect to diagnosis of lung pathologies.

7A.19 (G) Describe spirometry. How are different pulmonary volumes measured using spirometry? What types of lung pathologies can be detected using spirometry?

Part IV—Heart-Lung Bypass System

Figure 7A.9 is a diagram of a heart-lung bypass machine. Refer to it for questions in Part IV.

7A.20 (E) Estimate the energy expended by the venous/arterial pump to move the blood through the system. Assume energy lost in interaction with the tissue and membrane oxygenator is 62 J/min. Assume that the blood pressure entering the body tissue is that found in the aorta and the blood pressure leaving the body tissue is that found in the vena cavae. Assume that the tubing is the same size throughout the system.

7A.21 (G) What salt(s) is contained in the cardioplegia line? How does it arrest the heart?

7A.22 (M, E) Note that there are two heater/cooler units (indicated by "hot/cold water") in the system. They are most often operated to cool the blood and the cardioplegia lines.

(a) Typically, the patient's blood needs to be cooled very quickly (in less than 5 min). Describe the parameters of the heater/cooler to achieve this, including temperature of circulating water, rate of heat transfer between the water and blood, and flow rate of circulating water. For your design, estimate the time to cool an average adult's blood to 28–30°C. This calculation should be more involved than Example 7A.2.

(b) To operate most efficiently, the flow lines in the system need to be primed with ~750 mL (otherwise, too much blood is outside of the body during the procedure). A solution of crystalloid (dextrose and Ringer lactate) or donated blood must be added to the system lines. What is the temperature of the crystalloid or donated blood that you add? How does this affect the time needed to cool the patient's blood to 28–30°C?

(c) The temperature of the cardioplegia line must also be 28–30°C. Describe the parameters of the heater/cooler to achieve this, including temperature of circulating water, rate of heat transfer between the water and blood, and flow rate of circulating water. Assume the cardioplegia packs are refrigerated prior to surgery.

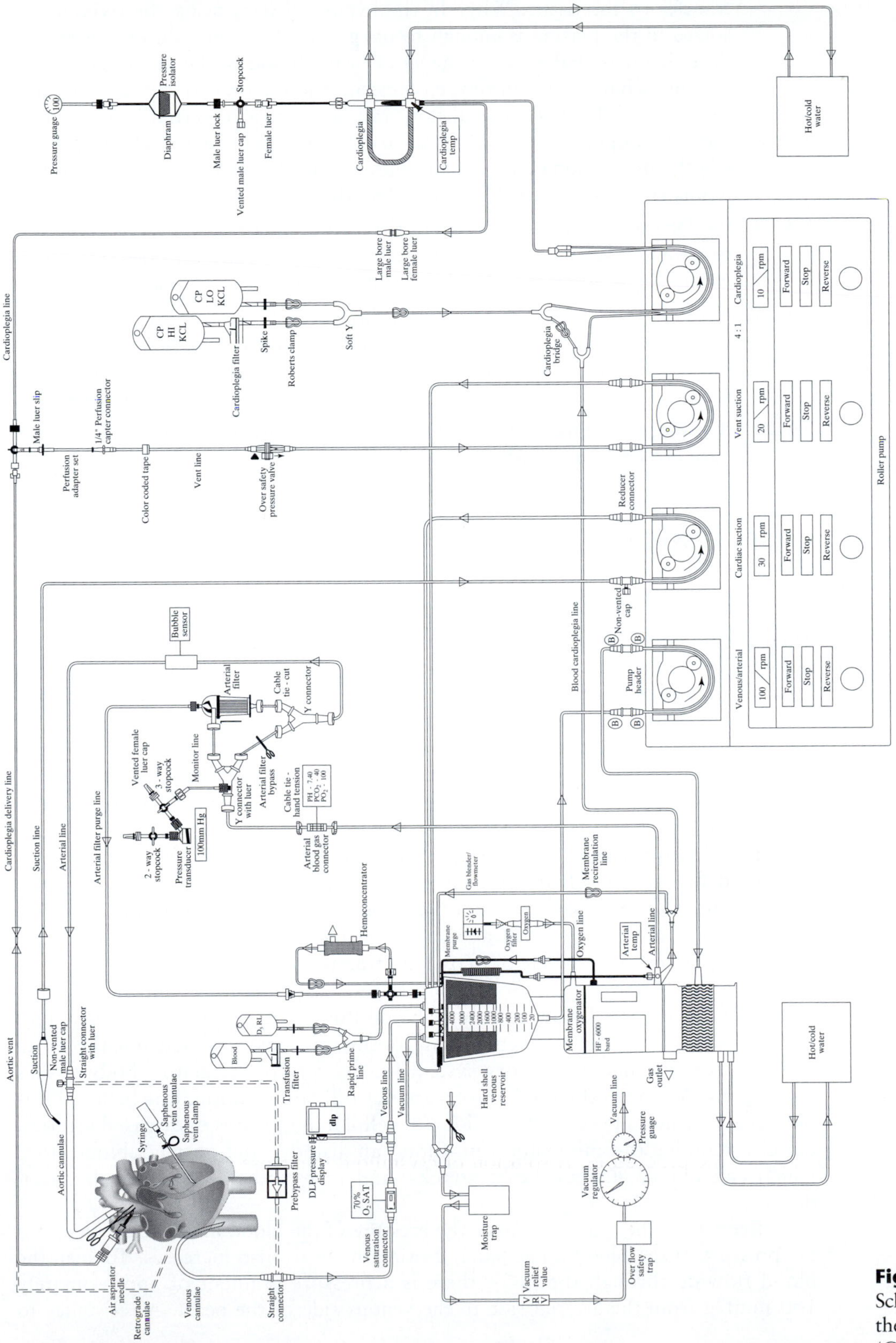

Figure 7A.9
Schematic flow diagram of the heart-lung machine. (Courtesy of Terry Crain.)

7A.23 (M) At this lower temperature, hemoglobin is not very useful in oxygen transfer to the tissue. Why? In the bypass system, using the oxygen dissolved in the plasma is enough to oxygenate the tissues during the procedure. Do mass balances on the gases in the blood in the main lines of the system. What change in oxygen concentration is needed in the membrane oxygenator? Estimate the airflow rate and oxygen partial pressure that must be supplied to the membrane oxygenator to make this change. (*Hint*: Check information on Q_{10} of enzymes in a biochemistry book. Take the solubility of O_2 to be 0.0023 mL/(mL blood).

7A.24 (G) Describe the general design of the membrane oxygenators used in these procedures. What important characteristics of the alveoli does this design mimic? How efficient is the transfer of oxygen into the blood?

7A.25 (G) Other than those mentioned in the text, list two important properties for the materials used in the pumps and two important properties for materials used in the membrane oxygenator. Explain. Would metals or polymeric materials (plastics) be more appropriate for these two applications?

7A.26 (G) What are three safety concerns for the heart-lung bypass device that have not been previously discussed?

Case Study B

Keeping the Beat: The Human Heart

The human heart (Figure 7B.1) is the vital organ that drives the circulatory system, pumping blood throughout the body's intricate capillary beds and tissue spaces. The heart is responsible for pumping oxygen-deficient blood through the pulmonic system (lungs) to exchange gases associated with cellular respiration, and for pumping oxygen-rich blood throughout the systemic circulatory system for delivery of necessary nutrients, gases, and waste products. How the heart pumps blood throughout its chambers and the body is a complex process controlled by muscular contractions timed by electrical signals and pressure gradients.

Blood flows along a certain pathway in the heart with each heartbeat. Oxygen-deficient blood from the systemic circulatory system collects in the inferior and superior venae cavae. From the venae cavae, the blood begins to fill the right atrium. Contraction of the atrium pushes blood through the tricuspid valve into the right ventricle, where a second contraction pumps the blood to the pulmonic system. After oxygenation, blood flows out of the lungs and fills the left atrium. When the left atrium contracts, blood flows through the mitral valve into the left ventricle, where the ventricular contraction pushes blood out the aorta and into the systemic circulatory system for delivery of oxygen to body tissues and organs. Although the mechanisms are not fully understood, it is believed this pulsatile flow helps prevent the buildup of cellular aggregates in diseased arteries, thereby avoiding thromboembolism. Clots can dislodge and obstruct blood flow, which can induce strokes, heart attacks, and many other cardiovascular accidents.

Every time the heart contracts, the pressure of the blood in the heart rises, and the pressure of the blood immediately exiting the heart also increases. To keep the blood flowing through the body, there is a pressure gradient of approximately 100 mmHg from the arterial side to the venous side of the heart. A resistance to

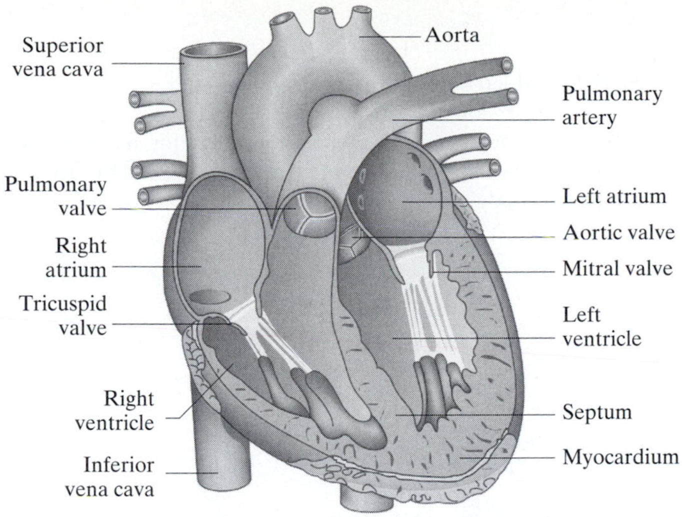

Figure 7B.1
The human heart.

blood flow, or peripheral resistance, is energy in the form of friction that is lost as blood comes in contact with the vessel walls during circulation.

The unique "lub-dub" sounds heard in the heartbeat correspond to the diastolic and systolic phases. While these phases can be used to describe where the blood is in the individual chambers at a particular period in time, diastole usually refers to the time period in which both ventricles relax simultaneously, while systole refers to the time period in which both ventricles simultaneously contract and eject their blood volume out. Diastole is usually twice as long as systole. At the end of diastole, atrial contraction occurs. While the atria simultaneously relax, the ventricles contract in systole. The volume of blood ejected in one contraction is the stroke volume, and the volume ejected in one minute (obtained by multiplying the number of beats per minute by the stroke volume) is the cardiac output. The average cardiac output for a healthy individual is 5 L/min.

EXAMPLE 7B.1 **Work Done by the Heart**

Problem: Find the work done by the right side and the left side of the heart in one hour. Assume the cardiac output is 5 L/min and no energy is lost due to friction as the blood contacts the chamber walls. Table 7B.1 gives the pressures at the various entrances and exits of the heart. (Adapted from Cooney DO, *Biomedical Engineering Principles: An Introduction to Fluid, Heat, and Mass Transport Processes*, 1976.)

TABLE 7B.1

Pressure of Vessels at Juncture with Heart

Side	Location	Pressure (mmHg)
Right	Venae cavae	0
	Pulmonary artery	15
Left	Pulmonary vein	6
	Aorta	97

Solution:

1. Assemble
   (a) Find: work done by the right and left sides of the heart in one hour.
   (b) Diagram: The right side of the heart contains deoxygenated blood that is just completing systemic circulation and has not yet been sent to the lungs for reoxygenation (Figure 7B.2). The left side of the heart contains oxygenated blood returning from the lungs, ready to be sent out to the body's organs and tissues.

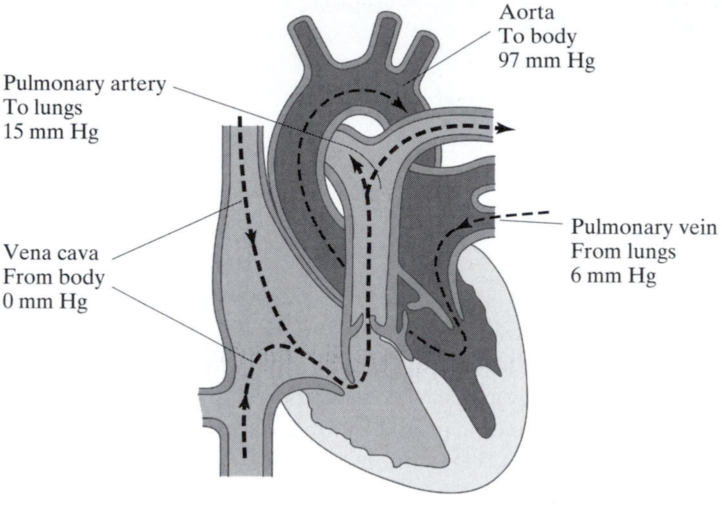

Figure 7B.2
Right and left sides of the heart system. Systemic side is shaded dark; pulmonic side is shaded light.

2. Analyze
   (a) Assume:
      - Only the heart does pumping work.
      - All blood vessels are at the same height relative to a fixed reference.
      - No energy loss due to friction or other causes.
      - Kinetic and internal energy changes across system are negligible.
      - Flow rate through each of four vessels listed is 5 L/min.
      - Density of blood is constant.
      - System is at steady-state.
   (b) Extra data:
      - The density of blood is 1.056 g/mL.
   (c) Variables, notations, units:
      - RS = right side of the heart
      - LS = left side of the heart
      - vc = vena cava
      - pa = pulmonary artery
      - pv = pulmonary vein
      - ao = aorta
      - Use J, L, mmHg, min.
   (d) Basis: We can use the given blood flow rate through the heart (5 L/min) as our basis:

$$\dot{m} = \rho\dot{V} = \left(\frac{1.056\text{ g}}{\text{mL}}\right)\left(\frac{5\text{ L}}{\text{min}}\right)\left(\frac{60\text{ min}}{\text{hr}}\right)\left(\frac{1000\text{ mL}}{\text{L}}\right) = 316{,}800\ \frac{\text{g}}{\text{hr}}$$

3. Calculate
   (a) Equations: Since the extended Bernoulli equation [6.11-9b] was used to solve Example 6.21, we use the differential energy conservation equation [4.3-5] for this problem:

$$\sum_i \dot{m}_i(\hat{E}_{P,i} + \hat{E}_{K,i} + \hat{U}_i) - \sum_j \dot{m}_j(\hat{E}_{P,j} + \hat{E}_{K,j} + \hat{U}_j) + \sum \dot{Q} + \sum \dot{W} = \frac{dE_T^{sys}}{dt}$$

   (b) Calculate:
   - Since the heart is at steady-state, total energy does not change over time; thus, the Accumulation term of the differential energy conservation equation becomes zero. Potential energy does not change in the system, so this term can be eliminated. While the kinetic energy of the blood does change across the system (since the velocity of blood changes), we can show this contribution is negligible, so this term is also eliminated (see Example 6.21). We assumed no energy loss due to friction or heat, so these terms become zero. Thus, the energy equation simplifies to:

$$\dot{W} = 0$$

     where $\dot{W}$ consists of both flow work and nonflow work, so:

$$\dot{W}_{\text{flow}} + \dot{W}_{\text{nonflow}} = 0$$

   - Flow work is the product of the pressure and specific volume, multiplied by the mass flow rate. Using the relationship between specific volume and density, we can substitute these variables for flow work:

$$\dot{W}_{\text{flow}} + \dot{W}_{\text{nonflow}} = \frac{\dot{m}}{\rho}(P_{\text{in}} - P_{\text{out}}) + \dot{W}_{\text{nonflow}} = 0$$

   - Using the given pressure differences, the blood density, and the basis mass flow rate, we can calculate the pumping (i.e., nonflow) work done by the right side of the heart ($\dot{W}_{\text{RS}}$), which pumps blood from the venae cavae ($P_{\text{vc}} = 0$ mmHg) to the pulmonary artery ($P_{\text{pa}} = 15$ mmHg):

$$\dot{W}_{\text{RS}} = \dot{W}_{\text{nonflow}} = \frac{-\dot{m}}{\rho}(P_{\text{vc}} - P_{\text{pa}}) = \left(\frac{-316{,}800\frac{\text{g}}{\text{hr}}}{1.056\frac{\text{g}}{\text{mL}}}\right)(0\text{ mmHg} - 15\text{ mmHg})$$

$$\dot{W}_{\text{RS}} = 4{,}500{,}000\frac{\text{mmHg}\cdot\text{mL}}{\text{hr}}\left(\frac{1.01325\times 10^5\text{ N}}{760\text{ mmHg}\cdot\text{m}^2}\right)\left(\frac{1\text{ m}^3}{1\times 10^6\text{ mL}}\right)(1\text{ hr})$$

$$\dot{W}_{\text{RS}} = 600\text{ N}\cdot\text{m} = 600\text{ J}$$

   - The same procedure is followed for the left side, which pumps blood from the pulmonary vein ($P_{\text{pv}} = 6$ mmHg) to the aorta ($P_{\text{ao}} = 97$ mmHg). The work in one hour equals 3640 J.

4. Finalize
   (a) Answer: In one hour, the right side of the heart performs 600 J of work and the left side 3640 J.
   (b) Check: The left side of the heart must perform much more work than the right, since the left side is responsible for pumping blood through the entire circulatory system. These number compare favorably with the work estimates in Example 6.21. The significantly greater work performed by the left side of the heart is the reason physicians frequently use left ventricular assist devices (LVADs) to alleviate the work requirements for patients who are awaiting a heart transplant. ■

Myocardial cells have a negative electrical resting potential, which means the charge inside the cell is negative compared to its surroundings. Periodically, this potential becomes positive and stimulates the cells to contract in a phenomenon called depolarization. During repolarization, the potential returns to its negative resting potential value, and the heart cells recover from the contraction. The transmission of the electrical impulse and the cycles of depolarization and repolarization are called an action potential. Myocardial cells possess automaticity, a property that causes them to fire automatically and periodically. However, in the normal heart, only one region demonstrates spontaneous electrical activity to synchronize the beating of the heart.

To correctly time the depolarization and repolarization of muscle tissue in the atria and ventricles, the heart's contractions are stimulated by an intrinsic electrical conduction system (Figure 7B.3). Specialized conductive fibers in the heart transmit the cardiac stimulus to all the myocardial cells in a specific order. The heart's pacemaker is called the sinoatrial (SA) node, located in the posterior wall of the right atrium near the opening of the superior vena cava. The electric signal travels quickly through myocardial gap junctions to depolarize and contract both atria, ejecting blood in the atria to

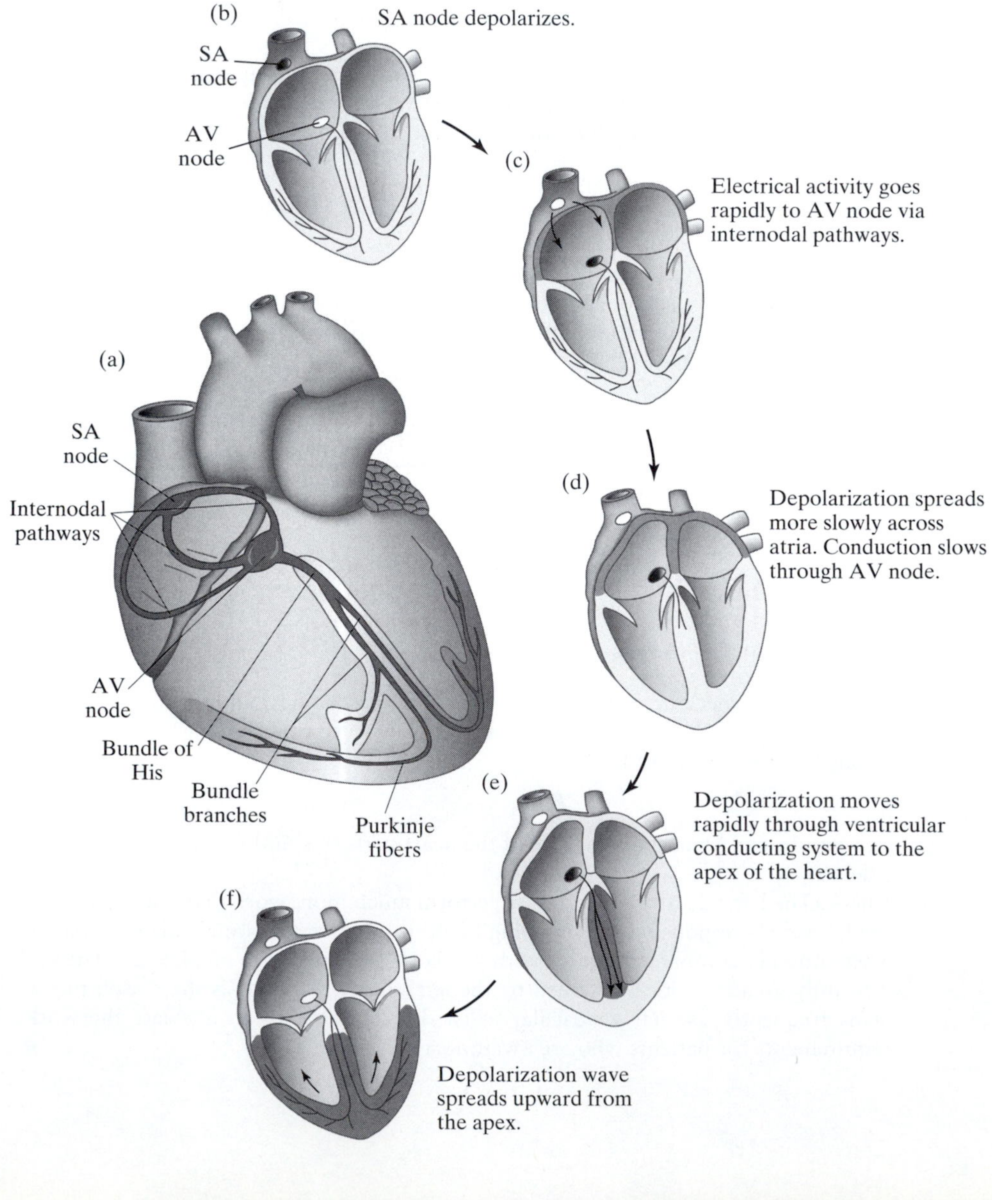
(b)
SA node depolarizes.
SA node
AV node
(c)
Electrical activity goes rapidly to AV node via internodal pathways.
(a)
SA node
Internodal pathways
AV node
Bundle of His
Bundle branches
Purkinje fibers
(d)
Depolarization spreads more slowly across atria. Conduction slows through AV node.
(e)
Depolarization moves rapidly through ventricular conducting system to the apex of the heart.
(f)
Depolarization wave spreads upward from the apex.

Figure 7B.3
Electrical conduction system of the human heart. (a) Parts of conduction system. (b)-(f) Process of depolarization. (Source: Silverthorn D.H *Human Physiology*, 2nd ed, Prentice Hall, 2001.)

the ventricles. The signal then slowly travels through the atria and septum wall to the atrioventricular (AV) node, located between the atria and the ventricles. This time delay allows the ventricles to fill up with blood. Once the electrical stimulus passes through the AV node, the impulse quickly continues down a conducting tissue called the Bundle of His, which divides into the left and right bundle branches and finally to the Purkinje fibers. Stimulating the Purkinje fibers causes simultaneous ventricular depolarization and contraction, allowing blood in the ventricles to be ejected to the pulmonary and systemic circulatory systems. If the SA node fails to fire a stimulus, the myocardial cells are capable of automatically initiating stimulation and functioning as a pacemaker. However, this usually results in a heart disorder, since action potential firing may not be coordinated.

EXAMPLE 7B.2 Charging and Discharging a Defibrillator

Problem: Ventricular fibrillation is a condition in which the individual muscle fibers in the ventricle do not act together (i.e., some are contracting and some are relaxing), inhibiting the ventricle from pumping blood efficiently. Without a quick remedy, a person suffering ventricular fibrillation will die. A biomedical device known as a defibrillator is used to send a large electrical shock through the heart, resynchronizing all of the muscle fibers. A defibrillator works by charging a capacitor using a battery and then releasing this buildup of charge into the body.

(a) Table 7B.2 tracks the process of charging the capacitor. Develop a mathematical equation that describes the accumulation of charge on the capacitor.
(b) The capacitor has a 30-μF rating. To what voltage is the capacitor charged? When applied, how much energy is released into the body?
(c) Table 7B.3 shows the voltage discharge of the capacitor as a function of time. These data can be fitted to the general formula as follows:

$$v(t) = \alpha(e^{-bt}) \sin(\beta t)$$

TABLE 7B.2

Charging of Capacitor for Defibrillator

Time (s)	Charge (C)
0.0	0
0.5	0.0410
1.0	0.0706
1.5	0.0917
2.0	0.1068
2.5	0.1177
3.0	0.1254
3.5	0.1308
4.0	0.1349
4.5	0.1375
5.0	0.1398
5.5	0.1413
6.0	0.1423
6.5	0.1432
7.0	0.1436
7.5	0.1440
8.0	0.1443
8.5	0.1445
9.0	0.1446
9.5	0.1447
10.0	0.1448

TABLE 7B.3

Voltage Discharge of Defibrillator

Time (s)	Voltage (V)
0	0
0.001	2857
0.002	4443
0.003	4990
0.004	4779
0.005	4080
0.006	3142
0.007	2147
0.008	1231
0.009	475
0.010	−87
0.011	−454
0.012	−647
0.013	−700
0.014	−658
0.015	−550
0.016	−415
0.017	−276
0.018	−151
0.019	−50
0.020	24

where v is voltage as a function of time t. Determine the parameters α, β, and h for the best fit of the data. Plot the data and the fitted curve. What is the peak voltage?

Solution:

(a) Plotting the given data in MATLAB, Microsoft Excel, or another program gives the graph shown in Figure 7B.4. The charge initially increases rapidly but then plateaus. The equation is of the form:

$$q = A(1 - e^{-kt})$$

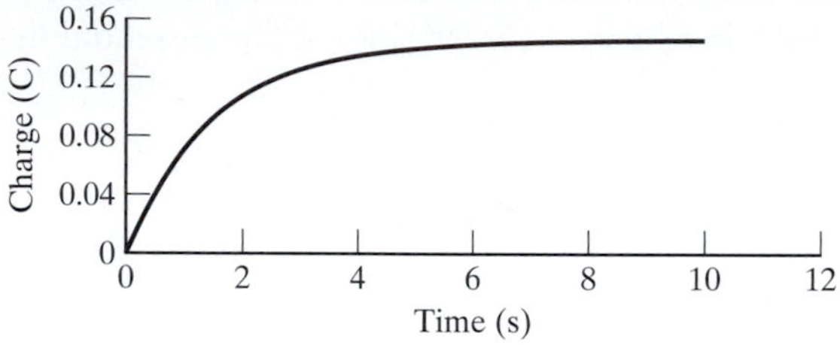

Figure 7B.4
Charging of a capacitor over time.

where q is charge and A and k are fitted parameters. The best fit of the graph returns the values $A = 0.145$ C and $k = 0.666$ 1/s. Thus, the correct mathematical model that describes the charging of the capacitor for a defibrillator is:

$$q = 0.145\left(1 - e^{-(0.666\ 1/\mathrm{s})t}\right)\ \mathrm{C}$$

(b) To find the voltage of the charged capacitor, we can use the relationship between capacitance and charge given in equation [5.7-7]. In the given data, the charge on the capacitor reaches 0.145 C after 10 s and the capacitance rating is 30 μF:

$$v_c = \frac{q}{\mathrm{C}} = \frac{0.145\ \mathrm{C}}{30 \times 10^{-6}\ \mathrm{F}} = 4830\ \mathrm{V}$$

All of the energy stored as charge is released to the body (the system). To count how much energy is released into the body when the defibrillator is discharged, we can use an algebraic energy conservation equation [4.3-15]. If we assume no energy leaves the system and no heat or work acts on the system, the conservation equation simplifies to:

$$E_{T,i} = E_{T,f}^{\mathrm{sys}} - E_{T,0}^{\mathrm{sys}}$$

Thus, the energy delivered to the body by the defibrillator is:

$$E_{T,i} = \frac{1}{2}Cv_c^2 = \frac{1}{2}(30 \times 10^{-6}\ \mathrm{F})(4830\ \mathrm{V})^2 = 350\ \mathrm{J} = E_{T,f}^{\mathrm{sys}} - E_{T,0}^{\mathrm{sys}}$$

Therefore, the accumulation of energy in the body is 350 J. This is enough to synchronize all of the muscle fibers.

(c) We can use the given equation as a template to model the data (Figure 7B.5). Using a program such as MATLAB reveals best fit values of $\alpha = 11{,}000$ V, $\beta = 320$ 1/s,

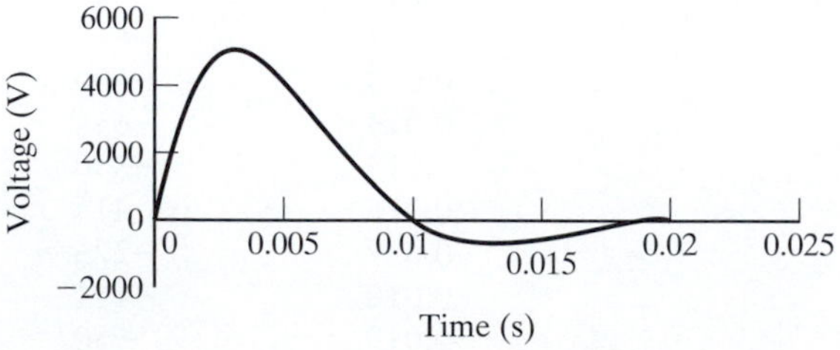

Figure 7B.5
Voltage discharge of a capacitor in a defibrillator over time.

and $h = 200$ 1/s. Thus, the full equation for voltage of the defibrillator that fits the data is:

$$v(t) = (11{,}000\text{ V})\left(e^{-\left(200\frac{1}{s}\right)t}\right)\sin\left(\left(320\frac{1}{s}\right)t\right)$$

According to the graphed data, the peak voltage occurs at 0.003 s at 4990 V. We can also determine the time at which the voltage is a maximum by taking the derivative of the equation for voltage and setting this equation equal to zero:

$$\frac{dv(t)}{dt} = (11{,}000\text{ V})\left(-\left(200\frac{1}{s}\right)e^{-\left(200\frac{1}{s}\right)t}\right)\sin\left(\left(320\frac{1}{s}\right)t\right)$$

$$+ (11{,}000\text{ V})e^{-\left(200\frac{1}{s}\right)t}\left(\left(320\frac{1}{s}\right)\cos\left(\left(320\frac{1}{s}\right)t\right)\right) = 0$$

$$(11{,}000\text{ V})\left(\left(200\frac{1}{s}\right)e^{-\left(200\frac{1}{s}\right)t}\right)\sin\left(\left(320\frac{1}{s}\right)t\right)$$

$$= (11{,}000\text{ V})e^{-\left(200\frac{1}{s}\right)t}\left(\left(320\frac{1}{s}\right)\cos\left(\left(320\frac{1}{s}\right)t\right)\right)$$

$$t = 0.00288\text{ s}$$

This calculated time is very close to the graphical value of 0.003 s. ■

Pressure gradients across the heart work in conjunction with the electrical activity of the heart to keep the blood flowing. When discussing pressures in the circulatory system, gauge pressures (not absolute pressures) are typically used. Consider the left side of the heart. During diastole, blood in the systemic arteries has a pressure of about 80 mmHg. When systole begins, the pressure inside the ventricle increases, causing the mitral valve to snap shut so that back flow of blood into the atrium is prevented. Pressure inside the left ventricle continues to increase until it becomes greater than pressure in the aorta (systolic pressure of about 120 mmHg), causing the aortic valve to open and ejection to begin. Diastole begins when the left ventricle pressure falls below the aortic pressure, and the aortic valve closes. The pressure in the aorta continues to plummet to 80 mmHg and the left ventricle pressure approaches 0 mmHg, while venous blood fills the atrium. When the left ventricle falls to a pressure below the atrial pressure, the mitral valve opens and blood rapidly fills the ventricle. The atrium then contracts to eject its blood volume, and the cardiac cycle repeats.

The circulatory system provides a path for the blood to travel to all of the body's tissues. The blood vessels constituting the circulatory system possess distinct properties specifically suited to their particular functions. The major blood vessels are the aorta, arteries, arterioles, capillaries, venules, veins, and venae cavae. Characteristics of some of these vessels are given in Table 7B.4.

When blood is ejected from the left ventricle, it enters the aorta, the largest artery in the body. The aorta has a thick smooth muscle layer and can sustain a high

TABLE 7B.4

Blood Vessel Characteristics

Vessel	Diameter (cm)	Velocity (cm/s)
Ascending aorta	2.0–3.2	63
Descending aorta	1.6–2.0	27
Arteries	0.2–0.6	20–50
Capillaries	0.0005–0.001	0.05–0.1
Veins	0.5–1.0	15–20
Venae cavae	2.0	11–16

mean pressure to accommodate the continuous pumping of blood from the heart. The aorta branches into arteries with strong vascular walls that transport blood rapidly. Arteries branch further into arterioles, which have strong, muscular walls that can close the arteriole completely or dilate it severalfold to alter blood flow to the capillaries in response to the tissues' needs. Arterioles pass blood to capillaries, which have walls just one endothelial cell thick to facilitate the exchange of fluid, nutrients, electrolytes, hormones, and gases between the blood and the interstitial fluid. Blood passes from capillaries to venules and next to veins. Since pressure in the venous system is very low (near 0 mmHg), venous walls are very thin. Nevertheless, veins have a relatively large elastin component, allowing responsive contraction or expansion to meet the needs of circulation. For example, when cardiac output increases during exercise, venous walls stretch to hold the increased volume of blood flow. Veins finally pass blood to the venae cavae for reentry to the heart, where it is transported to the lungs for reoxygenation before entering the systemic circulation once again.

All of the body's blood supply travels through the aorta once per cycle. After supplying the coronary arteries with the oxygen necessary to nourish the heart itself, the aorta extends upward toward the neck to feed the branches carrying blood to the head and arms. Branches of the aorta called the right and left carotids bring blood to the eyes and the brain, while the arms receive blood from the right and left subclavian arteries. The aorta also branches and bends downward, directing blood into the chest's arterial system. Blood travels through an opening in the diaphragm called the aortic hiatus and into an extensive arterial network in the abdomen to supply oxygenated blood to the liver, stomach, kidneys, intestines, gonads, and other organs. Two iliac arteries bring blood to the legs.

As the body's most integral distributor, blood facilitates transport of nutrients, gases, and waste products. Constituting 45% of the blood volume, erythrocytes (red blood cells) transport oxygen and carbon dioxide to and from the tissues. The other 55% of whole blood is comprised of plasma that carries vitamins and minerals. Plasma consists of water (92%), albumin, and fibrinogen proteins (6%), as well as various carbohydrates, hormones, ions, and wastes. Blood concentrations of ions, which are necessary to maintain electrical activity in the body, are often measured as an equivalent (eq), which is equal to the molarity of an ion times the number of charges the ion carries. Ions in the blood include sodium (135–145 meq/L), chloride (100–108 meq/L), calcium (4.3–5.3 meq/L), and potassium (3.5–5 meq/L).

EXAMPLE 7B.3 **Blood Velocities in the Aortic Arch**

Problem: Immediately after leaving the heart, the aorta arches downward, a region commonly called the aortic arch (Figure 7B.6). Major arteries branch off the aortic arch. The mass flow rates and pressures in the aorta and aortic arch are higher than in any other vessels

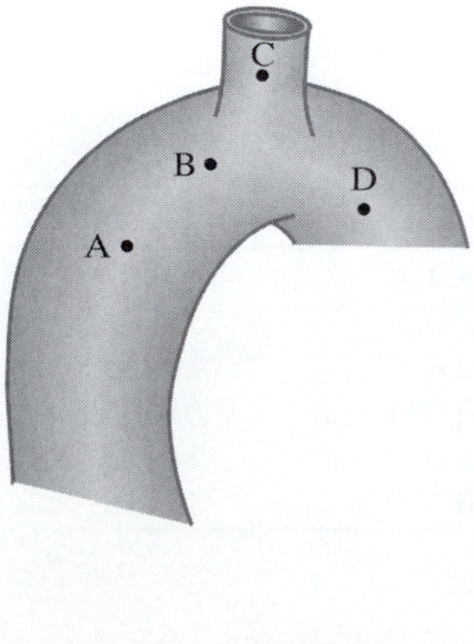

Figure 7B.6
Points in the aortic arch.

TABLE 7B.5A

Setup for Flow Patterns in Aortic Arch

Location	Velocity (cm/s)	Pressure (mmHg)	Diameter (cm)
A	35	97	2.5
B		97	—
C	40		0.75
D			2.1

in the body. Assume the system is at steady-state, velocity profiles are uniform at all points, and there is no resistance to flow. Calculate the appropriate pressure and velocity at each location in the vessel to complete Table 7B.5A.

Solution:

1. Assemble
   (a) Find: pressure and velocity at each location in the aortic arch.
   (b) Diagram: The diagram of the system is given in Figure 7B.6.
   (c) Table: A table setup is given in Table 7B.5A.
2. Analyze
   (a) Assume:
   - The system is at steady-state (i.e., no blood accumulates in system).
   - No leaks.
   - Pressure at each location is constant (i.e., pressure is not pulsatile).
   - Blood density is constant.
   - Uniform velocity profiles throughout system.
   - No changes in potential energy.
   - No frictional losses (i.e., no resistance to flow).
   - No work is done on the system.
   - All vessels can be modeled as cylinders.

   (b) Extra data:
   - The density of whole blood is 1.056 g/cm^3.

   (c) Variables, notations, units:
   - Use cm, s, mmHg, g.

   (d) Basis: Using the given diameter, density, and velocity, we can calculate a mass flow rate at point A in the vessel to use as our basis:

$$\dot{m}_A = Av\rho = \frac{\pi}{4}D^2 v\rho = \frac{\pi}{4}(2.5\text{ cm})^2\left(35\frac{\text{cm}}{\text{s}}\right)\left(1.056\frac{\text{g}}{\text{cm}^3}\right) = 181\ \frac{\text{g}}{\text{s}}$$

3. Calculate
   (a) Equations: We can use the mass conservation equation [3.3-10] to keep track of the mass in the system:

$$\sum_i \dot{m}_i - \sum_j \dot{m}_j = \frac{dm^{\text{sys}}_{\text{acc}}}{dt}$$

   Velocity, pressure, and elevation of two points along the path of a fluid in steady-state flow can be related by the Bernoulli equation [6.11-11]:

$$(gh_i - gh_j) + \left(\frac{1}{2}v_i^2 - \frac{1}{2}v_j^2\right) + \frac{1}{\rho}(P_i - P_j) = 0$$

   (b) Calculate:
   - We can first simplify the mass conservation equation such that Accumulation equals zero, since the system is at steady-state:

$$\sum_i \dot{m}_i - \sum_j \dot{m}_j = 0$$

We can then set up a mass balance across regions A, C, and D. Since the diameter and velocity at A and C are given, as well as the diameter at D, the velocity at D can be calculated:

$$\sum_i \dot{m}_i - \sum_j \dot{m}_j = \dot{m}_A - \dot{m}_C - \dot{m}_D$$

$$= 181\frac{\text{g}}{\text{s}} - \frac{\pi}{4}(0.75\text{ cm})^2\left(40\frac{\text{cm}}{\text{s}}\right)\left(1.056\frac{\text{g}}{\text{cm}^3}\right)$$

$$-\frac{\pi}{4}(2.1\text{ cm})^2(v_D)\left(1.056\frac{\text{g}}{\text{cm}^3}\right) = 0$$

$$v_D = 44.3\frac{\text{cm}}{\text{s}}$$

- Since we assume no changes in potential energy, we can simplify the Bernoulli equation and then rearrange it to find the velocity at B:

$$\left(\frac{1}{2}v_A^2 - \frac{1}{2}v_B^2\right) + \frac{1}{\rho}(P_A - P_B) = 0$$

$$\frac{v_A^2}{2} + \frac{P_A}{\rho} = \frac{v_B^2}{2} + \frac{P_B}{\rho}$$

Substituting in the given pressures ($P_A = P_B = 97$ mmHg), we determine the velocity at B is the same as the velocity at A, which is 35 cm/s.

- Looking at positions B and C, we can now apply the same simplified Bernoulli equation to find the pressure at C:

$$\frac{v_B^2}{2} + \frac{P_B}{\rho} = \frac{v_C^2}{2} + \frac{P_C}{\rho}$$

$$\frac{\left(\frac{35\text{ cm}}{\text{s}}\right)^2}{2} + \frac{97\text{ mmHg}\left(\frac{1.01325\text{ dynes}}{760\text{ mmHg}}\right)\left(\frac{\frac{\text{g}}{\text{cm}^2\cdot\text{s}}}{\text{dynes}}\right)}{\frac{1.056\text{ g}}{\text{cm}^3}}$$

$$= \frac{\left(\frac{40\text{ cm}}{\text{s}}\right)^2}{2} + \frac{P_C}{\left(\frac{1.056\text{ g}}{\text{cm}^3}\right)}$$

$$P_C = \left(\frac{129{,}125\text{ g}}{\text{cm}^2\cdot\text{s}}\right)\left(\frac{1\text{ dyne}}{\frac{\text{g}}{\text{cm}^2\cdot\text{s}}}\right)\left(\frac{760\text{ mmHg}}{1.01325\times10^6\text{ dynes}}\right) = 96.9\text{ mmHg}$$

Applying this same calculation to positions B and D gives a pressure of 96.7 mmHg at D.

4. Finalize
   (a) Answer: The answers are given in Table 7B.5B.
   (b) Check: We see very slight pressure drops with relatively large velocity increases. For example, when blood travels from B to C, the velocity increases by 5 cm/s, but the pressure drops only 0.1 mmHg. Significant pressure drops occur with increased arterial bifurcations. ■

In the past century, the prevalence of coronary heart disease has risen, making it the number-one cause of death in the United States. Diet changes and modern conveniences have led to a widespread decrease in physical activity and a corresponding increase in clogged blood vessels, heart attacks, and strokes. In 2002, nearly

TABLE 7B.5B

Flow Patterns in Aortic Arch

Location	Velocity (cm/s)	Pressure (mmHg)	Diameter (cm)
A	35	97	2.5
B	35	97	—
C	40	96.9	0.75
D	44.3	96.7	2.1

70.1 million Americans—approximately 1 out of every 4 persons—had at least one cardiovascular disease (CVD), which can lead to high blood pressure, heart attack, congestive heart failure, and stroke.[1]

In 1948, researchers enlisted more than 5000 middle-aged patients with no signs of heart disease for biennial examinations in the 30-year Framingham Heart Study, and the children of these patients were enrolled in the Framingham Offspring Study in 1971. These two unprecedented studies allowed physicians to compile priceless profiles on predicting heart disease. Two major risk factors determined include high cholesterol and high blood pressure. Symptoms of heart disease are varied, but frequently involve inadequate blood flow caused by blocked arteries, resulting in pain in the chest, arms, neck, and or back after physical exertion. Other discomforts following physical activity may include shortness of breath, nausea, fainting, and sweating. Whereas intense angina pectoris (chest pain) usually indicates a heart attack in men, women are more likely to experience nausea or vomiting during a heart attack and may feel no chest pain at all, a phenomena known as a "silent" heart attack.

Today, the causes of heart disease are well known, and some are easily preventable. Certain factors are unavoidable, such as age, gender, and heredity. However, several major risk factors can be controlled by sustaining a healthy lifestyle. The best ways to prevent heart disease are to eat healthily, exercise regularly, and avoid smoking.

With this knowledge, the field of medicine and biomedical engineering has also developed new technology to delay mortality and increase quality of life. In 1952, the first successful open heart surgery was performed, but it was the invention of the cardiopulmonary bypass (CPB) that allowed more time-consuming, risky open-chest surgeries to be performed. Since then several impressive firsts became possible: the first implantation of a mechanical device to assist a diseased heart (1965), the first whole heart transplant from one person to another (1967), the first implantation of a total artificial heart (1969), and the first implantation of a permanent total artificial heart (1982). The field of cardiology is constantly expanding to meet the growing number and needs of heart disease patients.

The design of heart assist devices is one of the most prominent examples of the interface between engineering and medicine. Patients with heart failure have hearts that simply fail to pump enough blood to satisfy the body's needs. For the more severe cases of heart failure, patients can be bridged to transplant, in which steps are taken to sustain the heart until a suitable donor is found, or patients can be treated with destination therapy, in which a long-term device is permanently implanted to replace heart function. One common class of devices is the left ventricular assist device (LVAD). This device mimics the function of the left ventricle, which is the side of the heart most likely to fail first (see Example 7B.1), but does not require removal of the native heart. Instead, the LVAD is implanted in the body, where it is usually connected to the heart by a tube that passes blood from the left atrium into the LVAD. Using pneumatically driven pumps or magnetically levitated turbines, blood is pumped into the systemic circulation from the LVAD's exit tube. The LVAD has an external computer controller and a power pack.

Two such LVADs designed for bridge-to-transplantation and destination therapy are produced by Thoratec (Pleasanton, CA). Both known as the HeartMate®, the two different left ventricular assist systems (LVAS)—the Implantable Pneumatic LVAS (HeartMate IP) and the Vented Electric LVAS (HeartMate VE, now XVE)—are approved for clinical use by the Food and Drug Administration (FDA). With the HeartMate, patients can greatly improve from very severe heart failure (classified as discomfort and symptoms that can occur even at complete rest) to mild heart failure (classified as suffering no symptoms from ordinary activity, such as walking up stairs). Meanwhile, they can undergo physical rehabilitation. In 1986, clinical trials on the HeartMate began at the Texas Heart Institute (THI) (Houston, TX), and the FDA approved the HeartMate for the market in 1994.

Both HeartMates feature blood chambers, drive lines, and inflow and outflow conduits. The conduits are made of a Dacron-fabric graft with a pig valve, which is connected to the native heart, and the blood chamber's textured surfaces reduce the risk of thromboembolism. Patients on the HeartMate IP have pneumatic pumps powered by a large drive console connected by a long cable, which controls the IP's 83-mL stroke volume and maximum pumping rate of 140 beats/min, making it capable of providing blood flow rates up to 12 L/min. While patients on the HeartMate IP can remain mobile, they cannot be discharged from the hospital. In contrast, the HeartMate XVE is electrically powered and features a smaller external console, as well as a portable battery pack that allows patients to be tether-free and ambulatory for up to 8 hours. The device has the same stroke volume as the HeartMate IP, but lower maximum pumping rates (up to 120 beats/min) for lower cardiac outputs (about 10 L/min). In 1991, the first patient to receive a HeartMate VE implant was sustained on the device for 505 days. Surgeons and engineers redesigned the original HeartMate VE to prevent further thrombus formation, decrease high pressures experienced by the pump's valves, and further improve the implantation technique and design to reduce infection. Thus, the newer generation HeartMate XVE can greatly increase the quality of life for a patient, making it suitable for destination therapy in addition to bridge-to-transplantation.

An alternative LVAD is the Jarvik 2000 (Figure 7B.7), developed in 1988. This nonpulsatile device with a valveless, electrically powered axial flow pump as its only

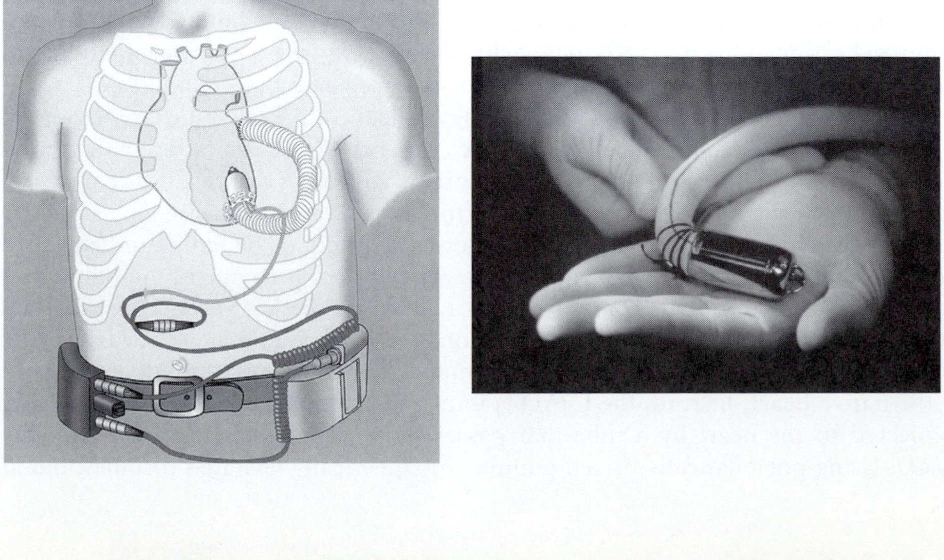

Figure 7B.7
A schematic placement diagram and close-up photo of the Jarvik 2000. (Reprinted with permission from The Texas Heart Institute.)

moving part maintains a continuous flow of oxygenated blood throughout the body. Being the size of a "C" battery, the Jarvik 2000 can be implanted directly into the patient's left ventricle. A waist pack with a small unit monitor controls the pump speed [outputs 5 L/min at adjustable speeds of 8000 to 12,000 revolutions per minute (rpms)] and battery life for the electrically powered motor. A percutaneous cable crosses the abdominal wall to deliver power to the rotating impeller, which is a magnet enclosed in a titanium shell. All blood-contacting surfaces are constructed of highly polished titanium. The monitor emits audible and visual signals to warn patients of potential complications. FDA-approved in 2000 for bridge-to-transplantation, the Jarvik 2000 can sustain patients with heart failure for over 200 days.

Despite the success of LVADs, heart transplantation still remains the most desirable option for prolonging the life of a patient with heart failure. In 1967, Dr. Christiaan Barnard performed the first heart transplantation, making history by defining a donor's death as brain death. Immediately, the procedure was reproduced by other surgeons, proving its viability. However, despite its success, at the end of any given year, nearly 4000 people who are eligible for much-needed heart transplants remain on the waiting list, and the severe shortage of donors (only about 2200 Americans receive one every year) and high risk of immune rejection still leave doctors in the same situation as before–looking for other, more permanent options.[2]

In 1969, Dr. Denton A. Cooley at THI and Dr. Domingo Liotta at the Baylor College of Medicine (Houston, TX) implanted the first total artificial heart (TAH) in a patient who could not have survived long on CPB. The Liotta TAH supported the patient for 64 hours, until a donor heart was obtained. The patient later died of pneumonia, but examination of the Liotta TAH revealed pristine parts, with no indication of thrombus on any of its smooth lining surfaces. Although the implantation was highly controversial, this experience demonstrated that patients could successfully be bridged to transplantation with mechanical circulatory support systems.

Subsequent designs, such as the most widely used Jarvik-7 [now known as the CardioWest (SynCardia Systems, Inc., Tucson, AZ)], improved on the earlier TAHs. In 1982, surgeons permanently implanted the Jarvik-7 into a patient, who survived 112 days with the device. Although others were implanted as destination therapy, the Jarvik-7's bulky external console and excessive maintenance costs diminished the quality of life for patients and caused the FDA to halt production of the device. In its place, ABIOMED (Danvers, MA) has created the AbioCor™ Implantable Replacement Heart (Figure 7B.8).

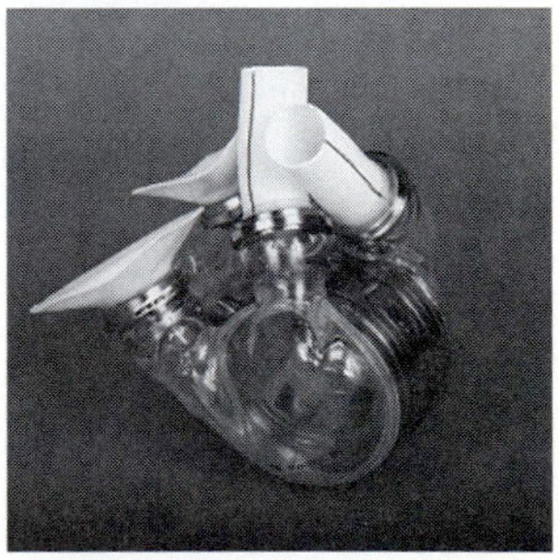

Figure 7B.8
The AbioCor™ by ABIOMED (Danvers, MA).

The AbioCor is the first device of its kind: a fully implantable pulsatile flow pump TAH that has undergone FDA-approved clinical trials. Weighing in at about 2 lb_f, the titanium/polymer-constructed AbioCor uses a motor that rotates at 4000–8000 rpms to hydraulically balance the fluid in the TAH and pump blood throughout the body. A transcutaneous energy transmission (TET) system transmits power from an external battery pack without piercing the skin, greatly reducing chances for infection. The external battery can last up to 4 hours and constantly recharges an internal emergency battery that lasts up to 20 minutes when not connected to the externed battery. The AbioCor's blood flow is monitored and controlled by an internal electronics package.

In 2001, an FDA-approved feasibility study of the AbioCor began as surgeons in Louisville, KY, performed the first implantation in a human patient.[3] Since then, other hospitals throughout the United States have implanted the TAH. In spring 2002, the initial five implantations were evaluated. Thrombus formation was found on its cagelike device on the inflow valve, and the AbioCor was remodeled to eliminate these struts. By May 2003, the AbioCor had been implanted into six more patients.

The results from the feasibility study are encouraging thus far. The AbioCor itself has been reliable and functions as intended. The TAH beats 150,000 times every day, and no pump failures or shortages in energy transmission have occurred.[3,4] The system has also been relatively easy to maintain at home. The pump is totally implantable, which has reduced infection considerably in comparison to other mechanical circulation devices.

Results from studies show that long-term survival with the AbioCor is still an engineering challenge. Thromboembolism is still the biggest problem, and proper anticoagulation protocols must be developed. If a new design using different materials to combat thrombosis is developed, engineers must make sure the materials are still compatible with the immune system and do not damage erythrocytes, without compromising mechanical reliability and stability. The device's current large size also rules out implantation in women, smaller men, and children, and its size is being modified. In addition, engineers and doctors must be sure that any total artificial heart does not decrease the quality of life for patients.

As technological advances continue to extend longevity, heart disease will also become more prevalent throughout the population, increasing the need to find ways to prevent heart failure. The obstacles experienced in the search for a way to prolong heart function indicate a persistent need for further research and improvements to a device that could improve length and quality of life for thousands of patients with heart disease.

References

1. American Heart Association. Heart disease and stroke statistics—2005 update. 2005. http://www.americanheart.org/downloadable/heart/1105390918119HDSStats2005Update.pdf> accessed January. (22, 2005).
2. United Network for Organ Sharing. 2003 U.S. Organ Procurement and Transplantation Network and the Scientific Registry of Transplant Recipients Annual Report. 2005. http://www.optn.org/AR2003/default.htm accessed January. (22, 2005).
3. Frazier OH, Dowling RD, Gray LA, Shah NA, Pool T, and Gregoric I. The total artificial heart: where we stand. *Cardiology* 2004,101:117–21.
4. Cooley D. The total artificial heart. *Nat Med* 2003, 9: 108–11.

Problems

Part I—Focus on the Heart

7B.1 (P) When the ventricles are pumping blood, the mitral and tricuspid valves are closed. This closure allows blood to flow out to the aorta and pulmonary artery, respectively. The diameter of the tricuspid valve is 29 mm, and that of the mitral valve is 31 mm.
- **(a)** Estimate the largest force acting on the mitral valve.
- **(b)** Estimate the largest force acting on the tricuspid valve. The peak pressure in the right ventricle is 25 mmHg.
- **(c)** During exercise, the pressure in the left ventricle can increase by 30%. What is the force on the valve during exercise?
- **(d)** Aortic stenosis is a condition in which the aortic valve decreases in diameter, which constricts flow into the aorta. This constriction results in a buildup of pressure in the ventricle as high as 300 mmHg. What is the force acting on the mitral valve in this disease state?

7B.2 (E) Estimate the number of calories that you consume each day. Chart the type of food, the number of servings, the number of calories per serving, and the total number of calories. Do not forget to include the soda (with caffeine) you are consuming to keep yourself up at night. Determine your basal metabolic rate (BMR) in kcal/day, assuming that BMR is 60% of caloric intake. Table 7B.6 shows the percent of BMR used by a variety of organs. Find the kcal/day that your heart uses.

TABLE 7B.6

Percent Basal Metabolic Rate for Key Organs

Organ	BMR (%)
Kidney (each)	3.85
Brain	16.0
Abdominal organs	33.6
Lungs (each)	2.2
Skeletal muscles	15.7
Heart	10.0
Other	12.6

7B.3 (M) To provide necessary nutrients to the heart, the heart has a coronary circulatory system. Determine the nominal oxygen consumption rate, or metabolic rate, of a tissue lying in the capillary bed in the heart, as shown in Figure 7B.9. Assume that the tissue space is well mixed with respect to oxygen, and that the blood at any point in a vessel is well mixed with respect to oxygen across the cross-sectional area of the vessel. The following variables are defined to help solve the problem: V is volume $[L^3]$, C_{O_2} is the concentration of dissolved oxygen $[NL^{-3}]$, $C_{Hb\text{-}O_2}$ is the concentration of oxygen bound to hemoglobin $[NL^{-3}]$, $\dot{V}$ is the blood flow rate to the tissue $[L^3t^{-1}]$, A is the surface area of capillary bed $[L^2]$, P is the pressure $[ML^{-1}t^{-2}]$, Γ is the metabolic rate of tissue $[NL^{-3}t^{-1}]$, ω is the permeability of blood in the capillary bed $[Lt^{-1}]$, AR is the designation for arterial side, VE is the designation for venous side, and T is the designation for tissue.

The volumetric flow rate out of the capillary bed into the tissue is given as ωA. The O_2 flow rate out of the capillary bed into the tissue can be modeled as $\omega A(C_{VE,O_2} - C_{T,O_2})$. The nominal pressure of oxygen on the arterial side is 95 mmHg and on the venous side 40 mmHg. The concentration of hemoglobin (Hb) in the blood is 2200 μM. On the arterial side, Hb is 96.6% saturated. On the venous side, He is 66.1% saturated. The Henry's law constant for O_2 is 0.74 mmHg/μM. Assume the mass of the heart tissue is 327 g and the blood flow rate to tissues ($\dot{V}$) is 225 mL/min.

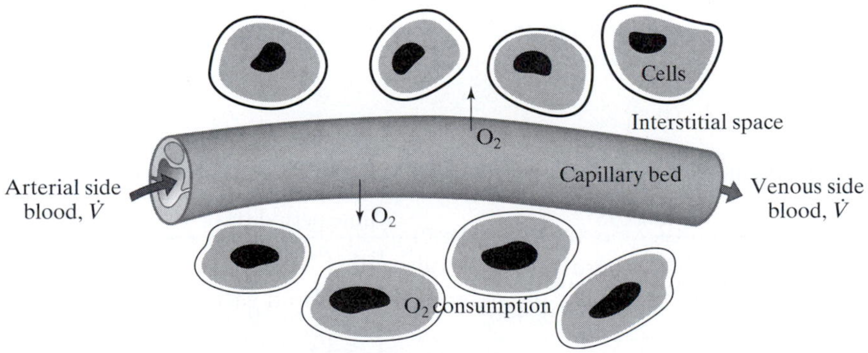

Figure 7B.9
Capillary bed in the heart.

Part II–Electrical Activity of the Heart

7B.4 (G) Describe how a cardiac impulse travels through the heart.
(a) Draw a diagram of the different portions of this conduction system and label each part. How long does it take for the impulse to travel from the SA node to the AV node, Bundle of His, bundle branches, and Purkinje fibers?
(b) Describe the anatomy of the AV bundle.
(c) Why is conduction faster in some parts of the heart when compared to others? Where is conduction the fastest, and what is the speed there? Where is the conduction slowest, and what is its speed there?

7B.5 (C) Intracellular and extracellular concentrations of ions present in cardiac cells are shown in Table 7B.7.

TABLE 7B.7

Ion Concentrations in Cardiac Cells

Ion	Extracellular concentration (mM)	Intracellular concentration (mM)	Conductance (S)
Na^+	145	10	3.0×10^{-6}
K^+	4	140	3.0×10^{-4}
Ca^{2+}	2	1×10^{-4}	3.0×10^{-6}

(a) Calculate the Nernst equilibrium potentials for each ion: Ca^{2+}, Na^+, and K^+.

(b) Remember that conductance is the inverse of resistance. Model the cell membrane as three parallel conductors, each representing one of the ions. What is the total or equivalent conductance?

(c) Assuming that the resting membrane potential is determined primarily by Ca^{2+}, Na^+, and K^+, calculate the resting membrane potential of the cell.

7B.6 (C) During an action potential, the conductances of Ca^{2+}, Na^+, and K^+ change according to the equations in Table 7B.8.

(a) Plot each ion's conductance for a 300-ms time period.

(b) Calculate and plot the resulting membrane potential for a 300-ms time period. Assume that the concentrations of the ions inside and outside the cell remain constant during the action potential. (Even though there is actually a flow of ions into and out of the cell during the action potential, it is not enough to change the concentrations. In other words, the Nernst equilibrium potentials for each ion do not change.) A program such as MATLAB may be useful.

(c) What is the effect of Ca^{2+} on the action potential? What would the action potential look like if Ca^{2+} were not involved?

TABLE 7B.8

Conductance Models for Ions Involved in Action Potentials

Ion	Conductance (S)	Time period (ms)
Na^+	3.0×10^{-6}	$0 < t \leq 10$
	$2.0 \times 10^{-4}t - 0.001997$	$10 < t \leq 15$
	$-2.0 \times 10^{-4}t + 0.004003$	$15 < t \leq 20$
	3.0×10^{-6}	$t > 20$
K^+	3.0×10^{-4}	$0 < t \leq 10$
	$3.0 \times 10^{-5}t$	$10 < t \leq 20$
	6.0×10^{-4}	$20 < t \leq 200$
	$-1.5 \times 10^{-5}t + 0.0036$	$200 < t \leq 220$
	3.0×10^{-4}	$t > 220$
Ca^{2+}	3.0×10^{-6}	$0 < t \leq 10$
	$4.97 \times 10^{-5}t + 4.94 \times 10^{-4}$	$10 < t \leq 20$
	5.0×10^{-4}	$20 < t \leq 150$
	$-9.94 \times 10^{-6}t + 0.001991$	$150 < t \leq 200$
	3.0×10^{-6}	$t > 200$

Part III—The Circulatory System

7B.7 (M, P) In delivering nutrient-rich blood to the body, large vessels split into two or more smaller vessels in progression from the aorta to the arterioles and finally the capillaries. In returning blood to the heart, the capillaries join to form venules and then finally the venae cavae. A typical diameter of each type of vessel and a typical blood velocity are given in Table 7B.9. Calculate the volumetric flow rate, mass flow rate, and Reynolds number of blood through each structure in the circulatory system. Postulate a reason for the extensive branching in the circulatory system and for the low Reynolds number flow.

TABLE 7B.9

Properties of Blood Vessels in Humans

Structure	Diameter (cm)	Blood velocity (cm/s)
Ascending aorta	2.6	63
Descending aorta	1.8	27
Main arterial branches	0.4	35
Arterioles	0.003	3
Capillaries	0.0006	0.05
Venules	0.002	2
Main venous branches	0.5	15
Venae cavae	2.0	14

7B.8 (P) Figure 7B.10 is a diagram of the aortic arch. The magnitude of the velocity of blood through the aortic arch is constant at 0.372 m/s. The direction of momentum depends on the position of the person. The volume of blood in the system is 49 cm^3. As drawn, the inlet flow has a y-direction component; the outlet flow has both x- and y-components. Imagine that the aortic arch is in your body and is oriented as shown when you are standing up. The magnitude and direction of the resultant force required to hold the aortic arch vessel in place depend on your position. Do not neglect the effects of gravity on the vessel.

(a) Assume that you are standing. Determine the resultant force ($\vec{F}_x$, $\vec{F}_y$, and $\vec{F}_z$ terms, as appropriate) that the body exerts on the aortic arch to keep it in place.

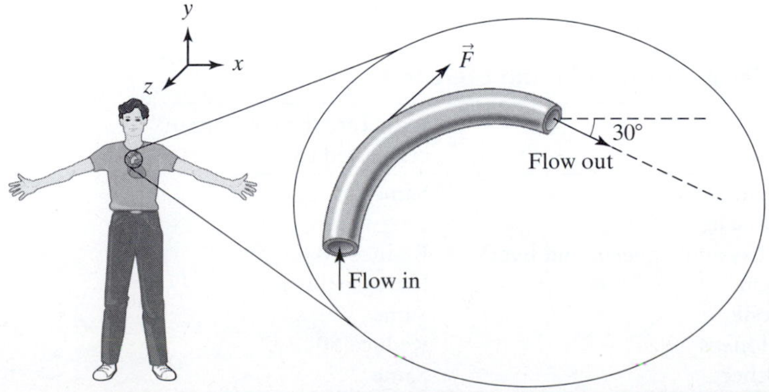

Figure 7B.10
Schematic of blood flow in aortic arch.

(b) When you lie down, the direction of the blood flow changes with respect to the shown x-, y-, and z-axes. (Do not change the orientation of the x-, y-, and z-axes.) Determine the resultant force ($\vec{F}_x$, $\vec{F}_y$, and $\vec{F}_z$ terms, as appropriate) that the body exerts on the aortic arch to keep it in place when you are supine.

7B.9 (P, C) There is a pressure drop from the left side to the right side of the heart.

(a) What is the magnitude of this pressure drop?

(b) As the blood moves through the body, the pressure drops due to vascular resistance, or peripheral resistance, which is the friction of the blood along the vessel wall. What is the numerical value of the peripheral resistance for the body? Report the answer in peripheral resistance units (PRU [mmHg·s/cm^3]).

(c) During exercise, a person's total peripheral resistance (TPR) drops. If a person has a TPR of 0.47 PRU, and his mean arterial pressure is 140 mmHg, what is his cardiac output?

(d) Discuss qualitatively and quantitatively (with formulas) how the resistance through various organs and parts of the body combines to give the TPR. Is the resistance seen through the kidneys greater or less than the total peripheral resistance? Why?

7B.10 (G) Only a fraction of the total cardiac output is seen by any organ. For example, the brain receives 0.7 L/min of blood flow, which is 14% of the total cardiac output of a resting person. Table 7B.10 shows the consumption of some other organs of the body.

(a) Complete Table 7B.10. Consult textbooks or journals as necessary.

(b) What is the flow rate of blood seen by the lungs?

During exercise, the cardiac output increases, allowing for the muscles to receive more nutrients. The distribution of blood through the body changes as shown in Table 7B.11.

TABLE 7B.10

Cardiac Output at Rest

Organ	Flow rate through organ (L/min)	Percent of cardiac output (%)
Brain	0.7	14
Muscles		18
GI system, spleen, and liver	1.35	
Skin	0.3	
Bone		5
Kidneys		
Other		

TABLE 7B.11

Cardiac Output During Exercise

Organ	Flow rate through organ compared to rest
Brain	Same
Muscles	
GI system, spleen, and liver	Reduce 50%
Skin	Increase fourfold
Bone	Same
Kidneys	Reduce 50%
Other	Same

(c) Assuming a cardiac output of 12.8 L/min, how much does the amount of blood going through the muscles change?

(d) What is the physiological reason that, during exercise, the amount of blood to the skin increases fourfold?

7B.11 (M, P) The narrowing of blood vessels is termed stenosis and is a very significant contributor to heart disease and stroke.

(a) List three risk factors for developing stenosis in an artery.

(b) The small artery in Figure 7B.11 is showing signs of stenosis. The following information is known about the flow at location A: diameter = 0.5 cm, systolic pressure = 110 mmHg, diastolic pressure = 70 mmHg, velocity = 10 cm/s. The following information is known about the flow at location B: diameter = 0.1 cm. Find the velocity, systolic pressure, and diastolic pressure at location B. State all your assumptions.

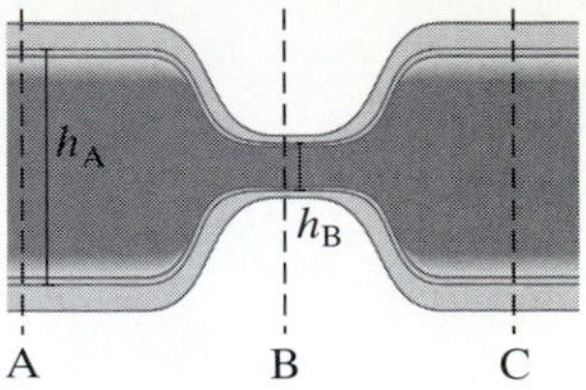

Figure 7B.11 Diagram of stenotic narrowing in an artery.

Another way to model this system is to look at the mean arterial pressure (MAP) instead of the systolic and diastolic pressures.

(c) Give a mathematical definition of the MAP in terms of systolic and diastolic pressures. Why is the MAP not the average of the systolic and diastolic pressures? Why do clinicians often use the MAP rather than systolic and/or diastolic pressures?

(d) Calculate the MAP at locations A and B based on the information given and the solutions calculated in part (b).

(e) Using the value for the MAP at location A (but not the MAP at location B) calculated in part (d) and the velocity and diameter information given in part (b), calculate the MAP at location B. Compare your answer to that in part (d).

(f) Assume a pressure drop of 0.1 mmHg due to the stenotic part of the vessel. Using the velocity and MAP of location A, calculate the velocity of the blood at location C.

7B.12 (M) When a person stands from a supine position, the pressure rises in vessels in the legs. This rise is caused by the added volume and weight of blood from the upper body. The increased pressure increases the amount of blood contained in the vessels for as long as the person stands. Assume a situation in which at $t = 0$, a person stands. The volume of blood leaving a section of a vein decreases and then returns to normal, as shown in Figure 7B.12. Before standing, the diameter of the vein is 0.5 cm and the length of the vein is 30 cm.

(a) What is the volume change in this part of the vein during 6 s?

(b) How much does the diameter of the vessel increase?

(c) In the condition known as varicose veins, this distension becomes permanent. In what groups of people is this condition most common? What complications arise from this condition?

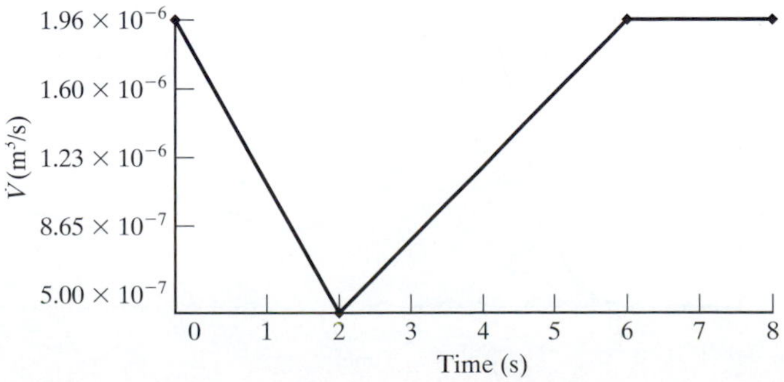

Figure 7B.12 Volumetric blood flow rate from vein over time when standing after being supine.

TABLE 7B.12

Concentration of Dye in Blood*

Time (s)	Concentration (mg/L)	Time (s)	Concentration (mg/L)
0.0	1.35	2.3	207.06
0.1	1.73	2.4	155.76
0.2	2.23	2.5	121.31
0.3	2.86	2.6	94.47
0.4	3.67	2.7	73.58
0.5	4.71	2.8	57.30
0.6	6.05	2.9	44.63
0.7	7.77	3.0	40.61
0.8	9.98	3.1	42.12
0.9	12.81	3.2	46.75
1.0	16.45	3.3	54.63
1.1	21.12	3.4	63.54
1.2	27.12	3.5	67.12
1.3	34.82	3.6	71.48
1.4	44.71	3.7	73.98
1.5	57.40	3.8	74.85
1.6	73.71	3.9	72.10
1.7	94.64	4.0	70.23
1.8	121.52	4.1	66.31
1.9	156.04	4.2	63.49
2.0	200.00	4.3	62.57
2.1	242.07	4.4	63.03
2.2	245.47	4.5	64.54

*Data are fabricated.

7B.13 (M) A method used to measure the flow rate of blood is the indicator–dilution method, where a dye is injected into the pulmonary artery and its concentration is monitored in the radial artery. Table 7B.12 shows the concentration of the dye in the radial artery over time. The first time that any dye is measured in the radial artery is arbitrarily defined to be $t = 0$. After the peak of dye concentration has passed, the curve enters an exponential decay region, which would continue to zero if no recirculation occurred. The patient was injected with 19.3 mg of dye into the pulmonary artery. Determine the average flow rate in the circulatory system of this patient. Assume that all of the dye stays in the circulatory system and that the dye does not react or decay. At times greater than 3 min, a phenomenon called "wash-out" occurs. Make sure you treat this region of data appropriately (Figure 7B.13). Try to incorporate the measured dye concentration in the Accumulation term.

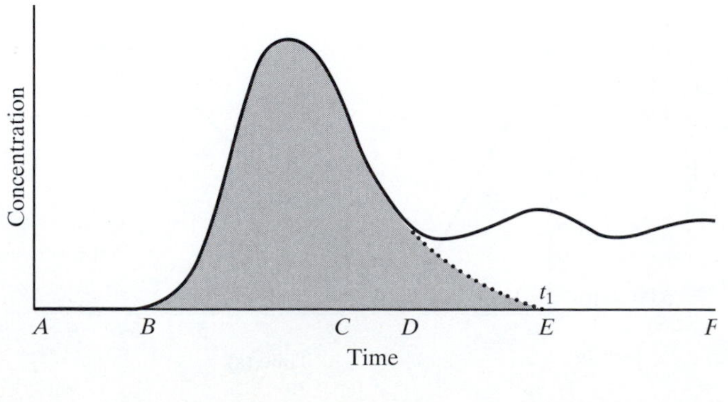

Figure 7B.13
Rapid-injection indicator-dilution curve. After the bolus is injected at time A, there is a transportation delay before the concentration begins rising at time B. After the peak is passed, the curve decays between C and D; it would continue along the dotted curve to t_1 if there were no recirculation. However, recirculation causes a second peak at E before the indicator becomes thoroughly mixed in the blood at F. (*Source*: Webster JG, *Medical Instrumentation: Application and Design*, 3d ed. New York: John Wiley & Sons, 1998.)

Part IV—Focus on Transport at the Capillary Level

7B.14 (G, P) Hematocrit is a term that quantifies the volume percent of blood that is comprised of red blood cells. In a healthy person, the average hematocrit for a female is 40. In some disease conditions, such as anemia and polycythemia, the hematocrit can be dramatically different. For example, the average hematocrit can reach as low as 15 in anemia and as high as 65 in polycythemia. When hematocrit changes, the viscosity of the blood changes as well.

(a) What is the viscosity of normal blood? Locate a figure or graph that plots viscosity as a function of hematocrit. Find the viscosity of blood at the other two stated disease conditions.

(b) Fit a mathematical model to the graph used to answer part (a). Find the viscosity of blood at the three hematocrit values (15, 40, and 65) using the mathematical model. Are your values significantly different?

(c) Calculate the Reynolds number in the aorta for the three hematocrit values. State the flow condition (laminar, transition, or turbulent) for each disease condition.

(d) What hematocrit level delivers the most O_2 to the tissues for a fixed pressure difference?

7B.15 (M) Hematocrit varies depending on location within the body. For example, the hematocrit values in the spleen and kidney are 80 and 20, respectively. Assume that two veins, one from each organ, join together. Given the information in Table 7B.13, what are the hematocrit and the velocity of the final, joined vein?

TABLE 7B.13

Hematocrit for Kidney and Spleen Veins

	Velocity (cm/s)	Diameter (mm)	Hematocrit (−)
Spleen vein	3	0.15	80
Kidney vein	5	0.20	20
Joined vein		0.21	

7B.16 (G, P) Along a capillary, pressures inside and outside the vessel act to keep the vessel open and intact and keep fluid flowing across the vessel wall. The interstitial fluid colloid osmotic pressure, capillary pressure, plasma colloid osmotic pressure, and interstitial fluid pressure all contribute to the net pressure. The following information is given: interstitial fluid colloid osmotic pressure is 8 mmHg, plasma colloid osmotic pressure is 28 mmHg, and interstitial fluid pressure is 3 mmHg.

(a) Give a short definition or explanation of the four types of pressure.

(b) Calculate the capillary pressure and the direction of pressure at the arterial end of a capillary. The net outward pressure on the arterial side is 13 mmHg.

(c) Calculate the capillary pressure and the direction of pressure at the venous end of a capillary. The net inward pressure at the venous end is 7 mmHg.

(d) Assume a mean capillary pressure of 17.3 mmHg. What are the net pressure and the direction of pressure along the length of the capillary?

(e) The average of the capillary pressures at the arterial and venous ends is 20 mmHg. Yet, the mean capillary pressure is 17.3 mmHg. How is mean capillary pressure calculated? Why is mean capillary pressure used in clinical situations?

(f) Where does the extra fluid in the tissue space go to maintain a steady-state volume in the tissue space?

7B.17 (M) Consider the case of a solute diffusing out of a capillary and into the tissue as shown in Figure 7B.14. Assume that the diffusion of a solute out of a capillary vessel and into the tissue follows Fick's law:

$$\dot{n} = PS(C - C_T)$$

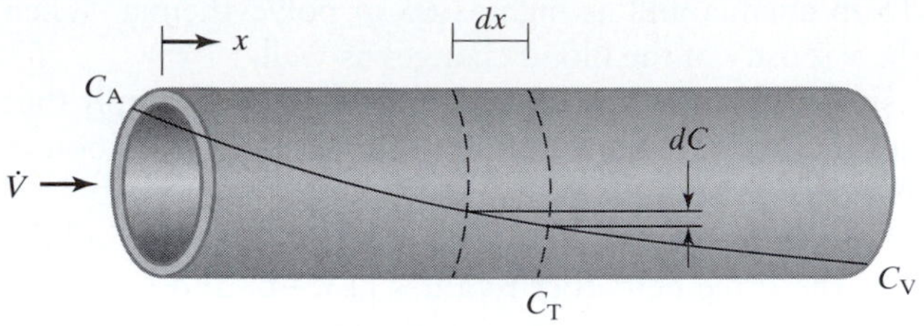

Figure 7B.14 Diffusion of solute from capillary bed.

where $\dot{n}$ is the solute flux out of capillary vessel [Nt^{-1}], P is the permeability coefficient of the solute [Lt^{-1}], S is the surface area of vessel segment [L^2], C is the concentration of solute in the capillary [NL^{-3}], C_T is the concentration of solute in the tissue [NL^{-3}], X is the length scale of vessel [L], $\dot{V}$ is the volumetric flow rate of blood [L^3t^{-1}], A is the designation for arterial side, V is for venous side, T is for tissue, v is the velocity of blood [Lt^{-1}], L is the length of vessel [L], and r is the radius of vessel [L].

(a) Define an expression for the concentration of the solute at the end of the capillary, C_V, in terms of C_A, C_T, P, r, L, and $\dot{V}$. (*Hint*: Set up a balance on a differential unit, dx. You may need to consider a differential concentration, dC.)

(b) Find the concentration at the end of the capillary, C_V, for the solute glucose. The following information is given: $C_A = 5$ μmol/mL, $v = 0.7$ mm/s, $L = 1.6$ cm, $r = 4$ μm, $P = 5.76 \times 10^{-5}$ cm/s. Assume that the concentration of the solute in the tissue is zero.

(c) For the solute described above, calculate the length of vessel required to lose 90%, 95%, and 99% of the solute to the tissue.

7B.18 (M) The lymphatic system plays an important role in collecting interstitial fluid and removing proteins and large particulate matter from the tissue space. The flow rate and conductivity of the solute vary along the vessel walls, as shown in Figure 7B.15. The flow rate across the vessel wall at any location is defined by:

$$\dot{V}_i = K_i(\Delta P_i)$$

where $\dot{V}_i$ is the flow rate of solute across the vessel wall at location i [L^3t^{-1}], K_i is the conductivity of solute across the vessel wall at location i [L^4tM^{-1}], and ΔP_i is the pressure difference across the barrier at location i [$ML^{-1}t^{-2}$].

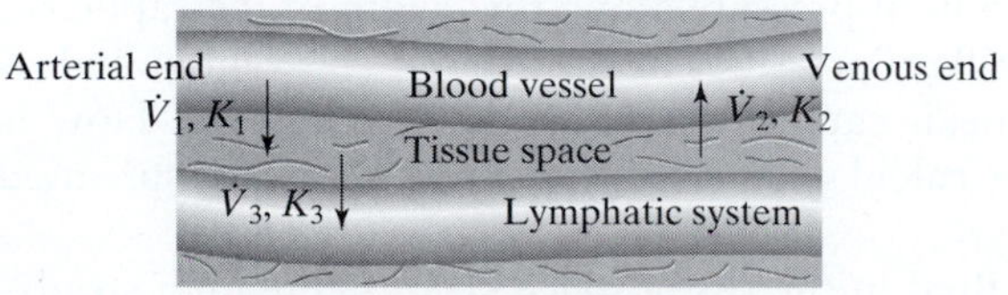

Figure 7B.15 Varying flow rates and conductivities of solute along vessel walls.

Derive an expression for the pressure in the tissue space in terms of K_1, K_2, K_3, P_1, P_2, and P_3. P_1 and P_2 are the pressures at the arterial and venous ends of the blood vessel, respectively; P_3 is the pressure of the lymphatic system. Although the pressure changes along the blood vessel, assume that the pressures in the tissue space and lymphatic system are constant.

7B.19 (M) While O_2 is being delivered from the blood vessels to the tissues, CO_2 is being removed from the tissues and carried away by the blood vessels. The partial pressures of CO_2 on the arterial and venous sides are 40 mmHg and 46 mmHg, respectively. The Henry's law constant for CO_2 is 17.575 mmHg/μM. Calculate the uptake rate of CO_2 from the tissue into the blood vessels.

Part V—Design of Heart Assist Devices

7B.20 (G) Consider the pulsatile LVAD (e.g., HeartMate®).

(a) List three advantages of this device as compared to other designs.
(b) List three disadvantages of this device as compared to other designs.
(c) How does the LVAD work? How many chambers does this device have? Describe the pumping mechanism. Include a schematic.
(d) Where are the device and other accessories implanted?
(e) How much blood can the LVAD circulate in one minute?
(f) How does the LVAD receive the energy it needs to pump continuously? Describe the accessory devices required for this function and how they work.

7B.21 (G) Consider the rotary pump (e.g., Jarvik).

(a) List three advantages of this device as compared to other designs.
(b) List three disadvantages of this device as compared to other designs.
(c) How does the rotary pump work? Describe the blood propulsion mechanism. Include a schematic.
(d) Where are the device and other accessories implanted?
(e) How much blood can the device circulate in one minute?
(f) How does the device receive the energy it needs to pump continuously? Describe the accessory devices required for this function and how they work.

7B.22 (G) Consider the total artificial heart (TAH) (e.g., AbioCor™).

(a) List three advantages of this device as compared to other designs.
(b) List three disadvantages of this device as compared to other designs.
(c) How does the TAH work? How many chambers does this device have? Describe the pumping mechanism. Include a schematic.
(d) Where are the artificial heart and other accessories implanted?
(e) How much blood can the TAH circulate in one minute?
(f) How does the TAH receive the energy it needs to pump continuously? Describe the accessory devices required for this function and how they work.

7B.23 (G) Pick one of the three devices described in Problems 7B.20–7B.22 List two or three improvements that could be made to the device and why the improvements are important. Describe technically how you could implement these ideas to create a better product.

7B.24 (G) After studying the TAHs in development and on the market, you decide to design and build your own TAH based on several novel ideas.

(a) List five very important and critical criteria to take into consideration when designing a TAH.
(b) List five important criteria that are not required but would be desired.
(c) List 8–10 specific technical specifications for a TAH. Examples of technical specifications include size, fluid output, stroke volume, energy requirements, strength of device, etc. State why you have chosen these values.
(d) List 2–3 novel ideas that you would incorporate into your design of a TAH.

7B.25 (G) Material selection is very important for any implantable device. What materials are currently used in implantable cardiac devices? What are the strengths and weaknesses of these materials? What improvements are possible?

7B.26 (G) Once you have the device built, you need to conduct extensive testing to insure safety and efficacy.
(a) List several major categories of tests that you need to conduct before the device is implanted.
(b) What animal models would you test your device in and why?
(c) You will use the results from your animal trials as support to start human clinical trials. What organization will you need to get approval from before you can start human trials?
(d) What major issues are addressed during human clinical trials?

7B.27 (G) An ideal artificial heart would be tissue-engineered from cardiac tissue of the person needing a replacement heart. However, researchers are not yet close to this goal.
(a) What research has been done to date? What parts of the heart are researchers focusing on in their attempts to construct tissue-engineered implants?
(b) you think that a full tissue-engineered heart will be created and implanted in a patient in your lifetime? Why or why not? What major technical obstacles must be overcome in pursuit of this goal?
(c) One source of cells for cardiac tissue is stem cells. Comment on the realistic promise that stem cells hold for building a tissue-engineered artificial heart. Comment on the ethical controversies surrounding stem cells and the current federal guidelines on the use of stem cells.

Case Study C

Better than Brita®: The Human Kidneys

The kidneys (Figure 7C.1) are two bean-shaped organs that regulate the amount of fluid in the body through the formation of urine. Located in the lower back near the rear wall of the abdomen on either side of the spine, each kidney is a mere 11 cm in length and weighs only 160 g. The kidneys are individually encased in a transparent, fibrous membrane called the renal capsule, which shields them against trauma and infection. The concave cavity of the kidney attaches to the renal artery and the renal vein, two of the body's crucial blood vessels, as well as to the ureter, the vessel that transports urine to the bladder.

Blood enters the kidneys through the renal artery at an average flow rate of 1.2 L/min (~25% of cardiac output). The artery branches into a network of smaller blood vessels called arterioles, eventually ending in tiny capillaries in the nephron, the functional unit of the kidney responsible for urine formation (Figure 7C.2). The kidneys have approximately one million nephrons to clean the blood through filtration, reabsorption, and secretion processes. Each nephron is composed of the

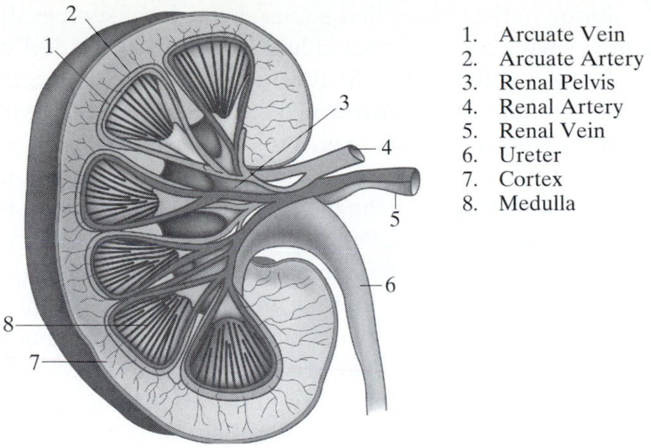

Figure 7C.1
A schematic drawing of a single kidney.

glomerulus surrounded by the Bowman's capsule, a renal tubule, and collecting duct. The glomerulus filters the blood, retaining red blood cells and proteins, and the filtered components—consisting of water and other low-molecular-weight molecules—are passed to the renal tubule. Composed of the convoluted tubules and the loop of Henle, the renal tubule mainly reabsorbs and secretes ions, water, and wastes. Semipermeable membranes surrounding the renal tubule selectively allow particles to pass back into the blood (reabsorption) or from the blood into the tubule (secretion). All material remaining in the filtrate after passage through the renal tubule accumulates in the collecting duct and exits the kidney as urine. Clean blood exits the kidneys' filtration system through the renal vein at a rate slightly under 1.2 L/min, while urine leaves the collecting duct at an approximate rate of 1.1 mL/min. Every day, approximately 180 L (about 50 gallons) of blood passes through the pair of kidneys, and about 1.5 L (1.3 quarts) of urine is produced.

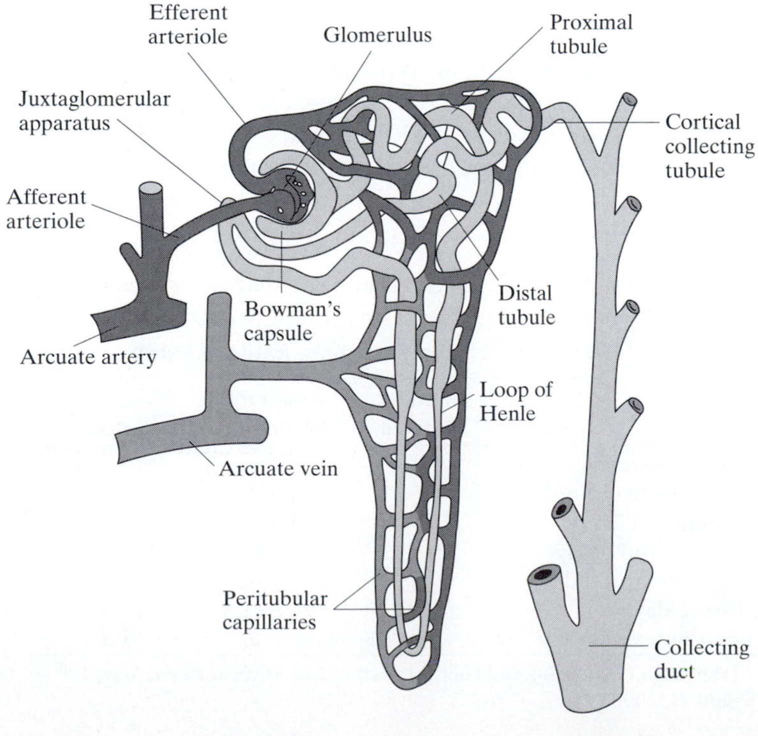

Figure 7C.2
A single nephron, the functional unit of the kidney. (*Source*: Guyton AC and Hall JE, *Textbook of Medical Physiology, 10th ed*. Philadelphia, WB Saunders, 2000).

Blood in the kidney first passes through the glomerulus, where filtrate fluid containing small, unbound molecules is passively filtered. A cuplike structure surrounding the glomerulus called the Bowman's capsule collects the filtrate. Molecules in the filtrate are selectively sorted in the capillary walls by size, where compounds with lower molecular weights (e.g., water, ions, urea) easily pass into the filtrate, while larger particles (particles with diameters greater than 8 nm, such as red blood cells and proteins) remain in the bloodstream. The charge of the molecules also interacts with the negatively charged proteoglycans in the membrane, with negatively charged particles like albumin being less easily filtered than neutrally and positively charged molecules. The filtered blood leaves the glomerulus through an arteriole, which branches into a network of blood vessels surrounding the renal tubule.

The primary role of the renal tubule is to selectively reabsorb molecules such that it leaves waste products behind for excretion. As the filtrate flows through the renal tubule, the network of blood vessels surrounding the tubule both passively and actively reabsorbs salts and virtually all of the nutrients, especially glucose and amino acids, which were filtered in the glomerulus. In response to the salt concentration differences between the nephron and the blood vessels, water is reabsorbed passively through osmosis. This important process, called tubular reabsorption, enables the body to selectively keep necessary substances while eliminating wastes. Overall, about 99% of the water, salt, and other nutrients are reabsorbed. In contrast, relatively small amounts of waste products, such as urea, uric acid, and creatinine, are reabsorbed; they are left behind in the filtrate.

In addition to reabsorbing valuable nutrients from the glomerular filtrate, a lesser role of the renal tubule is tubular secretion. Unwanted substances from the capillaries surrounding the nephron are passed into the filtrate. Such substances include ammonium, hydrogen, and potassium ions, as well as organic ions, which can be derived from foreign chemicals or the natural by-products of the body's metabolic processes. The renal tubule eventually empties its waste products into a collecting duct, and the collecting ducts from many nephrons collectively feed into the ureter. The ureters (one from each kidney) empty the liquid waste into the bladder for storage until it is excreted out the urethra for removal from the body. Urine usually consists of the major end products of metabolism (urea, creatinine, uric acid) and other wastes (sulfates, phenols), as well as any excess ions (Na^+, Cl^-, K^+) (Table 7C.1).

How the kidneys are functioning can be quantitatively determined by the glomerular filtration rate (GFR), defined as the volume of blood filtered per unit time. In a normal individual, the GFR is about 10% of renal blood flow. Many

TABLE 7C.1

Filtration, Reabsorption, and Excretion, Rates of Different Substance*

	Amount Filtered	Amount Absorbed	Amount Excreted	% of Filtered Load Reabsorded
Glucose (g/day)	180	180	0	100
Bicarbonate (mEq/day)	4,320	4,318	2	>99.9
Sodium (mE/day)	25,560	25,410	150	99.4
Chloride (mEq/day)	19,440	19,260	180	99.1
Potassium (mEq/day)	756	664	92	87.8
Urea (g/day)	46.8	23.4	23.4	50
Creatinine (g/day)	1.8	0	1.8	0

*Table from Guyton AC and Hall JE, *Textbook of Medical Physiology*, 10th ed. Philadelphia: Saunders, 2000.

different stimuli can influence and change the GFR. For example, when the arterioles experience a decrease in pressure—and therefore a decrease in renal blood flow—increasing levels of the vasoconstrictor hormone angiotensin II causes the renal arterioles to constrict, which increases the pressure on the glomerulus and thus increases the GFR. Although minor fluctuations in the sympathetic nervous system change the GFR relatively little, strong effects on the renal sympathetic nerves can greatly affect GFR. For example, a severe hemorrhage can lead to the release of epinephrine and norepinephrine, resulting in constriction of the renal arterioles, which leads to a decrease in GFR. GFR can also increase with hormones that act as vasodilators, such as endothelial-derived nitric oxide, prostaglandins, and bradykinin. Other factors, such as a high-protein diet and increased blood glucose, can also increase renal blood flow and GFR.

EXAMPLE 7C.1 **Glomerular Filtration Rate**

Problem: Inulin is a polysaccharide that passes freely through the glomerulus and is neither secreted nor reabsorbed by the nephron, making it ideal for determining GFR. Inulin is infused into an individual until a steady-state concentration of 0.1 g/(100 mL blood) is established. Over a 2 hr period, 180 mL of urine is collected with an average inulin concentration of 0.08 g/mL. Find the GFR of the individual. Assume that throughout the test, the inulin level in the blood remains at the steady-state concentration given. Assume no inulin is metabolized by the body and is only excreted in the urine.

Solution:

1. Assemble
   (a) Find: GFR.
   (b) Diagram: The system is shown in Figure 7C.3.
2. Analyze
   (a) Assume:
      - Inulin does not react.
      - No stimuli change the inulin levels (i.e., inulin concentration in blood does not change and is at steady-state).
      - All inulin in the blood that enters the kidney is excreted in the urine.
   (b) Extra data: No extra data are needed.
   (c) Variables, notations, units:
      - Use mL, min, g.
   (d) Basis: Given the concentration of urine in the 2 hr time period, we can calculate the mass of inulin for a basis:

$$m_{\text{urine,inulin}} = C_{\text{urine,inulin}} V_{\text{urine}} = \left(\frac{0.08 \text{ g inulin}}{\text{mL urine}}\right)(180 \text{ mL urine}) = 14.4 \text{ g}$$

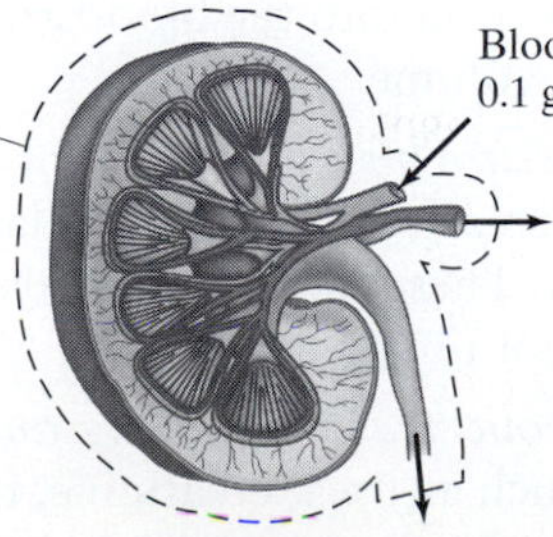

Figure 7C.3
System diagram of inulin flow in the kidney.

3. Calculate
   (a) Equations: Because we calculate the GFR using the mass of inulin excreted over a definite time period, we can use the algebraic mass accounting equation [3.3-3]. With our assumption that inulin does not react, we can eliminate the Generation and Consumption terms, leaving the algebraic mass conservation equation [3.3-9]. Since we also assume the concentration is at steady-state, the Accumulation term is zero. So for any component in a stream, we use equation [3.6-10]:

$$\sum_i m_{i,s} - \sum_j m_{j,s} = 0$$

   (b) Calculate:
   - Using the algebraic mass conservation equation we can write an equation specific to our system:

$$\sum_i m_{i,s} - \sum_j m_{j,s} = m_{\text{blood in,inulin}} - m_{\text{urine,inulin}} - m_{\text{blood out,inulin}} = 0$$

   - We assume all inulin is excreted in urine, so no inulin can flow out of the kidney in the blood. We can then use the given inulin concentrations to determine the volume of blood flowing in:

$$\begin{aligned} m_{\text{blood in,inulin}} - m_{\text{urine,inulin}} &= C_{\text{blood in,inulin}} V_{\text{blood in}} - m_{\text{urine,inulin}} = 0 \\ &= \left(\frac{0.1 \text{ g inulin}}{100 \text{ mL blood}}\right) V_{\text{blood in}} - 14.4 \text{ g} = 0 \\ V_{\text{blood in}} &= 14{,}400 \text{ mL} \end{aligned}$$

   - GFR is the volume of blood filtered per unit time, so using the 2-hr time period, we calculate the GFR to be:

$$\text{GFR} = \left(\frac{14{,}400 \text{ mL}}{2 \text{ hr}}\right)\left(\frac{1 \text{ hr}}{60 \text{ min}}\right) = 120 \ \frac{\text{mL}}{\text{min}}$$

4. Finalize
   (a) Answer: The individual's GFR is 120 mL/min.
   (b) Check: Checking this number against literature reveals the calculated GFR is nearly the same. We also know that GFR is about 10% of renal blood flow, which is about 1.2 L/min, so we expect the GFR to be about 120 mL/min. ■

In addition to forming urine, the kidneys are also responsible for performing several other vital functions to maintain homeostasis, including:

1. *Controlling body fluid volume*: One substance that influences fluid volume is antidiuretic hormone (ADH), which is released into the bloodstream when concentrations of salts and other substances become too high. ADH increases permeability to water in the renal tubules and collecting ducts, causing increased water reabsorption into the bloodstream. In contrast, diuretics are used to excrete extra fluid volume.
2. *Regulating blood pressure by regulating blood plasma volume*: The kidneys can secrete substances, such as renin, that stimulate the release of vasoactive factors, such as angiotensin II. These factors can cause arterial vessels to constrict or dilate for short periods of time.
3. *Concentrating metabolic waste products and foreign chemicals*: The kidneys concentrate wastes, such as urea, creatinine, drugs, and food additives, for elimination from the body.

4. *Regulating electrolytes in the plasma*: The kidneys can retain or excrete ions based on a person's diet. Such ion concentrations influence how much water is reabsorbed by the nephrons.
5. *Regulating acid-base balance*: In conjunction with the lungs and other fluid buffers in the body, the kidneys adjust the body's pH by controlling excretion of acids and fluid buffers. Improper hydrogen ion concentrations result in acidosis or alkalosis, which impair the central nervous system. If the blood pH drops below 7.35, the kidneys move the excess hydrogen ions into the urine through tubular secretion.
6. *Activating gluconeogenesis*: Like the liver, the kidneys can also synthesize glucose from amino acids when the body experiences prolonged fasting.
7. *Helping control the rate of red blood cell formation*: Erythropoietin is crucial to red blood cell production. Under hypoxic conditions, the kidneys secrete erythropoietin. The kidneys also process vitamin D, converting it to an active form that stimulates bone development.

When any of the kidneys' fundamental roles malfunction, maintaining homeostasis becomes difficult and kidney disease can develop. Kidney disease can range from a mild infection to life-threatening kidney failure. Additionally, having hypertension can cause kidney disease or having some types of kidney disease can cause hypertension, creating a vicious cycle in which kidney disease exacerbates hypertension, which in turn damages the kidneys further. Diabetes mellitus can also lead to kidney damage because high levels of blood glucose can damage the small blood vessels in the kidneys.

Severe kidney diseases can be separated into two different categories: acute renal failure or chronic renal failure. In acute renal failure, the kidneys unexpectedly stop working, but recovery to normalcy is possible. The most significant effect of acute renal failure is the retention of fluids, metabolic waste products, and electrolytes, leading to edema and hypertension. High concentrations of certain ions (e.g., potassium, hydrogen) can lead to severe imbalances, causing conditions such as hyperkalemia and metabolic acidosis. If left untreated, severe cases can cause patients to die in 8 to 14 days. In contrast, chronic renal failure is characterized by the gradual, irreversible loss of kidney function as nephrons progressively begin to fail. Remaining nephrons can increase function to excrete electrolytes and fluid normally to a certain extent, but metabolic waste products (e.g., creatinine, urea) are not easily reabsorbed and are lost at a rate proportional to the GFR. Wastes that cannot be adequately filtered can consequently accumulate in the blood and tissues at toxic levels, a condition known as uremia (which means literally "urine in the blood"), and can eventually lead to death.

One renal condition is renal tubular acidosis (RTA), in which the kidneys fail to excrete enough hydrogen ions or reabsorb enough bicarbonate or both. RTA displays impaired ability to transport ions, particularly hydrogen and bicarbonate, across the renal tubule or collecting duct, causing blood pH to decrease ($pH < 7.41$), thus possibly changing urine pH.[1] Three forms of RTA exist: Type I, Type II, and Type IV. All types of RTA can be inherited or can result from strain on the nephrons, such as in kidney transplantation. Type I affects the distal tubule and is characterized by low potassium levels in the blood, which can lead to kidney stones. For Type I, urine pH does not fall below 5.5. Type II affects the proximal tubules and is traced to an assortment of disorders, including vitamin D deficiency, thyroid problems, and fructose intolerance. It can also appear as a side effect of certain drugs, such as acetazolamide and outdated tetracycline. The urine pH ranges between 5.5 and 7 for Type II. Type IV affects the distal renal tubule but is classified by high potassium levels in the blood and normal urine pH. Low levels of the hormone aldosterone, which regulates sodium, potassium, and chloride ions, can cause Type IV, leading to heart problems (e.g., arrhythmias). If treated early enough, permanent kidney failure can be prevented.

EXAMPLE 7C.2 Renal Tubular Acidosis

Problem: Renal tubular acidosis (RTA) is a kidney abnormality that results in a decreased blood pH (< 7.41). In this problem, we will focus on Type I RTA, which is the impaired ability to secrete hydrogen ions in the distal tubule, creating a urine pH that is usually greater than 5.5.

An acid load test where ammonium chloride (NH_4Cl) is taken orally is used to confirm Type I RTA. Ammonium chloride acts as a hydrogen donor, thus decreasing blood pH. In an individual with normal kidney function, the kidney clears excess hydrogen ions, and the urine pH drops to less than 5.2 in approximately 3–6 hr.[1] However, in an individual with Type I RTA, urine pH remains greater than 6 during the same time duration.

Suppose a woman complains of symptoms associated with RTA. Test results show pH values of 7.0 and 8.5 for her blood and urine, respectively. Assuming she has normal kidney function, what quantity of ammonium chloride would decrease the urine pH to 5.2? How many moles of hydrogen ions accumulate in the bladder as the urine pH drops from 8.5 to 5.2? The dissociation constant (K_a) for NH_4^+ is 5.6×10^{-10} M.

Solution:

1. Assemble
   (a) Find: amount of NH_4Cl needed to decrease urine pH from 8.5 to 5.2 in an individual with normal kidney function.
   (b) Diagram: The system is shown in Figure 7C.4.

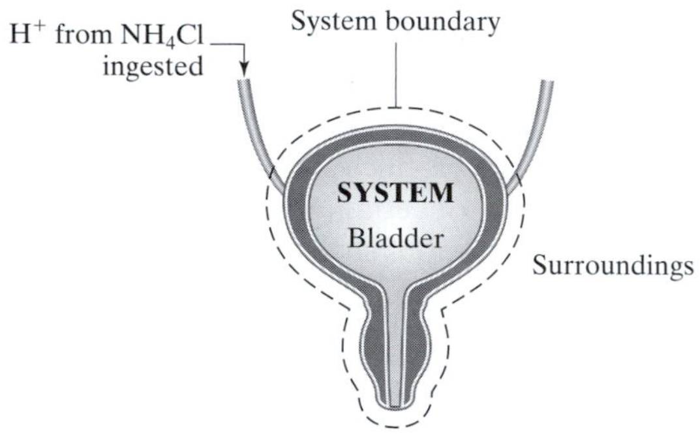

Figure 7C.4
System diagram of ammonium chloride flow in the bladder.

2. Analyze
   (a) Assume:
   - The woman has normal kidney function (i.e., no Type I RTA).
   - NH_4Cl completely dissociates (100%) into NH_4^+ and Cl^-.
   - The increase in H^+ ions and the change in pH in the urine results exclusively from H^+ ions from NH_4Cl dissociation.
   - Urine is produced at a constant rate throughout the day.
   - Changes in H^+ and NH_3 concentrations are modeled by their direct addition to a fixed volume of urine in the bladder.
   - No NH_4^+ or NH_3 is present in the bladder before ingestion of NH_4Cl.

   (b) Extra data:
   - Urine output ranges from 1 to 1.5 L/day.

   (c) Variables, notations, units:
   - Use mg, hr.

   (d) Basis: The basis is the initial amount of H^+ ions in the bladder (calcuation below).
   (e) The equilibrium dissociation reactions are:

$$NH_4Cl \rightleftharpoons NH_4^+ + Cl^-$$

$$NH_4^+ \rightleftharpoons NH_3 + H^+$$

3. Calculate
   (a) Equations: Because we are interested in the pH change over a discrete time interval, we can use an algebraic molar accounting equation [3.3-4] to count the number of charged hydrogen ions:

$$\sum_i n_{i,\mathrm{H}^+} - \sum_j n_{j,\mathrm{H}^+} + n_{\mathrm{gen,H}^+} - n_{\mathrm{con,H}^+} = n^{\mathrm{sys}}_{\mathrm{H}^+,\mathrm{f}} - n^{\mathrm{sys}}_{\mathrm{H}^+,0}$$

   To relate pH and hydrogen ion concentrations, equation [5.9-4] is necessary:

$$\mathrm{pH} = -\log[\mathrm{H}^+]$$

   (b) Calculate:
   - We assume the NH_4Cl dissociates into its ions in the bloodstream prior to entering the bladder. Since no charge is generated or consumed in the bladder, we can simplify the accounting equation such that Generation and Consumption terms are zero. Since we assume the woman does not urinate in the time interval, the Output term is also zero, simplifying the equation to:

$$n_{\mathrm{in,H}^+} = n^{\mathrm{sys}}_{\mathrm{acc,H}^+} = n^{\mathrm{sys}}_{\mathrm{H}^+,\mathrm{f}} - n^{\mathrm{sys}}_{\mathrm{H}^+,0}$$

   - Using the given initial pH of the urine (8.5), we can determine the hydrogen ion concentration in the system of the bladder:

$$(\mathrm{pH})^{\mathrm{bladder}}_0 = -\log[\mathrm{H}^+]^{\mathrm{bladder}}_0$$

$$[\mathrm{H}^+]^{\mathrm{bladder}}_0 = 10^{-(\mathrm{pH})_0^{\mathrm{bladder}}} = 10^{-8.5} = 3.16 \times 10^{-9}\ \mathrm{M}$$

   Solving for the hydrogen ion concentration at the final condition (pH = 5.2) in the same manner gives 6.298×10^{-6} M. Since her urine output is about 1.25 L/day, the amount of urine that accumulates in the bladder in 4 hr is:

$$V = \left(\frac{1.25\ \mathrm{L}}{\mathrm{day}}\right)\left(\frac{1\ \mathrm{day}}{24\ \mathrm{hr}}\right)(4\ \mathrm{hr}) = 0.208\ \mathrm{L}$$

   Thus, the initial amount of hydrogen ions in the bladder is:

$$n^{\mathrm{bladder}}_{\mathrm{H}^+,0} = [\mathrm{H}^+]^{\mathrm{bladder}}_0 V = \left(3.16 \times 10^{-9}\frac{\mathrm{mol}}{\mathrm{L}}\right)(0.208\ \mathrm{L}) = 6.58 \times 10^{-10}\ \mathrm{mol}$$

   Solving for the hydrogen ion amount at the final condition in the same way gives 1.31×10^{-6} mol. Thus, the change in the amount of hydrogen ions to drop urine pH from 8.5 to 5.2 is:

$$n^{\mathrm{bladder}}_{\mathrm{acc,H}^+} = n^{\mathrm{bladder}}_{\mathrm{H}^+,\mathrm{f}} - n^{\mathrm{bladder}}_{\mathrm{H}^+,0}$$

$$= 1.31 \times 10^{-6}\ \mathrm{mol} - 6.58 \times 10^{-10}\ \mathrm{mol} = 1.31 \times 10^{-6}\ \mathrm{mol}$$

   - Since we assume that no NH_3 is present prior to ingestion of NH_4Cl and that NH_4^+ dissociates to H^+ and NH_3 in a 1:1 ratio (i.e., 1 mole of NH_3 is produced for every mole of H^+), the amount of NH_3 present in the urine must equal the amount of H^+ that accumulates:

$$n^{\mathrm{bladder}}_{\mathrm{acc,NH_3}} = n^{\mathrm{bladder}}_{\mathrm{acc,H}^+} = 1.31 \times 10^{-6}\ \mathrm{mol}$$

$$C^{\mathrm{bladder}}_{\mathrm{acc,NH_3}} = \frac{1.31 \times 10^{-6}\ \mathrm{mol}}{0.208\ \mathrm{L}} = 6.298 \times 10^{-6}\ \mathrm{M}$$

   - Using the calculated NH_3 and H^+ concentrations (at pH 5.2) and the given dissociation constant for NH_4^+, we can figure out the molar concentration of NH_4^+:

$$K_a = \frac{[H^+][NH_3]}{[NH_4^+]}$$

$$[NH_4^+] = \frac{[H^+][NH_3]}{K_a} = \frac{(6.298 \times 10^{-6}\ M)(6.298 \times 10^{-6}\ M)}{5.6 \times 10^{-10}\ M} = 0.0708\ M$$

- Using the urine volume during the 4-hr time period required for the pH to drop to 5.2, we can figure out how many moles of NH_4^+ are needed:

$$n_{NH_4^+}^{bladder} = [NH_4^+]V = \left(0.0708\ \frac{mol}{L}\right)(0.208\ L) = 0.0147\ mol$$

- We assume NH_4Cl dissociates completely, so every mole of NH_4Cl ingested yields one mole of NH_4^+. The mass of ammonium chloride that must be consumed is calculated using its molecular weight:

$$m_{NH_4Cl} = n_{NH_4Cl}M_{NH_4Cl} = (0.0147\ mol)\left(53.49\ \frac{g}{mol}\right)\left(\frac{1000\ mg}{1\ g}\right) = 788\ mg$$

4. Finalize
   (a) Answer: To lower urine pH from 8.5 to 5.2 in a normally functioning kidney during 4 hr, 788 mg of ammonium chloride must be ingested.
   (b) Check: According to values found in literature, a person with a normal kidney would need 100 mg/kg of NH_4Cl (~6800 mg NH_4Cl) to lower urine pH from 8.5 to 5.2 in 4 hr.[1] In this problem, we made simplifying assumptions regarding ammonium chloride that may have been too simple, such as the absence of NH_4^+ or NH_3 in the bladder before ingestion of NH_4Cl. One probable reason for the greater dosage than what was calculated could be buffering reagents in the blood and the urine. ■

When the kidneys and nephrons deteriorate to the point where patients with chronic renal failure can no longer function, patients must be placed on dialysis treatment or put on a waiting list for a kidney transplant, and are classified as having reached a stage called end-stage renal disease (ESRD). In 1978, the prevalence of ESRD in the United States was 42,000.[2] By 1991, the number increased fivefold to 200,000, and it doubled again in the last decade, reaching 400,000 in 2001. Although the aging baby boomer generation and prolonged longevity contribute to these increasing numbers, the largest growth comes from the mounting number of patients with diabetes, the number-one cause of ESRD and an indicator of rising obesity trends. Technology may have improved since 1978, but still newer methods and equipment are needed to effectively care for patients with kidney malfunctions while balancing safety and costs.

As it is for patients with failing hearts, the best option for patients with ESRD is an organ transplant. Kidney transplants are the most common of all transplant operations and have excellent success rates. However, although kidneys can be contributed by living donors, since the kidney is one of the few organs that can singly compensate for the work of two, the number of available donors is not increasing to match the need. As of the end of 2002, the number on the waiting list nationwide totaled more than 50,000 patients, and the average waiting time to transplant for the 25% of new candidates deemed most in need ranged from 107 days to 341 days.[3] The urgent need for alternatives to perform in place of the kidney is readily evident. One commonly used alternative for treating kidney failure is kidney dialysis, a procedure in which blood is circulated through a machine that removes wastes and excess fluid from the bloodstream. Some patients use dialysis for a short time while their kidneys recover from injury or disease. Others must use dialysis for their entire lives or until a kidney transplant becomes available. Several methods of artificial kidney dialysis are used in clinical situations, with hemodialysis being the most prevalent.

Artificial kidney dialysis attempts to mimic the chemical principles the kidneys use naturally to maintain the chemical composition of the blood. The concepts of polarity and concentration gradients are central to diffusion across semipermeable membranes, the mechanism of the dialysis process for both natural and artificial kidneys. In the United States, the most used hemodialysis machine is the hollow-fiber membrane device. The main unit, called a dialyzer or artificial kidney, is constructed of a cylindrical container that holds 10,000 to 20,000 parallel hollow fibers made of semipermeable cellulose membrane (Figure 7C.5). In hemodialysis, blood with high levels of waste circulates out of the body connected to the dialyzer, which cleans the blood and pumps it back to the body. Since hemodialysis requires three sessions every week for 3–5 hr each, an arterio-venous fistula is surgically created in the patient's forearm by sewing together an artery and a vein (Figure 7C.6); the rapid blood flow from the artery enlarges the vein for more efficient dialysis, as well as making repeated needle insertions into the vein easier. The blood is then pumped through the hollow-fiber membrane chamber before returning into the patient's

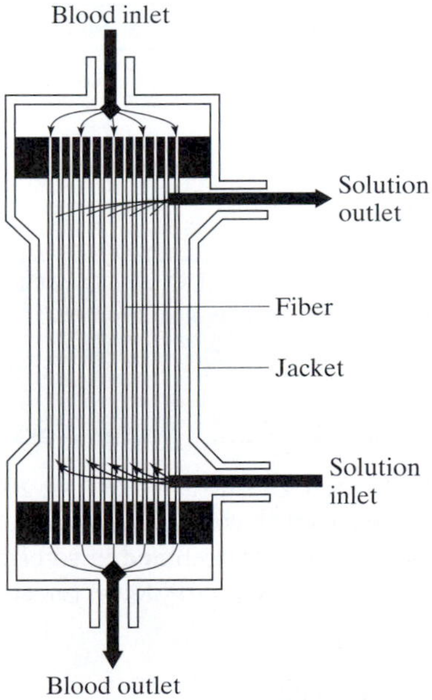

Figure 7C.5
Structure of a typical hollow fiber dialyzer. (*Source*: National Kidney and Urologic Diseases Information Clearinghouse, "Treatment methods for kidney failure: Hemodialysis," National Institute of Diabetes and Digestive and Kidney Diseases, National Institutes of Health.)

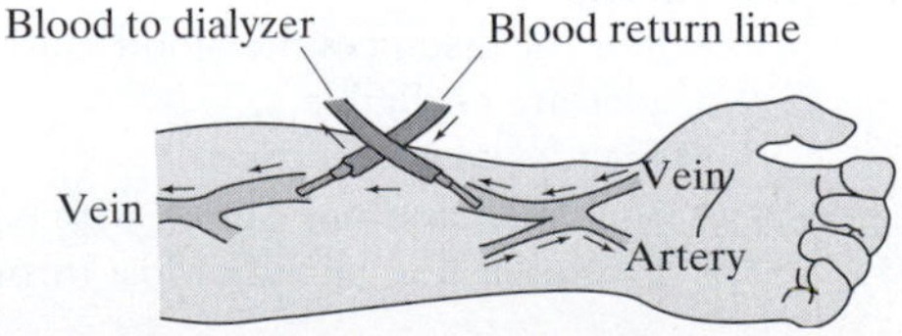

Figure 7C.6
Diagram of the arterio-venous fistula used in a hemodialysis machine. (*Source*: Kidney Dialysis Foundation, "Dialysis: Related care.")

vein. In the chamber, the semipermeable membranes allow solutes in the blood that meet the molecular weight restrictions (e.g., urea, glucose, Na^+, Cl^-) to pass through the walls of the tubing while retaining the larger proteins and cells.

To create the concentration gradients necessary for diffusion of these molecules to occur, the membranes are immersed in dialysate. A solution of purified water that contains concentrations of the extracted solutes near or slightly lower than the desired concentrations in the blood, the dialysate usually has a pH between 7 and 7.8. In the hollow fiber container, the blood and dialysate flow in opposite directions. This countercurrent exchange design allows more toxins to diffuse from the blood in the artificial kidney. In this manner of diffusion, excess unwanted substances in the blood are removed from the body into the dialysate (Table 7C.2). For maintenance of low-molecular-weight species vital to the blood, the species concentration in the dialysate is equal such that the two solutions are in dynamic equilibrium. The flow rates of both the blood and the dialysate control the rate at which fluid and wastes are exchanged between the two streams. Usually a volume of 120 L of dialysis fluid is needed to clean the blood of a single patient. Blood is considered clean when levels of creatinine, which cannot be completely filtered even in the natural kidney, reach 1 mg/dL.

TABLE 7C.2

Typical Dialysate Composition*

Component	Concentration (g/L)	Component ion	Concentration (meq/L)
NaCl	5.8	Na^+	132
$NaHCO_3$	4.5	K^+	2.0
KCl	0.15	Cl^-	105
$CaCl_2$	0.18	HCO_3^-	33
$MgCl_2$	0.15	Ca^{2+}	2.5
$C_6H_{12}O_6$	2.0	Mg^{2+}	1.5

*Table from Cooney DO, *Biomedical Engineering Principles: An Introduction to Fluid, Heat, and Mass Transport Processes*. New York: Marcel Dekker, 1976.

EXAMPLE 7C.3 Pump Work During Hemodialysis

Problem: During hemodialysis, 20 J/min of energy is lost to frictional resistances in the tubing. How much work must the pump do to move the blood from the patient to the machine and back to the patient? Blood exits the body through an artery and returns through a vein located in the patient's arm. Blood volumetric flow rate through the dialysis machine is 300 mL/min.

Solution:

1. Assemble
   (a) Find: work required to pump blood from patient to dialysis machine and back.
   (b) Diagram: The system is shown in Figure 7C.7.
2. Analyze
   (a) Assume:
   - For the distances the blood must travel, the changes in potential and kinetic energies are negligible.
   - Steady-state operation.
   - One inlet and one outlet, with equal tube diameters.
   - Energy is not gained or lost through any means other than friction.

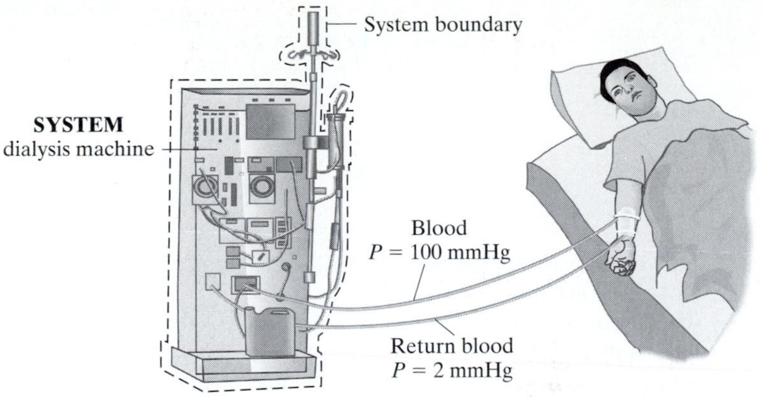

Figure 7C.7 Blood flow between patient and dialysis machine.

(b) Extra data:
- The density of whole blood is 1.056 g/mL.
- Venous gauge pressure is 2 mmHg.
- Arterial gauge pressure is 100 mmHg.

(c) Variables, notations, units:
- A = blood that exits the body through an artery and enters the machine
- V = blood that exits the machine and returns to the body through a vein
- Use J, min, mL, mmHg.

(d) Basis: Given the blood volumetric flow rate, we can find the blood mass flow rate to use as our basis:

$$\dot{m}_{\text{blood}} = \rho_{\text{blood}}\dot{V}_{\text{blood}} = \left(\frac{1.056\text{ g}}{\text{mL}}\right)\left(\frac{300\text{ mL}}{\text{min}}\right) = 316.8\ \frac{\text{g}}{\text{min}}$$

3. Calculate

(a) Equations: Since we are interested in the amount of work the dialysis pump does, we can use the mechanical energy accounting equation [6.10-8]:

$$\dot{m}(gh_i - gh_j) + \dot{m}\left(\frac{1}{2}v_i^2 - \frac{1}{2}v_j^2\right) + \frac{\dot{m}}{\rho}(P_i - P_j) + \sum \dot{W}_{\text{shaft}} - \sum \dot{f} = 0$$

(b) Calculate:
- Because we assume the potential and kinetic energy changes are negligible, we can reduce the mechanical energy accounting equation to:

$$\frac{\dot{m}}{\rho}(P_i - P_j) + \sum \dot{W}_{\text{shaft}} - \sum \dot{f} = \frac{\dot{m}_{\text{blood}}}{\rho_{\text{blood}}}(P_{\text{A}} - P_{\text{V}}) + \sum \dot{W}_{\text{pump}} - \sum \dot{f}_{\text{tubing}} = 0$$

Rearranging this equation and substituting the given values yields the work the dialysis pump must do to circulate the blood:

$$\sum \dot{W}_{\text{pump}} = \sum \dot{f}_{\text{tubing}} - \frac{\dot{m}_{\text{blood}}}{\rho_{\text{blood}}}(P_{\text{A}} - P_{\text{V}})$$

$$\sum \dot{W}_{\text{pump}} = 20\frac{\text{J}}{\text{min}} - \frac{316.8\frac{\text{g}}{\text{min}}}{\left(1.056\frac{\text{g}}{\text{cm}^3}\right)}(100\text{ mmHg} - 2\text{ mmHg})$$

$$\times \left(\frac{101{,}325 \dfrac{\text{J}}{\text{m}^3}}{760 \text{ mmHg}} \right) \left(\frac{1 \text{ m}}{100 \text{ cm}} \right)^3$$

$$\sum \dot{W}_{\text{pump}} = 16 \, \frac{\text{J}}{\text{min}}$$

4. Finalize
   (a) Answer: The pump must do 16 J/min of work to circulate the blood from the body to the machine and back.
   (b) Check: As seen by the pressure difference in the lines entering and exiting the dialysis machine, pressure drops across the dialysis machine, and energy is added to the system. However, there is a friction loss, which is greater that the energy gain in the pressure drop, so net nonflow work on the system is required. ■

Projected numbers of patients who will need treatment for ESRD in 2030 amount to 1.3 million diabetics and 945,000 nondiabetics—totaling more than 2.2 million patients.[2] Medicare's current expenditures for patients with kidney malfunctions alone equal about 6.4% ($22.8 billion) of their budget and continue to increase each year. Although hemodialysis is the best alternative treatment method available in response to the severe donor shortage, it is an expensive treatment that has many design issues which challenge engineers to further improve the machine without compromising cost efficiency. Some of the most serious issues include:

1. *Dehydration*: Regulating fluid volume is vital to maintaining blood pressure and enabling cellular transport.
2. *Infection*: This is common in patients undergoing hemodialysis, since multiple components of the machine must be properly disinfected, and repeated percutaneous injections, as with the needle insertions, increase the chance of microbes and other agents entering the body.
3. *Fluid flows*: How long each patient undergoes dialysis is determined by the area of tubing and how fast the dialysate and blood lines must flow to clean the blood while still maintaining blood pressure.
4. *Biocompatibility*: All materials used that come in contact with blood must not put the patient at risk
5. *Filration of waster:* Toxins of intermediate molecular weight do not easily cross the artificial kidney membrane, yet their bulildup in the blood is toxic for the patient.
6. H*ormone replacement:* Many of the hormones and fluids filtered out in the dialysate must be replaced, such as aldosterone. For example, injecting erythropoietin is difficult and hemoglobin substitutes often quickly deteriorate or clot, making fewer red blood cells available for oxygen and resulting in anemia and fatigue.
7. *Purifying water for dialysate*: Dialysate itself must be purified and kept from being stagnant, since removal of chlorine from the water makes it vulnerable to bacteria growth.
8. *Disposability and disinfection*: While the machine is reusable, components that come into contact with blood, such as the dialyzer and tubing, must be kept separate for each patient. Some of these components are disposable, such as the dialysate solution, and some can be disinfected. Often, however, the time and

money it takes to properly disinfect each component outweighs the benefits of avoiding disposal.

This ever-increasing prevalence of organ failure—not only the kidneys, but also the heart, lungs, liver, pancreas, and others—in the United States demands the attention and expertise of skilled bioengineers. As we learn more and more about how our bodies fail, we must step up and conceive innovative technologies to cope with the stresses we place on our body in order to improve our quality of lives and extend longevity.

References

1. "Genitourinary disorders." Beers MH and Berkow R, eds. *The Merck Manual of Diagnosis and Therapy*, 17th ed. Whitehouse Station, NJ: Merck Research Laboratories, 1999.
2. U.S. Renal Data System. "USRDS 2003 annual data report: Atlas of end-stage renal disease in the United States." National Institutes of Health, National Institutes of Diabetes and Digestive and Kidney Diseases. Bethesda, MD, 2003.
3. United Network for Organ Sharing. "2003 U.S. Organ Procurement and Transplantation Network and the Scientific Registry of Transplant Recipients Annual Report." 2005. http://www.optn.org/AR2003/default.htm (accessed January 22, 2005).
4. Fleck C and Braunlich H. "Kidney function after unilateral nephrectomy." *Exp Pathol* 1984, 25:3–18.

Problems

Part I—Kidney Function

7C.1 (G) Draw and label a diagram of the kidney and the major inlets and outlets to the kidney. What are the fluid flow rates in these major inlet and outlet streams?

7C.2 (G) Identify five important functions of the kidney. Discuss how each is critical in maintaining homeostasis in the body.

7C.3 (M) Competitive athletes consume large quantities of water to remain hydrated during strenuous exercise. Suppose two competitive runners need to rehydrate themselves as quickly as possible after completing a grueling marathon. People typically prefer cold beverages after exercise. However, a cool beverage may cause the esophagus to constrict, leading to a slower rate of water consumption and rehydration. Suppose two runners want to drink a volume, V. One decides to drink ice water and the other decides to drink water at room temperature. Derive an equation for the maximum water drinking time for a constricted esophagus in terms of the maximum water drinking time of an unconstricted esophagus, the radius of the esophagus in the unconstricted state, and the radius of the esophagus in the constricted state. Assume that the linear velocity of the liquid is the same in the constricted and unconstricted states. Assume that the radius of the esophagus decreases by 10% when cool beverages are consumed. Estimate and compare the consumption times for a constricted and unconstricted esophagus.

7C.4 (M) John empties his bladder after a long morning of working outside. He drinks 24 fl oz. of water, then eats three slices of pizza and drinks two beverages over the next 2 hours. Table 7C.3 summarizes his consumption. On a splendid day with moderate temperature and low humidity, John perspires at a rate of 0.1 mL/(m^2·min). After 2 hours John goes again to empty his bladder. Determine the amount of liquid excreted to return John to the state he was at prior to eating and drinking. Assume that the total amount of liquid in his body does not change. How do the air temperature and humidity affect the perspiration rate, and hence the total volume of liquid excreted to return John to his earlier state?

TABLE 7C.3

Summary of John's Meal Consumption

Product	Weight or volume	% Water content
Water	24 fl oz.	100
Pizza, 3 slices	0.5 kg/slice	20
Coke	8 fl oz.	100
Sprite	4 fl oz.	100

7C.5 (G) Different beverages have different effects on the kidneys. For example, some students drink coffee to help them endure long nights of studying for bioengineering tests. Water and coffee have very similar physical properties (e.g., density, viscosity), but they differ in terms of dictating kidney function: coffee is a diuretic and water is not. Explain how a diuretic works and what an individual taking a diuretic should expect in terms of kidney function. What happens at the cellular and biochemical level when an individual consumes a diuretic?

7C.6 (G) The kidney determines which molecules in the blood are retained in the body and which are excreted in the urine. Filtration selectively allows molecules to transfer between blood and excretion flow. Size and charge dictate the filtration of molecules in the kidney. Name five different constituents present in blood. Identify whether the constituent is primarily filtered (enters the excretion flow) or is primarily not filtered (is retained in the blood). Your list should include both types of constituents.

7C.7 (M) We have established that inulin is an ideal molecule for determining GFR (Example 7C.1). Suppose inulin is infused into an individual until a steady-state concentration of 0.1 g/(100 mL blood) is obtained. After steady-state is achieved, the inulin infusion is halted. With a catheter and sampling devices, you are able to take the instantaneous inulin concentration in the urine as well as the mixing-cup average of the inulin in the urine. (A mixing-cup average is the average concentration of all the samples collected up to a specified time.) Over a 2-hr period, 180 mL of urine is collected with an average inulin concentration of 0.08 g/mL.

(a) Determine the length of time until the instantaneous concentration of inulin in the blood drops to one-tenth of its original concentration.

(b) Use an integral mass balance to show that the total mass of inulin excreted in the urine is equivalent to the mass of inulin initially in the blood.

(c) Derive an equation describing the mixing-cup inulin concentration as a function of time, blood volume, GFR, collected urine volume, and other variables you find necessary.

(d) At what time does the mixing-cup inulin concentration equal the instantaneous inulin concentration?

(e) At what time does the mixing-cup inulin concentration equal two times the instantaneous inulin concentration?

7C.8 (M) Kidney removal (nephrectomy) has become common during the last decades. Once a partial nephrectomy (removal of one kidney) has been performed, the remaining kidney is required to compensate by increasing many of its activities. In particular, the remaining kidney typically increases its GFR to 75% of the original GFR function of both kidneys.[4] Determine the length of time until the instantaneous concentration of inulin in the blood drops to one-tenth its original concentration in a patient who has undergone a partial nephrectomy. How does this compare with the time calculated for an individual with normal kidney function? Use the data and calculations presented in Problem 7C.7.

Part II—Modeling the Nephron

In this part, you develop a sophisticated model of the nephron. Engineering principles and processes should drive the selection of the units in your model nephron. Major chemical constituents should be identified and tracked through the nephron. Understanding how chemical constituents are processed illuminates how the kidney works.

7C.9 (G) The nephron is the basic functional unit of the kidney. Draw and label a diagram of the nephron, including its major functional units. Describe the role of each major functional unit.

7C.10 (G) Model the nephron as a multi-unit system containing 6–10 units. For example, the Bowman's capsule could be a unit. Identify the primary engineering concept (e.g., filtration, reabsorption, etc.) that occurs in each of the units in your model nephron. Discuss the primary feature or characteristic that drove the selection of each unit.

7C.11 (G) Draw appropriate streams to connect the units. Determine the flow rate in each stream. (You may need to gather physiological data from books and journals.)

7C.12 (M) Identify 8–10 major chemical components in the blood that are processed in the nephron. You must consider water, bicarbonate, sodium, and urea. Develop mass balances on each of the chemical constituents. Determine the concentration of each component in each stream. (Again, you will need to gather data from books and journals. Using a computer may be helpful.) Present your data in a concise way, such as a table.

7C.13 (M) Can the chemical components from Problem 7C.12 be grouped into classes of compounds based on their patterns of movement through the nephron? If so, describe.

7C.14 (M) Condense the 6- to 10-unit model of the nephron into only 2 to 4 units. Describe each unit and the engineering concept(s) occurring in it. Justify your selection in light of the conclusions from Problem 7C.13.

Part III—Kidney Diseases and the Hemodialysis Machine

7C.15 (C) Renal tubular acidosis (RTA) is a kidney abnormality that results in a decreased blood pH ($pH < 7.41$) and a change in urine pH. Type II RTA is the decreased reabsorption of bicarbonate ions in the proximal tubule. A bicarbonate titration test helps identify Type II RTA. Sodium bicarbonate ($NaHCO_3$) is infused into an individual orally or by IV to raise the bicarbonate concentration in the blood. In an individual with normal kidney function, bicarbonate acts as a H^+ sink, and blood pH increases. In an individual with Type II RTA, bicarbonate appears in urine quickly followed by a slower rise in blood pH and blood bicarbonate concentration.

Joanne walks into your office complaining of symptoms associated with RTA. You immediately check the pH of her blood and urine. Test results show pH values of 7.0 and 8.5 for blood and urine, respectively. Assume that Joanne weighs 150 lb_f and that she excretes urine at a rate of 1–1.5 L/day.

(a) Identify the pathophysiology and symptoms of Type II RTA.

(b) Determine the amount of sodium bicarbonate needed to raise Joanne's blood pH to normal (7.41), assuming that she does not have Type II RTA. Assume that 0.014 meq/min of bicarbonate leaves in the urine.

(c) Determine the amount of sodium bicarbonate needed to raise Joanne's blood pH to normal (7.41), assuming that she has Type II RTA. In an individual with normal kidney function, 0.48 meq/min of bicarbonate is reabsorbed in the proximal tubules. However, in an individual with Type II RTA, 0.48 meq/min of bicarbonate leaves in the urine.

(d) How do the amounts computed for parts (b) and (c) compare with clinical literature values? Discuss any similarities or differences.

7C.16 (M) Consider a patient undergoing hemodialysis (Figure 7C.8). The following equation describes the concentration of creatinine in blood after dialysis, C_{Bo}, as follows:

$$C_{Bo} = C_{Bi} \exp\left(\frac{-KA}{\dot{V}_B}\right)$$

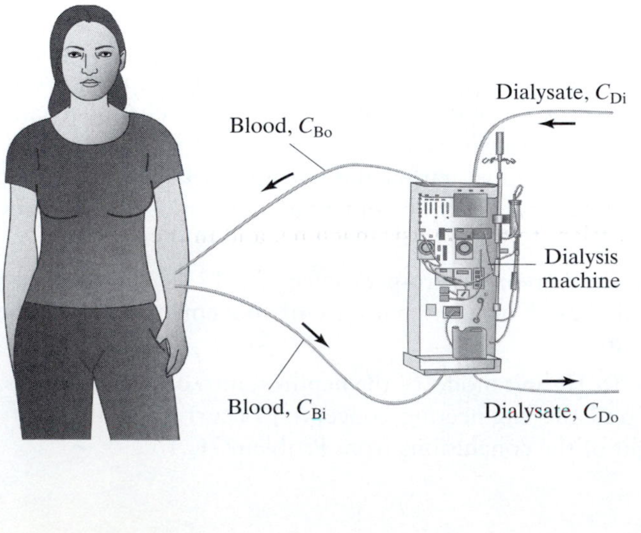

Figure 7C.8 Dialysate and blood flow in hemodialysis.

with variable definitions as follows: C_{Bo} is the concentration of creatinine in the bloodstream leaving the dialysis machine (i.e., prior to entering the patient), C_{Bi} is the concentration of creatinine in the bloodstream leaving the patient (i.e., prior to entering the dialysis machine), K is the mass transfer coefficient of creatinine through the artificial kidney, A is the surface area for exchange in the artificial kidney, and $\dot{V}_B$ is the volumetric flow rate of blood out of the patient, through the machine, and back into the patient.

Set up a dynamic balance on creatinine mass flow rate. Develop an equation for the creatinine concentration in the patient's blood as a function of time, and graph the result. The patient begins dialysis with a blood creatinine concentration of 10 mg/dL. In addition to the variables listed above, the solution may include the following variables: t is the time of dialysis machine operation, V is the volume of blood in body, and C_{Bi}^0 is the initial concentration of creatinine in the bloodstream leaving the patient (i.e., prior to entering the dialysis machine).

7C.17 (M) Dialysis technicians operate the machine until the blood concentration of creatinine reaches 1 mg/dL. Why don't the technicians run the machine until all creatinine is removed?

7C.18 (M) Consider the concentration of creatinine in the dialysate fluid over the course of dialysis. The following equation describes the mass of creatinine filtered from the blood to the dialysate in the dialysis unit as follows:

$$\dot{W} = KA\left[\frac{(C_{Bi} - C_{Di}) - (C_{Bo} - C_{Do})}{\ln\left(\dfrac{(C_{Bi} - C_{Di})}{(C_{Bo} - C_{Do})}\right)}\right]$$

with variable definitions as follows: $\dot{W}$ is the rate of mass of creatinine removed from the blood in the dialysis machine, C_{Di} is the concentration of creatinine in the dialysate stream entering the dialysis machine, and C_{Do} is the concentration of creatinine in the dialysate stream leaving the dialysis machine.

Using the information developed in Problem 7C.16, calculate the concentration of creatinine in the waste dialysate leaving the dialysis unit, C_{Do}, as a function of time. A numerical solution is required. In addition to the variables listed above, the solution may include $\dot{V}_D$, the volumetric flow rate of dialysate.

7C.19 (M) Based on the results from Problem 7C.18, graph and describe how C_{Bo} and C_{Do} change as a result of the operational parameters $\dot{V}_B$ and $\dot{V}_D$.

7C.20 (M) Based on the results from Problem 7C.18, graph and describe how C_{Bo} and C_{Do} change as a function of surface area (A).

7C.21 (P) Calculate the Reynolds numbers in the needle in the patient's arm, in the tubes connecting the patient to the dialysis machine, and in the hollow-fiber membrane tubes. Is the flow laminar or turbulent in each region? A 15-gauge needle is inserted into the patient's arm. Specific measurements on the machine and tubing are listed in Table 7C.4.

7C.22 (E) The dialysate fluid that exchanges with the blood in the device must be at nearly the same temperature as the blood in order to avoid heating or cooling the blood before it reenters an individual. A two-stage heating

TABLE 7C.4

Artificial Kidney Machine Components

Cobe–Centrysystem 3	
Blood flow rate	200–300 mL/min
Dialysate flow rate	500 mL/min
Toray artificial kidney filter	
Number of inner tubes	11,000
Diameter of inner tubes	225 μm
Surface area	2.1 m^2
Tubing length	13.5 cm
Patient tubing	
Dialysate tubing diameter	0.5 in
Blood tubing diameter	0.375 in

system is used to heat the dialysate fluid from room temperature to 38°C (Figure 7C.9). The temperature of the fresh dialysate increases when it passes through a heat exchanger (stage I) with the waste (or processed) dialysate. Both streams leave the exchanger at the same temperature. Stage II involves a heater, which warms the fresh dialysate up to 38°C prior to entering the hollow-fiber reactor. Blood from the patient is at 37°C and must be returned to the patient no cooler than 36.5°C. The hollow-fiber reactor is not insulated and loses 1000 cal/min of heat. How much heat (energy) must be added to the system by the stage II heater? Also, determine the temperature of the waste dialysate.

7C.23 (G) Two nonreusable parts of the dialysis machine are the artificial kidney and the tubing between machine and the patient. What are two important properties of the tubing and materials that contact the blood and dialysate that should be considered in their design? What other factors should be considered (e.g., cost) in selecting materials for these parts?

7C.24 (G) Identify three kidney functions that are currently lacking in hemodialysis machines. Brainstorm possible solutions to these problems. Select one possible solution and describe it in detail. Include references to support your ideas.

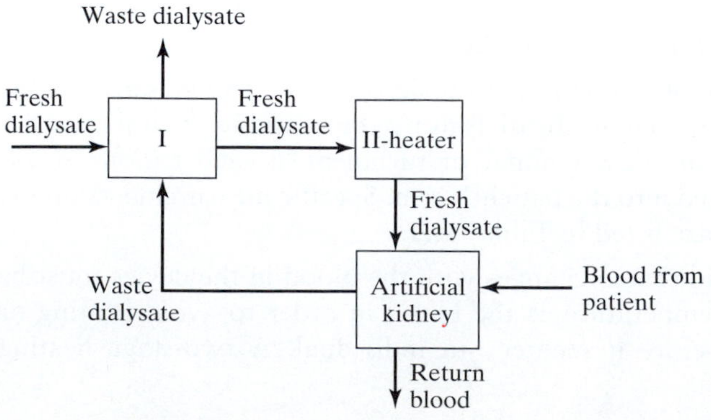

Figure 7C.9
Two-stage heating system used to heat dialysate fluid from room temperature to 38°C.

7C.25 (G) When designing a hemodialysis machine, you can control the volumetric flow rate of blood and of dialysate. How do issues such as time of operation, cost, heater requirements, pump requirements, and volume of dialysate affect your design of these controllable variables?

7C.26 (G) You are in the middle of developing a new and improved artificial kidney machine. What specific safety concerns do you need to address? What federal and/or state agencies regulate and oversee the safety of these devices?

7C.27 (G) As a technician operating the dialysis machine, you are concerned about environmental, health, and safety issues. Can you pour the waste dialysate fluid down the drain? What other operational concerns do you have? What state and/or federal agencies oversee the use and maintenance of these devices?

APPENDIX A List of Symbols

Symbol	Meaning
a	Acceleration [Lt^{-2}]
A	Area and cross-sectional area [L^2]
C	Mass concentration [$L^{-3}M$]
C	Molar concentration [$L^{-3}N$]
C	Capacitance [$L^{-2}M^{-1}t^4T^2$]
C_0	Initial mass concentration [$L^{-3}M$]
C_0	Initial molar concentration [$L^{-3}N$]
C_p	Heat capacity at constant pressure [$L^2t^{-2}T^{-1}$]
C_v	Heat capacity at constant volume [$L^2t^{-2}T^{-1}$]
D	Diameter [L]
E_T	Total energy [L^2Mt^{-2}]
$\dot{E}_T$	Rate of total energy [L^2Mt^{-3}]
E_E	Electrical energy [L^2Mt^{-2}]
$\dot{E}_E$	Rate of electrical energy [L^2Mt^{-3}]
E_K	Kinetic energy [L^2Mt^{-2}]
$\dot{E}_K$	Rate of kinetic energy [L^2Mt^{-3}]
$\hat{E}_K$	Specific kinetic energy [$L^2Mt^{-2}N^{-1}$]
E_P	Potential energy [L^2Mt^{-2}]
$\dot{E}_P$	Rate of potential energy [L^2Mt^{-3}]
$\hat{E}_P$	Specific potential energy [$L^2Mt^{-2}N^{-1}$]
EMF	Equilibrium potential [$L^2Mt^{-3}T^{-1}$]
f	Fractional conversion [−]
f	Frequency [t^{-1}]
f	Rotational frequency [t^{-1}]
$\dot{f}$	Frictional losses [L^2Mt^{-3}]
F	Force [LMt^{-2}]
F_R	Resultant force [LMt^2]
g	Acceleration due to gravity [Lt^{-2}]
g_c	Conversion factor for force units [−]
$\dot{G}_{elec}$	Rate of electrical energy generated [L^2Mt^{-3}]
h	Height [L]
h	Heat transfer coefficient [$Mt^{-3}T^{-1}$]
H	Enthalpy [L^2Mt^{-2}]
$\hat{H}$	Specific enthalpy [$LM^2t^{-2}N^{-1}$]
$\dot{H}$	Rate of enthalpy [L^2Mt^{-3}]
H	Hematocrit [−]
H	Henry's law constant [L^2t^{-2}]
H_M	Molal humidity [−]
H_P	Percent humidity [−]
H_R	Relative humidity [−]
i	Current [I]
i	Inlet index [−]
j	Outlet index [−]
k	Inlet index [−]
k	Thermal conductivity [$LMt^{-3}T^{-1}$]
L	Angular momentum [L^2Mt^{-1}]
L	Inductance [$L^2Mt^{-2}I^{-2}$]
$\dot{L}$	Rate of angular momentum [L^2Mt^{-2}]
l	Length [L]
m	Mass [M]
m_A	Mass of constituent A [M]
$\dot{m}$	Mass flow rate [Mt^{-1}]
m_A	Mass flow rate of constituent A [Mt^{-1}]
M	Molecular weight or molar mass [MN^{-1}]
M_{av}	Average molecular weight [MN^{-1}]
n	Number of moles [N]
n_A	Number of moles of constituent A [N]
$\dot{n}$	Molar flow rate [$t^{-1}N$]
p	Linear momentum [LMt^{-1}]
$\dot{p}$	Rate of linear momentum [LMt^{-2}]
P	Power [L^2Mt^{-3}]
P	Pressure, vapor pressure [$L^{-1}Mt^{-2}$]
P_i	Partial pressure [$L^{-1}Mt^{-2}$]
P_i^*	Saturated vapor pressure [$L^{-1}Mt^{-2}$]
P_0	Ambient pressure [$L^{-1}Mt^{-2}$]
q	Charge [tI]
$\dot{q}$	Rate of charge [I]
q_+	Positive charge [tI]
q_-	Negative charge [tI]
Q	Heat [L^2Mt^{-2}]

$\dot{Q}$	Rate of heat [L^2Mt^{-3}]
r	Position vector [L]
r	Radius [L]
R	Ideal gas constant [$L^2Mt^{-2}T^{-1}N^{-1}$]
R	Rate of reaction [Mt^{-1}]
R	Resistance [$L^2Mt^{-3}I^{-2}$]
Re	Reynolds number [−]
RQ	Respiratory quotient [−]
S_M	Molal saturation [−]
S_P	Percent saturation [−]
S_R	Relative saturation [−]
SG	Specific gravity [−]
t	Time [t]
t_0	Initial time [t]
t_f	Final time [t]
T	Period [t]
T	Temperature [T]
U	Internal energy [L^2Mt^{-2}]
$\dot{U}$	Rate of internal energy [L^2Mt^{-3}]
$\hat{U}$	Specific internal energy [$L^2Mt^{-2}N^{-1}$]
v	Velocity [Lt^{-1}]
v	Voltage [$L^2Mt^{-3}I^{-1}$]
V	Volume [L^3]
$\dot{V}$	Volumetric flow rate [L^3t^{-1}]
$\hat{V}$	Specific volume [L^3N^{-1}]
w_A	Mass fraction of component A [−]
W	Weight [LMt^{-2}]
W	Work [L^2Mt^{-2}]
$\dot{W}$	Rate of work [L^2Mt^{-2}]
$\dot{W}_{elec}$	Rate of electrical energy consumed [L^2Mt^{-2}]
$\dot{W}_{flow}$	Rate of flow work [L^2Mt^{-2}]
$\dot{W}_{shaft}$	Rate of shaft work [L^2Mt^{-2}]
x	Direction in space [L]
x_A	Mole fraction of component A [−]
$\bar{x}$	Mean value [depends on system]
y	Direction in space [L]
z	Direction in space [L]
z	Height above a reference plane [L]
μ	Fluid viscosity [$L^{-1}Mt^{-1}$]
ρ	Density [$L^{-3}M$]
ρ_{ref}	Reference density [$L^{-3}M$]
σ	Stoichiometric coefficient of a compound [−]
σ	Standard deviation [depends on system]
Σ	Summation [−]
τ	Torque [L^2Mt^{-2}]
ψ	Extensive property [depends on system]
$\dot{\psi}$	Rate of extensive property [ψt^{-1}]
ω	Angular velocity [t^{-1}]

APPENDIX

B Factors for Unit Conversion

Factors for Unit Conversion

Quantity	Equivalent Values
Mass	1 kg = 1000 g = 0.001 metric ton = 2.20462 lb_m = 35.27392 oz 1 lb_m = 16 oz = 5 × 10^{-4} ton = 453.593 g = 0.453593 kg
Length	1 m = 100 cm = 1000 mm = 10^6 microns (μm) = 39.37 in = 3.2808 ft = 1.0936 yd = 0.0006214 mile 1 ft = 12 in = 1/3 yd = 0.3048 m = 30.48 cm
Volume	1 m^3 = 1000 L = 10^6 cm^3 = 10^6 mL = 35.3145 ft^3 = 220.83 imperial gallons = 264.17 gal = 1056.68 qt 1 ft^3 = 1728 in^3 = 7.4805 gal = 0.028317 m^3 = 28.317 L = 28,317 cm^3
Force	1 N = 1 kg·m/s^2 = 10^5 dynes = 10^5 g·cm/s^2 = 0.22481 lb_f 1 lb_f = 32.174 lb_m·ft/s^2 = 4.4482 N = 4.4482 × 10^5 dynes
Pressure	1 atm = 1.01325 × 10^5 N/m^2 (Pa) = 101.325 kPa = 1.01325 bar = 1.01325 × 10^6 $dynes/cm^2$ = 760 mmHg at 0°C (torr) = 10.333 m H_2O at 4°C = 14.696 lb_f/in^2 (psi) = 33.9 ft H_2O at 4°C = 29.921 in Hg at 0°C
Energy	1 J = 1 N·m = 10^7 ergs = 10^7 dyne·cm = 2.778 × 10^{-7} kW·hr = 0.23901 cal = 0.7376 ft-lb_f = 9.486 × 10^{-4} Btu
Power	1 W = 1 J/s = 0.23901 cal/s = 0.7376 ft·lb_f/s = 9.486 × 10^{-4} Btu/s = 1.341 × 10^{-3} hp

Example: The factor to convert grams to lb_m is $\left(\frac{2.20462\ lb_m}{1000\ g}\right)$.

APPENDIX C

Periodic Table of the Elements

The Nonmetals — Transition Elements — The Active Metals

IA	IIA	IIIB	IVB	VB	VIB	VIIB		VIIIB		IB	IIB	IIIA	IVA	VA	VIA	VIIA	Noble Gases
1 **H** 1.00794																	2 **He** 4.00260
3 **Li** 6.941	4 **Be** 9.01218											5 **B** 10.81	6 **C** 12.011	7 **N** 14.0067	8 **O** 15.9994	9 **F** 18.9984	10 **Ne** 20.1797
11 **Na** 22.98977	12 **Mg** 24.305											13 **Al** 26.98154	14 **Si** 28.0855	15 **P** 30.9738	16 **S** 32.066	17 **Cl** 35.4527	18 **Ar** 39.948
19 **K** 39.0983	20 **Ca** 40.078	21 **Sc** 44.9559	22 **Ti** 47.88	23 **V** 50.9415	24 **Cr** 51.996	25 **Mn** 54.9380	26 **Fe** 55.847	27 **Co** 58.9332	28 **Ni** 58.69	29 **Cu** 63.546	30 **Zn** 65.39	31 **Ga** 69.72	32 **Ge** 72.59	33 **As** 74.9216	34 **Se** 78.96	35 **Br** 79.904	36 **Kr** 83.80
37 **Rb** 85.4678	38 **Sr** 87.62	39 **Y** 88.9059	40 **Zr** 91.224	41 **Nb** 92.9064	42 **Mo** 95.94	43 **Tc** (98)	44 **Ru** 101.07	45 **Rh** 102.9055	46 **Pd** 106.42	47 **Ag** 107.8682	48 **Cd** 112.41	49 **In** 114.82	50 **Sn** 118.710	51 **Sb** 121.75	52 **Te** 127.60	53 **I** 126.9045	54 **Xe** 131.29
55 **Cs** 132.9054	56 **Ba** 137.33	57* **La** 138.9055	72 **Hf** 178.49	73 **Ta** 180.9479	74 **W** 183.85	75 **Re** 186.207	76 **Os** 190.2	77 **Ir** 192.22	78 **Pt** 195.08	79 **Au** 196.9665	80 **Hg** 200.59	81 **Ti** 204.383	82 **Pb** 207.2	83 **Bi** 208.9804	84 **Po** (209)	85 **At** (210)	86 **Rn** (~222)
87 **Fr** (223)	88 **Ra** 226.0254	89† **Ac** 227.0278	104 **Unq** (261)	105 **Unp** (262)	106 **Unh** (263)												

Inner Transition Metals

	58	59	60	61	62	63	64	65	66	67	68	69	70	71
*Lanthanides	58 **Ce** 140.12	59 **Pr** 140.9077	60 **Nd** 144.24	61 **Pm** (145)	62 **Sm** 150.36	63 **Eu** 151.965	64 **Gd** 157.25	65 **Tb** 158.9254	66 **Dy** 162.50	67 **Ho** 164.9304	68 **Er** 167.26	69 **Tm** 168.9342	70 **Yb** 173.04	71 **Lu** 174.967
†Actinides	90 **Th** 232.0381	91 **Pa** 231.036	92 **U** 238.0289	93 **Np** 237.048	94 **Pu** (244)	95 **Am** (243)	96 **Cm** (247)	97 **Bk** (247)	98 **Cf** (251)	99 **Es** (252)	100 **Fm** (257)	101 **Md** (258)	102 **No** (259)	103 **Lr** (262)

Tables of Biological Data

APPENDIX D

TABLE D.1

Molecular Weight of Common Biological Molecules

Biological molecule	Molecular weight (Da)
Water	18
Amino acid, average	135
Amino acid, range	57.06 (glycine)–186.21 (tryptophan)
Glucose	180
Cholesterol	387
Lipid membrane, average	600
DNA base pair, average	610
Integral membrane protein, average	60,000
Protein, range	5000–3,000,000
Hemoglobin	64,000
DNA in a haploid nucleus	1.83×10^{12}

TABLE D.2

Standard Man Biophysical Values*

Height	1.73 m or 5 ft 8 in
Mass	68 kg or 150 lb_m
Body surface area†	1.7 m^2
Body core temperature	37.0°C
Mean skin temperature	34.2°C
Heat capacity	0.86 kcal(kg · °C)
Percent body fat	12%
Body fluids	41.0 L (60% of body weight)
Basal metabolism	40 kcal(m^2 · hr) or 72 kcal/hr
Blood volume	5 L
Resting cardiac output	5 L/min
Systemic blood pressure	120/80 mmHg
Resting heart rate	65 beats/min

*Table modified from Cooney DO, *Biomedical Engineering Principles: An Introduction to Fluid, Heat, and Mass Transport Processes.* New York: Marcel Dekker, 1976.

†Data from Guyton AC and Hall JE. *Textbook of Medical Physiology,* 10th ed. Philadelphia: Saunders, 2000.

TABLE D.3

Standard Man Lung Values at Biological Temperature and Pressure*

Total lung capacity	6.0 L
Vital capacity	4.2 L
Ventilation rate	6.0 L/min
Alveolar ventilation rate	4.2 L/min
Tidal volume	500 mL
Dead space	150 mL
Breathing frequency	12 breaths/min
Pulmonary capillary blood volume	75 mL
O_2 consumption	284 mL/min
CO_2 production	227 mL/min
Respiratory quotient	0.80

*Table modified from Cooney DO, *Biomedical Engineering Principles: An Introduction to Fluid, Heat, and Mass Transport Processes.* New York: Marcel Dekker, 1976.

TABLE D.4

Content of a 70-kg Adult Man*

Constituent	Mass (g)
Water	41,400
Fat	12,600
Protein	12,600
Calcium	1,160
Phosphate	670
Carbohydrate	300
Potassium	150
Sulfur	112
Chlorine	85
Sodium	63
Other	860

*Table from Cooney DO, *Biomedical Engineering Principles: An Introduction to Fluid, Heat, and Mass Transport Processes.* New York: Marcel Dekker, 1976.

TABLE D.5

Average Daily Water Balance*

Water intake (mL)		Water excretion (mL)	
Drinking water	1200	Urine	1400
Water content of food	1000	Insensible water loss through skin	350
Water from oxidation	300	Insensible water loss through lungs	350
		Sweat	200
		Stool	200
Total	2500	Total	2500

*Table from Cooney DO, *Biomedical Engineering Principles: An Introduction to Fluid, Heat, and Mass Transport Processes.* New York: Marcel Dekker, 1976.

TABLE D.6

Physical Properties of Human Blood (Normal Adult Mean Values)*

Whole blood	Volume	5 L
	Density	1.056 g/cm^3
	pH	7.41
	Viscosity at normal hematocrit (37°C)	3.0 cP
	Venous hematocrit	
	Male	0.47
	Female	0.42
	Whole blood volume	78 mL/(kg body weight)
Plasma or serum	Volume	~3 L
	pH	7.3–7.5
	Viscosity (37°C)	1.2 cP
	Density	1.0239 g/cm^3
Erythrocytes	Volume	~2 L
	Density	1.098 g/cm^3
	Count	
	Male	5.4×10^9/(mL whole blood)
	Female	4.8×10^9/(mL whole blood)
	Corpuscular volume	87 μm^3
	Diameter	8.4 μm
	Hemoglobin concentration	0.335 g/(mL erythrocyte)
Leukocytes	Count	7.4×10^6/(mL whole blood)
	Diameter	7–20 μm
Platelets	Count	2.8×10^8/(mL whole blood)
	Diameter	2–5 μm
Dissolved gas concentrations at STP	Arterial O_2 content	0.195 mL O_2/(mL blood)
	Arterial CO_2 content	0.492 mL CO_2/(mL blood)
	Venous O_2 content	0.145 mL O_2/(mL blood)
	Venous CO_2 content	0.532 mL CO_2/(mL blood)

STP = standard temperature and pressure, 0°C and 1 atm.
*Table modified from Cooney DO, *Biomedical Engineering Principles: An Introduction to Fluid, Heat, and Mass Transport Processes*. New York: Marcel Dekker, 1976.

TABLE D.7

Approximate Blood Distribution in Vascular Bed of a Hypothetical Man*†

Pulmonary circulation	Volume (mL)	Systemic circulation	Volume (mL)
Pulmonary arteries	400	Aorta	100
Pulmonary capillaries	60	Systemic arteries	450
Venules	140	Systemic capillaries	300
Pulmonary veins	700	Venules	200
Total pulmonary system	1300	Systemic veins	2050
		Total systemic vessels	3100
Heart	250		
		Unaccounted (probably extra blood in reservoirs of liver and spleen)	550

*Hypothetical man is assumed to be age 30, weight 63 kg, height 178 cm, blood volume 5.2 L.
†Table from Cooney DO, *Biomedical Engineering Principles: An Introduction to Fluid, Heat, and Mass Transport Processes*. New York: Marcel Dekker, 1976.

TABLE D.8

Blood Flow to Different Organs and Tissues Under Basal Conditions*

Organ or tissue	Blood flow rate (mL/min)
Brain	700
Heart	150
Bronchial	150
Kidneys	1100
Liver, total	1350
Portal	1050
Arterial	300
Muscle (inactive state)	750
Bone	250
Skin (cool weather)	300
Thyroid gland	50
Adrenal glands	25
Other tissues	175
Total	5000

*Table modified from Cooney DO, *Biomedical Engineering Principles: An Introduction to Fluid, Heat, and Mass Transport Processes.* New York: Marcel Dekker, 1976.

TABLE D.9

Systemic Circulation of a Man*

Structure	Diameter (cm)	Blood velocity (cm/s)	Tube Reynolds number
Ascending aorta	2.0–3.2	63	3600–5800
Descending aorta	1.6–2.0	27	1200–1500
Large arteries	0.2–0.6	20–50	110–850
Capillaries	0.0005–0.001	0.05–0.1	0.0007–0.003
Large veins	0.5–1.0	15–20	210–570
Venae cavae	2.0	11–16	630–900

*Table from Cooney DO, *Biomedical Engineering Principles: An Introduction to Fluid, Heat, and Mass Transport Processes.* New York: Marcel Dekker, 1976.

TABLE D.10

Metabolism of Different Classes of Foods*

	Carbohydrate	Lipid	Protein
Liters O_2 used/g	0.81	1.96	0.94
Liters CO_2 produced/g	0.81	1.39	0.75
Respiratory quotient	1.00	0.71	0.80
Heat of reaction (kcal/g)	4.1	9.3	4.5

*Table from Cooney DO, *Biomedical Engineering Principles: An Introduction to Fluid, Heat, and Mass Transport Processes.* New York: Marcel Dekker, 1976.

TABLE D.11

Osmolar‡ Substances in Extracellular and Intracellular Fluids*

Constituent	Plasma ($mOsm/L\ H_2O$)	Interstitial ($mOsm/L\ H_2O$)	Intracellular ($mOsm/L\ H_2O$)
Na^+	142	139	14
K^+	4.2	4.0	140
Ca^{2+}	1.3	1.2	—
Mg^{2+}	0.8	0.7	20
Cl^-	108	108	4
HCO_3^-	24	28.3	10
$H_2PO_4^-$, HPO_4^{2-}	2	2	11
SO_4^{2-}	0.5	0.5	1
Phosphocreatine	—	—	45
Carnosine	—	—	14
Amino acids	2	2	8
Creatine	0.2	0.2	9
Lactate	1.2	1.2	1.5
Adenosine triphosphate (ATP)	—	—	5
Hexose monophosphate	—	—	3.7
Glucose	5.6	5.6	—
Protein	1.2	0.2	4
Urea	4	4	4
Other	4.8	3.9	10
Total constituent	301.8	300.8	301.2
	Plasma (mmHg)	Interstitial (mmHg)	Intracellular (mmHg)
P_{O_2}†	35	—	20
P_{CO_2}†	46	—	50
Total osmotic pressure (37°C)	5443	5423	5423
	Plasma	Interstitial	Intracellular
pH†	7.4	7.35	7.0

‡One Osmol is equal to one gram-molecular weight of undissociated solute.

*Table modified from Guyton AC and Hall JE, *Textbook of Medical Physiology*, 10th ed. Philadelphia: Saunders, 2000.

†Data from Cooney DO, *Biomedical Engineering Principles: An Introduction to Fluid, Heat, and Mass Transport Processes*. New York: Marcel Dekker, 1976.

TABLE D.12

Cell Structures and Functions

Cell structure or organelle	Function
Cell membrane	• Fluid lipid bilayer that provides protective barrier • Regulates chemical traffic into and out of cell
Cytoplasm	• Fluid portion containing dissolved proteins, electrolytes, glucose, fat globules, secretory vesicles, and organelles
Nucleus	• Directs cellular reproduction and metabolic activities • Contains the DNA that determines the characteristics of the cell's proteins
Nucleolus	• Involved in synthesis of rRNA and precursors of ribosomes
Granular (rough) endoplasmic reticulum (ER)	• Packages and transports proteins produced by ribosomes
Agranular (smooth) endoplasmic reticulum (ER)	• Synthesis of lipid substances and other enzymatic processes
Ribosomes	• Site of protein synthesis • Can be free in cytoplasm or attached to rough ER membrane
Golgi apparatus	• Functions in storage, modification, and packaging of secretory products • Major director of macromolecular transport within cells • Site of synthesis of polysaccharides from simple sugars and attachment of polysaccharides to lipids and proteins
Lysosomes	• Storage vesicles for hydrolytic enzymes for digestion of damaged cellular structures, ingested food particles, and unwanted matter such as bacteria
Peroxisomes	• Contain oxidative enzymes to catalyze condensation reactions such as the detoxification of alcohol and the oxidation of hydrogen peroxide
Mitochondria	• Site of cellular respiration, the chemical reactions by which energy is extracted from nutrients and made available for energy-demanding cellular functions like metabolism
Microfilaments	• Serves as a structural component of the cytoskeleton • Involved in cell movement such as muscle contraction and intracellular transport of vesicles
Microtubules	• Provide general structure to cell • Move chromosomes to locations of new nuclei during cell division • Define pathways to be followed by secretory vesicles
Centrioles	• Focus of the microtubule spindle during cell division
Cilia and flagella	• Functions in moving the cell or in moving liquids or small particles across the surface of the cell

TABLE D.13

Physical Cell Properties*	
Mammalian cell composition by weight (except fat cells):	
Water	70%
Proteins	18%
Miscellaneous small metabolites	3%
Phospholipids	3%
Other lipids	2%
Polysaccharides	2%
Inorganic ions (Na^+, K^+, Mg^{2+}, Ca^{2+}, Cl^-, etc.)	1%
RNA	1.1%
DNA	0.25%
Cell membrane composition by mass:	
Proteins	55%
Phospholipids	25%
Cholesterol	13%
Other lipids	4%
Carbohydrates	3%
Cell membrane thickness, range	7.5–10 nm
Cell diameter, range	10–20 μm
Cell volume, approximate	4×10^{-9} cm^3
Number of chromosomes	46 (23 pairs)
Nuclear volume, approximate	2.4×10^{-10} cm^3
Number of different types of proteins	10,000
Mass of proteins	0.18 ng
Mass of DNA	0.0025 ng
Number of cell surface receptors	500–100,000

*Data from Alberts B, Johnson A, Lewis J, et al., *Molecular Biology of the Cell,* 4th ed. New York: Garland Science, 2002.

TABLE D.14

Molecular DNA	
Number of base pairs per chromosome, average*	150 million
Length of a typical gene†	3000 base pairs
Number of bases	4 (A, C, G, T)
Number of bases in the human genome†	3 billion
Number of genes in the human genome‡	20,000–25,000
Portion of genome containing protein-coding sequences (exons) of genes†	10%
Number of active genes per cell§	10,000–20,000

*Data from Alberts B, Johnson A, Lewis J, et al., *Molecular Biology of the Cell,* 4th ed. New York: Garland Science, 2002.
†Casey D., *DOE Human Genome Program: Primer on Molecular Genetics.* U.S. Department of Energy. June 1992. http://www.ornl.gov/sci/techresources/Human_Genome/publicat/primer/primer.pdf (accessed August 7, 2005).
‡"How many genes are in the human genome?" Human Genome Project Information, http://www.ornl.gov/sci/techresources/Human_Genome/faq/genenumber.shtml (last modified October 27, 2004; accessed August 7, 2005).
§Szallasi Z., "Genetic network analysis: From the bench to computers and back," Second International Conference on Systems Biology, November 4, 2001. http://www.chip.org/people/zszallasi/ICSB2001+Tutorial.pdf (accessed August 7, 2005).

APPENDIX E Thermodynamic Data

TABLE E.1

Heat Capacities

Compound	State	Temperature unit	a	$b \times 10^2$	$c \times 10^5$	$d \times 10^9$	Temperature range (units of T)
Acetone	g	°C	71.96	20.10	−12.78	34.76	0–1200
Air	g	°C	28.94	0.4147	0.3191	−1.965	0–1500
	g	K	28.09	0.1965	0.4799	−1.965	273–1800
Ammonia	g	°C	35.15	2.954	0.4421	−6.686	0–1200
Calcium hydroxide	c	K	89.5				276–373
Carbon dioxide	g	°C	36.11	4.233	−2.887	7.464	0–1500
Ethanol	*l*	°C	103.1				0
	l	°C	158.8				100
	g	°C	61.34	15.72	−8.749	19.83	0–1200
Formaldehyde	g	°C	34.28	4.268	0.000	−8.694	0–1200
Hydrogen	g	°C	28.84	0.00765	0.3288	−0.8698	0–1500
Hydrogen chloride	g	°C	29.13	−0.1341	0.9715	−4.335	0–1200
Hydrogen sulphide	g	°C	33.51	1.547	0.3012	−3.292	0–1500
Methane	g	°C	34.31	5.469	0.3661	−11.00	0–1200
	g	K	19.87	5.021	1.268	−11.00	273–1500
Methanol	*l*	°C	75.86				0
	l	°C	82.59				40
	g	°C	42.93	8.301	−1.87	−8.03	0–700
Nitric acid	*l*	°C	110.0				25
Nitrogen	g	°C	29.00	0.2199	0.5723	−2.871	0–1500
Oxygen	g	°C	29.10	1.158	−0.6076	1.311	0–1500
Sulphur							
(rhombic)	c	K	15.2	2.68			273–368
(monoclinic)	c	K	18.3	1.84			368–392
Sulphuric acid	*l*	°C	139.1	15.59			10–45
Sulphur dioxide	g	°C	38.91	3.904	−3.104	8.606	0–1500
Water	*l*	°C	75.4				0–100
	g	°C	33.46	0.6880	0.7604	−3.593	0–1500

Data from Felder RM and Rousseau RW, *Elementary Principles of Chemical Processes*. New York: John Wiley & Sons, 1978.

Table from Doran PM, *Bioprocess Engineering Principles*. London: Academic Press, 1995.

C_p (J/(mol · °C)) $= a + bT + cT^2 + dT^3$

Example. For acetone gas between 0°C and 1200°C:

C_p (J/(mol · °C)) $= 71.96 + (20.10 \times 10^{-2})T - (12.78 \times 10^{-5})T^2 + (34.76 \times 10^{-9})T^3$, where T is in °C.

Note that some equations require T in K, as indicated.

State: g = gas; *l* = liquid; c = crystal.

TABLE E.2

Mean Heat Capacities of Gases

T (°C)	C_p (J/mol · °C)					
	Air	O_2	N_2	H_2	CO_2	H_2O
0	29.06	29.24	29.12	28.61	35.96	33.48
18	29.07	29.28	29.12	28.69	36.43	33.51
25	29.07	29.30	29.12	28.72	36.47	33.52
100	29.14	29.53	29.14	28.98	38.17	33.73
200	29.29	29.93	29.23	29.10	40.12	34.10
300	29.51	30.44	29.38	29.15	41.85	34.54
400	29.78	30.88	29.60	29.22	43.35	35.05
500	30.08	31.33	29.87	29.28	44.69	35.59

Data from Himmelblau DM, *Basic Principles and Calculations in Chemical Engineering*, 3d ed. Englewood Cliffs, NJ: Prentice-Hall, 1974.

Table from Doran PM, *Bioprocess Engineering Principles*. London: Academic Press, 1995. *Reference state*: $T_{ref} = 0°C$; $P_{ref} = 1$ atm.

TABLE E.3

Specific Heats of Organic Liquids

Compound	Formula	Temperature (°C)	C_p (cal/(g · °C))
Acetic acid	$C_2H_4O_2$	26–95	0.522
Acetone	C_3H_6O	3–22.6	0.514
		0	0.506
		24.2–49.4	0.538
Acetonitrile	C_2H_3N	21–76	0.541
Benzaldehyde	C_7H_6O	22–172	0.428
Butyl alcohol (*n*-)	$C_4H_{10}O$	2.3	0.526
		19.2	0.563
		21–115	0.687
		30	0.582
Butyric acid (*n*-)	$C_4H_8O_2$	0	0.444
		40	0.501
		20–100	0.515
Carbon tetrachloride	CCl_4	0	0.198
		20	0.201
		30	0.200
Chloroform	$CHCl_3$	0	0.232
		15	0.226
		30	0.234
Cresol (*o*-)	C_7H_8O	0–20	0.497
Cresol (*m*-)	C_7H_8O	21–197	0.551
		0–20	0.477
Dichloroacetic acid	$C_2H_2Cl_2O_2$	21–106	0.349
		21–196	0.348
Diethylamine	$C_4H_{11}N$	22.5	0.516
Diethyl malonate	$C_7H_{12}O_4$	20	0.431
Diethyl oxalate	$C_6H_{10}O_4$	20	0.431
Diethyl succinate	$C_8H_{14}O_4$	20	0.450
Dipropyl malonate	$C_9H_{16}O_4$	20	0.431

TABLE E.3 *(Continued)*

Specific Heats of Organic Liquids

Compound	Formula	Temperature (°C)	C_p (cal/(g · °C))
Dipropyl oxalate (*n*-)	$C_8H_{14}O_4$	20	0.431
Dipropyl succinate	$C_{10}H_{18}O_4$	20	0.450
Ethanol	C_2H_6O	0–98	0.680
Ether	$C_4H_{10}O$	−5	0.525
		0	0.521
		30	0.545
		80	0.687
		120	0.800
		140	0.819
		180	1.037
Ethyl acetate	$C_4H_8O_2$	20	0.457
		20	0.476
Ethylene glycol	$C_2H_6O_2$	−11.1	0.535
		0	0.542
		2.5	0.550
		5.1	0.554
		14.9	0.569
		19.9	0.573
Formic acid	CH_2O_2	0	0.436
		15.5	0.509
		20–100	0.524
Furfural	$C_5H_4O_2$	0	0.367
		20–100	0.416
Glycerol	$C_3H_8O_3$	15–50	0.576
Hexadecane (*n*-)	$C_{16}H_{34}$	0–50	0.496
Isobutyl acetate	$C_6H_{12}O_2$	20	0.459
Isobutyl alcohol	$C_4H_{10}O$	21–109	0.716
		30	0.603
Isobutyl succinate	$C_{12}H_{22}O_4$	0	0.442
Isobutyric acid	$C_4H_8O_2$	20	0.450
Lauric acid	$C_{12}H_{24}O_2$	40–100	0.572
		57	0.515
Methanol	CH_4O	5–10	0.590
		15–20	0.601
Methyl butyl ketone	$C_6H_{12}O$	21–127	0.553
Methyl ethyl ketone	C_4H_8O	20–78	0.549
Methyl formate	$C_2H_4O_2$	13–29	0.516
Methyl propionate	$C_4H_8O_2$	20	0.459
Palmitic acid	$C_{16}H_{32}O_2$	65–104	0.653
Propionic acid	$C_3H_6O_2$	0	0.444
		20–137	0.560
Propyl acetate (*n*-)	$C_5H_{10}O_2$	20	0.459
Propyl butyrate	$C_7H_{14}O_2$	20	0.459

(Continued)

TABLE E.3 *(Continued)*

Specific Heats of Organic Liquids

Compound	Formula	Temperature (°C)	C_p (cal/(g · °C))
Propyl formate (*n*-)	$C_4H_8O_2$	20	0.459
Pyridine	C_5H_5N	20	0.405
		21–108	0.431
		0–20	0.395
Quinoline	C_9H_7N	0–20	0.352
Salicylaldehyde	$C_7H_6O_2$	18	0.382
Stearic acid	$C_{18}H_{36}O_2$	75–137	0.550

Data from Perry RH, Green DW, Maloney JO, eds. *Chemical Engineers' Handbook*, 6th ed. New York: McGraw-Hill, 1984.

Table from Doran PM, *Bioprocess Engineering Principles*. London: Academic Press, 1995.

TABLE E.4

Normal Melting Points and Boiling Points, and Standard Heats of Phase Change

Compound	Molecular weight	Melting temperature (°C)	$\Delta\hat{H}_M$ at melting point (kJ/mol)	Normal boiling point (°C)	$\Delta\hat{H}_V$ at vaporization (boiling) point (kJ/mol)
Acetaldehyde	44.05	−123.7		20.2	25.1
Acetic acid	60.05	16.6	12.09	118.2	24.39
Acetone	58.08	−95.0	5.69	56.0	30.2
Ammonia	17.03	−77.8	5.653	−33.43	23.351
Benzaldehyde	106.12	−26.0		179.0	38.40
Carbon dioxide	44.01	−56.6	8.33	(sublimates at −78°C)	
Chloroform	119.39	−63.7		61.0	
Ethanol	46.07	−114.6	5.021	78.5	38.58
Formaldehyde	30.03	−92		−19.3	24.48
Formic acid	46.03	8.30	12.68	100.5	22.25
Glycerol	92.09	18.20	18.30	290.0	
Hydrogen	2.016	−259.19	0.12	−252.76	0.904
Hydrogen chloride	36.47	−114.2	1.99	−85.0	16.1
Hydrogen sulphide	34.08	−85.5	2.38	−60.3	18.67
Methane	16.04	−182.5	0.94	−161.5	8.179
Methanol	32.04	−97.9	3.167	64.7	35.27
Nitric acid	63.02	−41.6	10.47	86	30.30
Nitrogen	28.02	−210.0	0.720	−195.8	5.577
Oxalic acid	90.04			(decomposes at 186°C)	
Oxygen	32.00	−218.75	0.444	−182.97	6.82
Phenol	94.11	42.5	11.43	181.4	
Phosphoric acid	98.00	42.3	10.54		
Sodium chloride	58.45	808	28.5	1465	170.7
Sodium hydroxide	40.00	319	8.34	1390	
Sulphur					
(rhombic)	256.53	113	10.04	444.6	83.7
(monoclinic)	256.53	119	14.17	444.6	83.7
Sulphur dioxide	64.07	−75.48	7.402	−10.02	24.91
Sulphuric acid	98.08	10.35	9.87	(decomposes at 340°C)	
Water	18.016	0.00	6.0095	100.00	40.656

Data from Felder RM and Rousseau RW, *Elementary Principles of Chemical Processes*. New York: John Wiley, 1978.

Table from Doran PM, *Bioprocess Engineering Principles*. London: Academic Press, 1995.
All thermodynamic data are at 1 atm.

TABLE E.5

Properties of Saturated Steam (SI Units): Temperature Table

		$\hat{V}$ (m^3/kg)		$\hat{U}$ (kJ/kg)		$\hat{H}$ (kJ/kg)		
T (°C)	P (bar)	Water	Steam	Water	Steam	Water	Evaporation ($\hat{H}_V$)	Steam
0.01	0.00611	0.001000	206.2	zero	2375.6	+0.0	2501.6	2501.6
2	0.00705	0.001000	179.9	8.4	2378.3	8.4	2496.8	2505.2
4	0.00813	0.001000	157.3	16.8	2381.1	16.8	2492.1	2508.9
6	0.00935	0.001000	137.8	25.2	2383.8	25.2	2487.4	2512.6
8	0.01072	0.001000	121.0	33.6	2386.6	33.6	2482.6	2516.2
10	0.01227	0.001000	106.4	42.0	2389.3	42.0	2477.9	2519.9
12	0.01401	0.001000	93.8	50.4	2392.1	50.4	2473.2	2523.6
14	0.01597	0.001001	82.9	58.8	2394.8	58.8	2468.5	2527.2
16	0.01817	0.001001	73.4	67.1	2397.6	67.1	2463.8	2530.9
18	0.02062	0.001001	65.1	75.5	2400.3	75.5	2459.0	2534.5
20	0.0234	0.001002	57.8	83.9	2403.0	83.9	2454.3	2538.2
22	0.0264	0.001002	51.5	92.2	2405.8	92.2	2449.6	2541.9
24	0.0298	0.001003	45.9	100.6	2408.5	100.6	2444.9	2545.5
25	0.0317	0.001003	43.4	104.8	2409.9	104.8	2442.5	2547.3
26	0.0336	0.001003	41.0	108.9	2411.2	108.9	2440.2	2549.1
28	0.0378	0.001004	36.7	117.3	2414.0	117.3	2435.4	2552.7
30	0.0424	0.001004	32.9	125.7	2416.7	125.7	2430.7	2556.4
32	0.0475	0.001005	29.6	134.0	2419.4	134.0	2425.9	2560.0
34	0.0532	0.001006	26.6	142.4	2422.1	142.4	2421.2	2563.6
36	0.0594	0.001006	24.0	150.7	2424.8	150.7	2416.4	2567.2
38	0.0662	0.001007	21.6	159.1	2427.5	159.1	2411.7	2570.8
40	0.0738	0.001008	19.55	167.4	2430.2	167.5	2406.9	2574.4
42	0.0820	0.001009	17.69	175.8	2432.9	175.8	2402.1	2577.9
44	0.0910	0.001009	16.04	184.2	2435.6	184.2	2397.3	2581.5
46	0.1009	0.001010	14.56	192.5	2438.3	192.5	2392.5	2585.1
48	0.1116	0.001011	13.23	200.9	2440.9	200.9	2387.7	2588.6
50	0.1234	0.001012	12.05	209.2	2443.6	209.3	2382.9	2592.2
52	0.1361	0.001013	10.98	217.7	2446	217.7	2377	2595
54	0.1500	0.001014	10.02	226.0	2449	226.0	2373	2599
56	0.1651	0.001015	9.158	234.4	2451	234.4	2368	2602
58	0.1815	0.001016	8.380	242.8	2454	242.8	2363	2606
60	0.1992	0.001017	7.678	251.1	2456	251.1	2358	2609
62	0.2184	0.001018	7.043	259.5	2459	259.5	2353	2613
64	0.2391	0.001019	6.468	267.9	2461	267.9	2348	2616
66	0.2615	0.001020	5.947	276.2	2464	276.2	2343	2619
68	0.2856	0.001022	5.475	284.6	2467	284.6	2338	2623
70	0.3117	0.001023	5.045	293.0	2469	293.0	2333	2626
72	0.3396	0.001024	4.655	301.4	2472	301.4	2329	2630
74	0.3696	0.001025	4.299	309.8	2474	309.8	2323	2633
76	0.4019	0.001026	3.975	318.2	2476	318.2	2318	2636
78	0.4365	0.001028	3.679	326.4	2479	326.4	2313	2639
80	0.4736	0.001029	3.408	334.8	2482	334.9	2308	2643
82	0.5133	0.001030	3.161	343.2	2484	343.3	2303	2646
84	0.5558	0.001032	2.934	351.6	2487	351.7	2298	2650
86	0.6011	0.001033	2.727	360.0	2489	360.1	2293	2653

TABLE E.5 *(Continued)*

Properties of Saturated Steam (SI Units): Temperature Table

		$\hat{V}$ (m³/kg)		$\hat{U}$ (kJ/kg)		$\hat{H}$ (kJ/kg)		
T (°C)	P (bar)	Water	Steam	Water	Steam	Water	Evaporation ($\hat{H}_V$)	Steam
88	0.6495	0.001034	2.536	368.4	2491	368.5	2288	2656
90	0.7011	0.001036	2.361	376.9	2493	377.0	2282	2659
92	0.7560	0.001037	2.200	385.3	2496	385.4	2277	2662
94	0.8145	0.001039	2.052	393.7	2499	393.8	2272	2666
96	0.8767	0.001040	1.915	401.1	2501	402.2	2267	2669
98	0.9429	0.001042	1.789	410.6	2504	410.7	2262	2673
100	1.0131	0.001044	1.673	419.0	2507	419.1	2257	2676
102	1.0876	0.001045	1.566	427.1	2509	427.5	2251	2679

Data from Haywood RW, *Thermodynamic Tables in SI (Metric) Units*. Cambridge University Press, 1968. Adapted with permission.
Table from Reklaitis GV, *Introduction to Material and Energy Balances*. New York: Wiley, 1983.
$\hat{V}$ = specific volume, $\hat{U}$ = specific internal energy, and $\hat{H}$ = specific enthalpy.

TABLE E.6

Properties of Saturated Steam (SI Units): Pressure Table

		$\hat{V}$ (m³/kg)		$\hat{U}$ (kJ/kg)		$\hat{H}$ (kJ/kg)		
P (bar)	T (°C)	Water	Steam	Water	Steam	Water	Evaporation ($\hat{H}_V$)	Steam
0.00611	0.01	0.001000	206.2	zero	2375.6	+0.0	2501.6	2501.6
0.008	3.8	0.001000	159.7	15.8	2380.7	15.8	2492.6	2508.5
0.010	7.0	0.001000	129.2	29.3	2385.2	29.3	2485.0	2514.4
0.012	9.7	0.001000	108.7	40.6	2388.9	40.6	2478.7	2519.3
0.014	12.0	0.001000	93.9	50.3	2392.0	50.3	2473.2	2523.5
0.016	14.0	0.001001	82.8	58.9	2394.8	58.9	2468.4	2527.3
0.018	15.9	0.001001	74.0	66.5	2397.4	66.5	2464.1	2530.6
0.020	17.5	0.001001	67.0	73.5	2399.6	73.5	2460.2	2533.6
0.022	19.0	0.001002	61.2	79.8	2401.7	79.8	2456.6	2536.4
0.024	20.4	0.001002	56.4	85.7	2403.6	85.7	2453.3	2539.0
0.026	21.7	0.001002	52.3	91.1	2405.4	91.1	2450.2	2541.3
0.028	23.0	0.001002	48.7	96.2	2407.1	96.2	2447.3	2543.6
0.030	24.1	0.001003	45.7	101.0	2408.6	101.0	2444.6	2545.6
0.035	26.7	0.001003	39.5	111.8	2412.2	111.8	2438.5	2550.4
0.040	29.0	0.001004	34.8	121.4	2415.3	121.4	2433.1	2554.5
0.045	31.0	0.001005	31.1	130.0	2418.1	130.0	2428.2	2558.2
0.050	32.9	0.001005	28.2	137.8	2420.6	137.8	2423.8	2561.6
0.060	36.2	0.001006	23.74	151.5	2425.1	151.5	2416.0	2567.5
0.070	39.0	0.001007	20.53	163.4	2428.9	163.4	2409.2	2572.6
0.080	41.5	0.001008	18.10	173.9	2432.3	173.9	2403.2	2577.1
0.090	43.8	0.001009	16.20	183.3	2435.3	183.3	2397.9	2581.1
0.10	45.8	0.001010	14.67	191.8	2438.0	191.8	2392.9	2584.8
0.11	47.7	0.001011	13.42	199.7	2440.5	199.7	2388.4	2588.1
0.12	49.4	0.001012	12.36	206.9	2442.8	206.9	2384.3	2591.2
0.13	51.1	0.001013	11.47	213.7	2445.0	213.7	2380.4	2594.0
0.14	52.6	0.001013	10.69	220.0	2447.0	220.0	2376.7	2596.7
0.15	54.0	0.001014	10.02	226.0	2448.9	226.0	2373.2	2599.2

(Continued)

TABLE E.6 *(Continued)*

Properties of Saturated Steam (SI Units): Pressure Table

		$\hat{V}$ (m^3/kg)		$\hat{U}$ (kJ/kg)		$\hat{H}$ (kJ/kg)		
P (bar)	T (°C)	Water	Steam	Water	Steam	Water	Evaporation ($\hat{H}_V$)	Steam
0.16	55.3	0.001015	9.43	231.6	2450.6	231.6	2370.0	2601.6
0.17	56.6	0.001015	8.91	236.9	2452.3	236.9	2366.9	2603.8
0.18	57.8	0.001016	8.45	242.0	2453.9	242.0	2363.9	2605.9
0.19	59.0	0.001017	8.03	246.8	2455.4	246.8	2361.1	2607.9
0.20	60.1	0.001017	7.65	251.5	2456.9	251.5	2358.4	2609.9
0.22	62.2	0.001018	7.00	260.1	2459.6	260.1	2353.3	2613.5
0.24	64.1	0.001019	6.45	268.2	2462.1	268.2	2348.6	2616.8
0.26	65.9	0.001020	5.98	275.6	2464.4	275.7	2344.2	2619.9
0.28	67.5	0.001021	5.58	282.7	2466.5	282.7	2340.0	2622.7
0.30	69.1	0.001022	5.23	289.3	2468.6	289.3	2336.1	2625.4
0.35	72.7	0.001025	4.53	304.3	2473.1	304.3	2327.2	2631.5
0.40	75.9	0.001027	3.99	317.6	2477.1	317.7	2319.2	2636.9
0.45	78.7	0.001028	3.58	329.6	2480.7	329.6	2312.0	2641.7
0.50	81.3	0.001030	3.24	340.5	2484.0	340.6	2305.4	2646.0
0.55	83.7	0.001032	2.96	350.6	2486.9	350.6	2299.3	2649.9
0.60	86.0	0.001033	2.73	359.9	2489.7	359.9	2293.6	2653.6
0.65	88.0	0.001035	2.53	368.5	2492.2	368.6	2288.3	2656.9
0.70	90.0	0.001036	2.36	376.7	2494.5	376.8	2283.3	2660.1
0.75	91.8	0.001037	2.22	384.4	2496.7	384.5	2278.6	2663.0
0.80	93.5	0.001039	2.087	391.6	2498.8	391.7	2274.1	2665.8
0.85	95.2	0.001040	1.972	398.5	2500.8	398.6	2269.8	2668.4
0.90	96.7	0.001041	1.869	405.1	2502.6	405.2	2265.6	2670.9
0.95	98.2	0.001042	1.777	411.4	2504.4	411.5	2261.7	2673.2
1.00	99.6	0.001043	1.694	417.4	2506.1	417.5	2257.9	2675.4
1.01325 (1 atm)	100.0	0.001044	1.673	419.0	2506.5	419.1	2256.9	2676.0
1.1	102.3	0.001046	1.549	428.7	2509.2	428.8	2250.8	2679.6
1.2	104.8	0.001048	1.428	439.2	2512.1	439.4	2244.1	2683.4
1.3	107.1	0.001049	1.325	449.1	2514.7	449.2	2237.8	2687.0
1.4	109.3	0.001051	1.236	458.3	2517.2	458.4	2231.9	2690.3
1.5	111.4	0.001053	1.159	467.0	2519.5	467.1	2226.2	2693.4
1.6	113.3	0.001055	1.091	475.2	2521.7	475.4	2220.9	2696.2
1.7	115.2	0.001056	1.031	483.0	2523.7	483.2	2215.7	2699.0
1.8	116.9	0.001058	0.977	490.5	2525.6	490.7	2210.8	2701.5
1.9	118.6	0.001059	0.929	497.6	2527.5	497.8	2206.1	2704.0
2.0	120.2	0.001061	0.885	504.5	2529.2	504.7	2201.6	2706.3
2.2	123.3	0.001064	0.810	517.4	2532.4	517.6	2193.0	2710.6
2.4	126.1	0.001066	0.746	529.4	2535.4	529.6	2184.9	2714.5
2.6	128.7	0.001069	0.693	540.6	2538.1	540.9	2177.3	2718.2
2.8	131.2	0.001071	0.646	551.1	2540.6	551.4	2170.1	2721.5
3.0	133.5	0.001074	0.606	561.1	2543.0	561.4	2163.2	2724.7
3.2	135.8	0.001076	0.570	570.6	2545.2	570.9	2156.7	2727.6
3.4	137.9	0.001078	0.538	579.6	2547.2	579.9	2150.4	2730.3
3.6	139.9	0.001080	0.510	588.1	2549.2	588.5	2144.4	2732.9
3.8	141.8	0.001082	0.485	596.4	2551.0	596.8	2138.6	2735.3
4.0	143.6	0.001084	0.462	604.2	2552.7	604.7	2133.0	2737.6
4.2	145.4	0.001086	0.442	611.8	2554.4	612.3	2127.5	2739.8
4.4	147.1	0.001088	0.423	619.1	2555.9	619.6	2122.3	2741.9
4.6	148.7	0.001089	0.405	626.2	2557.4	626.7	2117.2	2743.9
4.8	150.3	0.001091	0.389	633.0	2558.8	633.5	2112.2	2745.7

TABLE E.6 *(Continued)*

Properties of Saturated Steam (SI Units): Pressure Table

		$\hat{V}$ (m³/kg)		$\hat{U}$ (kJ/kg)		$\hat{H}$ (kJ/kg)		
P (bar)	*T* (°C)	Water	Steam	Water	Steam	Water	Evaporation ($\hat{H}_V$)	Steam
5.0	151.8	0.001093	0.375	639.6	2560.2	640.1	2107.4	2747.5
5.5	155.5	0.001097	0.342	655.2	2563.3	655.8	2095.9	2751.7
6.0	158.8	0.001101	0.315	669.8	2566.2	670.4	2085.0	2755.5
6.5	162.0	0.001105	0.292	683.4	2568.7	684.1	2074.7	2758.9
7.0	165.0	0.001108	0.273	696.3	2571.1	697.1	2064.9	2762.0
7.5	167.8	0.001112	0.2554	708.5	2573.3	709.3	2055.5	2764.8
8.0	170.4	0.001115	0.2403	720.0	2575.5	720.9	2046.5	2767.5
8.5	172.9	0.001118	0.2268	731.1	2577.1	732.0	2037.9	2769.9
9.0	175.4	0.001121	0.2148	741.6	2578.8	742.6	2029.5	2772.1
9.5	177.7	0.001124	0.2040	751.8	2580.4	752.8	2021.4	2774.2
10.0	179.9	0.001127	0.1943	761.5	2581.9	762.6	2013.6	2776.2
10.5	182.0	0.001130	0.1855	770.8	2583.3	772.0	2005.9	2778.0
11.0	184.1	0.001133	0.1774	779.9	2584.5	781.1	1998.5	2779.7
11.5	186.0	0.001136	0.1700	788.6	2585.8	789.9	1991.3	2781.3
12.0	188.0	0.001139	0.1632	797.1	2586.9	798.4	1984.3	2782.7
12.5	189.8	0.001141	0.1569	805.3	2588.0	806.7	1977.4	2784.1
13.0	191.6	0.001144	0.1511	813.2	2589.0	814.7	1970.7	2785.4
14	195.0	0.001149	0.1407	828.5	2590.8	830.1	1957.7	2787.8
15	198.3	0.001154	0.1317	842.9	2592.4	844.7	1945.2	2789.9
16	201.4	0.001159	0.1237	856.7	2593.8	858.6	1933.2	2791.7
17	204.3	0.001163	0.1166	869.9	2595.1	871.8	1921.5	2793.4
18	207.1	0.001168	0.1103	882.5	2596.3	884.6	1910.3	2794.8
19	209.8	0.001172	0.1047	894.6	2597.3	896.8	1899.3	2796.1
20	212.4	0.001177	0.0995	906.2	2598.2	908.6	1888.6	2797.2
21	214.9	0.001181	0.0949	917.5	2598.9	920.0	1878.2	2798.2
22	217.2	0.001185	0.0907	928.3	2599.6	931.0	1868.1	2799.1
23	219.6	0.001189	0.0868	938.9	2600.2	941.6	1858.2	2799.8
24	221.8	0.001193	0.0832	949.1	2600.7	951.9	1848.5	2800.4
25	223.9	0.001197	0.0799	959.0	2601.2	962.0	1839.0	2800.9
26	226.0	0.001201	0.0769	968.6	2601.5	971.7	1829.6	2801.4
27	228.1	0.001205	0.0740	978.0	2601.8	981.2	1820.5	2801.7
28	230.0	0.001209	0.0714	987.1	2602.1	990.5	1811.5	2802.0
29	232.0	0.001213	0.0689	996.0	2602.3	999.5	1802.6	2802.2
30	233.8	0.001216	0.0666	1004.7	2602.4	1008.4	1793.9	2802.3
32	237.4	0.001224	0.0624	1021.5	2602.5	1025.4	1776.9	2802.3
34	240.9	0.001231	0.0587	1037.6	2602.5	1041.8	1760.3	2802.1
36	244.2	0.001238	0.0554	1053.1	2602.2	1057.6	1744.2	2801.7
38	247.3	0.001245	0.0524	1068.0	2601.9	1072.7	1728.4	2801.1
40	250.3	0.001252	0.0497	1082.4	2601.3	1087.4	1712.9	2800.3
42	253.2	0.001259	0.0473	1096.3	2600.7	1101.6	1697.8	2799.4
44	256.0	0.001266	0.0451	1109.8	2599.9	1115.4	1682.9	2798.3
46	258.8	0.001272	0.0430	1122.9	2599.1	1128.8	1668.3	2797.1
48	261.4	0.001279	0.0412	1135.6	2598.1	1141.8	1653.9	2795.7
50	263.9	0.001286	0.0394	1148.0	2597.0	1154.5	1639.7	2794.2
52	266.4	0.001292	0.0378	1160.1	2595.9	1166.8	1625.7	2792.6
54	268.8	0.001299	0.0363	1171.9	2594.6	1178.9	1611.9	2790.8
56	271.1	0.001306	0.0349	1183.5	2593.3	1190.8	1598.2	2789.0
58	273.3	0.001312	0.0337	1194.7	2591.9	1202.3	1584.7	2787.0
60	275.6	0.001319	0.0324	1205.8	2590.4	1213.7	1571.3	2785.0
62	277.7	0.001325	0.0313	1216.6	2588.8	1224.8	1558.0	2782.9
64	279.8	0.001332	0.0302	1227.2	2587.2	1235.7	1544.9	2780.6

(Continued)

TABLE E.6 *(Continued)*

Properties of Saturated Steam (SI Units): Pressure Table

		$\hat{V}$ (m^3/kg)		$\hat{U}$ (kJ/kg)		$\hat{H}$ (kJ/kg)		
P (bar)	T (°C)	Water	Steam	Water	Steam	Water	Evaporation ($\hat{H}_V$)	Steam
66	281.8	0.001338	0.0292	1237.6	2585.5	1246.5	1531.9	2778.3
68	283.8	0.001345	0.0283	1247.9	2583.7	1257.0	1518.9	2775.9
70	285.8	0.001351	0.0274	1258.0	2581.8	1267.4	1506.0	2773.5
72	287.7	0.001358	0.0265	1267.9	2579.9	1277.6	1493.3	2770.9
74	289.6	0.001364	0.0257	1277.6	2578.0	1287.7	1480.5	2768.3
76	291.4	0.001371	0.0249	1287.2	2575.9	1297.6	1467.9	2765.5
78	293.2	0.001378	0.0242	1296.7	2573.8	1307.4	1455.3	2762.8
80	295.0	0.001384	0.0235	1306.0	2571.7	1317.1	1442.8	2759.9
82	296.7	0.001391	0.0229	1315.2	2569.5	1326.6	1430.3	2757.0
84	298.4	0.001398	0.0222	1324.3	2567.2	1336.1	1417.9	2754.0
86	300.1	0.001404	0.0216	1333.3	2564.9	1345.4	1405.5	2750.9
88	301.7	0.001411	0.0210	1342.2	2562.6	1354.6	1393.2	2747.8
90	303.3	0.001418	0.02050	1351.0	2560.1	1363.7	1380.9	2744.6
92	304.9	0.001425	0.01996	1359.7	2557.7	1372.8	1368.6	2741.4
94	306.4	0.001432	0.01945	1368.2	2555.2	1381.7	1356.3	2738.0
96	308.0	0.001439	0.01897	1376.7	2552.6	1390.6	1344.1	2734.7
98	309.5	0.001446	0.01849	1385.2	2550.0	1399.3	1331.9	2731.2
100	311.0	0.001453	0.01804	1393.5	2547.3	1408.0	1319.7	2727.7
105	314.6	0.001470	0.01698	1414.1	2540.4	1429.5	1289.2	2718.7
110	318.0	0.001489	0.01601	1434.2	2533.2	1450.6	1258.7	2709.3
115	321.4	0.001507	0.01511	1454.0	2525.7	1471.3	1228.2	2699.5
120	324.6	0.001527	0.01428	1473.4	2517.8	1491.8	1197.4	2689.2
125	327.8	0.001547	0.01351	1492.7	2509.4	1512.0	1166.4	2678.4
130	330.8	0.001567	0.01280	1511.6	2500.6	1532.0	1135.0	2667.0
135	333.8	0.001588	0.01213	1530.4	2491.3	1551.9	1103.1	2655.0
140	336.6	0.001611	0.01150	1549.1	2481.4	1571.6	1070.7	2642.4
145	339.4	0.001634	0.01090	1567.5	2471.0	1591.3	1037.7	2629.1
150	342.1	0.001658	0.01034	1586.1	2459.9	1611.0	1004.0	2615.0
155	344.8	0.001683	0.00981	1604.6	2448.2	1630.7	969.6	2600.3
160	347.3	0.001710	0.00931	1623.2	2436.0	1650.5	934.3	2584.9
165	349.8	0.001739	0.00883	1641.8	2423.1	1670.5	898.3	2568.8
170	352.3	0.001770	0.00837	1661.6	2409.3	1691.7	859.9	2551.6
175	354.6	0.001803	0.00793	1681.8	2394.6	1713.3	820.0	2533.3
180	357.0	0.001840	0.00750	1701.7	2378.9	1734.8	779.1	2513.9
185	359.2	0.001881	0.00708	1721.7	2362.1	1756.5	736.6	2493.1
190	361.4	0.001926	0.00668	1742.1	2343.8	1778.7	692.0	2470.6
195	363.6	0.001977	0.00628	1763.2	2323.6	1801.8	644.2	2446.0
200	365.7	0.00204	0.00588	1785.7	2300.8	1826.5	591.9	2418.4
205	367.8	0.00211	0.00546	1810.7	2274.4	1853.9	532.5	2386.4
210	369.8	0.00220	0.00502	1840.0	2242.1	1886.3	461.3	2347.6
215	371.8	0.00234	0.00451	1878.6	2198.1	1928.9	366.2	2295.2
220	373.7	0.00267	0.00373	1952	2114	2011	185	2196
221.2 (critical point)	374.15	0.00317	0.00317	2038	2038	2108	0	2108

Data from Haywood RW, *Thermodynamic Tables in SI (Metric) Units*. Cambridge University Press, 1968. Adapted with permission.
Table from Reklaitis GV, *Introduction to Material and Energy Balances*. New York: Wiley, 1983.
$\hat{V}$ = specific volume, $\hat{U}$ = specific internal energy, and $\hat{H}$ = specific enthalpy.

TABLE E.7

Thermodynamic Data for Organic Compounds (all values are for 298 K and 1 bar)

	M (g/mol)	$\Delta\hat{H}_f^o$ (kJ/mol)	C_p (J/(mol·K))	$\Delta\hat{H}_c^o$ (kJ/mol)
C(s) (graphite)	12.011	0	8.527	−393.51
C(s) (diamond)	12.011	+1.895	6.113	−395.40
CO_2(g)	44.010	−393.51	37.11	
Hydrocarbons				
CH_4(g), methane	16.04	−74.81	35.31	−890
CH_3(g), methyl	15.04	+145.69	38.70	
C_2H_2(g), ethyne	26.04	+226.73	43.93	−1300
C_2H_4(g), ethene	28.05	+52.26	43.56	−1411
C_2H_6(g), ethane	30.07	−84.68	52.63	−1560
C_3H_6(g), propene	42.08	+20.42	63.89	−2058
C_3H_6(g), cyclopropane	42.08	+53.30	55.94	−2091
C_3H_8(g), propane	44.10	−103.85	73.5	−2220
C_4H_8(g), 1-butene	56.11	−0.13	85.65	−2717
C_4H_8(g), *cis*-2-butene	56.11	−6.99	78.91	−2710
C_4H_8(g), *trans*-2-butene	56.11	−11.17	87.82	−2707
C_4H_{10}(g), butane	58.13	−126.15	97.45	−2878
C_5H_{12}(g), pentane	72.15	−146.44	120.2	−3537
C_5H_{12}(l)	72.15	−173.1		
C_6H_6(l), benzene	78.12	+49.0	136.1	−3268
C_6H_6(g)	78.12	+82.93	81.67	−3302
C_6H_{12}(l), cyclohexane	84.16	−156	156.5	−3920
C_6H_{14}(l), hexane	86.18	−198.7		−4163
$C_6H_5CH_3$(g), toluene	92.14	+50.0	103.6	−3953
C_7H_{16}(l), heptane	100.21	−224.4	224.3	
C_8H_{18}(l), octane	114.23	−249.9		−5471
C_8H_{18}(l), iso-octane	114.23	−255.1		−5461
$C_{10}H_8$(s), naphthalene	128.18	+78.53		−5157
Alcohols and phenols				
CH_3OH(l), methanol	32.04	−238.66	81.6	−726
CH_3OH(g)	32.04	−200.66	43.89	−764
C_2H_5OH(l), ethanol	46.07	−277.69	111.46	−1368
C_2H_5OH(g)	46.07	−235.10	65.44	−1409
C_6H_5OH(s), phenol	94.12	−165.0		−3054
Carboxylic acids, hydroxy acids, and esters				
HCOOH(l), formic	46.03	−424.72	99.04	−255
CH_3COOH(l), acetic	60.05	−484.5	124.3	−875
CH_3COOH(aq)	60.05	−485.76		
$CH_3CO_2^-$(aq)	59.05	−486.01	−6.3	
$(COOH)_2$(s), oxalic	90.04	−827.2	117	−254
C_6H_5COOH(s), benzoic	122.13	−385.1	146.8	−3227
$CH_3CH(OH)COOH$(s), lactic	90.08	−694.0		−1344
$CH_3COOC_2H_5$(l), ethyl acetate	88.11	−479.0	170.1	−2231
Alkanals and alkanones				
HCHO(g), methanol	30.03	−108.57	35.40	−571

(Continued)

TABLE E.7 *(Continued)*

Thermodynamic Data for Organic Compounds (all values are for 298 K and 1 bar)				
	M (g/mol)	$\Delta\hat{H}_f^o$ (kJ/mol)	C_p (J/(mol·K))	$\Delta\hat{H}_c^o$ (kJ/mol)
$CH_3CHO(l)$, ethanol	44.05	−192.30		−1166
$CH_3CHO(g)$	44.05	−166.19	57.3	−1192
$CH_3COCH_3(l)$, propanone	58.08	−248.1	124.7	−1790
Sugars				
$C_6H_{12}O_6(s)$, α-D-glucose	180.16	−1274		−2808
$C_6H_{12}O_6(s)$, β-D-glucose	180.16	−1268		
$C_6H_{12}O_6(s)$, β-D-fructose	180.16	−1266		−2810
$C_{12}H_{22}O_{11}(s)$, sucrose	342.30	−2222		−5645
Nitrogen compounds				
$CO(NH_2)_2(s)$, urea	60.06	−333.51	93.14	−632
$CH_3NH_2(g)$, methylamine	31.06	−22.97	53.1	−1085
$C_6H_5NH_2(l)$, aniline	93.13	+31.1		−3393
$CH_2(NH_2)COOH(s)$, glycine	75.07	−532.9	99.2	−969

Table from Atkins P, *Physical Chemistry*, 6th ed. New York: W. H. Freeman, 1998.

TABLE E.8

Thermodynamic Data (all values relate to 298 K and 1 bar)			
	M (g/mol)	$\Delta\hat{H}_f^o$ (kJ/mol)	C_p (J/(K·mol))
Aluminum			
Al(s)	26.98	0	24.35
Al(l)	26.98	+10.56	24.21
Al(g)	26.98	+326.4	21.38
$Al^{3+}(g)$	26.98	+5483.17	
$Al^{3+}(aq)$	26.98	−531	
$Al_2O_3(s, \alpha)$	101.96	−1675.7	79.04
$AlCl_3(s)$	133.24	−704.2	91.84
Argon			
Ar(g)	39.95	0	20.786
Antimony			
Sb(s)	121.75	0	25.23
$SbH_3(g)$	124.77	+145.11	41.05
Arsenic			
As(s, α)	74.92	0	24.64
As(g)	74.92	+302.5	20.79
$As_4(g)$	299.69	+143.9	
$AsH_3(g)$	77.95	+66.44	38.07
Barium			
Ba(s)	137.34	0	28.07
Ba(g)	137.34	+180	20.79
$Ba^{2+}(aq)$	137.34	−537.64	
BaO(s)	153.34	−553.5	47.78

TABLE E.8 *(Continued)*

Thermodynamic Data (all values relate to 298 K and 1 bar)

	M (g/mol)	ΔH_f° (kJ/mol)	C_p (J/(K·mol))
Barium (Continued)			
$BaCl_2$(s)	208.25	−858.6	75.14
Beryllium			
Be(s)	9.01	0	16.44
Be(g)	9.01	+324.3	20.79
Bismuth			
Bi(s)	208.98	0	25.52
Bi(g)	208.98	+207.1	20.79
Bromine			
Br_2(l)	159.82	0	75.689
Br_2(g)	159.82	+30.907	36.02
Br(g)	79.91	+111.88	20.786
Br^-(g)	79.91	−219.07	
Br^-(aq)	79.91	−121.55	−141.8
HBr(g)	90.92	−36.40	29.142
Cadmium			
Cd(s, γ)	112.40	0	25.98
Cd(g)	112.40	+112.01	20.79
Cd^{2+}(aq)	112.40	−75.90	
CdO(s)	128.40	−258.2	43.43
$CdCO_3$(s)	172.41	−750.6	
Cesium			
Cs(s)	132.91	0	32.17
Cs(g)	132.91	+76.06	20.79
Cs^+(aq)	132.91	−258.28	−10.5
Calcium			
Ca(s)	40.08	0	25.31
Ca(g)	40.08	+178.2	20.786
Ca^{2+}(aq)	40.08	−542.83	
CaO(s)	56.08	−635.09	42.80
$CaCO_3$(s) (calcite)	100.09	−1206.9	81.88
$CaCO_3$(s) (aragonite)	100.09	−1207.1	81.25
CaF_2(s)	78.08	−1219.6	67.03
$CaCl_2$(s)	110.99	−795.8	72.59
$CaBr_2$(s)	199.90	−682.8	
Carbon			
C(s) (graphite)	12.011	0	8.527
C(s) (diamond)	12.011	+1.895	6.113
C(g)	12.011	+716.68	20.838
C_2(g)	24.022	+831.90	43.21
CO(g)	28.011	−110.53	29.14
CO_2(g)	44.010	−393.51	37.11
CO_2(aq)	44.010	−413.80	
H_2CO_3(aq)	62.03	−699.65	
HCO_3^-(aq)	61.02	−691.99	
CO_3^{2-}(aq)	60.01	−677.14	
CCl_4(*l*)	153.82	−135.44	131.75
CS_2(*l*)	76.14	+89.70	75.7
HCN(g)	27.03	+135.1	35.86
HCN(*l*)	27.03	+108.87	70.63
CN^-(aq)	26.02	+150.6	

(Continued)

TABLE E.8 *(Continued)*

Thermodynamic data (All values relate to 298 K and 1 bar)

	M (g/mol)	$\Delta \hat{H}_f^o$ (kJ/mol)	C_p (J/(K·mol))
Chlorine			
$Cl_2(g)$	70.91	0	33.91
$Cl(g)$	35.45	+121.68	21.840
$Cl^-(g)$	35.45	−233.13	
$Cl^-(aq)$	35.45	−167.16	−136.4
$HCl(g)$	36.46	−92.31	29.12
$HCl(aq)$	36.46	−167.16	−136.4
Chromium			
$Cr(s)$	52.00	0	23.35
$Cr(g)$	52.00	+396.6	20.79
$CrO_4^{2-}(aq)$	115.99	−881.15	
$Cr_2O_7^{2-}(aq)$	215.99	−1490.3	
Copper			
$Cu(s)$	63.54	0	24.44
$Cu(g)$	63.54	+338.32	20.79
$Cu^+(aq)$	63.54	+71.67	
$Cu^{2+}(aq)$	63.54	+64.77	
$Cu_2O(s)$	143.08	−168.6	63.64
$CuO(s)$	79.54	−157.3	42.30
$CuSO_4(s)$	159.60	−771.36	100.0
$CuSO_4 \cdot H_2O(s)$	177.62	−1085.8	134
$CuSO_4 \cdot 5\,H_2O(s)$	249.68	−2279.7	280
Deuterium			
$D_2(g)$	4.028	0	29.20
$HD(g)$	3.022	+0.318	29.196
$D_2O(g)$	20.028	−249.20	34.27
$D_2O(l)$	20.028	−294.60	84.35
$HDO(g)$	19.022	−245.30	33.81
$HDO(l)$	19.022	−289.89	
Fluorine			
$F_2(g)$	38.00	0	31.30
$F(g)$	19.00	+78.99	22.74
$F^-(aq)$	19.00	−332.63	−106.7
$HF(g)$	20.01	−271.1	29.13
Gold			
$Au(s)$	196.97	0	25.42
$Au(g)$	196.97	+366.1	20.79
Helium			
$He(g)$	4.003	0	20.786
Hydrogen			
$H_2(g)$	2.016	0	28.824
$H(g)$	1.008	+217.97	20.784
$H^+(aq)$	1.008	0	0
$H^+(g)$	1.008	+1536.20	
$H_2O(l)$	18.015	−285.83	75.291
$H_2O(g)$	18.015	−241.82	33.58
$H_2O_2(l)$	34.015	−187.78	89.1
Iodine			
$I_2(s)$	253.81	0	54.44
$I_2(g)$	253.81	+62.44	36.90

TABLE E.8 *(Continued)*

Thermodynamic data (All values relate to 298 K and 1 bar)

	M (g/mol)	$\Delta\hat{H}_f^\circ$ (kJ/mol)	C_p (J/(K·mol))
Iodine (continued)			
$I(g)$	126.90	+106.84	20.786
$I^-(aq)$	126.90	−55.19	−142.3
$HI(g)$	127.91	+26.48	29.158
Iron			
$Fe(s)$	55.85	0	25.10
$Fe(g)$	55.85	+416.3	25.68
$Fe^{2+}(aq)$	55.85	−89.1	
$Fe^{3+}(aq)$	55.85	−48.5	
$Fe_3O_4(s)$ (magnetite)	231.54	−1118.4	143.43
$Fe_2O_3(s)$ (hematite)	159.69	−824.2	103.85
$FeS(s, \alpha)$	87.91	−100.0	50.54
$FeS_2(s)$	119.98	−178.2	62.17
Krypton			
$Kr(g)$	83.80	0	20.786
Lead			
$Pb(s)$	207.19	0	26.44
$Pb(g)$	207.19	+195.0	20.79
$Pb^{2+}(aq)$	207.19	−1.7	
PbO(s, yellow)	223.19	−217.32	45.77
PbO(s, red)	223.19	−218.99	45.81
$PbO_2(s)$	239.19	−277.4	64.64
Lithium			
$Li(s)$	6.94	0	24.77
$Li(g)$	6.94	+159.37	20.79
$Li^+(aq)$	6.94	−278.49	68.6
Magnesium			
$Mg(s)$	24.31	0	24.89
$Mg(g)$	24.31	+147.70	20.786
$Mg^{2+}(aq)$	24.31	−466.85	
$MgO(s)$	40.31	−601.70	37.15
$MgCO_3(s)$	84.32	−1095.8	75.52
$MgCl_2(s)$	95.22	−641.32	71.38
Mercury			
$Hg(l)$	200.59	0	27.983
$Hg(g)$	200.59	+61.32	20.786
$Hg^{2+}(aq)$	200.59	+171.1	
$Hg_2^{2+}(aq)$	401.18	+172.4	
$HgO(s)$	216.59	−90.83	44.06
$Hg_2Cl_2(s)$	472.09	−265.22	102
$HgCl_2(s)$	271.50	−224.3	
HgS(s, black)	232.65	−53.6	
Neon			
$Ne(g)$	20.18	0	20.786
Nitrogen			
$N_2(g)$	28.013	0	29.125
$N(g)$	14.007	+472.70	20.786
$NO(g)$	30.01	+90.25	29.844
$N_2O(g)$	44.01	+82.05	38.45
$NO_2(g)$	46.01	+33.18	37.20
$N_2O_4(g)$	92.01	+9.16	77.28
$N_2O_5(s)$	108.01	−43.1	143.1

(Continued)

TABLE E.8 *(Continued)*

Thermodynamic Data (all values relate to 298 K and 1 bar)

	M (g/mol)	$\Delta \hat{H}_f^o$ (kJ/mol)	C_p (J/(K·mol))
N_2O_5(g)	108.01	+11.3	84.5
HNO_3(l)	63.01	−174.10	109.87
HNO_3(aq)	63.01	−207.36	−86.6
NO_3^-(aq)	62.01	−205.0	−86.6
NH_3(g)	17.03	−46.11	35.06
NH_3(aq)	17.03	−80.29	
NH_4^+(aq)	18.04	−132.51	79.9
NH_2OH(s)	33.03	−114.2	
HN_3(l)	43.03	+264.0	43.68
HN_3(g)	43.03	+294.1	98.87
N_2H_4(l)	32.05	+50.63	139.3
NH_4NO_3(s)	80.04	−365.56	84.1
NH_4Cl(s)	53.49	−314.43	
Oxygen			
O_2(g)	31.999	0	29.355
O(g)	15.999	+249.17	21.912
O_3(g)	47.998	+142.7	39.20
OH^-(aq)	17.007	−229.99	−148.5
Phosphorus			
P(s, wh)	30.97	0	23.840
P(g)	30.97	+314.64	20.786
P_2(g)	61.95	+144.3	32.05
P_4(g)	123.90	+58.91	67.15
PH_3(g)	34.00	+5.4	37.11
PCl_3(g)	137.33	−287.0	71.84
PCl_3(l)	137.33	−319.7	
PCl_5(g)	208.24	−374.9	112.8
PCl_5(s)	208.24	−443.5	
H_3PO_3(s)	82.00	−964.4	
H_3PO_3(aq)	82.00	−964.8	
H_3PO_4(s)	94.97	−1279.0	106.06
H_3PO_4(l)	94.97	−1266.9	
H_3PO_4(aq)	94.97	−1277.4	
PO_4^{3-}(aq)	94.97	−1277.4	
P_4O_{10}(s)	283.89	−2984.0	211.71
P_4O_6(s)	219.89	−1640.1	
Potassium			
K(s)	39.10	0	29.58
K(g)	39.10	+89.24	20.786
K^+(g)	39.10	+514.26	
K^+(aq)	39.10	−252.38	21.8
KOH(s)	56.11	−424.76	64.9
KF(s)	58.10	−576.27	49.04
KCl(s)	74.56	−436.75	51.30
KBr(s)	119.01	−393.80	52.30
KI(s)	166.01	−327.90	52.93
Silicon			
Si(s)	28.09	0	20.00
Si(g)	28.09	+455.6	22.25
SiO_2(s, α)	60.09	−910.94	44.43
Silver			

TABLE E.8 *(Continued)*

Thermodynamic Data (all values relate to 298 K and 1 bar)			
	M (g/mol)	$\Delta\hat{H}_f^\circ$ (kJ/mol)	C_p (J/(K·mol))
Ag^+(aq)	107.87	+105.58	21.8
AgBr(s)	187.78	−100.37	52.38
AgCl(s)	143.32	−127.07	50.79
Ag_2O(s)	231.74	−31.05	65.86
$AgNO_3$(s)	169.88	−129.39	93.05
Sodium			
Na(s)	22.99	0	28.24
Na(g)	22.99	+107.32	20.79
Na^+(aq)	22.99	−240.12	46.4
NaOH(s)	40.00	−425.61	59.54
NaCl(s)	58.44	−411.15	50.50
NaBr(s)	102.90	−361.06	51.38
NaI(s)	149.89	−287.78	52.09
Sulfur			
S(s, α) (rhombic)	32.06	0	22.64
S(s, β) (monoclinic)	32.06	+0.33	23.6
S(g)	32.06	+278.81	23.673
S_2(g)	64.13	+128.37	32.47
S^{2-}(aq)	32.06	+33.1	
SO_2(g)	64.06	−296.83	39.87
SO_3(g)	80.06	−395.72	50.67
H_2SO_4(l)	98.08	−813.99	138.9
H_2SO_4(aq)	98.08	−909.27	−293
SO_4^{2-}(aq)	96.06	−909.27	−293
HSO_4^-(aq)	97.07	−887.34	−84
H_2S(g)	34.08	−20.63	34.23
H_2S(aq)	34.08	−39.7	
HS^-(aq)	33.072	−17.6	
SF_6(g)	146.05	−1209	97.28
Tin			
Sn(s, β)	118.69	0	26.99
Sn(g)	118.69	+302.1	20.26
Sn^{2+}(aq)	118.69	−8.8	
SnO(s)	134.69	−285.8	44.31
SnO_2(s)	150.69	−580.7	52.59
Xenon			
Xe(g)	131.30	0	20.786
Zinc			
Zn(s)	65.37	0	25.40
Zn(g)	65.37	+130.73	20.79
Zn^{2+}(aq)	65.37	−153.89	46
ZnO(s)	81.37	−348.28	40.25

Table from Atkins P, *Physical Chemistry*, 6th ed. New York: W. H. Freeman, 1998.

TABLE E.9

Heats of Combustion

Compound	Formula	Molecular weight, M (g/mol)	State	Heat of combustion $\Delta\hat{H}_c^o$ (kJ/mol)
Acetaldehyde	C_2H_4O	44.053	*l*	−1166.9
			g	−1192.5
Acetic acid	$C_2H_4O_2$	60.053	*l*	−874.2
			g	−925.9
Acetone	C_3H_6O	58.080	*l*	−1789.9
			g	−1820.7
Acetylene	C_2H_2	26.038	g	−1301.1
Adenine	$C_5H_5N_5$	135.128	c	−2778.1
			g	−2886.9
Alanine (D-)	$C_3H_7O_2N$	89.094	c	−1619.7
Alanine (L-)	$C_3H_7O_2N$	89.094	c	−1576.9
			g	−1715.0
Ammonia	NH_3	17.03	g	−382.6
Ammonium ion	NH_4^+			−383
Arginine (D-)	$C_6H_{14}O_2N_4$	174.203	c	−3738.4
Asparagine (L-)	$C_4H_8O_3N_2$	132.119	c	−1928.0
Aspartic acid (L-)	$C_4H_7O_4N$	133.104	c	−1601.1
Benzaldehyde	C_7H_6O	106.124	*l*	−3525.1
			g	−3575.4
Biomass	$CH_{1.8}O_{0.5}N_{0.2}$	25.9	s	−552
Butanoic acid	$C_4H_8O_2$	88.106	*l*	−2183.6
			g	−2241.6
1-Butanol	$C_4H_{10}O$	74.123	*l*	−2675.9
			g	−2728.2
2-Butanol	$C_4H_{10}O$	74.123	*l*	−2660.6
			g	−2710.3
Butyric acid	$C_4H_8O_2$	88.106	*l*	−2183.6
			g	−2241.6
Caffeine	$C_8H_{10}O_2N_4$		s	−4246.5*
Carbon	C	12.011	c	−393.5
Carbon monoxide	CO	28.010	g	−283.0
Citric acid	$C_6H_8O_7$		s	−1962.0
Codeine	$C_{18}H_{21}O_3N.H_2O$		s	−9745.7*
Cytosine	$C_4H_5ON_3$	111.103	c	−2067.3
Ethane	C_2H_6	30.070	g	−1560.7
Ethanol	C_2H_6O	46.069	*l*	−1366.8
			g	−1409.4
Ethylene	C_2H_4	28.054	g	−1411.2
Ethylene glycol	$C_2H_6O_2$	62.068	*l*	−1189.2
			g	−1257.0
Formaldehyde	CH_2O	30.026	g	−570.7
Formic acid	CH_2O_2	46.026	*l*	−254.6
			g	−300.7
Fructose (D-)	$C_6H_{12}O_6$		s	−2813.7
Fumaric acid	$C_4H_4O_4$	116.073	c	−1334.0
Galactose (D-)	$C_6H_{12}O_6$		s	−2805.7
Glucose (D-)	$C_6H_{12}O_6$		s	−2805.0
Glutamic acid (L-)	$C_5H_9O_4N$	147.131	c	−2244.1
Glutamine (L-)	$C_5H_{10}O_3N_2$	146.146	c	−2570.3

TABLE E.9 *(Continued)*

Heats of Combustion

Compound	Formula	Molecular weight, M (g/mol)	State	Heat of combustion $\Delta \hat{H}_c^o$ (kJ/mol)
Glutaric acid	$C_5H_8O_4$	132.116	c	−2150.9
Glycerol	$C_3H_8O_3$	92.095	*l*	−1655.4
			g	−1741.2
Glycine	$C_2H_5O_2N$	75.067	c	−973.1
Glycogen	$(C_6H_{10}O_5)_x$per kg		s	−17530.1*
Guanine	$C_5H_5ON_5$	151.128	c	−2498.2
Hexadecane	$C_{16}H_{34}$	226.446	*l*	−10699.2
			g	−10780.5
Hexadecanoic acid	$C_{16}H_{32}O_2$	256.429	c	−9977.9
			l	−10031.3
			g	−10132.3
Histidine (L-)	$C_6H_9O_2N_3$	155.157	c	−3180.6
Hydrogen	H_2	2.016	g	−285.8
Hydrogen sulphide	H_2S	34.08		−562.6
Inositol	$C_6H_{12}O_6$		s	−2772.2*
Isoleucine (L-)	$C_6H_{13}O_2N$	131.175	c	−3581.1
Isoquinoline	C_9H_7N	129.161	*l*	−4686.5
Lactic acid (D, L-)	$C_3H_6O_3$		*l*	−1368.3
Lactose	$C_{12}H_{22}O_{11}$		s	−5652.5
Leucine (D-)	$C_6H_{13}O_2N$	131.175	c	−3581.7
Leucine (L-)	$C_6H_{13}O_2N$	131.175	c	−3581.6
Lysine	$C_6H_{14}O_2N_2$	146.189	c	−3683.2
Malic acid (L-)	$C_4H_6O_5$		s	−1328.8
Malonic acid	$C_3H_4O_4$		s	−861.8
Maltose	$C_{12}H_{22}O_{11}$		s	−5649.5
Mannitol (D-)	$C_6H_{14}O_6$		s	−3046.5*
Methane	CH_4	16.043	g	−890.8
Methanol	CH_4O	32.042	*l*	−726.1
			g	−763.7
Morphine	$C_{17}H_{19}O_3N.H_2O$		s	−8986.6*
Nicotine	$C_{10}H_{14}N_2$		*l*	−5977.8*
Oleic acid	$C_{18}H_{34}O_2$		*l*	−11126.5
Oxalic acid	$C_2H_2O_4$	90.036	c	−251.1
Papaverine	$C_{20}H_{21}O_4N$		s	−10375.8*
Pentane	C_5H_{12}	72.150	*l*	−3509.0
			g	−3535.6
Phenylalanine (L-)	$C_9H_{11}O_2N$	165.192	c	−4646.8
Phthalic acid	$C_8H_6O_4$	166.133	c	−3223.6
Proline (L-)	$C_5H_9O_2N$	115.132	c	−2741.6
Propane	C_3H_8	44.097	g	−2219.2
1-Propanol	C_3H_8O	60.096	*l*	−2021.3
			g	−2068.8
2-Propanol	C_3H_8O	60.096	*l*	−2005.8
			g	−2051.1
Propionic acid	$C_3H_6O_2$	74.079	*l*	−1527.3
			g	−1584.5

(Continued)

TABLE E.9 *(Continued)*

Heats of Combustion

Compound	Formula	Molecular weight, M (g/mol)	State	Heat of combustion $\Delta\hat{H}_c^o$ (kJ/mol)
1,2-Propylene glycol	$C_3H_8O_2$	76.095	l	−1838.2
			g	−1902.6
1,3-Propylene glycol	$C_3H_8O_2$	76.095	l	−1859.0
			g	−1931.8
Pyridine	C_5H_5N	79.101	l	−2782.3
			g	−2822.5
Pyrimidine	$C_4H_4N_2$	80.089	l	−2291.6
			g	−2341.6
Salicylic acid	$C_7H_6O_3$	138.123	c	−3022.2
			g	−3117.3
Serine (L-)	$C_3H_7O_3N$	105.094	c	−1448.2
Starch	$(C_6H_{10}O_5)_x$per kg		s	−17496.6*
Succinic acid	$C_4H_6O_4$	118.089	c	−1491.0
Sucrose	$C_{12}H_{22}O_{11}$		s	−5644.9
Thebaine	$C_{19}H_{21}O_3N$		s	−10221.7*
Threonine (L-)	$C_4H_9O_3N$	119.120	c	−2053.1
Thymine	$C_5H_6O_2N_2$	126.115	c	−2362.2
Tryptophan (L-)	$C_{11}H_{12}O_2N_2$	204.229	c	−5628.3
Tyrosine (L-)	$C_9H_{11}O_3N$	181.191	c	−4428.6
Uracil	$C_4H_4O_2N_2$	112.088	c	−1716.3
			g	−1842.8
Urea	CH_4ON_2	60.056	c	−631.6
			g	−719.4
Valine (L-)	$C_5H_{11}O_2N$	117.148	c	−2921.7
			g	−3084.5
Xanthine	$C_5H_4O_2N_4$	152.113	c	−2159.6
Xylose	$C_5H_{10}O_5$		s	−2340.5

Data From *Handbook of Chemistry and Physics*, 73d ed. Boca Raton, FL: CRC Press, 1992, *Handbook of Chemistry and Physics*, 57th ed. Boca Raton, FL: CRC Press, 1976, and Felder RM and Rousseau RW, *Elementary Principles of Chemical Processes*. New York: John Wiley, 1978.

Table from Doran PM, *Bioprocess Engineering Principles*. London: Academic Press, 1995.

Reference conditions: 1 atm and 25°C or 20°C; values marked with an asterisk* refer to 20°C.

Products of combustion are taken to be CO_2 (g), H_2O (l), and N_2 (g); therefore, $\Delta\hat{H}_c^\circ = 0$ for CO_2(g), H_2O(l), and N_2(g).

State: g = gas; l = liquid; c = crystal; s = solid.

Index